HANDBOOK OF
QUALITY INTEGRATED
CIRCUIT MANUFACTURING

HANDBOOK OF QUALITY INTEGRATED CIRCUIT MANUFACTURING

ROBERT ZORICH

ACADEMIC PRESS, INC.
Harcourt Brace Jovanovich, Publishers
San Diego New York Boston
London Sydney Tokyo Toronto

ACADEMIC PRESS, INC.
San Diego, California 92101

United Kingdom Edition published by
Academic Press Limited
24–28 Oval Road, London NW1 7DX

Library of Congress Cataloging-in-Publication Data

Zorich, Robert.
Handbook of quality integrated circuit manufacturing / Robert
Zorich.
p. cm.
Includes bibliographical references and index.
ISBN 0-12-781870-7 (alk. paper)
1. Integrated circuits--Wafer-scale integration--Design and construction--Handbooks, manuals, etc. I. Title.
TK7874.Z67 1990
621.381'5--dc20 90-697
CIP

PRINTED IN THE UNITED STATES OF AMERICA
91 92 93 9 8 7 6 5 4 3 2 1

To my loving wife Cathleen,
who brought me the joy of Michelle and Jonathan.

CONTENTS

CHAPTER 3

METHODS OF SEMICONDUCTOR ANALYSIS

CHAPTER 4

CLEANROOM FACILITIES AND PROCEDURES

CHAPTER 5

WAFER CLEANING TECHNIQUES

CHAPTER 6

PHOTOLITHOGRAPHY AND ETCH PROCESSES

CHAPTER 7

THE DIFFUSION PROCESSES

CHAPTER 8

LOW-PRESSURE CHEMICAL VAPOR DEPOSITION PROCESSES

PREFACE

Since the 1970s, the use of microelectronics has been increasing at an incredible rate. Ideas and applications that were considered science fiction just a few years ago have now become commonplace. Very few products had integrated circuits (ICs) designed into them, and those that did were either very expensive, very specialized, or both. A quasi-programmable calculator or 8-bit refrigerator-size minicomputer (often with no more than 4K of usable RAM) were the closest devices the average person could compare with what is now the common personal computer. Mainframe computers were fed stacks of punched cards; there were no floppy disks or drives. TVs, stereos, and other consumer goods often were solid-state, but consisted of discrete components, not ICs. They consumed a lot of power by today's standards and had limited features. Few kitchen or household appliances had electronics in their designs, and microwave ovens were just beginning to appear. The use of IC electronics in critical areas of vehicles was considered risky, if not outright dangerous. (After all, who knew when the electronics would fail? In fact, common relay failures occurred all the time.)

Today, ICs of all types exist in everything from popcorn makers and microwave ovens to home entertainment centers, personal computers, cameras, and cars. The military and aerospace industries use electronics extensively; applications range from aircraft that could not fly without electronics, to intelligent self-guided munitions, and to sophisticated satellites that can transmit live pictures with unbelievable resolution. Integrated circuits now control the operations of nearly all new vehicles. The use of microprocessors and microcontrollers has spread to virtually every niche within the American business scene. In fact, the proliferation of miniaturized circuitry has allowed for entire industries to form that did not previously exist.

The increase in processing power and the widespread distribution of this computing power has driven the semiconductor market toward ever-higher density, ever-faster microprocessors and memory chips, and ever-increasing varieties of peripheral devices. Computer application designs have been greatly reduced in sophistication, such as computer interconnection, the reduction of modems and local area networks to a few or in some cases one IC, high-density storage devices like CD-ROMs, and affordable high-resolution color graphics that previously were very difficult and expensive to design and manufacture. New IC design methodologies have allowed the systems designer to customize circuitry and have prototypes within weeks of conception.

The impact of these devices is very real and apparent, whereas the effort required to construct the IC is usually unknown to the average person. To the customer or end-user there appears to be little difference in the chips themselves, since what is usually seen is the hermetically sealed package. Those who have had exposure to chips know that the circuits are visible only with a microscope. (Even a human hair is gigantic compared to the circuitry placed on the surface of a chip.) You will see the motor of your car in action (or at least see the effects of its action), but you will never see the "motor" of your PC in action. At best, you will notice that it gets warm.

The various techniques used to construct ICs are also unknown to the users of the chips. The diverse applications of ICs require a wide variety of processing techniques to allow them to work properly. And though it can be fairly easy to design and construct systems using ICs, the design manufacture of the chips themselves can be a long, drawn-out process.

The ICs that go into these technological wonders are manufactured in plants that are often 1000 or more times cleaner than a hospital operating room. (Fortunately, human beings are more tolerant of contamination than integrated circuits.) Workers are allowed into these factories only while wearing special cleanroom suits. The equipment is often fully automated, requiring only a small amount of human interaction during operation. However, this equipment can be difficult to maintain as its tolerances are narrow, and high precision must be maintained. Extremely harsh chemicals are used to remove any contamination that may appear on the substrate materials. Water is purified (deionized) to the point that it acts as an acid, is undrinkable, and can dissolve many common metals and plastics. The purity of the chemicals, air, and water is very difficult to maintain.

Added to the difficulties of cleanroom manufacturing is the fact that circuit designs vary greatly, change often, and can include anywhere from a few thousand transistors to several million transistors per circuit. There are often more than 200 to 300 chips per silicon wafer, which results in the construction of several hundred million transistors per silicon wafer. Defects on any of these circuits can render the chips unusable. At the high average selling price of some

ICs, the loss of a single wafer can mean a loss of thousands of dollars in revenue.

The sophistication of the processes used in the manufacture of ICs is quite remarkable and has been continually evolving since the 1960s. Each generation of ICs has new challenges and new complications, often resulting in massive procedural changes and equipment modifications. Improvements in technology have greatly increased the complexity and difficulty of manufacturing the devices and have caused the "rules of the game" to change rapidly, sometimes before the old rules have been thoroughly understood. Ultimately, one thing is guaranteed for the future: the technology will continue to evolve rapidly. New technologies once again will stretch the limits of manufacturing ICs.

Unfortunately, even as the industry requires sound and experienced decision makers with problem solving skills to further develop its technology, the schools of the United States graduate fewer engineers. In addition, unlike the typical Japanese engineer, a U.S. manufacturing or process engineer does not stay with one job particularly long and often moves rapidly into research and development or middle management positions. If these trends and the practice of intercompany secrecy in processing continue (which hopefully will be reduced slightly with the advent of Sematech, the government and industry sponsored IC consortium), a significant amount of basic information and understanding of fabrication processing will be lost. Thus each new generation of engineers has to undergo a steep learning curve for almost a year just to learn one of the many processing areas. These individuals do not have time to become adequately prepared when change occurs. Projects collapse or are put many weeks behind schedule because engineers leave a company to join another, or processes "go south" when the owner is on vacation. It is imperative that the experience of the manufacturing engineers in this country not become lost in the scramble to get ahead. While the U.S. semiconductor industry is very strong, competition from foreign lands is growing. Every effort should be made to improve the efficiency of the manufacturing operation. This includes reducing the amount of time spent by new personnel learning a job and helping prevent more experienced engineers from chasing problems down blind alleys.

It is in this light that this book has been written. It is designed to be a well-rounded source of information for experienced process engineers, novice process engineers, manufacturers, maintenance, and, of significant importance, equipment manufacturers who can use it as a basic tutorial for the details of their job. It is also a reference for the many other individuals, various group managers, and production and maintenance personnel who need to understand the technical nature of the decisions that they are being asked to make. These individuals often hear that costs are high and that everything must be done correctly, but often there is little cross-fertilization of ideas, so that many individuals do not have a well-rounded view of high-quality wafer processing. This handbook will also be very useful to the fabrication equipment manufacturers; it may clarify some of the requests they receive and contribute to their ability to develop new products which may find more rapid acceptance in the wafer fabrication environment.

This book does not duplicate textbooks on semiconductor physics and theory, but focuses on the primary problems of the operating manufacturing area. It is the intent of this book to encourage the reader to find suitable solutions to manufacturing problems or to seek directions to find a solution. After all, who wants to waste time by running a complete analysis of a clean station if it turns out that only a filter change is required? This simple example illustrates my point that there is enough to do in the semiconductor industry without reinventing solutions to problems that have been solved previously. The book is predicated on the concept that quality must come first in order to produce chips profitably. It focuses on manufacturing techniques with the following industry priorities in mind: defect analysis and elimination, equipment downtime reduction and control, yield enhancement, inventory management and throughput time reduction, and statistical process control in concert with management by exception and preventative quality control techniques.

Although there is a large amount of specific technical data presented, in some cases statements are made which of necessity are somewhat general and are based more on observation than on hard scientific data. These observations may not hold true for all situations or for all devices and technologies. For example, the requirements of a 1.2-μm CMOS technology vary significantly from those of a 3-μm bipolar technology, even though basic analysis techniques remain the same. However, since this is a book about problem solving techniques, rules of thumb, and methods for improving integrated circuit yield, it is useful to have ideas for possible directions of action in order to solve particularly difficult problems, even if the ideas have not been rigorously defined. The general logic of a technique may be applicable even if the particulars of a process have changed. In any event, before any idea is implemented, it should be tested thoroughly, even if it seems simple. (A simple change can lead to more serious problems later.) In general, my philosophy has been to cover many of the support and operations aspects of wafer manufacturing and not just the engineering or technical aspects.

There are several types of wafer fabrication areas, each with different needs. For instance, there is the large scale manufacturing area which requires a high monthly output of ICs. These factories build RAM chips, microprocessors, EPROMs, and other support ICs. They require both adequate hardware and the facilities to meet the needs of production and must have procedures in place to maximize output in order to remain profitable in a very competitive industry. Another type of wafer fabrication area is the small company supplier that manufactures custom chips (chips for specific markets) or is a captive supplier for a larger company. They may have limited resources and may not have the most adequate facilities. These companies may derive significant benefit from the procedures described in this book. A third type of wafer fabrication area is the research and development group. They may not be as sensitive to the effects of contamination as others because they usually only need to get a sufficient number of chips or sample to obtain reasonable data for their tests. This book can give the research or technology development engineer some guidelines for how the chips are manufactured when the projects they are working on are developed into products or production processes.

In all cases, it is assumed that the reader has had some exposure to semiconductor technology, either in a semiconductor physics class or as experience in the fabrication process. It is recommended that the reader has available to him or her one or more of the textbooks listed in the bibliography. The book by Muller and Kamins and the several books by Sze are all highly recommended.

This book consists of the following chapters:

1. *Overview of Semiconductor Manufacturing.* This chapter describes some of the basic tenets of the book and sets up basic definitions for many terms and practices to be described throughout the rest of the book. In addition, several integrated circuit elements are described, along with a description of a wafer manufacturing process.
2. *Yield Optimization.* This chapter discusses the impact of die yield and wafer (line) yield changes. Low-yield analysis techniques and isometric yield prediction curves are defined, along with inventory analysis and control techniques, management by exception techniques, and so on.
3. *Methods of Semiconductor Analysis.* Finding and identifying contamination and other problems that can affect yields can be extremely complex. This chapter addresses the issues of methods for particulate detection, examples of typical particle contamination patterns, and electrical tests for determining chemical contamination levels.
4. *Cleanroom Facilities and Procedures.* The basic starting point for developing any wafer fabrication process is the setup of the cleanroom manufacturing space. This chapter covers the various types of cleanroom layouts and the issues critical for the operation of the cleanroom, which include airflow considerations, pressure balancing, and chemical and piping compatibility issues. It concludes with a discussion on the safety requirements for the fabrication area.
5. *Wafer Cleaning Techniques.* This chapter covers wafer cleaning, methods of evaluating the cleaning process, and options to

improve the sequence. Cleaning solutions are discussed along with related safety and environmental issues.

6. *Photolithographic and Etch Processes.* The first section describes the lithography process. Wafer geometry is determined here and the precision of the operation determines most of the characteristics of the final product. Contamination, maintenance, and safety issues are covered. The second section covers the etching processes and the very precise process that must be accomplished to define the circuits that have been created during the lithographic steps. The various types of etching processes and the chemical by-products are discussed.
7. *Diffusion Processes.* This chapter covers wafer oxidation processes, dopant redistribution processes, anneals, alloys, and so on. Because of the high temperatures involved in these processes, not only are uniformity and particulate issues important, but chemical contamination must be reduced to an absolute minimum. The maintenance and safety issues inherent to the diffusion processes are discussed.
8. *Low-Pressure Chemical Vapor Deposition Processes.* This chapter covers the various chemical vapor deposition steps that are used in the semiconductor industry. The problems associated with these types of deposition techniques are discussed. The polysilicon, silicon dioxide, and silicon nitride deposition steps are considered, and potential sources of contamination and the effects of the anneal processes on the wafers are covered.
9. *Ion Implantation Processes.* This chapter covers the ion implant steps used to dope regions of the substrate or films with controlled amounts of impurities. This is done to control the resistivity of the materials in question. This chapter covers other aspects as well as the equipment used to provide the ion implantation and the effects of particle and chemical contamination on the wafers at this step.
10. *Metallization Processes.* This chapter covers both physical vapor deposition and, to a lesser degree, chemical vapor deposition techniques. Developing good contacts and maintaining high integrity metal lines are discussed as well as reliability problems with electromigration and processing problems such as hillock and void formation. Different types of metallization equipment are also covered.
11. *IC Testing.* This chapter describes the circuit testing process, in-fab electrical test procedures, wafer sort procedures, and low-yield analysis techniques. Reasonable limits of contamination, methods of detection, and impact of defects are covered.
12. *Statistical Quality Control.* In this chapter statistical process control is defined, and methods of implementation are outlined. The use of SPC in a management by exception environment to

detect trends, predict yield and output, and to schedule maintenance work is covered.

13. *The People Who Make the Chips.* This chapter covers the typical organization of a wafer fabrication area and describes the various roles of the groups within it. The prime responsibilities of each group are outlined and some ideas for resolving the numerous conflicts are given.
14. *Conclusion.* The final chapter covers the future of the semiconductor industry and discusses the impact that fab personnel make on the manufacturing process. The methods of implementation of the various process changes described in the book are reviewed, and the need for the involvment of all, from management to operators, is stressed.

Appendix A: *Chemical Toxicity and Flammability.* This appendix lists many of the chemicals used in wafer manufacturing and the guidelines to their toxicity and flammability.

Appendix B: *Statistical Experimental Design.* This appendix outlines the fundamentals for engineering experiment design. Due to the high costs and complexity of an IC manufacturing process, experiments must be designed to provide a maximum of information with a minimum of impact to the rest of the manufacturing sequence or to the output of the fab.

As a senior member of the engineering staff at a major semiconductor manufacturing firm I am constantly looking for ways to improve semiconductor processing technology. I therefore welcome the reader's insights, suggestions, or questions, which may be directed to me at 24331 Muirlands, #4357, El Toro, California 92630.

Robert Zorich

CHAPTER 1

OVERVIEW OF SEMICONDUCTOR MANUFACTURING

Semiconductor manufacturing is one of the most challenging and complicated of the new technologies that have developed in the last half of this century. The complexity of the designs of modern integrated circuits is matched by the complexity of manufacturing the devices. The specialized conditions inside the factories are far cleaner than operating rooms. Operators routinely handle multimillion dollar equipment and extremely hazardous chemicals while completely encased in protective clothing (with the exception of safety gear, this is to protect the wafers, not the people). The equipment is constantly being pushed to its operational limits as the marketplace demands denser and denser circuitry.

As a result, the rapidly evolving needs of the semiconductor industry require a thorough discussion of manufacturing techniques. While there has been a great deal published about the expectations of the wafer fabrication areas (or "fab areas"), most of it has been published by research and development engineers and scientists rather than experienced on-line process engineers. This is not meant as a slight to either party, as the job of a process engineer is to keep a wafer fab running, not to publish papers, whereas the opposite is true of the R&D scientists and engineers.

However, keeping a semiconductor manufacturing process running is an art as well as a science (sometimes viewed more as black magic than any other kind of art) and, to maximize the efficiency of the fab area, the process engineer must balance a large number of tasks, priorities, and responsibilities. He or she usually bears a significant portion of the brunt of the problem when wafers are rejected (scrapped) or when equipment is shut down for whatever reason. The consequences of mistakes, even small ones, can be great, both in cost of lost product and in terms of potential safety problems. It is of benefit to know the plant, the facilities, and the equipment well, even if it is not the main work area, to detect potential areas of danger or of defect generation. It is also important to understand the ultimate goals of the manufacturing organization, and how they differ from the goals of the process engineering organization.

The pressure on the processing people can be quite high at times, even when the risks are limited through the implementation of proper procedures. The rewards are usually very good, allowing career opportunities to move into management at a younger average age than in most other industries. In fact, as a result of this mobility, fab personnel (especially engineers) tend to move out of wafer fab processing very early in their careers. The situation in the United States is much different from that in many foreign countries, where longevity on the job is considered an attribute. Putting further pressure on the staffs of the fab manufacturing groups is the fact that the universities in the United States have been graduating fewer semiconductor engineers than are required by the industry. Given these factors, it is not surprising that there is a critical shortage of qualified process engineers in the United States. Unfortunately, the motivation for moving out of wafer fab engineering is stronger than the motivation for going into the profession. As a result of the high demand for experienced personnel, the turnover rate among the semiconductor firms has been very high. These issues have the unfortunate consequence of diluting the amount of information that is available to effectively operate a wafer fabrication plant.

Manufacturing integrated circuits is a very complex task, requiring interactions between a number of groups. The number and types of these groups may vary by fab area size, number of products manufactured, and many other factors. However, most fab areas have the following areas of responsibility. First, the manufacturing group is chartered to make enough chips to pay for everybody else. The manufacturing group consists of the operators who run the various processes and the process technicians who perform most preventive maintenance, as well as the supervisors and support personnel, such as the training department. The maintenance groups are typically separated into plant and line maintenance groups. The plant maintenance personnel are assigned to all facilities up to their delivery points in the fab, and the line maintenance group services the equipment in the fab area itself. Fab engineering is usually broken up into two areas, process engineering and product or yield engineering. Process engineers analyze the equipment and the processes themselves, making adjustments to the various parameters and verifying that specifications are being followed. The product or yield engineers control the individual product lines that are manufactured in the fab, following up on die yield changes, and implementing the changes in the process to optimize the yields. There are often equipment engineers who evaluate and purchase new items for the fab to use and work with facilities engineers on the fab's designs. Last but not least is the research and development group, who designs the chips, builds prototypes, and debugs the parts prior to the transfer to manufacturing. The transfer of a technology from an R&D environment is an intricate procedure and sometimes requires individuals or small groups to facilitate the transfer. Controlling all of this activity is the plant manager, who generally reports to a vice president.

While we will be discussing the relationships between the various groups in depth, you will see that I view the overall goal of the process engineering group to be "independent technical experts." This is an ideal which will rarely be attained, and then only for short periods. By this, I mean that the engineering group should have the fabrication area running so well within tolerance that the production and maintenance groups can run without engineering assistance (a rare occurrence in today's operating environments). However, engineers should be prepared to address problems that the other groups may come up with when faced with extraordinary downtime. In addition, the engineers are responsible for making improvements in the operation of the fab area in order to permit the ever-increasing complexity of circuit design. It should be noted that a reduction in linewidth of about one-half requires an improvement in overall factory performance of nearly an order of magnitude. Note that, due to the high cost of building new wafer fabrication space, it becomes imperative to be able to upgrade existing conditions to those required by higher technologies. The engineering staff should be able to focus more on these issues and less on day-to-day fire fighting as a general rule.

Nevertheless, the process engineer must know and understand a variety of techniques, and why they are practiced. They can be used to reduce the confusion that is felt when there are major problems with no apparent resolution. This is especially true of those who are new to the industry, or new to their areas, and need some general guidelines and directions. This book attempts to meet some of these needs.

In this chapter, we will describe the basic elements of an integrated circuit. First, we will describe the construction of a few of the basic building blocks of the chip and from there we will discuss some of the concerns of the wafer manufacturer.

1.1 THE SEMICONDUCTOR MANUFACTURING PROCESS

The complexity of the construction of today's device designs remains one of the miracles of the technological age. Most people, even many in the integrated circuit and associated industries, do not understand just how complex and critical is the process for manufacturing integrated circuits. There are discussions of the merits of building low-particulate equipment and of types and styles of automated handling systems. In too many cases, the equipment selections have been based on what looks good to individuals who have not had to run the systems and do not understand some of the fundamental issues driving the fabrication process. One of the goals of this book is to give the reader an appreciation of the complexity and precision required in the wafer fabrication process. We must start with a generalized discussion of the construction of some semiconductor devices. This discussion is not meant to be an all-inclusive look at wafer fabrication process sequences, but mostly as a guide to the various films and different structures that are used in some of the most common devices. We will build a transistor and two common types of memory cell.

First, we'll build a transistor. This type of transistor is called an *n*-type (or *n*-channel) enhancement mode transistor. There are four basic types of MOS transistors, both *p*- and *n*-type, and both enhancement and depletion mode. The essential differences of these technologies are given in Table 1-1. The combination of *p*-type and *n*-type transistors and circuitry is the basis of CMOS technology, which has the advantage of being able to operate with much lower power, and requires very little power to remain in a standby condition. There are also speed and other design benefits in the use of CMOS. We will discuss CMOS in more detail later.

Several definitions must be given at this time. These are terms that are used in some manner in every fab area. While there are variations in these definitions from company to company and sometimes even from fab to fab, we will attempt to keep our definitions consistent throughout the book. To start with, we will call the silicon pieces *wafers*. Some companies are known to call them *slices*. Groups of up to 25 to 50 wafers are placed in Teflon cassettes. They are each identified as *lots* or *runs* and have a particular *lot* or *run number*. There is little consistency in these terms, so expect to hear both used at various times. (Be careful not to confuse this type of "run" with an equipment "run.") Today, lots usually have 25 wafers, although they sometimes consist of 50 wafers. The main reason for the reduction is that better control can be maintained over lot integrity with smaller lots and, in addition, there will be less impact to fab output if the lot is somehow misprocessed.

Wafers are identified individually early in the manufacturing process. Lot numbers, wafer serial numbers, and other optional information are all cut into the surface of the wafer with a laser shortly after the first nitride deposition step. This is shown in Figure 1-1, which describes the basic outlines of the silicon wafer. The lot number will sometimes incorporate the device type, and identifying numbers often involve the work week and the lot started that week or month. For instance, lot number 4452 might be lot 52, started in work week 44.

TABLE 1-1
Essential Differences in MOS Transistor Technology (Silicon-based)

Type of transistor	Schematic	State	Side view
p-channel depletion	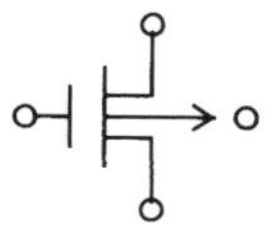 Bulk substrate	Normally on	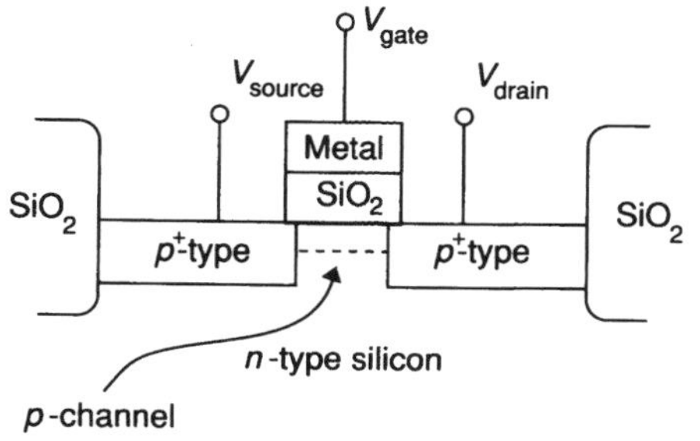 V_{gate}, V_{source}, V_{drain}, Metal, SiO_2, SiO_2, p^+-type, p^+-type, SiO_2, *n*-type silicon, *p*-channel
p-channel enhancement	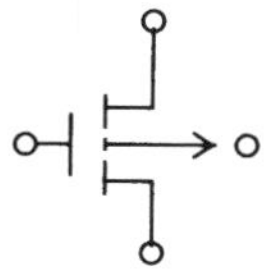 Bulk substrate	Normally off	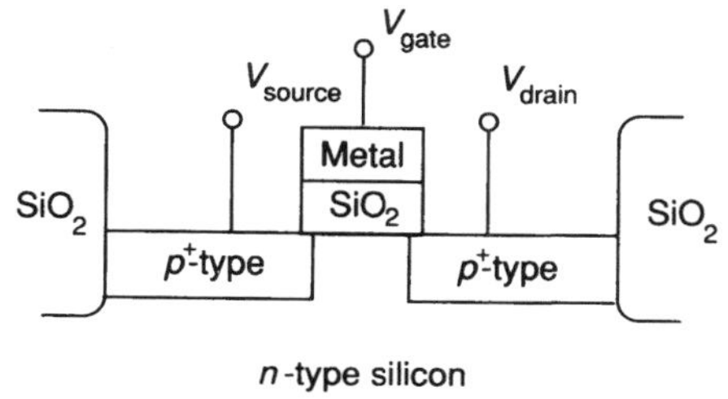 V_{gate}, V_{source}, V_{drain}, Metal, SiO_2, SiO_2, p^+-type, p^+-type, SiO_2, *n*-type silicon
n-channel depletion	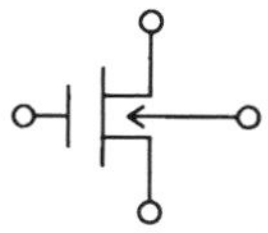 Bulk substrate	Normally on	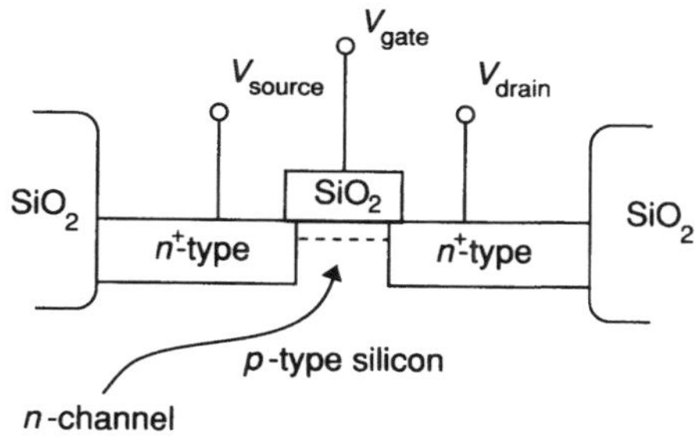 V_{gate}, V_{source}, V_{drain}, SiO_2, SiO_2, n^+-type, n^+-type, SiO_2, *p*-type silicon, *n*-channel
n-channel enhancement	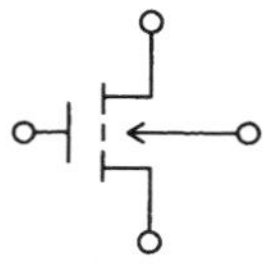 Bulk substrate	Normally off	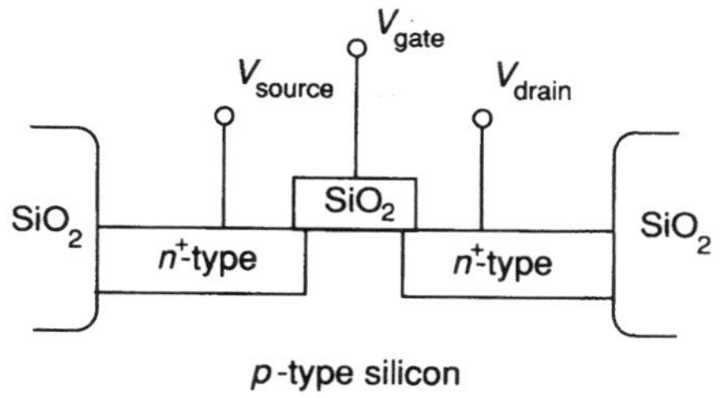 V_{gate}, V_{source}, V_{drain}, SiO_2, SiO_2, n^+-type, n^+-type, SiO_2, *p*-type silicon

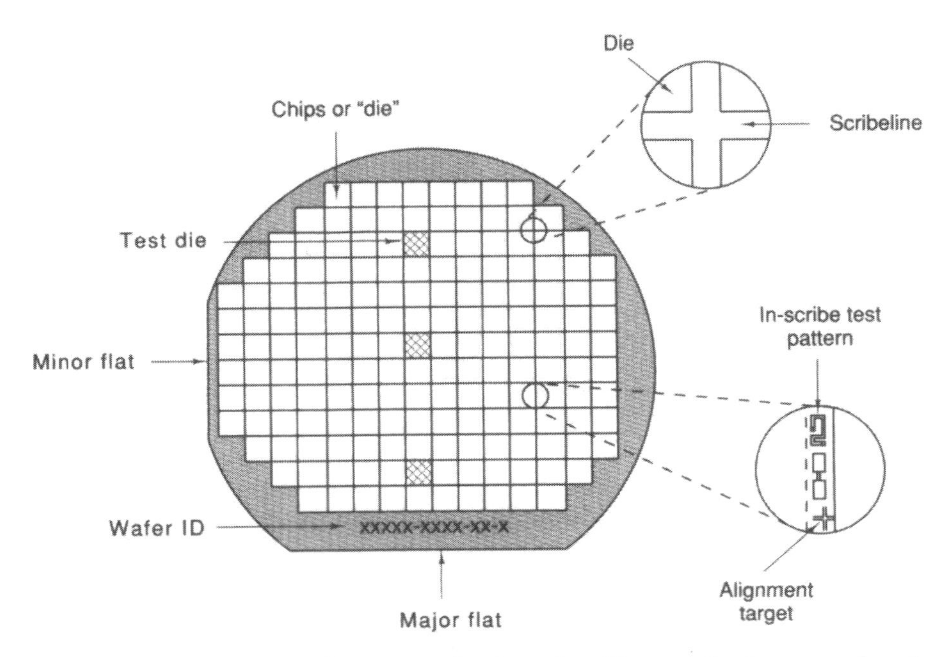

Figure 1-1. The Silicon Wafer In production, the silicon wafer has several main features, including the chips or die, a major and minor flat, an identifier, and a number of tests and alignment patterns in the scribelines.

Each wafer will usually be identified with a wafer number from one (1) to the maximum number of wafers. For example, a particular wafer ID may be 4452–1. It is very convenient to number lots in this way, as process control and wafer tracking become much easier. Wafer numbers usually have no relation to their position within the storage cassette, which is referred to as *slot position* or *slot number*. There is often some confusion about slot positions and slot numbers because, historically, many operators were taught that slot position number one was at the front of the cassette. Unfortunately, standard cassettes are manufactured with wafer position 1 being defined as the last position of the cassette, to facilitate robotics and other types of automated equipment.

When a lot of wafers is processed through a machine, it will be called a *run*, although in many wafer fabs this is called a *pass*. In the case of diffusion furnaces, there may be a number of lots making up one run or pass. In the case of a single wafer processing machine such as a wafer stepper, one run will be considered the completion of one lot of wafers. When discussing process control charts, the phrase *run number* will reference a particular time that a machine was used. We will see that, in most processes, the ability to control a process is determined through observation of the statistical results of test wafers, which are unpatterned, lower quality silicon wafers with known characteristics.

The basic starting material for the construction of NMOS parts is *p*-type silicon, which allow the *n*-type channels to be easily and clearly defined. The silicon is doped with boron to make it *p*-type, and comes with a range of dopant concentrations. This is measured by changes in the bulk resistivity of the wafer, or the resistivity through the wafer. This is to be distinguished from *sheet resistivity*, which is the resistivity along the surface of the wafer, and which is the type of resistivity usually discussed in the context of diffusion or implant operations. The silicon itself will be in the form of a thin wafer with specified dimensions, usually four to eight inches in diameter. Standards have been established for the bowing, flatness, thickness, shape of the edges, front and backside polishing techniques, and impurity concentrations of substances like oxygen, carbon, heavier metals, and so on. In addition, there are certain features of the wafer defined, such as the size of the major flat, size and location of the minor flat, and so on. In many cases, wafer fabrication houses purchase silicon substrates with a layer of epitaxial silicon on top of the standard substrate. The epitaxial silicon or *epi* is merely a very pure, controlled, single-crystal silicon deposited on the standard silicon substrate. This process is done to prevent substrate impurities or crystal damage from propagating into the surface zone, where the integrated circuit itself is constructed, thus improving yields. The cost of these wafers is significantly higher than for standard silicon. There are methods of processing that achieve similar ends without the expense of using epitaxial films.

The first step is the initial oxidation. This first oxide is used primarily to give the photolithographic people something to put a pattern on, and to protect the substrate from damage during subsequent operations. This first oxidation is usually from 500 to 1000 Å thick. The primary attributes for this film are that it remains clean and that it has a uniform thickness. Wafers must remain clean at this stage because the wafer is at its most sensitive, without having even the field isolation regions defined. Any particle delivered at this point will create a site for uneven growth of nitride in the next step, which then grows into a much larger particle. Defects created at this point often turn up as problems much later in the fabrication process, and at that point are impossible to repair. Good uniformity is required to allow the photolithographic equipment to be focused as clearly as possible.

The second step is the deposition of a layer of silicon nitride. The initial oxide and first nitride steps are shown in Figure 1-2. This layer is usually grown to a thickness of 1000 to 1500 Å. Although this layer of nitride is seldom used for any permanent structure on the chip and is ultimately removed entirely, it serves the critical purpose of protecting the active regions of the circuit during growth of the field isolation oxide. This film must have good uniformity and must remain particle-free. A particle here can result in pinholes in the nitride after the etch process, damaging the protection of the active regions and possibly

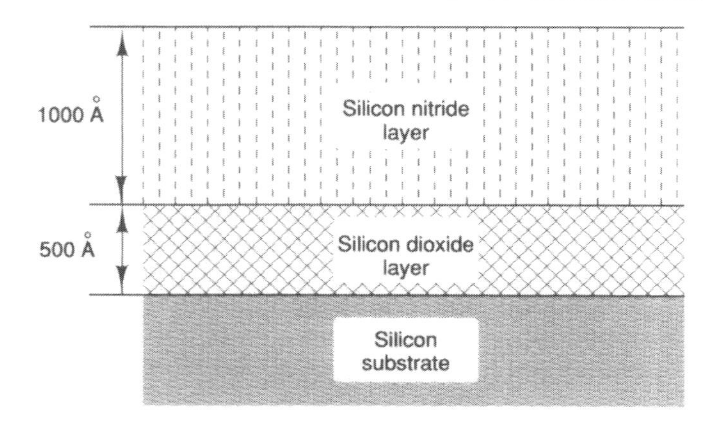

Figure 1-2. Initial Oxide and Nitride Layers The first layers placed on the wafer are the initial silicon dioxide and silicon nitride films. In this example, they are 500Å and 1000Å, respectively.

damaging the device operation itself. Finally, since nitride is so hard and impervious to chemical attack, particles that are delivered to the surface of the wafer after the nitride process is complete can usually be removed by cleaning.

It is at this point that the wafers are usually identified with a laser scriber. Using this machine before this point can be disastrous, as an incredible number of particles and slag are created. A certain amount of this material is produced on even the best of the laser scribing machines. When performed after the nitride has been deposited, there is less damage to the underlying substrates, and it is much easier to clean the slag from the wafer.

The next step is the field definition masking step. This involves depositing a layer of photoresist on the wafer and placing a pattern on the wafer (imaging). This pattern defines the areas which will contain active parts of the circuit from each other and from the areas that will be covered with the field oxides. After the pattern has been placed on the wafer, the nitride is removed from the wafer over the field isolation areas. To protect them, the nitride is left over the active areas of the chips. In many cases, there is an implant step at this point of a *p*-type dopant (boron), which is used to enhance the substrate typing. After this implant, the first oxide layer is removed from the wafers, and they are completely cleaned. Figure 1-3 shows a diagram of the field implant with a thin silicon oxide and silicon nitride film defining the edges of the implant region.

The next step is field oxidation. This thermally grown silicon oxide film is about 5000 to 10,000 Å thick. During this process, the method of oxidation becomes readily apparent. Unlike a deposited film, silicon oxide forms by the combustion of silicon. This consumes the silicon, displacing it with SiO_2, which is much less dense. As a result, the field oxide grows with a characteristic shape, called a bird's

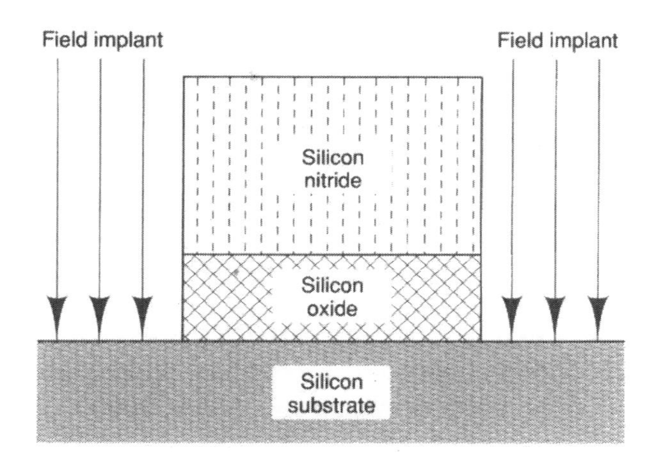

Figure 1-3. Field Implantation and Definition The initial oxide and nitride films are removed from the field isolation regions, and are left in place to protect the future transistor.

beak, due to the way the film looks at its corner, as shown in Figure 1-4. The silicon nitride prevents the oxidation of the silicon underneath, and thus isolates the areas that will later have circuits placed on them. Finally, the field oxidation process will consume a significant amount of the boron that is near the surface of the substrate, so there may be some need for replenishment. This is usually done at a subsequent ion implant step, called the threshold adjust implant.

At this point, there are a couple of different directions that the process can go. If it is a CMOS process, there will be separate regions

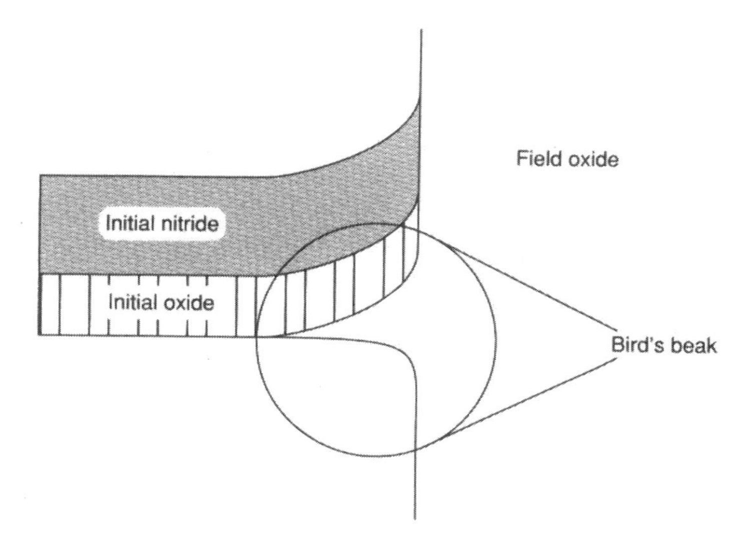

Figure 1-4. Bird's Beak When the field oxidation layer is grown, the oxidant diffuses under the nitride layer, bending it up, and forming a structure called the bird's beak.

on the wafer covered with nitride that will be used for the *n*-type and *p*-type devices. In this case, there will be one or more steps that involve a photomask step, an implant step of either phosphorus for *n*-type wells or boron for *p*-type wells, followed with very long anneal processes to drive the dopants to the required depth. This is usually done without further oxidation of the surface of the wafer. In addition, there must not be any area of exposed silicon during anneal steps, as substrate damage can easily occur and can propagate throughout the wafer. The drive processes must have the ultimate in chemical cleanliness, as these processes are often the highest temperature steps in the process. Any chemical contaminant that gets into the furnace or onto the wafers will be driven into the wafer very easily. Metallic contaminants are often the villains here. The additional steps required to make a CMOS device make the technology considerably more complex but they are not very interesting in this context, and we will not discuss them here. For our purposes in this section, we will be describing the construction of fairly simple NMOS devices so that the main themes can be shown. For reference, a pair of simple transistors, constructed using CMOS technology, are shown in Figure 1-5.

In our standard NMOS transistor, we now have a space defined for the transistor that is isolated on either side by the field oxidation, as seen in Figure 1-6. First, we grow a thin layer of silicon dioxide on this surface. This is called the gate oxide, and is typically from 250 to 450 Å thick. This oxide is sometimes as thin as 150 Å or less. For example, electrically erasable PROMs use thicknesses of around 85 to 120 Å in certain locations. The gate oxide quality is critically important to the performance of the integrated circuit. For instance, if it is too thick, the device may operate too slowly or require too much power to reach specification targets. If a particle or other contaminant is trapped in the oxide, the oxide will deteriorate rapidly and break down easily, causing early chip failure.

The next process is the deposition of polysilicon on top of the wafer. This film is made up of many individual silicon crystals, each one oriented at random. This film can be doped to a number of levels, depending on whether the polysilicon layer is to be used for a resistor line or whether it will be used for a storage cell or other type of device. Any contamination deposited on the wafer surface between the gate oxidation step and prior to the polysilicon deposition process will usually result in a fairly serious defect, and can often be associated with die yield loss later in the process. In the case of the transistor we are describing, the polysilicon will be used to control the "gate" action of the transistor. A polysilicon film structure is shown as-deposited in Figure 1-7.

The wafers go through a phosphorus deposition process following polysilicon deposition. This deposition changes both the conductivity of the polysilicon layer and the crystal structure of the polysilicon itself. The exact dopant concentration of the film is critical as

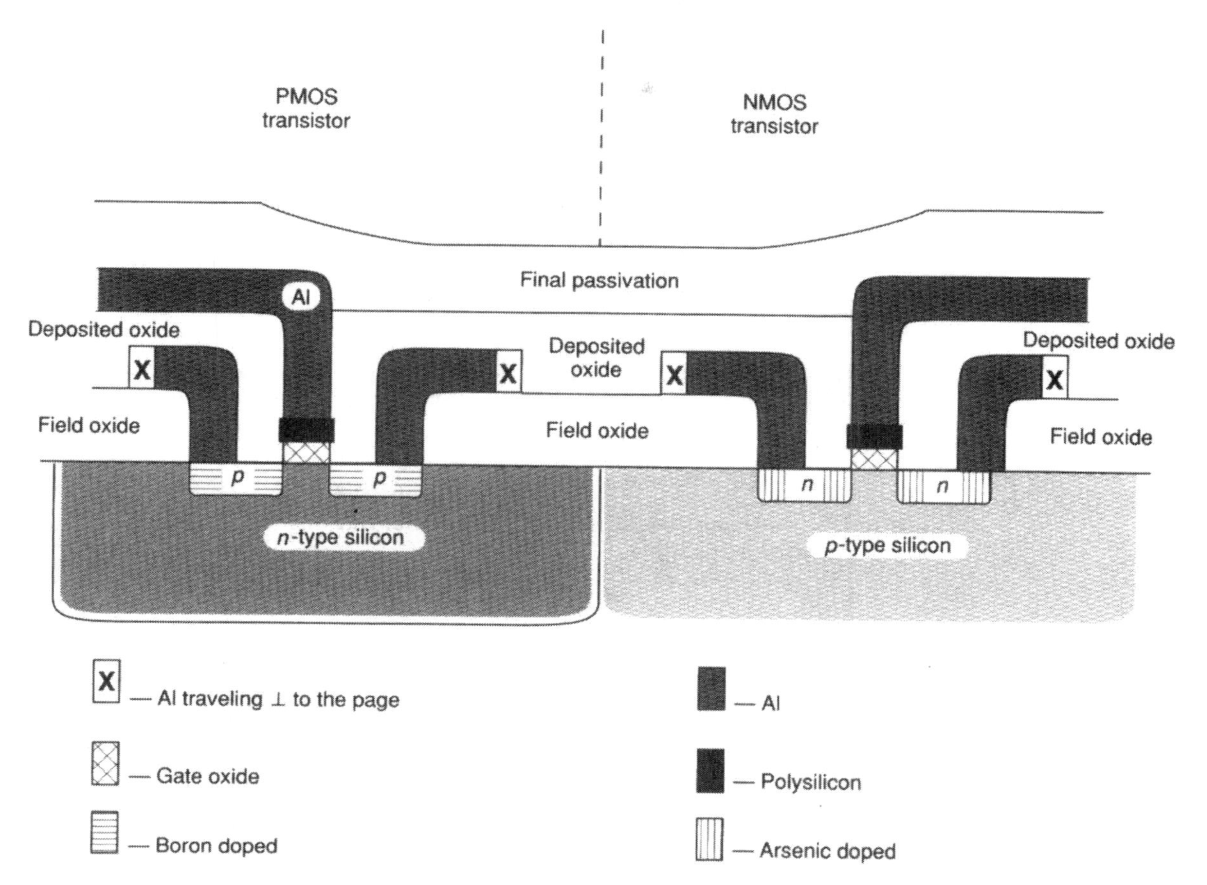

Figure 1-5. CMOS Structures CMOS devices contain structures in both NMOS and PMOS. This figure shows two transistors built side by side in CMOS. They connect to the outside world by lines traveling perpendicular to the page.

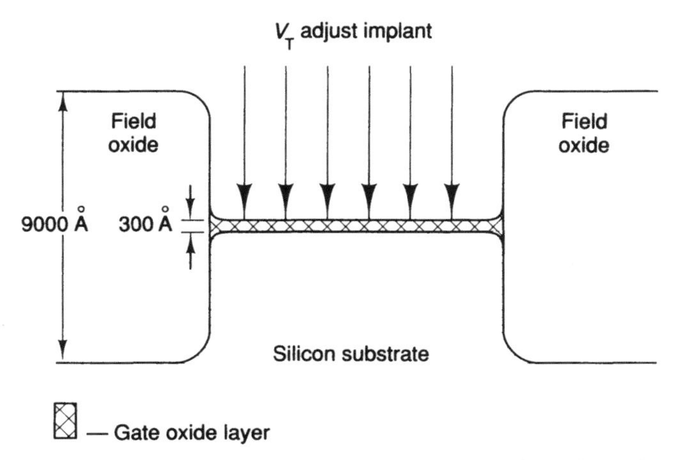

Figure 1-6. Field Oxidation Cut-Out Regions The initial oxide and nitride are removed and active regions of the circuit are exposed. They then receive an implant and a thin layer of gate oxide is grown from the silicon.

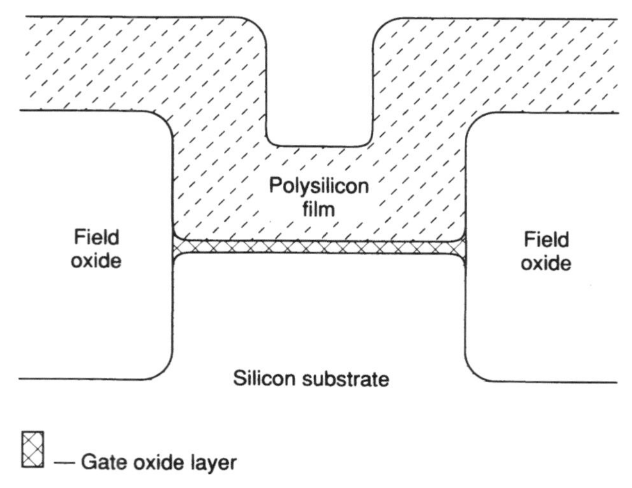

Figure 1-7. As-Deposited Polysilicon Film After gate oxidation, the entire structure is covered with polysilicon.

changes in this process can result in changes in integrated circuit (IC) processing speed, power consumption, and, ultimately, die yield. During the phosphorus deposition process, a layer of oxide grows on top of the polysilicon; it is usually of poor quality and is removed.

At this point, the substrate gate features have been defined and the gate structure itself constructed and safely buried under a layer of doped polysilicon. While still extremely sensitive, the wafers have now passed one major hurdle. In the case of CMOS designs, and of most memory chips, a second deposition process sequence occurs at this point in the process, involving patterning of the current layers, growth or deposition of an interlayer dielectric film (an insulator such as oxide or nitride), and the deposition of another polysilicon film or an advanced metal film, such as tungsten silicide. Often very complex structures are constructed at this level, allowing for quite a degree of three-dimensional design. Our very simple device requires only a single-layer polysilicon structure, and thus is easy to describe. Suffice it to say that few devices are built with as few conductive layers as our transistor.

Now we have a wafer that is uniformly coated with polysilicon. The wafer must now be sent to the photolithographic area, where the polysilicon film is patterned and etched. As shown in Figure 1-8, we can see that the top portion of the gate is first defined at the polysilicon etch process, which clears away the excess poly. This step may be followed by an oxide etch, which removes the oxide from source and drain regions of the transistor. Sometimes this etch is performed at a later point in the process, depending on the method of poly/metal interface structure that is required. In any event, it is necessary to remove any thermally grown oxide from the areas to be used for contacts prior to the start of the metallization process, as the deposited

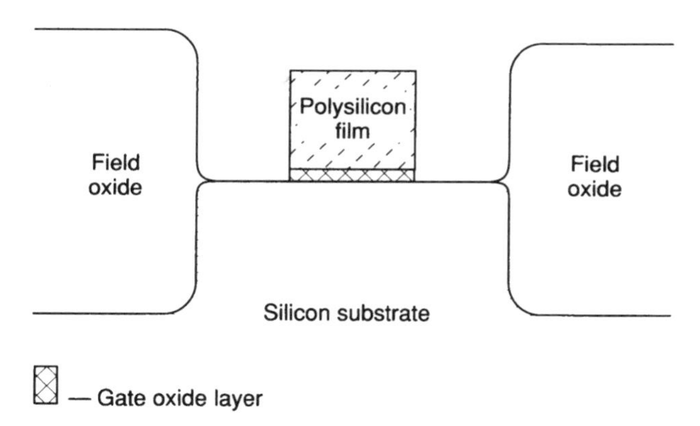

Figure 1-8. The Self-Aligned Silicon Gate The polysilicon is etched to define the top of the gate structure and the SiO_2 is then cleared to open the source and drain regions. This procedure self-aligns the gate position within the transistor.

dielectric films etch at a much higher, and incompatible, rate than the thermally grown silicon oxide films.

With many devices, it is necessary to cap the final polysilicon layers with a layer of thermally grown silicon dioxide. This layer of oxide comes at the expense of the layer of polysilicon film, as the growth of the oxide consumes some of the polysilicon. If this is required, there would not have been an oxide etch in the previous step, and there would be an oxidation step, followed by a masking step, culminating with an oxide etch step, in order to define the source and drain regions. For our simple device, this extra oxidation sequence is not required.

Now that the source and drain areas of the transistor have been defined and exposed, it is time for us to make these regions electrically active. This is done by coating the wafers with a fairly thick layer of photoresist, and then implanting the surface of the wafers with an *n*-type dopant. The usual dopant is arsenic. After the implant is complete, the photoresist is removed and the wafers are annealed at high temperature. This causes the arsenic to diffuse into the regions of interest with characteristic well shapes, which are seen in Figure 1-9. The distance between the wells is called the effective channel length (L_{eff}) of the transistor. This is to be contrasted to gate width, which is the width of the polysilicon gate. The annealing process not only drives the arsenic to its proper locations, it also frees all of the available electrons in the arsenic for use in the device. Any oxide that may be grown in the source and drain regions must be removed prior to any further processing.

It is now time to cover the substrate devices with the first layer of protective glass. This structure is shown in Figure 1-10. This glass is deposited using a variety of deposition techniques, and is much less dense than thermally grown silicon dioxide. Chemically, it is the same, SiO_2. The glass can be doped with a variety of agents. This is not

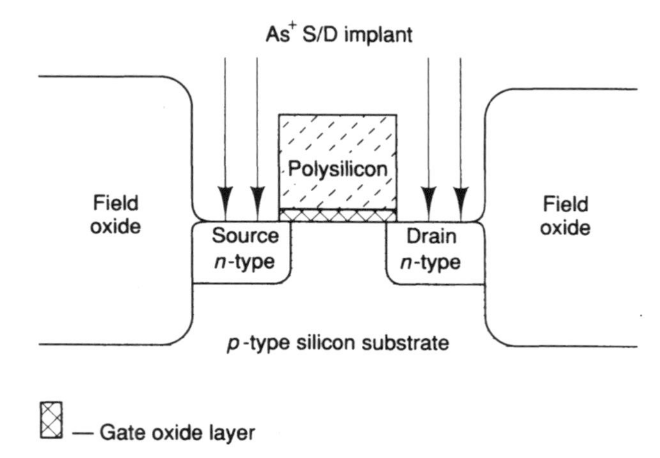

Figure 1-9. Arsenic-Doped Source / Drain Regions Arsenic is implanted into the source and drain parts of the transistor, creating conductive regions within the substrate.

to change its electrical properties, as the glass is an insulator, but instead is used to cause the glass to melt at lower temperatures and to flow more smoothly over steps. The typical dopants used are phosphorus and boron, usually in the forms of P_2O_5 and B_2O_3, along with other combinations. The concentrations used are quite high, ranging

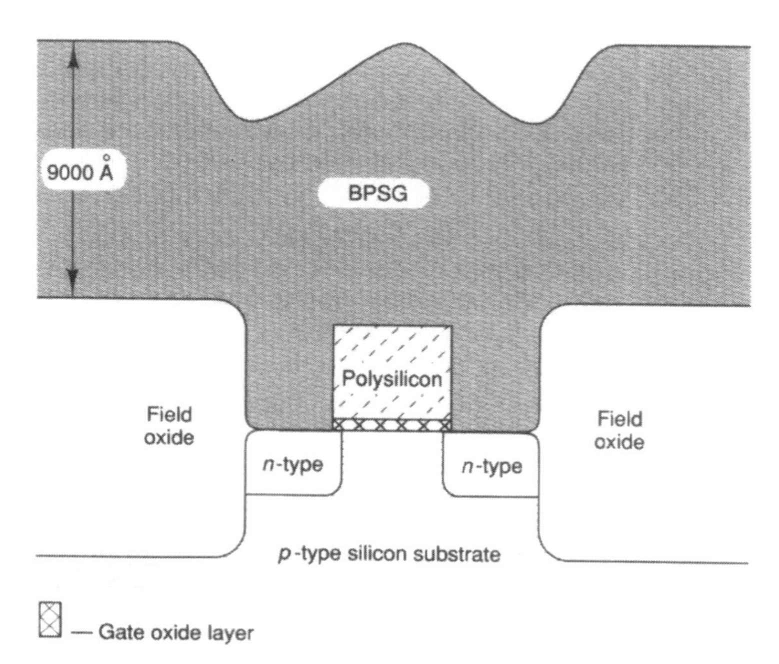

Figure 1-10. BPSG-Coated Transistor Now that the basic transistor structure is complete, it is coated with a layer of deposited silicon dioxide or BPSG for insulation and protection from the environment.

as high as 6 or 7 weight percent of dopant in the film. After the glass is deposited, it must be annealed, where it liquifies somewhat, flowing down over the steps uniformly, and becomes smooth and more dense. For our device, we will assume that we are using a film of BPSG (or boro-phospho-silicate glass), a deposited glass that contains both boron and phosphorus.

After this layer is put on, contact holes must be punched through to the source, drain, and gate regions to allow the device to work. This is done by placing the appropriate image onto the BPSG layer and then etching the oxide down to the surface of the wafer. This is usually done with a plasma reactor, as these contact holes must be very precise. Two of the crucial aspects of contact etching are that all of the oxide must be removed cleanly from the surface of the silicon, and there must not be any sideways (or lateral) etching (called *undercut*), as that will cause the contact to become too wide and could cause a short between the source and the gate or the gate and the drain areas. A view of the contact area is shown in Figure 1-11.

Once the contact holes are etched and the resist removed, the first metal layer is placed on the wafer. In most cases, a number of metal layers will be placed onto the wafer, at least two and often three, but we will use only one. This metal layer is usually aluminum, and may be mixed with about 1 to 1.5% silicon, and possibly some copper. This metal must be cleanly deposited into the holes with no cracking or gaps left between the metal and the surrounding parts. It is very critical that a good contact be made in all of the proper locations. After the metal has been deposited, it is tested to verify that it is of the correct conductivity. The next step is to alloy the wafer to guarantee good contact between the substrate and the aluminum at the contact point. This is usually done at a fairly low temperature (not greater than 425°C), in an atmosphere of some 15% or more of hydrogen.

At this point the metal pattern is placed onto the wafers. This allows the metal lines to be hooked up to other devices on the chip or

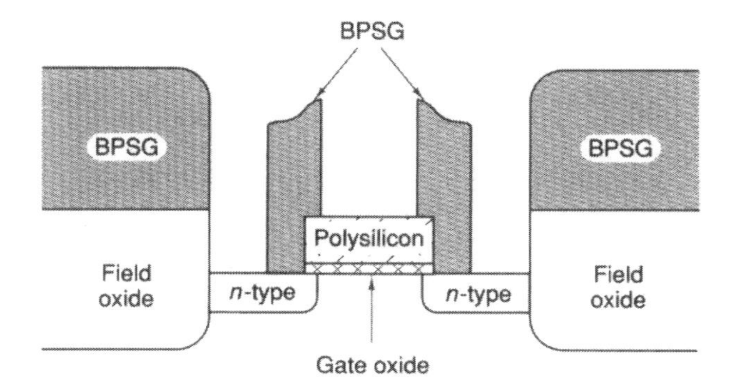

Figure 1-11. Open Transistor Contacts To connect the transistor to the outside world, contacts are cut into the BPSG film.

to contact pads to connect to the outside world. In our design, we are assuming that the lead for the source of the transistor comes in from the left side of the page, and that the drain lead goes out toward the right. The gate lead itself is a line that is perpendicular to the plane of the page that is connected to a device above or below the level of the page that controls the switching action of the transistor. The gap between the aluminum lines is critical and must be maintained or short circuits may occur that will cause the chip to fail. The aluminum deposition step is described in Figure 1-12.

Now that we have built the transistor, we come to the protective layers of the device. There are a number of films placed onto the surface of wafers to protect against various damaging components of the environment. First, another layer of glass is placed onto the surface. Sometimes this glass will be doped but, in many cases, the fab area will use multiple oxide films with the last oxide layer being undoped. A film of silicon nitride is usually deposited on top of these glass layers. This is a very hard substance that will protect the chips from scratches, chemicals, and other damage. Sometimes a layer of polyimide is placed on top of this structure to help reduce potential radiation damage (e.g., for parts used in spacecraft). After these final layers, often called passivation layers, are placed on the wafers, the bonding pads must be etched out so that the device can contact the outside world. This final appearance is shown in Figure 1-13.

Finally, the devices are tested for both electrical performance and overall yield. Bad chips are usually marked with a red dot after testing, and these chips are later thrown away. There may be lifetime tests or other unusual electrical tests performed on the devices while they are still in wafer form since, after this point, wafers will be broken up and each chip will have to be handled individually. Wafers that have more

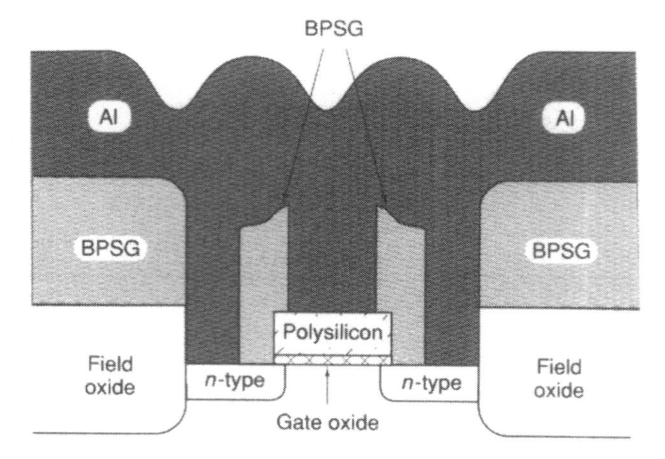

Figure 1-12. Transistor after First Metal A layer of aluminum-silicon alloy is deposited over the open contacts to form the connections.

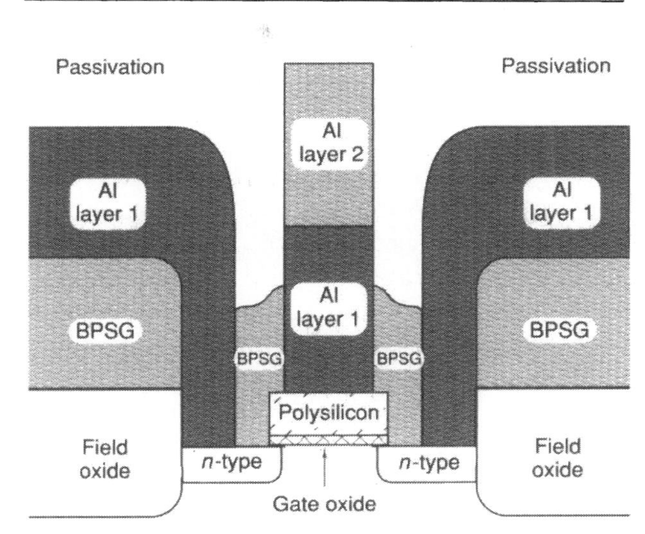

Figure 1-13. A Complete NMOS Transistor The first aluminum layer is etched, then another passivation and aluminum layer is added, and this simple transistor is complete.

than a certain number of failures are often removed at this point and sent back to engineering for analysis. Testing wafers that have failed often gives insights into ways to improve the overall process and fab yield.

The wafers that have survived up to this point are now broken up into their separate chips. This is done by carefully sawing (usually called *dicing*) the wafers apart. Chips with red dots on them are thrown away and the rest are carefully taken to the packaging areas, where the chips are glued into small plastic packages (sometimes the packages are made of other materials), and then tiny wires are soldered onto the bonding pads of the chips and onto the pins of the package. This is a tedious and delicate process that up until a few years was done entirely by hand. A significant amount of this "die-bonding" is still done by hand, although most of the advanced manufacturers have automated their die-bonding operations. After the chips are placed securely into the packages, they are sealed and the chips undergo final test. Those that survive this step are stamped with a logo and shipped to the customers. The rest are thrown away.

It is clear from the number of steps involved and the number of ways that wafers can be damaged that huge losses can be incurred during the manufacturing process. This is precisely why the motivation for manufacturing high-quality wafers is so high.

Now that we have built this transistor, just what does it do? Probably everyone has heard the word transistor, but few know how one works. A transistor is first and foremost a switch. In the case of the chip we have designed, when there is no power to the gate, the switch is off and no electricity can flow through the switch. When there is power applied to the gate, the switch is on and electricity can flow through the transistor. Specifically, what happens is that power is placed into the source region on the left where it is stopped by the edges of the well. The drain region is connected to another device, let's say a small light-emitting diode (LED) light. While there is no power applied to the gate region of the transistor, no power can travel to the light, so it remains off. When a positive voltage is applied to the gate, however, electrons are pulled up to the surface of the silicon at the gate oxide interface (hence the term *enhancement mode transistor*). This allows a small channel of *n*-type material to form, which will allow current to flow though. Thus, electrons from the source region can traverse across to the drain region, where they travel out of the device to the LED, which then lights up. The minimum voltage that must be placed onto the gate region to cause this reaction to occur is called the *threshold voltage*, and is a critical term in discussing integrated circuits. The transistor we have built is very simple, but if we were able to place thousands of them on a chip in the correct sequences, we would be able to build very complex logical devices, including microprocessors. With some minor changes in design, it is possible to build transistors that will act as power amplifiers by allowing current to flow from the gate to the drain of the transistor. While we will not spend much time on this application, it is important to note its value to the electronics world.

While we have designed and constructed the transistor, which is one of the basic building blocks of semiconductor technology, it is clear that there are many other types of structures and devices, ranging from random access memory (RAM) cells to electrically programmable read-only memory (EPROM) cells to complex transistor-capacitor-resistor networks that can provide virtually any type of data manipulation and storage. Virtually all of the major electronic components that are designed or constructed in the "regular" world can be reproduced in the micro world, with some ingenuity.

As an example of the structure of a more complex device than our transistor, we will now take a quick look at a completed EPROM cell, as in Figure 1-14. We can see, looking from the bottom of the device up, that there are some extra wells that come in under the device in the substrate. These wells are often used to sense the state of the EPROM. For instance, if the cell is charged, the transistor will be in the "on" state, and can be counted as a bit of "1." (In real life, the EPROMs use inverse logic, so that if a charge is on the cell, the bit is "0.")

Above the substrate we see the gate oxide, with an unusual block of polysilicon (poly I) above it. This block of polysilicon is the memory cell itself, with either the existence or lack of a stored charge representing a bit condition. This polysilicon gains the charge through "erasing" the chip with ultraviolet light (this is why the designers use reverse logic when designing EPROM chips). When the UV photons

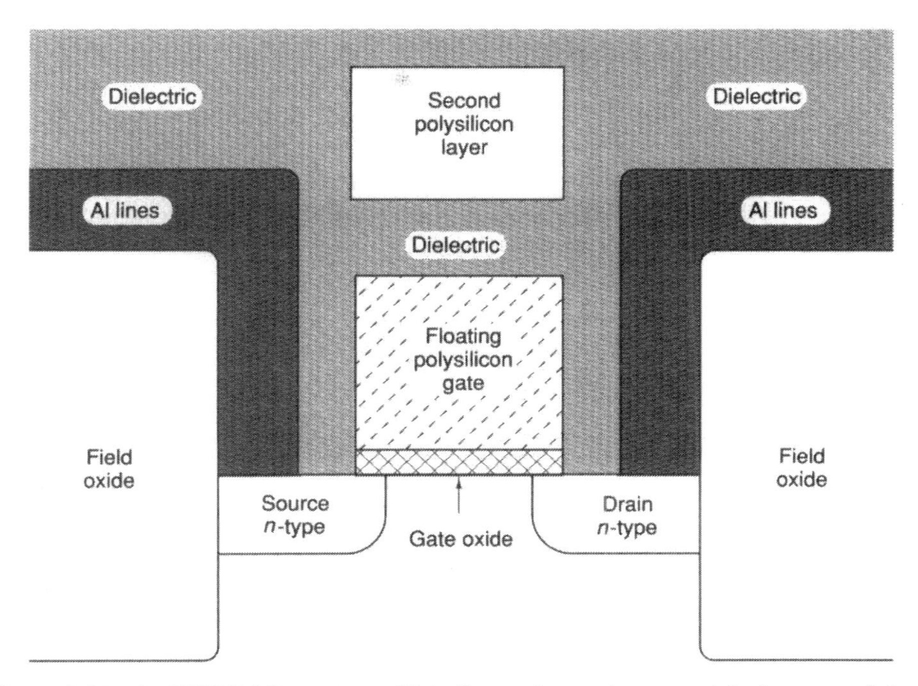

Figure 1-14. An EPROM Structure This figure shows the essential elements of the floating gate EPROM cell. Charge is developed on the floating gate by exposure to UV light. This charge can be drained off the gate by applying a strong voltage to the drain region. Memory state is determined by whether or not the transistor under the cell is on or off. In this figure, the transistor will be on when there is a charge on the floating gate.

strike the polysilicon, electrons are freed; they are then trapped in the polysilicon layer. After the light is removed, these charges remain trapped for up to an estimated 100 years. To program the EPROM, a high voltage is placed across the memory cell, allowing the electrons to be removed from the polysilicon and the structure to go to ground state.

The high voltage is applied from above (often partly in conjunction with the drain of the sense transistor) in the next layer of polysilicon. The voltage pushes the electrons into the drain of the underlying transistor, which causes damage to that oxide. For the first several dozen to several hundred programming cycles this is not a problem, and in most cases, EPROMs are programmed only a few dozen times. (In fact, as a percentage of sales, most are programmed only once.) After this, however, the probability of failures increases rapidly.

The second layer (poly II) must be separated from poly I by a strong insulator, and often a combination of thermally grown silicon oxide and deposited silicon nitride is used to separate the two polysilicon films. This portion of the device can used to boost the output (and thus the longevity) of the memory cell, and is the primary contact that the memory cell has with the outside world. This second layer of polysilicon is covered entirely in dielectric films, although these films do not need to be of the strength of the interpoly dielectric films. Incidentally, the second poly film is sometimes replaced with one of a number of metal-silicon combinations, such as titanium silicide. In some designs, there is yet a third polysilicon or silicide film used, often for memory cell addressing purposes.

After the second poly film is put down, there is a layer of BPSG and then two layers of metal, separated by BPSG or other types of glass insulators. These lines are usually made of aluminum, and are typically used to control the function and addressing on the chip. The memory cells are placed onto the chip in a two-dimensional array. Two wires are required to track down the exact bit, the row address line and the column address line. These are represented by metal I and metal II. In some cases, additional circuitry will require a third metal layer to be deposited on the wafer. Finally, the last protective passivation layers are deposited on the chip.

Typically, devices do not have more than five or six conductive layers on them, simply because the ability to manufacture the devices becomes incredibly complex. However, the researchers and designers are even getting around this, depositing another layer of epitaxial silicon on top of the existing circuits and then building another layer of circuitry on that. This is not yet production-worthy, but could eventually lead to new and incredibly dense and complex parts.

The other devices that are manufactured use similar techniques to achieve operation. For example, we have the dynamic RAM (DRAM) chip, which is used in large quantities in nearly every computer system. This memory cell consists of a capacitor that has a charge stored on it. See Figure 1-15 for a description of the DRAM memory cell. The existence of the charge is detected in much the same way as the charge is detected in the EPROM cell. The problem with a DRAM cell is that all of the capacitors on a chip must be recharged several dozen times a second. For a one-megabit DRAM, this means that over one million capacitors must be recharged per chip per second. Obviously, this takes significant amounts of time and power, and also tends to limit the maximum data transfer rates of the system.

The static RAM (SRAM) chip is simply a chip that has several transistors designed into a loop that, once put into a certain state, will remain in that state until it is intentionally changed or until the power is turned off. This type of design allows for the cell to be accessed rapidly, requires no refreshing of capacitors, and requires only a little power. The disadvantage of the static RAM design is that it takes up much more space per cell and thus cannot be designed to hold as much memory as a DRAM. These chips usually respond very quickly, sometimes as rapidly as 15 nanoseconds.

The electrically erasable PROM (EEPROM) is essentially an EPROM cell that does not use ultraviolet light to charge the polysilicon cell. Instead, the gate oxide has a region in it that is cut down to

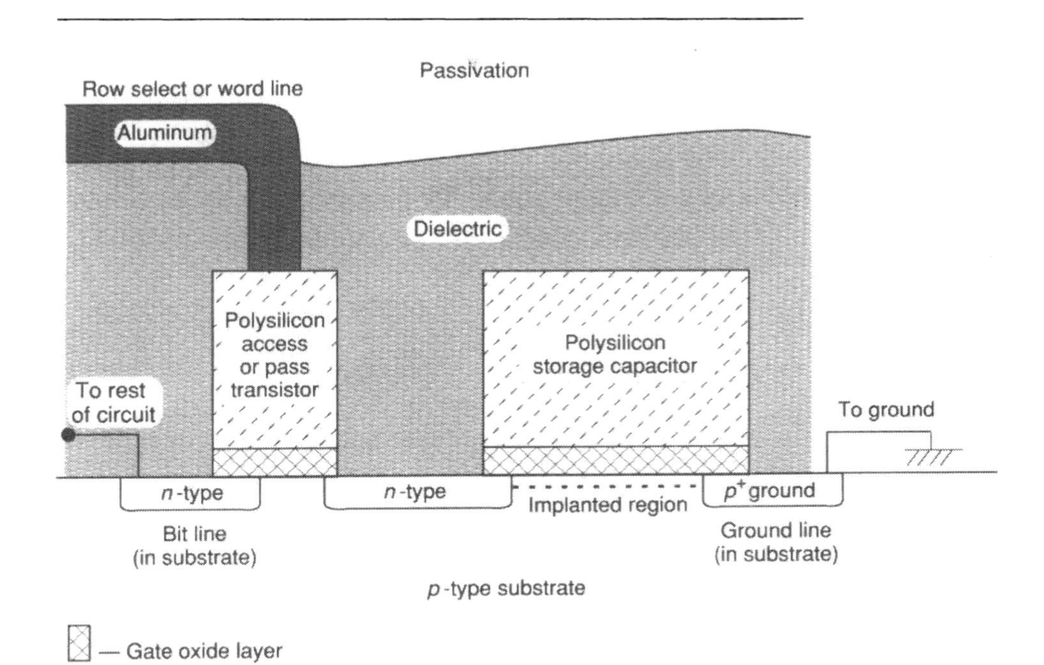

Figure 1-15. A DRAM Structure Shown is a simple single-transistor dynamic RAM memory cell. Charge is placed on the capacitor where it momentarily holds the depletion region in an inverted state. This is detected and decoded as a bit. The charge drains off quickly, so it must be replenished repeatedly every 100 milliseconds or so.

around 100 Å, which allows for a certain amount of quantum mechanical tunneling of electrons to take place for charging and discharging the poly I layer. While this does cause some damage to the oxide layer, it is far less damaging than the technique of hot electron injection used in EPROMs. The EEPROM can be erased and reprogrammed up to 100,000 or more times. EEPROM cells can be used for a number of other functions also, including control of programmable logic devices and others.

Microprocessors and their peripheral chips are mostly combinations of these building blocks, with transistors for control, static RAM cells for registers, and so on. The genius of the microprocessor revolution has been in the design and implementation of relatively few basic building blocks.

As can be seen, the design and construction of integrated circuits can be rather complex. However, when broken down into constituent parts, and when the proper framework for manufacture in the fabrication area is employed, a certain amount of control can be obtained over the wide number of variables. Attaining and maintaining this control can permit the wafer manufacturer to build chips with some hope for profit and success.

1.2 BASIC REQUIREMENTS AND DEFINITIONS

There are a variety of problems that are inherent in nearly all aspects of the semiconductor industry. They include items such as defect generation and propagation; control of purity of all of the chemicals, gases, and air used in the fabrication area; and the degradation of wafer quality and yield as a function of time—putting time limits on how long wafers can sit between steps, for instance. As a result, significant areas of focus for the manufacturing operation include defect reduction, throughput time improvement, equipment downtime reduction, and effective inventory management. These tasks are usually distributed among several groups, but it is important for everyone to have an understanding of all of the various roles. For now, we will focus on the more technical aspects of the engineer's role, and will discuss some of the basics of manufacturing quality control at the end of the section.

1.2.1 Linewidths

One of the most "visible" definitions in terms of impact to the industry, and receiving the most publicity, is that of linewidth. This is the source of the widely publicized phrase "x.x micron technology." The phrase has been distorted in recent years, usually for marketing reasons, to include "gate width" or "channel length" measurements. *Linewidth* quite literally is the width of the connecting conductor lines on the circuit, as shown in Figure 1-16. An integrated circuit's technology base is defined by the minimum feature sizes on the chip,

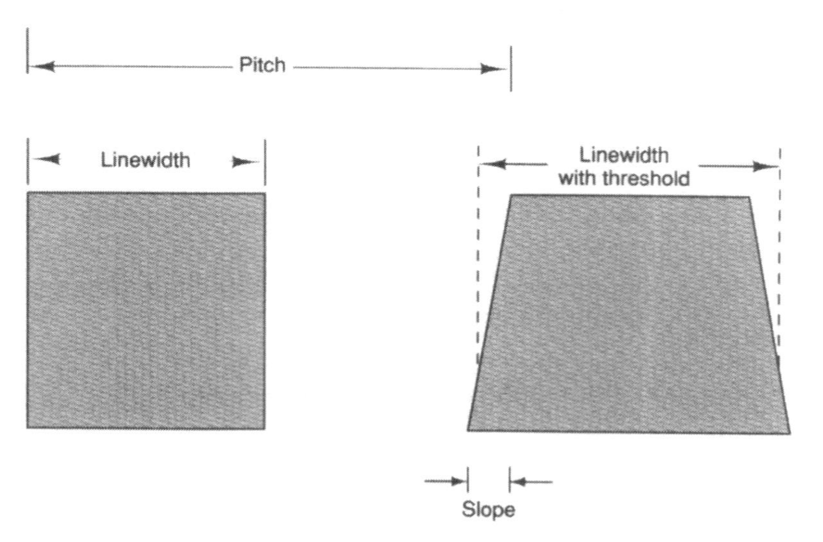

Figure 1-16. Linewidth Definitions Two 1-μm lines are shown, one with vertical sidewalls and one with positively sloped edges.

which are often found on the uppermost metal lines (but can be found in other structures as well). The minimum feature size is incorporated into the technology design specifications (or design rules) that are created by the research group when the device is created. The design rules usually keep most spaces, lines, and other support circuits about 33% larger than the minimum feature sizes. Minimum linewidth is defined by the photolithographic group, which is limited by the resolution of the imaging devices available to them. Modern step-and-repeat devices can obtain resolutions to about 0.5-μm linewidth, although a variety of instruments are appearing on the market that are capable of producing 0.35-μm lines.

As this book is being written (1990), the most common linewidths are between 1.0 and 2.0 μm, although there are many devices produced on both sides of that range. In the 1970s, integrated circuit technology was in the 10-μm linewidth and up range, which reverted to about a 2.0 to 2.5 μm "standard" through the early and mid-1980s, and has now progressed to the 1.0 to 2.0 μm scale. By 1993, it is expected that 0.35-μm linewidths will represent the leading edge of manufacturing technology. Reduction in linewidth scale is occurring even more rapidly than previously, because a wide variety of new devices are now possible at the higher chip densities, and there is a large-scale migration of older parts to modern technology, due to the massive improvement in die yield available on the tighter geometries. To give an idea of the increases in density available as a result of the decrease in linewidth, a very simple look at the increase going from a 5-μm to a 1-μm linewidth results in an increase of the number of contacts (Fig. 1-17) available by nine times.

One of the most critical components of the linewidth decision is that of how much power must be run through the line. This gets into the technology being used (i.e., CMOS, bipolar, etc.), and the material

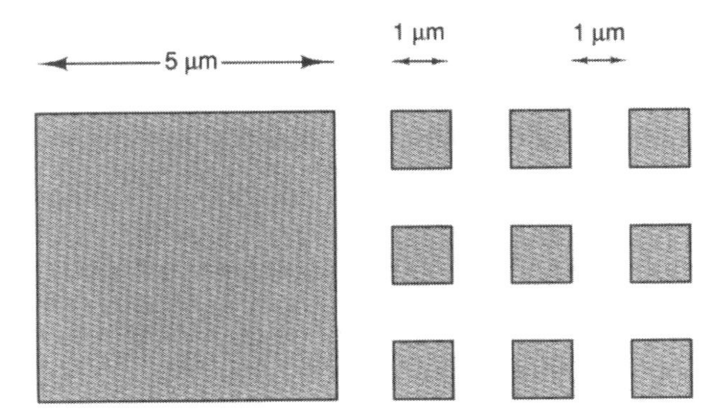

Figure 1-17. Contact Density as a Function of Reduced Linewidth Nine square 1-μm contacts can be placed in the area of one square 5-μm contact.

being used to carry the current. An excess amount of power sent to an element of the chip that cannot handle it is likely to result in failure of that element. Unfortunately, if a line is carrying that much power, it is probably important; therefore, a failure may well mean the failure of the entire circuit. However, as technology has improved, the power required to run the various components (transistors, capacitors, and the like) on the chips has been reduced at a rate that keeps the designers ahead of the various materials' current-carrying capabilities. These requirements have led to some exotic metallization processes. There have been a few times in the last 10 years where one group or another has stated that we cannot go below a certain linewidth due to the structures of the materials and, so far, wafer fabrication scientists have found new ways to cheat these predictions!

However, there are real practical limits to the minimal linewidth issue. The primary issues are: What is the maximum size at which a chip can be produced profitably, and will it cost more to shrink this IC down to the next level (due to increase in manufacturing and equipment costs) than the increase in density will make back in a reasonable time frame? At some point the answer to the second question will become obvious. For instance, it makes as little sense to build 19-cent transistor-transistor logic (TTL) parts on 0.5-μm technology, as it does to try to build the 80486 or 68040 at 5-μm levels (assuming it was possible), since, in the first case, it could cost several hundred million dollars to produce the TTL parts, which, at a profit margin of 1 cent per chip, would require sales in excess of 10 to 15 billion ICs just to pay for the plant. In the second case, the microprocessors at 5-μm technology would allow only a couple of ICs per wafer, and would result in chip costs of $5,000 to $10,000 for each microprossor just to break even in manufacturing.

Clearly there is a spacing requirement around each contact; in Figure 1-17 it is 1 μm. Whenever a line size is specified, there must be a space defined around that line. While that may seem obvious, at the microscale of these devices line-spacing issues become very critical. A deviation of a few tenths of a micrometer can cause an unexpected electric field distortion which can lead to an increase in the failure rates of the dielectric insulating two adjacent lines. The problem is pointed out in Figure 1-18. The spacing requirements of an integrated circuit are as tightly defined and as potentially critical as the line sizes.

Notice also that, if round contacts are used, a higher density may be possible. Square contacts, while still common, are often avoided as sizes decrease, due to the space restrictions, manufacturability issues (such as contact hole etching), and due to somewhat unpredictable electric field generation at the corners of the contacts. Sharp points in a structure can lead to localized increases of electric fields which can lead to wear-out mechanisms that will affect the long-term reliability of the devices.

The term channel length has sometimes been used in sometimes

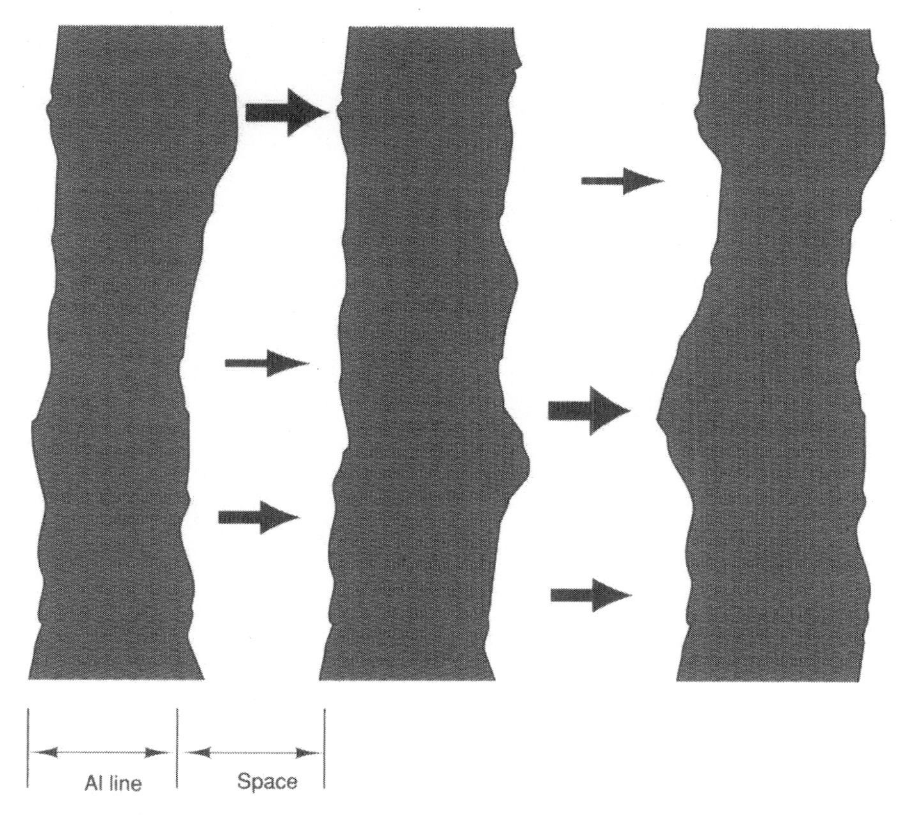

Figure 1-18. Variations in Electric Fields due to Rough Lines Variations and roughness of the current-carrying lines can cause variations in electric field densities across the intervening spaces. These variations are exacerbated by defects in the dielectric between the lines.

misleading ways. Some manufacturers have described their technology as "x.x micron" technology based on channel length as opposed to linewidth for the definition. This has been done primarily to impress potential customers with the technology used in the devices. Some of these manufacturers have given relatively good arguments in favor of the use of this terminology. Nevertheless, it is wise to be aware of the differences in the terminology and to read the fine print. The *channel length* is the distance under the gate regions, between the source and drain regions of the transistors on the devices. The main impact of shorter channel length is that the device will typically respond faster. A description of channel length is shown in Figure 1-19.

The specific way a technology is designed, which linewidth is specified, what shape and type of distribution for contact, line and hole spacing requirements, and so on, are included in what we have referred to as the design rules. While there is often no general reason why some of these rules could not be violated, they are built in as a rigor to thoroughly guarantee device reliability and repeatable device performance. The design rules also incorporate specific allowances for

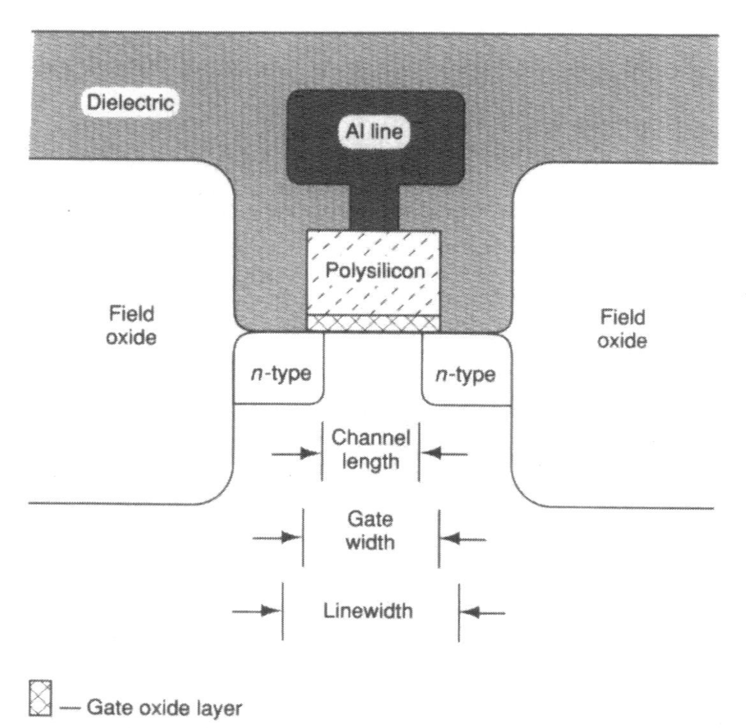

Figure 1-19. Definitions of Some Key Parameters We can see that the channel length, gate width, and linewidth definitions are all quite distinct. While the channel length is shorter than the gate width, there are no restrictions on the relative sizes of the gate width and minimum linewidth.

film thicknesses, power handling requirements, redundant circuitry, and others. As noted, these rules are built into the technology when it is designed in research. This raises an issue sometimes when process technology has gone beyond the technology of a certain design, or if the design expected certain processes to exist that have not become manufacturable. In these instances, exceptions may be made to the rules, as much as the designs will allow. Very often, violations of the design rules will result in significantly reduced yields and device performance.

1.2.2 Film Types and Characteristics

The semiconductor device itself is made up of a number of layers of varying types of thin films. These films may be dielectric in nature, used for insulation or capacitive storage, or they may be conductive, allowing for free or controlled current flow. As many as 10 or 12 layers of various films may be piled up on a particular device, although that would make for a complicated structure. A variety of problems exist in film control. They are amplified by the continued reduction in designed film thicknesses for the transistor gates, and for use in other structures. In many cases, the gate oxide layers are only 100 to 150 Å thick. Consider that silicon dioxide has an average bond length of about 1.5 Å, and you get the idea that we are not talking about many layers of molecules. Some technologies use even thinner films. At the other end of the spectrum are the thicker films, which must handle the insulation load, and which must be capable of being smoothed enough to allow proper deposition of succeeding layers. Since most deposited films are rough and bumpy as-deposited, this can become a serious issue. The variety of films are displayed in Table 1-2.

Requirements are placed on all of the conductive metal layers that are deposited, including control of film grain structure, uniformity of resistivity, film performance when built into complex topographies (such as the requirement that films do not crack when deposited over high steps), and electrical performance over time, as typified by phenomena such as electromigration in aluminum films.

All of these films have a variety of common problems, including those of defect and particulate generation. Film creation or diffusion

TABLE 1-2
Films Used in the Manufacture of Integrated Circuits

Film type[a]	Use	Process	Maximum temperature[b] (°C)	Oxidize[c]	Layers[d]	Doping[e]
SiO_2	Gate dielectric	Grown	>1400	—	2–4	No
	Insulation	CVD sputter spin-on	>1400 (densifies)	—	2–4	No
Si_3N_4	Insulation Passivation	CVD	>1400	No	1–3	No
Poly Si	Capacitors	CVD	>1000 (changes structure)	Yes	2–3	Yes
amorphous	Conductors Gates			Yes	2–3	Yes
Aluminum Al alloys	Conductors	Sputter	~450	No	2–3	No
Tungsten	Vias, contacts Barrier metal	CVD	>1400	No	1–2	No
Silicides	Conductors Vias	CVD Sputter	>1400	Some	1–3	Yes
SiO_xN_y	Passivation	CVD	>1400	No	1–2	No
Polyimide	Passivation Reduce radiation	Spin-on	200–300	No	1	No

[a] Assumes typical multipoly/multimetal RAM and microprocessor IC.
[b] Maximum temperatures limited to 1400°C, simply because few processes exceed this value.
[c] Oxidize means that the film itself will oxidize (some permit diffusion).
[d] Layers means number of times film type is used during manufacture of the typical IC.
[e] Doping means that the material is electrically changed through addition of impurities.

processes, especially the film deposition technologies, have the maximum potential for defect contribution. After all, chemicals are being delivered through space to the surface of the wafer, where a particular chemical reaction must take place. Any impurity in the delivery of the process materials will probably result in the creation of a defect somewhere on the surface of the wafer. Film deposition materials also have the nasty habit of finding contaminants left behind on the wafers from previous steps, upon which they deposit more rapidly, forming certain types of defects. One of the most common of these problems is "haze," usually traced to residual moisture on the surface of a wafer.

Particles can result in a number of different failure modes, including immediate failures on power-up (also called infant mortality), when the particle or defect causes either a short or an open circuit somewhere on the device. Other failures can be stuck bits or other memory array-type failures. Additionally, small particles can lead to reduced lifetime and increased numbers of reliability failures as a result of slow degradation of the electrical properties of the films.

Finally, defects in the deposited films, especially the protective coatings or "passivation" layers, can allow the penetration of the elements (air and water) into the device structure, particularly moisture, which can react with the various chemicals in the device and cause breakdowns.

As technology has progressed, the allowance for defects has decreased by several orders of magnitude. The tolerances that were allowed 10 years ago would not allow many devices to even be constructed with today's designs.

The tolerance for film uniformity has also tightened considerably. Whereas 10 years ago a gate oxide could range within a 50 or 60 Å window, now the tolerances being specified to equipment vendors are typically ±2 to 3% at three standard deviations from the average at a target thickness below 200 Å. This gives a maximum range allowable of only 8 to 12 Å. This requires that the process be well understood and characterized, and that the conditions of the system are maintained at peak cleanliness and control at all times. The many variables of the process must be kept in balance at all times in order to maintain this degree of control. In the case of the CVD operations, this will be seen to be a delicate balancing act.

At the higher circuit switching speeds and clock rates in computer systems being produced today, it is imperative that the integrated circuits have increased lifetime. A chip's average lifetime has been increased by many times already, and will have to be further increased in the future. This is in spite of the requirements of thinner films and tighter linewidths and spacing. The density and purity of thin films, especially the gate oxides which are undergoing the most stress during the high speed cycling, must be controlled to a degree of precision that is far tighter than the requirements of a few years ago.

The overall effect of these forces is to make process development very difficult.

The complexity of the processes in the film deposition and diffusion area results in a variety of unique problems that must be considered in detail prior to the creation of the device. These limitations give rise to specifications within the design rules that give predicted target ranges and tolerances for critical aspects for the films, such as film conformality requirements, resistivity, and current handling expectations.

1.2.3 Circuit Design Evolution and the Goals of the Semiconductor Industry

The demand for speed has pushed microcomputer technology since the start, and continues to do so today. The most common memory chips, for instance, cannot even hope to keep up with the microprocessors that drive them. This has led to the creation of a market for high-speed instruction cache controller chips to give near "zero-wait state" performance to the latest computers. This push for increased speed has its inherent problems, including higher intrinsic power consumption at the higher clock speeds, higher operating temperatures, and consequent problems with reliability of device performance. Small changes in the manufacturing process can induce major changes in the device properties, which may add significantly to the cost of manufacturing. For example, a film analyzer that allows wafers to be processed with out-of-tolerance readings may result in a wafer with low device yield, or with a batch of material that operates at an unacceptably slow rate. While these parts may function, they could be salable only at a reduced price. These kinds of surprises are usually discovered too late, and usually have an impact on far too many wafers. The response from top management to this situation is usually not very positive, to say the least.

To give the reader an idea of the impact of the increase in clock rates, in 1980 the Apple II was driven by a 6502 with about a 2-MHz clock rate. The early IMSAI microcomputer (the one used in the movie "War Games") was a 2.5-MHz 8080-based machine that later was upgraded to a 4-MHz machine. The memory chips in those days sometimes used two or more wait states and were typically capable of retrieving data at a rate no better than 500 nanoseconds. The clock speed in modern high-speed computers at the time of this writing is around 25 MHz, and will probably go to 32 MHz or higher in the near future. This means that the CPU can access data in as little as 40 nanoseconds. The most common type of dynamic RAM IC right now runs at 100 nanoseconds, with 70 nanosecond parts available at premium prices. This requires a minimum of one "wait state" to allow the memory chips to catch up with the CPU. Higher speed but lower density and more expensive static RAM chips with speeds of about 30 nanoseconds are used in caches to help reduce this wait-state requirement. The challenge in manufacturing integrated circuits at these high rates of speed is that tolerances must be held tightly in all aspects of circuit processing.

This leads directly to the next challenge of the semiconductor industry, and that is the issue of increasing density. Ten years ago, a 2048-bit static RAM chip or 4096-bit dynamic RAM chip was considered high technology. Now the same size (or even smaller) package holds over one million bits of information, with 4- to 64-megabit chips already in development or pilot production. This race to improve the density of memory and other integrated circuitry will not slow down, and in fact is only intensifying with each generation of devices. This has had a direct impact on the requirements for defect, particulate, and quality control in the wafer fabs. The more extreme the density requirements of the technology, the more extreme the requirements for clean manufacturing conditions, and the higher the consequent cost of the wafer fab will be, in both operating costs and initial construction costs.

The increase in the density of the devices does not necessarily relate to changes in size of the average integrated circuit, although the ability to shrink an existing design to reduce manufacturing costs is a significant driving force in the industry. In fact, however, the cleaner requirements of newer wafer fabs have allowed larger chips to be built profitably. Extreme examples of this are the attempts at wafer scale chip manufacturing. Even some of the more common chips, particularly the custom or ASIC (application-specific integrated circuit) chips can be quite large, sometimes nearly as large as a small postage stamp. These parts become very expensive due to the reduced number of chips available on each wafer.

When line geometries on existing devices are reduced to a new set of design rules, there is usually a significant boost to the overall die production. This can sometimes range as high as 40%, depending on the particular integrated circuit. A side benefit of die-size reduction is that for a given defect level there will be a proportionately larger number of die surviving through to the packaging process. Usually, there is money to be made shrinking the design of an integrated circuit up to the point where a major investment must be made in the wafer fab to manufacture the new chip. At that point, the design is usually old enough that a large market share or high selling price no longer exists to justify the expense of adding new equipment and further shrinking of the design ceases.

Another goal of the semiconductor industry is the quest for ever lower power dissipation on integrated circuits. There are a number of reasons for this. One of the primary reasons is technical—excess power dissipation on very small devices can cause localized destruction of the integrated circuit structure. Excess power can also cause currents to be induced in neighboring circuit traces. This leads to problems with crosstalk and other undesirable effects. Other problems that are amplified with excess power consumption can include things such as increases in parasitic transistor activity. Finally, lower overall power consumption by each chip allows systems to be built with smaller and less expensive power supplies, useful in laptop computers, calculators, portable tape decks, and so on. The power consumption of a device is largely determined by the technology with which it is constructed. For example, CMOS chips consume less power than their NMOS cousins.

1.2.4 Manufacturing and Reliability Aspects

Device reliability must be one of the most crucial concerns for the manufacturing process. Integrated circuit failures in the field are unacceptable, and in many instances very dangerous. (Next time you fly, think about the chips that went into the plane. Would you want a radar altimeter chip giving incorrect readings as you approach the field in a storm? Or any other circuit failure for that matter?) In addition, any chips that are to go into critical applications are usually specified to be able to maintain more rigorous standards, such as military specification (Mil-Spec) quality levels. If batches of chips cannot pass these tests, they can no longer be sold as high-reliability parts and there is a consequent reduction in their average selling price. In most cases, reliability is related to the wear-out mechanisms inherent in the various films of the device, and also is closely related to the contamination present during the manufacture of the wafers. The issues of contamination become increasingly important as process geometries are reduced and chip densities increase, as very tiny amounts of contaminants may cause significant device problems.

Bear in mind that integrated circuits are usually specified with reliability measured in failure rate after a certain number of hours of operation. While the industry has improved on chip reliability, the amount of effort required just to stay in place is very significant. To illustrate this, consider that since the increase in clock speeds from 2.5 to 25 MHz has meant that devices must undergo a tenfold increase in cycles per year, and since the lifetime of integrated circuits has been expected to at least double, the total number of cycles that the various gates must undergo has increased by 20 to 25 times. Now consider that the total number of transistors on a chip has increased in 10 years from around 50,000 to over 1,000,000 (an increase of 20 times). The result is a requirement for a real increase in chip reliability of over 400 times that of the technology of 10 years ago. In reality, integrated circuit reliability has probably improved much more than this, and circuit failure is not usually a major concern on known high-quality parts, at least for quite a while after purchase.

There are two interrelated special cases of high-reliability integrated circuits. One is the Mil-Spec part, the other is the radiation-hardened (or Rad-Hard) devices. The military specification part is typically a standard part that has been produced and tested under the strictest of conditions and under a series of specifications designed by the military. These chips do not tend to be the circuits that are at the state of the art, as the military has a more serious need for reliable parts than for high-tech parts. This does not, however, mean that the military does not use the highest tech parts in their systems, just that the preference is for the tried and true.

Radiation-hardened chips are almost always produced under the same level of scrutiny as the Mil-Spec devices, even if they are to be used for nonmilitary purposes. These types of chips are typically used in spacecraft, reactor systems, and other devices that operate in or near an environment of high radiation. The major differences in these chips as compared to other Mil-Spec devices is that they receive special processing to make them more immune to the effects of radiation. These types of processes include heavier implants to strengthen diode barriers, special annealing and alloying cycles, and special passivation sequences to absorb the energy and reduce damage.

Two more of the common issues that arise in wafer fabrication are the issues of die per wafer yield and wafer line yield. *Die yield* is defined as the number of working chips that can be made on a wafer. (Integrated circuits are called "dice" or singular, die, when in raw form.) This number typically ranges from 40 to 90% of the available chips. A yield of 40% is pretty bad, but if the chip is very large or has a very high average selling price, it may be that a 40% die yield is acceptable. If yields decline much below this, the wafers become interesting only as engineering experiments. Most merchant manufacturers hope for die yields of 70 to 85% on their wafers, and few can sustain die yields above 90% on modern technologies. One of the top priorities of the engineer is to develop methods to maximize and stabilize die yield.

Steady increases in die yields are preferred, as it is important from a business standpoint to be able to accurately predict yield. Having yield stability at a somewhat lower die yield may prove more profitable than having unstable yields that have occasional wafers or lots with higher yields. Yield instability can complicate forecasting factory output, and can result in delays to customers, or in the delivery of inferior parts to customers.

Often, process changes can be implemented to improve die yield by a few die per wafer. This can often be very profitable, as it does not take much in the way of a die yield increase to pay for a procedural change or for a new piece of equipment.

As an example of the impact that changes in die per wafer can make, consider the following: An integrated circuit may be produced on a wafer that can contain perhaps 400 chips. Of these, there may be

240 working ICs. Assuming that the average selling price of each chip is about $5.00, each wafer would have a potential value of around $1200. Each lot usually contains 25 wafers (although some fab areas use 50-wafer lots); therefore each lot contains up to 10,000 ICs, out of which 6000 are functional. Each lot is thus worth approximately $30,000. Each 10 die per wafer increase, then, results in a $1250 increase in potential revenue per lot. A medium semiconductor fab can move at least 5000 wafers, or 200 lots per week. Therefore, each 10 die per wafer increase can result in an increase of potential revenue of around $250,000 per week! Not only that, but this is often nearly free additional revenue, as the actual manufacturing cost changes may be minimal. In many cases, incremental increases in die per wafer yield can be used to justify equipment purchases and upgrades.

One of the primary reasons that wafer sizes have been increasing over the last several years is the increased die available per wafer. Since it presumably costs about the same to make a six-inch as a four-inch wafer, the number of integrated circuits manufactured can be greatly increased for a given level of operational expense. This benefit is somewhat offset by the much higher prices being commanded by equipment compatible with larger wafer sizes. Depending on the conversion costs and on whether the die yield rates can be kept constant during the transition, there can be a substantial increase in potential die on a wafer. Ten years ago, the standard size of a silicon wafer was four inches (having just come up from three inches). Today's five-inch wafer has 56% more surface area available to production than does the four-inch wafer; the six-inch wafer has 2.25 times the area of the four-inch wafer; and an eight-inch substrate holds nearly four times the surface area. Design restrictions may prevent use of all of this area, but the impact on die per wafer is clearly enormous. Now instead of a lot holding 6000 integrated circuits, it can hold up to 24,000 integrated circuits, making each lot worth $120,000 in potential revenue (assuming the earlier example).

Clearly, if a small increase in die wafer yield can help the potential revenue of a fab area, increases in line yield can help the bottom line also. Line yield is defined as the number of wafers that survive each process step. Scratches, breakage, surface contamination, haze, and other defects can cause wafers to be rejected at any step in the process. It is not unusual for a lot that started at 25 wafers to end up with 15 or fewer wafers. Using the above example, a five-wafer line yield loss for the average lot would result in a net loss of 1200 working chips per lot, or $6,000 per lot, projecting to a weekly loss of potential revenue of over $300,000. Obviously, any improvement in this area can result in enormous benefit, again with minimal impact to the expense level of a wafer fab. Clearly, line yield becomes even more critical as wafer size increases.

Incidentally, so that these numbers can be viewed in proper perspective, the cost of wafer fab operations is often in the range of

$250,000 to $350,000 per day. Chemicals, wafers, and equipment are very expensive in the semiconductor industry.

Finally, in light of the daily costs of wafer fab operations, another major issue for the manufacturing and process engineer is that of equipment uptime. An instrument that has excessive maintenance requirements or does not operate within specification tolerances will cause numerous problems, from being the source of chronic factory inventory imbalances, to die and line yield degradation and reliability problems. Every effort must be made to identify, control, and predict equipment maintenance issues as soon as they arise. While equipment preventive maintenance can be expensive, time-consuming, and inconvenient, there is little that wrecks a fab area's output more rapidly than unplanned equipment downtime. This is a classic case of either paying a dollar now, or paying the dollar (plus lots of interest) later.

1.2.5 Sources of Process Problems

Problems with semiconductor devices usually fall into one of three categories: (1) particulate contamination, (2) chemical contamination, and (3) mechanical damage. All three can result in die yield or line yield losses, depending on the severity of the problem. Many of these problems have predictable causes, but the source of the problem can be obscured. This can result in a considerable amount of fruitless effort. Occasionally there are problems that creep in when a process or a piece of equipment is upgraded. A common situation is the case where one problem exhibits itself in a number of different ways, depending on the exact sequence of processing. For instance, particles that are deposited in a nitride process may be trapped on the surface and reappear later as etch defects. These etch defects may look very similar to defects that could be produced during photoresist deposition. It is at this stage in particular that experience comes into play, as the source of the problem must be clearly identified and eliminated. It is especially easy to treat the symptoms of problems and not treat the main problem. It is crucial at this point to solicit input from the fab operators, as they are often the only ones with the missing pieces of the puzzle.

Each of the three classes of defects introduces its own type of detection problem. Some defects are easy to detect; for example, scratches on the surface of the wafer. This type of mechanical damage, which is lethal to integrated circuits, can be seen with the naked eye, although sometimes requiring special bright lights for observing the full impact of the scratch. Particulate matter can be detected by sight and by machine, although there are no machines today that accurately measure particles that exist on any arbitrary pattern. Accurate particle counts done by machine require smooth, bare substrates to operate correctly. Companies have widely varying standards as to what is considered acceptable for particulate defect levels, although it is well known that excessive particles cause circuit failure. Defect detection and wafer testing procedures will be covered in more detail in chapter three.

Sources of particulate contamination cover the spectrum of possibilities. Humans are one of the primary causes of particles. Skin, hair, saliva, smoke, salt, dust, pollen, you name it, a human drags it around with him or her. All manufacturing areas require clean-room suits, with varying degrees of cover. Some require complete coverage, even to the extent of wearing a filtered air supply, while others may require only gowns and hats. With narrower geometries and larger wafers, wearing the proper gear is more critical. Equipment can also create and store particles for later distribution. Moving parts rubbing on surfaces, static electricity buildup, corrosion, improper handling or cleaning of parts, and a wide variety of other items can cause equipment to become a particle generation source. Finally, the process gases and chemicals in which the wafers are immersed must be kept perfectly clean, or they will become sources of particles. Gas distribution and water distribution systems are both notorious for becoming particulate sources, sometimes a long time after installation. Filters must be used and maintained properly, as filter failures can cause gigantic increases in defects in an extremely short time.

Chemical contamination comes in a variety of forms, including trapped charged contaminants in films, which have an increased probability of causing dielectric breakdown and other problems; atoms of highly mobile elements that become redistributed while the chip is operating, which changes performance through time; and dopant cross contamination, which can change the resistivity or type of the substrate. Chemical contamination contributes greatly to reliability failures since the problems are sometimes subtle and can permit wafers to pass the final tests without any obvious flaws.

Chemical contamination is the hardest type of contamination to detect. It often requires specialized analytical equipment, or must be detected through electrical tests. Occasionally, the defects will appear at the in-line electrical tests, while others may not show up until sort. In other cases, the contamination may be confined to particular areas of the wafers. Another unfortunate feature of chemical contamination problems is that they can manifest themselves in a variety of ways, often adding to the difficulties of evaluating the problem. Major cases of chemical contamination occur when the wafers are misprocessed or when wafers have gone through a contaminated solution. These problems usually show up very quickly, and the wafers can be rejected. In the worst case, problems are not discovered until the products have been sold and are failing.

Some of the most common sources for chemical contamination are the gases and chemicals through which the wafers are passed. Care must be taken to identify what wafers are run through each piece of equipment, as dopants from one type of wafer may be leached or etched from the surface of the wafers and become dissolved in the chemical solution or deposited on the sidewalls of the vessel. Wafers that are sensitive to the particular dopant used in the first process should not be processed in that vessel unless it has been very thoroughly cleaned. The best practice is to restrict the use of the vessel or solution to a particular process.

Other sources of chemical contamination can include wafer misprocessing, such as implantation of the wrong element (not terribly uncommon, unfortunately) and improper handling.

Mechanical damage covers items from crystal defects to surface scratches. All can be fatal to the integrated circuits affected, but most defects of this nature produce only localized effects. Scratches are usually caused by equipment alignment being out of tolerance or by mishandling by a human, and are almost always lethal to the affected chips. Nearby chips can be affected by the particles that are created from the action of scratching.

Silicon crystal defects can be caused by stress, improper processing, dopant atoms within the crystal structure, surface defects, and silicon growth defects. Silicon manufacturers have gone to great length and expense to supply defect-free silicon, but there are many situations where fab processes can create and magnify defects. Crystal defects can be difficult to isolate, unless they lead to a predictable problem such as excessive substrate diode leakage. Once identified, they are usually easy to observe by using any of a wide variety of methods, after which the culprit process can usually be tracked down. Sometimes substantial crystal-defect problems are created by the sequence of the process. As a result, we can sometimes make substantial changes in the process or in the substrate selection to produce improvements. Some of these techniques reduce defects and minimize their impact on a wafer. These techniques should be considered whenever a new process is being designed.

Also in the mechanical defect category are items such as etch nonuniformities and mask defects, haze creation in the CVD processes, and so on. These types of defects can usually be readily identified and can be traced, and the yield impact can be gauged. Obviously, since many mechanical defects are caused by fairly gross problems, the results of fixing these kinds of problems are usually immediate and very beneficial to the output of the fab area.

1.2.6 Defect and Contamination Control Requirements

While defects are recognized as public enemy number one in the semiconductor industry, it is generally accepted that complete removal of particles, once deposited, is nearly impossible. While part of the problem is technical, the rapidly evolving industry standards themselves produce requirements that are beyond those that are attainable with the current generation of processing equipment. For instance, as linewidths shrink, film thicknesses are reduced. Both of these areas are sensitive to particles, and with each decrease in dimension there is a consequent decrease in tolerance in allowable particle sizes and quantities. Unfortunately, it is nearly as complex to control small particles on these structures as it is to build the structures in the first place.

There are two basic regimes for maximum particle size that must be considered. Each process technology must be evaluated as to which type will be more lethal and the process specifications set appropriately. The types are maximum particle size as a function of linewidth, and particle size as a function of film thickness.

In the linewidth case, there are at this time two general rules of thumb, neither of which has been thoroughly tested, at least with published yield impact reports. The first assumption, which is more prevalent in the United States, is that the maximum allowable particle size can be no more than one-fourth to one-half the minimum linewidth. It is assumed that particles larger than this can cause short circuits through intervening dielectric films. The other view, more prevalent in Japan, is that this threshold is no greater than one-tenth the minimum linewidth. The difference in these numbers is measured in terms of manufacturing costs, as it is much more costly to build chips at the higher levels of cleanliness.

The other primary issue when dealing with particles is their impact as a function of film thickness. Since even the very smallest particles detectable are very large in comparison to thin gate oxidation thicknesses, it must be assumed that all particles are capable of causing damage to the critical transistor regions. Particle-size restrictions are of crucial importance when dealing with interlevel dielectrics and metal layers. The existence of a particle in or on top of these films is very likely to cause problems with short circuits, open circuits, and related failure modes. Since the smallest particles that can be detected are so much larger than the typical film thicknesses involved, the main line of defense at these levels is to control the particles at their sources, and attempt to attain cleanliness levels beyond the minimum requirements. This is one of the reasons that the tighter one-tenth minimum linewidth rule is used. By monitoring the defects at those lower levels, the probability of controlling defect densities for larger particles increases.

Chemical contamination levels must be kept to a very low level. In the early 1980s, it was said that 1 sodium atom in 10,000 silicon atoms could significantly reduce device performance. Today that number has been reduced to one in 1.0E + 8 to 1.0E + 9, and almost every alien element is forbidden from the device for various reasons. It is currently required to use electrical testing to determine chemical impurity levels that are acceptable. The use of high-tech chemical analysis equipment as standard fabrication test equipment has increased.

1.2.7 Quality Management Techniques

The implementation of proper fab management techniques can allow improved control of defect levels, uptime, and device yield and reliability. These techniques include procedures for controlling wafer inventory, equipment preventative maintenance routines, implementation of statistical process control methods, and effective low-yield analysis methodology. It has been found that total defects, especially those in the category of particle contamination, increase at a rate directly proportional to the time that the wafers sit idle. A batch of wafers that takes twice as long to process through the fab area as an otherwise equivalent batch will show a significant reduction in die yield, sometimes as high as 25%. The yield impact can be even more severe if there are delays after certain processes. The gate region of a transistor or the tunneling region of an EEPROM is especially susceptible to damage if it is exposed to the environment for a short time.

As a result, the so-called Just-in-Time (JIT) methods are being implemented. The JIT techniques are used so that material is always flowing to the next step, neither too early for the next step to be ready nor too late, but "just in time." These systems are used in the wafer manufacturing environments to reduce and maintain controllable wafer inventories with reasonable throughput times. Controls on wafers in process (usually called WIP) such as these have been known to improve fab throughput times from 40 and 50 days to under 20 days. This allows nearly twice the number of wafers to be processed in the fab area in an equivalent time, and simultaneously increases yield. These methods do impose restrictions on the maintenance and engineering staffs, as there is much less margin for error. Maintenance items must be done on schedule, and engineering runs become more valuable. Excess inventory is removed, and errors become prominent immediately.

When linked with carefully and properly implemented statistical process control methods, yields and throughput can be maximized simultaneously. Statistical process control (SPC) is used to implement Management-by-Exception control techniques. In these, the operator records pertinent data about each process run, and if a statistically significant event (an exception) occurs, the process is shut down. At this point, an engineer will be called in to make an appropriate adjustment, or a process technician may clean the quartzware in a furnace, or whatever is the appropriate action for the exception. After the situation is resolved, a report of the event is made. After thorough studies are made on a particular technology, limits are placed on the critical parameters of a device in order to maintain control on the process. While this is a very common practice, sometimes the statistics are abused, and specification tolerances are widened in the mistaken assumption that fewer discrepancies on the yield report show a fab that is running within tighter control. This is a mistaken belief. As yields and fab control improve, the engineering target windows should automatically tighten as a result of the improved control, which in turn should lead to even more precise control on the process.

If batches of wafers then fall outside of yield expectations, it will become necessary to perform effective low-yield analysis. If specification windows are allowed to become progressively looser, then effective low-yield analysis becomes impossible. These techniques must be fairly specific, but will allow analysis of failures, and hopefully will produce methods for preventing the failure modes from recurring in the future. Associated with this, but somewhat more costly to implement, is high-yield analysis. Looking in detail at your highest yielding wafers may indicate what parameters to look at to improve yields. This kind of analysis can lead to rapid yield gains after the key critical factors have been identified.

In summary, it is clear that the primary focus of the engineering staff must be to design quality control into the process. The manufacturing group must attempt to meet those specifications in every possible way to ensure that quality workmanship is attained. The maintenance crew must strive to keep the equipment up and in good condition, which is a constant struggle in itself. The goal should be to control the process in as proactive a way as is possible or practical. We will come back to the common and important themes of defect control, inventory management, downtime control, yield enhancement, and quality manufacturing throughout the book.

CHAPTER 2

YIELD OPTIMIZATION

In the preceding chapter, we discussed the general impact that defects in a wide variety of locations could make on die and line yields. In doing so, we covered a simple example demonstrating the impact on revenue of a die yield increase. Clearly, relatively small reductions in defect densities can make significant improvements in overall factory profitability. Conversely, small increases in defect densities can decrease profitability and can lead to no end of trouble. In this chapter, we will discuss optimization of yields and methods for measuring defects. We will also discuss methods for implementing changes in the manufacturing process that will control defects and control the situations that allow defects to be created.

Obviously, a number of advantages can be gained from the implementation of defect-reduction plans in semiconductor manufacturing environments, including die per wafer yield increases, line yield increases, reliability improvement and field failure rate reduction, and improved performance and ability to command premium average selling prices. The term *defects* in this case can cover all classes of contamination, including particulate contamination, chemical contamination, and mechanical damage, and, in some cases, can even include uniformity issues.

These advantages take on differing amounts of significance to each product line as a result of several factors such as wafer size, die size, device complexity, and the inherent sensitivity of any certain technology to each particular type of contamination. As discussed in the previous chapter, reductions in linewidths have resulted in severely restricted particle-size requirements in comparison to those of a few years ago. As a result, linewidth restrictions are probably the first and most crucial aspect to keep in mind when setting up a defect-free environment. Other major aspects related to the linewidth requirements that must be considered include maximum environmental particle-size tolerance, the tolerance limits on mask defects and etch artifacts, and the quality of deposited films.

The second priority to consider when developing the plans for a cleanroom would probably be the choice of substrate material and processing requirements for that substrate. For example, gallium arsenide substrates are much more fragile than silicon substrates and are easier to damage. Also, certain films are more sensitive to scratches than others; for example, polysilicon films are softer and therefore more prone to scratches than silicon nitride films. Additionally, there are chemical contamination issues that are more sensitive in some areas of the fab than in others. This is true for sodium, which is often used as the basis for many photoresist developers, but which is absolutely intolerable in a high-temperature environment, such as a diffusion furnace. Knowledge of these aspects of the technology can allow the engineer to build in appropriate preventative measures as required.

Another factor to consider when outlining strategies for defect reduction programs is the impact of die size increases. As the integrated circuits take up more of the real estate of the wafer, there are two factors to consider. One is that the number of chips per wafer produced will be reduced. The other attribute to keep in mind is that the defect densities must be greatly reduced in order to get reasonable yields. We will quantify this as we discuss defect density versus yield models in detail, but this has a significant impact on the process. The costs of making large integrated circuits are much higher than those of comparable chips of smaller surface area. The main reasons for these cost increases are the increased purity requirements and the reduction in the chips produced per wafer.

A major factor to consider when developing any strategy for defect control is the impact that those controls will have on the manufacturing organization. Sudden, drastic changes in policy will not always show positive results if there is less than full cooperation by all parties. It is sometimes better to have a slower, more evolutionary change in contamination consciousness. However, the full impact on profitability of the defect reduction programs will not be felt unless there is a complete commitment by all parties. Even an aggressive team will find that there are practical limits to the rate of improvement. The result of a drastic increase in requirements can be increases in paperwork, as more material is "suddenly" out of the specification limits (or "discrepant"), and significant increases in downtime as equipment is repaired or upgraded to meet the new requirements. A middle road must be reached which allows rapid improvement without wrecking the manufacturability of the production line. Specific techniques on upgrading a fab area will be discussed further throughout the book.

For now, it is important to keep this basic conflict in mind when discussing improvement projects: Each generation of integrated circuit design needs tighter requirements than can be attained with existing equipment and procedures and tends to evolve much more rapidly than the equipment. This results in almost continual change, which reduces productivity and can have short-term negative yield impacts. Thus, the very act of trying to improve the fab area's productivity could have a negative impact on it. Ultimately, a fine balancing act is required to keep the rate of improvement greater than the rate of regression due to confusion and errors brought on by changes in procedures, and all available means of information control should be employed to maintain the balance, including the statistical process control and management-by-exception methods noted before.

2.1 BASIC MANUFACTURING COSTS

To get a feel for the costs involved in the economics of running a wafer fabrication area, we will consider the case of a medium-sized wafer fab. This is not designed to be a complete list of all of the expenses of the fab, but is meant for an example only. Specific calculations must be made for each manufacturing environment. Table 2-1 shows several types of semiconductor process, the typical cleanroom requirements for each technology, and the approximate costs per wafer for each.

Our fab area will be designed to handle about 15,000 wafers per month (defined as wafers in process [or production] or WIP) and will be assumed to be a class 10 clean room. See chapter 4 for discussions on the various classes of clean room. The substrate material will be silicon, six inches in diameter. We will further assume that the factory runs 24 hours a day, five days a week, or 120 hours per week. (Many factories work full time seven days a week, but we will be conservative in this area.) In a four-week month, there are 480 hours to produce the

TABLE 2-1
Basic Technology Requirements

Device	Minimum linewidth	Room class	Wafer size	Die size	Typical die available	Typical wafer cost[a]
TTL logic	3–5 μm	1000	4 in. A	<10 mm^2	1000+	$100–125
DRAM						
64k–256k	<2 μm	100	4–6 in.	20 mm^2	400–900	$200–600
1M–4M	<1 μm	1–10	6–8 in.	75 mm^2	270–480	$1000–5000
Microprocessors						
8-bit	<2 μm	10–100	4–6 in.	20 mm^2	400–900	$250–750
32-bit	<1 μm	1–10	6–8 in.	95 mm^2	185–330	$3000–10,000

[a] Costs include manufacturing, packaging, test, design, and overhead costs.

15,000 wafers, therefore meaning that the average output of the fab area is 31.25 wafers per hour. This is shown in Table 2-2, which also gives a summary of the cost estimates we will be making below.

If the technology uses 10 photolithographic layers, there must be at least 312.5 wafers produced per hour through the photolithographic area. This defines the minimum number of steppers or aligners that must be purchased to meet the fab area's requirements. If each stepper can produce 50 wafers per hour, at least six will be required to run the operation. However, assuming 20% downtime (which implies excellent control—often downtime can exceed this by a long way), at least two additional steppers will be required to operate at peak efficiency. We now have an approximate lower bound on the number of steppers required for the fab. In our case we will need eight steppers that can run at 50 wafers per hour. Many steppers are unable to reach these speeds; typical rates are 20 to 30 wafers per hour. Part of the stepper

TABLE 2-2
Typical Wafer Manufacturing Costs

Assumptions:
Wafers out: 15,000/month (small fab ~10,000 ft^2 cleanroom)
Hours available per month: 480
Shipments per hour: 31.25 required to meet target
Approximately class 10 to class 100 fab; ~5 in. wafers

This estimate will take into account only the "big ticket" items, and is therefore estimated to be no more than 70% of the actual costs of a factory of this size. Many fabs will have costs that exceed these values by many times. This will especially be the case in very large volume and high cleanliness (class 1) areas.

This is meant as an illustration and does not represent the actual costs of running a fab. Specific details about fab costs vary widely from fab to fab.

Minimum monthly costs:

Site costs (building)		$ 520,000
Facilities, chemicals		$ 400,000
Raw wafer costs		$ 500,000
Quartzware, etc.		$ 80,000
Reticles/masks		$ 65,000
Sundries		$ 55,000
Salaries/benefits		$ 750,000
Subtotal		$2,370,000
(We assumed this is <70% of total costs, so add. . .)		$1,185,000
Total		$3,555,000
Average cost per wafer	$237	
Operations cost per hour	$7,406.25	
per day	$177,750	

rate is a function of wafer size and that must be considered. A fab area producing 15,000 six-inch wafers per month will need more than twice as many steppers as a fab running 15,000 four-inch wafers per month.

Similarly, if there are 15 diffusion and thin film processes required for the technology, and each process can produce 33 wafers per hour (a typical diffusion process may contain around 100 wafers per run, each taking about three hours to complete), there will be a capacity requirement 468.75 "moves" per hour, meaning at least 17 or 18 diffusion or thin-film systems. In reality, due to chemical cross-contamination constraints (i.e., requirements that wafers that have been processed with conflicting dopant types are not processed together), it will be necessary to have more than this. For this type of fab area, 24 to 30 different diffusion and CVD systems would normally be installed.

This logic will also hold up for all of the various clean, implant, etch, and other steps that are required. As a result, we can start to get a lower bound on the requirements for cleanroom space. We will assume that all of the equipment is *chase-mounted*, which means that the equipment is mounted to the wall of the cleanroom, so that the machine facilities are in the less expensive aisles behind the cleanroom, and only the front panels extend into the cleanroom. There will be at least 60 major pieces of equipment involved plus a large array of other equipment used to analyze the major items, or to analyze the wafers, or to do any one of a number of tasks. If each piece of equipment takes up six feet of space on a wall, and a person requires eight square feet to work, plus table space, and so on, then at least 10,000 square feet of class 10 fab area space would be required, and even this would be a very tight fit. (Remember, though, the equipment is chase-mounted, and the "chase area" space has yet to be determined.) Assuming that this could be done comfortably, the cost to build the clean room alone (that is to put up walls and install the laminar air flow systems in the ceiling) exceeds $2,250,000, assuming that a very conservative figure of $225 per square foot is used for determining cleanroom costs. To this must be added facilitization, such as water delivery, chemical delivery, exhaust, power, and so on.

As a result, many fabrication areas cost in excess of $100 million when completed, counting in the costs of all of the equipment, and so on. For example, a fab area being built in Japan to construct four-megabit DRAM devices in mass production is estimated to cost $300 million. We will assume that our fab will cost a mere $50 million, and has an expected useful lifetime of 10 years. That means that (without interest), ultimately, the costs of construction and the consequent amortization of the site will average out to $420,000 a month. Add to this another estimated $100,000 of associated office and laboratory space and interest payments, and the total site costs are around $520,000 a month.

Fab operations require a lot of electricity, water, and other facilities, as fans, lights, furnaces, and so on must be on continuously. This will run into the hundreds of thousands of dollars per month. We will assume for this exercise that the cost of electricity for the area is $80,000 per month. Water costs for generating the ultrapure water required may run in the range of $50,000 a month, while toxic waste removal could exceed $50,000 a month. Toxic waste removal costs are increasing rapidly, as more stringent laws go into effect and toxic waste sites fill up. Chemical costs at $20 a gallon (which is also very conservative, depending on the chemical and purity), and usage rates of up to 500 gallons per day add up to chemical use of around $200,000 per month. Ultrapure gas use is lower, but the unit costs are greater, so that there are will be as many as 30 bottles of gas at costs of around $3500 apiece used on the premises at any time. Since they last up to four months, this implies a usage rate of gases of $25,000 per month. Therefore, the costs of the facility requirements alone add up to over $400,000 per month.

For a wafer fab to produce 15,000 wafers per month, it must consume a larger number than that per month, dependent on the "line yield" loss of the fab. If 1000 wafers per month fail for some reason, it must be assumed that the fab will start 16,000 wafers per month to attain an output of 15,000 wafers. Therefore, wafer costs, at $30 per wafer, contribute a healthy $480,000 per month expense to the fab. The fab may use 1500 additional test wafers a month at $15 per wafer, for a total of $22,500 a month. The costs for quartzware for diffusion furnaces, as well as other related consumables, can easily exceed $50,000 per month. Cleanroom supplies are very expensive, each notebook costing around $10, and the average bunnysuit about $600 to $700 complete. These costs can run another $30,000 or more per month. Masks (also called reticles), used to project the desired pattern on the wafers, cost around $5000 apiece, and must be replaced occasionally, and must be amortized across the board. If there are 10 photolithographic steps, there are 10 masking layers. In a high-volume fab area, there may be dozens of mask sets, but we will assume only 15 sets (which is three sets of reticles, for each of five different integrated circuits, a very small number of reticles). With 10 reticles per set, at $5000 per reticle, and 15 sets of reticles, there is a total investment of $750,000 in reticles, with an average lifetime of perhaps 12 months, which results in a per month cost on reticle sets of $62,500. Therefore, adding all of this up, the costs of the various items used in the manufacture of the wafers is about $700,000 a month.

Finally, there are the costs of salaries and benefits for the employees. A fab of this size, with maintenance, production, engineering, and management all counted in, requires at least 200 people, probably more. But assuming an average annual salary of $30,000 (an arbitrary round number, not necessarily representative of industry salary standards), plus average benefit and overhead costs of around 50% of that figure, or $15,000, the average annual cost of personnel exceeds $9 million. This means that salaries and benefits add up to another $750,000 per month in our hypothetical fab area.

We can see that, even given this crude estimate of the costs of the fab area, and not counting in extra costs, such as new equipment purchases, tools, spare parts, backup parts, and so on, the costs of running this wafer fab exceed $2.5 million per month ($2,500,000). Adding in unexpected extras could easily add 30 to 40%, so our basic estimate for the fab monthly expenses could go up to $3.4 million. Finally, bear in mind that these are not fully tested and operational integrated circuits. Wafer fabs seldom have the chip packaging plants associated with them, so they must usually ship the wafers out in wafer form. There is still a significant amount of work to be done on the wafers, including dicing, die attachment, wire bonding, packaging, and final testing. For now, however, we will not include these costs. So, if we assume a monthly cost of $3,400,000, those 15,000 wafers will cost an average of around $225 apiece. This is a fairly reasonable number, although a little lower than that usually achieved by IC manufacturers. In actual practice, a wafer may cost as much as $250 to $300 to produce on silicon. Gallium arsenide is much more expensive, and the average processing costs at the equivalent point in manufacture can exceed $1000 per wafer. Wafers produced in an R&D lab, or in a small pilot production line can be as much as five times this expensive to build.

The costs of the final assembly and test of the integrated circuits has increased just as the complexity of chips has increased. While these costs used to be around 33% of the cost of the chips, testing costs have increased so that it is not uncommon to spend 55 to 60% or more of manufacturing cost on testing and packaging, bringing the per wafer cost for manufactured parts to around $450 to $800 per wafer.

Another way of viewing this expenditure rate is in terms of costs per day. Clearly, if our operation costs $3.4 million per month, and there are 20 days of operation per month, our fab would cost $170,000 per day to run. Again, this number turns out to be on the conservative side, since real fabs of nearly equivalent volume cost as much as $250,000 a day. A very large wafer fabrication area can easily cost much more than that to operate. Yield variations and unexpected and excessive equipment downtime will have a tremendous impact on the cost of operations of the fab area.

Now that we have reached this point, we can start to discuss the breakeven point for die yields. This is a very fluid number, depending heavily on the state of the fab, the soundness of the chip design, and the fluctuations of the marketplace. Its continual variations cause frustrations for many managers but, if viewed statistically, breakeven point analysis can be used to determine and control trends. A summary of the breakeven analysis for various chips is given in Table 2-3.

Let's assume there are two types of integrated circuits of different

TABLE 2-3
Manufacturing Cost Breakeven Analysis[a]

IC Type	Die size (mm^2)	Die yield (%)	Wafer size	Available die	Die manufactured per month	Average cost per die($)[b]
TTL logic	6	70	4	1310	13,744,500	0.25
Small RAM/ microprocessor	20	70	4	390	4,123,350	0.82
Large RAM	65	50	6	130	1,950,000	1.74
Large microprocessor	95	50	6	90	1,350,000	2.52
Large microprocessor	95	50	8	170	2,550,000	1.33
Large microprocessor	95	<25	6	30	450,000	7.56

[a] Assumes 15,000 wafers manufactured per month and $3.4 million / month operating costs.
[b] The cost per die values are very good indicators of the impact of larger wafers and of reduced die yield. Controlling the price of dice is critical in maintaining a competitive market position.

sizes manufactured in this fab area. Since the manufacturing costs of a wafer are related more closely to the number of steps involved in the process than to any other factor, the wafers themselves will cost about the same as long as they use similar technology. We will assume for this example that the second chip has sides that are exactly twice as long as the first chip; in other words it covers four times as much surface area on the wafer. If the first type of chip has 200 die per wafer available, then the second type of chip would have no more than 50 die per wafer. If we then assume the die yields on each device are 60%, the first part will yield 120 working parts, and the second chip 30 working parts. Now, if total wafer costs are $400 per wafer, the first chip will have to command an average selling price of more than $3.33 per chip to make money. Taking into account the marketer's pricing rule of thumb—marking up the price three times—means you will have to command prices near $10 per chip to make reasonable profits. The same numbers worked out for the second chip show that the average selling price would have to exceed $13.33 to break even, and $40 to make good business sense. It is rare for a fab to be able to command sales prices that are three times the costs, due to the well-known erosion of chip prices. This is one of the basic driving forces behind the continual improvement in yield and quality required.

From here, it becomes mostly a matter of scale. Per wafer costs for finished-product integrated circuits usually fall in the ranges discussed here regardless of size of the operation. Price increases are usually found in the cleanliness requirements needed for the technology. If our fab had been built to class 1 standards, it would have cost more than twice the original $50 million estimate. In addition, a wafer fab requiring output of 5 million DRAMs per month, with a chip the size of the first (smaller) chip used in the example above, would have to produce at least 2.75 times the number of wafers. (Our fab can produce 1.8 million chips a month at 120 die/wafer.) It is easy to see then that the cost of a class 1 cleanroom with the capacity to make 5 million four-megabit DRAMS would cost around $275 million. Costs scale up in a fairly linear fashion, as increases in costs of some items will be offset due to a relatively lower increase in the requirements for other items that can be purchased in cheaper bulk forms.

2.2 FINANCIAL IMPACT OF YIELD CHANGES

The financial impact of changes in a fab area can be gauged by one of four key indicators: the die yield—number of chips per wafer; the line yield—number of wafers surviving the process; downtime—percentage of time the factory produces output; and reliability—ability of chips to perform their assigned tasks without failing. The costs of problems in these areas are covered in Table 2-4.

We have assumed above that the wafer fab is able to manufacture chips at a rate of 60% die yield. This is equivalent to saying that we are

TABLE 2-4
Financial Impact of Fab Manufacturing Problems

Problem	Definition	Costs	Impact[a]	Potential revenue lost[b]	Impact[c]	Potential revenue lost
Die yield	Number of working die per wafer	Per wafer 1 IC = $3.33 $100	$15,000/mo per 1 DPW	$50,000/mo per 1 DPW	Up to $1.5M per 1 DPW	Up to $7.5M per 1 DPW
Line yield	Number of wafers that survive	Per wafer 1 wafer = $225 $1,500	$33,750/mo per 1% loss	$100,000 per 1% loss	$750,000 per 1% loss	$3.75M per 1% loss
Downtime	Time when no manufacturing because equipment is unavailable	Per hour 1 hr = $7,500	$168,000/mo per 5% down	$300,000/mo per 5% down	$3.75M/ mo per 5% down	$18.75M/mo per 5% down
Reliability	Ability of IC to perform to spec for time	Lost time; lost prestige	Several month delay	Based on size of product line	Same	Same
Total per month (Does not count reliability costs)			$216,750	$450,000	$6,000,000	$30,000,000

[a] Impact includes requirements to make up lost factory output.
[b] First column of Impact/revenue lost assumes IC with selling price of $3.33 (256k DRAM).
[c] Second column of Impact/revenue lost assumes IC with selling price of $100 (80386 costs ~ $200).

throwing away 40% of the chips on any given wafer. We have also assumed that most wafers with problems are thrown away during manufacture, and that all equipment stays up at least as much as it needs to in order to run a perfect Just-in-Time inventory control system. However, as in many situations in life, seldom does the reality of an event match the theoretical optimum. Die yields vary, often far beyond their expected range, operators drop wafers or run them through an incorrect process, equipment breaks down when least expected and most required, and so on. This all adds to the cost of manufacturing a wafer. Murphy would have loved a wafer fab.

First, we will look again at die yield. If each chip must sell for $3.33 to break even with "perfect" manufacturing costs, at 15,000 wafers produced per month, a *one* die per wafer decrease on all material will result in a revenue loss of $50,000. While the optimist would state that there would be a real influence to the bottom line if we could make the 40% failures into operable chips, realistically there will be some upper limit to the amount of yield gain that may be possible. A sudden 5 or 10% increase in die yields (5 to 12 die per wafer, or 75,000 to 180,000 additional chips per month) may produce so much additional work that the fab line cannot sustain the output. This would include problems in testing, sorting, and packaging the individual chips. In any event, as pointed out earlier, a steady, controlled, and predictable growth is preferable to rapid, uncontrolled or unpredictable growth.

As can be seen, if a new piece of equipment can result in a verifiable die per wafer increase, this can be used to justify the purchase of the new piece of equipment. In business terms, this called an ROI, or return on investment analysis. For example, if a new CVD machine produces chips at a rate four die per wafer higher than the standard, and the machine costs $400,000, the payback period is two months. (Four die/wafer will result in a $200,000 increase in revenue per month.) If six of the machines are required for fab operations, the payback period is one year. Usually, payback periods of machines that improve die yields can be measured in months, which is very rapid compared to most industries. This has a direct payback in the long-term profitability of the company.

Another big impact on the profitability of the semiconductor house is the loss of wafers during process, also called line yield losses. A wafer that is thrown away results in two obvious losses: the loss of the work performed up to that point (often several hundred dollars worth), plus the loss of potential revenue. The subtle problems that also occur include: For every wafer that is thrown out, you miss the scheduled output by over 100 chips, which can throw manufacturing plans off a bit, especially if there are a large number of wafers lost, and there is a consequent upward shift in the breakeven point for each wafer that has survived the process.

Taking a look at the numbers, we can see that the loss of one wafer per week will cause an impact of over $1500 per month. Most of the time, losses are much greater than that, often on the order of 50 to 100 losses per week, which adds up to $75,000 to $150,000 per month. When considering line yield, it is important to note the cumulative effects of wafers lost per operation. There are a very large number of operations a wafer will go through, often exceeding 100 individual steps. If there is one wafer lost per run every four steps in a 25-wafer lot's history, then clearly no wafers would remain in the lot at the end of the process. Even losses of 0.1% per step result in losses of 10% of the wafers after 100 steps. This means a loss in wafers of 1500 per month or 375 wafers per week. Clearly, 1500 wafers will result in a lot of loss, no matter how the loss is viewed. In terms of lost time and resources, the cost would be (at $200 per wafer), over $300,000 per month, and in terms of lost revenue, there would be over 180,000 chips not manufactured or a total of nearly $600,000 of revenue that would not be obtained. Since line yield losses are usually easier to fix than die yield problems, fab personnel often go after these kinds of issues first.

Next, there are costs associated with equipment downtime. This is in addition to the repair costs themselves. It is for these reasons that it is imperative to purchase enough systems to get the job done assuming a "reasonable" amount of downtime for the particular issue. Reasonable can range from 1 to 25% downtime, depending on the complexity and sensitivity of the machine. Certainly, the lower the downtime requirements, the better. The costs associated with insufficient equipment backup and insufficient spare part stocks become obvious very quickly. If there is a machine that "gates" or "controls" the inventory (i.e., there are always bottlenecks at a certain step), when that machine goes down there will be trouble.

Two points: First, notice I used the word *when* the machine goes down—whenever a machine is in the critical path like this, it is used more and is therefore more likely to have a failure; second, Murphy's laws have very powerful effects in wafer fabrication houses—it is almost universally true that when a machine is most needed, that is when it will decide to "shut itself down." If this machine that is in the critical path goes down and the work builds up, it will have a serious impact on the wafer production schedule. This requires additional time to reach the desired output goals. If the line stops completely at that step, the costs become the full daily operation costs for the wafer fab—$170,000 a day or more. If overtime is required to make up for the loss, you can tack on an additional large amount for the additional salaries (including overtime pay), overhead, facilities, and so on that will be required. One day of overtime will cost around $240,000. A few screw-ups in a month could cost the wafer manufacturer hundreds of thousands if not millions in lost revenue (and has been known to happen many times). Even if the problem is not as catastrophic as the shutdown of the entire fab line, the lost productivity can still easily add up to tens or hundreds of thousands of dollars of loss in a short time.

There is another type of yield loss which is somewhat more subtle, and does not immediately appear in the manufacturing area. This is in the area of reliability failures. Samples are pulled from lots and subjected to extremely nasty tests to see how the chips will fare in the real world. They are almost always run to complete failure and then analyzed for the failure mode. If a new failure mode suddenly crops up, or if an unusually large number of chips fail, or if they fail at an earlier stage of the testing than usual, the production of this device may be stopped. Sometimes no further shipments can occur until the problem is resolved, or else prices, specifications, or other items will have to be adjusted. Parts failing reliability qualifications cannot be sold as prime parts. If the problem is so serious that shipments have to stop, the fab output could be set back several months (the time it takes to run experiments to find the problem, run material through the problem area for samples, and then run the reliability tests on these new samples). This is why it is so crucial to build quality in at each step. A several-month delay in a product line can mean (and has meant) the demise of an entire company.

Integrated circuit reliability is one area that cannot be taken lightly. Since the problems are usually much more subtle, they are almost invisible to fab operations. As a result, they are much more easily avoided or swept under the rug than line or die yield losses. Reliability failures must be addressed immediately and aggressively or the problem will probably get worse. While there are cases of problems solving themselves (more likely that someone acted on the problem without realizing it or reporting it), most problems only get worse until they become catastrophic. The time delays involved in reliability problem solving prohibit errors and force tight restrictions on the engineers and on the manufacturing area as a whole.

2.3 AREAS FOR YIELD ENHANCEMENTS

Clearly, improvements in die per wafer and in line yields can result in significant improvements in the output and revenue of a fab area. These improvements often take a significant amount of effort, time, and engineering, and may involve so many changes that incorporating the improvement results in lost cost effectiveness, or they may cause sufficient confusion that yield losses elsewhere will increase (which can sometimes offset any potential gain). In any event, fab management people are always looking for easier ways to increase the number of chips that can be produced in their factories. This is particularly crucial in factories that have limited floor space or insufficient capitalization to allow expansion. These increases come in four basic areas: more efficient chip design (using fewer transistors to achieve the same function, for instance), decreases in chip size, increases in wafer size, and reductions in throughput time (the time it takes to manufacture the wafers).

Chip design improvements can sometimes result in increased output, but the designers are usually more concerned with scaling the current design down to the next geometry level, or incorporating their ideas into the next generation of chips. The main problem with chip redesign efforts is that sometimes a significant amount of effort can go into a design with only a marginal increase in output or functionality. There will be cases, however, where a design was coded into an ASIC chip that is later made into a standard integrated circuit. In this case, the redesign effort could prove to be very profitable, since application specific designs and other semicustom circuit solutions are not particularly efficient from a design standpoint. A carefully designed single purpose chip can be made much smaller than its ASIC equivalent.

Decreasing chip area has some advantages, especially if the chip is very large to begin with, or is at a larger linewidth design rule than the fab can handle. (In other words, a fab may be able to run 1.5-μm chips, but its designs may be using a 2.0- to 2.5-μm design rule.) While the relationships are not exact, the decrease in chip size will usually result in an increase in chip density of about the same amount. In other words, a shrink of 10% should result in an almost 10% increase in die available. This is somewhat limited by the exact layout used on the wafer. There comes a point where further shrinking of the part makes no sense. This usually occurs when significant modifications are required in the fab to manufacture the new part, for example, requiring the upgrade from projection aligners for photolithographic use to stepper systems. Typically, a chip can be shrunk by about 30% from the first design until major changes have to be made. Again, this is very dependent on the design of the device.

Increasing wafer size is a common method of increasing available chips. Going from a four-inch to a five-inch wafer results in an increase in wafer surface area of 56% (from 78.54 to 122.72 cm^2). Going from a four-inch to a six-inch wafer results in an increase of 125% (from 78.54 to 176.71 cm^2). Thus, the surface area on a six inch wafer is more than double that of a four-inch wafer and more than twice the number of chips can be placed on the wafer. Changing wafer sizes in a fab can cause a significant amount of trouble and expense, depending on which changes are made. Upgrading from four-inch to five-inch wafers requires fewer process and equipment changes than does the change from four to six inches, since most of the equipment that is used on four-inch wafers can be used for five-inch wafers, which is not usually the case for six-inch wafers. Changes in wafer sizes for a particular device or technology often occur when a new wafer fabrication area is built. Sometimes fab upgrades can be accomplished within an existing fab with careful planning and an understanding that there will probably be a major impact on the fab's productivity during the transition.

The last method of increasing factory output is through techniques for reducing throughput time. Buildup of inventory at bottleneck steps and allowing inventory to sit around at other steps have been shown to have numerous detrimental effects on both yield and productivity, ranging from increased defect densities on the wafers to wafers that require reworks due to problems created by excess elapsed time. There are also detrimental psychological effects on the production personnel when wafers are allowed to sit around. While throughput time reduction techniques may not always seem to have an obvious relationship to fab output, there are a number of reasons why they can be used effectively. First, atmospheric gases and moisture can diffuse into films on the wafer. For example, atmospheric moisture can cause BPSG films to lose boron in the form of boric acid crystals that form on the surface of the wafer. Atmospheric moisture and oxygen can cause thin gate oxides to become extremely leaky and incapable of holding charges, and can cause general deterioration of photoresists and other films. These problems can be reduced or eliminated with reductions in throughput time.

Another way that increased throughput time can reduce output is through increased defect density. Particles tend to accumulate steadily on wafers within any storage medium. The rate of accumulation is reduced when the wafers are kept within ultraclean environments and in static-free boxes, but there are still particles that are deposited on the surface of the wafers, jarred free from boats, backs of other wafers, sides of the boxes, and so on. This occurs when the boxes are moved around (usually to make space for more inventory or to keep track of the lot numbers). These particles ultimately add up to a decrease in die yield, usually a few die per wafer on the average. These and other problems that are associated with stationary inventory result in die yield and reliability failures, along with occasional line yield losses.

Another justification for decreasing throughput time is that the same techniques required to reduce the throughput time are associated with inventory reduction. The combination can be used to improve the efficiency of the factory itself, thus allowing the factory to maintain a slightly larger maximum capacity per month. An improvement of 5% in factory capacity means a potential increase of 750 wafers produced per month or 90,000 chips per month. This translates to additional revenue of nearly $300,000 per month.

We can see when we add it all up that there are ways of improving factory output anywhere from 5 or 10% up to two or more times, as summarized in Table 2-5. Much of the decision on how to increase the total number of chips manufactured is based on economics. The throughput time reduction techniques come nearly free of charge and can result in significant increases in income, especially if the factory has not had inventory and statistical process controls installed. At the other end of the spectrum, output can be enhanced very significantly

TABLE 2-5
Potential Yield Improvement Techniques[a]

Area of improvement	Cost	Complexity	Timeliness	Effectiveness
Enhance quality[b]	Low	Low–Medium	Excellent	Good
Better circuit	High	High	Long lead	Fair
Shrink circuit	Medium	Medium–High	Long lead	Good
Increase wafer size	High	Medium–High	Long lead	Very good
Better throughput	Low	Low	Good	Good

[a] There are a number of methods for improving yields. As usual, the most profitable methods are those that have the least cost and complexity associated with them, while at the same time providing maximum impact on the yields.
[b] Quality enhancements include improvements in handling, attention to detail, and good process control.

through changing wafer sizes. In some cases, such as the transition from four- to five-inch wafers, the relatively low expense may justify upgrading an older fab, although all the changes that are required could still cost millions of dollars. The cost to upgrade to six-inch wafers is so great that in most instances it has proved more cost-effective to build a new fabrication area. The costs that are involved to upgrade to eight-inch wafers or build an eight-inch wafer line are very high, as most of the equipment on the market has not been constructed for eight-inch substrates. However, as of this date, a few manufacturers have announced plans for eight-inch wafer fabs and others will soon be following their lead. There is currently a debate in the industry as to whether the expansion of wafers beyond eight inches will be truly cost-effective, although technically it is possible to process larger substrates. The transition to the much larger substrates will probably not occur until the advent of "individual wafer processing," which implies a fully automated environment. This is due in part to the complexity of manufacturing the parts on a surface of that size, and to other items such as the weight of batches of the large (for example, 12-inch) wafers.

Some processes may prove difficult to run as wafer substrates change, and some process parameters may change as a result of new restrictions. For example, CVD processes must often run in a different pressure regime to create effective, uniform films. Other changes may be introduced when a device produced in one fab area is transferred to a new fab area. These may result in minor parametric changes to the devices, which may require some adjustment of the process targets. While this is interesting from the engineer's or scientist's point of view, it can be a nightmare for manufacturing managers. The time required to run the experiments necessary to requalify a process can sometimes be much longer than forecast, especially in a new fab where there is less familiarity with the equipment and procedures. There may also be a reduction in die yield during the transition. This can affect revenues sharply, and sometimes can have a lingering impact on operations. Finally, there may be some customers who will require either more extensive testing or requalification of the new circuits after changes are made. This is especially true of military or Mil-Spec customers.

While all this is not meant to discourage fabrication area upgrades, it is designed to make the reader aware of the many consequences and time sinks that can result during a fab area upgrade or transferring a product to a new fabrication line. Any time such a major effort is undertaken, it must be done with the proper long-term strategy, with plenty of time built into the schedule, and with expectations that there will be periods of significant reductions in factory output. Basically, you must plan for the unexpected.

2.4 INVENTORY MANAGEMENT TECHNIQUES

We have discussed the concept of inventory management techniques a number of times up to this point, without too much description of what that term meant. While the ideas presented here are not necessarily the only methods that can achieve the goals of maximum output at minimum cost, they do represent tracks that can serve as a basis for a customized system in any factory.

First, we must define the basic goal of the wafer manufacturing operation. This is, simply, to build as many of the most high-tech integrated circuits possible at the lowest possible cost, with the highest reliability possible. While this goal is easily stated, it is by no means a simple goal to attain. It follows from this that the ideal fabrication operation should make the integrated circuits as rapidly as possible once the wafers are introduced into the fab area. Since these wafers are on paper only "inventory" or "costs of operations," they have no value to the company until they are completed and ready for shipment. Clearly, the time a wafer sits stationary costs money. As a result, it is important to make sure that each process is prepared for the wafers that are being delivered to it. While this can be done easily with a very few wafers hand-carried from place to place, the complexities introduced by going to mass production with this kind of system are expensive, in terms of both resources and manpower. That is where the inventory management techniques come in.

Fortunately, there are a number of computerized fab area inventory management systems that can assist the implementation of these techniques into wafer fabs. They can give the entire organization the information necessary to control the manufacturing process. Most often, they are used merely as instruments of observation but they are increasingly being relied upon to "tell" the managers what areas are in need of attention. Like any tool, the reports generated must be used properly to be effective. This requires that the reports be kept current, or be produced in real time. The following are some of the reports that are recommended to control wafer fab operations:

1. Inventory located at each step, which can be placed on a bar graph to visually display line balance, as shown in Figure 2-1. By observing the total number of wafers in inventory at all times, it becomes obvious when bottlenecks appear in the production area. This may indicate that more training is required, or that more equipment is required.
2. Total time at each step. While this is often just set at some arbitrary limit, by observing the average throughput times of the various steps and comparing them to the expected throughput times, judgments may be made as to where additional resources need to be applied, or where there is some underlying problem. Figure 2-2 describes a typical time at process step report.
3. Wafer discrepancies by step. By observing where the wafers are out of specification ranges, resources can be applied to small problem

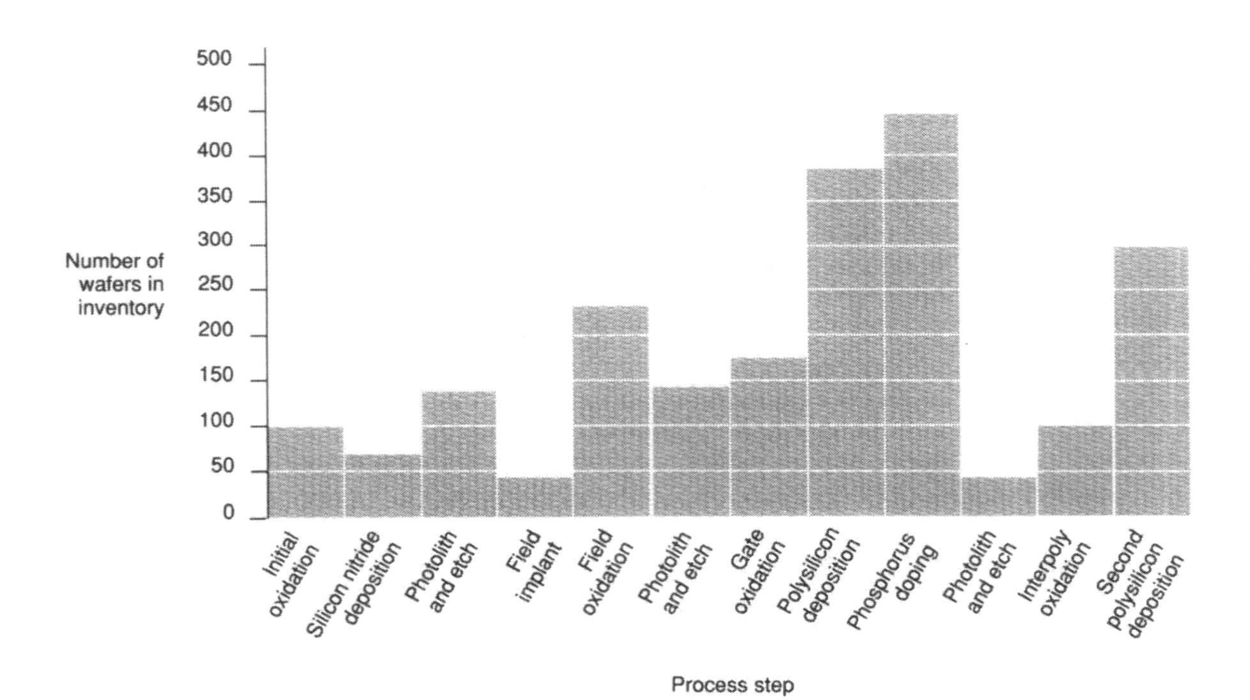

Figure 2-1. Inventory Control Chart This partial inventory control chart demonstrates how the chart can indicate problem areas in the fab. The first several process are fairly well balanced. The field oxidation process is a long (8–10 hr) process, so can often show higher inventory levels. However, a problem clearly exists at the polysilicon deposition and/or phosphorus deposition steps. This has caused excessive inventory buildup at those steps, as well as starvation of succeeding steps.

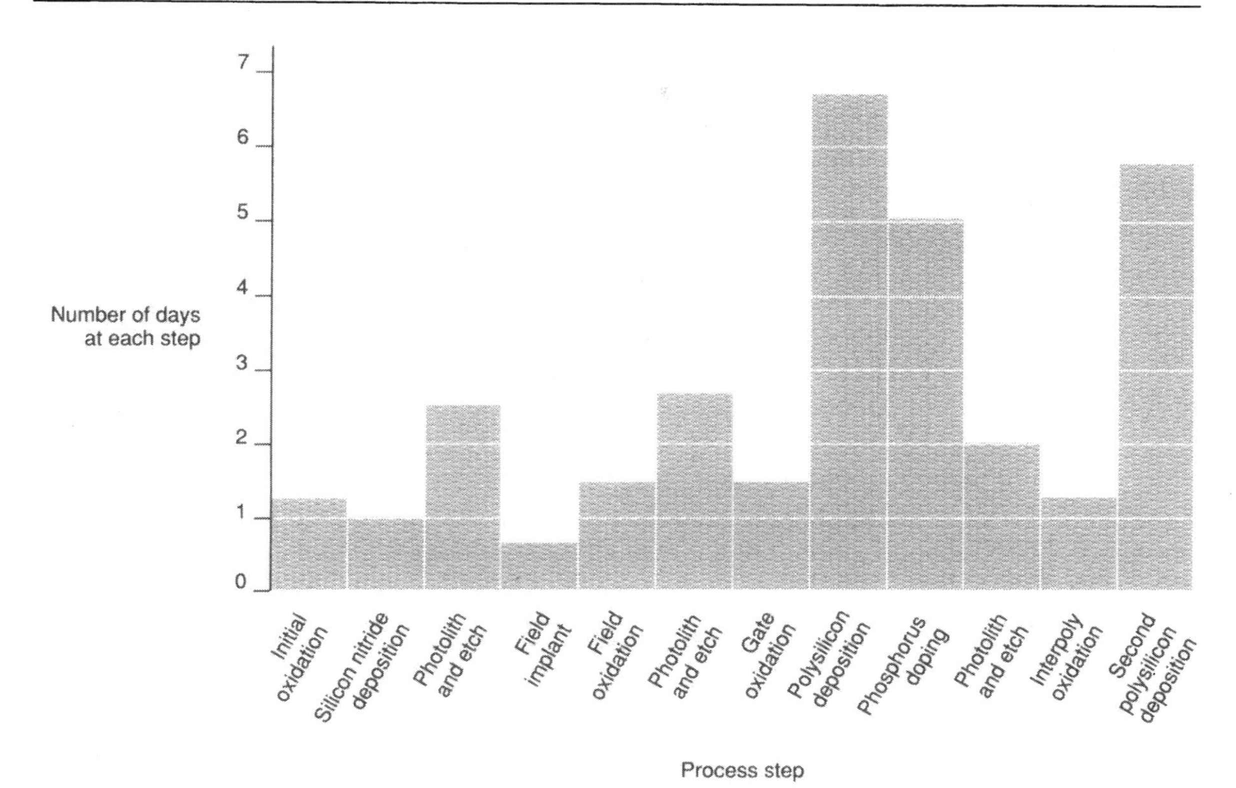

Figure 2-2. The Time at Process Chart The time at each process can be monitored and analyzed with this chart. Again, problem areas stand out clearly. An optimized fab area will have most of the process step throughput times at under a day.

areas before they become major problem areas. This report is displayed in Figure 2-3.

4. Wafer scrap reports. These reports produce information similar to the discrepant material reports, but highlight line yield and maintenance issues, whereas the discrepant reports tend to indicate engineering issues, such as process drift. The wafer scrap report is described in Figure 2-4.
5. Regular cumulative trend reports on discrepancies and wafer losses, as shown in Figure 2-5. This will tend to show up long-term trends, or other areas of weakness within the process that require attention.

Viewing inventory as a function of process step can create a significant benefit when trying to control line balance. To sustain a positive wafer flow, and keep throughput times to a minimum, it is important to prevent buildups from occurring at any step. This can occur if a bottleneck exists due to insufficient equipment, excessive downtime, or low productivity. This is often seen in the case of the CVD processes. They often become bottlenecks, as the steps that

Discrepant Material Report

Step	Number of wafers	Run numbers	High readings	Low readings	High particles	Other parameters
Initial oxidation						
Nitride deposition						
1st mask						
Field implant						
Field oxidation						
⋮	⋮	⋮	⋮	⋮	⋮	⋮

Discrepant Material Trend Report

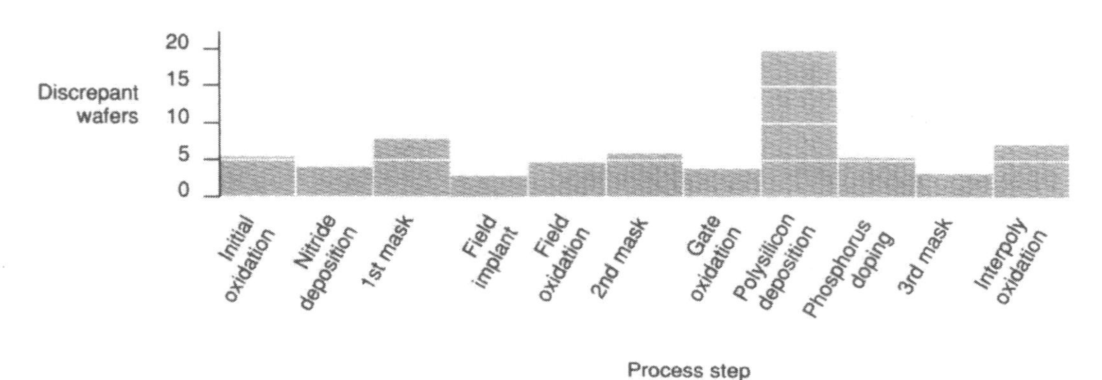

Figure 2-3. Discrepant Material Report Material that falls outside of the specification ranges is recorded on the discrepant material report. This report is used to summarize the number of wafers that have had problems at each step, and a rough idea of what the problems were. The discrepant material trend report can be used to suggest areas where the most significant improvement can be made.

precede them are relatively easy, and have less inherent downtime. For instance, four gate oxidation furnaces can produce around 500 wafers in three hours. Four polysilicon furnaces can produce about the same number of wafers in that time. However, the tubes for the polysilicon furnaces require four times as many cleans, and can sometimes produce only partial loads, due to haze or other film-quality problems. In this case, continued processing of wafers at gate oxidation can more than completely bury the four polysilicon deposition furnaces. This results in excess inventory piling up prior to the polysilicon deposition process. Since the gate oxide quality is very critical for the reliability of the devices, excessive delays before the deposition process

Scrap Material Report

Step	Number of wafers	Run numbers	Haze	Particles	Scratches
Initial oxidation					
Nitride deposition					
1st mask					
Field implant					
Field oxidation					
⋮	⋮	⋮	⋮	⋮	⋮

Scrap Material Trend Report

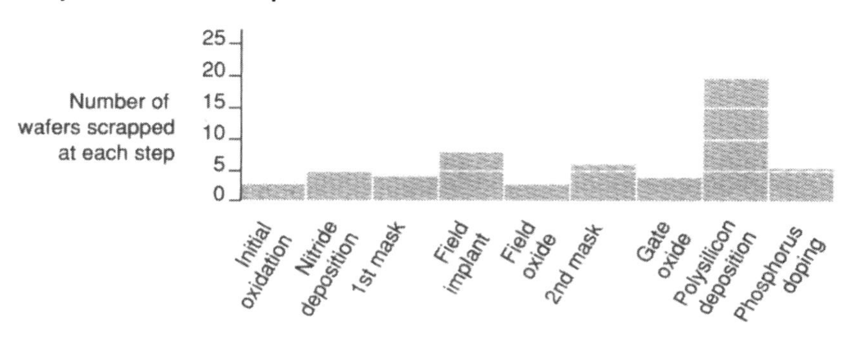

Figure 2-4. Wafer Scrap Reports The wafer scrap reports are similar to the discrepant material reports, although the problems reported here are more severe and often equipment-related. The trend report can be used to identify areas of improvement.

can cause significant yield impact. Other pieces of equipment that can become bottlenecks are ion implanters, etchers, and steppers.

Other problems can occur from allowing imbalances to exist in the manufacturing line, including damage to delicate equipment after the excessive, hard use required to reduce the buildup after the problem causing the bottleneck is solved. This puts strain on other facilities, sometimes causing breakdowns, slowing production further. Another problem is that excess pressure is placed on the manufacturing group to reduce the inventory. This can result in increased misprocessing or in a reduction of quality due to less time being available for proper inspections, fatigue, careless handling, and so on. Also, the line imbalance can become so great that the original problem is lost in the confusion surrounding the pileup. Finally, the presence of imbalances in the fab line act much as friction, preventing free flow of the wafers through the wafer fabrication area and reducing the overall fab productivity significantly.

Discrepant Material Cumulative Trend Report

Step	Work week 1	2	3	4	5	6	7	8	....
Initial oxidation									
Nitride deposition									
1st mask									
Field implant									
Field oxidation									
2nd mask									
Gate oxidation									
⋮	⋮	⋮	⋮	⋮	⋮	⋮	⋮	⋮	

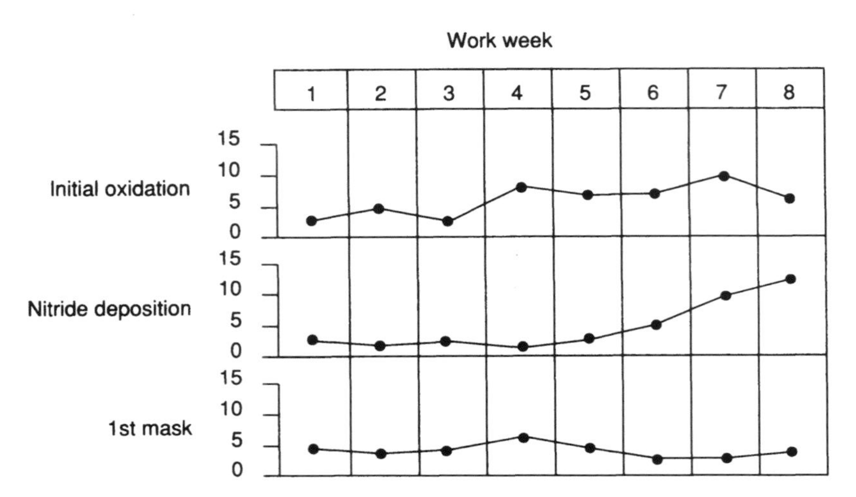

Figure 2-5. Discrepant and Scrapped Material Cumulative Trend Reports This report allows tracking of discrepant material trends. This type of report can permit visibility into subtle problems, and can be used for either discrepant material or wafer scrap reports. For instance, a sudden increase in work week 7 might be correlated to the increase in problems seen in the Initial Ox step that week.

One of the primary weapons in the war on inventory buildup is the concept of Just-in-Time (JIT) manufacturing. Basically, this means that the materials being processed in one place come out of the step and are immediately prepared for processing in the subsequent step, which has been timed to be completely ready for use when the wafers have been prepared. Hence the concept Just-in-Time—all steps are completed just in time to be processed in the next step. This has been applied to a number of other industries and has been applied quite successfully at a number of semiconductor manufacturing houses. This is actually more of a discipline than a procedure and is a very difficult discipline to maintain. The concepts involved in JIT manufacturing are often difficult for production supervisors who must show wafer activity no matter what. Sometimes, it may be necessary to stop production at a certain point in the process in order to guarantee process integrity and control. It is important that supervisors understand that "paper moves" (which are made to make the productivity to look good) can cost the fab a tremendous amount in lost productivity. As mentioned earlier, major problems can occur where the bottlenecks exist, most often when large amounts of material are piled up there. If a problem exists at a certain point, material from that point on back must be stopped at a safe location and the production line kept balanced until the problem is resolved. (Ideally, the processes would be so tightly controlled that a failure in a process would translate back to changes in the start rate and no changes would be seen in overall line balance.) It is incumbent upon the maintenance crew to maintain sufficient supplies, spare parts, and expertise to keep unscheduled downtime to a minimum. The strategy as to which operations to shut down in case of problems should be made beforehand so that problem areas have contingency plans or alternative methods for continued manufacturing. This could include "commandeering" of other equipment, contracting the work out to another fab, or working with a vendor.

Usually the first items to be addressed when considering line imbalance issues are the chronic maintenance and setup problems, such as tube cleaning and implant source changing. Building *realistic* schedules for preventive maintenance and then observing the actual ability to meet those schedules will force a certain discipline on the entire organization. Ultimately, this will benefit the entire project.

Once line balance is being observed and JIT techniques are being employed, the overall fab inventory can be reduced. At first, this is relatively easy as the various records will point to the problem areas and the largest problems can be hit. This alone can result in significant throughput time reductions, often initially in the range of 20 to 30%, with very large reductions in inventory possible as the bottlenecks are identified and removed. After most of the larger obstacles have been removed from the production line, and the line inventory has reached a rough balance of inventory, we need to reduce the overall fab inventory. This means reducing the inventory evenly at each step. This can be done through reducing wafer starts, which is ineffective over the long haul, but can be useful in the early stages of an inventory reduction plan, to accustom personnel to the concept of low-inventory management. A more effective method for reducing inventory is to keep the wafer starts constant and improve the productivity of the operation. This can be done by reducing the overall maintenance time required (which usually means increased preventive maintenance and much care and effort); increasing the throughput of the operation (such as increasing the load size in a diffusion furnace); and time and motion studies of the operation itself, including observing and optimizing such items as wafer cleaning times, loading and unloading times, measurement and analysis times, and inspections. Each of these items should be evaluated and if time can be saved by new methods, then by all means the new methods should be employed (as long as the quality of manufacture is still guaranteed). Once again, the inventory control graphs can be used to pinpoint chronic areas of buildup and alleviate the pressure points as required.

One of the immediate outcomes of this reduction in inventory is that the throughput times will drop dramatically. Naturally, there is some absolute minimum time that it takes for a wafer to be processed, which is about 8 to 12 days, depending on the technology. Realistically, there will be some delay but throughput times of 15 to 20 days for standard material are not unrealistic. Since it often takes as many as 40 or more days to get material processed through wafer fabs without inventory management, it's clear that there is a lot of room for improvement. Clearly, if the fab that we have designed was capable of producing 15,000 wafers a month with a throughput time of 36 days, and now is able to do that same amount in 24 days, then the effective capacity will have increased 50%: from 15,000 to over 22,000 wafers/month. Thus, reductions in throughput time result directly in improvements in overall productivity.

One of the benefits of reduced inventory is that each lot may be tracked more closely, so that fewer inspections may be necessary. In fact, some of the inspection steps should be reduced or even removed since inspecting wafers has often been shown to increase defect densities. It must be kept in mind that it takes careful control and proper statistical proof to remove inspection steps. Quality must never be compromised in the structure of this system. In fact it must be an integral part of the entire project, and is an inherent part of the JIT concept. If there are failures in quality techniques, there will eventually be delays, during which other material that has been prepared for processing will have to wait for the problem to be resolved.

Thus, we have seen that there are several main rules for increasing the fab output and improving chip quality through inventory management. They are:

1. Attain and maintain line inventory balance.
2. Reduce excess inventory buildups at all bottlenecks.
3. Identify and permanently relieve bottlenecks.
4. When line is stable, reduce overall inventory uniformly.
5. Minimize step cycle times, and remove excess steps.
6. When throughput times come down, there will be excess capacity in the fab area, and therefore it will be reasonable to increase the

wafer starts. This must be done carefully to prevent overloading the fab line, and then all other techniques must be closely followed.

As noted, to guarantee that the maintenance group can sustain the equipment, adequate supplies of spare parts, extra gas bottles, chemicals, and all other requirements of the wafer fab must be maintained. This requires also that all of the parts and gases are of appropriate quality for the operation. This is a form of JIT that is often called "ship to stock," and assumes that the manufacturer of a supply has done sufficient quality control on the product, so that the user does not have to provide the testing. It is important to maintain good relations with all of the vendors (and from the equipment suppliers' viewpoint, the customers), in order to be able to set up and qualify ship-to-stock programs. Ultimately, ship-to-stock programs save both vendor and user quite a bit of money, as the user does not have to supply incoming quality assurance (QA) people, and the vendor gets fewer returns. The key to all of this is the guarantee of quality. A great deal must be taken on faith and reputation, but many companies, especially in Japan, consider these methods to be standard practice, and consider incoming QA as redundant.

Finally, efforts must be made to inform and instruct the people who operate the fab about the key details of their jobs, and to lead the personnel in a manner that will inspire them to the near perfection required in the fab. Everyone must be conscious of two key details that can make or break a wafer fab: performance to specification and quality workmanship or craftsmanship. Performance to specification is often interpreted to mean that, as long as the process is in spec, everything is all right. Therefore to reduce apparent problems, specs are sometimes widened. This is a mistake and a perversion of the original goal. The specification ranges originally picked for a particular process are restricted by the limitations of the equipment used in manufacture. Although some processes are more sensitive than others, it is important to identify the optimum process window for each process. This can be done using production wafers but requires a significant amount of data and computing power. Since we are attempting to reach the optimum in yield and reliability specifications, everyone from operator to engineer should attempt to improve the control of each process. As this real control is demonstrated on the process control charts, new and sometimes tighter specification windows can be created, if required. Ultimately, the spec windows would be equal to those values that have no impact on yields, even by a single die per wafer. This will result in even more control over fab yield and the predictability of die yields, and will result in less discrepant material or losses, therefore increasing line yields.

In the real world, specification windows are often set arbitrarily or by equipment limitations, which are usually wider than the optimum values. While for any one process this may make only a one or two die per wafer loss, there can be problems that are based on combinations of marginal material. A device that is sensitive to low gate oxide thicknesses and to thick polysilicon depositions may be in for trouble if the gate process is marginally in spec on the thin side and the poly layer is marginally too thick. Clearly, identifying those exact limits is very crucial to the success of a product line.

Finally, there is the issue of craftsmanship. This is an ancient concept that seems to have been largely forgotten in the mechanized, modern world. However, especially in the IC business, care and concern while manufacturing the devices causes substantial increases in yield, especially line yield, and can impact long-term reliability in positive ways. An operator trained to work within this framework will have pride in his or her work, and will very likely be more productive. As usual, there is far more that can be written about this subject than there is space to discuss it, but again, I must emphasize that performance to specification, tied with steady reduction in specification tolerances, coupled with master craftsmanship in production, can make a very significant impact on the productivity and profitability of a wafer fab, without requiring any capital investment.

2.5 MANAGEMENT BY EXCEPTION

As the name implies, this technique is used to react to exceptions or problems that have arisen. Even though many fab areas run in a fire-fighting mode much of the time, in general the operations of a fabrication area should be pretty smooth from day to day if all of the proper methods of fab control are used (good maintenance, proper work flow and balance, etc.). The only time the management team should become involved in direct operations is in the event of a serious problem not already covered by guidelines. This management method presupposes that appropriate process control techniques are in place so that, when situations arise, it is clear that there is a problem, exactly what the problem is, how it will impact operations (or how much it has already impacted them), and how quickly the problem can be resolved. It also assumes that proper records of past similar occurrences are kept and properly tracked in specification form.

One of the most important tools for maintaining a management by exception system is the statistical process control system. This is also called "Statistical Quality Control," "SPC," and "SQC," as well as a variety of other names. These systems have been installed in most fab areas in the last several years, and used with varying degrees of success. The typical method of implementation is to set up a data entry system, usually consisting of operators filling in test data on a log sheet, including the calculation of the average, range from highest to lowest parametric value, and sometimes standard deviations.

Next the data are placed on a process control chart, which is simply a two-part, time-dependent graph, with the run number on the x axis, and the average and range values on the y axis (sometimes standard deviations are used in place of the range values). A sample process control chart is shown in Figure 2-6. There are usually two sets of limits defined on the process control chart. The outermost is the specification limit. This is the limit beyond which the process will produce wafers with detectable die yield loss, due to any of a number of reasons. Usually, specification limits are ±5 to 10%, and yield or performance will suffer if the wafers are processed outside of this range. Within this range are the engineering control limits. These are used to identify trends in the process. Since all processes vary from run to run, it is important to be able to analyze that variation when it starts to occur, but before it becomes excessive. The control limits are used as flags to the engineer that there has been some sort of significant drift in the process. This drift can then be compensated for before

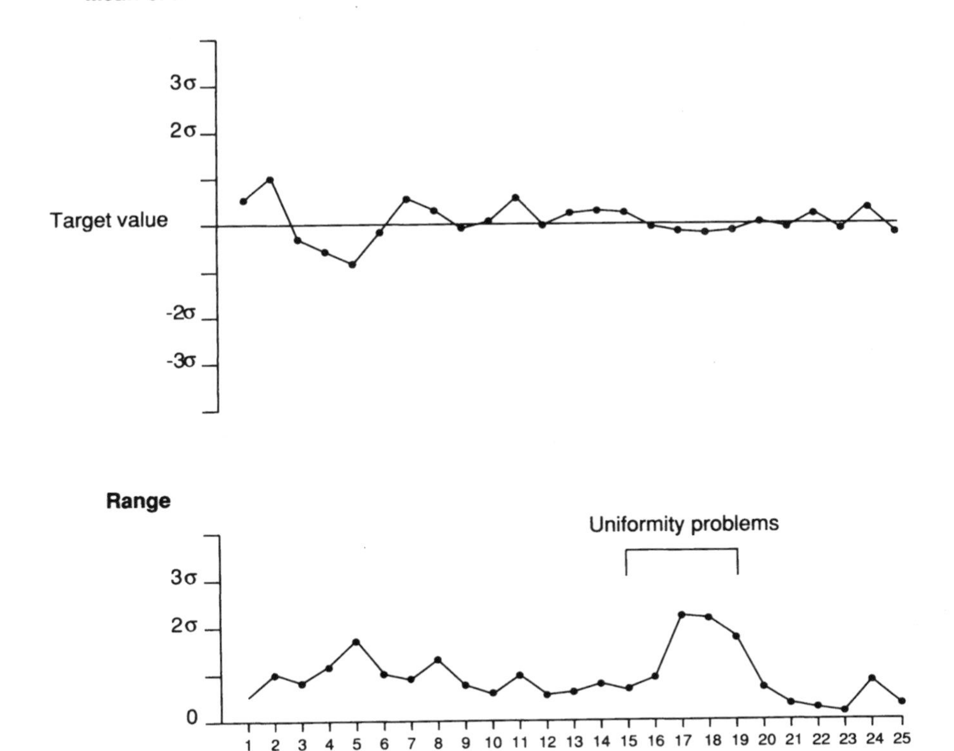

Figure 2-6. Example of a Statistical Process Control Chart A basic tool for monitoring process control is the process control chart. The chart can be used in nearly every process, and provides information as to when preventive maintenance or engineering work must be performed. Nominal 2σ and 3σ values are used for warning limits and specification limits, respectively.

wafers exceed the specification limits. As mentioned earlier, it is important to analyze the impact of the process parameters as a function of yield in order to determine what true specification and control limits should be.

We can see that management by exception procedures are easily described in a flow chart, as seen in Figure 2-7. The procedure begins with an excursion of the process beyond the control or specification limits. The problem must be detected as close to its source as possible, and identified as rapidly as possible. A belated discovery that a problem has occurred in the process may have jeopardized both the affected wafers and other wafers or equipment. Additional material may

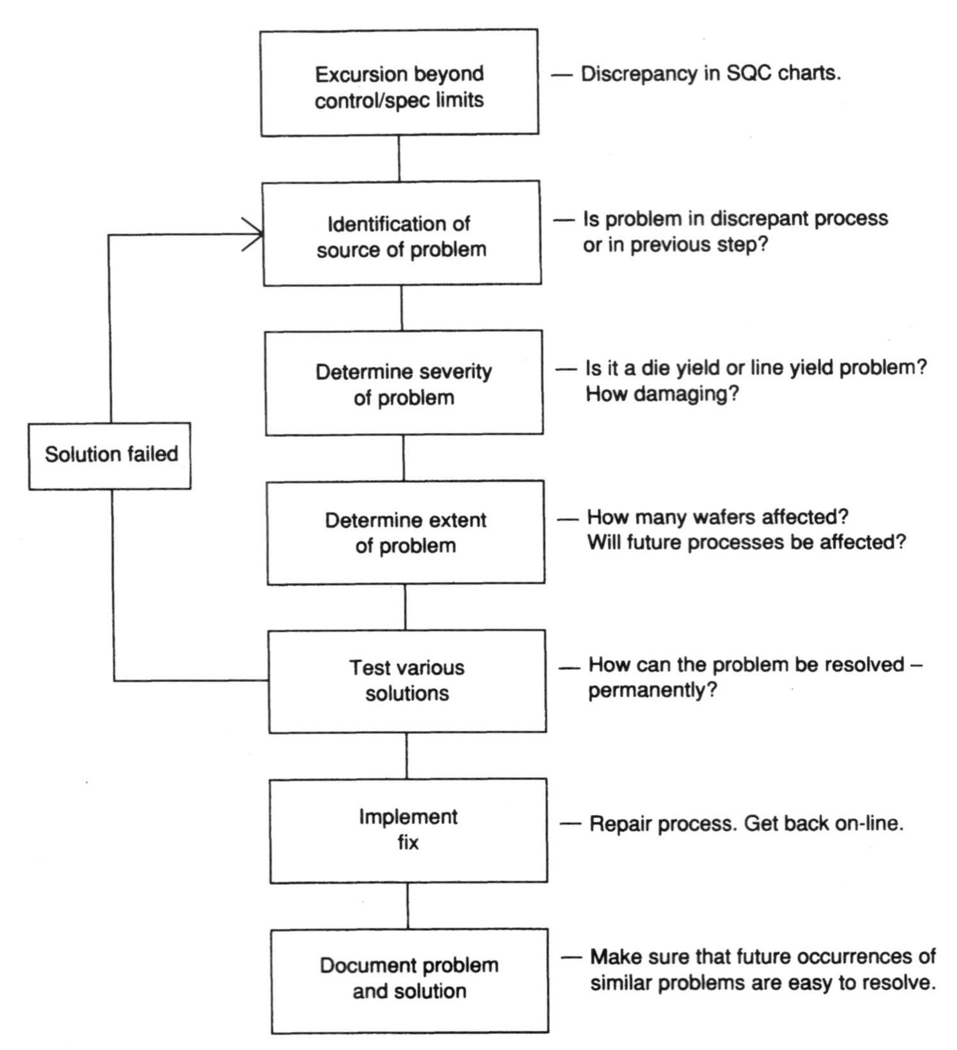

Figure 2-7. Management-by-Exception Response Flowchart When a problem occurs in a wafer fab, certain steps should always be followed in order to completely resolve the issue and prevent future downtime due to repeated occurrences of the same problem.

become damaged in the same way if processed prior to discovery of the damaged wafers. An example would be a situation where a chemical solution such as a developer receives an incomplete rinse, which appears as an electrical test or oxidation thickness problem. With suitable process control, it may be possible to detect that a problem has occurred when the film thickness varies. Then, the control and maintenance logs of equipment can be examined. The timing could then be correlated to show that the thickness problem occurred when the developer started receiving incomplete rinsing. Sometimes, there will be no clear evidence of a problem until many steps later, in which case, clever thought and careful experimentation with a large number of steps (and therefore a large number of variables) will be required. One of the most dangerous pitfalls of these searches is that simple problems or basic fundamentals can be overlooked in the search of a seemingly complex problem. Careful monitoring of all steps and of all process control reports, including losses, help to identify problems.

After the problem and its source have been identified, we must determine the seriousness of the problem. This is also used in determining the extent of the problem. If the problem is small, it may not be necessary to spend a lot of time resolving it, and the time can be better spent on more serious issues. If, on the other hand, there is a significant loss reported (which does happen occasionally, in a "yield crash"), other priorities may have to wait while the problem is resolved. Usually, by the time a problem reaches yield crash magnitude, several problems may have developed that are contributing to the overall yield problem. In addition, there may be so much pressure to do *something* (and so much may have already been lost), that drastic actions may be taken which do not alleviate the real problem and may in fact aggravate the problem even more. Opening up gas systems to replace mass flow controllers when a contamination problem is within a nearby clean station is a common and often disastrous example. Also important in determining the severity of the problem is the determination as to what type of loss is occurring. Responses for a line yield problem will differ from those required for a reliability or a major die yield problem.

The next step in resolving the problem is to analyze its extent. This involves determining how many wafers may be affected by the problem, which consists of determining when the problem occurred, and when the last wafers suspected of being affected were processed. In many cases, finding the source of the problem also pinpoints the timing of the problem. If the start date cannot be isolated, it may be necessary to search through electrical test, wafer sort, or final test results to identify the amount of time that the process has been affected. Usually, the main concern is for the wafers currently in process. If there are a number of wafers in the fab that have received the suspect process, they should be identified, so that the impact to the chip output can be quantified. Additional starts may be made to compensate for the loss, if required. In addition, all wafers upstream of the problem process should be stopped at appropriate "safe" places and should not receive further processing until the situation is resolved. As you can see, a problem of a serious magnitude will have a significant impact on the JIT system described above. It is imperative that appropriate controls be in place to prevent a problem from reaching a serious enough level to justify shutting the line down. It is for this reason that control limits should be steadily tightened in order to reduce the amount of variability permitted and allow predictive efforts to be used to prevent problems.

The fourth step of the procedure is to determine and implement the solution. These are items very specific to the equipment involved and the problem encountered. In some cases, engineering experiments may be needed which could stop production of specific products for several weeks or months. In other cases, only specific groups of processes will need to be tested, and the testing performed in a few days. In some cases, the fix may be relatively simple, but will involve some design and equipment modification. In the case of the developer solution that caused electrical and oxidation problems, a possible solution could be a simple device to rinse both sides of the wafer simultaneously. Obviously, the amount of time required to implement a solution must be minimized, but there must be enough time allotted to allow the work to be completed properly or else the problem may not be resolved or may even be made worse.

Finally, the last step, and often the most overlooked, is the documentation of the problem. All too often, this consists of writing down a list of material affected and issuing a short report to wrap up the problem before going on to the next one. Then, when a similar problem occurs a year later, there is no clear record (if there is any record at all) of the original problem. If the original engineers are still there, the amount of time required to discover the problem will be reduced, but if they are not, the entire problem will have to be reanalyzed. A specification or a central file should be created, with a common problem/solution form that can be used after a problem has been resolved, so that the problems can be referenced later. This simple procedure can save many hours or more of effort, even if the problems are not exactly the same, and the information only points in general directions.

So, we can see that there are essentially five steps in the implementation of management by exception techniques:

1. Detection and identification of problem.
2. Determine severity of the problem.
3. Determine extent of problem.
4. Determine solution, and implement.
5. Documentation of the problem.

Effective management by exception must have the support and involvement of all groups equally. The commitment must be made from the production group to be willing to stop production for short periods when small problems are discovered before they become large problems; from the engineering group, who must be willing to stop whatever they are doing to respond to the exception conditions; and finally from the maintenance group, who must simultaneously juggle preventive maintenance requirements and immediate equipment repair requirements. Only through effective data-gathering techniques can this support be maintained. If the same small problems occur repeatedly with no clear attempt to implement long-term solutions, the personnels' motivation will deteriorate and the techniques will be destined to failure. The individuals who are operating the fabrication area must strive for perfection in their work, and be observant of even minor changes, and management must be willing to respond to violations of the process control efforts at all levels. Visible exceptions can include the occasional visits to the clean rooms by managers and VIPs for tours. If not handled properly, these tours not only bring in contamination, but can have a significant, detrimental impact on productivity as well.

2.6 SPECIFICATION WRITING

There is a school of thought that believes that if common sense is allowed in specifications, and if less detail is placed into specifications, there will be less confusion, less discrepant material, and thus fewer problems. This has seldom proved to be true. In fact, whenever common sense is expected to be used, discussion can and will arise as to the "correct" method of operation of the specific step. Eventually, this discussion will lead to someone getting incorrect information and misprocessing wafers. In addition, the extreme cleanliness and contamination requirements of a wafer fabrication area do not constitute a natural condition. Therefore, common sense may not make any sense at all. For instance, wiping the surface of a wafer to dry it or dust it off is the worst thing one can do, and even using an air gun can add more contamination than is being removed. Finally, there may be individuals who are new to the industry, who may not understand what they are trying to accomplish, or who do not speak common languages well enough to understand all of the rules and details. Many, if not most, of the problems in a wafer fab can be traced to poor or insufficient specification of the task.

These problems can be avoided through certain specification writing techniques. They are completely opposite to the common sense school techniques. A sample of this specification writing style is given in Appendix A. This technique is often called the "cook-book" style of specification writing. Specifications must be as precise as the technology they are trying to specify. If the operators must spend time to do a job right, the engineer should verify that the procedure is right and spend the time documenting the process exactly. Every detail should be observed and individually placed into the specification in order and as required. While creating the specification for a process, the engineer may discover ways of improving the process or may discover subtle problems in it. By observing the process in action, and verifying the exact sequence, the probability of writing an impossible situation into the specification is reduced. Writing an impossible situation into a specification has often occurred and is almost always associated with the fact that the individual writing the specification has never attempted or successfully completed the task that he or she is specifying. In my mind, the mark of the truly excellent engineer is that the individual has performed nearly every task that he or she requires others to do. This not only keeps confusion down, it allows the engineer to have some understanding of the problems faced by the various other groups in the wafer fab.

When writing specifications, nothing should be left to chance. In a sense, this is similar to programming a computer in that every option must be thought out or the computer may do something unpredictable. If there is an event that can have multiple outcomes, all of the potential outcomes must be covered, and if there is any possibility that there are yet more potential outcomes, contingency plans should be written into the specification to give a clear idea of what to do in the event of an unexpected result.

All key tasks and actions should be identified and outlined in the specification. If it is very long or especially complicated, an index may be in order. Subtasks can be organized within each of the major tasks, and may not have to be indexed unless they are especially complex. It is important to make sure that all subtasks of a step are included in the specification. Using the computer program analogy again, if a step is left out or if the description is incorrect, there will be a significant probability that an unpredictable event will occur.

Specifications should be written in the style of a specialized cookbook. That is, all of the ingredients are laid out, the process conditions are outlined, the exact sequence of events is described, and the process is tested to make sure it came out "done." The sequence should be in the order that each event happens, including subtasks. In some cases, this can make writing the specification somewhat difficult, requiring some creative writing. However, if the task can be performed it can be described, if not in writing, then through pictures and diagrams. In the case of duplicate or repetitive procedures, a common specification or section of a spec can be referred to. This is acceptable practice as long as it is done sparingly. The writer must be careful not to write something that requires the operator to jump from page to page in the specification manual.

In general, specifications should be written in a modular format so that changes in one procedure that are common to a number of processes need to be made once only. The advantages of this are several, including the prevention of mistakes (forgetting to change one specification out of many is a common error), consistency from process to process, and ease of paperwork and training. From the writer's point of view, this is preferable as it makes the writing of a specification much simpler.

Finally, in order to obtain contracts from the military and certain other customers, specifications must conform to certain standards and must cover certain points. If the specifications are written with the above-mentioned details in mind, there is a much higher probability that the specifications will pass a military spec audit (which occur periodically). In addition to specifications, military and other high-reliability customers usually require certain tests to be performed at key steps, with clear documentation for each lot. This must be clearly defined and outlined in the specs. Many fab areas have trouble with military audits as a result of inadequate specifications and incomplete lot audit trails. It is much easier to write the specifications with these goals in mind before starting into the audit cycles, as opposed to being forced to write them "under the gun."

2.7 LOW-YIELD ANALYSIS

Up to this point, we have spoken about yield improvements through mostly proactive methods. However, there are many times that problems do not show up on process control reports. Instead they show up at the sort steps as wafers with reduced yields. Sometimes the loss can be very significant; for instance, the loss of 50 to 75% of the expected yield. When these wafers show up, they are not usually processed. If there are only a few wafers with poor yield, they are usually removed and the rest of the wafers shipped on. If there are very many wafers in a lot with low yield, the entire lot will probably be held for low-yield analysis.

There are a number of steps that are followed while examining low-yielding wafers. They are outlined in the flow chart given in Figure 2-8. The first task is to identify the failure mechanism. Often, certain types of failures are identified as particular types of problems, and can be classified as such. After the failure mode is identified, the lot history should be analyzed. Look for something unusual at any step that could have caused the problem. This might include a discrepant material report, meaning that the run exceeded its specification limits. It might be a high particle count at some step, or other underlying problems that, summed together, caused some parameter to fall out of line. Usually, over 50% of the problems will have been determined through the first checks of the lot history.

The next step is to start electrical testing if required. Often, the test die or scribeline test patterns are available for tests. They may include oxide breakdown, contact resistance, continuity, and other tests. In some cases, special test modes are designed into the chip.

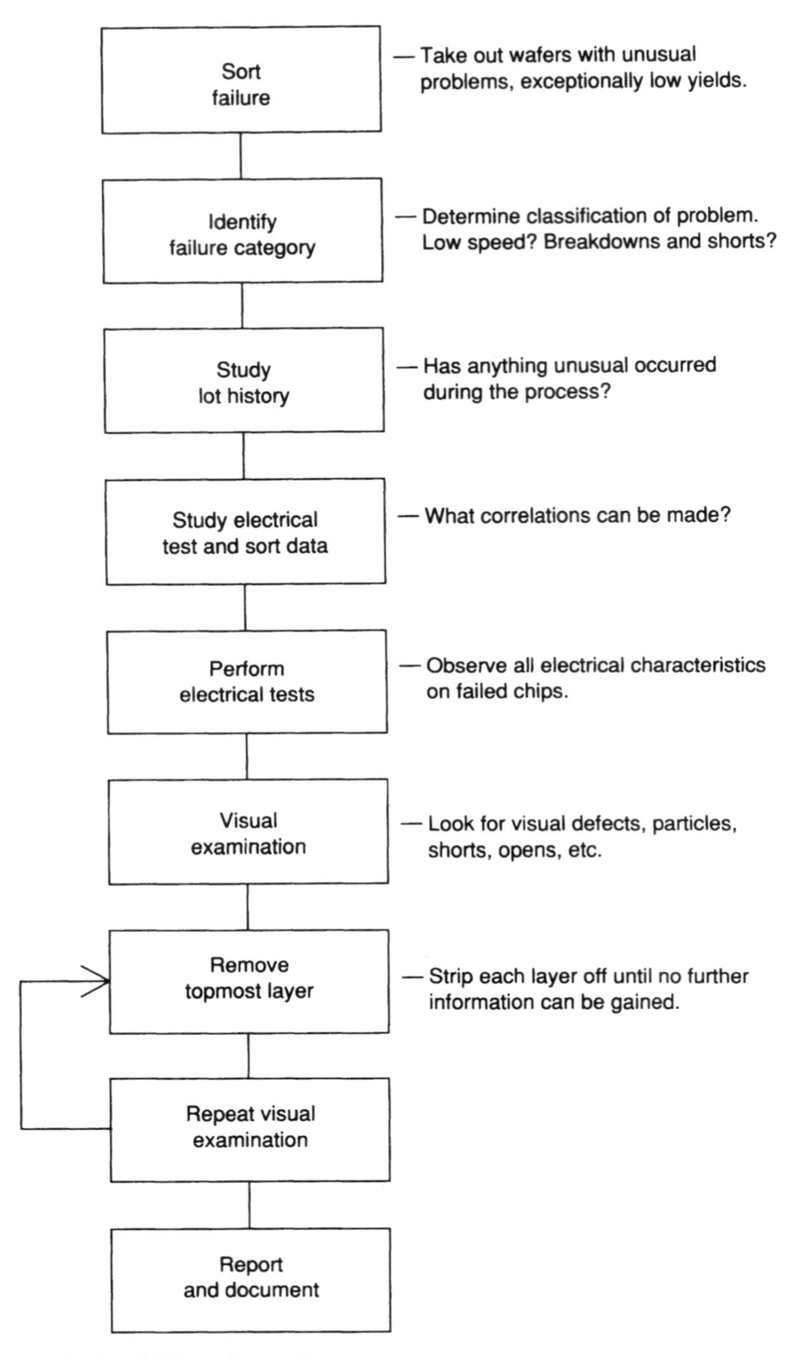

Figure 2-8. Low-Yield Analysis Flowchart Low-yield analysis should be performed carefully to reduce error. The raw data available on any batch of wafers should be examined carefully before damaging or testing the wafers. Similar techniques should be used for studying high-yielding material for high-yield analysis.

These modes can be utilized with probers programmed to run the chip in these states.

If the problem has not yet been determined, the next step is to strip back the circuitry layer by layer, in an attempt to see the defect. This must be done very carefully to avoid causing damage to the circuit that could be misinterpreted as the defect in question. After each layer is etched off, the wafer is thoroughly examined in those areas of the chips that have failed. Often, particles and other subtle problems are discovered while undergoing this kind of inspection.

After the problem area has been found, it is crucial that the low-yield analysis (LYA) engineer report the findings to the appropriate process engineer, so that the problems found at LYA can be addressed directly in the process. The removal of these types of defects will result in fewer overall wafer losses at the sorting operation, and will usually improve the yields on wafers that were not affected to the point of significant loss.

Unfortunately, even the best of LYA methods cannot identify all problem wafers. A certain percentage of failures, often in the range of 20% or so, cannot be positively identified. A typical range of problems found is shown in Figure 2-9. There can be a number of reasons for this, from "invisible" particulate defects (actually just too small to be

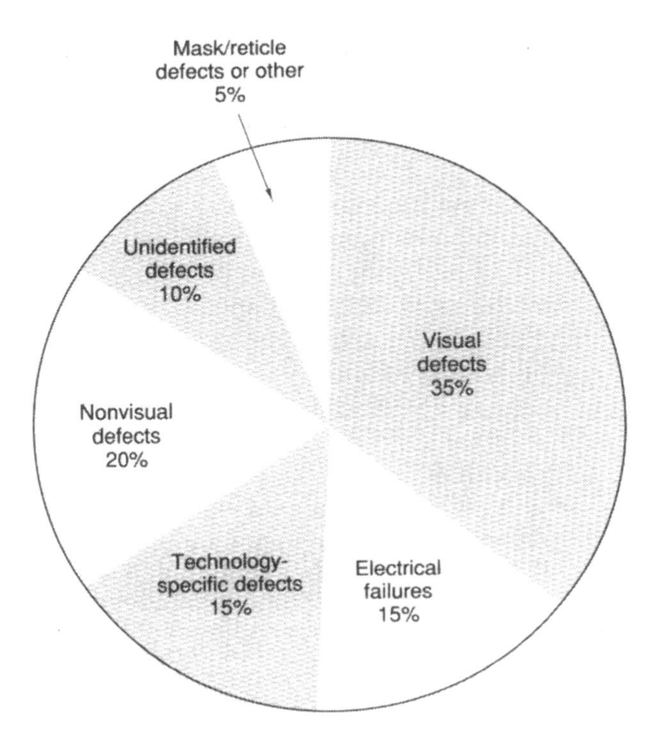

Figure 2-9. Distribution of Yield Failures This pie chart demonstrates the typical failure modes that occur in the wafer fab LYA procedure.

found using the relatively crude methods used for removing the various layers on the wafer), to unexpected chemical contamination. In any event, the number of problems that can be found using LYA methodology can result in a drastic increase in the output and performance of the wafer fab.

2.8 HOUSEKEEPING

To control yields at the highest possible level, housekeeping becomes a constant chore, with something happening almost continually, whether it is mopping the floor or cleaning the furnaces. Housekeeping specifications must be clearly written and scrupulously followed by all personnel to guarantee the maintenance of the clean environment. Schedules for all tasks must be set up, and individuals identified to perform each of the specific jobs. Especially important is the idea of consistency. A manager who goes through a fabrication area in improper cleanroom attire can cause as much damage to fab housekeeping efforts as a chemical spill, in the sense that the personnel will not feel the same commitment to perfection if they think the rules apply to them only. The rules must apply equally to everyone.

It is important that the upper management levels understand and support the housecleaning efforts of the fabrication area, and not view them as additional headaches. These techniques can improve the workers' pride in their work, thereby improving productivity and yields. In addition to the support, it is important that the rules are applied consistently in all areas. Just as it undermines efforts for individuals to appear exempt from the rules, it can also be detrimental to allow inconsistency in cleanliness, depending upon which part of the fab area one is in.

Finally, there are many times when there are so many problems with fab defects that special defect-reduction teams must be named. These teams should have representation from all groups at all levels. Operators should nominate a representative or two (since they are the ones who handle the wafers the most); maintenance, management, engineering, and other groups that have vested interests in the fab area should all attend. The ideas given by any group should be welcomed, and all ideas discussed and, if found to be effective, implemented. Often, the individuals who work on the fab line itself have the most insight as to solving a particular issue. Engineers and managers must never belittle these ideas, as many times there are small things that can be done which will result in major improvements. Remember that some of the larger pitfalls of hunting for problems (or for resolving them) occur as a result of trying to design much more complex solutions than are really required by the problems at hand.

2.9 CONCLUSIONS ON YIELD

In this chapter, we have covered a lot of territory, involving a large number of issues. In many respects, this chapter has been about the costs of uncontrolled yield and uncontrollable defects, and about a number of management techniques to bring order to the fab. The large effects of the lack of quality control were investigated and found to impact the bottom line in the range of hundreds of thousands of dollars. Just-In-Time, management by exception, precision specification writing, statistical process control methods, and low-yield analysis techniques are all used to reduce the effects of defects and to eliminate sources of defects of many kinds. The general rules for all of these ideas are simple, however. They involve allowing all of the personnel to do the very best they can, and to inspire them to strive for perfection. The tools they need to make their jobs easy must be put in place. The fab management team must be willing to listen to the on-line personnel and to properly support the new ideas. The requirements that are getting consistently tighter, and the parameters that are continually becoming more critical, make it essential that every possible technique be tried to keep the process together. It takes a team effort, and the costs of failure are very high.

CHAPTER 3

METHODS OF SEMICONDUCTOR ANALYSIS

We have discussed the major impact that defects have on die yield, reliability, and wafer line yield. Clearly, if there is to be any progress made on the reduction of the various types of contamination, the fab area personnel must have the ability to detect and identify that contamination. The contamination and other defects can be classified into three basic categories: particulate, chemical, and mechanical. The methods of detection of these three items are quite different from one another and will each require a separate discussion. Typically, particulate testing is accomplished by observing the medium to be tested with a laser or high-intensity lamp. Chemical contamination is tested through a variety of electrical tests and through specialized analytical gear. Mechanical testing is done through visual observation, sometimes involving chemical preprocessing, and also through some very specific procedures. The data to be gathered must include the following: determination of size and distribution of sizes of the defects (the size of particle, length of scratch, or surface area affected in the case of chemical contamination), determination of the nature of the defects (in other words, is the defect a hole in a film or a bump on top of the film; is the contamination haze or small particles?), determination of the extent of the problem (defects per wafer, impurity concentrations), and, finally, what the distribution of the defects is (do they cover the wafer uniformly, are they distributed in any particular pattern).

Detection of the defects mentioned so far can be a difficult task. For example, typical maximum particle sizes allowed in wafer fabs can be dozens of times smaller than the human hair, usually 0.2 to 1 μm in diameter. The particles are often invisible to the most thorough visual inspection with a microscope. Chemical contamination can be even more difficult to detect and analyze, as potentially damaging concentrations of the contaminants are often on the order of parts per billion, and may reside only in a small layer right at the surface of the wafer (which of course is where the integrated circuit is). Furthermore, each generation of integrated circuits is becoming more sensitive to chemical contamination. Another concern when dealing with chemical impurities in the wafers is the existence of atoms that are not necessarily contaminants, but which are detrimental to the performance of the films or other structures in which these atoms lie. For example, hydrogen trapped in a silicon nitride film can cause the film structure to undergo stress, which can result in film cracks or other problems. Some of these problems can be very subtle. Often, problems of a chemical nature emerge only after many hours of operation of a device. As a result, continuous full-cycle testing of various crucial parts of the circuits is necessary in order to show evidence of a problem.

Naturally, since the detection of the defects is so difficult, the price of the analytical gear has gone up as a direct function of the requirements of the industry. A good machine to measure 0.2-μm particles on the surface of a wafer costs well over $150,000, and a typical wafer fab will need two or three, at least, to meet output requirements. This is yet another case where a cost versus performance decision is required. While the ideal situation would call for the fab area to have one of these devices per major operation, that would probably be prohibitively expensive. If too few of the machines are purchased, problems with throughput can arise, with productivity suffering as fab personnel are required to wait for machine time, or must move to other areas of the wafer fab to use the machine. Data can be compromised, as it can prove difficult to distinguish the source of particles that may have appeared on the wafers during a process from those that may have been added by the box in which wafers were carried to the inspection machine. In addition, the use of larger wafers and the use of very small geometries increases the costs of the measurement machines and often increases the time required for contaminant measurements. Finally, due to the sensitivity of the machines, regular preventive maintenance is a requirement. As with any state-of-the-art instrument, there are often additional unexpected failures, such as laser burnouts or photomultiplier drifts, that will require periodic repair or recalibration. Therefore, the maintenance requirements must be analyzed when deciding how to purchase test gear. As usual, a fine balance must be met between the needs of the wafer fabrication process and the output (therefore revenue) requirements while making any decision to purchase large quantities of test gear. Table 3-1 lists a variety of types of equipment used for measuring chemical, particle, and mechanical defect levels of silicon wafers.

TABLE 3-1
Semiconductor Analytical Equipment

Equipment	Technique	Parameter monitored
Particle counters		
Surface	Laser/white light	Particles, haze, scratches, stress
Fluid		Particles, bubbles
Airborne		Particles
C–V Probe	Capacitance–voltage	V_{FB}, Q_{SS}, film ionic contamination, dielectric constant, others
Film thickness analyzers		
	Spectrophotometer	Film thickness, refractive index, optical constants
	Ellipsometer	
	Prism coupler	
	Acoustooptic	
	Profilometer	
	Eddy current (for metal films only)	
Linewidth analyzers		
	Optical	Linewidth, pitch, registration
	SEM	
Chemical analysis test gear		
	FTIR	Film stoichiometry, bond structure
	UV–VIS	UV transmissivity, other for EPROMs
	X-ray diffraction/Fluorescence	Dopant concentration in films
	Chromatography	Chemical content, boron in BPSG
Film stress reflectance test		
	interferometry	Wafer bowing translated to stress
Inspection		Visual inspection for defects
Optical	Electronic/optical	Automated defect analysis
	SEM	High-resolution defect inspections
Resistivity	4-point probe	Dopant concentration, resistivity
	IR spectral analysis	

3.1 PARTICLE DETECTION METHODS

To properly test for the various types of particle and chemical contamination, we will need to identify the key kinds of equipment that serve the needs of the fab. We will also need to devise test sequences to isolate the problem areas. In this section, we'll focus on the particle counting devices, their general techniques of operation, and the analysis of the data that are retrieved.

3.1.1 Equipment for Particle Detection

The detection of particles is a very active field of study, so much so that whatever is written here could seem dated by the time this book is published. However, the basic methods of detection are fairly well established at this point. The basic idea is to scan the sample, whether a gas, a liquid, or a wafer, with lasers or other high-intensity light sources. Particles will show up as bright, tiny point sources of light, which are picked up by sensitive photodetectors and analyzed by a computer. These data are used to tell the operator the size, count, and other attributes of the particles in the sample. The white-light machines have more limitations as far as minimum detection size capabilities and sometimes lack some of the more exotic features, but can be used for many tasks and are relatively inexpensive.

Airborne-Particle Detection

Airborne-particle counters are relatively easy to construct, and are relatively less expensive than some of the other types of detectors. In most cases, the exact location and type of the particle detected is irrelevant or is not directly measurable by the detector, and the determination of size and number of particles is the main requirement. This requires that the laser sources and detectors remain stable and that the air flow through the machine is carefully calibrated and maintained.

As shown in Figure 3-1, the instrument brings an air sample through the laser chamber at a fixed velocity. The laser scans through this airstream and results in a reflection when it hits a particle. The detector then picks up the reflection, the intensity of which varies as a function of the size and reflectivity of the particle. When a reflection is detected, the instrument registers the particle's reflectance, calculates

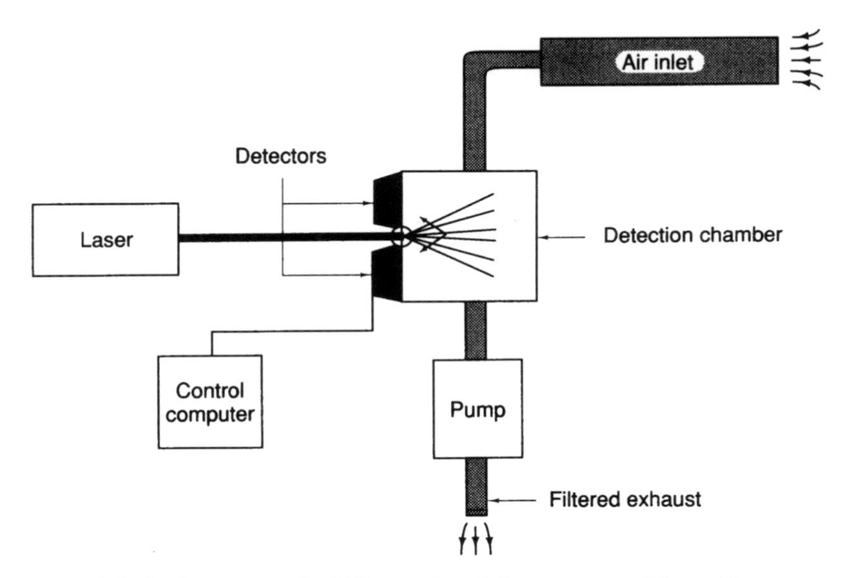

Figure 3-1. Block Diagram of Airborne-Particle Counter The airborne-particle counter draws cleanroom air into the detection chamber at a calibrated rate. Particles drawn into the chamber with the air are illuminated by a laser, from which reflections can be obtained and analyzed.

its size, and reports that back to the user. There are some drawbacks to this technique of measuring airborne particles. First is how one guarantees that the air sample has not been contaminated within the tubing or in the machine itself. It is nearly always true that if you shake the inlet line of the machine you will generate a significantly higher count than if the inlet tubing is held still. The contaminants may build up until they fall off in a large clump, showing up to the user as a sudden huge increase in particles, followed with instabilities in readings for some time after the occurrence. The next drawback is that placing the inlet too near or too far from people can significantly impact the validity of the test results, producing a certain amount of ambiguity. Finally, the machine must make assumptions about the reflectivity of the particle which may not always hold true.

The amount of subjectivity and ambiguity that results from the use of this type of machine makes the term "Class X" cleanroom more difficult to define, even with precise numerical industry guidelines specified (see the chapter on cleanroom design). One can easily find ways of modifying the tests in subtle and usually quasi-legitimate ways, so that the results come out more or less favorably, depending on the requirements. Obviously, this is not a recommended practice, as the important thing is to find out what the defect levels seen by the wafers are, not produce impressive reports. In general, if one really wants to know the conditions in which the wafers are being processed, the readings should be taken as near the station under consideration as possible during operation, whereas if one wants to know the background of the room only, then the samples should be taken in the work-station areas, but with no one in the area, and the room allowed to stabilize for a while, typically around 30 minutes. These data should be taken at regular intervals in order to build up a database of expected values before initiating any major action based on the readings. Be aware that there is a significant amount of uncertainty in any individual reading.

Liquid-Borne Particle Detectors

The automatic detection of liquid-borne particles is very similar in many respects to that of airborne-particle counting. As shown in Figure 3-2, these devices pump a measured quantity of liquid through a chamber which is illuminated by a bright light or laser, and the reflections of the particles are picked up by photodetection circuitry. This, however, is a very rapidly changing field with serious technical difficulties. These difficulties have limited the capabilities of liquid-borne particle detectors and have allowed continued use of the older but better characterized manual counting methods. While there are a great many technical problems that have had an impact on the development of liquid-borne particle counters, the advent of more powerful, faster computers and more reliable detection methods have provided a basis for dramatic improvements.

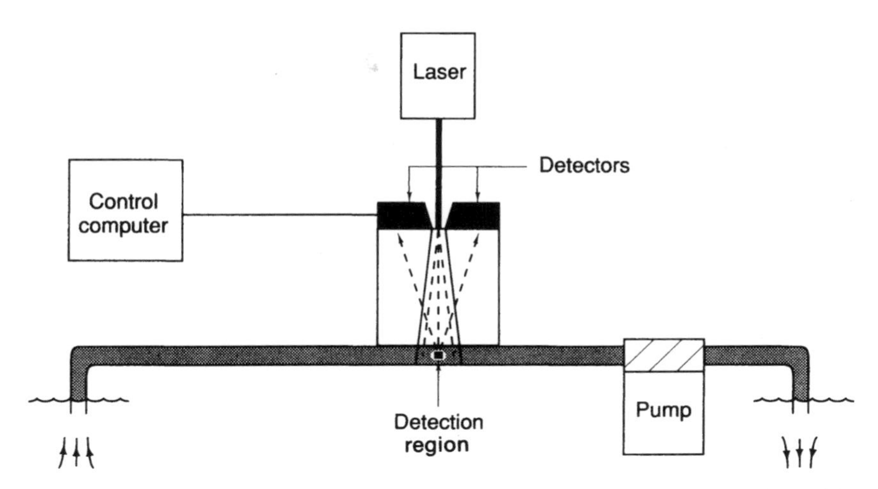

Figure 3-2. Block Diagram of Liquid-Borne Particle Counter The liquid-borne particle counter is very similar to the airborne-particle counter, drawing in the liquid and illuminating it with a laser. Particles show up as reflections in the stream of liquid.

One of the difficulties in measuring liquid-borne particles is that liquids, unlike most gases, refract and absorb light, often to a very significant degree. Different chemicals will be sensitive to different wavelengths of light, so a laser system may not work for all chemicals (while the red He-Ne laser is transparent to most common wafer fabrication solutions, there can be problems with some chemicals). Also, as chemicals are dissolved into the solution, color and other attributes of the chemical may change, without any indications of large particles. If this change in characteristics is too great, the calibration of the machine may vary. Some sort of reference beam may be needed to measure changes in transmitted light intensities or color. Also, the machine must control the flow and consistency of the fluid very carefully. The fluid should always remain in a smooth laminar regime or turbulence may induce false readings.

Another problem with liquid-borne particle counters is the formation of microbubbles in the solution. We will come up against microbubbles in a number of areas, and will find that there are many problems created by the existence of these very elusive little creatures. In many cases, pumps, restrictions, nozzles, valves, and other items cause a large amount of turbulence in the fluid stream. If these turbulent regions find a pocket of air, or in some cases, other types of fluids or contaminants, a bubble can form. This is called cavitation. The surface tension of a microbubble is very high, which tends to hold the bubble together for an inordinate amount of time. However, it does eventually break up, sometimes leading to confusing and contradictory results. Their transient nature makes these microbubbles hard to quantify, and makes finding the source a complex task. They can range from less than a *micron* to several *microns* in size. Since these bubbles will scatter light in a manner similar to that of particles immersed in the solution, their existence within the particle counter will disturb the final readings. Of course, since the bubbles are not particles, they can reflect light back in some unexpected ways, which can lead one to the conclusion that the "particles" or bubbles are of drastically different size than they actually are. In other words, it may be possible that a 1-μm bubble reflects light the same way as a 2-μm particle and will be interpreted by the detector as the larger particle. The unique characteristics of bubbles can allow some degree of identification capability, but because this takes a significant amount of processing power, this remains a significant problem area. Trying to find the source of microbubbles can be a difficult and unrewarding job. As a result, the only reliable way to keep from having microbubbles cause disturbed readings is to not allow them into the system in the first place.

Finally, there is one additional problem that must be considered in the case of fluid-borne particle detection. Since the sample fluid must flow through the system at a known rate, the flow meter portion of the instrument must know the type of fluid coming through the system, in order to account for viscosity and heat-carrying capacity of the varying solutions. If the system believes the type of fluid is different from what is actually being used, it will miscount the amount of the fluid. This can result in grossly distorted results.

Clearly, there is room for innovation in the liquid-borne particle counting market. While the problems are not insurmountable, a successful liquid-borne particle counter will be a complicated machine. It will have to be able to measure any fluid with arbitrary characteristics, and be able to discern between bubbles, particles, waves, and other anomalies in the fluid. There are already a variety of devices on the market that can measure particles in specific fluids, and by the time this is published the number of instruments available will have undoubtedly increased.

Wafer Surface-Particle Counters

Ultimately, it makes little difference if the room air is clean, the fluids used in the fab are clean, and the rest of the fab area is spotless, if the wafers themselves end up dirty. Therefore, the key indicator for determining the impact of contamination on production wafers is to measure the surface of wafers processed in the same manner as the production wafers. This is accomplished with surface-particle counter devices. The reason that test wafers are required is that, at this point in time, there is no generally accepted method for detecting particles on top of fully patterned wafers, although there are several companies working very hard on the problem.

While it is relatively easy in concept to measure particles on a smooth, bare silicon surface, largely through the use of holographic and other optical effects, the issue becomes very complex indeed when one wishes to measure particles on a patterned wafer. There are thousands of different patterns and techniques of putting those patterns on the wafers, and many different films on top of the substrates that absorb light in different ways. A design feature may have the appearance of a defect on one set of designs, and a defect on another design may appear to the machine to be a design feature . The first instruments used to measure in-pattern defects were designed for masks and reticles, since there is only one type of pattern on each.

There are a number of instruments used to investigate the contamination present on the surface of blank test wafers. These instruments use either a laser or bright light to look for reflections of particles off the surface of the wafer. The computer breaks up the surface of the wafer into a set of pixels in two dimensions for simplicity's sake and speed of computation. As shown in Figure 3-3, the laser scans along a predetermined path while the wafer is moved under or through the beam. The amount the sample moves is carefully gauged in order to control the reliability of the readings. The light is reflected from particles into a photodetector above the scanning area. The signal is picked up and sent into an amplifier. This signal is analyzed, and from its characteristics and timing, the exact location of the particle within the scanned area can be determined and some characteristics of the particle can be discerned. For instance, a large particle will produce a bright, sharp reflection and will return more light than a smaller particle. Figure 3-4 shows some sample waveforms for various particles on a wafer surface. The defect may occupy more than one pixel or may also be detected on a subsequent scan. Haze can also cover more than one pixel, and can appear as large areas of diffuse reflection. By examining any phase shifts and other features in the reflected light signature, it is possible to determine whether the particle is on top of a film, is embedded in the film, or is buried under the film.

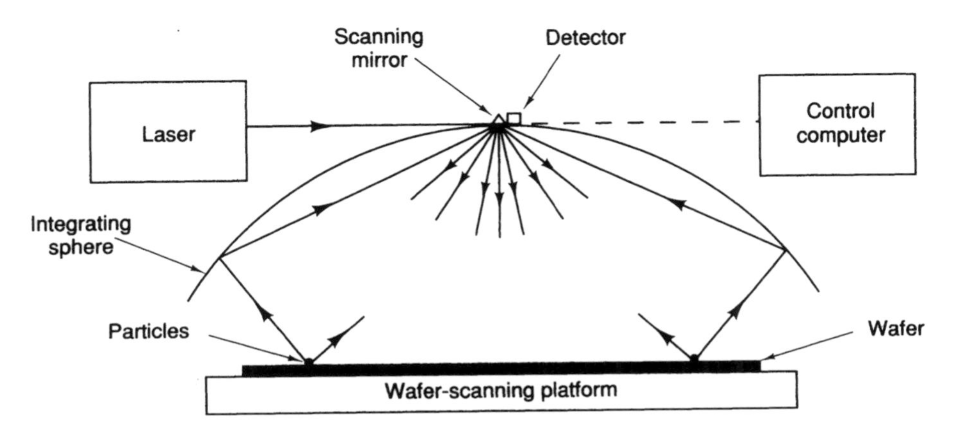

Figure 3-3. Block Diagram of Surface-Particle Counters The surface-particle counter operates by scanning the wafer under a laser. Particles on the surface of the wafer will reflect light back to the detector in a characteristic fashion. The reflected light is gathered and intensified through use of the integrating sphere.

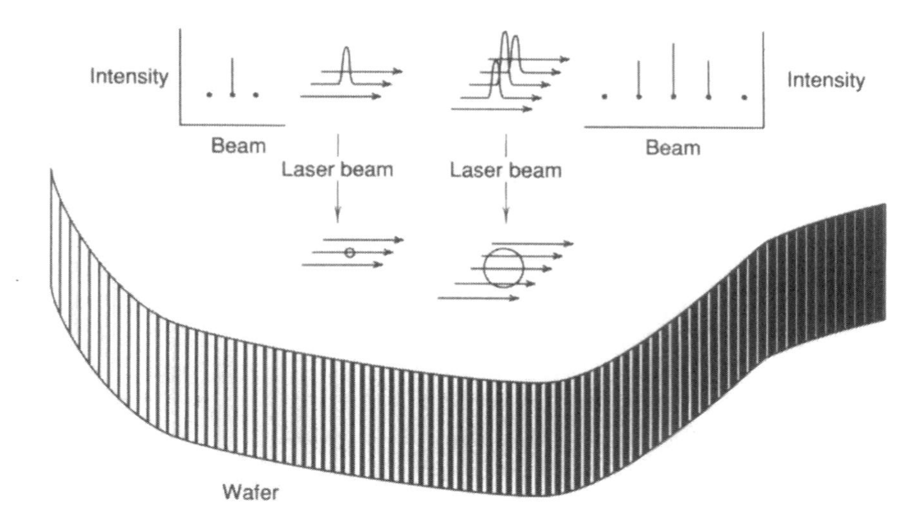

Figure 3-4. Particle-Counter Responses The particles on a wafer will reflect light back in predictable fashions, including total intensity of the reflection and the number of scans in which the defect is detected. These parameters can be used to identify both the size and type of the defect.

There are some regions of film thicknesses where the light from the red He–Ne laser gets mostly absorbed. A dark blue (around 1000 Å) silicon nitride (Si_3N_4) film absorbs a significant amount of the red light, which can cause the particle counter to see an unusually large number of defects. This response is because the smaller particles reflect more light relative to the substrate with a dark film, thus being more easily identified. Obviously, if any substance or substrate is being tested that absorbs the particular wavelength of light that is being used to perform the measurement, the detector must be properly calibrated to allow for that absorption prior to its use. This is usually a simple task, and in many cases the most common films are preprogrammed into the machines.

The surface-borne particle detectors work by moving the wafer under a beam scanning what is essentially a straight narrow line across the wafer. The beam spot size on these machine is usually in the range of 10 to 50 μm. Smaller spot sizes produce more accurate readings, but involve more information to process. Therefore, either more time or a more powerful CPU is required to analyze the data. Computer algorithms used to do this kind of work are becoming very complex, and the quantities of data being analyzed are also increasing rapidly as a result of the latest methods, which we multiple overscan techniques (similar to the techniques used by CD players). Usually, there is enough overscan involved so that the laser beam will actually illuminate all points on the wafer two or three times from slightly different angles. This allows the computer to more precisely analyze the surface. The instrument will search for specific patterns while the data is being analyzed. For instance, if a series of pixels in a row show defects, that may indicate a scratch. In another case, several pixels in an area may be covered with defects that scatter the light in a certain fashion. This may be interpreted as haze on the surface. The amount of haze can then be determined by the amount of light that is scattered and the amount of diffusion of the light from the substrate. Finally, the overscan abilities also allow the machine to determine the shape of the particles to some degree as well as their size.

There are some problems with this technique. The first has been pointed out already, which is that this technique has limitations when used with patterned wafers. This limitation will be overcome in the fairly near future. One method allows the computer to analyze the topography and determine the uppermost pattern. Generally, it will only be necessary to look at a variety of die, determine the "average" expected pattern, and then test each chip to verify that there is no deviation. It then looks at any deviations from that pattern in more detail to analyze the anomaly. From this look, a particular signature would be generated which could be compared to expected defects. This type of machine could be programmed to look for all types of defects, including mask defects, once the basic algorithms were created.

Another difficulty with this detection method is that it is relatively substrate-dependent. The changes in substrate reflectivity range from the almost totally absorbing nitride films just mentioned to the very highly reflective surface of a wafer with aluminum deposited on its surface. This requires a thorough recalibration and gain or sensitivity adjustment of the photodetection circuitry. In many cases, the machines can self-calibrate by reading a set of data, either from read-only memory (ROM) or disk files. The instruments that are delivered with a variety of preset film types will usually cover substances such as bare silicon, several film thicknesses of silicon nitride, several film thicknesses of silicon oxide, and some types of metal films and substrates.

The basic calibration method assumes that initial calibration of the system is done with calibrated latex spheres on top of a polished, totally reflecting surface. Incidentally, the latex spheres are the ones manufactured on the space shuttle, as they are exceptionally uniform in characteristics and size. In this procedure, wafers with each of the various types of films have a calibrated number of latex spheres deposited on them. Then the threshold at which these spheres are detected is specified, and the machine is checked and calibrated by adjusting either the light intensity or the photodetector gain to fall within reasonable ranges. At this point, the settings are then stored and the exact number of particles on the wafers noted. This data is used whenever one of the specified films is measured.

One film on which it is particularly difficult to measure particle counts is polysilicon. The film itself is deposited in relatively large grains that vary in size from a few tenths of a micron to several microns. These grains produce a scatter that makes it nearly impossible to identify particles much less than 4 to 5 μm in size. This problem can also occur with otherwise bumpy films, such as some BPSG films. In these cases, a surface particle counter has only a limited amount of usefulness, but can still be used to determine gross problems or haze.

Wafer surface-particle detection is one of the main tools in the fight against particles on wafers, as this is the only method that gives even a close approximation to the conditions seen by the production wafers. The results are very objective, and are many times better than the visual techniques used for so many years. Particles can be seen down to an incredibly small size and accurately mapped, so that process trends can be discovered.

Observations about Particle Contamination

When particles are measured by any of the preceding methods, certain distinct aspects will be noticed. The first and most obvious is that the number of particles increases nearly exponentially as the minimum size of the particles observed is decreased. This is not a terribly surprising result, confirming the basic intuition that, if a few particles are visible at any size, then more will exist that we cannot see. What may be surprising is the rate of increase of the particle counts. A wafer with five or six 2-μm particles may have dozens of 1-μm and hundreds of 0.5-μm particles. A typical defect distribution curve is shown in Figure 3-5. The slope of the curve can change drastically as a result of the type of processing the wafer has received. Chemical vapor deposition processes of all types will produce a profusion of "particles" below a certain size. These are likely to be associated with the grains or bumps of the film as-deposited. Other processes may produce relatively more large particles due to the production of the particles as a result of some sort of abrasion or other mechanical problems.

In general, a good rule of thumb is that very large particles are probably generated due to equipment-related problems, whereas very small particles are often process-related. This may not always hold true, but is a basis for the start of a search for a defect source. The reason for this rule of thumb is that particles generated by abrasion tend to be in the 3- to 5-μm and above range, and the processes tend to create smaller particles (see the chapter on CVD processing), usually under 1 μm. The particles in the intermediate ranges often cause a fair amount of consternation, as they can be formed in a variety of ways, and are often "composite" particles, or particles that start off small and slowly accrete material until they become much larger. It is also important to view the pattern of the defects on the wafer when making determinations on particle sources. For instance, the clustering of particles at two or three points at the edge of the wafer may indicate that the particles are coming from diffusion quartz boats, or a radial distribution around the center of the wafer may indicate a spinner problem.

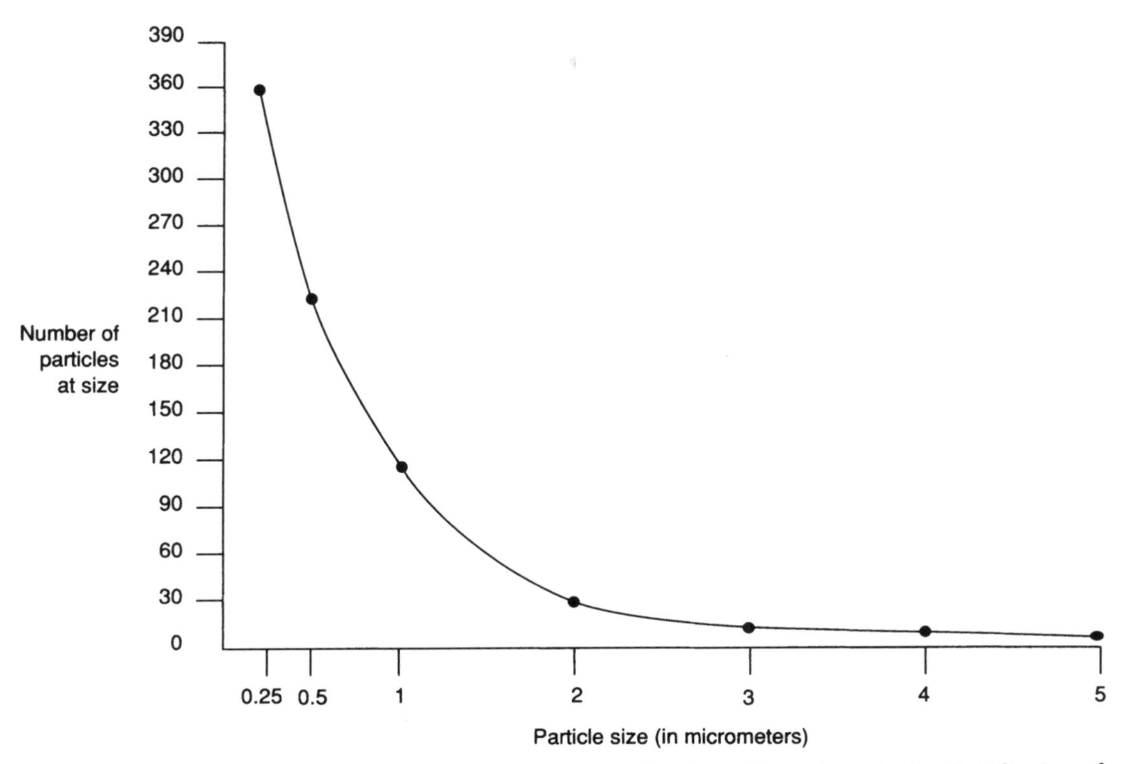

Figure 3-5. Typical Distribution of Particle Sizes Even though the distribution of defects on different wafersis never the same,they will generally exhibit the attributes of the sample defect size vs. distribution curve.

Another feature of small defects is that, as particle size decreases, the force required to remove the particles increases. This is primarily due to the fact that, as particle sizes shrink, the charge/mass ratio and surface-area/mass ratio increase. This is described as van der Vaal force. This causes attractive forces on the wafer surface to be relatively stronger for small particles than for those of a larger particle. The binding force as a function of particle size is demonstrated in Figure 3-6. The adhesion of these particles is increased even more if the surface is bare, etched silicon. The free, floating bonds of the silicon will attach themselves to the particles with many times the force with which they will be attached to oxidized silicon.

3.1.2 Methods of Particulate Testing

There are a number of methods for using the various particle detectors to determine the source of particle contamination in a wafer fab. The steps are similar in many of the cases, and in any event the concepts used to carry out the tests are generally consistent. One of the important facets of the testing program is that the individual steps must be repeated enough times to give a reasonable level of statistical validity.

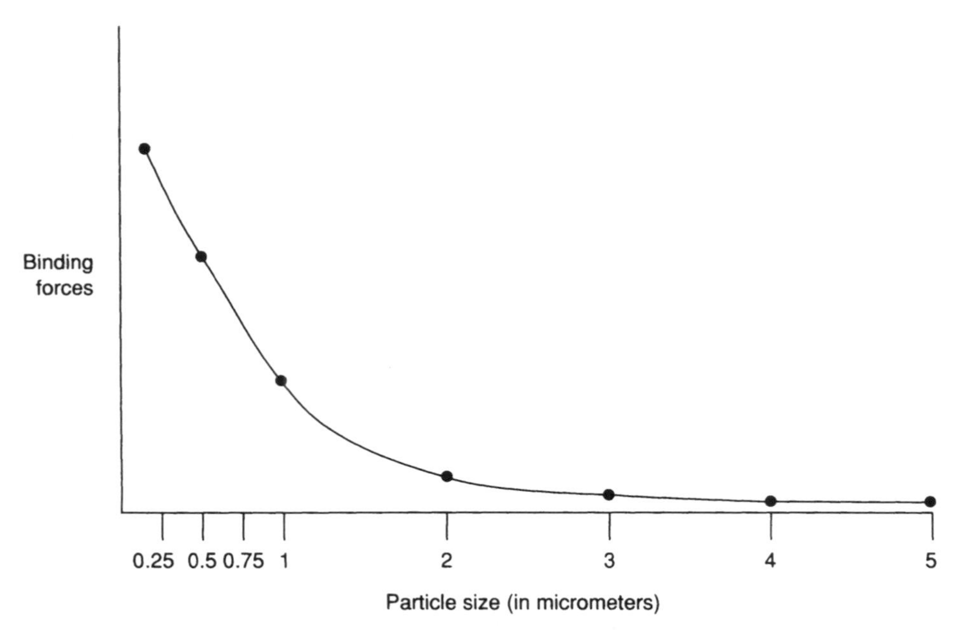

Figure 3-6. Binding Force as a Function of Wafer Size As particle size decreases, the forces that bind the particles to the wafers increase nearly exponentially. This is why the smallest particles must be prevented from adhering to the wafers in the first place.

This type of testing is prone to wide variations in the data. Variations from area to area in a wafer fab and from day to day within any area can be as high as 200 to 300%. In fact, one of the main goals of the particle reduction efforts in a wafer fab is not only to be able to attain a certain level of cleanliness, but to maintain it from day to day. One of the dangers in embarking on a defect reduction program is not doing a sufficient job in creating a basic database of fab area performance prior to introducing changes. If the data is not clear, improper changes may be made, which could prove to be detrimental to the clean-up program. In the long run, just keep in mind that the end result is to achieve consistency and scientific control in all fab area cleanliness experiments, and there will be fewer problems and fewer blind alleys.

Airborne Contaminant Levels

Since the wafers will almost always be transported or moved through the air of the clean room, it is important that this air remain clean at all times. The methods for keeping the fab area's air clean are discussed in the chapter on cleanroom design; however, we will discuss the basics of the testing procedures here. The tests for determining these conditions are actually quite straightforward. Some need to be done on a regular basis, such as daily or once a shift, while others may need to be performed only occasionally in order to verify that proper conditions are being maintained. The most common tests to be run in the fab area are the air-flow velocity test, the wafer layout test, and the airborne particle count test. These tests can help identify equipment problems and gross failures and can be used to track and identify the more subtle sources of contamination. The instruments used for measuring air velocity and airborne particulate are portable, whereas the layout tests require the use of a surface particle counter. The portable units are sometimes hand-held, and are quite easy to use.

Now we get into the area of work-station particulate testing. Each of the three basic tests—the air velocity test, the airborne particle test, and the wafer layout test—should be run routinely, and precise logs and process control charts kept up on the results. Most of these tests should be run daily if not every shift, and at the very least, spot checks should be done on all shifts. The exact locations of the tests should be clearly defined, so that there is no ambiguity about the results. In general, the tests should be run with the room in as close to standard running condition as is possible. Since the ideal case is to monitor the operation in progress, the individual performing the tests should find suitable locations for obtaining accurate data without interfering in the operation of the fab.

When the factory clean-up plans start to go into effect, it will be necessary to run tests on the room with the operation in process and with the room empty. This will give the engineer the information as to the contribution of the activity of operators in the room as well as the background count due to the environment itself, and will allow for the construction of an accurate contaminant dispersal model. The data generated from early sets of tests should be rechecked periodically to verify that new procedures, equipment, and personnel introduced to the room are not causing room cleanliness to decline.

The work-station airborne-particle tests are quite simple to perform. The technician merely stands with the measurement device near the area to be tested and takes a sample using the instructions accompanying the measurement equipment. In particular, it is important to keep the probes pointed away from the person taking the sample, as the interaction of the technician with the room may contribute an unpredictable amount to changes in air velocity or particulate contribution. In the case of the air velocity meter, results will be generated almost immediately. In the case of the airborne-particle counter, each sample usually takes one minute. There are sometimes "tricks" to get the most stable readings from the particle measurement machines, such as shaking the inlet line before starting the tests to remove any excess buildup on the tube walls that could come off and cause erroneous readings. If unusual readings are recorded, it is often a good idea to thoroughly shake out the line and wait until the counter restabilizes (usually 10 to 20 minutes), then re-run the tests. After the results are obtained, they should be recorded in the process control log immediately.

During the initial phases of the fab analysis, samples should be taken in a number of areas of a process room. If there is a wafer cleaning area or system in the room, that would be a prime location to perform a test. Another excellent location to evaluate would be a load station, or staging area. Another location that should be tested for airborne particles is around any test or inspection gear that may used with production material. Several samples should be taken in each position, and the contribution from a number of directions should be calculated. A reasonable test sequence for these early tests would be to take five samples from the north direction, turn to the east, take five more samples, and so on through the south and west directions. Each of these sets of readings from each of the positions would then be averaged and recorded. Each of these four average readings could then be averaged again and recorded as a more general single data point.

The full set of data is useful to the fab engineer in the event that a particle problem relating to the room environment is found. This data may point to some evidence of its source. The data are also useful for day-to-day trend analysis of work area cleanliness. Since these tests should take place in each room or work station of the fabrication area, clearly there are a lot of tests to be performed. Therefore, it is highly recommended that the routine become computerized as rapidly as possible. After there is a sufficient history on the various areas in the fab, the number of tests can be reduced to just the main areas of the fab, with fewer readings in only one direction.

Incidentally, the units of measure of the airborne-particle counters are usually given in particles per cubic foot, as required by regulations, and as such are directly related back to the room cleanliness standards. The standard is designed around the number of particles of given sizes that can be found in the air at any point in time. The characteristic curves for the cleanroom standards are shown in Figure 3-7. It is clear that the standards recognize the effects of the increase in particle count as size decreases. In part, the lower size standards are defined since, if these levels cannot be obtained, it is unlikely that the levels of larger particles will be reached.

The wafer layout test procedure is somewhat more complex, and certainly far less standardized throughout the industry than the airborne-particle test, but can provide a significant amount of information if done correctly. Each test itself is simple and is performed as follows. The number of particles on a wafer are counted using a surface-particle counter. The wafer is then placed onto a tray in a predetermined location in the fab area, typically near a work station. It is left there for a certain amount of time, usually one hour. After that, the wafer is removed from the tray and remeasured in the surface-particle counter. As shown in Figure 3-8, in that time you will expect a certain amount of air to flow over the surface of the wafer. A certain percentage of particles that have been free in the air and that have actually gone past the wafer will have landed on the surface of the wafer and adhered. In this case, the changes in the defect levels of the wafer from day to day will indicate changes in the room quality levels. While the

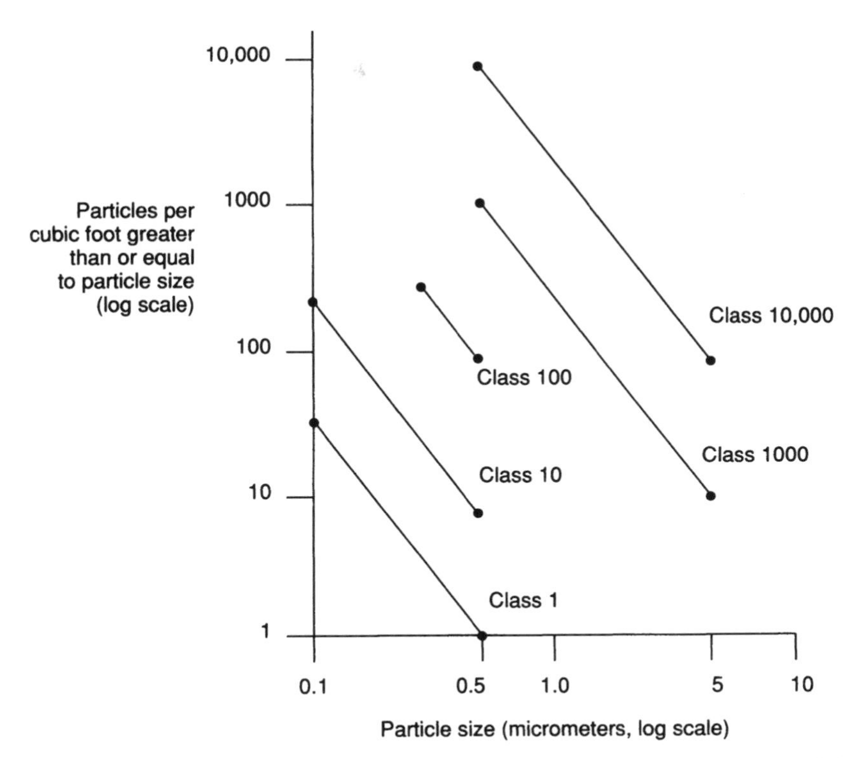

Figure 3-7. Cleanroom Definition Curves The federal government has defined cleanroom standards, as shown in this graph. The standard particle size is defined as 0.5 μm, as measured by a light-scattering monitoring method (as described in the text).

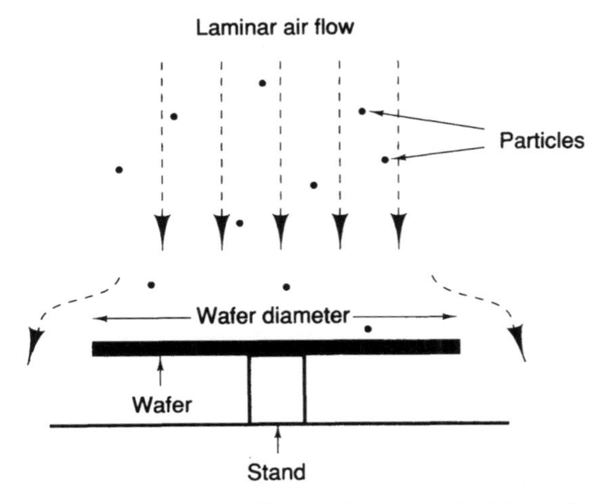

Figure 3-8. The Wafer Layout Test The wafer layout test is performed by placing a wafer under the air flow of a work area. Since the air will be moving at a steady velocity, the number of particles in the air can be estimated by the number that land on the wafer. If the air velocity is 5 feet/minute and the wafer area is 0.2 square feet, 1 cubic foot of air passes over the wafer each minute. Although some particles will not adhere to the surface of the wafer, a correlation can be made between the number of particles seen on the surface of the wafer and airborne-particle counts performed nearby.

results from this test can vary due to a fairly large number of variables not associated with the room cleanliness, the wafer layout test is considered to be one of the most sensitive for determining what the wafers themselves will actually pick up. This test places an object of equivalent properties directly into the region of interest for a long enough time that some real information can be obtained. If there are particles on these wafers, there are probably particles on the production wafers.

One of the problems with this test is that it is not standardized, and by definition may not be able to be completely standardized. For example, how one defines the exact location to place the layout wafer in a variety of different fabrication areas can be critical. It has been shown that there is an impact on the amount of particulate deposited as a function of location in the room environment. (For example, the particles deposited on a layout wafer will increase if placed too near a wall.) However, as a general rule of thumb, when using five- or six-inch wafers, one to three particles per hour landing on the wafers is a very good result. If the numbers are almost always zero, the test period should be lengthened to half of a shift or a shift. With particle counts of greater than five particles per hour, there is room for improvement, but the problems are usually traceable to personnel or handling sources. Counts above 10 to 15 are often caused by equipment failures, or by too many personnel in the cleanroom. Very often problems with high counts are associated with maintenance work in the room in question. This phenomenon should be observed closely. If it is determined that certain procedures cause particle counts to go up in the work-station area, the operation should be shut down during the times that these procedures are being performed, or the procedures should be modified to prevent contamination. These types of tests should be run as often as airborne particle tests (once a day, at least), and preferably in conjunction with those tests. The results from these layout tests should be monitored as closely as possible and should be tracked on process control charts.

Finally, there is one other type of airborne test that can be run to simulate airflow patterns within the cleanroom. This simple test is excellent for conceptual testing of air-flow dynamics and particle trapping sites in the clean room, especially in regions around equipment. The test is commonly called the smoke test. This must not be done when wafers are present in the room, and in fact should be done only when a major change has taken place in the room, such as the addition of new equipment. It is also highly recommended that liquid nitrogen, steam, or other modern, noncontaminating, and easy to clean "smoke" sources be used. Older sources burn oil, which will thoroughly mess up your clean, new fab area. The main areas to test are the interfaces between the laminar flow hoods, the exhaust areas, flows over tables and around equipment, at doors and entrances to other areas, and finally around clean sinks in all directions as these sinks have their own exhausts and will set up regions of influence between themselves and the room. The desired effect is to have the smoke travel in the straightest, least perturbed line possible. If there are regions where it is impossible to remove turbulence, such as the interface region between the fab air exhaust and a clean-sink air exhaust, the exhaust flows from the sink should be optimized to reduce the area of this turbulence to a minimum. Another item to be on the lookout for is leakage of air either into or out of the cleanroom. While leakage out is not usually a problem, leakage coming in probably will also allow contamination to enter the cleanroom. Corners and wall connections should be checked for fit using these methods. Finally, if there is some area of inherent turbulence that cannot be further reduced, the placement of equipment, a wafer storage area, or a work area in that space is discouraged. Use it for a logbook station instead.

Liquid-Borne Chemical Testing

As mentioned, the issues involved with the development of liquid-borne particle tests cause the complexity of the test procedures to increase and the control and repeatability of the test results to decrease. As a result of the problems with in-situ methods of particle counting in fluids, most of the tests that have been devised for isolating particles in liquid solutions have been manual, involving the use of test wafers to simulate the production wafer environment.

These tests can be time-consuming and laborious due to the fact that each solution and system must be tested independently. While it is possible to combine the tests into easier to accomplish test matrices, generally the tests must be done in a precise sequence and must cover all of the various possibilities or the results may be skewed, if not completely incorrect. It is important to identify the particulate source quickly and precisely, as intrusion into the clean system based on incorrect information could cause a new particle source to be created in addition to time lost in failing to solve the original problem. It is important to maintain the proper sequence of the tests in a clean station and to verify that the tests are performed properly in a manner similar to that of production wafer cleans. In many cases, the tests should be performed from the last operation (the dry step) and progress forward through the other steps in the clean so that all of the subcomponents of the clean are evaluated. Finally, as always, any test procedure must be repeated in a statistically significant number of samples, which are required due to the wide uncertainty associated with these types of tests.

There are some clean stations that have been set up for in-situ testing of the process fluid. To use the liquid-borne particle counters that are currently on the market, they must be installed directly into the piping loop. The monitoring stations are specific to one step in the process, and while the same techniques can be translated from one chemical bath to another, the equipment must be set up for that particular solution. In these systems, it is important to identify the sources of the particles that are seen in the detector. It is also important to determine whether there are certain times that give more reasonable results than others. For example, testing the solution immediately after wafers have been placed in the solution may not be particularly worthwhile if all of the particles dissolve in the solution within the span of the clean sequence. However, if the particles have not been removed before the next batch of wafers is prepared for cleaning, there will be a potential for adding contamination. In fact, the increase in particles through time in an unfiltered acid bath is often the main reason that the acids or other chemicals are thrown away. Often, the chemical strength of the acid has not been reduced by any significant amount even though there may be an unacceptable number of particles. However, it is much easier to filter some solutions than others.

The clean station with in-situ particle monitors should be tested in a fashion similar to the following. First, the system should be cleaned thoroughly and all acids and filters changed as required. For initial tests, you should use new acids and filters. Then the system should be monitored while operating until the particle counts stabilize. The time will vary, depending on the initial level of contamination of the tank/chemical/filter combination and on the rate of filtration of the chemical. In any event, it will probably take about half an hour to an hour to stabilize unless there is a problem. If possible, when first being evaluated the system should be allowed to recirculate but remain unused except for test runs for several days before any conclusions are drawn. The performance of most filtered acid systems will usually improve very slowly with time over a period of several days to a week. This test sequence needs to be done once only and provides information that will indicate the lowest baseline value at which the individual tank can operate.

After the acid has reached stability, a series of tests should be run to determine the impact that the cassettes (or boats) and wafers have on the cleaning solution. First, 5 to 10 empty Teflon® cleaning boats should be placed into the solution one at a time, and the amount of increase in particulate noted, along with the amount of time required for the particle counts to reduce to a stability. Then remove the boat and try the next one. This will give information on how much particulate is carried around on the boats, as well as indicating of the recovery time of the chemical bath. After this, the boats should be filled about half full with wafers and the test sequence repeated. Then the boat should be completely filled and the test sequence repeated. This data can be used as a good baseline for controlling the process. For instance, you now have information on the acid recovery rate as a function of number of wafers introduced to the solution. You can use this information to decide if the amount of time that the wafers are submerged in the solution is sufficient.

After the initial test stage is complete, it will not usually be necessary to perform these extensive test procedures again. Monitoring the fluid particle levels at the end of the cleaning cycles should be sufficient for most needs. This data should be stored on a process control log. This data can be used later to tie defect and yield degradation problems to problems associated with the cleaning process. Finally, it may be necessary to repeat these tests to verify that all is under control after major changes have been implemented; for example, the replacement of an etch module.

As has been mentioned, there are only a few automated defect monitoring clean stations, and they have tended to be developed for very specific processes. As a result, there are a vast number of other types of tests used to verify chemical solution cleanliness. In this next section, we will be discussing the use of manual methods involving the use of test wafers to measure the quality of the clean. This is in many respects a more valid method, since the true measure of the clean process is the impact on the wafers themselves. However, this can be an expensive proposition, as problems with trying to reclaim wafers will prevent the repeated use of the test wafers. In general, the tests should not be performed with wafers that start with more than 20 to 30 particles on the wafers. Any more than this and the numbers will be skewed by the exchange of particles from the wafer surface back into the solution, and from the fact that some of the existing particles may simply move around on the wafer surface. While the test wafers must be handled carefully, they can sometimes be reclaimed for later use, at least a few times. The technique usually involves a strong sulfuric peroxide clean or an ammonia chemical scrubbing. At the high cost of test wafers, even a few reclaims can save a lot of money. Bear in mind that these tests have a larger margin of error than many of the tests described in this chapter due to the mechanics of the handling involved.

A typical wet-station testing sequence would consist of the following. First, a batch of 10 to 25 wafers would be measured on a wafer surface-particle monitoring device. The number and pattern of any particles would be noted for each wafer. Now, approximately three to five wafers would be placed in a clean Teflon® cleaning boat, and these wafers then placed into the spin rinse drier and the operation started. Note that, for greater precision, a test of the wafer transferring operation could also be run. Transferring of wafers, even with utmost control and delicacy, can sometimes add a few particles. In monitoring the production area, however, an excessive number of tests can lose cost effectiveness by taking up too much time or consuming too many test wafers. Therefore, the specification control limits are usually set so that the average amount of contaminant added is figured into the limits. In any event, the wafers should be processed in the spin drier first, and those wafers set aside. The next three to five wafers should be processed through the last dump rinse system, and then through the spin drier. These wafers should also be set aside. After this, the last acid clean sink should be tested with three to five more fresh wafers, followed with a rinse in the previously used rinse tank and drying in the spin drier. This should continue until all of the sinks have been tested. After this, all of the wafers should be remeasured on the surface-particle monitoring machine.

At this point, you will have a compilation of the data from each of the various steps. What is required is to remove the cumulative particulate addition from each step from the previous step. This is clarified in Table 3-2. Thus, in order to know the particles generated in the last rinse sink it is necessary to average the readings from the drier-only test and also to average the numbers from the rinse-sink/spin-drier combination, and then to find the difference between the two: net particles = total clean − drier contribution. In each case, the count from each step in the clean can be calculated by taking the number from that operation and subtracting the average number from the previous operation. This is why it is so important to make sure that your database is statistically sound and that the tests are carried out correctly, or small errors can be introduced that are later magnified. In any event, if this data has been collected properly, you will soon have a list of the net particles added (or removed) from each of the clean sinks in the process station.

If there is a particular problem with one area of the clean, this sink can be tested independently, by putting wafers into that clean sink, then rinsing and drying only. The test described above is more indicative of the total process clean, but will not always identify particular areas of problems.

When these wafers are observed, certain things will stand out immediately. First, certain acid cleans always tend to show more

TABLE 3-2
Example Results from Clean-Sink Testing[a]

Test	Item	Total count	Net count
1	Spin rinse drier	2	2
2	Final dump rinse[b]	6	4
3	$HCl:H_2O_2$[b]	12	6
4	Dump rinse	34	22
5	HF	37	3
6	Dump rinse	40	3
7	Sulfuric peroxide	42	2

[a] When analyzing a particle problem in the clean sink, a careful sequence of experiments must be performed in order to isolate the source of the contamination.

[b] Problem areas are the post-HF dump rinse and the $HCl:H_2O_2$ solution.

particulate problems than others. This is true in the case of hydrofluoric acid, for instance. In this case, particles added when the HF etch step is included in the test procedure may not be indicative of particles suspended in the acid. As noted, the freshly etched surface of the wafers attracts particles much more rapidly than typical wafers, which have a thin layer of silicon oxide on the surface. Even if the HF acid is clean, any particles suspended within the rinse water will be attracted to and will attach themselves to the bare silicon surface. The effects of each must be isolated.

A simple test for determining whether the particles are added by the HF or by the water is as follows: Take two wafers (or two statistically sound groups of wafers, although usually the results from this test are so striking that statistical deviations are irrelevant) and place them into two wafer cleaning boats. Immerse them into the HF acid simultaneously, then remove them and place one into the rinse tank and continue the process through to the drying step. Place the other wafer in a safe place to air dry. (This will not take long as the water and acid will not adhere to the bare wafer surface.) Place the two wafers into the surface-particle counter. If there are a large number of particles on both sets of wafers, there is contamination in the HF and possibly in the water. If the air-dried wafer is dirty, and the rinsed wafer is clean, the HF is probably dirty, and the water is very clean. In the vast majority of the cases, however, the finding is that the rinsed wafer has more particles on it than the air-dried wafer. This implies that the water has more particles in it than the acid. This should not be surprising since many semiconductor vendors filter their HF acid, but do not use point-of-use deionized (DI) water filtration.

Another observation that might be noted when analyzing the test wafers is that there sometimes appear to be layers of contamination on the wafers. This is shown in Figure 3-9. This is usually indicative of a solution that has received insufficient agitation or stirring. These solutions include $HCl:H_2O_2$ and phosphoric acid/water combinations. The various chemicals do not mix well, and may be of sufficiently different density to separate, and are often mixed at fairly high temperatures, which can accelerate that separation. The particles in the solutions will tend to be trapped at the interface between the chemicals; they are then transferred to the wafers in the form of a line or level on the wafers.

It should be clear from the preceding discussion that wafer cleaning is an area where a significant amount of work needs to be done. Characterization and control of the fluids in the wafer fab can be a very difficult task, but does have its rewards, as yields are usually enhanced by a very significant amount if the clean operations are controlled. As an aside, the wafer cleaning operation is typically one of the areas in a wafer fabrication area where the most significant problems can occur, and usually is one of the least monitored and controlled of the fab area processes. This small lapse of focus often

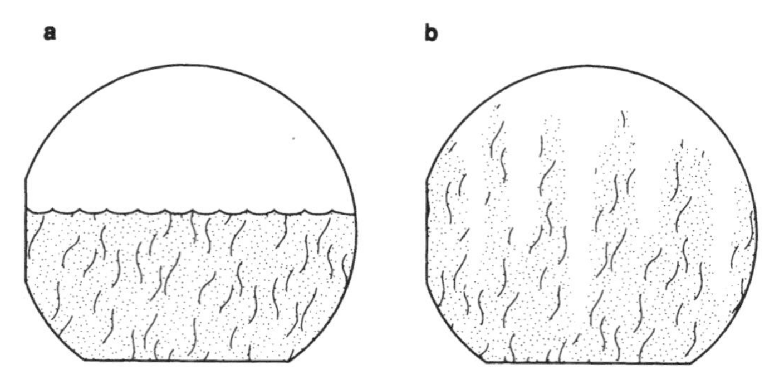

Figure 3-9. Chemically Streaked Wafer Patterns Two common problems that occur due to the use of mixed chemical baths are shown here. (a) Two chemicals of different densities are used. If they are sufficiently miscible, separation can occur if constant agitation is not applied. Particles in the mixture will tend to be trapped at the fluid interface, with streaks below the line. (b) The two chemicals are of similar density, so that the streaks appear more uniformly across the surface of the wafer. While more common, this problem is more easily addressed through the use of filtered chemical baths and magnetic stirrers.

leads to significant impacts when the inherent problems of the clean operations rear up their various ugly heads and there is no information as to what happened, why, or how the problem can be avoided in the future. Perhaps it is because of the perceived lack of glamour in the clean stations and operations, but often too little attention is paid to the clean processes until there is problem. Nevertheless, the requirements for a good, controlled, and reproducible clean cannot be emphasized enough. It is important that consistent, statistically sound methods and a systematic approach be employed in the attempt to eliminate particles and other contaminants from being introduced from the wafer cleaning process before they can be propagated throughout the factory.

Surface-Particle Counting Methods

This is in many ways the easiest of the testing procedures, in that the mechanics of the tests themselves are very easy to describe and make a good deal of intuitive sense. Essentially, the wafers are observed before and after a process is performed on them, and the difference between these two readings is taken to be the quantity of particles added for that step at that defect size range. There are, however, limits to this concept, and certainly the tests must be performed in controlled ways or the results will be unsound.

We have performed a number of tests using the surface-borne particle counter; they give the general idea of the specific tasks that must be performed. These tasks include a number of key items to keep in mind when designing wafer particle tests. One of these items includes running a sufficient number of test wafers in a test to calculate reasonably valid standard deviations with the data. The results of this type of test can be skewed by a number of factors. For example, many of the particle counters on the market are orientation-sensitive. While some of the machines have attempted to reduce this type of problem, it still can become significant, especially when testing CVD films. In the CVD systems, there are often subtle grain-structure and orientation changes that can cause the wafer to appear quite different when viewed from various angles.

To maintain statistical significance it is necessary to build baseline data on the quantities of particles added through all the subcomponents of a process. This holds true no matter what the process, whether diffusion, photolithography, or any other. These subcomponents include the steps that are necessary in order to carry out the testing sequence. In other words, the acts of taking the test wafers from the storage box, placing them onto the particle counter, and then placing them into a wafer transfer box for transport around the fab should be analyzed individually as part of the initial testing process. This contamination must be controlled also, as this kind of false information can cause a significant amount of effort to be spent trying to find and resolve a nonexistent problem. Any time an unusually high particle count is obtained, a second test should be run to verify the results of the first run prior to major action being taken.

To test any process with wafer particle-counting techniques, it is necessary to break the process up into its subcomponents, just as the wafer transport was broken up. As we can see from Table 3-3, the LPCVD process testing may include the quartzware loading sequence, the push and pull cycles, the pumpdown and backfill cycles, and finally the full process cycle. Careful consideration must be given

TABLE 3-3
Example for LPCVD Particle Testing[a]

Test	Item	Total count	Net count
1	Load/unload	8	8
2	Standby gases	12	4
3	Pumpdown/backfill[b]	20	8
4	O_2 on	21	1
5	Grow LTO (SiH_4, O_2 on)	23	2
6	Grow PSG	26	3
7	Grow BPSG[b]	38	12

[a] Finding contaminant sources is similar in LPCVD processes to the procedures described in Table 3-2. In this process, however, contaminants that are present on the surface of the wafer, but are too small to be noticed, may become highlighted and much more apparent after the deposition step.

[b] Problem areas are the loaders, the pumping and backfill sequences, and the boron source gas.

while creating the tests to make sure that all of the major variables are accounted for. For example, if the gas system tests and the pump cycle tests are checked simultaneously, particles that are observed may have come from a number of sources within the rapidly changing process chamber environment. If sufficient control is maintained during the testing procedure, it may be possible to discern potential problems as a result of the patterns of particles that are deposited on the wafers.

3.2 CHEMICAL CONTAMINATION ANALYSIS METHODS

There are a number of methods for testing the impurity levels of semiconductor devices. Many of these methods can be used to test for quantities of desired dopants as well as for measuring the impurity levels of the devices. This capability can give these instruments some marketing advantages. However, if a particular contaminant is sufficiently lethal, the purchase of specialized test equipment can often be justified. In this section, we will cover the methods for testing for these contaminants, without going into great detail on their interpretation, which will be left for the appropriate chapters.

3.2.1 C–V Testing

One of the more common tests that is performed in a wafer fabrication area is the C–V test. This stands for capacitance–voltage test, and is a direct measure of the quality and purity of a dielectric film. This test has become the standard procedure for guaranteeing the quality of the diffusion and other high-temperature equipment and C–V failures can result in equipment being temporarily shut down or cleaned. The procedure itself can yield far more data than is usually used on a day-to-day basis, and a careful examination of C–V results can provide the engineer with a clearer understanding of the quality of the process. Some of the factors that can be calculated from C–V tests include dielectric constants, mobile ion densities, fixed (trapped) ion densities, surface state charges, flatband and inversion threshold voltages, and substrate dopant levels.

Although I will not attempt to duplicate all of the excellent work available on the subject of C–V testing, I will discuss the basics of the testing procedures as used in a wafer fab. There will be a further discussion of C–V results in the electrical test and diffusion processing chapters, and much more information can be found in the bibliography.

Capacitance–voltage qualification is usually done on all equipment that is used in high-temperature operations. The various operations that typically receive C–V qualification are listed in Table 3-4. This has typically meant only the diffusion area but, increasingly, even lower temperature processes, such as metallization and poly-

TABLE 3-4
Processes Requiring C–V Qualification

Process	C–V wafer type	Attributes tested
Initial oxide	Grown	Chemical contamination
Field oxide	Grown	Contamination
Gate, source, drain oxides	Grown	Surface conditions, film quality, contamination
Anneals	Known good oxide	Cross contamination
Polysilicon	Known good oxide	Tube quality, chances of hazing
Metal deposition	Known good oxide	Work function, alloy quality, metal purity
Silicon nitride	Deposited	Surface, states, dielectric constant, film quality

silicon deposition steps, are being tested for their intrinsic C–V characteristics. In addition, in many cases C–V qualification is required on clean stations and other room temperature process equipment. The main reason that the higher temperature equipment is usually the first tested is simply due to the fact that contamination moves much more freely in the higher temperature environment, both between pieces of gear and the wafers and between wafers, as well as within the semiconductor structure itself. The primary contaminants that are monitored are sodium, which is mostly delivered to the wafers through human contact, and phosphorus or other process dopants that have been delivered to the wafers through improper cleaning or handling.

The test procedure itself is somewhat involved and requires cooperation from a number of groups within the fabrication area. The operators must take new wafers (these can sometimes be of test wafer quality, although sensitive process lines may require the use of prime wafer quality) and run them through a standard process clean cycle followed with a special-process sequence through each furnace. The wafers themselves can be *p*-type or *n*-type, and are usually the same type as the base substrate of the technology being produced. In most cases, the substrate is *p*-type. In any event, the substrate bulk resistivity should be between 20 and 40 ohm·cms. Wafers that have resistivity above that can exhibit strange behavior, such as switching from *p*-type to *n*-type if the oxygen concentration is too high. Wafers with very low resistivity can give inconsistent results and should be avoided. If more precision is required, substrate oxygen and other dopant concentrations should be controlled or monitored.

The structure of the C–V test cell is shown in Figure 3-10. We will describe the construction of this structure in detail. This sequence is shown in Table 3-5. There are often two wafers processed through

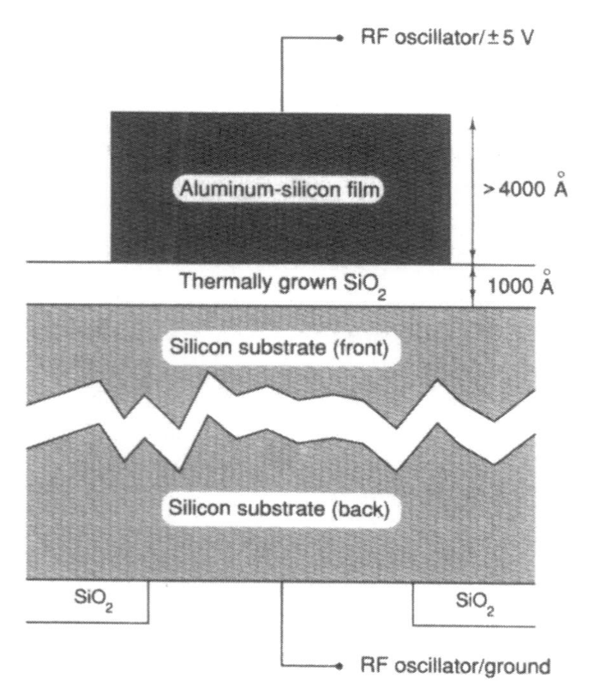

Figure 3-10. The C–V Test Structure The C–V test structure consists of a spot of aluminum about 5 mm in diameter on top of a silicon dioxide film. This film and a cleared area on the back of the wafer are attached to the C–V plotter electronics.

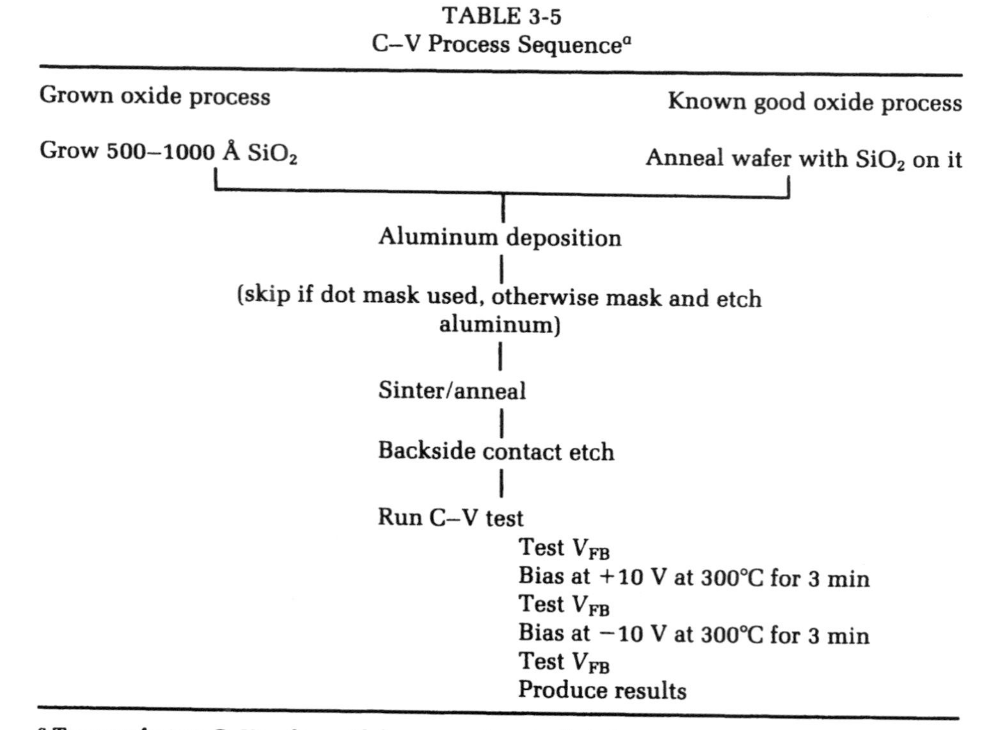

TABLE 3-5
C–V Process Sequence[a]

Grown oxide process	Known good oxide process
Grow 500–1000 Å SiO_2	Anneal wafer with SiO_2 on it

Aluminum deposition

(skip if dot mask used, otherwise mask and etch aluminum)

Sinter/anneal

Backside contact etch

Run C–V test
- Test V_{FB}
- Bias at +10 V at 300°C for 3 min
- Test V_{FB}
- Bias at −10 V at 300°C for 3 min
- Test V_{FB}
- Produce results

[a] To manufacture C–V wafers with known and controlled characteristics, a specific sequence of events should be followed. While not the only possible sequence, the procedure shown here will result in consistent, reasonable results.

each furnace to ensure that there will be no problem due to breakage or other contamination. In the processes that typically grow an oxide film, such as field or gate oxidation steps, the C–V process will grow the dielectric oxide film. Usually this oxide is 600- to 1000-Å thick. In processes that are not typically used for oxidation, such as anneal tubes, wafers are used that have had oxide grown on them previously. These known good oxide (KGO) wafers are processed in batches and are then qualified by running one or two of the wafers through the entire C–V process and analyzing their C–V characteristics while holding the others to one side immediately after oxidation. After this set of wafers has been qualified, they are placed into nonoxidizing furnaces and allowed to anneal at processing temperatures for half an hour to an hour. At this point there will be two types of wafers produced out of the diffusion area, one that has just had oxide grown on it, the other having just had an anneal cycle.

These wafers are then prepared for C–V contact preparation. This involves placing aluminum dots all over the surface of the wafer. These dots are about 1 to 1½ μm thick and 1 to 2 mm in diameter. Larger dots are sometimes used, but they usually give poor results. The aluminum must be relatively uniform in thickness and should not be thinner than about 5000 Å. Thinner films can cause measurement problems and are easily damaged by the C–V probes. The aluminum quality is very critical to the success and reproducibility of a C–V result. Sometimes it is thought that since the C–V tests use high frequencies, there is little sensitivity to the aluminum, but experience shows that to be a false assumption.

The C–V dots can be produced in one of two ways. The more precise and cleaner method is to deposit the aluminum uniformly over the surface of the wafer and then produce the dots through a photomask operation. The other technique is to mount an aluminum mask in front of the oxidized wafer inside the aluminum deposition system and allow the aluminum to be deposited through the holes in the plate. This process tends to produce inconsistent aluminum thickness and the mask itself can contaminate the wafer, thus invalidating the results. However, this second method results in much faster throughput times, and is a common method used in wafer fabs.

After the wafers have the dots placed on their surfaces, it is necessary to anneal the aluminum at around 400°C in a hydrogen ambient for about 30 minutes. This is done in the standard alloy furnace in the fab area. After the anneal process has been completed, there is a need for a backside contact on the wafer. This is produced by dabbing HF acid on the back of the wafer with a swab until the oxide is cleared from the surface. The wafer must then be rinsed and dried. Another way to remove the oxide from the backside of the wafers is to place straight (49%) HF acid into a plastic bottle, and then hold the back of the wafer in direct contact with the open container. The HF fumes that are present within the bottle will rapidly etch the oxide, leaving an excellent clean surface. Since the HF will not react with silicon, there is no chance that HF will be transmitted to the wafer surface so no rinse or dry is required. This method is quick and easy, but can be slightly hazardous if the fab personnel are clumsy at handling chemicals.

Now the basic C–V structure has been constructed, and the wafer is placed on the C–V stage, the probes placed on contact on the aluminum dots, and the test process started. Usually, three dots are measured and the best two out of three readings are recorded. Often there will be minor problems that cause one area of a test wafer to give poor results, so that no action should be taken on the results of one reading or one wafer.

The test itself involves scanning a range of voltages, usually from −5 to +5 volts. The polarity of this voltage is varied over the capacitor at around 1 MHz. While the voltages are being scanned, the capacitance of the structure is analyzed. A sample C–V curve is shown in Figure 3-11, and should be referred to throughout the discussion of C–V analysis. The exact position where all of the energy bands balance out is called the flatband point, and is used as a point of reference for C–V testing. Usually, the absolute value of the flatband voltage is not very critical as long as it is within a reasonable range, as it can vary as a function of many variables, including the oxide thickness and

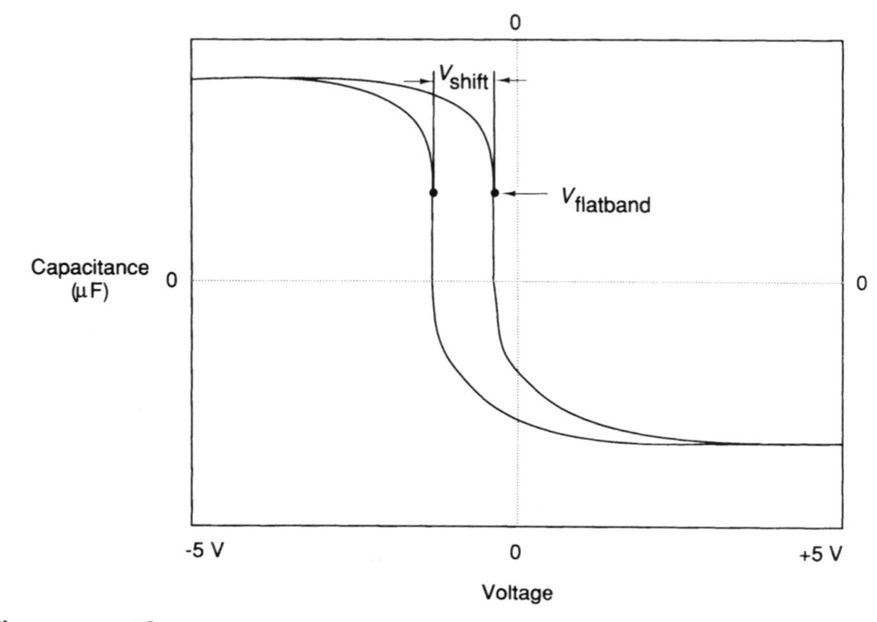

Figure 3-11. The Typical Capacitance–Voltage Curve The C–V plot is a measure of the quality of the dielectric. Threshold voltage, dielectric constant, and total capacitance can be determined by examining the flatband voltage and capacitance values. Chemical contamination of the dielectric film is determined by measuring the shift in flatband voltage after high-temperature processing.

metal quality. A typical flatband value for a 1000-Å SiO_2 film is around −1.0 volts and can vary as much as 0.2 to 0.3 volts.

The wafer is next processed through a heating cycle and the C–V plot is run again. Specifically, after the wafer has had the first C–V plot calculated, it is heated to around 300°C, and a constant 10 volts is placed across the capacitor. At this temperature, ions trapped within the oxide will move more freely and, in the presence of the electric field, will tend to move to one or the other of the oxide interfaces. The wafer is kept under these conditions for three minutes, which is an arbitrary but reasonable standard time. The C–V characteristics are then calculated again. If there were contaminating ions in the oxide film that moved, there will be a shift in the flatband point, which is then relayed back to the operator. Sometimes the high-temperature stress test is performed twice with the polarity on the capacitor reversed, and the C–V data taken three times. The amount of flatband voltage shift is then monitored as a measure of the amount of contaminant in the film. Usually, the maximum shift allowed is in the range of 0.1 volts, although some devices are more sensitive to contamination than this.

A number of traits can be identified through analysis of the C–V chart itself. For instance, the existence of a long tail at the base of the lower end of the C–V curve indicates that the plotter is driving the substrate into a condition called *deep depletion*. What this means is that the wafer surface is becoming further charged after the time it should have stopped. This occurs when there is an excess in the depletion of electrons from deeper in the substrate. This can be caused by unknown crystal contaminants or excessive damage to the silicon crystal, among others.

Another test that is often performed while evaluating the C–V characteristics of a wafer is the measurement of Q_{ss} and Q_{ox}. They both relate to the charges that have been trapped in the oxide or at the silicon/SiO_2 layer during the oxidation process. Q_{ss} refers to the charges at the interface, and are typically kept below a value of 1×10^{11}. They are controlled by oxidizing in chlorinated steam environments, and by preventing drastic temperature changes. In a sense, they can be viewed as "static charges," and can be controlled with procedures that will prevent static charges at the macro level. Q_{ox}, on the other hand, refers to immobile charges trapped within the silicon dioxide layer. These types of charges can affect the threshold voltages of MOS transistors, and the capacitance and charge storage capabilities of floating gates and capacitors.

There are a number of other pieces of data that may be obtained as a result of running C–V tests. They will be analyzed as required in the chapter on electrical tests, but some of these tests include information such as the number of surface state charges, trapped charges, substrate doping and depletion levels, and other data. The C–V plot machine is a very useful analytical tool.

Sometimes the C–V plotter is used to test the dielectric strength of thin oxides. This can work quite well up to a certain point, usually about 100 volts. Unfortunately, the typical C–V plotter does not have a sufficiently high voltage probe to be able to measure the breakdown of thick oxides and nitrides. The C–V plotter can be used for many gate dielectric applications. Usually, the breakdown test is implemented as a function separate from the C–V plot program. We will come back to the issue of gate breakdowns later.

As with all other fab area information, the C–V data should be kept in process control logs. It will be possible to detect more subtle problems that could be affecting the fab area by observing the C–V logs on a regular basis. For instance, if a bad batch of quartz cleaning chemicals has been brought into the factory, it may be noticed by seeing that a large number of furnaces that had been cleaned on a certain date have also been failing their C–V qualifications.

3.2.2 Optical Spectra Tests

I have placed a fairly large number of otherwise divergent tests into this category because they all have something in common, that is, they measure the interactions that occur between various wavelengths of light and the wafers. In many instances these tests can distinguish between the types and quantities of the various dopants.

The first type of test we will consider is the FTIR (Fourier transmissive infrared) test. Silicon crystals are transparent to infrared radiation at certain wavelengths. Since most of the dopant species and films absorb infrared, some very strongly, the nature of the films can be studied in detail. In the FTIR test, a wafer is placed into the processing chamber, where an infrared laser is directed perpendicular to the surface of the wafer. A block diagram of the FTIR is depicted in Figure 3-12. The intensity of the transmitted light at each frequency is measured, and compared to a sample spectrum. In a typical situation, the reference sample would have been the spectrum of the same wafer prior to receiving a process. For instance, as seen in Figure 3-13a, if a strong absorption peak is seen at around 1065 cm^{-1}, there is a coating of SiO_2 on the wafer. Since the absorption of the infrared is strongly influenced by absorption of certain molecular bonds, it follows that the FTIR can be used to study the film characteristics in more detail. For instance, a steep narrow peak may indicate a more pure and uniform structure than a broad and diffuse peak. This is shown in Figure 3-13b. Some of the items that can be analyzed with an FTIR machine in a semiconductor environment include the hydrogen content of deposited silicon oxide and silicon nitride films, measuring the boron or phosphorus content of BPSG films, analysis of deposited film structures, and purity.

Another optical technique for measuring chemical purity is to measure the fluorescence of the films or surface when exposed to

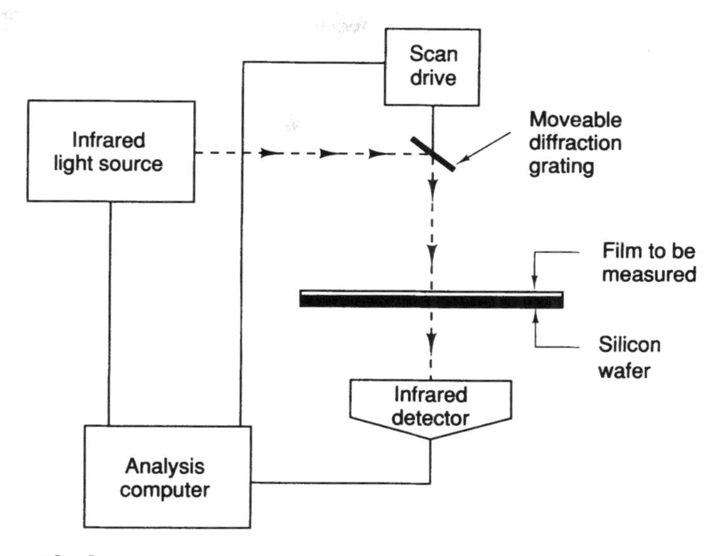

Figure 3-12. Block Diagram of FTIR Analysis System The FTIR is used to analyze the absorption of IR photons as a function of wavelength. This method permits determination of film constituents such as boron, phosphorus, and hydrogen.

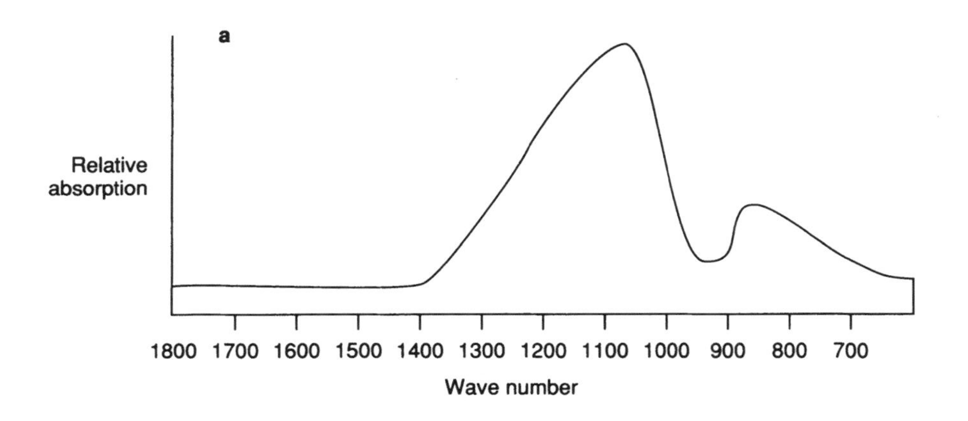

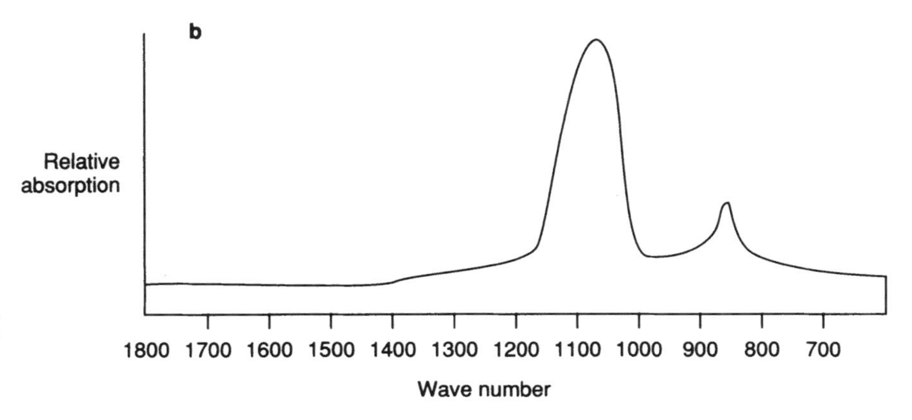

Figure 3-13. FTIR Analysis of SiO_2 Film Film structures and density can be determined through observation of the FTIR spectra of the film. Narrower absorption bands can indicate a reduction in the number of chemical bonds, which is typical of denser, purer films. (a) Broad absorption peak for SiO_2; (b) narrow absorption peak for SiO_2 using FTIR.

certain wavelengths of light. Some of these devices use ultraviolet light and others use X rays to induce fluorescence. Just as with the FTIR, the device scans through a range of frequencies, and compares fluorescence spectra against known standards to identify the dopants. These devices work quite well for discerning and quantifying a number of compounds. There are, however, limitations in certain areas. For instance, both the IR absorbance and the various fluorescence spectra of boron- and phosphorus-based dopant compounds in BPSG films are very similar, so similar in fact that neither of these methods is very reliable for determining dopant content of BPSG films. We will see that there are other chemical methods for determining these.

Ultraviolet spectrophotometers are sometimes used in a wafer fab to measure contaminant levels, especially hydrocarbons, and are also used by EPROM manufacturers to test the transparency of their protective layer films. They are often used to measure the reflectivity of metal films, which then can be translated back to changes in grain structure, and so on. Ultraviolet spectrophotometers are less common in the fab environment, but can be useful in some circumstances. New applications for these devices include the measurements of very thin oxides under 100 Å, and for measuring dielectric films on polysilicon and aluminum.

3.2.3 Electrical Test Procedures

Ultimately, the true test for quality of manufacture of an integrated circuit is whether the devices function properly when completed. For this to occur, all of the structures on the chip must be operating within the correct ranges. Since it costs so much to manufacture a wafer, it is important to find out as early as possible if it has been contaminated or otherwise damaged. As a result, there are a series of electrical tests performed in the fab during the manufacturing stage. These are usually performed after the first metal deposition is completed, but there are instances where electrical testing is performed immediately after the polysilicon deposition and patterning is complete. There are also a series of tests performed, along with die sort, after the chips are completed. We will devote an entire chapter to electrical testing, so we will cover only the fundamentals of in-fab electrical testing here.

Essentially, all that can be evaluated in these procedures are test patterns. Since the wafers are not completed, they are very sensitive. Any attempt to test a working chip at this point will probably result in its destruction. The test patterns are sometimes actual small chip-sized areas set aside on the wafer for test purposes. However, with the advent of steppers, this has become uneconomical, and the test patterns are now placed in the scribe lines. This makes testing the wafers quite challenging as it is important to run these tests without creating any particles or otherwise affecting the exposed integrated circuit.

The test patterns themselves are usually composed of a number of different structures. They will depend on the devices being constructed, but will often cover areas such as transistor gain, dielectric breakdowns of the various films, contact resistance, special types of C–V tests to determine the quality of very thin dielectrics of known areas, current-carrying capacity and resistance of the thin-film resistors and conductors, and diode breakdown and leakage. Usually, the tests themselves are carried out automatically, with operator intervention required only for probe placement and process control.

3.2.4 Chemical Processing Techniques

There are a number of chemical processing techniques for determining film and wafer impurity concentrations. Some of these can be quite simple, while others involve a more elaborate procedure. In these tests, the rate of reaction is usually the primary indicator of quantity of impurity, although in most cases, the tests will not directly tell you what the impurity is. For manufacturing purposes, the expected contaminant or dopant must be known beforehand or the test time can become excessive.

First, there is the standard chemical chromatography test. This is essentially done the same as in chemistry class, but in a more controlled fashion. A wafer with a film to be sampled, for instance BPSG, is broken up into small pieces. A controlled-size piece is measured for film thickness and then the film is dissolved in a carrier solution, usually a weak HF solution. This solution is then drawn up through tubes, where the components can be identified by the location of the contaminant in the tube. A rough estimate is made of the quantity of the dopants in the film from the quantity deposited at each level in the tube. While this procedure is less accurate than the optical methods discussed earlier, it does have the advantage that the boron and phosphorus contents of the BPSG film can be readily identified. When the ratio of the two components is used in conjunction with an FTIR spectrum of the total concentration of the constituent impurities of the film, a fairly accurate estimation can be made of the true quantities of the various dopants. Combinations of techniques like this can often be used to clarify ambiguous situations.

Another type of test commonly employed to verify that a film is of the proper consistency is to run an etch test on the film. If the etch rates of a substance in a chemical solution are characterized under given conditions, this information can be used to predict total concentration ranges for a process, and to provide an easy method to ensure that there have been no significant changes in the process. Again in the case of BPSG, the BPSG glass will etch in weak HF solutions. The etch rate is dependent on the phosphorus and boron concentrations, and is also dependent on the relative ratio of these dopants, as well as the etch bath temperature and concentration. If a variety of test films with different concentrations of phosphorus and boron are etched and the data analyzed, a test sequence can be specified in which the operator takes a test wafer from each run and measures it for thickness, etches it for a fixed time, and then measures a second time, thus obtaining its etch rate. While there is not an accurate unit for measure, the procedure can act as an indicator of a potential problem. This type of test is particularly useful for those companies that cannot afford the several hundred thousand dollars worth of test gear required to do the analysis themselves, or have the time or money to pay an analytical lab for the services. Etch rate tests also have another purpose, in that they are commonly used to verify that a chemical is up to full strength. For instance, an HF bath is tested by dipping a wafer with a known quantity of thermally grown silicon oxide, and measuring the amount removed in certain time span.

3.2.5 Other Testing Techniques

There are a wide variety of more exotic tests that are occasionally performed on wafers. These include Auger analysis, and other surface analysis techniques, such as secondary ion mass spectrometry (SIMS) and electron spectroscopy for chemical analysis (ESCA), as well as even more esoteric tests, such as neutron absorption analysis to obtain more precise measurements of the boron in a wafer or in a film. Various other attributes that can be monitored with these instruments include grain size and structure characterizations, both surface and embedded contaminant identification, and analysis of the surface structure and morphology. A device called the deep level transient spectrometer (DLTS) is another new technique used for finding and identifying metallic contamination trapped in the substrate. While interesting to pursue, the techniques and results are beyond the scope of this book. The reader is referred to one of the many articles and texts on these subjects as outlined in the bibliography.

3.3 TESTING FOR MECHANICAL DAMAGE

A wafer that has undergone some sort of mechanical damage can often go unnoticed in the manufacturing environment. This can result in subtle problems that significantly impact yields. For example, a nitride passivation process that has drifted off target can put the film under stress. This stress can result in microcracks that cause a failure at a later time. There are a wide number of potential problem areas and an equally wide array of testing procedures used to quantify the problems.

Some common problems for integrated circuits are those of latch-up, parasitic transistors, leaky diodes, and other unusual effects caused by the propagation of silicon crystal defects during processing. Some of these types of problems are shown in Figure 3-14a. While

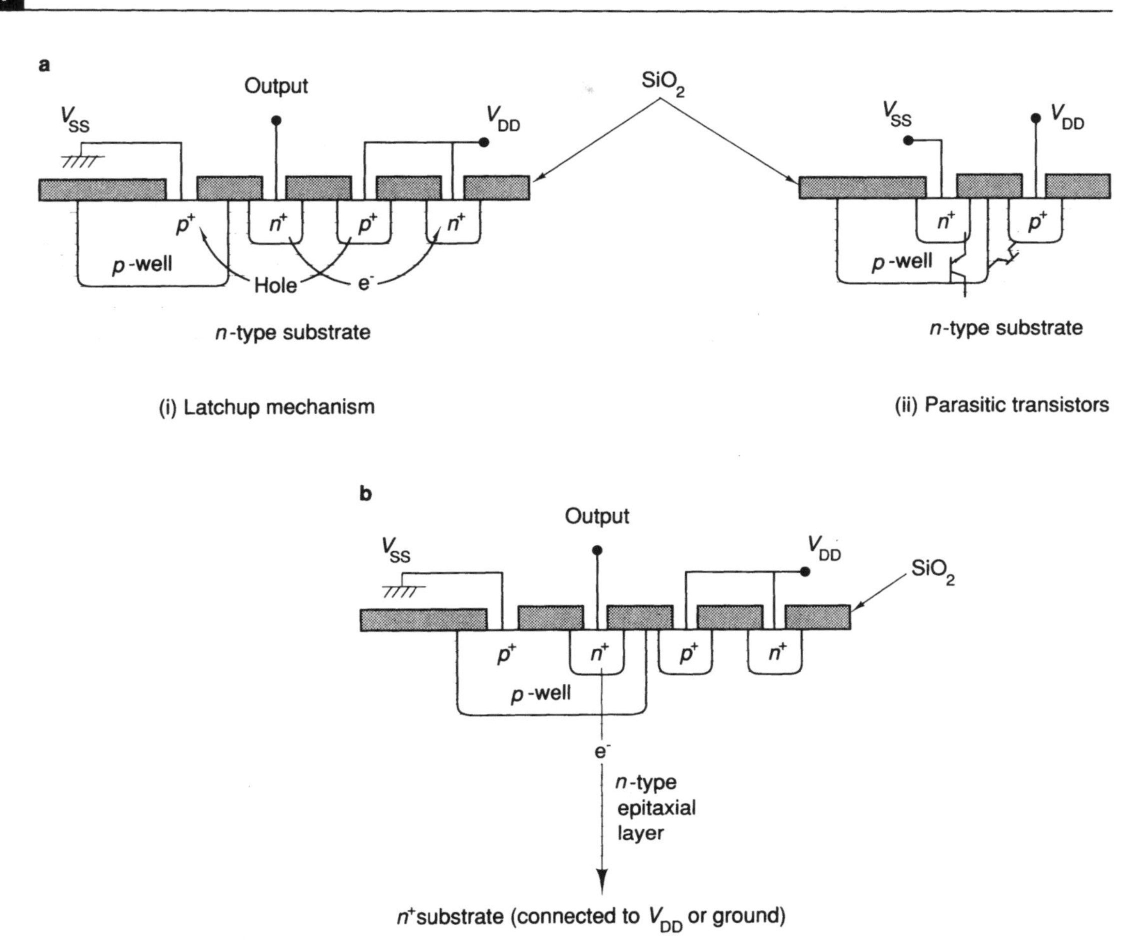

Figure 3-14. Latch-up Mechanisms and Prevention In a CMOS circuit, a phenomenon called latch-up can be caused when a voltage surge is transmitted to the IC through an I/O line. This can permit electrons to escape the n^+ region, especially if the silicon has crystal damage so that the diode junctions are not sharp. These electrons can then migrate to other portions of the circuit. This activity forces holes to move back from p^+ region, eventually forming a self-perpetuating circuit. In Figure 3-14b, the use of epitaxial silicon allows the electrons to be collected before latch-up can occur. Other methods for reducing latch-up include special implants and "guard rings" around the I/O lines to prevent excess electron injection into the substrate.

there have been many manufacturing methods proposed to avoid this problem, few manufacturers actually process their wafers in such a way as to actively influence the formation and propagation characteristics of crystal defects. The techniques for producing defect-free zones are not complex, but do add to the manufacturing time. In most cases, the manufacturers use more expensive epitaxial silicon to reduce these effects. This is a silicon wafer produced with a layer of single-crystal silicon deposited on top of it (see Figure 3-14b). This film effectively stops the propagation of defects at the epitaxial/substrate interface. This reduction in device problems can improve junction isolation for diode arrays (a common feature with high-density RAM, ROM, and ASIC devices). It should be noted that crystal defects that form in the epitaxial layer will propagate through the layer, although the interface between the epitaxial layer and the bulk substrate will act as a barrier to limit damage.

There is also a common requirement to measure the junction depths of a doped region precisely. As a result, there may be factors requiring a physical confirmation of the effects of processing on the substrate. Thus, there is a class of devices used to groove an area into the surface of the wafers a few microns, and run etching and staining solutions over the surface of the groove. The result is an area of stained silicon where there is damage or dopant. This area can be precisely measured and the exact depth of the diffusion junction or the extent of defect propagation can be determined.

Another issue we have mentioned is the problem of stress on a wafer. This type of test is usually performed optically with a device that measures the deflection of the wafer surface as a result of processing. In one case a very bright light source is placed near the wafer and the reflectivity measured. The wafer is then processed and the measurement repeated. If the reflectivity is higher than it was prior to processing, it is assumed that the wafer has bowed closer to the light source, and if the reflectivity is less, then the wafer will have bowed away from the light source. This is shown in Figure 3-15. This bowing is then used to calculate the stress on the wafer. While this test is fast and easy, it can be prone to some error, especially if the process being tested affects the reflectivity of the back of the wafer. Another method for testing stress involves the use of interferometric techniques. While much more accurate, these methods are more complicated, and the equipment more difficult to use. Therefore they are typically used only in laboratories and not in manufacturing areas. Innovations just being introduced include devices that scan the wafer with a laser, measuring the deflection across virtually the entire wafer before and after processing to calculate the induced stress.

Another mechanical failure mode is that of pinholes in films. While this is not now as common a problem as it used to be, and is largely a function of the cleanliness of a process, there are still concerns about this very destructive problem. The test for this problem is quite easy, but requires a specialized setup. A test wafer from the process to be analyzed is placed in a small dish on top of an electrode. The dish is filled with isopropyl alcohol and power is applied to the electrode. When a sufficient voltage is applied, wherever there is a contact with the wafer surface through the oxide (i.e., at a pinhole), bubbles will form. These little sources of bubbles are then counted and recorded. Usually, a good map of the particles that existed on a wafer prior to the deposition or growth of a film will correspond fairly closely to the number and distribution of the pinholes.

A very common source of wafer failures from mechanical causes

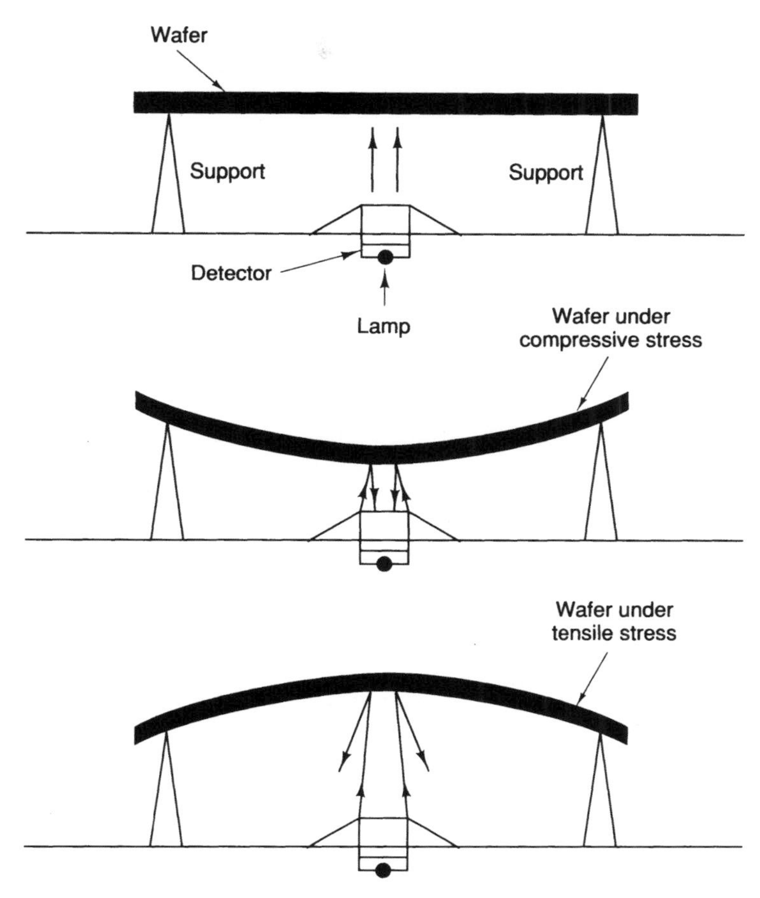

Figure 3-15. Stress Gauge Test The stress test is performed by measuring the change in reflectance after processing. When the wafer bows toward the detector, the total reflectance increases, and when the wafer is bowed away from the detector, the reflectance decreases. This amount of change can be calibrated and the stress of the wafer estimated in dynes/cm^2.

is a scratch on the wafer surface itself. Scratches on the wafers are often the bane of a wafer fabrication line, as the solutions for reducing scratches often appear very simple, but result in minimal improvement, if not actually becoming an additional problem. These scratches can be delivered from machines and from humans, and in both cases the only true solution is to require and enforce discipline by all involved. Usually, machines cause scratches due to some misalignment, or from some unexpected vibration. Robot support arms are famous for these types of problems. Humans cause scratches in most cases by trying to move too fast or through carelessness. The scratch is never intentional and in fact may not even seem noticeable to the individual involved, but the films are very thin and often very soft and can scratch easily. The best method for determining that these scratches exist is the bright light or long-wave UV visual inspection. In this test, a wafer is held at a high angle to the light source. The scratch becomes clearly visible and its size can be measured. There are machines on the market that allow the operator to visually inspect and sort all of the wafers in a run without handling them. Many fabrication areas still rely on the tried and true method of operator inspection, although that carries with it the risk of further contamination or damage to the wafers.

Usually, the wafer's scratch pattern can be a clue as to whether the scratch is machine- or human-related, and sometimes can give secondary clues as to the source of the scratch itself, if it has not yet been identified. For example, a wavy or curved scratch (Figure 3-16a) can often be attributed to poor handling, where an operator has reached between wafers and come into contact with the front of a wafer with the wand. A very straight gouge is usually related to a track mechanism, say, a laser scriber, or the wafer transfer system of a stepper (Figure 3-16b). A circular cut can often be attributed to problems in a spinner or in a wafer scrubber, although there is little equipment used in wafer fabs these days that comes into contact with the wafers in this fashion (Figure 3-16c). In all of these cases, the wafer could have been scratched as a result of some underlying equipment problem that can make the problem very subtle and intermittent. For instance, a lead screw may be slightly warped, and at one point

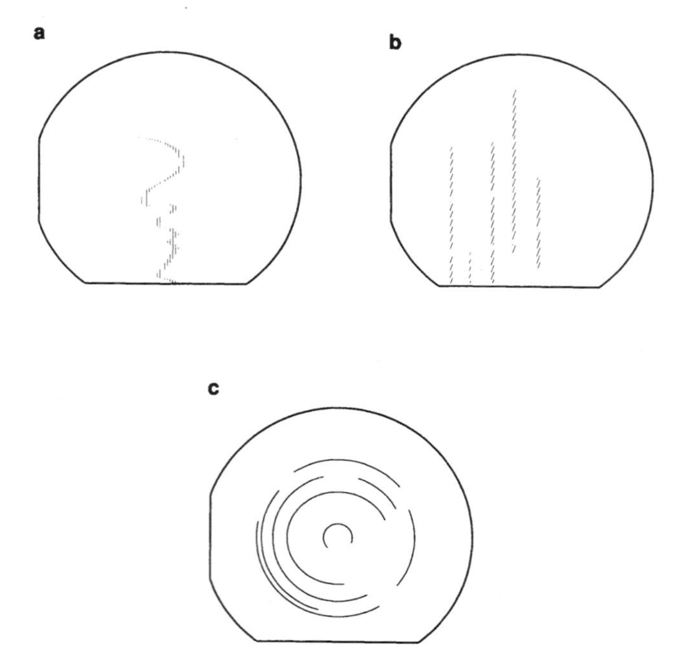

Figure 3-16. Wafer Scratch Types Although hard and fast rules are hard to come by, these figures show some typical scratch patterns produced by human mishandling, wafer transfer misalignments, and spinner misalignments. Note in each case that large areas of the wafers are often affected. (a) Vacuum wand scratch; (b) machine-induced scratch; (c) rotational machine-induced scratch.

changes the orientation of the wafer, and in doing so it contacts the edge of a cable that is improperly positioned in the chamber. This is why it is so important to observe the fabrication equipment in such minute detail, and take care when purchasing equipment to verify that no such sources exist. We will discuss wafer scratches in much more depth throughout the book.

Finally, there are certain types of mechanical damage that can be observed only at high magnification, using a scanning electron microscope (SEM). They include structural defects such as cracks and particles as shown in Figure 3-17, as well as other phenomena. There are often process-related problems that can be identified in this manner, such as the spiking of aluminum through the silicon contact interface and into the substrate below. Film conformality, surface roughness, and other related phenomena can be characterized through SEM inspection analysis (Figure 3-18). Lower voltage SEMs can be used to measure linewidth and perform other inspection tasks.

3.4 OVERALL OBSERVATIONS ON SEMICONDUCTOR CONTAMINATION

As we have seen, there are a significant number of problems that can occur in a wafer fabrication area that can impact the wafer yields. This chapter has covered some of the most common and most obvious problems in order to illustrate the methods by which the various test techniques are applied. Clearly, there are underlying themes to all of the methods outlined, and they really boil down to: Use the scientific method and your instincts. There are a huge number of variables in any wafer fabrication process, too many to control at any one time. This requires that a statistical approach be taken to all decisions. However, if each process and its subcomponents are broken down and controlled tightly enough, the probabilities of failure can be more tightly controlled and, in the event of a failure, the source can be more rapidly isolated and eliminated.

The best way to attain these goals is through the maintenance of complete, thorough records and through a complete series of process control applications. While this may seem like unnecessary effort by some, controlling the fab area manufacturing quality before problems occur and fixing them as soon as they do occur is much simpler than being forced to analyze a problem that has shown up as a sort yield failure, which may have been occurring in the fab for weeks or months and is now obscured by other factors that have masked the original problem.

Proper process control requires that a significant amount of data be obtained and kept current in order to validate the statistics used to control the fab area. Specification limits and control limits, as well as process optimization, can all be obtained through analysis of this data.

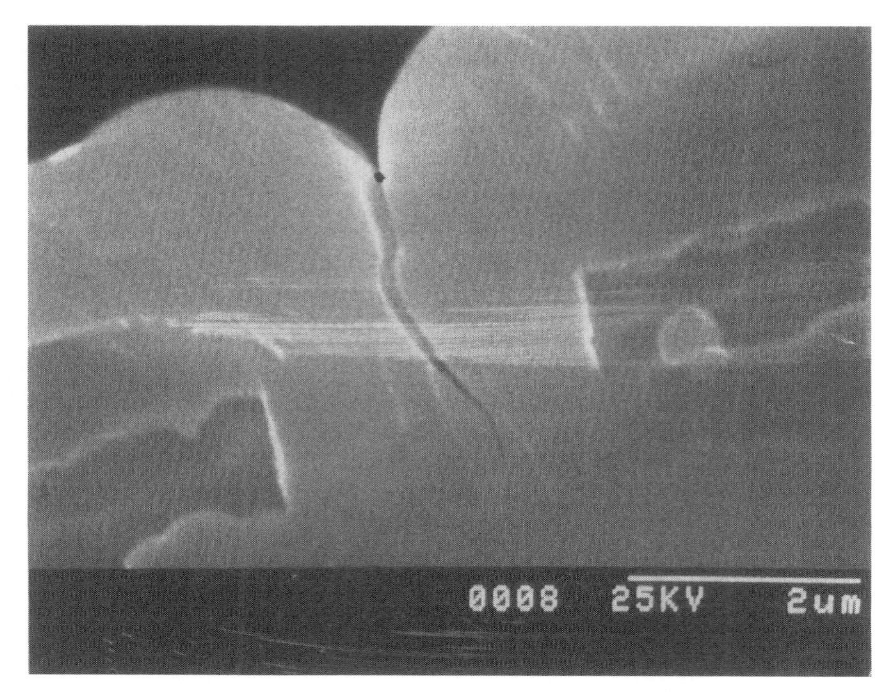

a

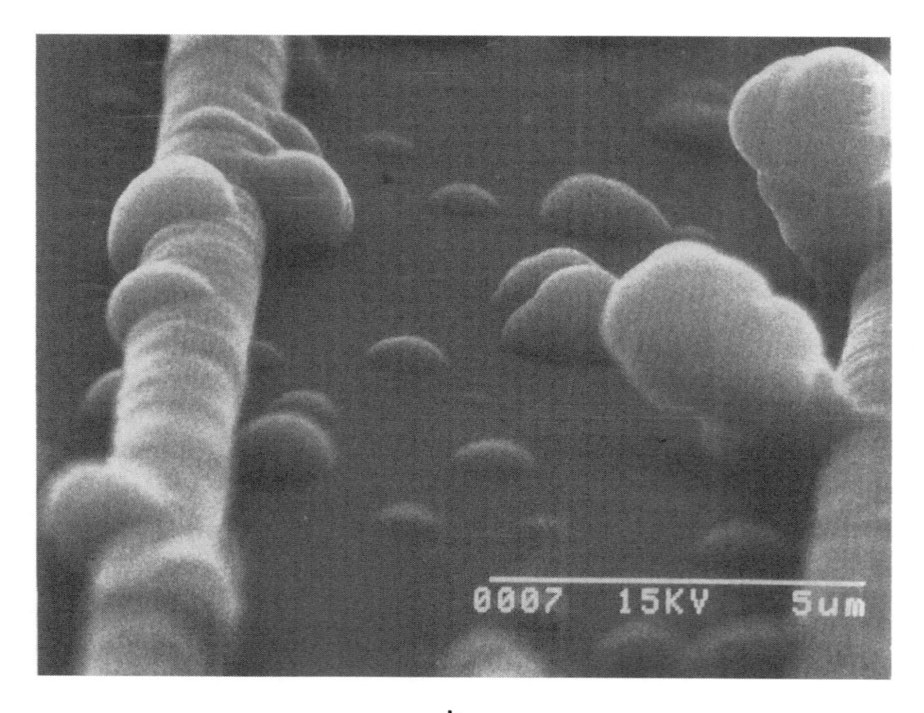

b

Figure 3-17. Defects in Deposited Films Scanning electron micrographs can be used to capture excellent images of numerous types of wafer defects and structures. Photographs reproduced with the permission of Greg Roche.

a

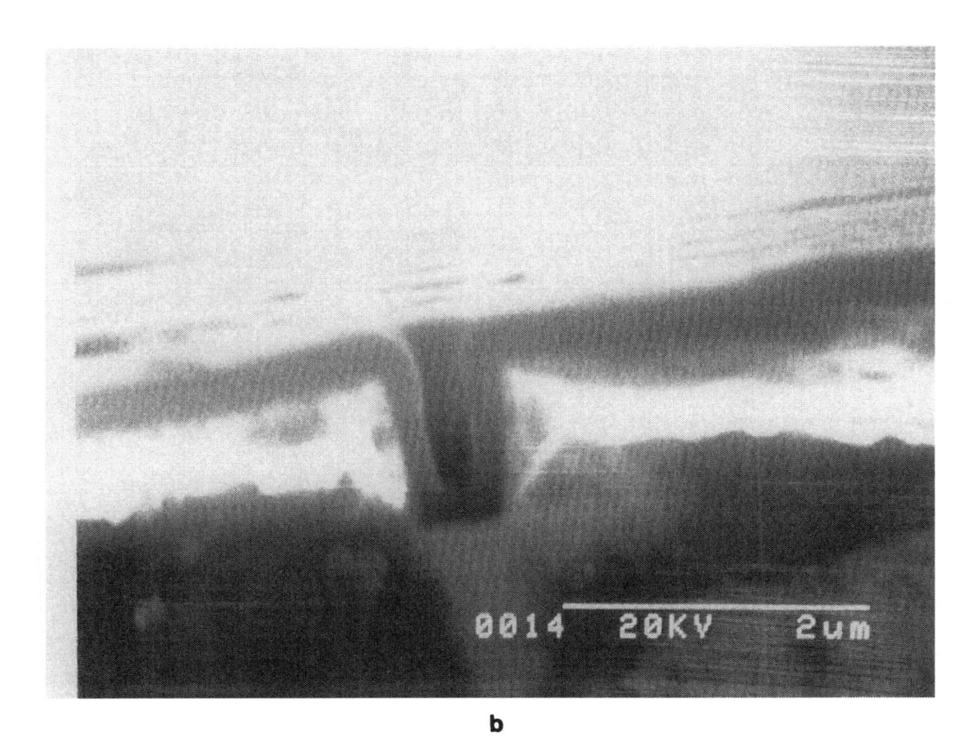

b

Figure 3-18. The Use of the Scanning Electron Micrograph for Observing IC Structures Structures such as array lines and contacts or trenches can be investigated and characterized in detail with help from the SEM. Photographs reproduced with the permission of Greg Roche.

While contamination by unpredicted sources is not desired on the surface of a wafer, certain types of desired dopants can sometimes appear to be contaminants. Minor process fluctuations can cause temporary changes in the characteristics of the films being observed. All of this is important when analyzing the results of these tests, in order to be able to discern the real problems from the underlying noise levels.

In obtaining this data, the process engineer has a method to compare process results with yield results, and discover areas where process enhancements may be made. The techniques of measurement and equipment described in this chapter can be used effectively to monitor virtually all contamination that is present in the manufacturing line, if used with the proper control techniques. It often turns out that improvements made in the areas discussed in this chapter can cause significant yield increases. The statistical quality control methods discussed in this section can reduce the amount of time required to identify problems and can prevent small problems from becoming large problems. In addition, the process control logs, along with properly archived sustaining logs and specifications, can allow the fab area personnel to build up a history and break out of the "we did something like this before" mode of problem solving.

The increase in productivity and improvements in yields that are a result of statistical process control programs can easily justify the extra effort and concentration required to implement them. Since these improvements can be directly related to improvements in profitability, it is clear that implementation of the testing methods we have discussed can significantly enhance the output of the semiconductor manufacturer.

CHAPTER 4

CLEANROOM FACILITIES AND PROCEDURES

In this chapter, we will discuss the cleanroom environment. The cleanliness level of this environment must be the first issue addressed, whether you are dealing with the construction of a new fab area, or upgrading an existing one. The environment must be free of chemical and particle contamination sources in a number of areas, including the air handling system, water supply system, and process gas handling systems. In addition, there must be plans in place for controlling the expected deluge of particles that will be delivered into the cleanroom from the moment the first piece of equipment is delivered. Thought must be given to the process sequence, and how the wafers will move from room to room and from place to place within any room. In addition, after the facilities have been optimized for performance and the basic levels of contamination of the environment are under control, it becomes imperative to install the appropriate procedures and restrictions to apply to the maintenance and operation of the equipment. The rules for personnel entering the cleanroom environment will be outlined. It will be seen that the maintenance of a high-purity wafer fabrication area is largely a matter of discipline, once the basics are set in place.

4.1 GENERALIZED CLEANROOM LAYOUT

Cleanroom design and construction are dominated by two overriding and mutually incompatible goals—cost and performance. To achieve the cleanliness required to manufacture integrated circuits profitably, the company must commit millions of dollars for the improvement of an existing cleanroom, or construct a new fab area from the ground up. The logical time to upgrade a fab area is when the cost of restoring and replacing the existing facility is less than the cost of new facilities. This is not an easy calculation. If the older manufacturing area has to cease operations for an extensive period of time to effect the upgrades, in addition to the large additional equipment costs, the decision to start construction of a new manufacturing area gains merit. This can be an exciting task, but very expensive. NEC reportedly is going to spend $300 million dollars on an automated manufacturing area to build four-megabit DRAM chips. Several American manufacturers are reported to have spent well over $200 million on their newest facilities. As a result, it is imperative that the design of the cleanroom be optimized for performance, not just for the initial setup but for planned and unplanned future expansion in the cleanroom. Thus, the fabs must be carefully designed and built by specialists in the area of cleanroom construction. In addition, with a cost that is so high, it becomes extremely crucial to keep the manufacturing area in as near to new condition as possible at all times in order to increase the lifespan of the area as much as possible, while minimizing shutdown periods for major overhauls.

As seen in Figure 4-1, the cleanroom itself consists of a number of different areas, all of which are important for achieving and maintaining the proper environment where wafers will be exposed. The different units will have different cleanliness and relative air pressure levels associated with each. The simple reason for the differences in air pressure between the various units is that particles have a tough time moving against air flow. (Although, surprisingly, particles can often be found coming into an area against the general air flow direction.)

Problems come up when trying to implement the different air pressure levels in the fab. A significant problem when balancing air pressure is verifying stability when various doors are opened. If the design of the fabrication area allows imbalances to occur, air pressures in the ultra-clean areas can be reversed, and contamination can easily stream in. In all cases, the various parts of the cleanrooms should not have abrupt interfaces to the nonclean areas, for example, a hallway that has a door which opens to the office areas, or a chase door that opens to a loading dock. While sometimes convenient, these interfaces between areas can completely wipe out benefits gained from many hours and thousands of dollars worth of work in a few minutes.

When the fab area is approached from the main office area, the first section that is encountered is the change room. Sometimes this is preceded by an anteroom that allows personnel storage locations for their personal belongings. In some cases, special fab area shoes are assigned to each individual and they are usually changed in the anteroom. An air shower may lead into the change room, and there will almost always be an air shower leading out of the change room into the

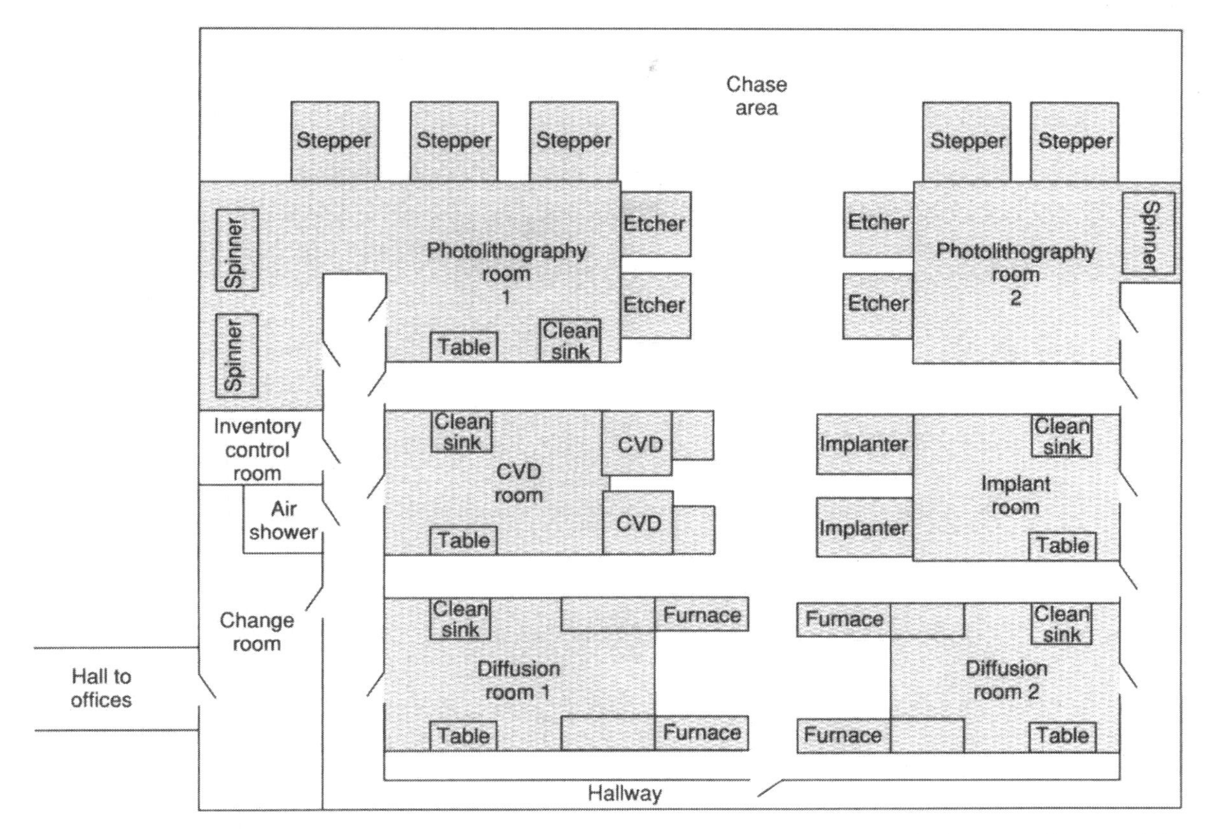

Figure 4-1. A Potential Wafer Fab Area Layout Here we have a small wafer fab layout utilizing the system of independent rooms for each function. Many different layouts are used by various manufacturers. The air pressure is highest in the processing rooms, with the hallway at the next pressure level, the chase area next, then the change room, and finally ending with the ambient air pressure in the main office areas. This arrangement minimizes the migration of particles from the outside world to the areas where the wafers are exposed to the environment.

cleaner areas of the fab. Actually, the name change room is usually somewhat of a misnomer, since people merely put their cleanroom suits over their regular clothes. In fact, street clothes such as jackets and so forth are not usually allowed in the change rooms. Nevertheless, the fab personnel are required to put on a variety of items that are meant to contain the contaminant sources of the human body. These items include full hoods, face covers that are worn over the mouth and nose, a cleanroom suit called a bunnysuit, gloves, and either special cleanroom shoes or shoe covers. In Figure 4-2, a cleanroom suit is shown, along with a chase-wall mounted CVD system.

The change room is usually the lowest pressure room in the fab area, but should be slightly higher in air pressure than the outside office areas or hallways. The change room should have adequate space for everyone to hang his or her garments freely. Cramming bunnysuits into lockers can increase particle shedding from the suits. The room

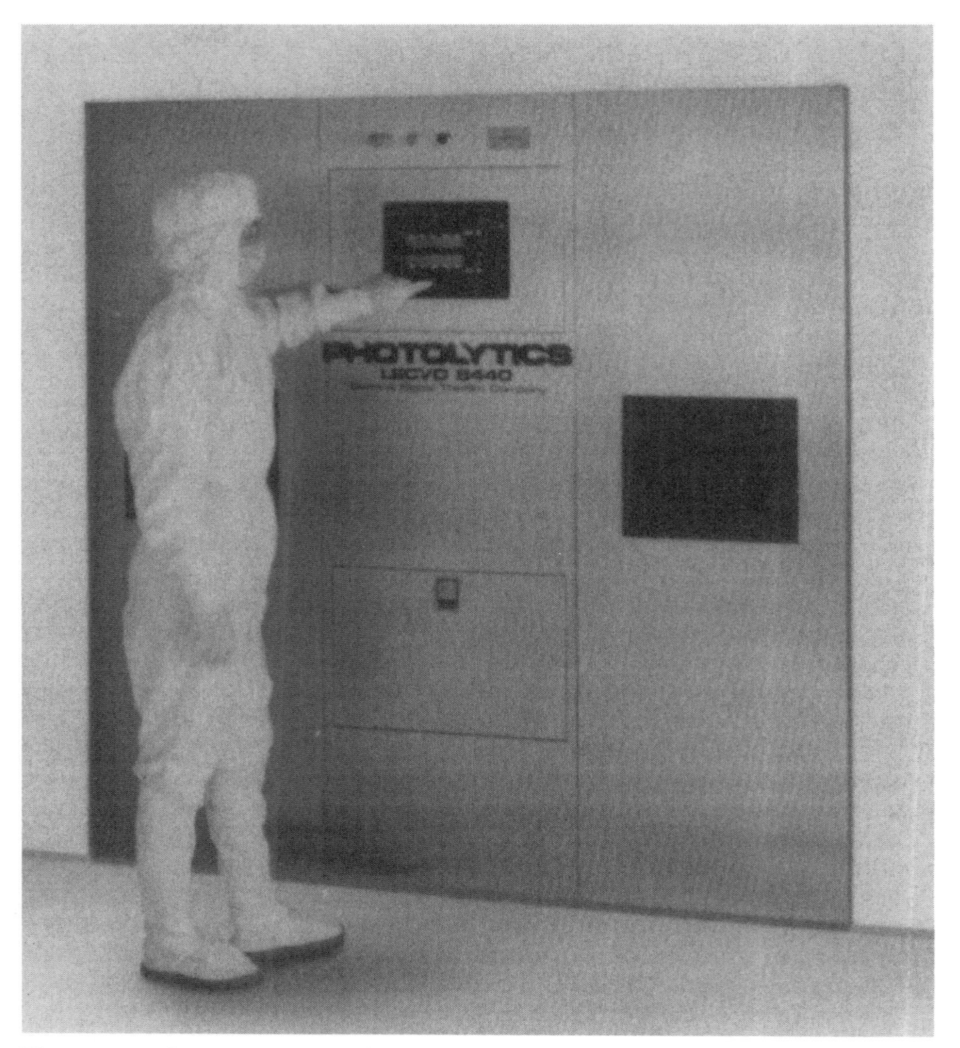

Figure 4-2. Cleanroom Attire with Chase-Mounted Process System As we can see, the cleanroom attire covers the operator entirely, preventing transmission of contamination. The wall-mounted access to the laser CVD reactor is typical of the most modern equipment design. Note the use of the "touchscreen" and "SMIF" (Standard Mechanical Interface) compatible load locks (behind the smoked plastic doors). Photograph courtesy of Photolytics, Inc.

itself should have some laminar flow in it, although it is not required to be as clean as the fab area itself. The change room will usually be one or possibly two classes below that of the clean area itself. If the clean area is class 10, the change room should be class 100 to class 1,000. There will then be an air shower between the fab main hallway and the change room. There usually will not be an air shower for the door leading back out of the fab area into the change room, but that door is often built to go only one way (out) to prevent people from using it to speed up entry into the fab area. The air shower itself should provide a strong flow of air and should provide room for at least three people at a time. For the air shower to be effective, it is important to be in the shower for a sufficient amount of time and to make sure the air covers the entire surface of the suit. While at first this may look unusual, soon everyone gets used to the "dance" in the air shower and it is done without thinking. The usual amount of time in an air shower is 30 to 60 seconds. Some air showers prevent entry into the fab area until the appropriate time has elapsed.

The next area of the fab is the main hallway. The pressure in this area must be kept above that of the change room and below the cleanroom. Since the main hallways are also the main thoroughfares, it is clear that the areas in the fab that will undergo the most drastic pressure changes will be the main hallways. As a result, the hallway should stay open to some larger area of equivalent pressure, or have a good exhaust to buffer these pressure changes. Often the pressure in the hallways is balanced by those of the chase area. The central fab hallway is often used for wafer transport, even in those fab areas that have pass-through windows from room to room. Since the hallways will not be extremely clean even under the best of conditions, wafer storage boxes should *never* be opened in the hallway. Even though it can be expensive to supply laminar air to the hallways, it is important to filter the air being sent into the hall and to pull the air out of the hallways and into the fab air return for proper filtration.

One of the biggest problems with maintaining cleanroom pressure stability is the requirement for doors. Opening a door can throw a lot of careful planning out the window if the door is in the wrong location, or if one localized area of pressure is way above or below that of other rooms. The imbalance may not be noticed at its source, but can show up elsewhere, where there might be a marginal balance in pressures with the doors closed. Room pressure balancing tests must be done in both doors-open and doors-closed conditions to be completely valid. In addition, a number of cleanrooms have determined that the standard swinging doors spread contamination themselves, especially in older buildings where they might be made of wood. There is also the possibility of dropping things while attempting to open one of these doors. As a result, a number of fab areas are installing electronic Star Trek–type doors which open by sliding out of the way when someone comes up to them. Critical processing equipment should never be placed near a door; paperwork areas, computer terminals, and other less critical procedures should be located nearest to the door.

The next area is the chase area, which is usually about one class above that of the cleanroom itself; it is where most of the maintenance work should take place. Tubes are cleaned, dried, stored, and transported for installation in the chase. The chemicals used in the fab are delivered to the process rooms in the chase area, where they are placed in pass-through storage cabinets. Often, other raw materials can be stored in the chase areas, such as the raw silicon wafers prior to the boxes being opened. In general, production wafers should not be exposed while in the chase area. It may be acceptable to allow test wafers into the chase area, if the areas that they are in are properly controlled. Reasons for allowing test wafers into the chase areas, can include test wafer reclaim, C–V wafer testing, SEM, or other complex or somewhat dirty test or analysis procedures. Since most of the equipment will be worked on in the chase area, it is important to maintain cleanliness in the processing chambers of the equipment being worked on, as well as in the surrounding cleanroom area. Wafers can easily be contaminated by problems that originate during the preventive maintenance procedures.

The main facilities of the fab area run through the chase area also. The maintenance requirements for these facilities can create conflicts in priorities at certain times. For instance, gloves are required in the cleanrooms even though use of the gloves may hamper the maintenance work in progress. It is much more difficult to perform delicate procedures and handle small items with gloves on than with the bare hands. This is especially true if the gloves are of improper sizes or are otherwise inadequate for fab area use. This can lead to a safety or quality problem if the maintenance crew is unable to properly secure gas and vacuum connections, or when high voltage equipment is being repaired. Handling objects is difficult and there is always a chance that a piece of a glove or other debris will end up stuck in a fitting or other crucial area.

There are also safety issues to bear in mind with the facilities in a chase area. Leaky gas lines, chemical spills, hot and broken quartzware, and toxic residues on equipment parts are all part of the job in a fab area, and while these and other accidents do not happen very often, nevertheless they do happen. The safety requirements for a fab area are paramount, and extra safety gear is often required. In some cases this gear can become contaminated and cause more harm than good. For example, this can be true when acid aprons are turned inside out and are later worn by unsuspecting personnel. Unfortunately, due to the nature of the work performed in the chase areas, accidents are more likely to occur there than elsewhere in the fab.

Another problem area in the chase area is the water and waste disposal system. The water is often brought in at or above the equipment height, while the drains are run through trenches in the fab floor. Water can leak through to the floor, sometimes at process equipment hookup points, and sometimes at joints. This is especially true of some piping materials. For instance, PVDF (a Teflon®-like plastic) plumbing is often more difficult to seal than other types of plastics. Any fluid leak has a number of consequences, some more obvious than others. Clearly, people could get chemical burns or damage clothing from leaks in drain lines, they could slip and fall, or the water or fluid could come into contact with a high-voltage electrical supply. Other more subtle effects include increased corrosion, the buildup of bacteria or mold spore sources, and health problems among workers and a potential hazard to the environment from chemical fumes and residues. It is important that leaks be completely cleaned up and that any leakage sources be repaired as soon as discovered.

There are a number of choices for materials for handling the DI water and the waste chemicals. They try to conform to the widest variety of specifications and requirements of the cleanroom; however, there are instances where there are inherent conflicts. For example, the need to have completely inert and stable substances such as PVDF used for plumbing means that it is difficult to find a material that can be used to bind the pipes together at joints. Cements cannot be used in many cases, as they may come into contact with the DI water that is being protected. In most cases, these cements will not hold up to the DI water as well as the piping materials, and will become the source for leaks and contamination. The use of threaded connectors without o-rings or other good gasket material does not work well due to the slickness of Teflon® materials used in the piping. Certain of these types of materials are well-known leak points in water systems. Some of the materials that are used react with DI water, such as some types of PVC, which deteriorates through time, causing particle counts to go out of control. Items like cast iron fittings will also react with DI water as well as fab chemicals, and should not be used in the drain lines of the cleanroom.

Finally, there is the manufacturing area itself. In older fab areas, there was only one main floor, but this has been given up for the more manageable independent room design. Now, there may be as many as 12 to 15 or more individual ultraclean rooms in the manufacturing area. They will usually be broken up by fundamental process and then even further by dopant type and process cleanliness desired. Thus the diffusion and thin-film processes are usually located near one another, as are the photomask and etch areas. However, there would usually be a separation between the clean gate oxidation processes and the dirty phosphorus deposition processes. The rooms may range anywhere from class 1 to class 100, which can sometimes even vary within one manufacturing area, depending on how critical the particular step is. The cleanroom itself usually has a very significant number of laminar flow hoods, often covering 100% of the ceiling area. Returns for the air-handling system are located either in the lower sidewalls of the cleanroom, or returns are built directly into the floor. These air-handling systems are capable of delivering up to 40 to 100 air changes per hour.

This high air flow is dictated by the factors discussed earlier in the effects of forces on small particles. Clearly, since gravity is of little effect, a room with still air will allow for particulate matter to build up in the air with the particles having very little relative motion after having been deposited in the atmosphere. It is necessary to push this air out of the way and into the filters as quickly as possible. The actual air flow required to maintain the cleanliness of the fab area is dictated by the number of particle-generating sources allowed in the room. A very clean setup will allow for lower volumes of air required to keep the room clean. There will be a rate below which the room will not maintain cleanliness but, economically, it makes sense to minimize the energy required to run the fab area. Moving hundreds of thousand of cubic feet of air an hour for 24 hours a day requires a tremendous amount of energy.

The air-flow dynamics within the work area itself should be studied and the equipment placed in such a way that turbulence is minimized and so that the wafers are protected from potential contamination as much as possible.

The amount of air moved is usually sufficient to replace the entire volume of air in a few minutes. The velocity of air required to attain predictable airchange rates is determined by the following equation:

$$\text{Air changes/hr} = (\text{Velocity} \cdot \text{VLF}_{\text{area}}) \;/\; (\text{Room volume})$$

In a 1000 sq. ft. room that is around nine feet high (leaving us with 9000 ft^3), with an air change rate of 30 changes per hour (a velocity of 4.5 feet/minute) means the air-handling systems produce 4500 ft^3 per minute (270,000 CFH). Usually, the air-handling system will recirculate large quantities of fab air, while bringing in limited quantities of outside air. This recirculation is performed due to the expense of purifying so much outside air. It much easier to recleanse the already purified air. It is important to understand the complex nature of the total air-handling system, as it will be handling well over a million cubic feet of air an hour, controlling the temperatures and humidity in several independent areas, and keeping all velocities at a reasonable level. There are many places where problems can arise and lots of equipment that requires maintenance. In addition, the exhaust for this quantity of air replacement must be strictly controlled, as imbalances in room exhaust will produce pressure imbalances within the fab. The recirculation of fab air presents problems of its own. Gas leaks in one portion of a building can be carried rapidly throughout the rest of the building. Localized concentrations of these fumes can exceed safe levels in areas that would otherwise be completely isolated from the source of the contamination.

In addition to the quantities of air required, the air must be cleaned through HEPA (high-efficiency particulate-absorbing) filters, which provide nearly 100% efficiency on all particles greater than 0.02 μm. The air must flow out of the filters uniformly at a fairly high and constant velocity in order to generate and maintain a laminar flow environment. Inconsistencies in the air flow can lead to areas of turbulence which will allow for the formation of airborne "particle traps" which will allow particles introduced into the air flow (such as from a human walking by) to become suspended for long periods of time. In addition, oscillations in the velocity of laminar air-flow hoods can lead to localized air pressure imbalances which can lead to turbulence in the air flow.

The filters used in the laminar flow hoods should be checked on a regular basis as holes, excess buildup, and otherwise damaged filters, or problems with the drive mechanisms of the blowers themselves will occur surprisingly often. This is not meant as criticism of the air-handling systems but should illustrate the difficulty of moving this much air at any time. When you consider that most of these machines are on 24 hours a day, seven days a week, they are very reliable, but even a medium fab (5000 sq. ft.) would have as many as 200 six-foot by four-foot laminar flow units in it. Even with an uptime of 99.99%, there is, on average, a failure requiring about one hour's worth of work on each hood every 59 weeks. With 200 laminar flow hoods in a factory, you could have as many as three or four of them failing per week. Incredibly, these workhorses can exceed these tough requirements and not fail at a high rate, but only when properly monitored and maintained.

In addition to all of these other requirements, the air must be kept at constant temperatures and humidity levels. While the photolithography areas are more sensitive to temperature and humidity changes than most of the other areas of the manufacturing area, effects of the conditions of the atmosphere are seen in a number of processes. Usually, it is not terribly critical what the exact temperature and humidity levels are, as long as the work area is comfortable, and does not change more than a degree or two or a couple of percent in humidity. Stability is the critical factor here.

However, keep in mind that even a Class 10 or Class 1 cleanroom is not perfect. In fact, in a Class 10 cleanroom (10 particles/cubic foot), using the previously mentioned dimensions and a flow of 50 air changes per hour and a nine-foot-high cleanroom, the air-flow velocity would be approximately 7.5 feet/minute. This means that approximately seven and one-half cubic feet of air containing 75 particles would impact on each one square foot surface perpendicular to the air flow each minute. Since the fab area must run 24 hours a day to remain clean, we can then estimate that each square foot (about the size of a floor tile) has contact with 108,000 0.5-μm particles per day, or 324,000 0.3-μm particles per day.

No matter how clean the air is that is delivered to the cleanroom, the room itself may generate far more particles than allowable by specification. There are many sources of these particles. They can be generated from the equipment, logbooks, humans, chemicals, or anything else that is brought into the cleanroom. Table tops can gather dust at a fairly high rate even in Class 10 areas. Small particles will be falling at a rate of some 40 particles per minute over every square foot in the cleanroom. This number will be exceeded in localized small areas for short periods of time after someone engages in activity in the area.

A significant contributor to the particle level of the cleanroom environment is the layout and setup of the equipment. If the equipment is laid out in such a way that wafers have to be carried around the room excessively in order to be processed, then there will be a greatly increased chance of contamination. Time and motion studies should be carried out prior to installation of equipment into a fab area. Also, the location and shapes of the equipment should be considered when designing the air flow. Placing a clean sink directly in front of an air exhaust instead of between two exhausts (if that option were available), would have detrimental effects on the air flow in the room. Similarly, contamination levels on wafers will be significantly increased if the load stations are near doors that experience much traffic.

There have been a number of definitions for the cleanroom environments in their various stages of development. They include the *as-built cleanroom*, which is a cleanroom that has all of the facilities installed but has not had any equipment installed in it yet. After the equipment is installed and the room is allowed to stabilize and no personnel are in the cleanroom and no activity is taking place, it is defined as an *at-rest cleanroom*. Finally, when the personnel are working in the fab and it is running in a normal manner, it is an *operational fab*. It usually takes about half an hour to an hour to completely clear the cleanroom of particles and to attain the at-rest levels. This time can vary depending on the amount of equipment in the room, how it is organized and kept up, and so on. There is often some residual for many hours, so that a room at rest for a day or two may have very clean air indeed.

Obviously, it is important to keep everything clean at all times in the room. This means wiping off table tops regularly, and inspecting the facilities to ensure that there are no chemical or water spills or leaks, or areas of corrosion. A common place to find corrosion is on metal timers used on acid sinks. This corrosion is transferred from the corroded timer to the operator's gloves, then to the Teflon® boats, and from there to the acids in the clean sink, and finally from there to the wafers. Another area for concern is rough edges that can snag cleanroom gear or gloves. Anything that is used to contact the wafer directly should be kept especially clean. Vacuum wands should have proper storage brackets, and in most fabs are hung from the ceiling to prevent them from touching a surface.

The key for maintaining the cleanliness of the room is discipline. All personnel must understand the importance of detailed workmanship in the cleanroom and must be actively enlisted in the struggle against contamination of all kinds. Maintaining a cleanroom is a difficult and often tedious chore, but nevertheless, the difference between a "World Class" cleanroom and any other fab is almost entirely due to a meticulous attitude. The "World Class" fab does not permit imperfections in the product or the manufacturing environment, as they understand that anything less will result in reduced success. To assist in the cleanroom discipline, each shift should designate a team or an individual to study the fab using a fab area quality control checklist (shown in Figure 4-3).

Figure 4-3. Quality Control Items The quality control checklist consists of a simple form identifying a piece of equipment or a particular area along with date, time, and an area for the items suggested above (plus others, as required). Each of the items should be checked and responsibility for correction of a problem assigned.

1. Quartzware

—All quartz handling pieces are in good condition
—Stored properly (in tube or on quartz sled)
—Clean (no fingerprints, haze, dust)
—Breakage (no chips or cracks)
—No devitrification, damaged slots, leaning wafers
—C–V tests have been completed and are up to date

2. Equipment Setup

—Tops of all surfaces, equipment clean
—Gas boxes and interiors of equipment clean
—No insulation or packing materials left in area
—Lead screws, robot arms clean
—Doors, vestibules, endcaps, other parts clean
—All thermocouples, gas and pressure gauge connections, electrical connections, etc. in place and secure
—All gas, fluid, vacuum, and air pressures are correct
—Optics aligned and cleaned
—Interiors of process chamber cleaned

3. Clean Deck Setup

—All tanks clean, no staining or damage apparent
—Dump rinses spray uniformly, fill completely to overflow, N_2 comes on, water dumps swiftly and drains completely
—Deck is clean, no stains, no standing fluids
—Exhaust is on and sufficient
—No corrosion anywhere
—Drier has correct pressures and exhaust settings; runs at correct temperature and speed
—Cleaning cassettes stored properly

4. General

—PM logs are up to date, all PMs performed to spec
—Tables, work stations clean
—Laminar flow hoods performing up to spec
—Wand tips in good condition, vacuum sufficient
—Excess paperwork removed
—Fire extinguishers up to date
—Utensils stored in correct locations, spare cleanroom supplies available
—All personnel in proper cleanroom and safety attire

There is yet another level of cleanliness that has been tested in ultraclean processing environments, and that is the nonintrusive fab, where the wafers are kept in controlled, hermetically sealed chambers at all times, and are never directly handled by humans. This is outlined in Figure 4-4. In theory, they keep the wafers at class 1 or better at all times and prevent picking up contamination during the occasional particle bursts that occur in the regular cleanroom environment. This requires the ultimate in automation techniques, with excellent equipment reliability and well-trained maintenance crews. Any misalignment could lead to damage. Improperly repaired equipment or failed sensors may allow the misprocessing of wafers. Since all test equipment must also be installed directly into these machines, the effort required to calibrate the test systems increases.

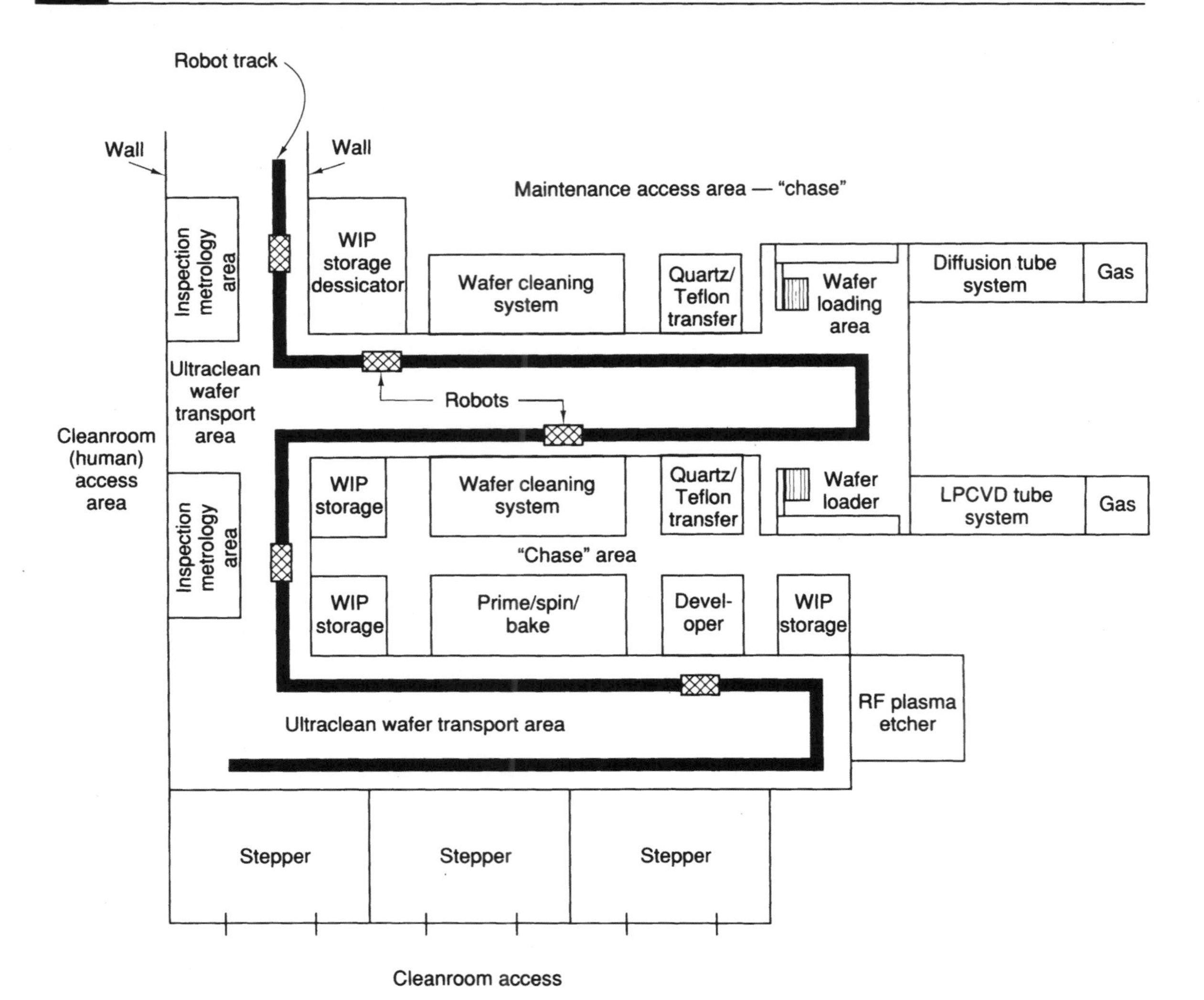

Figure 4-4. Automated Wafer Fabrication Automated wafer fabrication areas must be laid out in such a fashion that the wafer-handling robots can follow tracks to move from process module to process module. Here we see diffusion/LPCVD and masking/ etch modules interconnected with each other and associated metrology equipment. The wafers travel in an ultraclean transport area which is never entered by humans. Human access to the equipment and to the wafers is always from behind walls, either from the chase area or from the main (human-accessed) cleanroom, and wafers are not handled directly.

These fab areas are very expensive and typically do not at this time match the price/performance levels of more standard fab designs. However, several Japanese and American companies have designed and constructed a number of these wafer fabs. The existing operations have been severely hampered by the lack of standardization and of truly automated equipment. This concept, called "hands-off," or "lights-out" automation by various manufacturers, will allow the operators to run the area from a keyboard in a control room, isolated from the cleanroom. In theory, only maintenance personnel would need to enter the chase area for maintenance work on the equipment. This level of automation requires cooperation between the various vendors and the semiconductor manufacturers, as well as a joint investment of resources that until recently has not existed within the United States.

4.2 BASIC MOTIONS OF SMALL PARTICLES

Since particles are the most pervasive and most difficult to control of the various types of defects, we have up to now spent a significant amount of our time discussing them. To do this, it is important to understand some of the basic issues that come up when dealing with these problems. We will discuss the "activity" of particles several times throughout this chapter. Since it makes little practical difference to the wafer whether a particle was created moments before it was deposited on the wafer surface, or whether it was delivered to the wafer from some outside source, we will call any motion of a particle that results in the particle landing and adhering to the wafer, an *activity*. In this light, the creation of a particle in the gas phase of an LPCVD reaction is an action, as is the deposition of skin flake that traversed from the scalp, to a hair, to a comb, to a hand, and finally to a glove before ending up on the wafer. Particles are generated from a number of sources and can often be isolated and new procedural developments or repairs effected. Once generated, the particles can continue to swirl around and remain suspended for very long periods of time in an inadequately controlled or turbulent environment. It is important to understand these features when discussing cleanroom design and rules so that appropriate measures to prevent contamination can be implemented at an early stage.

4.2.1 The Effects of the Forces of Nature on Particles

Since we have defined a number of potential sources of particles and a number of mechanisms for their transport into the fabrication area, we can now look at what happens to those particles when they are in the environment. When at this supermicro scale, the effects of everyday forces often do not work quite the same as they do at our scale. For instance, the effects of gravity are severely limited, and often will not be observable in the presence of other forces. Essentially, the forces we will look at will include the effects of gravity, electrostatic forces, magnetic forces, "wind" or the effects of moving gases, and energy absorption from heat sources, such as the high-temperature diffusion processes. The forces are outlined in the list in Table 4-1.

First, we must describe our sample particles. They can include almost anything, but if the processing has been carefully done, the odds are that most of the particles on the wafer will be silicon dioxide. In some cases, there will be other types and we will look at them as

TABLE 4-1
Forces Acting on Small Particles

Force	Direction	Magnitude
Diffusion	Random	$<0.5\ \mu m$—larger than gravity $>0.5\ \mu m$—smaller than gravity, but comparable
Gravity	To ground	$\approx 10^{-14}$ N on 1 μm SiO_2 particle. More important effect in vacuum than at atmospheric pressure
Electrostatic	To "electrode"	$\approx 10^{-8}$ N on 1 μm SiO_2 particle. Can be used to control airborne particles (electrophoresis).
Air flow	With air flow	$\approx 10^{-12}$ to 10^{-13} N on 1 μm SiO_2 particle in typical VLF air flows. Much stronger at higher air velocity.
Convection	Toward heat source, very turbulent	Convection currents can be stronger than air flow above strong heat source.
Mechanical	Random	Can be caused through vibration from operating a piece of eqiupment or from secondary sources, for instance, vibrations from the laminar flow hoods vibrating a wall, which then sheds particles.

required. We will assume the particles are cubes 1 μm to a side, giving a volume of 1 μm^3. We will assume that the density of the silicon dioxide particle is the same as that of solid silicon dioxide. Now since the density of silicon dioxide is 2.27 g/cm^3, the mass of this particle is approximately 2.27×10^{-15} kg. As a result, the force of gravity on the particle can be calculated as 2.23×10^{14} Newtons. Note that many airborne particles are hydrocarbons by nature and therefore often lighter than SiO_2, thus reducing the effects of gravity even more.

If the particle has a charge on it due to friction or other reasons, of 1×10^{-10} coulombs and if the wafer has an electrostatic potential of as little of 100 volts (which is a very small potential; electrostatic potentials as high as 15,000 volts have been recorded on Teflon® wafer holders), and if the particle passes within one centimeter of the wafer, then the force exerted on the particle is about 9×10^{-7} N, or at least several hundred thousand to several millions of times the force exerted by gravity. Clearly, much smaller static charges will still result in relatively strong attractive forces. Since magnetic fields are not often encountered or desirable in a wafer fab, we will not calculate the effect of magnetic force on the wafers, but, if calculated, it would likely be much less than electrostatic forces, but may be much stronger than gravitational forces, depending on the materials involved. Clearly, magnetic forces would have no effect on our idealized particle unless there was a significant amount of iron or other magnetic material trapped in the silicon oxide structure.

If a particle is immersed in a moving sea of air, which we can assume to be mostly nitrogen and oxygen, and if the molecules are moving at an average of 10 centimeters per second, each gas molecule hitting the particle will impart a momentum of about 5×10^{-25} g·cm/s to the particle. Assuming we are at standard temperature and pressure, there are about 2.6×10^{10} gas molecules per cubic micron of space. If they are all moving at 10 centimeters per second and impact with the particle, they will impart 1.41×10^{-14} g·cm/s of momentum to the particle. If these particles hit the particle for 0.1 seconds, then 2.6×10^{14} molecules will have the opportunity to contact the particle. If there were 100% transference of energy from the gas molecules to the particle, the gas would have imparted 1.4×10^{-10} g·cm/s of momentum to the particle, or, in other words, will have accelerated the particle to 10 cm/s, which is the limit imposed by the gas stream (actually, if there were 100% transference of energy, the particle would be at full speed in 0.016 seconds). Obviously, the energy transfer is not that efficient, but this should be all right for our discussion. Looked at in another way, these results lead us to the conclusion that, with an acceleration of 62.5 m/s^2, a particle with a mass of 2.27×10^{-15} kg is acting under a force equal to 1.42×10^{-13} N, or about 6 to 10 times that of gravity. Thus, it is clear that even mild, light air flows can have a significant impact on particle migration, even though electrostatic fields still exert far more influence. This is easily confirmed by blowing an N_2 gun across the surface of a wafer measuring 1-μm particles before and after.

The adhesive forces, which are primarily electrostatic, will cause particles to be trapped on the surface of the wafer out of the high-speed gas stream, even though the particles have a significant amount of momentum at impact. The strength of electrostatic fields has led to the development of electrophoresis devices for cleanroom air scrubbing. This instrument sits in the corner, and is charged to a high electrostatic potential. The resulting field allows particles to be drawn to the device, where they can be easily removed from the cleanroom. They work with amazing efficiency.

Finally, whenever wafers are heated and cooled, there is a significant amount of activity. Static charges increase dramatically. Particles gather energy and vibrate at a higher rate. As a result, they will be gathered to the wafers and widely redistributed over their surfaces. Without going into great detail, note that the magnitude of energy imparted in a high-temperature diffusion process (say, at 950°C) causes particles to move around more than the lower temperatures of alloy-type steps (400°C). In addition, convection currents around the heat sources can lift particles from surfaces, bringing them near the wafers. Gases moving across the surface of the wafers while they are being inserted or removed from a processing chamber are typical sources of particles.

4.2.2 The Human Contribution

First, everyone must be made aware that humans are the number one source of non-silicute particulate contamination in a wafer fab. They eat, drink, smoke, wash, go outside, change clothes, comb hair, wipe faces, scratch, and any one of a hundred different, normal human activities. All of these actions are particle creators, or particle movers. The act of a human entering the fab, no matter how clean, will allow particles to ride in on the bunnysuit (this is the reason for an air shower, although if not performed correctly, they can add many times the amount of particulate that originally existed), to come in on the shoes, even with overshoes or booties (the reason for tacky mats), and on every other surface of the body. Scratching can remove small skin flakes (a very common contaminant in a fab) and leave them in the air where you were standing. The flakes may also adhere to the surface of the gloves until dislodged later. Smoking a cigarette causes particles to be exhaled in very heavy quantities for the first 15 minutes or so, with the distribution of smoke particles reducing to levels approaching that of nonsmokers only after many hours. If the face mask of a smoker is examined you will often see staining from these particles. As lethal as smoking is to humans, it is nothing compared to the effects it has on integrated circuits.

But, even if every possible measure is taken to ensure the cleanliness of the humans (short of enclosing them in a hermetically sealed spacesuit, which could itself get dusty), the human will still produce particles. Just walking around can produce millions of particles. High-speed photographic techniques (sometimes shown as training movies in wafer fabs) have shown that there is a near-perpetual plume of particles rising up off the human (due to the human's warmth). This plume can extend for several feet from the human and exists even with people in cleanroom suits. This plume and the areas of the suit that can have leaks for particle escape are shown in Figure 4-5. Many of the particles can be trapped by the suit if it is of the correct type and has proper ventilation and seals. The various particle counts produced by the different types of suits are given in Table 4-2.

There is another problem with human behavior that can cause wafer problems. This involves wafer handling. Obviously dropping wafers on the floor, smashing them into a table or into a quartz boat, and any number of other moves can cause damage and generate or propagate particles on wafers. Handling the wafer storage boxes roughly can also cause particles to dislodge from the sides of the box and land on the wafer surfaces. This problem increases if cycle times are long and JIT methods are not employed, so that the wafers sit around in *idle* inventory for extended periods of time. The reason I have emphasized idle is that these wafers will seldom be placed in a cabinet and left alone. Usually, they will be moved around, as operators and inventory management individuals look at the boxes, or at the

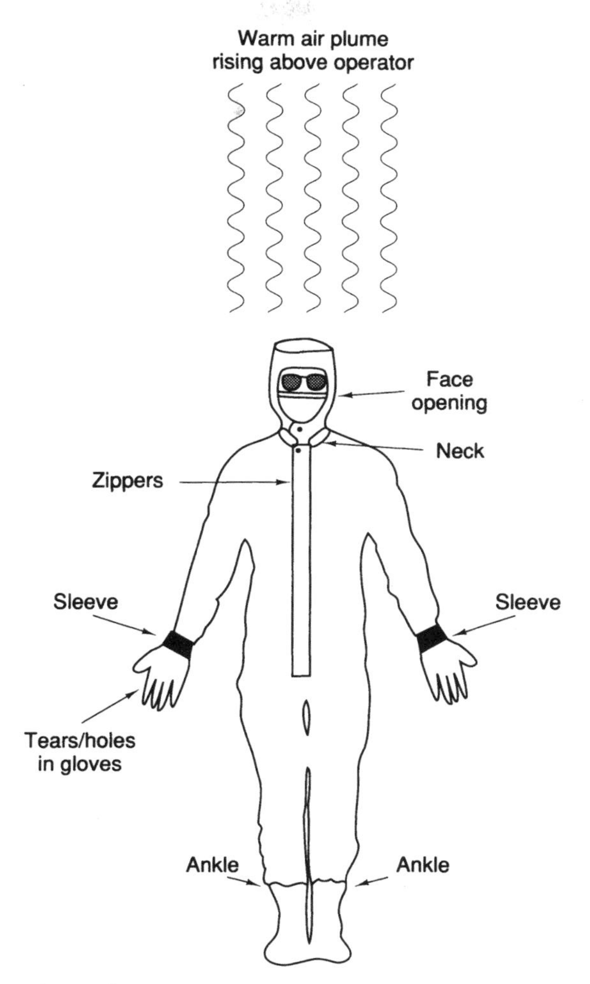

Figure 4-5. Location of Particle Leaks/Sources in Bunnysuits Even though cleanroom gear is designed to contain and control particles, leaks and openings can be significant sources of contamination. They occur at the sleeve, neck, and ankle openings, and are much worse if the material of the bunnysuit does not "breathe," allowing pressures to build up and push particles out of the suit as the operator moves around. Finally, the warm air rising above the person will contain disproportionate amount of particulate, especially if the hood and face covers are not designed properly.

wafers themselves. This extra, undocumented handling can add hundreds of particles to the wafers and seriously impact yields.

Therefore, it is clear that human behavior is directly related to one or more paths of activity of particles. We can create them ourselves through simple acts such as breathing, carry them into the clean area on our suits, and sometimes create them through our actions, such as misprocessing wafers or handling them roughly.

4.2.3 Water- and Chemical-Borne Contamination

Humans are not the only sources of particles, or the only vehicles for their delivery to the wafer surface. What is probably the second most

TABLE 4-2
Particle Counts Due to Cleanroom Suits[a]

Type of suit	Count[b]
Street clothes	400
Regular bunnysuit	100
"Goretex" style	5

[a] Cleanroom attire is a crucial barrier to fab contamination. Different materials produce drastically different results. Typically, the attire that "breathes" the exhausting particles is the best. The "Goretex" suits permit air to flow through the fabric while trapping even the smallest particles.
[b] Approximate 0.5 μm particles on wafer surface, after reaching over wafers for one-minute intervals.

common source of contamination and particles is the fab area's water supply. Despite the attempts to keep it pure, usually measured as a function of the resistivity of the water, deionized water can pick up a significant amount of contamination from a number of sources. Dead areas in the water lines can and will lead to bacteria buildup. The use of certain types of piping materials can lead to particulate generation, as the plastics in use may be damaged by the corrosive nature of the water. Chemical contamination can be introduced through the residual processing chemicals or from leaks or other problems within the water system, as well as from impurities in the chemicals used to clean the wafers.

One of the most serious problem areas in wet chemical processing is the case where wafers have been etched in hydrofluoric acid until all of the silicon oxide has been stripped off. This creates a highly active wafer surface, as the silicon crystal will leave a row of "dangling" bonds at the surface. These bonds provide a very strong attractive force for particles and chemicals in the cleaning solutions, and when a particle in the rinse water comes near the surface it will almost invariably be deposited on the surface of the wafers, and usually in the worst possible locations, as these types of etches are often used to clear off gate regions and other active areas of the circuit. In fact, there are methods of HF etch testing which can be used to quantify the amount of particulate that is being generated in the various dump rinser assemblies in the fab area. One of the serious consequences of adding particles or chemical contaminants at this stage is that it is nearly impossible to completely remove these defects once introduced, due largely to the strength of the bonding. Coupled with the probable location of the particles at the silicon interface, one can see that there is a significant probability of these types of particles affecting die yield.

Wafer cleaning chemicals and their storage containers and cabinets must be kept clean and free of corrosion. The effects of small amounts of impurities in the acids and solvents (which are often metallic in nature), can be devastating. A batch of bad chemicals that goes undetected can contaminate significant percentages of material in the fab. The problems often are hard to trace, since chemicals are used and then disposed of so rapidly in the industrial environment. In most cases, the chemicals have had specifications guaranteed by the vendors, and as long as the highest grades of chemicals are purchased, there will seldom be a problem, unless the chemicals are contaminated on-site. The quality of new vendors of chemicals should be tested through the use of engineering split lots, observing for yield variations. Finally, the bottles themselves usually come double-bagged and should be left in the bags until appropriate times for delivery to the various storage areas or use area within the fab area. Modern chemical bottles are made of unbreakable plastic, which is a real boon, as anyone who has ever seen a glass chemical bottle dropped will testify.

If acid storage containers are reused or left to sit around, or are placed in a dirty area, contamination will spread to the chemicals, the tanks the chemicals are used in, and ultimately the wafers themselves. Therefore, proper storage and handling procedures must be taught to all personnel who handle chemicals. There is something disheartening about pouring fresh chemicals into a clean tank and seeing "stuff" floating around in the solution. This is especially true when you realize the strength of these solutions, and the fact that these must be hardy particles indeed to be visible at all. The next question that immediately comes to mind is, How many are there that cannot be seen? One should always be on the lookout for the corrosion that can build up around the cabinets (such as hinges and door latches) and other metallic parts in the clean deck (such as timers). This corrosion can take place so slowly that no one notices it until it starts getting very bad.

Another basic issue when dealing with the impact of contamination is sometimes overlooked because it is such a simple concept. This is the fact that the wafers will be processed through the water and chemicals a significant number of times, and in a wide variety of processing systems throughout the processing cycle of the wafers. As a result, there will be a very large cumulative effect of contamination on yield even in a very clean fab area. A 10-mask layer, double-poly/single-metal type process will result in the wafers being placed into fluids anywhere from 60 to 95 times. If we are using six-inch wafers and add one or two particles each time the wafers are processed in a solution (a fairly typical number for a very clean fab area, working out to about 10 particles per cleaning step, although there are many factories that have numbers that are much higher, sometimes in the hundreds per cleaning step), we add as many as 125 particles, or at a defect density rate of 0.7 defects/square centimeter. These particles may become sites for later problems such as gate breakdowns and LPCVD asperities, and can cause a lot of grief due to the subtle nature of the problem and the significant amount of effort that can be required to clean up a water distribution or wafer-cleaning system once contaminated.

Clearly, the water and chemical transport systems, the vessels in which the wafers are processed, and the purity of the chemicals themselves can significantly impact the wafer yields due to both the intimate nature of the liquid surrounding the wafers and to the sheer number of occurrences of liquid immersion in a wafer fabrication process.

4.2.4 Gas System and Air-Flow Contributions

Since the wafers will almost always be surrounded by some sort of gas (at least until we can start processing in space), the conditions of that gas stream will be critical in determining the cleanliness of the final end product. In this section, I am combining the areas of process gases and the atmospheric ambient gases, since a number of basic principles apply to both, and we will discuss the specifics of each in more detail in their individual sections.

One of the primary contributors to gas-borne particles is turbulence. While this does not usually in itself create particles, it can aggravate a number of situations. One is that it will allow particles to be suspended in the gas stream near the wafers longer than they would ordinarily be, increasing the probability of contact with the wafers. The second is that the particles that are suspended in turbulent eddies can increase greatly in quantity, so that there will be localized regions where the cleanliness of the gas stream is far below standard. Third, a turbulent gas stream will dislodge dust and other particles from the surfaces of equipment, lights, and so on at an unpredictable rate. In the cases of LPCVD processes, excessive turbulence can lead to in-gas formation of particles (called gas-phase nucleation). These particles deposit onto the wafer and act as localized "hot spots," causing more reaction to occur on them and thereby causing the particle to grow to an excessive size.

One of the sources of turbulence in a gas system, and a very likely place for particles to hide in, is dead space in the tubing. *Dead space* is defined here as any area which cannot be directly purged out, and can include any kind of port, manifold, valve, or line that is off the main flow line. This dead space will have a slow circulation of the process gas through it and often the decrease in velocity, and the extreme local turbulence at the opening to the dead space, will allow particles to settle in these areas. When gases are later turned on or off, the shock waves and pressure changes that accompany the action can dislodge these particles, which then can be pulled into the gas stream and from there to the wafer surface.

Other items can contribute to small-particle propagation in a gas stream; for example, the use of filters in inappropriate locations or flexible tubing in a gas system. The main purpose of flexible tubing is to provide simple attachment of repaired parts or for movable process sections, with the first reason being the most common. While this may sometimes be necessary, it is not a recommended practice, and with careful design, systems can be constructed that use all or mostly straight tubing, and use flexible tubing only at joints that absolutely must flex. The inappropriate placing of filters is often done out of a desire to keep the system clean. However, there are often more problems caused by this inappropriate placing. For example, when filters are placed in locations such as downstream of mass flow controllers, or in areas where there will be rapid and continual changes in flow and pressures, such as between an upstream valve and a mass flow controller, dramatic increases occur in localized turbulence, providing a trap for particles on both sides of the filter, and stressing the filter membrane, sometimes causing premature failure and massive particulate generation. The use of filters in these regions can also causes pressure drops that may put the mass flow controller under its minimum pressure range. While there are proper places for filters to be placed, the use of filters in these other places is usually the result of an attempt to keep errors (such as poor fittings or welds, or improper purging procedures) from propagating as particles throughout the gas system. The use of extra parts, which could themselves create problems, should be discouraged in these cases, and the factory specifications on equipment and procedures should be sufficiently controlled and the workers sufficiently trained so that errors like these will be rare and will show up during qualification tests.

Another area where filtration problems can cause particulate propagation is in the cleanroom air-handling mechanisms. If the motors in these mechanisms fail, a belt breaks (many of the devices use belt-driven fans), or if the filter somehow gets a hole in it or has a region of the filter plug up, there will be additional particles moved into the cleanroom or there will be turbulence in the air stream over everyone's head. In any event, the cleanliness of the room will have been compromised. Bear in mind that all filters, whether for air-handling systems or for gas-flow systems or, for that matter, chemical-flow systems, must be replaced periodically. If they are not eventually replaced, they will start deteriorating and shedding particles or they will plug and cause pressure or flow problems downstream.

There is one last problem area when dealing with gas streams and this is the issue of exhaust flows. While many people assume that having high exhaust flow is sufficient, they may not realize that the exhausts must be properly balanced to be effective. In the case of an atmospheric diffusion furnace, the exhaust flow can drastically affect the uniformity of the oxide or the dopant. This exhaust pressure can change fairly drastically from furnace to furnace. In LPCVD and etch processing, the vacuum exhaust can be critical to the various aspects of process control, including uniformities and defect levels. In the cleanroom, insufficient exhaust will cause turbulence in the air stream, whereas too much exhaust can cause the room to go to a negative pressure with respect to other rooms, which may lead to the importing of contamination. Exhaust systems are also areas where large quantities of contamination exist. These areas must be watched for corrosion, buildup of process gas by-products, and other contamination. It should never be assumed that contamination in the exhaust is "OK" since it is not in the immediate vicinity of wafers. Exhaust problems and contamination in the exhaust stacks often work back into the process tube in unpredictable and detrimental ways.

4.2.5 Mechanical Contributions

There are clearly many areas that can cause particles from a mechanical point of view. Every time the wafer is moved, placed onto a piece of equipment, placed into a boat, grabbed by a robot arm, and so on, the wafer stands a chance of mechanical damage of some kind. This damage can range from the addition of a couple of particles to the severe scratching of the wafer surface or breakage of the wafer. Damage can even be caused from problems generated in one portion of a machine, then transported in some way to the wafer in another, sometimes isolated, area.

Within this category of defects, the most obvious is wafer scratching. Although easy to see, the defect can be hard to trace if it is intermittent. In this case, we are assuming the scratches to be equipment- and not operator-caused. There should be a flag raised with each engineer and operator whenever he or she sees a machine that moves a wafer in between two plates or through a slot. While this is often required, the robot or track that carries the wafer plus all of the equipment around the wafer must be constructed to tight tolerances, and must be very rigid. Any amount of play or vibration could eventually lead to a failure that could cause wafer damage. Indeed, in many cases, intermittent scratches are caused by robot arms being placed in marginally aligned positions. These are manifested as random scratches depending on the exact position the wafer handler is in on that each particular trip. Depending on the type of layout of the machine in question, a scratch on one wafer may even allow particles to propagate to the next wafer in the line, thus causing more yield problems. Furthermore, even if a scratch is small enough to allow the wafer to be further processed, the number of particles generated during the action of the scratch may still have a significant impact to the yield of the wafer not directly affected by the scratch.

The previously mentioned problems of corrosion are not unique to the wet decks. Many chemicals are corrosive in a fab area environment; for example, the exhaust stream of a diffusion furnace contains steam and hydrochloric acid which causes significant damage to the exhaust system of the diffusion furnace. Other processes have similar chemical incompatibilities that must be observed. This includes the effects of the residual "goo" that builds up around the area from various processes. This can range from splattered photoresist to residual phosphorus compounds in diffusion furnaces. Both of these problems spread both particle and chemical contamination far and wide and can have unpredictable effects at processes far removed from the original problem area. One final related area to watch out for is shredded plastic and Teflon® parts. With time, almost all of the plastic in a factory will deteriorate if the parts rub against anything. These types of problems cause primarily particulate damage, and are not generally sources of chemical contamination. Roughened or burred plastic surfaces can cause mechanical damage, such as light scratching of soft films such as photoresist.

One final area of defect generation and propagation is in the use of moving parts, especially lead screws, in a wafer fab. In general, any moving part will have some friction, and eventually will generate some particles. With modern materials, this is becoming less of an issue, but it is still important to consider. The problems become especially acute for rotating objects, such as lead screws. These objects create huge numbers of particles and are often situated near wafers or in front of laminar flow elements on automatic load diffusion furnaces. In addition, these parts are often hidden away, and can gather large quantities of contamination before cleaning and repair can be effected. One should never use anodized aluminum or other coated materials when specifying moving parts, as the coatings seldom can hold up to the stresses of everyday use. Moving parts also tend to generate vibration which can cause uneven wear, and can shake particles off of any surface on which they may have landed. In general, the attempt should be made to minimize moving parts in all systems, and all moving systems should be kept in as close to perfect alignment as possible.

Another good rule of thumb is that a part which makes excessive noise when moving is likely to be a particle generator (after all, even violin strings must vibrate to make sound; most sounds are the result of friction). This is especially true if a part has a new sound that it has not had before. This is a clear sign of some minor misalignment. This detail is often overlooked by equipment vendors simply because they are not used to the perfectionist attitude prevalent in the fab areas. The use of small clues such as these helps to prevent problems and becomes even more important as the machinery ages and tolerances change.

4.3 CLEANROOM ACTIVITY REQUIREMENTS

The quality and integrity of a cleanroom must be monitored continually and the people who work in the fab areas must be continually striving to keep the fab in its clean condition. Since activity in the fab area is the basis of the vast majority of contaminant sources, it is important that this activity be controlled. This is done through a combination of procedures, special clothing, and frequent disposal of used items, as well as frequent cleaning of all surfaces within the cleanroom. We will discuss these items in this section.

4.3.1 Cleanroom Clothing

The clothing that is worn inside the cleanroom is of critical importance. There are many types of cleanroom suits, and they need to be evaluated carefully for their cost/performance ratios prior to purchase by a manufacturing area. There are a number of components that should be worn in any cleanroom, and they should be put on in a strict sequence. These components are quickly outlined in Table 4-3. The cleanroom clothing is the primary method of containing the human generation of particles. While none of the garments have completely eliminated the human contribution to contaminant generation, they have certainly come close, especially when compared to 10 years ago, when clean lab coats were considered acceptable in many operations. Figure 4-6 shows the components of the bunnysuit in detail.

While I will be discussing the most common practices involved in cleanroom gowning, it should be noted that different manufacturers have very different requirements in gowning. Some will require that you wash your hands in special soap prior to starting the gowning procedure, in order to remove skin oils from the hands. Others require that the gloves are worn during the gowning procedure and that these gloves are thrown away after putting on the cleanroom gear and new gloves put on. Many fabs have shoe scrubbers for removing dust from shoes prior to putting on the lint-free shoe covers. The cleanest fabs

TABLE 4-3
Cleanroom Attire/Gowning Sequence[a]

Fab shoes	Comfortable leather shoes, never leave clean area
Hairnet	Thin, lint-free, contains hair, large particles
Shoe covers	Thin, Tyvek (plastic) cover for fab shoes
Change-room gloves	Lightweight, disposable gloves; short usage
Cleanroom boots	High traction, high-top nylon/Goretex boots
Beard/face cover	Thin, lint-free paper; holds facial hair
Hood	Nylon/Goretex; covers head, shoulders, most of face
Bunnysuit	Nylon/Goretex suit; complete body coverage
Filter pack	Some fabs may require a filtered breathing pack
Fab gloves	Remove change-room gloves; heavier, better gloves

[a] Fab may have additional requirements for entry into the cleanroom. For instance, special hand soaps are available for washing hands before changing, and sometimes shoes are "brushed" prior to use of the shoe covers. Multiple air showers are also required prior to entry into the cleanroom and often the change room.

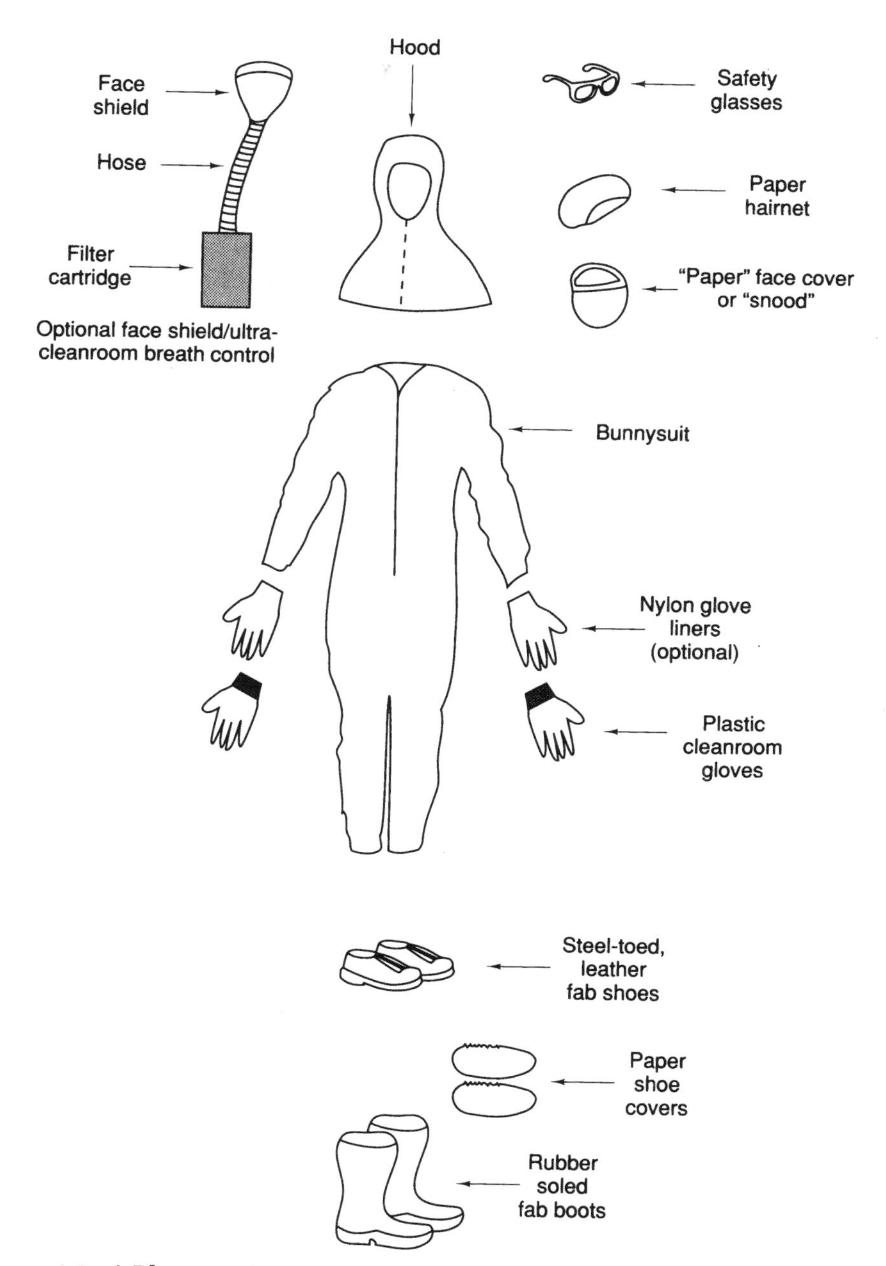

Figure 4-6. A Blow-up of Cleanroom Suit Components The cleanroom suit consists of a number of components, which are designed to contain and protect every square inch of the body. Critical particle-generating areas (such as the head, hands, and feet) are covered with multiple layers of protection. Ultraclean fabs require full coverage of the face with a spacesuit-like filtered-air faceshield.

will require all of this and more, although I will not go into detail about every practice.

We will start with the shoe. This is almost always the first thing that should be covered, as you do not want to contaminate the rest of your cleanroom gear by dragging your shoes through it all. Some fab areas dedicate shoes to the cleanroom, expecting that the investment in shoes will prevent dirt from shoes that have been worn outside from ever entering the fab area. This argument has a great deal of validity for very sensitive fabrication lines. However, it is absolutely required that the fab area shoes be cleaned on a regular basis or they will gather and spread dust and other contamination throughout the fab area. There is another problem with dedicated fab area shoes, which is that the fab area clothes must make special allowances for covering the ankle. An alternative is to wear high-top shoes.

A problem when selecting dedicated cleanroom shoes is the selection of material. While canvas-type shoes are washable, they will have a high probability of generating particles and other types of lint. They are also dangerous in the event of a chemical spill, as they will absorb the chemicals readily. Leather shoes are not as likely to produce particles and will be safer in the event of an chemical spill. These shoes will be more expensive and may not be as comfortable for all-day wear as some of the other types. Goretex shoes can be purchased but they are also expensive. The Goretex is fairly strong but can be torn, and will not provide much protection from chemical spills. The selection of a comfortable shoe is a difficult problem for the fab manufacturer, since the wrong shoe can result in a significant amount of grousing, and consequent loss of productivity. Unfortunately, everyone's shoe size and style are quite different. requiring very nearly customized shoe purchases.

There are other solutions available on the market, however. For example, one of the best solutions is pairs of nylon shoe covers, or boots, that have rubberized soles and reach up to just below the knees, where they are bound with Velcro® strips, and can be cleaned with the rest of the fab area garments. They allow the fab personnel to wear their own shoes under the boots, or the fab can require fab area shoes. These nylon boots eliminate particles escaping from ankles and socks and can help prevent the billowing of particles out of the bottom of the bunnysuits.

There are a variety of other, more temporary solutions to cleanroom shoes. For visitors, there are plastic and even lint-free paper boots that cover only the shoes themselves. They are sometimes used by manufacturing areas but rarely for any more than a temporary basis. They are disposed of immediately after use as they are not washable. In fact, in most cases, the boots do not survive long enough to even be useful for any extended time. It is best not to use these types of boots at all in a Class 10 or better cleanroom.

The second item that is worn is the head cover. The first item put on is the hairnet. These are lightweight nonlinting paper nets that contain the hair. This is done to isolate the head from the interior of the cleanroom suit. The hair and the scalp are major culprits in terms of particle production and therefore require more protection. The paper is usually able to trap particles down to about 1 μm, but this is very dependent upon the condition of the hairnet. These hairnets are disposable and should be thrown away after being removed, not reused. In some cases, it may be necessary to pull the hairnet down all the way over the ears, while in others that will not be necessary. This depends on the type of bunnysuit and hood used, as well as the length of the hair being covered. It is important that all of the hair is within the hairnet. This is especially true of the hair on the forehead. It is difficult to keep the hair in front under the hairnet and in the hood, but it is critical to do so, for leakage here will usually result in huge increases of particle generation from the person.

After the hairnet comes the "beard cover" or "snood." This is usually a lightweight lint-free paper or nylon cover that is worn on the face over the mouth, nose, and skin. Usually, the paper type have a string that goes around the head to hold them on, and the nylon types snap onto the cleanroom suit hood. The face shield is to prevent the particles from the breath and facial skin from getting out into the environment. Originally, they were used to prevent people with beards from contaminating the wafers. Interestingly enough, it has since been shown that a well-groomed beard produces fewer particles than bare skin, and that it is the breath that produces the most particles. The snood must be kept clean and undamaged, or it will do no good. As these items are usually disposable, they should be changed often. There are some types of face covers that can be cleaned with the other cleanroom garments but this practice is not recommended for sanitary purposes.

Over the hairnet is the cleanroom hood. This is a nylon hood that covers the entire head, with a low front over the forehead, and a section to cover the chin and neck. Usually the hood has long sections on all sides to tuck into the bunnysuit. Any snaps on these hoods (or on any cleanroom gear) should be fastened completely, and the hood placed on the head so that it is comfortable. It is very important that all of this gear fits properly. The proper fit for a hood is to be slightly tight, with no gaps between the edge of the hood and the face when standing and looking straight ahead. A hood that is too loose will allow particles to escape in large quantities, while a hood that is too tight will be uncomfortable and therefore will probably not be worn correctly.

Now that the cleanroom and the other gear are protected from the contamination coming from the shoes and the head, it is time to put on the main part of the cleanroom clothes, the bunnysuit, which gets its name from its obvious resemblance to a child's rabbit costume. (At $300 to $1000 each, they are definitely not children's costumes!) This is a one-piece nylon or Goretex suit, typically very lightweight, but tightly woven to prevent particles from escaping. Usually the arms and legs are bound with snaps or elastic and the front is sealed tightly. This is actually the source of a problem, in that when someone sits down or otherwise moves around in the bunnysuit, it becomes possible to greatly pressurize the suit, causing bursts of particles coming out of any opening. Many of the most modern suits have special Goretex patches for pressure equalization while controlling particle output both in quantity and location of discharge. The bunnysuits are definitely not disposable, but should be replaced and cleaned two to three times a week. The bunnysuit should be worn over the hood flaps, containing the hood exits as much as possible, but should be worn under the nylon boots. The boots should come up over the legs of the bunnysuit to trap particles that may be coming out of the bunnysuit at the ankle. There are lint-free paper bunnysuits that can be used for temporary purposes; however, they are generally not very durable or comfortable for everyday use.

The next thing that is put on are the gloves. There are two types of gloves that can be put on at this point. The first type are made of light nylon and are meant to be worn under the plastic gloves. The primary purpose for these gloves is to absorb moisture from the hands while working. There is some controversy over whether these gloves should be allowed. My feeling is that they should be, as they will trap and control contamination better than the plastic gloves alone, which can tear easily and allow built up drops of perspiration to escape. A second advantage is that it is less difficult to damage the plastic gloves with your fingernails. The main disadvantage to wearing these gloves is that it is more difficult to handle objects while working. Some people feel that the gloves hold moisture and make their hands feel clammy and therefore uncomfortable. Actually, it makes little difference whether the nylon gloves are worn or not, with the exception of the individual's comfort.

On top of the nylon gloves will be worn lightweight plastic gloves. These are usually quite thin and tight, much like surgeon's gloves, allowing for easy handling of various objects. These types of gloves are often dusted with talcum powder at manufacture. It is important that these types of gloves be avoided as the powder can spread to wafers. In addition, the gloves can come with long-sleeve portions or short sleeves. The best type to use are the long-sleeve type, with the sleeves extended up over the ends of the sleeves to the bunnysuit. The gloves should be used to attempt to control particles from coming out of the sleeves of the bunnysuits. One of the problems with these lightweight gloves is that they can be torn or damaged quite easily. They can also be contaminated by simple tasks such as scratching your nose. Plastic gloves should be changed regularly, as often as desired. The longest time that should pass between glove changes is four hours and that assumes that the person wearing the gloves that long is very careful with them. Changing gloves once an hour is reasonable in very clean manufacturing environments.

The last thing that is normally worn are safety glasses. These glasses are highly recommended and in most places legally required. Certainly with the amount of toxic and dangerous chemicals that are used in the fab, the use of safety glasses is at least prudent. There have been many instances where the use of safety glasses has saved the sight of a fab worker when something has splattered or burst into a person's face. I have personally seen $POCl_3$ bottles explode, hitting an individual in the face with $POCl_3$ ($POCl_3$ is also known as phosphorus oxychloride and is used as a phosphorus doping source. It is an acid which reacts violently with water and is very toxic and volatile.) I have also seen hydrogen explosions inside quartz reactor chambers, which have shattered all of the silicon into tiny fragments that have hit individuals. In all of these cases, the individuals could have been blinded had they not been wearing their safety glasses that day. Fortunately, most accidents like this are carefully guarded against, but even the best fail-safes have been known to fail.

In some manufacturing areas that are attempting to reach the ultimate in cleanliness, new types of head gear are used that look more like a space suit helmet than cleanroom gear. These units consist of clear plastic bubble shields which are placed over the entire face and which are attached by hose to a filter box which is carried on a belt. They prevent all exposure of the human surface to the wafer fab. Usually they are worn in conjunction with one-piece bunnysuits that incorporate hoods and boots as one unit. The disadvantages of these suits is the discomfort and difficulty communicating with other individuals (already a problem in standard cleanroom gear inside the usually noisy fab area), in addition to cost. There are many who believe that containing the humans is not the answer for controlling wafer environments, but instead expect that totally automated manufacturing is the solution. While this solution brings its own set of problems into the process, including high cost and maintainability issues, it will probably be the type of fab that will be most common in 15 to 20 years or less. At this time, the software and hardware are not quite at the level of integration required for this to be accomplished in a cost-effective manner.

4.3.2 Cleanroom Equipment and Supplies

Many pieces of gear are required for the operation of a cleanroom. These items range from vacuum wands and wafer flat finders to storage racks and other cleanroom supplies. Many companies have catalogs that list a wide variety of cleanroom supplies. Most of these products are excellent for the job that they are rated for, but it is important to verify that the article matches the job intended. For example, many cleanroom towels cannot be used with both acids and solvents. Some can even catch fire spontaneously in contact with the wrong substances. Chemical-handling gloves are other types of items that must be carefully monitored for safety. In addition, there may be convenience factors involved when purchasing cleanroom supplies; for example, some types of cleanroom paper are difficult to write on with certain other types of pens. This often results in smearing of inks (itself a source of contamination) or other problems. Price is often one of the main considerations here, as many of the items discussed have both low-tech and high-tech solutions. Needless to say, the high-tech solutions are more expensive, and sometimes are not as useful as the less sophisticated equivalents.

One of the most commonly used items in a wafer fab is the vacuum wand. This device is used to carry the wafers from place to place without touching the front surfaces of the wafers. There are many types of these devices, but we can narrow the choices down fairly quickly. First, the wand should be Teflon®-tipped, unless the wafer must be removed from something while still very hot. Generally, Teflon® vacuum wands cannot be used if the temperature of the wafers is much over 300°C. In the rare instances where a wafer must be handled at higher temperatures, metal vacuum wand tips are available. They are not recommended, as they can transfer metallic contamination from the tip to the back of the wafers. This contamination may be able to diffuse through the wafer to the circuits, or, more likely, will be lifted off the back of the wafer and redeposited on an adjoining wafer in a batch process or to chemical cleaning baths or other process equipment. The vacuum wand tips must be cleaned often, with care going into the cleaning of the vacuum opening, so as to remove all particles from this area without causing any plugging of the vacuum. Thus, they should be easily cleaned or replaced. Vacuum wand tips are typically cleaned by rinsing with isopropyl alcohol. A significant feature to observe when determining a type of vacuum wand tip is the area of the wafer exposed to the vacuum. Logically enough, the larger the area of the vacuum, the stronger the holding force. In addition, the relative area of the wand surface is important, in that it needs to be large enough to support the wafer and prevent it from falling off in the event of excessive torque. Also, the wand tip should be as thin as is practical in order to prevent damaging closely spaced wafers and for ease of use with fab equipment.

There are two basic types of wand control types, as shown in Figure 4-7. In one case, the wand is normally on, meaning that vacuum is on until the release button is pressed. This is often the most convenient configuration but requires that the wand be equipped with a manual method to turn it off when it is not in use. The other type has the vacuum normally off, so the operator is required to press the wand button to apply vacuum. While there may be some advantage to this setup, there is one major disadvantage, which is that if the button is released the wafer will fall off. Thus, a distraction or other interruption could result in a wafer falling to the floor. Vacuum wands should be supported from the ceiling HEPA filters, and are usually attached with a coiled plastic hose. These hoses should be long enough to reach all work surfaces, but not long enough to hang with the tip touching any surface. There should be a sufficient number and distribution of vacuum wands to allow the operators to perform their tasks without

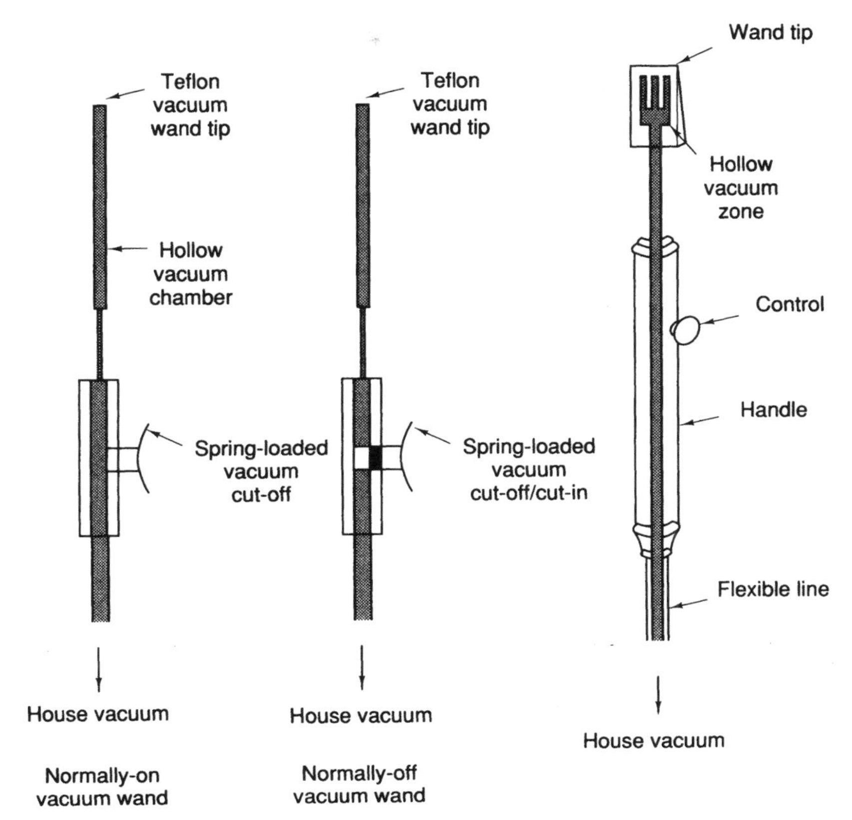

Figure 4-7. Vacuum Wand When wafers are handled in the cleanroom, small pencil-like devices called vacuum wands are used. They consist of a handle with a small push button that controls the flow of the vacuum to the wand tip. The tip is typically a wide, flat, and very thin piece of Teflon® that is placed onto the back of the wafer and vacuum is then applied. The wand control can be purchased in either a normally-on or normally-off mode. Most fabs use the normally-on variety to prevent the wafers from coming off of the wand before the control is pressed.

having to stretch the wand hoses excessively. In general, no matter how thin the vacuum wand tip, no one should ever reach between wafers, as this practice has a high chance of causing scratches and other defects.

Some operations require the use of tweezers for wafer placement. While very convenient, this practice is strongly discouraged due to wafer handling contamination that is prevalent with the use of tweezers. Tweezers produce a characteristic contaminant pattern, where the clips of the tweezers are seen on the front of the wafer. The use of tweezers may be necessary in some CVD operations where the wafers are loaded back to back or with platens that are designed without a method for pickup (e.g., C–V plotter hot chucks). The tweezers should be coated with Teflon® or another inert material, and should be wide enough for the appropriate wafer size. Extreme care should be exercised when tweezers are used, as the contact with the front surfaces of wafers is a well-known cause of particles and other damage. It is also very easy to cause other types of handling problems such as scratches when handling wafers with tweezers. Most operations that have historically required tweezers can be accomplished with slight modifications, such as the back-space-back CVD boat designs. See Figure 4-8 for the two cassette designs. They allow two wafers to be placed very close to one another but far enough apart that reaching between them with a wand is easy. If tweezers are allowed in a wafer fabrication area for a specific process, they should be restricted in some way (such as with a cord), as easy availability will increase the temptation to use them in other areas of the fab.

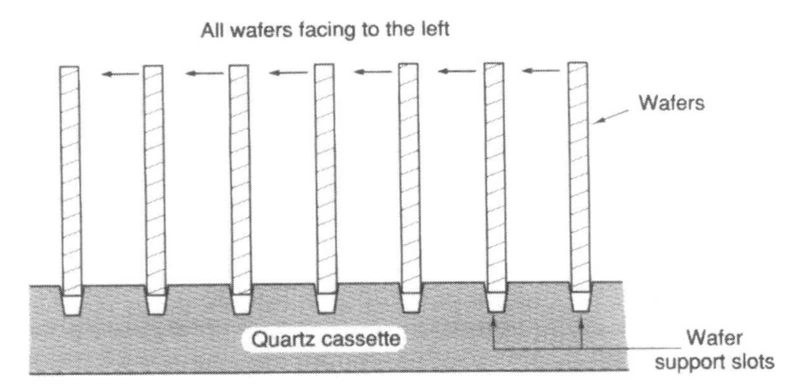

(a) Typical diffusion quartz cassette

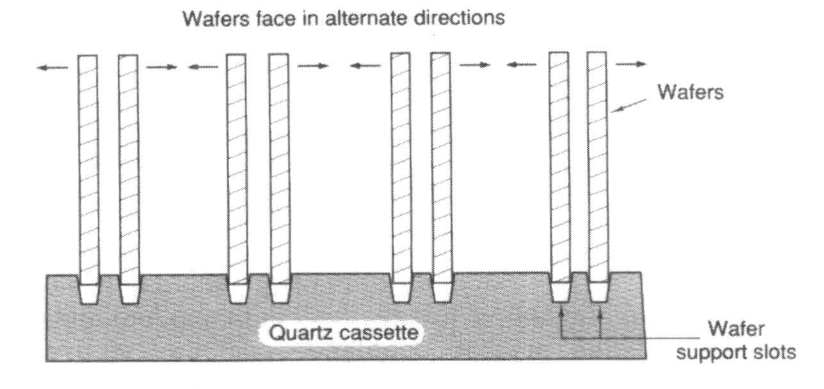

(b) Back-to-back quartz cassette

Figure 4-8. Two Types of Quartz Cassette Slots (a) Most cassettes used in wafer manufacturing have uniformly spaced slots, with enough room to get a hold on the wafers with a vacuum wand. For most operations, this is the cleanest method; the cassettes permit good uniformity. (b) Some CVD operations, particularly BPSG, use a back-to-back arrangement, making it impossible to use a vacuum wand to load or unload the wafers. While good for uniformity and throughput, the use of tweezers to load and unload wafers for these processes adds to the overall defect levels.

Many fab operations require knowledge of or control of the orientation of the wafer. Typically, this is done with a device known as a flat finder. Most modern manufacturing equipment is built with flat finders in the equipment, typically accomplished through the use of a rotating stage and a photo detector cell arrangement, as shown in Figure 4-9a. However, there are still many applications that require the use of a manual flat finder. As shown in Figure 4-9b, these are usually plastic or Teflon® objects with a stage area to place a cassette in and a rotating rod at the bottom.

The cassette is placed on the flat finder and the rod is rotated. This rod is situated at a height that allows the wafers to turn until the flats are at the bottom of the flat finder, at which point they will no longer touch the rod, and will quit turning. This allows all of the flats to be aligned to the bottom of the cassette. If this is acceptable, the cassette can be removed at this point. If the flats must be aligned to some other angle such as pointing straight up, a mechanism is provided so that, when a bar is pressed, the height of the rod is changed and the wafers can all be turned together to the correct angle. The key to the use of these devices is to turn the rod steadily. Moving it slowly can allow for somewhat better precision for flat alignment, but flats can be found quickly as long as the motion used to rotate the wafers in the cassette is smooth. Jerky motions often result in wafer misalignments.

Particle addition due to the use of flat finders is minimal as long as the machines are kept clean. Since the flat finders come into intimate contact with the edge of the wafers, and are one of the few systems in the fab that move wafers by their edges, it is imperative that they remain clean at all times. Some of the devices have been manufactured with black material on the key parts, such as the rod, that makes seeing any particle contamination easy. Another feature that is available is a motorized control. While slightly more convenient to use (in that they are generally hands-off), the reliability of the machines has sometimes been an issue. Manual flat finders require no power supplies and have virtually no maintenance requirements, other than cleaning.

Of course, when wafers are not actually being processed they must be stored somewhere. There are essentially three options available here. The first, and least recommended, is to place the boxes in stacks on top of a table. Since many of the boxes that are used are stackable, this seems like a logical thing to do. However, the amount of contamination that is generated by continually rubbing the boxes together is incredibly high, and there is the additional problem that, whenever a certain lot is required, many lots will have be shuffled around to find it, thus increasing the handling of the boxes significantly. The next option, practiced in many fabrication areas, is to place the wafer storage boxes on wire racks. This helps to prevent the boxes from rubbing on one another, and helps in the storage of the wafers in process (WIP). However, the wire racks are still not the

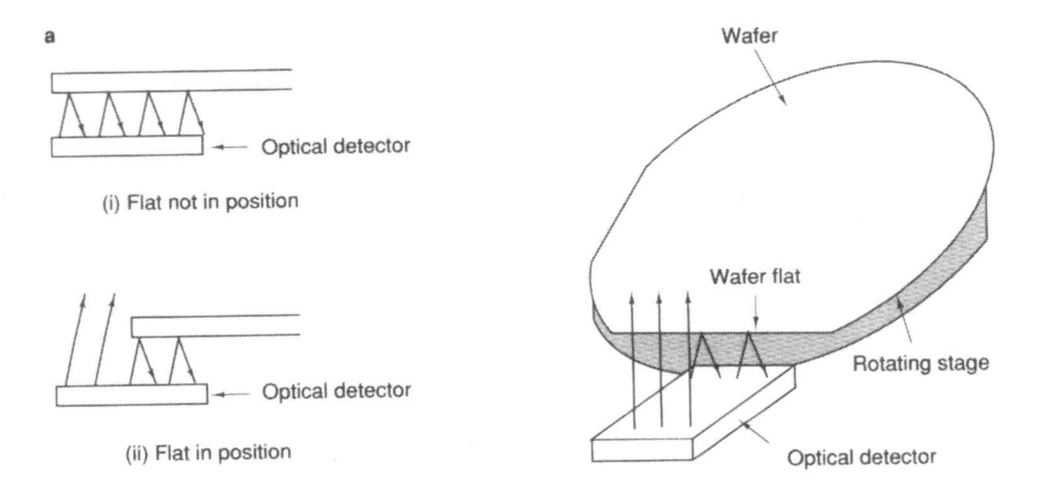

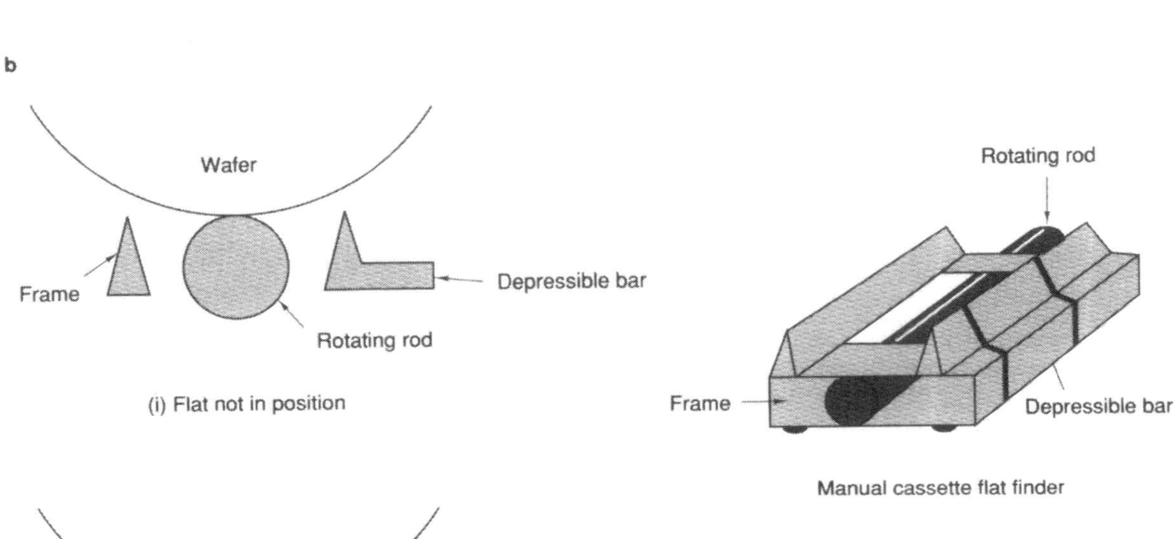

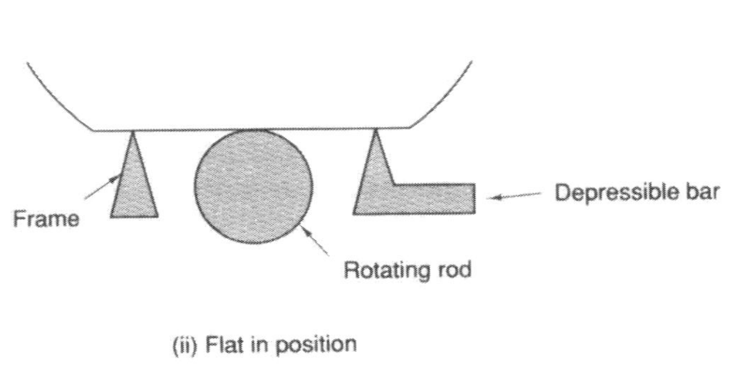

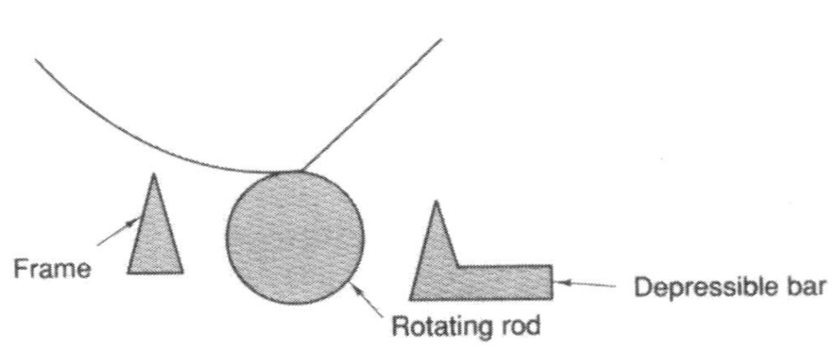

(iii) Flat being moved to alternate angle

ultimate solution. Any particles generated on racks above the position of other boxes will fall on the boxes below. In addition, the continual activity in the staging area will allow particle levels to build up. These particles can become trapped in the turbulent air surrounding the wire racks and the storage boxes. The biggest advantage of the wire WIP storage racks is their low cost.

The next step up for WIP storage is the "desiccator box." These are plastic cases that contain a large number of individual boxes, each of which is closed and purged with filtered nitrogen. They work exceptionally well for storing wafer boxes, and prevent any undesired contamination from entering the individual storage area, except that which is brought in with the wafers. This last item is a key point. For example, if the films on the wafers oxidize rapidly, the fact that the cabinet is nitrogen-purged will make little difference, as the ambient gases inside the box will be air—the same air that was in the box when it was closed. This can be somewhat helped by using vented storage boxes, although that option may not be desirable for other reasons. The main disadvantage to the use of desiccator boxes is the cost and floor-space requirements, which are the highest of these options by far. However, in an ultraclean area, this method is usually chosen, since the yield enhancement available with the cleaner storage can be shown to outweigh the costs of the desiccator boxes.

Since computers have become so pervasive in many industries, it is amazing to see how resistant fab area people have been to computerization. This tide has finally turned and more and more facilities are implementing proper process control with integrated computer systems. Nevertheless, without computerization, it will be almost impossible to rid the fab of the deluge of paper that is generated in the manufacturing process. There are control charts and processed wafer logs for every piece of equipment. Maintenance logs must be stored nearby, and a pass-down log kept in each room in order to pass on key information about that day's activities and problems. Finally, copies of the specifications relevant to the processes in a specific area are usually kept nearby for quick reference. All of these documents add tremendously to the amount of particle generation in the room as they are handled repeatedly by many different individuals, and are large and flat and leave a nice surface for particles to gather on. Modern

Figure 4-9. The Manual Full-Cassette Flat Finder Flat alignment is important in most semiconductor applications. There are a number of techniques for finding the wafer flat, two of which are shown here. (a) The on-stage flat finder in which the wafer is placed onto a rotating stage. The wafer is then rotated over an optical detector, which can determine when the flat passes over it. (b) The manual full-cassette flat finder in which the cassettes are placed on a small frame, and the rotating rod is turned under the wafers. When the flats are located (over the rotating rod), the wafers will be raised slightly off of the bar and will stop rotating. Pressing the bar and rotating the rod permits the wafers to be oriented at any angle.

cleanroom paper produces very little lint or other particles, often remaining particle-free even after being torn. Tests should be performed on cleanroom paper should be performed to guarantee that it meets all requirements prior to its introduction to the cleanroom. It should be tested for airborne-particle generation while held still vertically, while shaken lightly, while shaken violently, while being torn, and after being torn. The reaction of the towel with various cleanroom chemicals should be tested, especially isopropyl alcohol, as well as the effectiveness of the paper with various inks and with the company's photocopiers. Notice that pencils are never allowed into a fab area, only ball point ink pens. Pencils and many types of magic markers shed particles. Some inks will smear on certain types of paper, and experiments should be run to discover the best combination. Bear in mind that cleanroom paper is expensive at 25 to 50 cents a sheet, so that maximum efficiency should be maintained on all use of the paper.

Having said all this, it is recommended that all efforts be made to achieve a paper-less fab area. In this situation, centralized computer systems running large database management systems can control the information for a wafer fabrication area far more efficiently than any manual system. These systems allow for more detailed recordkeeping, with all information on each piece of equipment and each type of circuit integrated into the system so as to allow easy access to this information for all groups: manufacturing, engineering, maintenance, and management. With this extra information made available, the process engineer can analyze the effects on line or die yield for subtle differences in process parameters. The manufacturing manager can keep track of line inventory at all times, helping to prevent bottlenecks, and making the scheduling of operations much easier. In addition to the ability to control the fab area information, implementing computerization at this level can reduce the amount of paper that is used, both reducing particle generation and transportation as well as the cost of the paper itself. As a last resort, the fab area can implement personal computer-based data collection systems, although this requires more effort both to design, implement, and operate on a daily basis. In these types of systems the personal computer is used as a "data concentrator," typically hooked up to a variety of measurement systems. For instance, in one lab that was run by the author, a variety of measurements from the analytical devices could be manipulated directly on the lab computer. These devices included a thin-film measurement device, stress gauge, particle counter, C–V plotter, and a variety of optical analytical equipment.

It is a recommendation for cleanroom operation that a nonimpact printer be used. The standard impact (dot matrix) printer can generate a significant amount of contamination if it is not completely enclosed at all times. If monochrome, 16-level, gray scale graphics are acceptable, a printer such as a laser printer or ink-jet/Desk-Jet type printer can be supplied. For use with color graphics, a color ink-jet printer is required. Expect to pay twice as much for color graphics capability. Plotters are also acceptable if maintained within some sort of controlled mechanism or kept isolated from the wafers. The activity of a plotter will produce a large number of particles, but not nearly as many as an impact printer. Ink-jet printers may have some particulate problems, which could be a serious issue if the printer does not support lint-free paper. A number of these types of printers require special paper with surface layers of "clay," that absorbs the ink properly. Use of other types of paper with these printers usually results in smears on the relatively nonabsorbing material used in regular and especially lint-free paper. The Desk-Jet series of printers will allow the use of plain paper and most lint-free paper. Color ink-jet inks have a tendency to smear, especially if the paper is of the wrong type, so the printouts should be allowed to dry for a moment before handling if this is seen to be a problem.

Towels, used to clean all types of equipment, are an important source of fabrication area contamination. However, as shown in Table 4-4, there are many different types of these towels, with some dramatically different effects when mixed with different materials. For example, older woven paper products were often incompatible with DI water, producing obnoxious odors and reacting rather violently when placed in contact with sulfuric acid. A number of older fab areas have stories of garbage cans igniting spontaneously. Although seldom the cause for anything more serious than the depositing of smoke particles on work surfaces, these small fires could have caused more serious problems.

There are numerous compatibility issues with today's towels also. Often, towels that are the most compatible with acids may not be particularly compatible with alkalines or solvents, and vice versa.

TABLE 4-4
Chemical Compatibility of Cleanroom Towels[a]

Material	Good	Poor	Abs	Part	Chem
Woven cotton fabric	WA, SA, HC		1	7	5
Woven polyester	WA, HC	SA	5	1	1
Nonwoven cellulose	HC	SA, WA	2	2	4
Cellulose/polyester	WA, HC	SA	3	4	2
Nylon	WA, HC	SA	7	5	6
Polyurethane foam	WA	SA, HC	8	3	7
Nonwoven polyester	WA, HC	SA	6	6	3

[a] Cleanroom towel materials react with various materials, and are therefore suited to different uses. In the table, SA stands for strong acids and oxidants, such as sulfuric peroxide; WA stands for weak acids, such as HF, and HC for hydrocarbon solvents. Good means that the material is generally acceptable for use, while poor means that the material is not recommended. "Abs" refers to the relative absorbance of the material, "Part" to the particle-shedding characteristics, and "Chem" to the chemical release characteristics.

Towels may dissolve or have other adverse reactions and should be tested thoroughly before use. Another issue that often crops up is the effects of the use of towels on the various surfaces. For instance, there are towels that do not lint under normal conditions (e.g., when wiping up a spill from a plastic surface, such as the surface of a wet sink), that will lint when wiped on something slightly roughened, such as a metal surface. Another possibility for contamination is the transfer of particles from a clean deck to a table top or other equipment. Tight control of the use of fab area towels to prevent the contamination of the various operations is highly recommended.

One of the more important substances that should be available at all times to the fab operators is a small plastic squeeze bottle of semiconductor-grade isopropyl alcohol. This is used to clean virtually all surfaces. Almost every surface in the fab area needs regular cleaning. This should include reaching into corners and vacuum cleaning as much equipment as possible. Any object that is at a level higher than that of the average wafer storage area or working area should be cleaned especially thoroughly. Generally, the main advantage to the use of an alcohol is that it allows particles to be swept up by the towel during the wiping process. The alcohol evaporates quickly, especially in the windy cleanroom, and is relatively harmless. It dissolves most substances and can remove almost everything from the surfaces being cleaned. Generally, the isopropyl alcohol is inert to most substances, with the exception of some types of plastic. If there is a question as to the compatibility of the alcohol with a substance, a test should be run in an unobtrusive corner.

There are problems with some of the types of alcohol other than isopropyl alcohol that limit their effectiveness in a wafer fabrication area. For instance, ethyl alcohol dries more slowly, does not wet the surface as well as isopropyl alcohol, and can leave some residue, especially on stainless steel. Methyl alcohol is poisonous, although the vapors are not as persistent and as potent as isopropyl alcohol fumes. Methyl alcohol often leaves residues on the surfaces of materials being cleaned, especially brushed finishes on stainless steel. Interestingly, when cleaning glass, methanol tends to work better, and does not leave a film on the glass, whereas isopropyl alcohol will. On the other hand, the isopropyl alcohol is better on the metal surfaces, and not as efficient on glass. This underlines the importance of testing the compatibilities of the various cleaning materials to be used for the cleanroom prior to implementation. Incidentally, wiping a surface with a dry towel will generally have little effect, except to stir up whatever dust is already present on the surface. Clearly, there are safety issues to be kept in mind while using the isopropyl alcohol, such as the incompatibilities with some acids, such as sulfuric acid, which can react explosively. In addition, the alcohol can catch fire easily; for example, if it comes into contact with hot furnaces. The flame is clear and will not be visible in the event of an accident.

It is important that there are sufficient storage places for all of the safety equipment required in a wafer fab. The acid-handling safety gear must be hung up after use, so hooks are usually provided for this purpose in an out-of-the-way area near the clean sinks. Special small closets are occasionally built for the storage of acid gear, which is an excellent idea. The hooks themselves should be made out of heavy plastic, since metal hooks may react with the residual chemicals or fumes over a long span of time. Other safety gear should also be immediately available but placed off to one side of the main work area. These items include fire extinguishers, first aid kits, chemical spill kits, blankets, and so on. The fire extinguishers often use Halon® for reduced contamination in the event of fire. However, Halon has been shown to be extremely damaging to the ozone layer, and market research should be done on the most-effective, nondamaging fire prevention system. There are likely to be many new innovations in fire protection before this book is published. The first aid kits should include items for first aid for chemical burns, especially HF acid. There should also be bandages and burn ointment, as two of the most serious safety problems in a fab on a regular basis, are cutting or burning oneself on quartzware.

Often, there will be a storage box that will have a variety of emergency test items. This will usually include objects (often called Drager tubes) that are used to sample the air for toxic fumes. These devices use a plunger to draw air through a tube which contains a chemical that will change color if specific toxic fumes are present. While they are cheap and easy to use, their usefulness is limited by the short range and limited air sample capabilities. They can be used in a contaminated environment one time only, as the chemical reaction that takes place is not reversible, at least not in a reasonable time frame. Nevertheless, if there is a question about the quality of the immediate environment, they can be very informative. Also stored in the emergency test box are strips of litmus paper for testing the nature of unknown liquid spills. These items are all required to make identification of unknown contaminants in the environment easier. If there is a serious accident, it will be difficult for fire department or emergency response teams to know how to respond without knowledge of what chemicals are present in the environment.

Finally, every fabrication area has a series of emergency showers. They are used in the rare event that chemicals are splattered all over an individual's clothes. Many of these showers are enclosed in little stalls with curtains to allow some privacy, since the contaminated clothes should be removed immediately. It is important to verify that all showers are set up with proper and sufficient drainage. Many older fabrication areas had the showers set up in corners of the fab without adequate drainage, in some cases allowing the drainage water to flow easily under high-voltage equipment, causing shorts and the possibility of electrocution. Most of the modern showers are drained properly.

In addition to the emergency showers, there are also a number of eye washes, in the event that chemicals are splashed into the eyes or face. They work very well and are easy to use, but the same issue with drainage applies here. Unfortunately, safety gear is sometimes placed into the wafer fab as an afterthought and not as a primary consideration. In general, however, the safety equipment is there to be used when necessary, and no hesitation should be felt in the event that there is a safety problem.

4.3.3 Cleanroom Procedures

We have spent much time discussing the various items that go into a cleanroom operation. However, what one learns through experience is that the main technique for keeping the cleanroom clean is through the discipline of the cleanroom personnel. Improvement in working habits can often make more of an impact against a particle problem than thousands of dollars of investment in new equipment. Indeed, one of the main issues that prevents some Class 100 cleanrooms from becoming Class 10 cleanrooms is that the environment is not kept under strict control. These areas may be covered with 100% laminar flow, but using insufficient filtration or old filters, or using cleanroom garments that may not be adequate for the purpose, and so on. With time, the conscientious cleanroom worker will start to "see" particle contamination sources, even though the particles themselves are invisible. While this may sound strange, most of the fab area engineers I have worked with in cleanrooms have reported the same perception.

Since humans are the number one source of contamination within the cleanroom environment, it is important to control their exposure to the wafers. In general, this means preventing production wafers from being handled unless the production equipment is up and ready to run and no individuals other than the operators are in the working space. Usually, there is a maximum limit to the number of people allowed within each room of the fab area, with the number varying depending on the size of the room. One person per 100 square feet is not an unusual cutoff, although even that may be excessive, depending on the layout of the production equipment. If, for example, there are only three individuals allowed into each room, and another individual needs to enter the room, either production is stopped and all wafers are covered, or someone must leave the room. Usually, a leeway of at least two individuals in addition to those required to run the equipment is allowed, so that supervisors, engineers, and maintenance personnel can get into the room. Sometimes there will be short-term waivers that exceed the maximum limit so that certain situations can be taken care of. For instance, this may be the case when a cleaning cycle is being performed on a production machine. These waivers are not recommended as they can become regular work habits that eventually undermine fab cleanliness efforts.

There are limits to the types of tasks that can be performed in the various locations inside of the cleanroom. For instance, the computer terminals and logbooks are usually kept separate from any work station where production wafers are exposed. This is because there is a high level of contamination associated with the logbooks and with computer terminals. Usually, there will be one area dedicated for wafer storage, one for wafer cleaning, one for loading and unloading, along with the paperwork and chemical and safety gear storage areas. The test wafer measurement area in the room is often off-limits to production wafers, but this depends on the particular setup of the room, the metrology requirements of the technology, and the ability of the measurement equipment to handle wafers in a clean-enough fashion.

Clearly, all maintenance work should be done in as controlled a fashion as possible. All tools should be cleaned and degreased prior to entering the fab area, and tool boxes must be kept clean. Dirty tools, especially those with machine oil on them, can thoroughly contaminate ultraclean process equipment in a matter of minutes. Care must be taken by the maintenance staff to be sure their equipment is always as clean as possible. Paper packages, such as those used to hold VCR gaskets or other spare parts, should be removed prior to being brought into the fab area. If an object is double-bagged, the outer bag should be removed in the change room and the inner bag removed in the chase area. In addition, care must be taken when handling the parts of the equipment. After disassembly, some of them may be toxic or dangerous in some other way, and should receive careful handling. This requires preparation. Many accidents have been caused by individuals taking on a task with insufficient information or preparation. All parts must be handled with gloves and not damaged. The sheer volume of parts in the wafer fab reduces the total impact of the dollar volume in repair costs to damage or errors in handling. Fab areas have been known to replace entire gas lines only to have them misinstalled and contaminated within the first few hours of operation. This can cost hundreds of thousands of dollars in labor, and hundreds of hours in downtime. All maintenance personnel must be clearly trained in the task that they are being asked to do, as the proper operation of the fab equipment is affected significantly by their performance.

A common problem in a cleanroom is the generation of particles as a result of outside vendor work. This is often the case when new equipment is being installed or some major problem has cropped up within one of the production systems, and the manufacturer's representatives are in the cleanroom. In these cases, an extra number of individuals are in the cleanroom, and while this may be acceptable, they often do not know or understand the significance of the tasks they are being asked to perform. For instance, if the individual is not used to wearing gloves while working on the gear and is now being forced to, he may try to remove them at the earliest opportunity. This is often a source of friction between the fab area personnel and the vendor. It is important that both sides understand the issue and the importance of the job at hand. Contamination cannot be propagated in a fab, and the vendor needs to know what each fab considers unacceptable contamination, since every wafer fab operates differently. It is important that the semiconductor manufacturer state the cleanroom rules clearly to all visitors, contractors, or vendors and have all appropriate required gear available. The manufacturer should not expect the vendors to carry around cleanroom suits, for instance. There may be occasions, however, when the gloves will prove to be an impediment to getting the work done. If this is the case, then specific waiver procedures should be followed and documented.

One of the key factors to bear in mind while controlling cleanroom contamination is that contamination often creeps in when standards are not applied uniformly. If some cleanroom regulations are not followed uniformly, often a general loss of discipline follows, which results in other cleanroom rules being violated. There are many ways that cleanroom standards can be violated, some of them giving a very bad message to employees. This can occur when managers, engineers, and other individuals do not wear their fab gear properly, or do not know or follow all of the procedures involved. Fab area visitors are often sources of contamination and controversy. These individuals can include vendors, customers, and other "big-wigs" brought in to inspect the facilities. While these public relations trips are often required, it is crucial that cleanroom rules are followed to the letter. This will be especially true if the visitor is subtly testing to see how thoroughly the cleanroom procedures are followed (which can sometimes happen). There is usually a negative impact on the morale of the fab personnel when cleanroom rules are inconsistently applied, usually resulting in a reduction of output and productivity.

4.4 CLEANROOM WRAP-UP

It is clear that cleanroom construction and operation is not a trivial matter and involves a complex interaction between several groups of people and processes, sometimes with quite different priorities and goals. The manufacturing operation must be run within a framework of compromises starting from the very basics of the construction of the fab. To build in the optimum of cleanliness requires an enormous amount of money, and upgrading a fab area can be done up to a certain point only, beyond which it becomes cheaper to construct a new cleanroom. Since it does cost so much to construct a cleanroom, it is imperative to squeeze every last chip out of the manufacturing line. Therefore, it is important that a common set of goals and rules apply to all individuals, and that the manufacturing personnel are given the tools and the facilities needed to allow them to accurately monitor the environment at all times.

CHAPTER 5

WAFER CLEANING TECHNIQUES

There has been a significant amount of discussion up to this point about preventing particles and contamination from reaching the wafer surface. While these techniques can help reduce the contaminant levels on the wafers, they will not prevent all of the contamination. Films such as photoresist must be removed entirely, without leaving any residues such as particles. In these cases, the wafers must go through a cleaning process of some sort. The cleaning processes may involve chemical immersion cleans, plasma cleans, or mechanical cleans. Each will be discussed in detail. While the intricacies of the etch process will be covered in detail in a succeeding chapter, a significant amount of the discussion here will also apply for the etch process. Those interested in the etch process will find useful information in this chapter.

As discussed earlier, the wafer cleaning process can be among the most critical as far as the propagation of contamination is concerned. Chemicals are contaminated easily, especially since most of them are purchased for the express purpose of absorbing contamination. A significant issue to keep in mind at this point is the problem of cross contamination, which occurs when chemicals have improper types of impurities dissolved in them. For example, a pre-gate oxidation cleaning system may have BPSG films etched in it, contaminating it with boron and phosphorus. Cross contamination can occur as a result of many factors, but the most common problem is simple misprocessing, such as placing the wafers with BPSG on them into the pre-gate oxidation clean. The chemicals, which are in direct contact with the wafers, can transmit contamination to all of the wafers in a lot. In fact, in serious cases of contamination, the chemicals could distribute the contamination to several lots worth of material. This problem can also exist within the "dry" or plasma cleaning processes. Contamination occurs here although there is a lower probability of serious damage. Plasma systems can generate their own forms of contamination in a number of ways, such as particles deposited on the wafers, or film residues, or unpredictable contaminant species.

As a result, we will cover the various cleaning processes in detail, from the equipment used to perform the clean, to solutions for wafer cleaning, to techniques for making the cleaning processes more efficient. Since the wafer cleaning process is so sensitive and can so easily turn from a cleaning process to a contaminating process, it should not come as a surprise that proper, careful handling procedures can be extremely critical. We will discuss many of the tricks used to create clean wafers. However, there are many methods that work only in certain circumstances, or that may not be the most efficient for a particular situation. As always, these items are given only as guidelines, and it is important to try to discern the method being used, modify it to fit your own circumstances, and then experiment to see if there is any benefit to be gained with a certain technique.

There are a variety of styles of chemical cleans available, from the old-fashioned manual wet decks, to sophisticated modern chemical processing chambers. We will go through all of these systems and their cleans in depth.

5.1 CHEMICAL CLEANING VESSELS AND SYSTEMS

Several major types of immersion vessels have been produced for cleaning wafers. They have been integrated into a large number of manual and automated wafer cleaning systems. These vessels and systems share a variety of features. They must contain the powerful chemicals in a safe way. They must allow uniform distribution of the cleaning solutions over the wafer surfaces, and they must perform all of these tasks without adding particles. Obviously, this isn't always easy to do. In the case of manual cleaning systems, using standard immersion vessels, handling techniques become very critical, while in automated systems, component reliability becomes the major concern.

Almost all of the cleaning vessels described are made out of one of a number of polymeric materials, including a variety of types of Teflon®, PVDF, PVA, and other similar materials. A list of these materials is given in Table 5-1. PVC piping has a tendency to dissolve in DI water and other chemicals and is not recommended for use. Polypropylene is too soft and reactive for use in the tanks themselves, but is often used to build the cleaning deck structure. The materials to be used in the tanks must not react with the chemicals in any way, and preferably *dewets* (the solution beads off of the surface) when the solution is drained from the tank. These tanks are usually made by pressing the material in a mold or milling it in one piece and are seldom made by

TABLE 5-1
Materials Used for Chemical Processing[a]

Material	Inert (at 20–65°C)	Reacts	Maximum temperature (°C)
ETFE	2,3,4,5,6,7,8,12,15,16	1,10,11,14	100
ABS	7,9,15	2,3,5,6,1,12,14	85
PVDF	2,4,16,5,6,7,8,9,10,14,15	3,1,11,12,13	130
PFA	All	None (of this list)	180
Polycarbonate	2,3,6,15	1,5,7,8,9,10,12,14	80
Polypropylene	3,16,5,4,6,7,1,9,10,15	1,2,8,11,12,14	70

ID	Chemical	ID	Chemical	ID	Chemical
1	Acetic	6	Phosphoric	11	Ether
2	Hydrochloric	7	Bases	12	Ketones
3	Sulfuric	8	Hydrocarbons	13	Esters
4	Hydrofluoric	9	Alcohol	14	Chlorinated solvents
5	Nitric	10	Amines	15	Freon
				16	Chromic acid

[a] Information derived from manufacturers' data. This list is not all-inclusive, and there is no guarantee that any particular combination is suited to a specific purpose

cementing together several sheets of Teflon®, as was the case 5 to 10 years ago, since particles gathered in the welds and sharp corners of the tanks, as seen in Figure 5-1, contributing to a general deterioration of bath cleanliness. Other contaminants leached from the cements used to join the edges. As a result, most cleaning tanks manufactured today have rounded, smooth corners, and are manufactured in one piece.

All the cleaning systems described must easily interface to the fab area facilities, including power, DI water, hazardous waste drain, rinse water drain, air handler intake, and fume exhaust systems. The safety of these systems is critically important as the chemicals used in them are highly toxic and often mixed with other chemicals and elevated in temperature to enhance their performance. It is very important to verify that systems are in place for all eventual possibilities, including acid leaks or spills, or fume exhaust-systems failures. In some cases, the fab area will have chemical distribution systems or large acid storage areas so that the chemicals can be stored and used in drum form. While this is an excellent idea, and can result in a significant cost savings in chemical use, the safety and environmental issues revolving around the chemical distribution systems have somewhat limited their acceptance. One of the biggest problems is the containment of the chemicals in the event of leakage or structural failure of some sort.

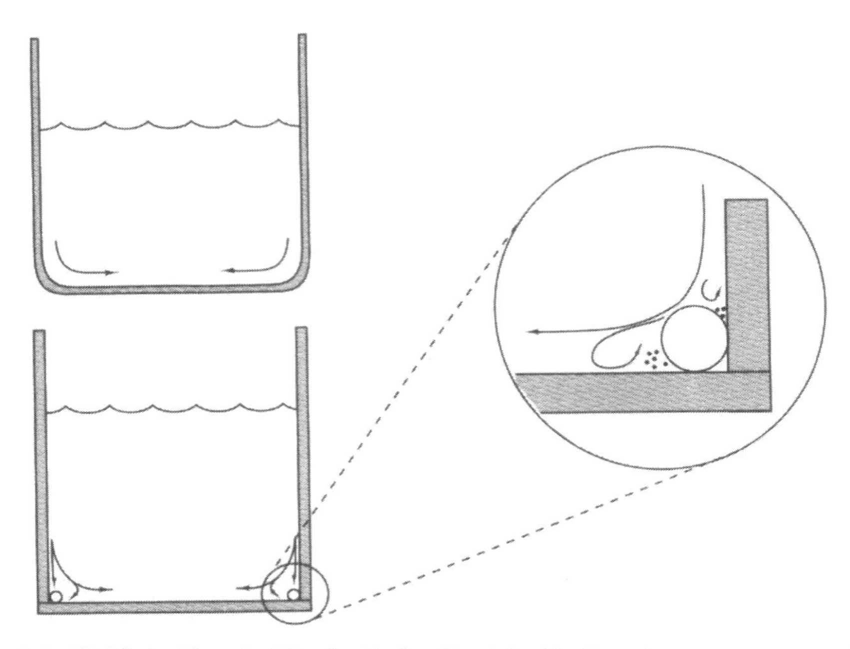

Figure 5-1. Welds in Chemical Tanks Gather Particles Easily As we can see, a tank with milled, smooth corners will not have regions of turbulence, so will not have the areas for particle entrapment as there are with welded tanks.

In almost all cases, when working around these types of cleaning systems, there are strict rules for wearing acid containment gear. This gear is provided for the operators' protection, and consists of heavy gloves, special gowns, air/sleeve guards, and face masks. Sometimes special shoes are required. This gear can protect the operator easily if used properly. However, most injury accidents occur when people are inadequately prepared. Not bothering to wear acid gear in order to save a few seconds is often a prelude to a trip to the nurse. Another case is when acid gear is improperly stored so that the chemicals get on the inside of the gear. This is obviously very dangerous, and something that is always somewhat disconcerting, since you can often do not know who else has worn the gear, and how well it may have been handled. Acid protective gear must be kept clean for the wafer's sake too, as it can get dirty easily, and is often ignored or taken for granted in the fab area. This allows particles and other contaminants to gather on the gear, which can then be dislodged into the cleaning solutions and from there directly to the surface of the wafers.

5.1.1 Immersion Cleaning Systems

We will start by discussing the various types of immersion vessels. The simplest is the standard heated tank, or hot pot. This item is seen in Figure 5-2. They are made out of quartz, ceramics, and plastic/Teflon® type materials. The heat can be provided by heater coils embedded in the walls of the vessels, or it can be provided by drop-in

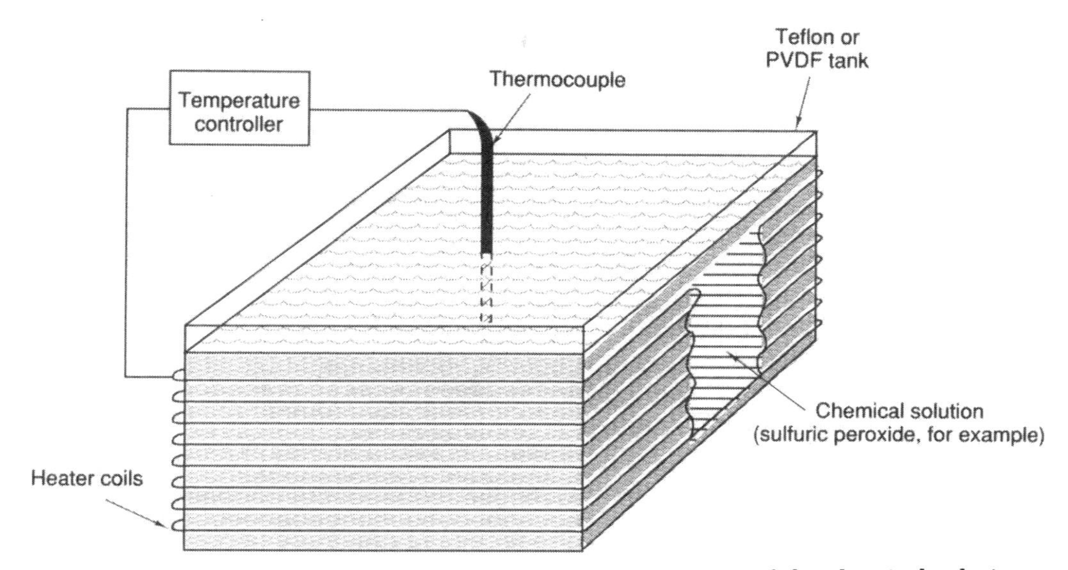

Figure 5-2. Schematic Representation of Hot Pot Many of the chemical solutions used in the wafer fab are kept at the proper temperature in the temperature-controlled chemical bath or hot pot.

immersion heaters. The drop-in immersion heaters are a nightmare from a cleanliness point of view, as they are difficult to clean, and can absorb contaminants in unpredictable ways. They are not recommended for use. The best of these tanks are made out of high-temperature polymeric materials that have embedded heating coils which can allow the tank to reach temperatures of around 180°C. They have smooth walls with rounded corners (square corners trap particles easily and are hard to clean). In some cases there are agitators or magnetic stirrers attached to these systems. The agitator is a device that hooks onto the cassette handler and moves the wafers into and out of the solution at a regular rate. The magnetic stirrer is simply a version of the device that you probably used in chemistry class. There are usually particular rates at which the magnetic stirrers can operate. The proper rate must be determined by experiment and will vary as a function of temperature and viscosity of the chemical and may be limited as well by the amount of particulate that it stirs up from the bottom. Hot pots are usually used for sulfuric peroxide, HCl solutions, and phosphoric acid solutions, as well as any number of other solutions as required. The hot pot is the basic, simple, heated chemical containment vessel.

The second type of vessel used in the wafer fab is the filtered recirculation etch module. This unit is shown in Figure 5-3. They are usually used with the less viscous solutions, including HF solutions. They can be used with ammonium peroxide solutions, as well as other solutions. Usually the filtration level is below 0.1 μm. The filters that can attain this level are very dense, resulting in a significant pressure drop across them. Recent developments have reduced the pressure

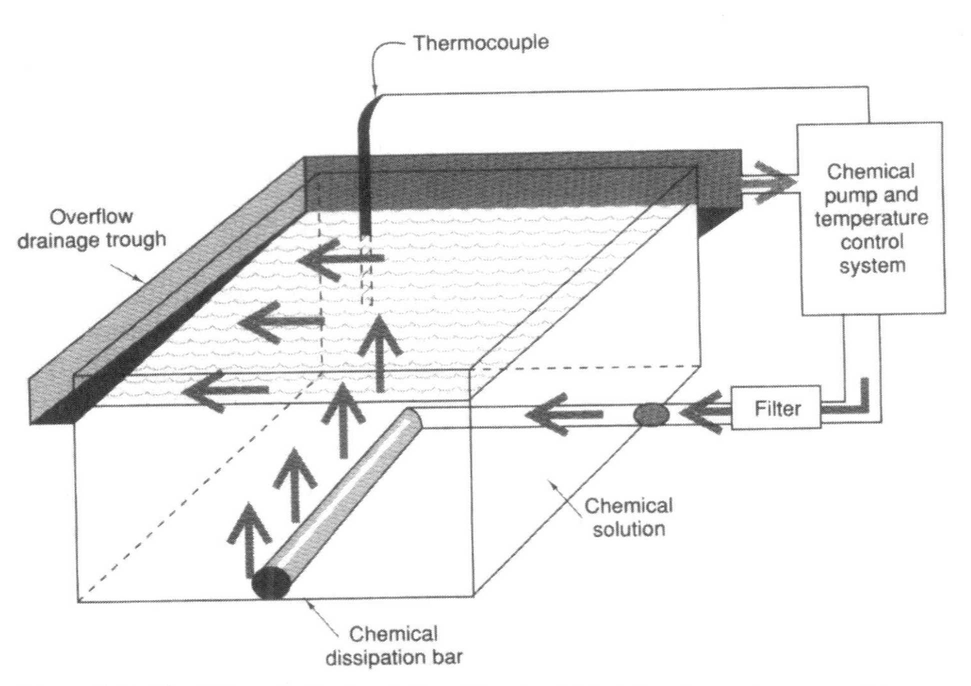

Figure 5-3. The Filtered, Recirculating Chemical Module Less viscous and lower temperature solutions are processes in a system that allows the system to be temperature controlled and filtered down to about 0.1%. Although not shown here, the main tank should overflow uniformly over all sides of the tank. All of the chemical solution should be pumped through the filters at least once per minute to ensure maximum cleanliness.

drop requirements, but a fairly strong pump motor is still required to produce enough flow through the filters to cause significant particle reduction. A major problem that occurs in filtered solutions is the formation of microbubbles. They can be formed in the pump area through tiny leaks or cavitation regions (creation of small vortices in localized regions) that form around moving parts. The use of air-operated bellows pumps exaggerates this problem. Another source of microbubbles can be the filters themselves, again through leakage or cavitation downstream of the filters. A lower pressure drop across the filter and a more consistent flow of solution through the filter will reduce the chance that microbubbles will form in the system. Usually, there are temperature controllers installed in these modules, but the temperatures are limited. Most of these systems cannot be run safely at temperatures above 40 to 50°C. However, it is possible, at high cost, to buy a filtered recirculating hot acid system, for recycling and filtering solutions such as sulfuric peroxide.

The third type of cleaning vessel is called the "dump rinse" (or "rinser"), as shown in Figure 5-4. This device is used to rinse wafers with deionized water. It has a drain on the bottom, and fills from the top through a series of spray nozzles. These nozzles are placed so that they spray over the wafers in the dump rinse tank uniformly. Some-

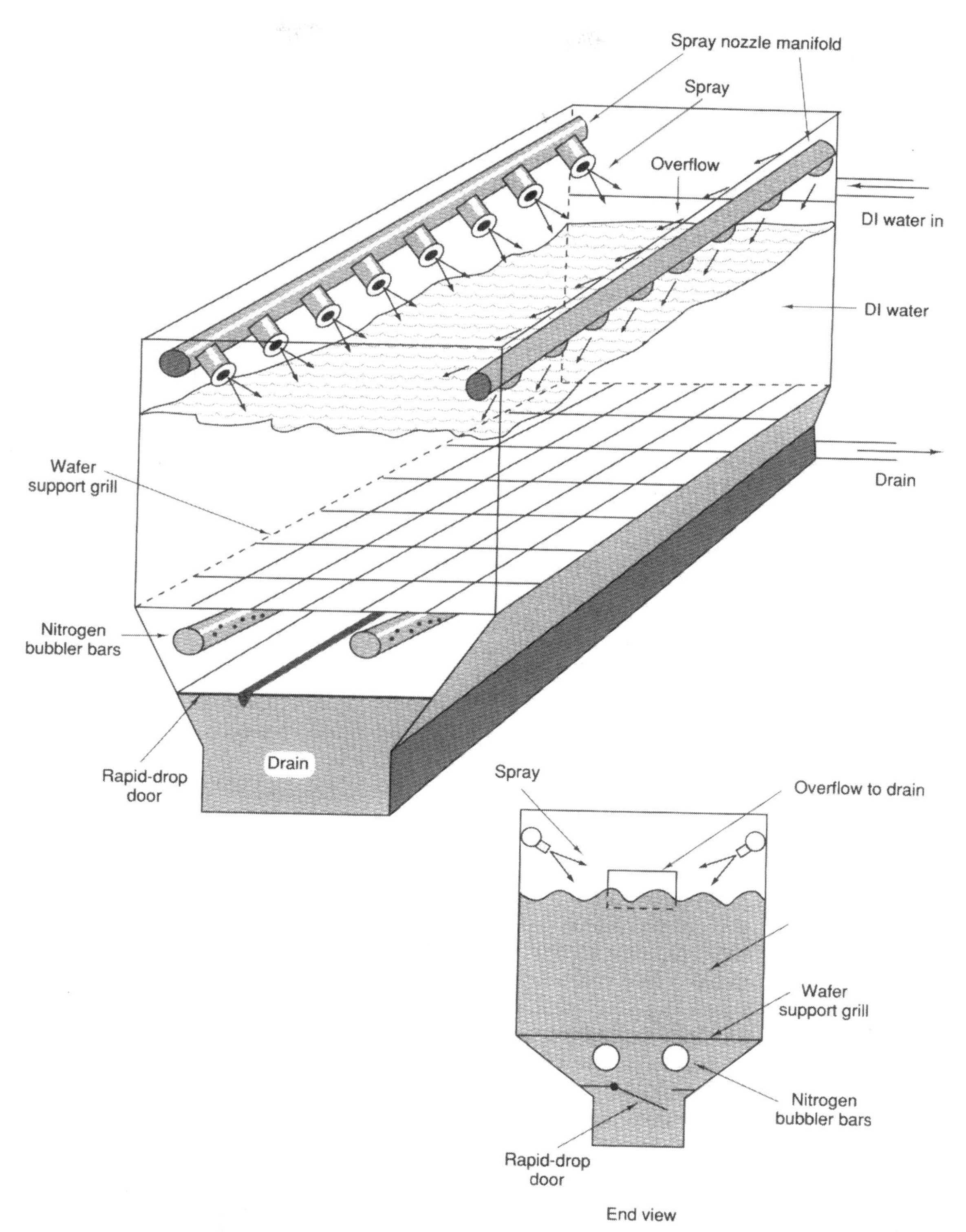

Figure 5-4. A Typical Dump-Rinse Module This tank is used to rinse excess chemicals from the wafers during the chemical clean process. Wafers are placed into the full tank. The spray nozzles are activated, the nitrogen is turned on, and the door opens to dump water from the tank. After it has emptied completely, the door closes, the tank fills to overflow, and the water is drained again. A typical dump rinse process consists of 6 to 10 of these cycles. The spray and nitrogen bubbling actions help to "scrub" the chemical residue from the surface of the wafers.

times there is some additional water fill from the bottom of the tank. In most cases, the water fills the vessel until it reaches a certain level, at which point the water is allowed to overflow into the drain, permitting contaminants that are drawn to the top to be drained off before the dumping cycle begins. After the tank has filled with water, and the water has been allowed to overflow for 5 to 15 seconds, the drain on the bottom of the tank opens and all of the water is rapidly drained out. It is important that this dump action take place rapidly and smoothly, as this type of action will produce the maximum cleanliness. Often, the dump rinses use filtered N_2 bubblers to help in the particulate removal activity. These bubbles are used to scrub the surface of the wafers clean and have been very effective in providing a mechanical source of particle removal energy. Some studies have shown that complete dumping of the water can create problems as the water surface moves past the wafers. Particles and other residue on the top of the water adhere to the surface of the wafers.

The last type of cleaning vessel is the wafer storage tank. It operates as a type of laminar flow unit, only, in this case, the fluid moves up in a vertical motion and then overflows into a drain. The base is usually set up to diffuse the water flow and allow uniform flow throughout the bath. Stagnant areas will result in localized buildup of contamination. These types of vessels are generally used to stage the wafers during the cleaning operation, for instance, while waiting for the drier to become available. In general, these vessels are avoided in large manufacturing areas due to the fact that they are easily contaminated. You will often find this type of facility in a research and development facility.

Wafers are dried in a device called a spin rinse drier, which is shown in Figure 5-5. They come in a variety of forms, from single cassette to batch processors. The cassette full of wafers is placed inside the chamber and secured to cassette holders. The wafers are spun in this chamber at 500 to 3000 rpm. During this time, deionized water is usually sprayed on the wafers for a minute or two or until the resistivity of the rinse water reaches a predetermined point. After this, the water is turned off, and hot filtered nitrogen is blown over the surface of the wafers, until they are completely dry. Usually, this takes about five minutes, but may take longer in batch processors. It is important that wafers are completely dry, otherwise many problems may occur, ranging from hazed CVD films to adhesion problems with photoresist and other spin-on films. It is sometimes surprising how many problems a little water can create in an associated processing step. Moisture remaining on the surface of the wafer can also attract and gather contaminants, leaving them as residues on the wafer when the moisture is finally driven off.

Single cassette driers can leave moisture on the wafers in some subtle ways. Since the wafers are spinning on axis, there is a point near the center of the wafers where there is relatively little motion and no

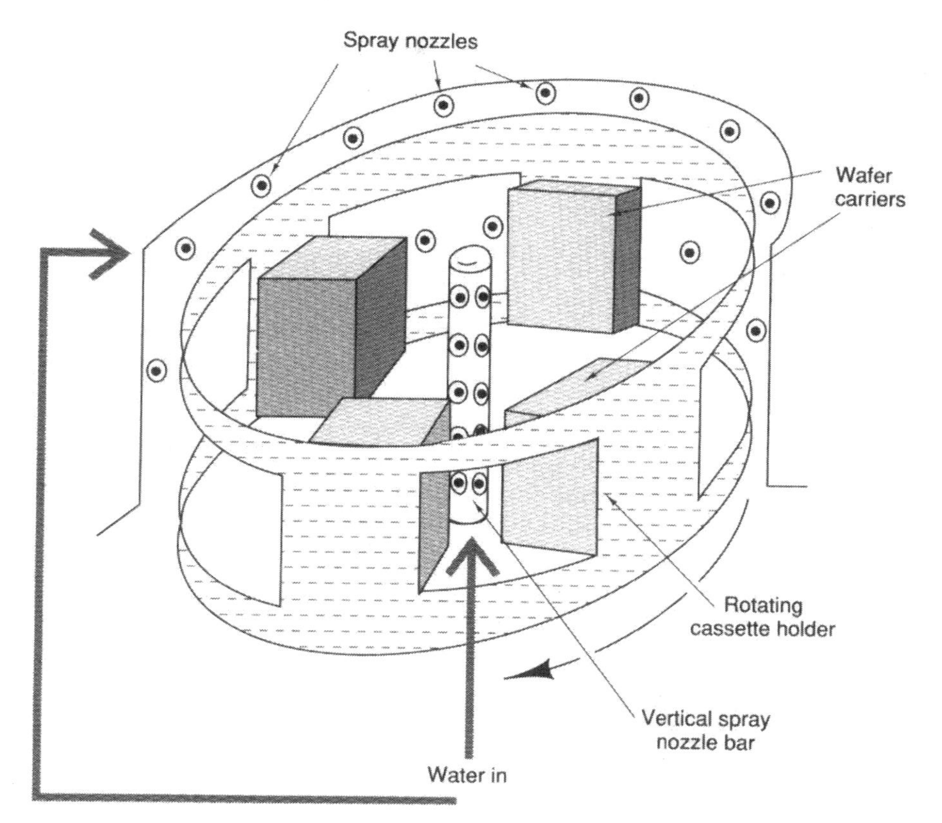

Figure 5-5. Schematic Representation of the Spin Rinse Drier The spin rinse drier is a device that can be used to remove all traces of chemicals and moisture from wafer surfaces. The wafers are placed in the wafer carriers and the cassette holder is spun at about 500 rpm while DI water is sprayed into the system. After a minute or so, the water is turned off, hot nitrogen is blown through the nozzles, and the wafers are accelerated to a rotation rate of 1500–3000 rpm. This process continues for up to 10 minutes so that all traces of moisture are removed.

significant centrifugal force. As a result, moisture that is residing in that area will not have as high a propensity to dry as other portions of the wafer. In addition, if the air flow through the drier is incorrect, usually due to improper setting of the inlet pressure, some wafers may not be dried in the center. When examining wafers with haze or other defects, always look for radial characteristics that would indicate this very common problem with a drier. (See Figure 5-6 for a description of this haze pattern.)

A characteristic problem with spin rinse driers is the buildup of static electricity within the drier. We have a rapidly moving piece of Teflon® or plastic in a hot, dry nitrogen ambient. This environment is perfect for the buildup of static charges. As a result, most driers are constructed with static eliminators as an option. These devices spray electrons into the drier chamber in an attempt to neutralize the charge buildup produced in the drier. These devices have some problems of

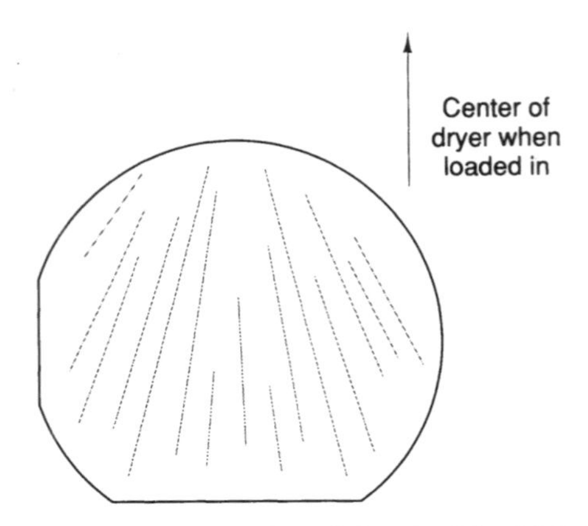

Figure 5-6. Typical Haze Pattern Produced by Residual Moisture When a spin rinse dryer does not dry the wafers sufficiently, haze can be produced at subsequent process steps, especially CVD operations. The source of the haze is readily identified if the haze appears in a pattern similar to that shown here.

their own, the most significant of which is that the static eliminators are themselves excellent particle generators.

Another problem with spin rinse driers is also due to the rapid rotation of the wafers in the drum of the drier. In this case, vortices form that can actually pull waste material from the exhaust drain into the drier. This is actually a common problem, but is often misdiagnosed. As shown in Figure 5-7, what happens is that there is a local region of pressure low enough that a channel can open through the swirling gases from some transfer which is to take place. Once this occurs, there will be circulation for a short time, allowing contamination to flow freely from the region of higher pressure (the drain) to the region of lower pressure (the vortex inside the drier). This problem can be avoided by verifying that the inlet line pressures to the drier are correct, that the exhaust lines are kept at a relatively low pressure compared to the drier, and (this is a critical one) that the use of a dump rinse does not flood the drain momentarily, stopping flow from the drier. There should be a trap installed on the bottom of the exhaust drain of the drain to help prevent this backstreaming.

A final problem related to the rotating drum of the spin rinse drier is the danger to the wafer in the event of an accident or minor problem. For instance, an unbalanced load can cause excess vibration, which can add particles, or even crack or break wafers. A broken wafer in the drier is a total disaster, as the pieces will fly out of the boat and shatter, with pieces hitting other wafers with relatively high velocity. This can cause secondary breakage and heavy contamination. The result of a few wafers broken can be the loss of an entire load, due to the large numbers of small particles that will spread everywhere.

Historically, the clean-sink setup has been the mainstay of the wafer cleaning process. These large units are typically made of a nonreactive plastic such as polypropylene. They have timers and other electronics and are hooked up to the DI water system, exhaust

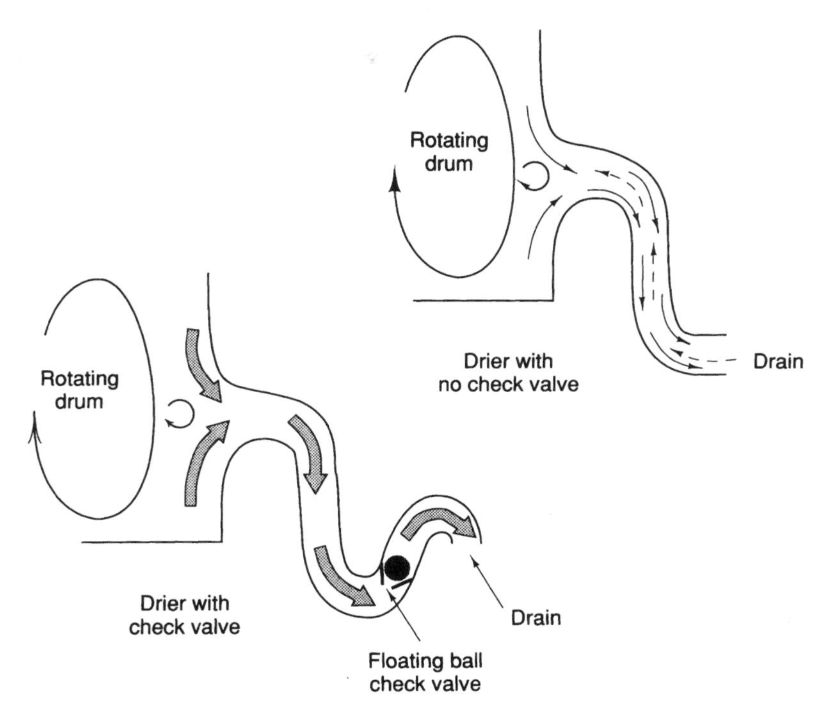

Figure 5-7. Drier Exhaust Drain Options The rotating drum of the spin rinse drier creates localized vortices within the drier. They produce low-pressure areas which permit gases and fumes from the drain to be drawn into the drier. This problem is combated by verifying that proper air pressures are maintained inside the drier, as well as through the use of check-valved and trapped drains.

system, and chemical distribution system, if available. The various cleaning modules are linked together in a manner consistent with the specific clean being performed at that step. For instance, a clean deck designed to run the "modified RCA clean" would have a sulfuric peroxide hot pot, followed with a dump rinse, followed with a filtered recirculating module for HF etch solutions, another hot pot for HCl peroxide and another for ammonium peroxide, each with its associated dump rinse, and a spin rinse drier. The modified RCA clean system is shown in Figure 5-8a. A simpler solution setup could be used for an etch process, as in Figure 5-8b. Typically, immersion clean systems are operated manually, with an operator moving the wafers from vessel to vessel. However, a number of robotic systems have been developed that can be integrated to the clean deck to perform this task with better consistency, but these systems are very large, expensive, and prone to mechanical problems.

5.1.2 Acid Processing Systems

The second most popular method of cleaning wafers is the rotating-drum batch processor. In this system, similar to the spin rinse drier, two to four cassettes are placed into special holders. The cassettes are then spun around at several hundred rpm while various chemicals are

a

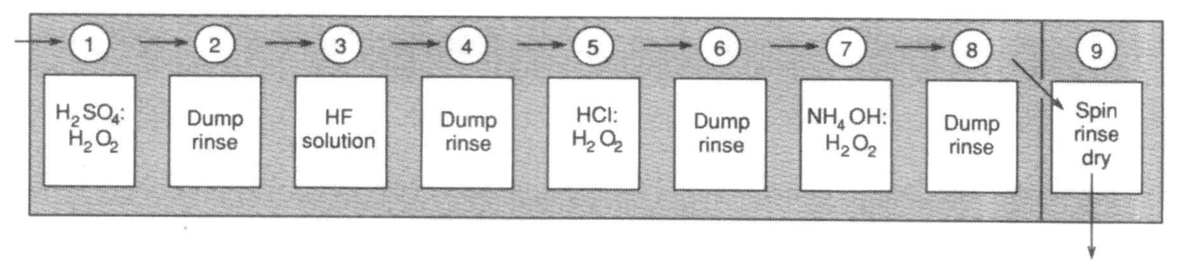

b

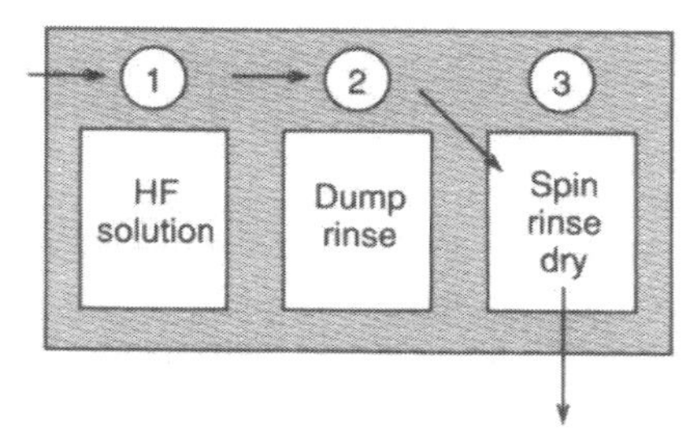

Figure 5-8. Block Layouts of Wafer Clean Sinks Chemical immersion cleans for wafers range from (a) the complex RCA-derivative cleaning sequences to (b) simple HF etch procedures. The exact procedure used depends on the cleaning effects desired.

sprayed over the surface of the wafers. In some cases, choline is used in this type of processor, in others, a more standard RCA type of clean is used. There were a number of problems with these machines in the past, although their reliability has improved dramatically in the last several years. The cleans produced with these types of systems when they are in good condition are excellent, removing even very small particles and providing excellent cleaning uniformity.

Chemical etching processes are not usually recommended in these types of processors, although in certain instances, such as removal of an entire film, the system can produce reasonable results. The major problem is trying to provide uniform etch results. The first issue is that the etchant is being atomized into a fine spray, which once in flight may or may not stay uniformly mixed and uniformly distributed in the chamber. The second issue is that of centrifugal force acting at the wafer/etchant interface. Clearly, if spin speeds become too high, there will be significant problems with the etch uniformity, not only across the wafer but across trough areas in the circuit itself, as shown in Figure 5-9. Despite the introduction of many more spray nozzles, careful work characterizing spin speeds, and other tricks, the spin acid processors are still not preferred for wet chemical etching. It should be noted after all this that most of the critical etch processes are performed by plasma etch systems, and that etch uniformity in the acid processor is not an issue in many fabs.

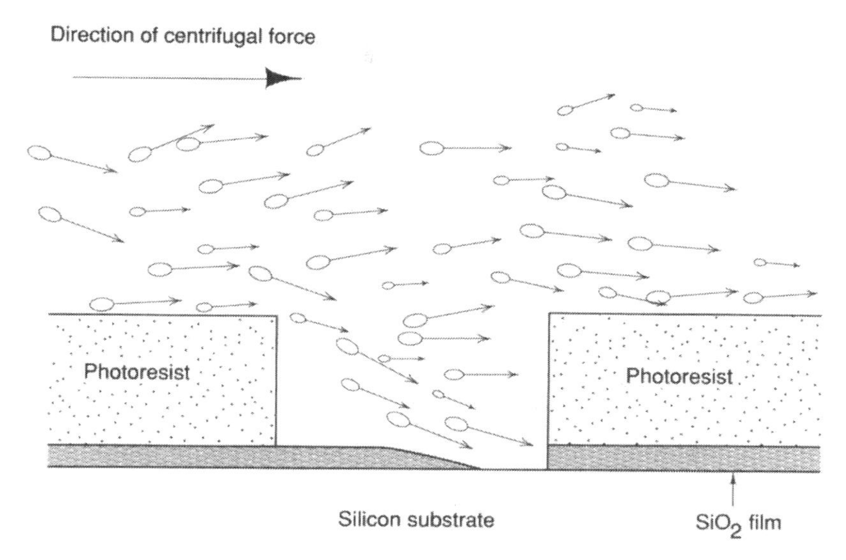

Figure 5-9. Spray Etch Nonuniformity In this extreme example, the etch-acid molecules are flying past the wafer at great speed (due to the high centrifugal forces associated with rapid rotation). Their velocities are great and directed in a nonrandom fashion, so that some areas requiring etch do not come into contact with the etch solution. Thus, the SiO_2 film is shielded from the HF by the photoresist on the left side of the opening, but has been etched and in fact is starting to attack the underlying oxides on the right side of the opening.

Naturally, these systems are quite complex and, as a result, have had quite a spotty record for reliability. Common issues include deterioration of the rotor pass-through seal, as this area undergoes high stress and heavy chemical attack, and is not always easily accessible for cleaning and preventive maintenance. These seals can fail in a number of ways, including leakage (allowing chemicals into the critical areas of the machine, including the drive motor), sustaining wear damage resulting in heavy particulate generation, and complete failure, which can damage everything within the acid processor. Improvement in the performance of the seals has been an area of intensive effort. Other areas of concern include the lid seals which can deteriorate or be damaged while loading and unloading wafers, as well as acid pumps and filtration systems which must handle large quantities of hot acids.

As mentioned in the discussion of spin rinse driers, there is the problem of batch wafer processing in this type of chamber. If there is problem or if a wafer, for whatever reason, breaks inside the spinning machine, the centrifugal force will pull the piece to the outside where it will be shattered by the rotor into tiny pieces. If this occurs, all of the wafers in the acid processor will be covered with small silicon chips of various sizes. While some wafers may be recovered and cleaned, in general, an accident like this usually results in a complete loss. This problem is somewhat more prevalent in the acid processors than in the spin rinse driers simply because the processing is more complex, and more time is spent in the chamber.

5.1.3 Other Cleaning Techniques

A fairly common cleaning process is called the megasonic clean. A block diagram of this system is shown in Figure 5-10. This is an outgrowth of a technology called ultrasonic cleaning, now commonly used to clean jewelry, tools, parts, and so on. In the system, the tanks are large enough to hold one or two cassettes of wafers. The cassettes are immersed in a solution and then passed over the megasonic generator. The term *megasonic* is used because the frequency of oscillation of the generator is much higher than that of ultrasonic cleaners. It was found that the lower frequency of the ultrasonic systems had insufficient energy to remove the smallest particles. There was also a tendency for the system to form standing waves, which allowed particles to migrate to certain low-energy boundaries, and become trapped on the wafers. This is seen in Figure 5-11. Therefore, instead of cleaning the wafers, the ultrasonic frequencies actually could make the particle problems worse. This is no longer the case, however, and megasonic cleaners are well known for their ability to remove particles. They are often implemented right after a laser scribing step, which will leave a significant amount of particulate slag.

There are three cleaning solutions that are typically used in megasonic cleaners. The first is simply deionized water, making the system

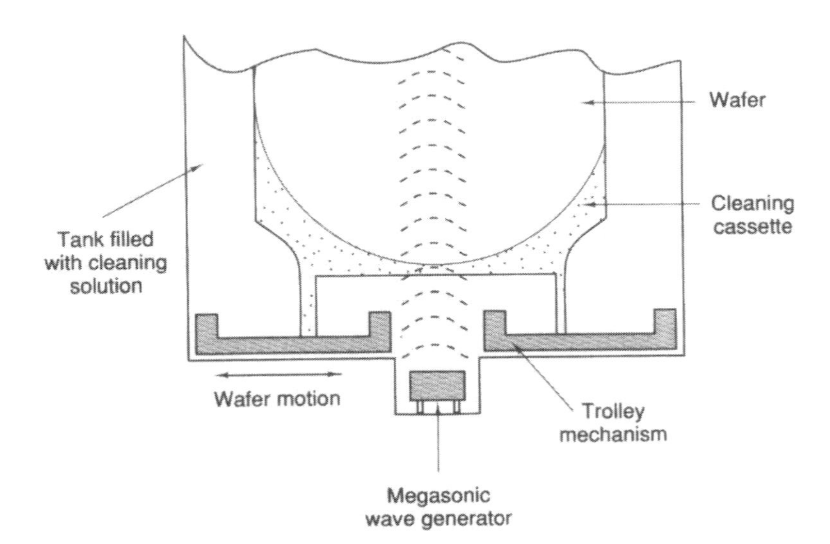

Figure 5-10. Functional Diagram of Megasonic Cleaner Here we can see a wafer immersed in the megasonic cleaner. The wafers are moved back and forth over the megasonic wave generator, which provides sufficient energy to lift particles off the surface of the wafer. The cleaning solutions range from straight DI water to ammonia, water, and hydrogen peroxide combinations.

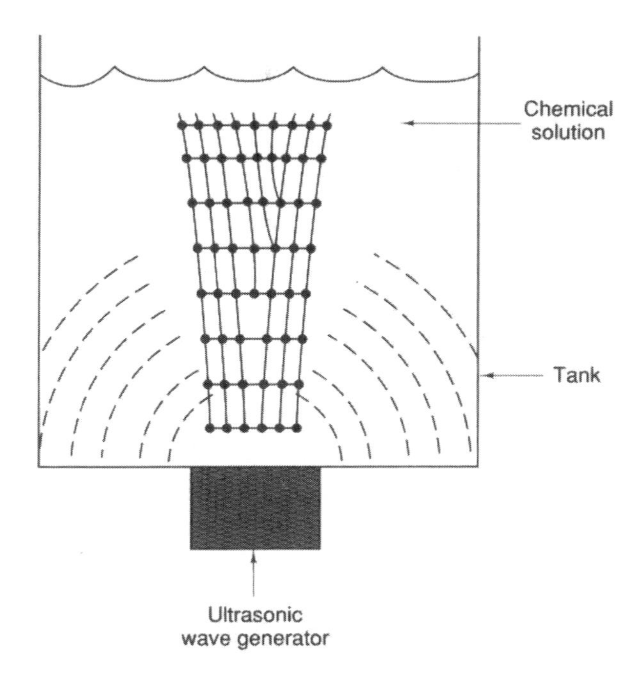

Figure 5-11. Standing Wave Patterns in Ultrasonic Cleaners Older ultrasonic cleaners had a number of subtle problems, including the formation of standing waves in the processing tanks. These standing waves do not permit the removal of particles, instead moving the particles around the surfaces of the wafers until they become trapped in the nodes of the vibration pattern. Thus, there are patterns of particles left on the wafer surface after cleaning.

essentially a powered rinse. The next most common item to place into the bath is hydrogen peroxide. The dissolution of the hydrogen peroxide in the megasonic "beam" provides a bubbling action which gives additional force to remove particles. The hydrogen peroxide also reacts slightly with some surface particles. Finally, the peroxide acts as an antibacterial agent, keeping the water in the tank clean and bacteria-free. The third type of solution used in the megasonic cleaner is a weak ammonia, or ammonia/hydrogen peroxide solution. This solution is used so that the surface-cleaning properties of the ammonia can further scrub the surface of the wafers. Straight ammonia can cause pitting of the substrate, so hydrogen peroxide is added to slightly oxidize and protect the substrate. In some cases, the megasonic cleaners have filtered recirculating modules attached to them to remove particles from the solutions. Typically, the solution is drained from the tank once per shift and replaced with fresh solution, although with careful replenishment and proper filtration, the solution may last for several days before requiring a change.

Another type of clean is called the scrub. This is done either through the use of brush contact or chemical spray. This is shown in Figure 5-12. Obviously, chemical spray scrubbers are greatly preferred over brush scrubbers, and neither is particularly popular these days.

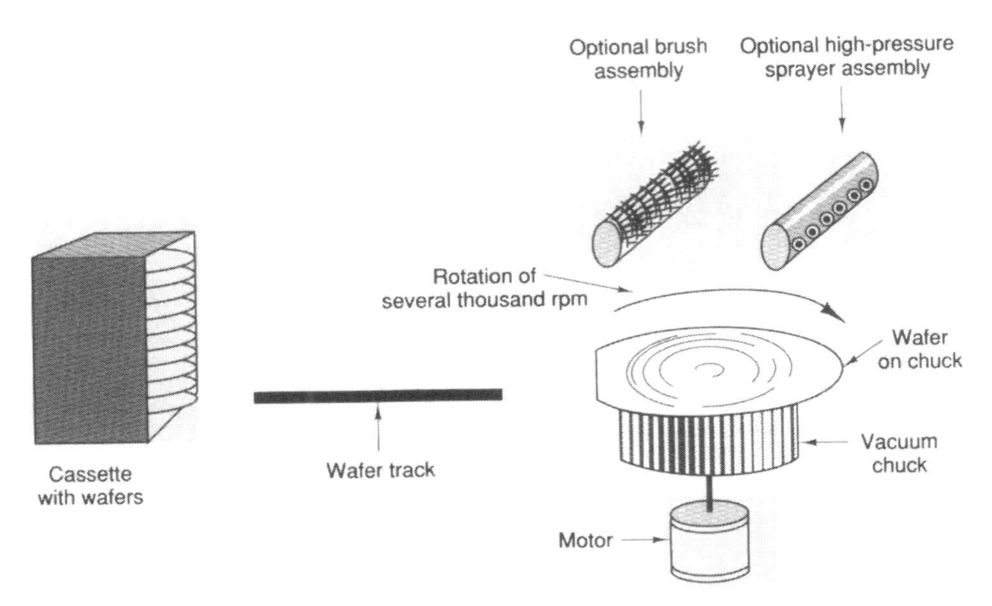

Figure 5-12. Wafer Scrubber Block Diagram The wafer scrubber is a device which allows the physical scrubbing of the surface of the wafer with either a soft brush or a high-pressure chemical jet. The wafer is rotated at 1000 to 3000 rpm, while the brush or spray is applied. This process is not recommended except in the rare cases of proven yield enhancement, since scratches and particles are both significant sources of yield loss using such procedures.

To prevent their drying out, brushes must be kept immersed in isopropanol at all times when not in use and even then the probability of damage, especially to modern ICs, is great. The high-pressure chemical scrubber can shoot small particles through the nozzle at high velocities onto the wafer surface. Although not quite as damaging to the wafer surface as brush scrubbing, defect density increases and yield degradations are common after the implementation of scrubber cleans. The difficulty in perfectly maintaining the scrubber can mean the loss of yield at some point in the future, even in those cases where a scrub can be shown to enhance yield. Typically, any yield increases that have been attributed to scrubber cleans have been associated with dirty LPCVD processes. The need for the scrubber cleans evaporates after the CVD processes are cleaned up and run properly.

A number of systems now inject ozone into the process chemicals instead of using hydrogen peroxide. The goal is essentially the same, which is to form a highly oxidizing bath to react with the hydrocarbons and photoresist residues. These systems have shown much promise, since it is relatively easy to control the ozone, and it is certainly much easier to purify and to keep pure than hydrogen peroxide. While unstable, it does not decompose at the rate that hydrogen peroxide will; plus, as it decomposes it will not cause dilution of the chemical bath. Particles will not be introduced to the system along with the chemical, as is possible with hydrogen peroxide. Finally,

H_2O_2 is easily contaminated with metals and other elements. This problem is eliminated with the use of ozone injection. Ozone is also used in plasma cleaning systems, and in the so-called afterglow plasma systems. We will discuss this further in Section 6.2, when we discuss the plasma processes. The use of ozone as a cleaning agent will certainly spread, and along with choline-based cleans may well dominate clean processing in the 1990s.

A promising new technology has recently been developed called the full-flow concept. In the system, shown in Figure 5-13, the acids are moved past stationary wafer cassettes for the appropriate intervals. As a result, there are never abrupt air/chemical interfaces, nonuniform sprays, or other problems that can occur with standard cleaning sys-

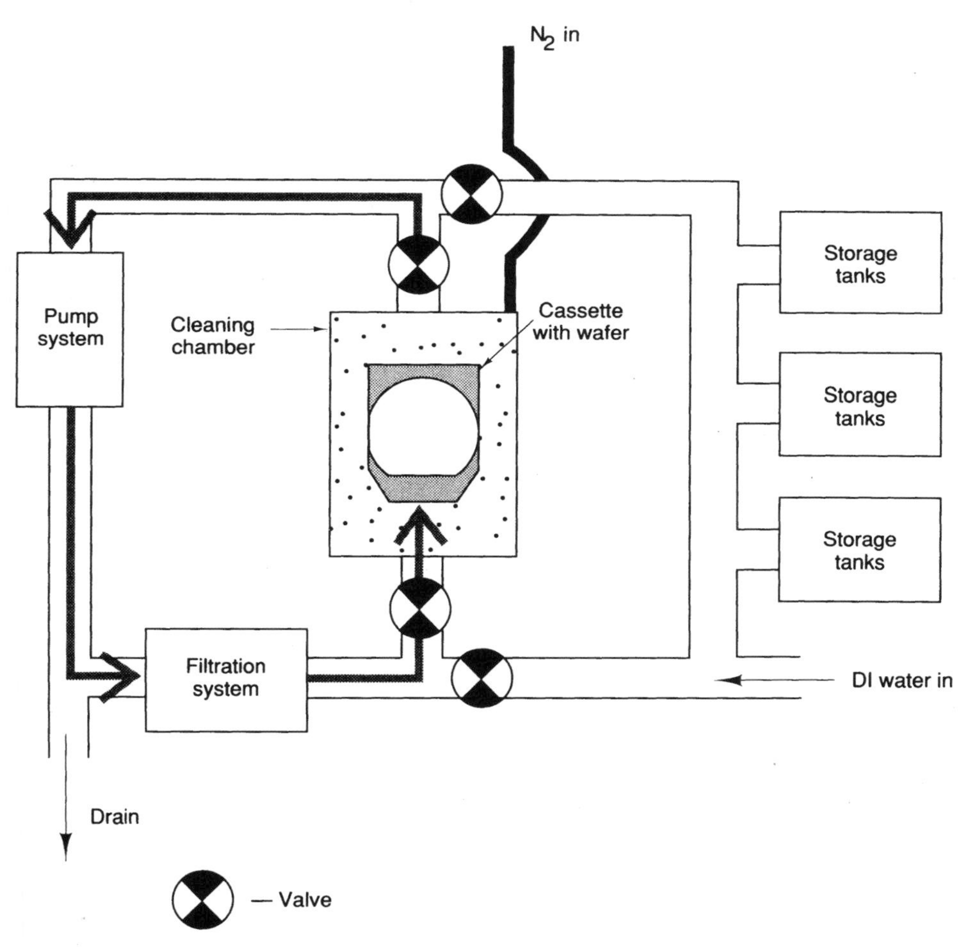

Figure 5-13. Stationary Wafer Cleaning System In this type of cleaning system, the wafers remain stationary and the chemicals are pumped past them. While having the advantage of no wafer motion, with consequent reduction in possibility of breakage and other damage, this system has the disadvantage of not being able to change chemical solutions rapidly enough, as well as a number of mechanical and contamination issues that must be resolved.

tems. The chemicals themselves can be replaced in the solution within a second or two, which is faster than an acid processor can clear itself of a chemical, and faster than a human or robot can move in a wet sink between the acid tank and the rinse tank. The actual usage rate of chemicals is drastically reduced as the chemicals are used much more efficiently, and can be recycled for longer periods of time. In addition, the reduction in wafer motion will result in a long-term reduction in line yield losses. The technical problems of this technology were not easy to overcome, due to the extreme cleanliness and precision requirements. Pumps have to handle diverse reagents at very large flow rates through Teflon® or PVDF plumbing. Fairly high pressures have to be obtained in order to filter the chemicals through the high-density filters. Nevertheless, the advantages of leaving the wafers still and processing wafers in this kind of controlled environment makes this technology one of the most promising to come along in a while.

5.2 CHEMICAL CLEANING SOLUTIONS

There are a wide variety of chemicals used in the manufacture of semiconductor devices. They have a range of uses, from photoresist and particle removal to controlled etching of the various films on the wafer and leaching of chemical contamination from the wafers. The chemicals that are commonly used in the wafer fab are listed in Table 5-2. Most of the chemical cleans practiced in a wafer fabrication area use combinations of chemicals, with the purpose of addressing as many of the contamination problems as possible. While there are a variety of names for the cleans used in wafer fabs, most of the common cleans are based on a clean procedure first described by W. Kern and commonly called the "RCA clean." There are actually a wide variety of cleans that are similar in nature to the RCA clean, and contain similar chemicals, but have variations in the sequence for various subtle reasons. It does not appear that there is one cleaning process which is perfect for all technologies and fab processes. It is up to the process engineer to determine the best combination of chemicals to use on the wafers at each step. This characterization can be a very complex and time-consuming task, so it is something that must be done carefully, and with as much statistical control as possible. A typical cleaning sequence that has been used with some success has been the following:

1. *Sulfuric peroxide* Removes residual photoresist and hydrocarbons from wafer surface. Removes metals and other contaminants from surface.
2. *HF etch* Removes oxide from areas of interest.
3. *HCl peroxide* Leaches metallic contaminant from the silicon substrate.

TABLE 5-2
Wafer Cleaning and Etching Chemicals

Chemical	Mixture	Temperature (°C)	Use
H_2O_2	Sulfuric peroxide	100–130	Oxidant; removes resist, hydrocarbons
	HCl peroxide	80–85	Oxidant, assists metal ion removal
	Ammonium peroxide	65–80	Oxidant, HC removal, surfactant
H_2SO_4	Sulfuric peroxide	100–130	Resist, metals, and hydrocarbons
HCl	$HCl:H_2O_2$	80–85	Metal ion leaching and removal
HF	HF, BOE	20–25	SiO_2 etch
H_2PO_3	Phosphoric : Water	120–140	Si_3N_4 etch
NH_4F	BOE	20–25	SiO_2 etch
NH_4OH	Ammonium peroxide	65–80	Surfactant, HC removal
Choline	Variety	20–80	Metal ion, HC removal, surfactant
HNO_3	Polysilicon etch	20–50	Oxidant for polysilicon etch
Acetic	Poly etch	25–50	Buffer for polysilicon etch
Chromic	Silicon etch[a]	20–30	Silicon substrate etch (for silicon defect analysis)

[a] These silicon etches go by a number of names, including Sirtl, Secco, and Schimmel etches, among others.

4. *Ammonium peroxide* Acid neutralization and removal. Surface clean and preparation. Removal of any residual hydrocarbons.

We will cover a variety of the cleaning combinations after we disucss the individual chemicals that will often be used.

5.2.1 Sulfuric Peroxide

This solution is a very strong oxidizing solution, often called the piranha clean due to its strong nature. Some wafer fabs have been known to impress newly hired employees with the dangers of the chemical by dipping a chicken leg into the solution. The meat of the chicken is dissolved to nothing in a few seconds, leaving a clean, white bone. This is a very impressive demonstration, as it clearly shows the new employee what will happen if he or she comes into contact with this acid.

Sulfuric peroxide is created by mixing pure sulfuric acid with a 30% H_2O_2/H_2O solution. This mixture causes a chemical reaction to occur, which heats the temperature of the solution to about 130° to

140°C, freeing oxygen and boiling the water in the solution. A highly active and unstable chemical, H_2SO_5, is formed which further strengthens the cleaning action. As a result, the solution bubbles and boils furiously. One of the keys to the solution's effectiveness is to use it while it is in this active state. Once the bubbles become small, and start to look more like an effervescence than a boiling action, it is time to recharge or change the solution. It should be noted that it is still difficult at the time that this is being written to run sulfuric peroxide through filtered recirculation systems. The solution is kept very hot, and is very reactive with the materials that make up the filtration systems. In addition, sulfuric acid is quite viscous and does not easily traverse filter membranes. There are a few systems that have overcome these problems to allow the acid to be continuously circulated through filters. They are not in widespread use at the time of this writing, however.

Sulfuric peroxide is formed in a variety of strengths, depending on the wafer fab. There is little difference in the results, as long as the solution is kept active. Often, the specified solutions are far too weak, and while they may remove the bulk of the photoresist, they may not be effective at removing all of the residual photoresist and contamination. The common solutions use three to five parts pure sulfuric acid to one part hydrogen peroxide, while the weakest practical solution uses about 300 to 500 ml of hydrogen peroxide in each tank of sulfuric peroxide. The 3 : 1 solution tends to react very quickly, taking the solution to over 130°C in just a few minutes. One of the problems with this solution is that the temperature is not very stable for about the first quarter to half an hour, after which the temperature can be maintained with a heated chemical bath. The solutions are typically maintained between 115° and 135°C. The higher temperature removes photoresist and other contaminants more readily, but results in a shortened lifetime for the solution. A four or five to one solution that is maintained at about 120 to 125°C will last about two to four hours before needing replenishment. The recharge, or replenishment, cycles can be repeated for several more hours before the solution needs to be replaced. Replenishment is done merely by adding more hydrogen peroxide to the solution. Usually, 150 to 250 ml are added during replenishment. This additional hydrogen peroxide can be added at a number of points in the production process, and is either added at regular timed intervals or prior to each one to three batches of wafers. The fixed time interval is about every 20 to 30 minutes, with weaker solutions requiring more replenishment. It must be remembered that the hydrogen peroxide solution being added is 70% water, and that as the solution goes flat, that is actually an indication that the peroxide has given up its spare oxygen, leaving only water. Therefore, the addition of hydrogen peroxide to the solution constantly dilutes the solution through time. As a result, each recharge cycle has a progressively shorter lifetime, until finally the recharge does not work, and there is little or no change in activity.

In order to reduce the instability of sulfuric peroxide, some companies inject ozone into the hot sulfuric acid to form H_2SO_5 and free oxygen. Some other companies have designed artificially stabilized H_2SO_5 solutions. All of these achieve the same end result, but standard sulfuric peroxide is still the preferred solution. Typically, there are about 50 wafers (two cassettes of up to 25 wafers) processed at a time. They are placed into the sulfuric peroxide bath for approximately 10 minutes. Some fab areas use a five-minute sulfuric clean time, but the problem with this length of time is that it is often not sufficient to remove all of the photoresist film, or may not have dissolved all of the resist into the acid, leaving particles remaining suspended in the solution that can be deposited onto the wafer surface. Ten minutes is usually enough time to do the job. However, this can depend on the processing that the photoresist has undergone. If the photoresist has received an implant, especially arsenic, it can become hardened, and can be very difficult to remove even with sulfuric peroxide. They can be removed by long cleans in sulfuric peroxide, sometimes up to 30 minutes in length. These cleans are preceded by a plasma strip, which oxidizes the hydrocarbons much as the sulfuric peroxide does. The plasma strip does not always remove all of the residuals, so it is followed with a sulfuric peroxide clean.

Since the sulfuric peroxide solution is such a good oxidant, it attacks all hydrocarbons in addition to photoresist, and as such can be used to remove some types of particles, skin flakes, grease, and other types of potential contaminants on the wafers. However, this extreme reactivity can have disadvantages. It is difficult to contain the solution, although glass and certain ceramics and plastics can do so. The chemicals will gather impurities from the environment very easily before and while being mixed to form sulfuric peroxide. Both of the chemicals, but especially the hydrogen peroxide, often have relatively high levels of metallic contaminants, which can cause severe problems if embedded into a critical structure. One of the more serious problems with contamination of the chemical solutions is not just that the wafers placed into that solution become contaminated, but that it is also possible to contaminate the cleaning tank itself. This can contaminate subsequent mixtures of the solution, even if those chemicals are clean and pure when poured.

Sulfuric peroxide cleans must be used with care when implemented into a wafer process step. It is important to understand the chemical relationship between the material exposed on the wafers and the chemicals prior to utilizing the sulfuric peroxide clean. In some cases, there will be little or no damage sustained by a film when using sulfuric peroxide, although in others there will be nearly complete destruction of the wafer with only a short exposure to the sulfuric acid solution. For instance, sulfuric peroxide is almost inert when it comes to silicon dioxide, so that the solution may be used on oxidized wafers with almost no fear of damage or change to the oxide on the wafer. In fact, it is sometimes possible to leach excess dopants or other contaminants from LPCVD glass films by immersion in high-temperature sulfuric peroxide cleans. Silicon nitride and silicon oxynitride films are also inert when placed in sulfuric peroxide solutions and as a result can be cleaned easily.

On the other hand, bare silicon wafers are extremely reactive in the oxidizing solution of sulfuric peroxide so that a bare silicon wafer or a wafer with polysilicon deposited on it will grow a thin layer of silicon dioxide on its surface. This oxide layer is about 25 to 40 Å thick, and is usually of very poor quality, with a significant amount of chemical contamination trapped within it. This chemical oxide can be found in any cleared area on the silicon surface, including gate oxidation regions, or other critical regions. The extremely reactive sulfuric peroxide baths also cause severe damage to patterned, but unoxidized polysilicon layers. The chemical reaction can be so violent that the lines may actually be lifted or torn right from the surface of the wafer.

Most metal films are highly reactive with sulfuric peroxide cleans. Aluminum films will dissolve in seconds in the solution, while some of the metal silicide films have a little better immunity to the acid. However, contact with these chemicals by one of these films will probably result in the total destruction of that film. In a few cases, it may be possible to redeposit the film onto the surface of the wafers, but this rarely works well, and will probably result in reduced die yield. Sulfuric peroxide cleans should not be used at any point in the process after metals have been deposited unless the metals are completely covered by an unreactive dielectric or other film.

In most cleaning processes, sulfuric peroxide is the first chemical solution that is used on the wafer, since it is used for photoresist removal and contamination reduction. This is done in order to remove the hydrocarbons that may have been introduced through handling or from other sources and to remove gross metallic contamination prior to the wafer going to any etch steps and before it enters the deionized water. Any chemical oxidation that occurs at this step is usually etched off in a consequent etch step.

5.2.2 Hydrofluoric Acid

Hydrofluoric acid is one of the most used acids in a wafer fabrication area, and may be found in variety of concentrations. These combinations are given in Table 5-3. This is a very potent acid, being a halogenic acid even more powerful than hydrochloric acid. Its most common use is to etch silicon dioxide, but it can be used to slowly etch silicon nitride, and certain other films. This acid is very strong even when heavily diluted, and can cause deep, severe burns. Unlike sulfuric peroxide which burns the skin severely and is instantly and painfully noticeable, hydrofluoric acid does not react quickly with the skin or soft tissues. Instead, it absorbs through the tissue to the bone, where it settles and starts reacting with the calcium to form calcium

TABLE 5-3
Hydrofluoric Acid Solutions[a]

Solution	Mixture	Etch rate (Å/s) (thermal SiO_2, 25°C)
"Straight" HF	49% HF in H_2O	>50
5 : 1	5 parts H_2O, 1 HF	10
10 : 1	10 parts H_2O, 1 HF	5
50 : 1	50 parts H_2O, 1 HF	<1
100 : 1	100 parts H_2O, 1 HF	<0.5
13 : 2 BOE[b]	13 parts NH_4F, 2 HF	≈ 25

[a] Hydrofluoric acid solutions are selected by the film that is to be etched. A partial etch on a gate oxide will be performed with a weak 50 : 1 or 100 : 1 HF solution, whereas thick films may be removed with 10 : 1 or 5 : 1 solutions, or even a buffered oxide etch (BOE) solution.
[b] BOE, buffered oxide etch, comes in a variety of concentrations, including 6 : 1 and 7 : 1 solutions.

fluoride. The only noticeable initial reaction is a slight reddening and itching of the skin several minutes after the exposure. A severe burn will be followed by aching and pain deep in the region of the burn. At this point, injections of neutralizer are required. In severe cases, the affected area may have to be operated on or amputated. Pure hydrofluoric acid can also produce extremely lethal fumes which can damage the lungs, sinuses, and other surrounding tissue. Pure hydrofluoric acid is so strong that a patch a few square centimeters in size can kill a human.

The purest form of hydrofluoric acid that is usually seen in a wafer fabrication area is a solution that is 49% (by weight) hydrogen fluoride dissolved in water. This is normally called straight HF, and is very strong and fumes easily. While it is possible to produce HF acid that is not in solution with water, it is difficult to handle, expensive, unstable, and, in almost all cases, unnecessary. Most hydrofluoric acid is delivered and used in diluted form. In some fabrication areas, a solution of five parts of deionized water to one of hydrofluoric acid is used, although most have converted to a more controllable solution such as 10 to 1 or even 50 to 1. These weak solutions are much easier to maintain, etch in a more uniform manner, and in general are safer to use than the stronger solutions. Sometimes hydrofluoric acid is mixed with ammonium fluoride and water solutions in concentrations of 5 and up to 15 parts ammonium fluoride solution to 1 part hydrofluoric acid. This solution is called buffered oxide etch, or more commonly BOE, and is used in place of the HF/water solutions for some etches. It tends to be a little more controlled and uniform than some of the stronger HF solutions while maintaining the high etch rates. Table 5-4 shows the etch rates of a variety of HF solutions on a variety of films. BOE solutions tend to be unstable and, if left open to the air, ammo-

TABLE 5-4
Wet Dielectric Etch Processes

Film	Etch rate (Å/s)	Solution
Grown SiO_2	5	10:1 HF
Dep SiO_2	6	100:1 HF
As-dep BPSG	27	6:1 BOE
PECVD SiO_2	7	100:1 HF
TEOS SiO_2	1	100:1 HF
Si_3N_4	1.5	6:1 BOE
SiO_xN_y	≈5	6:1 BOE[a]

[a] Etch rate varies as a function of O/N ratio.

nium fluoride precipitates will start to form in the tank, which will contaminate the etch solution. As a result, most of the advanced fab areas try to avoid the use of straight BOE, with the occasional exception of a noncritical step that must be etched relatively quickly. Diluted BOE solutions tend to be more stable and can be used effectively. BOE can be difficult to filter below 0.2 μm, because the NH_4F molecule tends to dissociate.

Notice that, since hydrofluoric acid dissolves glass readily, it is not practical to store the HF in glass. It is necessary to use plastic or Teflon® bottles to store the hydrofluoric acid. They are usually much stronger and safer than glass bottles and are often recommended for the storage of other chemicals.

An advantage of the weak hydrofluoric acid solutions is that they can be filtered with most standard water filters (as long as the filter membrane material is compatible with the acid). There are many acid recirculation systems that can clean and purify the hydrofluoric acid baths. They were developed partly because for many years it was believed that the particles picked up during HF particle tests were from the HF, and not from the DI water itself. Now, the filtration of HF is one of the most well known and well controlled of any of the wet cleaning processes. The filtration systems that are used with HF are excellent, producing low counts from after about one-half an hour (about two to three acid recirculations through the filters) to about a week. In fact, the HF acid needs to be replaced more as a result of changes in etch rate due to consumption of the HF than to contamination of the acid. Of course, metallic contamination can be a slight problem, especially if the sulfuric acid bath is not kept clean. In general, the filtered HF recirculation systems show results after a week that are as good as or better than the particle counts that were found shortly after the chemicals were poured.

One of the problems with the filtration systems that has been addressed by some of the systems manufacturers is that of microbubbles. We have discussed microbubbles already, but the filtration systems are among the most serious places for microbubbles to occur. In the case of a solution like sulfuric peroxide, the bubbling action enhances the process. However, with HF, it is important that the etch process is carried out in a very controlled fashion. If the etch bath produces bubbles there will be localized, unpredictable areas where the etch did not take place, since the bubbles will prevent the etch from taking place. If at a critical location, this problem could cause a significant change in the yield of the wafers. There are a couple of methods for reducing microbubbles, one of the most successful of which is the use of a special delivery rod that is porous and allows the chemical to move through it, but tends to break up the bubbles. Microbubbles are the worst in systems in which the solution is brought directly into the etch tank from the motor assembly. There should always be some sort of diffusing system to slow down, trap, or break the microbubbles. Air pumps generally should be carefully selected in order to reduce the possibilities of cavitation. The best plan is to purchase systems that are designed to prevent the formation of microbubbles in the first place.

As noted, hydrofluoric acid etches the silicon dioxide films that are not generally attacked by other acids. The silicon dioxide films will have varying etch rates, depending on whether the film is grown or deposited, and whether the film is doped or undoped. Doping a glass film with phosphorus and boron changes the etch rate of a film in a predictable enough fashion that etch rate can be used as a rough measure of the content of the phosphorus and the boron. Figure 5-14 shows the change in etch rate as a function of phosphorus and boron. The slowest etch rates are for the thermally grown films, as they are the densest films. Thermally grown silicon dioxide grown from polysilicon is somewhat less dense than silicon dioxide grown from single-crystal silicon. As a result, the etch rate of the polysilicon oxide will be somewhat higher than that of single-crystal silicon oxide. Undoped deposited silicon dioxide has a much higher etch rate than either of these, often three to five times higher, with dopants in the film causing even higher etch rates. Of course, the dopants in the film will be removed from the wafer and dissolved into the solution. In some cases, these films may not react uniformly and may lift off the wafers in a solid form, in which case they will act as particles and will usually end up being trapped in the filters. Acid baths that have had doped glass films etched in them are generally considered contaminated and are not used for most other film etches.

Hydrofluoric acid also can be used to etch silicon nitride films. This requires a somewhat stronger solution of HF, and is much slower than the etching of silicon dioxide. In most cases, the etch is too slow to be useful in a manufacturing area. Hydrofluoric acid can also be used to etch silicon oxynitride. Although the etch rates are between

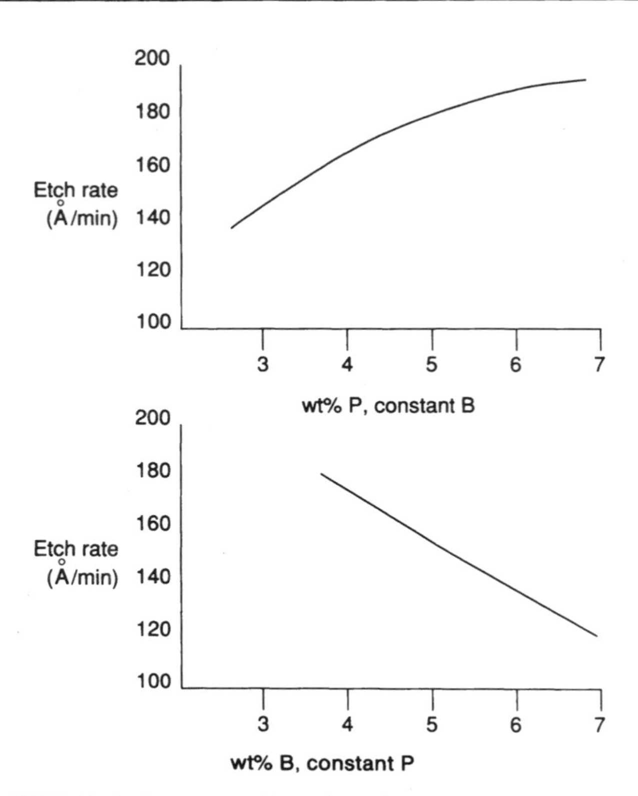

Figure 5-14. BPSG Etch Rates as a Function of Dopant Concentration BPSG etch rates vary with dopant concentration. The etch rates also may vary with the amount of annealing that is performed on the wafer. These trends permit the use of these etch rates as monitors for dopant concentration, especially for boron, which is difficult to monitor in the presence of phosphorus.

those of silicon nitride and those of silicon oxide, in general, the oxynitride films etch very slowly in hydrofluoric acid.

Fortunately, there are a number of substances which will not react with hydrofluoric acid such as silicon, in both the single-crystal and polycrystalline form, which is an extremely useful feature. This allows the process to etch down to the silicon surface and stop, with a complete cessation of activity once the acid reaches the substrate. In fact, the easiest way to ascertain that a silicon dioxide film has been completely removed is through visual observation of the phenomenon called "sheeting" or "dewetting." The water and acid solution will cling to the surface of the oxidized wafer, so that if the wafer is extracted from the acid solution, it continues to be covered with solution. When all of the silicon dioxide film has been removed, the silicon surface becomes hydrophobic, and will not support any moisture on the surface. The solution then rapidly clears off of the surface. The visual difference is quite striking, and is often used to define the end point of an etch: "Etch in HF until wafers dewet" is a common phrase used in a rework or test procedure. It used to be possible to etch the wafers and see contacts and other features become visible on production wafers. However, modern contact holes and feature sizes have become so small that the visual difference between the completely etched and the unetched surface is negligible. Therefore, most modern HF etches are for fixed times only, and assume that the correct oxide thickness is in place. Sometimes, for the most critical steps, the oxide thickness is measured prior to the etch process and then the etching time is calculated appropriately. Notice that as gate oxidations have been reduced to under 100 Å, HF etches on bare silicon have been shown to cause damage resulting in parametric changes.

Another substance which typically does not react strongly in a hydrofluoric acid solution is photoresist. There are a number of operations which can make use of this phenomena. For example, there can be etches through field oxidation steps to open up areas for an implant, or there may be a need to cut contacts or bonding pads through a deposited oxide layer. Some photoresists work better than others in this type of application, and most resists must be baked until hardened and can handle only fairly weak solutions of hydrofluoric acid.

One problem with wet etches is that they are by nature isotropic. By this, we mean that the etch does not cut straight down, but also spreads sideways into the adjoining reactive films. This is demonstrated in Figure 5-15. This undercutting can cause severe problems with most technologies. Typically, this undercutting is a less severe problem if the consequent step is an oxidation step, or if the film above the etched film will be removed (as in the case of photoresist or early layers of silicon nitride). This problem will be most severe when the succeeding steps involve deposition steps, as the space where the undercut has taken place can remain unfilled after the deposition step. This can cause a physical or electrical collapse of that layer, resulting in a failure at that step. Usually, if a problem like this becomes severe enough to affect one chip, it will affect a large number of chips at once. One of the main reasons that plasma etches have been researched so thoroughly and used so extensively is the ability of the etch process to provide anisotropic etches. Hydrofluoric acid etches are not assumed to etch sideways at the same rate that they etch down, so that the undercut amount can be calculated fairly easily. While this must be characterized for each oxide film by SEM analysis, SiO_2 films will typically undercut at about one-half to two-thirds the rate at which they will etch.

It is very important not to mix an oxidant into an HF etch bath. The mixture of HF with an oxidant produces a silicon etch (we will discuss silicon etches in more detail later). The oxidant will attack the silicon surface, which has been etched clean by the HF. The oxidant reacts with the silicon to form silicon dioxide which will immediately be attacked by the HF, which clears the surface for further reaction with the oxidant chemical. The cycle will continue until all of the silicon has been consumed. Needless to say, the reaction would very quickly destroy polysilicon films. One difficulty in identifying this

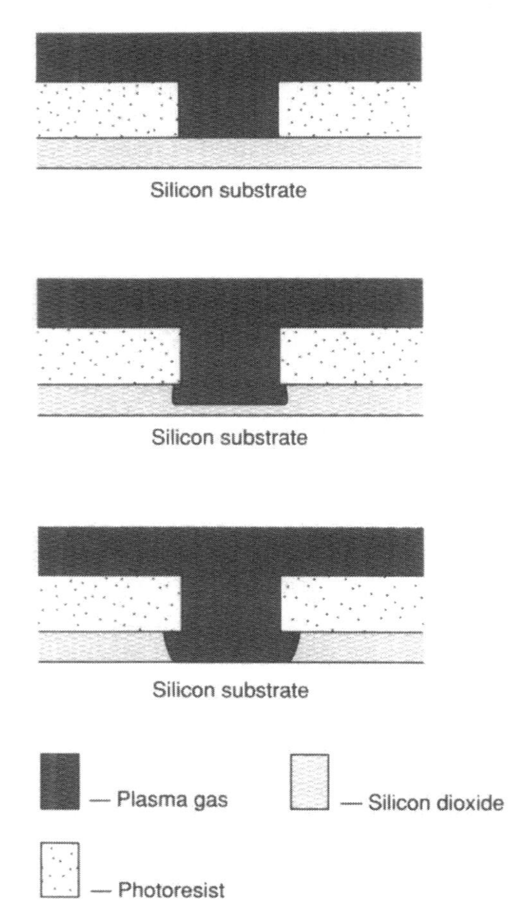

Figure 5-15. Undercutting of Photoresist As can be seen, etch processes can attack films under the etch blocking film. This undercutting can cause severe problems with yield.

problem immediately is that there is often no way of telling that the solution produced is a silicon etch. This will be especially true of hydrogen peroxide/hydrofluoric acid mixtures, since there is no visible reaction between the chemicals.

5.2.3 Hydrochloric Peroxide

Hydrochloric acid is used in the wafer fabrication area in a number of cleans, although it is less common than either sulfuric or hydrofluoric acids. Typically, the hydrochloric acid is mixed with hydrogen peroxide to form a potent oxidant and getter solution. The solution will react with metallic substances on the wafers and can leach metallic impurities from the surface of the wafer. Usually, the hydrochloric peroxide is used in a clean sequence along with sulfuric peroxide and hydrofluoric acid. This solution is very strong, but somewhat less dangerous than the sulfuric peroxide and HF solutions. It is easy to tell when one is burned, and there are no nasty surprises waiting several hours later as in the case of HF. The solution will not cause as much skin damage as the sulfuric peroxide. However, it is still an extremely toxic and dangerous chemical in its own right.

Both hydrochloric acid and hydrogen peroxide can trap metallic contamination easily, making it easy to leach the metals from the wafers, but also making it difficult to avoid having an excess of metallic impurities in the solution. Since the HCl : H_2O_2 solution is often the last solution in the clean sequence, it is important not to contaminate the wafers at the crucial point. Thus, it is important to keep the solutions and the baths clean at all times. Metallic contaminants drawn from the wafers or from the environment can also become trapped in the vessel lining and be transmitted to wafers in other lots. This is a particularly difficult situation since the main point of the hydrochloric clean is metallic removal.

Hydrochloric peroxide is also an oxidizing agent, although not nearly as strong as sulfuric peroxide. It will allow a thin (25 to 35 Å) film of chemical SiO_2 to be grown on the surface of the wafer. This oxide layer will be of very poor quality and will usually contain a high concentration of the metals that have been dissolved into the solution. It may be possible that device performance could be affected by excess metal atoms in the SiO_2. This will occur most often with thin oxidation steps, such as gate or EEPROM tunneling devices. If this is a problem, this step should be done prior to the HF etch step, and following the sulfuric peroxide step. Since the hydrochloric peroxide solution is not as effective at removing the hydrocarbon-based contaminants, this clean should follow the sulfuric peroxide step, and not precede it. If problems are seen with the specific sequence of chemical baths, it may be necessary to modify the process, for instance, adding an etch step after the HCl peroxide bath.

Hydrochloric peroxide solutions are prepared by adding hydrogen peroxide to the HCl with about one part of hydrogen peroxide to three or four parts hydrochloric acid. The solution is then kept at about 80 to 85°C. This is done to reach an optimum with respect to both the lifetime and the strength of the acid. Clearly, higher temperatures will cause the reactivity of the solution to increase, but this will come at the expense of the hydrogen peroxide, which will degrade much faster. The optimum temperature has the solution at a point of near-effervescence, not at the stage of heavy boiling action, as with sulfuric peroxide. After, HCl peroxide solutions are diluted to 5 parts water for each part HCl and H_2O_2. This solution is somewhat easier to handle, but not quite as effective.

One of the problems with the hydrochloric acid/hydrogen peroxide mixture is that the mixture tends to separate, with the lighter hydrogen peroxide floating to the top of the tank. If left undisturbed long enough, the solution will break up into two separate phases like oil and water. Not only is this situation highly ineffective as a clean, but will also cause an increase in the contaminant levels of the wafers. Contaminants, especially particles, will tend to gather at the interface between the two chemicals, which will reach concentrations that allow easy deposition of the contaminants on the wafers. Therefore, the hydrochloric peroxide mixture must be stirred at all times. This can be done by placing a fairly strong magnetic stirrer into the tank (which should be Teflon®-coated, and cleaned and observed frequently for any damage), or by stirring the solution immediately prior to placing the wafers into the tank. The wafers should be kept continuously agitated after that to ensure the best operation of the clean. This can be done with one of the many agitators available on the market. Usually, though, a good initial stirring will keep the mixture from separating while the wafers are in the vessel. Dilute HCl peroxide solutions are not as prone to this as the pure chemicals.

Hydrochloric peroxide mixtures will last somewhat longer than sulfuric peroxide solutions before going flat, but will still need a recharge after a few hours of use. Hydrochloric peroxide can accept only one or two recharges before requiring replacement. A larger quantity of hydrogen peroxide is required to recharge the hydrochloric peroxide than to recharge the sulfuric peroxide, sometimes requiring 500 ml or more per cleaning vessel. Typical cleaning times are 5 minutes per batch of two cassettes, although 10-minute cleans are possible in baths with good agitation.

Since the hydrochloric peroxide solution does not have a particularly high viscosity, it traverses filtration mechanisms quite easily, and is in a reasonable temperature range for filtration. As a result, the solution can be used in acid recirculation systems similar to those used for HF solutions. It is important to verify that the filter materials are compatible with hydrochloric acid to prevent their dissolving or otherwise contaminating the wafers. A high enough recirculation rate can help mix the solution at the same time that it is being filtered. Since the hydrochloric acid steps are usually embedded in the middle of the clean process and are not used to attack substances on the wafer (as in an etch process), there should not be many particles added to the bath due to the wafers being placed into it. Most particles will enter the vessel as a result of environmental factors. Chemical contamination may be delivered to the solution with the wafers, however. Hopefully, this amount will be minimal and all the acid is doing is pulling out residual traces of contamination. Large quantities of contamination require the acid bath to be changed. However, tests for acid contamination are not generally done on a regular basis.

Hydrochloric peroxide reacts with a number of the various films used in the semiconductor manufacturing process. The solution will attack many of the metal films to varying degrees. With the typical aluminum alloy films, immersing wafers with bare metal exposed into a solution of hydrochloric peroxide will result in the destruction of the film. As in the case of sulfuric acid, there may be instances where this film can be redeposited, although there is usually yield loss associated with this (typically only first-layer metals can be redeposited; after this point contacts are seriously damaged and redeposition cannot repair the damage). The hydrochloric acid itself will attack the silicon substrate, etching the silicon slowly, but quite destructively. One of the biggest problems with HCl etching of silicon is that the HCl will attack areas of crystal damage first, taking defects such as crystal slip and turning them into shallow etch pits. The etching phenomenon is one of the prime reasons for the addition of the hydrogen peroxide to the solution, as the peroxide will oxidize the surface faster than the HCl will etch it. It will therefore prevent damage to the substrate. The poor quality of this oxide has already been discussed. Hydrocarbon films such as photoresist will react with hydrochloric peroxide solution, but in most cases the reaction will not be as efficient as that of sulfuric peroxide. The end result will be that the acid may become contaminated with hydrocarbon "goo," and the wafers will have an incompletely removed mess on their surfaces.

Fortunately, hydrochloric peroxide does not attack silicon oxide or silicon nitride films and reacts with polysilicon in a manner similar to that of silicon, forming a thin layer of oxide on the surface of the film. As a result, this solution can be used freely on wafers through most of the early steps of the process, which is where the most sensitivity to metallic contamination occurs. Generally, all diffusion and thin-film/implant steps can utilize the HCl peroxide solution as long as the photoresist has been removed first, and there is no deposited metal on the surface of the wafer. As always, there are exceptions to the general rules. Some processes, especially very thin oxidation steps, have electrical failure modes associated with chlorine entrapment in the film.

5.2.4 Ammonium Peroxide

There are a number of clean processes which incorporate a strong solution of ammonium peroxide. This solution is used to remove hydrocarbons and other residual traces of acids or contaminants. As such it is often placed last in the clean sequence after all of the other acid cleans. The ammonium peroxide will also act as a surface cleaner, just as Windex® or other ammonia-based window cleaners are used to create clean surfaces on window glass. Ammonium peroxide is considered relatively safe, as long as it kept contained. The fumes from the solution can be very powerful and can easily overcome operators. However, coming into contact with the solution, while not pleasant, will not cause the burns and tissue damage associated with the previous chemicals.

Ammonium peroxide solutions range in strength from 2 : 1 ammonia to hydrogen peroxide to 5 : 1 or higher. The exact mixture will depend on the required task of the cleaning solution and also on the type of substrate that is being exposed to the solution. Straight ammonia can damage the silicon substrate, which is why hydrogen peroxide is added to the solution. Since there is hydrogen peroxide in the solution, care must be taken whenever a substrate that is easily oxidized is placed in the solution. A very thin layer of aluminum may be at risk in a solution that is rich in hydrogen peroxide. Fortunately, solutions that have a low concentration of peroxide will be able to remove many types of grease particles and other hydrocarbons, although they are generally not strong enough to remove baked-on photoresist. Typically, ammonium peroxide solutions are kept at around 65 to 80°C. Remember that ammonium hydroxide is fairly unstable, so that as the temperature is drastically raised the lifetime of the solution will be reduced. This effect will become quite noticeable near the boiling point of the water, where the solution may give off all of its ammonia content in as little as an hour. The hydrogen peroxide also deteriorates with increasing temperature, so clearly an ammonium peroxide solution is a short-lived and volatile commodity that must be maintained constantly. Obviously, a tankful of hot water is not going to clean anything very effectively!

Since the viscosity of the solution is low, ammonium peroxide solutions can be used in filtered recirculating acid tanks. The various components of the system should be vented to prevent undue pressurization of sections of the cleaning module as a result of decomposition of the ammonia and hydrogen peroxide solutions. This could occur in areas such as the filter housing or fluid pump. A buildup of pressure could cause failure of the mechanisms that move the chemicals or could even result in a small explosion. Another problem that can occur as a result of the use of filtration systems with ammonium peroxide solutions is the creation of microbubbles in the solution. As usual, they can cause problems with the consistency of cleans, possibly preventing localized areas of the circuits from receiving correct processing. Another problem with microbubbles is that particles that exist within the solution will congregate at the bubble interface. When the bubble comes into contact with a wafer and breaks, all of the particles that are attached to the bubble will be left on the surface of the wafer. To a degree, ammonium peroxide will form bubbles naturally. These bubbles are typically larger than the microbubbles formed by the system, and to some degree can sweep the smaller bubbles up, somewhat alleviating the microbubble problem.

Ammonium peroxide combines with a number of hydrocarbon compounds, allowing the use of the solution in situations similar to those encountered with sulfuric peroxide. In fact, some combinations, including the originally published RCA clean recipe, entirely replace the sulfuric peroxide solution with an ammonium peroxide solution. One of the advantages of this solution over the sulfuric acid solution is that it is not nearly as aggressive and will not damage existing circuitry. For example, the use of ammonia-based solutions may be acceptable on aluminum films (the use of hydrogen peroxide is not particularly recommended for this purpose, as it may create an Al_2O_3 film that could be detrimental to device performance). The ammonium peroxide solutions are also safe on both oxidized and bare silicon substrates. The use of ammonium peroxide for removal of thicker hydrocarbon films, such as photoresist removal, is not usually practiced, since the ammonia will not dissolve the photoresist as effectively as sulfuric acid will. The use of the ammonium peroxide as an initial chemical clean will be acceptable in the event that plasma photoresist stripping is used to remove most of the photoresist. The ammonium peroxide solution may still not be strong enough to remove residual resists after ion implant processes.

The ammonium peroxide solutions can react with and neutralize remnants of the acids used in the rest of the cleaning process and, as a result, the ammonium peroxide solution is often placed last in the cleaning sequence. The ability of the ammonia to perform excellent surface cleaning also makes it an excellent candidate for the final cleaning step. The surface cleaning and neutralization properties of the ammonia solutions also result in its use in chemical scrubbers which spray the ammonia solution onto a wafer spinning at high speed. Ammonia-based solutions are finding more widespread use in the wafer fabs due to their relatively low toxicity and lack of metallic element content. For example, despite continual vigilance against sodium contamination of wafers, until recently the most common photoresist developing solutions used sodium hydroxide as the primary agent. Many of them have switched to ammonia or other mixtures as a basis for the solution.

5.2.5 Deionized Water

One of the most common fluids in which the wafers are placed is deionized water. This is used to rinse almost every other solution in the fab off the wafers after the solution has performed its task. The water itself is highly purified, and is easily contaminated. Contamination in the water, especially if the source of the contamination is within the water delivery lines, can lead to widespread contamination and destruction of the wafers.

Deionized water contamination is usually measured by testing the resistivity of the water. Completely pure water has a resistivity in the range of 18 to 21 megohms. Even very small amounts of contamination will allow this value to slip. Usually, a fab area will cease operation at a step if the resistivity of the water at that step drops below 10 to 12 megohms. The resistivity of the water can be measured at a number of points. In fact, the optimum setup would be to measure the water as it leaves the recycling plant, again as it enters the fab, and again at each clean sink. This will help to isolate any problems that may be found in the water system. This setup is shown in Figure 5-16.

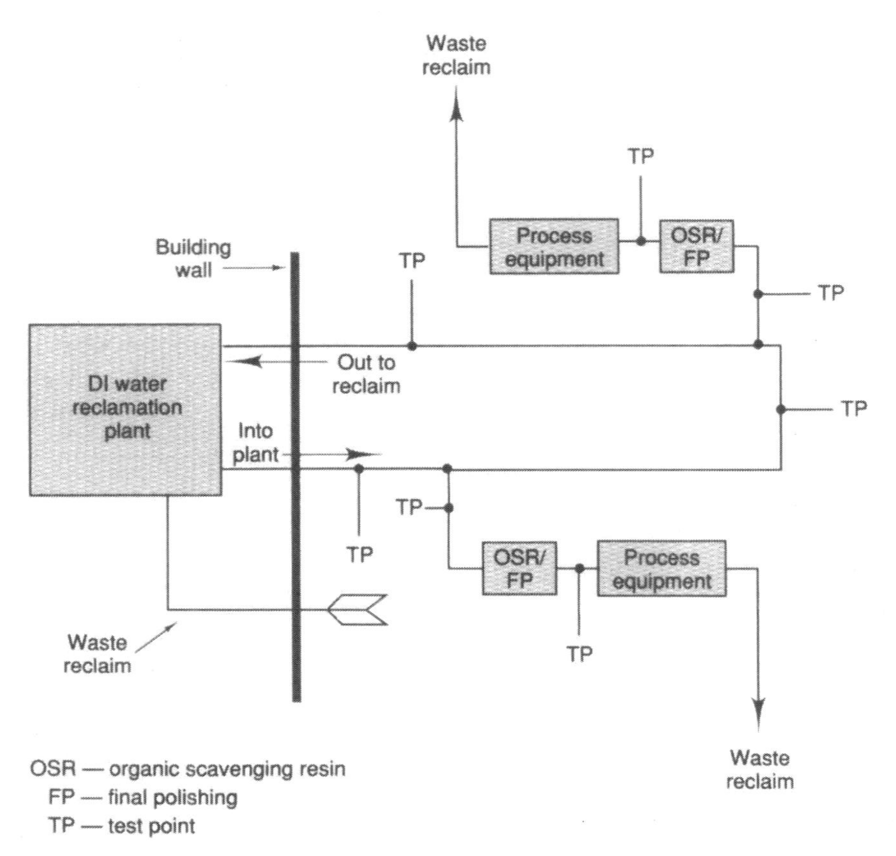

Figure 5-16. Fab Water Supply System Each cleaning system will be linked onto the main water supply line, with tests located upstream of the equipment to monitor the water quality. Many fabs incorporate final water purification (polishing) systems at the point of use of the water. These systems include organic absorbing systems, heaters, filters, and other purifiers.

Deionized water does not stay pure for very long, and will absorb contamination from the atmosphere and surrounding vessels if left to sit stagnant for even a few minutes. As a result, DI water is always kept circulating, and is continually being recharged and repurified at the DI water reclaim plant that every fab has. Usually, the water is kept trickling through all of the various gear. Spray nozzles are constructed with return flow lines, and dump rinser tanks usually are left with a small amount of water dribbling through the nozzles. The rinses are allowed to overflow into the drain continuously, to ensure that the water in the rinser tanks does not become stagnant.

Some of the most common contaminants in the water are bacteria, which can multiply rapidly. Left unchecked, the bacteria will form colonies in the water lines. These colonies will break off in clumps, which then float free in the water, and can be delivered to the wafers. These particles range in size from a few microns to very large and visible particles (tens or hundreds of microns). Fortunately, the adhesion of these clumps to the surface of the wafers is fairly low, so that they can usually be rinsed off fairly easily. Stagnant dump rinser tanks are especially susceptible to bacteria growth due to a low level of nutrients (chemical residues) within the water. This is especially true of tanks used to rinse sulfuric acid residue, since it gathers hydrocarbons. This is amplified if the sulfuric acid is being used to strip photoresist. In an attempt to both remove and kill floating clumps of bacteria and to remove organic "food" for the bacteria to consume, many factories incorporate organic scavenging systems prior to the point where the water enters the individual wafer cleaning systems, as shown in Figure 5-16. This is in addition to the purification performed in the DI water plant.

Metals, a variety of hydrocarbons, and particles delivered from the water system can all be found in the water in varying concentrations. They can reach lethal levels in local regions of the water system. These contaminants can be difficult to trace, especially if the source is some transitory condition. Unfortunately, it is then difficult to treat the problems directly. As a result, there have been a number of point-of-use filtration and purification systems developed that can be installed at each clean station.

Deionized water is often heated to around 80°C. This water will rapidly remove all traces of chemicals from other cleaning solutions. It will also leach any number of impurities trapped in the surface of the wafers. Hot deionized water is very reactive and will destroy most materials in a short time, so hot DI is usually produced at the point of use, and is mixed with colder water before it is dumped into drains. Usually, all the plumbing of the hot DI system must be made of an expensive material like PVDF in order to prevent rapid deterioration.

There is usually a deionized water recovery plant associated with each wafer fabrication area. In some cases, the plant is very sophisticated, allowing large quantities of city water to be added to the existing DI water and then purifying the entire solution. In most cases, however, the water is continually recirculated through the system and outside water is added in small quantities only as required. In smaller systems the water is sent through fixed beds of chemical which purify or polish the water. The chemicals in these beds must be replaced periodically as they deteriorate. The lifetime of these beds and of the water quality is significantly less than that of the larger systems.

5.2.6 Choline, Solvents, and Other Chemicals

There are a number of solvents used in the wafer fab. Many of the older types of solvents have proved dangerous over the years and are no longer used. They include chemicals such as trichloroethane (TCA) and trichloroethylene (TCE), which are highly carcinogenic and have low vapor pressure, thus producing toxic fumes. The solvents are used for a number of reasons, from thinning liquid photoresist to cleaning and degreasing equipment and wafers. The solvents may also be used for more detailed process steps such as photoresist edge bead removal.

In most cases, the solvents attack hydrocarbons, and generally do not attack metals. This, however, is not true of choline, which can be used to leach metal ions from wafer surfaces. Choline has a number of other interesting properties, which allows it to be used as a basis for a cleaning solution that can perform the metallic removal function of HCl peroxide and photoresist development *and* removal, and it performs an excellent surface clean, replacing the ammonium peroxide solutions. The various properties are obtained by adjusting the concentration of choline and through the use of proper mixtures of other substances. As a result, choline can conceivably be used to replace a number of toxic chemicals in the fab.

Choline is a fairly safe chemical. In fact, some choline compounds are required for proper health. However, the fumes can be very strong, and can be overwhelming, although not particularly dangerous (at least there are no known long-term health risks). Any use of choline should be within properly ventilated areas. Many of the other solvents produce strong and sometimes toxic odors. In most cases the toxicity of the chemical is low enough that short exposures will not cause harm. However, in general the best rule to follow is to avoid exposure to unknown solvents, and especially unknown mixtures of solvents. In all cases, specify the best air exhaust and purification facilities possible.

Other solutions used in wafer fabrication include chemicals such as phosphoric acid, nitric acid, and chromic acid, which are three rather lethal substances, especially in the applications in which they are used. The phosphoric acid is typically used as a nitride removal chemical, and is usually processed in a continually agitated bath, mixed with water, and heated to 85 to 180°C. Obviously, heating the solution over 100°C is a dangerous proposition, and must be done only in well-agitated vessels. Separation of the water from the phosphoric acid can result in the splattering or boiling over of the water in the solution. Sometimes this splattering can have a considerable amount of force behind it. Another problem with phosphoric acid is that it produces strong fumes that permeate the area around the clean station. The fumes are not particularly strong-smelling, but will deaden the sense of smell (as will many fab chemicals), making detection difficult. Even faint fumes of phosphoric acid can lead to increased incidence of bronchitis and other respiratory ailments.

Chromic acid solutions are often used in the testing of integrated circuits, to detect certain substrate defects. Chromic acid solutions are typically used for silicon etches, and can etch into crystal deformities at an enhanced rate, allowing these solutions to be used to find problems with surface pitting, crystal dislocations, and so on. The chromic acid solutions usually have some HF in them to help the etch process (the solutions actually work as a combination oxidizer and oxide remover). They also usually include acetic acid as a buffer chemical. This combination is very dangerous and toxic, exposing the user to heavy-metal poisoning, HF burns, and respiratory illness. In addition, the chromic acid is not allowed to be poured down the drains in most municipalities in the United States, and must be handled carefully and disposed of properly with other toxic waste.

Nitric acid is not used as much today as it was in years past, when it was mixed with acetic acid and hydrofluoric acid to form a polysilicon etch. Nitric acid is difficult to purify, easily stains things with which it comes in contact, and causes respiratory problems. There will probably be even fewer uses for nitric acid in the future.

5.3 ENGINEERING THE WAFER CLEANING PROCESSES

As you can see, the wafer cleaning process has a large number of variables, and an even wider area of potential contamination sources. The process engineer must develop a procedure that allows the operator to utilize these cleans in an effective and efficient way. This may range from requirements on how to agitate wafers in a manual immersion clean to the simpler procedures to prevent contamination from entering the chambers of acid processing systems. The specifications should cover all of the varieties of cleans that are going to be performed in the piece of equipment. There should also be procedures in place for tests to detect sources of problems.

First, we must develop an operations procedure. Since each system and fab is different we will not cover all of the details of the specific cleans, but we will go over the guidelines that should be used to come up with specifications. The procedure for cleaning the wafers should be designed to be as inherently clean as possible. For example, the cleaning system should be laid out so that the operator does not have to reach over the open vessels to perform any tasks. This practice allows contamination to fall from the protective gear into the vessels. Another possibility for introduction of contamination includes handling wet items and then being required to handle metallic parts of the clean system. This eventually corrodes the metal pieces, which then can contaminate the chemical vessels.

The sequence of events of the process should be carefully analyzed. If there are a number of steps to be performed that take different amounts of time to complete, the exact sequence and timing of each step should be noted and these values optimized. It will not do to have a process become bottlenecked in an unusual condition, forcing wafers to be trapped in an acid. This could be especially lethal in the event of HF etch processes. It is also important to make sure that the operation can be relatively easy for the operator to perform without excess labor. Finally, the specifications must be clear enough to permit a high probability of success without error. The final specifications should include all of the details that are necessary to produce the best possible cleans.

Therefore, we can see that there are several features that should be present in the operating procedures. First, the operation should be clearly analyzed and the exact sequence outlined. This should be done while keeping in mind the needs of the operators. This also includes identifying the ergonomic issues involved. Next, the procedure should be set up so that the cleans are done in a straight-through sequence. This is to prevent backtracking of the wafers through the process, which will add time and increase the probability of cross-contamination problems. Finally, handling issues must be identified and the actual mechanics of the handling identified to prevent contamination.

A set of procedures must be put into place to monitor the cleaning process. Both particulate and chemical contamination must be monitored, as well as verification of the correct solutions. The procedures for particle testing will usually involve running a set of test wafers through the cleaning process. The particle counts are measured both before and after each process. If the particle counts exceed the expected levels, further production is stopped and preventive maintenance applied. The system is then retested. The shutdown levels and procedures must be clearly defined and contingency plans for problems laid out ahead of time. A set of tests should be performed to verify chemical contamination levels in the cleaning system. This contamination is found using the capacitance–voltage test. This is performed by running test wafers through the clean process and then processing the wafers through a known clean diffusion furnace. After this, the oxide layer is tested for its capacitive properties. If the oxide shows an abnormally high amount of contamination, the cleaning system is shut down and the source of contamination cleaned out. It is important to verify that the etch steps are achieving good uniformities and etch rates. This etch rate test is performed by taking wafers with a known amount of oxide on them, etching them in the system for a fixed time, and then measuring the change in oxide thickness. If the concentration of the solution has changed, there will be a change in the etch rate or uniformity of the etch across the wafer. If the system falls out of tolerance, the etch solution must be changed. Finally, there are other tests that must be performed occasionally, for instance, to guarantee the reactivity of hydrogen peroxide or other solutions of the cleans. It is important that these procedures are clearly defined and the results and criteria for shutting down and fixing the systems also clearly specified. The systems should be monitored on a regular basis, and specific actions taken when problems are discovered. All procedures describing the actions to take should also be clearly specified.

Finally, there must be procedures to cover how to analyze and respond to the unexpected problems that come up. This can occur, for example, when an unknown source of haze has appeared, or when test wafers repeatedly fail qualification tests despite maintenance work. In these cases, it may be necessary to run special tests analyzing each subcomponent of the cleaning system. In these cases, each vessel and chemical must be isolated and the ultimate source of the problem identified. While these kinds of tests are usually performed by the engineering staff and are not explicitly specified, there really is no reason why these tests cannot be documented and specified just as any other set of procedures are specified.

5.4 WAFER CLEANING SYSTEM CONCLUSIONS

Clearly, a wide variety of chemical cleans can be performed on the wafers. In this chapter we have covered some of the most important points about the equipment used to clean the wafers and the chemicals used in these systems. The toxicity of these chemicals was discussed as well as the need for protective gear. One of the primary points that one must remember when developing cleaning processes for a wafer fab is that there is a very large amount of variability inherent in them. All efforts should be taken to ensure that the process variables have been reduced as much as possible, that the procedure is easy to follow, and that the actions that must be performed are carefully orchestrated and controlled to prevent inadvertent contamination of the wafers. Preventive maintenance methods, specific system cleaning techniques, and tight control of the chemicals used in the systems must be clearly defined in the cleaning specifications. Assuming that these ends can be met, the wafer cleaning practices will result in excellent chemical and particulate removal effects, which will enhance the yields of the integrated circuits being produced in the fab.

CHAPTER 6

PHOTOLITHOGRAPHY AND ETCH PROCESSES

Photolithography is probably the most important of the techniques required to manufacture integrated circuits. The photolithography processes define the geometries that must be manufactured, and are the limiting processes in the case of minimum feature size and usually in the case of equipment throughput. Certainly, the equipment used to perform the photolithography and etch processes is among the most sophisticated and complex manufacturing equipment used in any industry.

The term *photolithography* comes from the Greek roots for light, stones, and images, so is literally the placing of an image on stone with light. As such, ancient peoples would instantly recognize the process as magic. However, the process is clearly not magical and, while complex in operation, is not conceptually very difficult. The photolithographic process consists of a number of subprocesses, which we will discuss in detail in this chapter. These subprocesses can be listed as follows:

1. Photoresist spin.
2. Photoresist soft bake.
3. Pattern alignment.
4. Image exposure.
5. Photoresist develop.
6. Film etching.
7. Photoresist removal.

Each of these subprocesses has its own set of difficulties, but the imaging and etching steps are the most difficult. It is important to remember that the linewidths used in IC manufacturing are reaching the limits of the optical systems we have available to us. For example, the depth of field on many common optics at 1000 x magnification is less than 1.0 μm, and the resolution of visible wavelengths of light is limited to a few tenths of a micron. This can make the placing of a precise submicron image on a tall stack of films rather complex. Specialized optics have been designed to allow greater depth of field, and optical exposure systems have been extended into the ultraviolet range in order to help reduce these problems. In addition, X-ray and electron beam lithographic techniques have been studied extensively and are coming out of the labs and into the manufacturing area.

Of course, it is necessary that the lines that are produced do not vary, either in overall average width, or in local width (i.e., in roughness). In addition, the lines produced on the films must be uniform on their vertical aspect. Poor slope control can lead to reliability problems in the event that a film cannot be deposited on top of the improper slope. It should be pointed out also that the minimum feature sizes allowed in any technology are usually about 30% less than the stated design rule limits. In other words, a process that is designed to 1.5 μm design rules may have certain features on the chips as small as about 1.0 μm. all of these issues must be defined and addressed in detail.

This chapter will be broken down into two major portions, the first on the image-making processes and its subprocesses. The second section will focus on the etch processes. While these two groups of processes are intimately linked, the processes and techniques required in each area are significantly different.

6.1 PHOTOLITHOGRAPHY IN THE WAFER FAB

Photolithography, also called masking, usually refers to the image-making portion of the integrated circuit manufacturing process. In this process, which is shown in Figure 6-1, the wafer is coated with the light-sensitive photoresist, which is exposed in a pattern defined by the mask, or reticle. The photoresist is developed and patterns for film removal, or ion implantation, are defined. After this, the etch or implant step takes place, and the resist is removed. This is a delicate process, and is sensitive to a number of parameters.

We will discuss the environmental considerations such as vibration control, humidity and temperature requirements, contamination issues at each of the various steps, image control and metrology requirements, as well as the mechanics of the photoresist deposition, alignment, exposure, and develop processes.

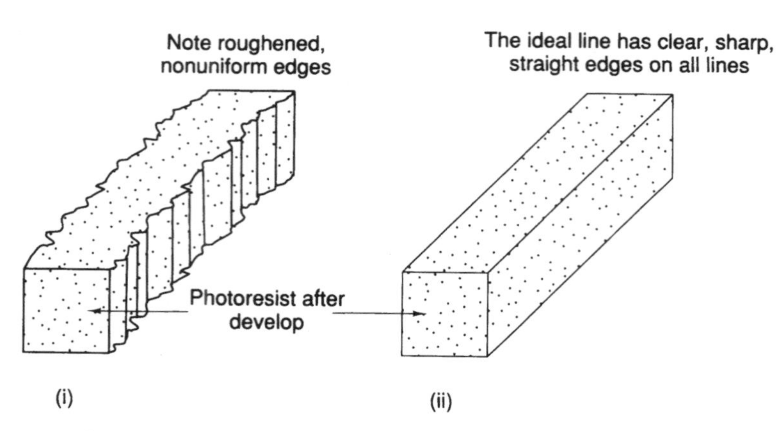

Figure 6-1. Photoresist Exposure Control Photoresist must be deposited, exposed, and developed in a uniform manner in order to produce straight, clean lines. For the semiconductor manufacturer to produce submicron geometries profitably, the ability to produce good images, and then to transfer these images to the underlying film, is most critical.

6.1.1 Photolithographic Environmental Considerations

Due to the sensitive nature of the photolithographic process, the environment can play a significant role in the success of the process. The factors that can affect it include lighting conditions, room temperature, humidity, and vibration. While they affect the process in different ways, the ultimate end result of instability in any of these factors is reduced die or line yields.

The first factor is that of room lighting. This has become less of a problem, as the photoresist formulations have become more and more wavelength-specific and less sensitive to visible light. In addition, many resist compounds are being developed that are sensitive only in the ultraviolet spectrum. In fabs using resist compounds that are sensitive to visible light, the lights in the rooms where the resist-coated wafers reside are all covered with yellow filters. This light will not react with the resist, and is analogous to the red light used in a photography darkroom. It is important to note that wafers with resist on them must be protected from standard room lighting or other sources of white light at all times. Subtle but very real sources of white light are inspection microscopes, surface-borne particle counters, and high-intensity inspection lamps. It is important to make sure that the product wafers are not inadvertently exposed to one of these light sources.

The second problem is that of controlled room temperature. This can be the source of a number of problems. In the first case, if the optics in a stepper or other alignment tool become too hot or too cold they may change position very slightly, but enough to cause pattern distortions. This problem is often addressed simply by enclosing the stepper in an environmental control chamber. While this is an expensive proposition, it makes sense if the cost of controlling the environment in the fab is excessive. Another problem is that the photoresist adhesion and viscosity may change slightly, since it is stored and used at room temperature and is not usually kept in temperature-controlled environments. In extreme cases, it may prove necessary to control the conditions of the photoresist. A third problem with unstable temperature control in the fab is that air at higher temperatures can contain a significantly larger quantity of water. This moisture will affect the priming of the wafers and can affect the adhesion of the photoresist. In all of these cases, changes in the temperature of the photolithography room can create a number of problems with the process.

A third problem, often related to that of temperature instability, is humidity instability. Very often, the same equipment failures that cause the temperature control to fail will cause the humidity control to fail. Variations in humidity can drastically change the adhesion of the photoresist films. These variations can also cause a deterioration in the amount of time that the wafers can sit between the various steps in the process. Drastic increases in humidity can also cause problems with condensation in optics and other equipment. Obviously, condensation on a lens will cause severe pattern distortions. Again, steppers are placed inside environmental control chambers in situations where this is an otherwise uncontrollable problem.

The last of the major environmental problems that can affect the photolithographic process is that of vibration. If the fab floor under the stepper vibrates, there will be an increased instability in line image, and proper focus will be difficult if not impossible to obtain, as seen in Figure 6-2. These problems will result in feature roughness, lack of good image resolution, and insufficient edge exposure, which can result in scumming and other resist development problems. Vibration is an especially severe problem when the steppers are placed upstairs in a multiple-story fab area. Modern earthquake-resistant buildings are also vibration-prone. It can also be a problem in a fab that is located near a freeway or other source of systematic vibration. Even small, otherwise imperceptible, vibrations can cause problems with focus and precise alignment. It must be kept in mind that there is a less than 10% tolerance for error with any of the photolithography processes, which can be exceeded with almost any vibration at today's submicron technologies. Steppers are often placed on vibration isolation tables which are placed on granite slabs or other solid, massive fixtures in order to reduce the effects of vibration.

6.1.2 Photoresist Deposition

The first step in the creation of a pattern on a silicon substrate is the deposition of photoresist. Photoresist is a substance that changes its chemical reactivity characteristics as a function of exposure to certain wavelengths of light. Older photoresists had fairly broad absorption

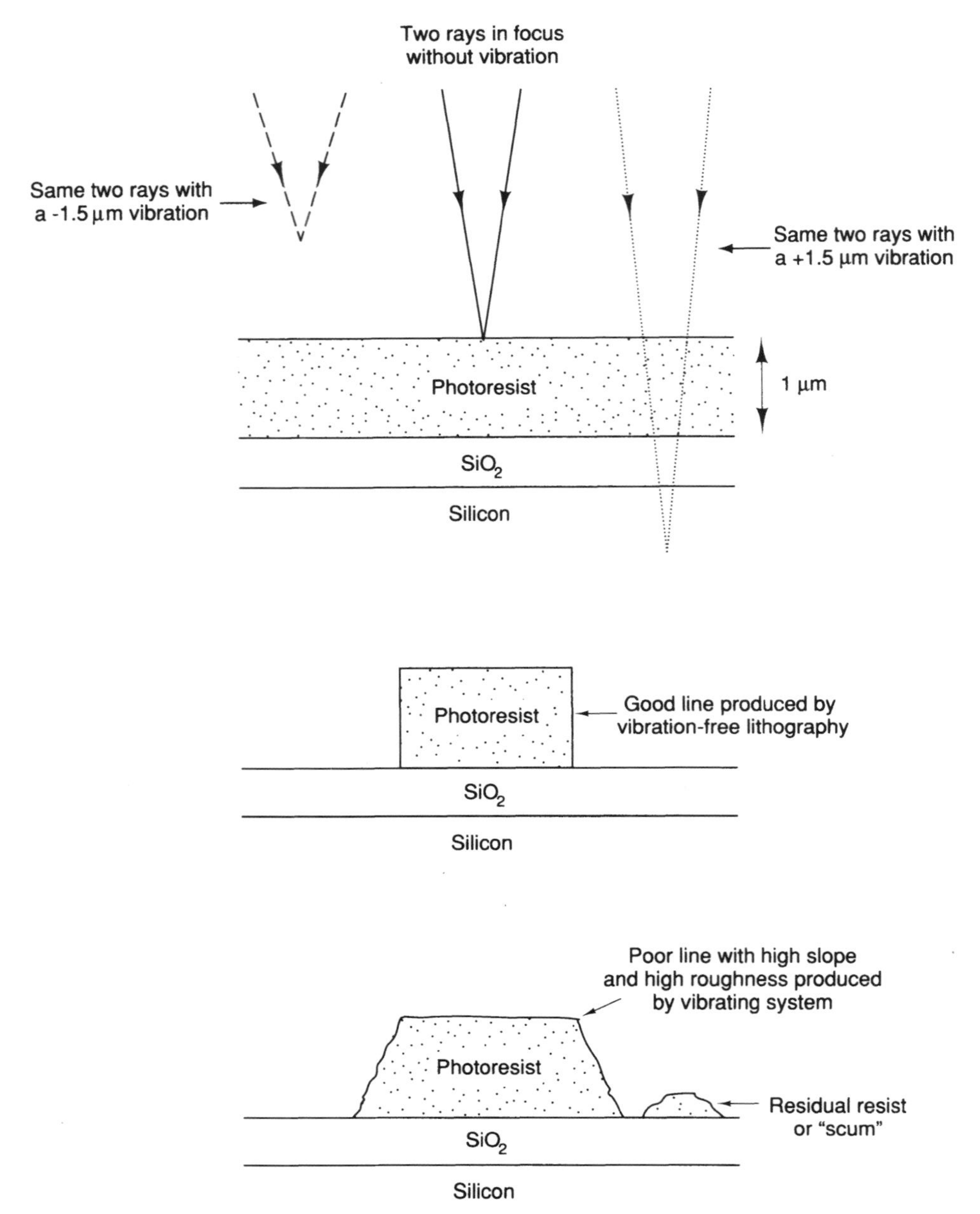

Figure 6-2. The Effects of Vibration on Photoresist Exposure Vibrations that occur during the exposure process will cause serious problems with linewidth control. They can include excess linewidth nonuniformity, lines that are too wide with nonvertical slopes, and anomalous exposure of the resist or "scumming."

and reaction spectra. However, it is preferable to keep the reaction range as specific as possible to prevent unexpected exposure of the resist, and to minimize problems due to chromatic aberration of the lens systems. Most of the photoresists available on the market are sensitive to fairly narrow ranges in the ultraviolet, defined by various mercury vapor and eximer laser emission lines, and often containing dyes and other chemicals to reduce the amount of reaction at unwanted wavelengths.

The photoresist itself can come in two forms, positive or negative. Positive photoresist is the most common type in use today. In the case of positive resist, the area of the resist that is exposed to light softens (made more reactive to the developer) and is removed in the developing process. This is done so that the exact pattern that is desired is transferred to the device, as seen in Figure 6-3a. In the case of negative resist, seen in Figure 6-3b, the section that must become

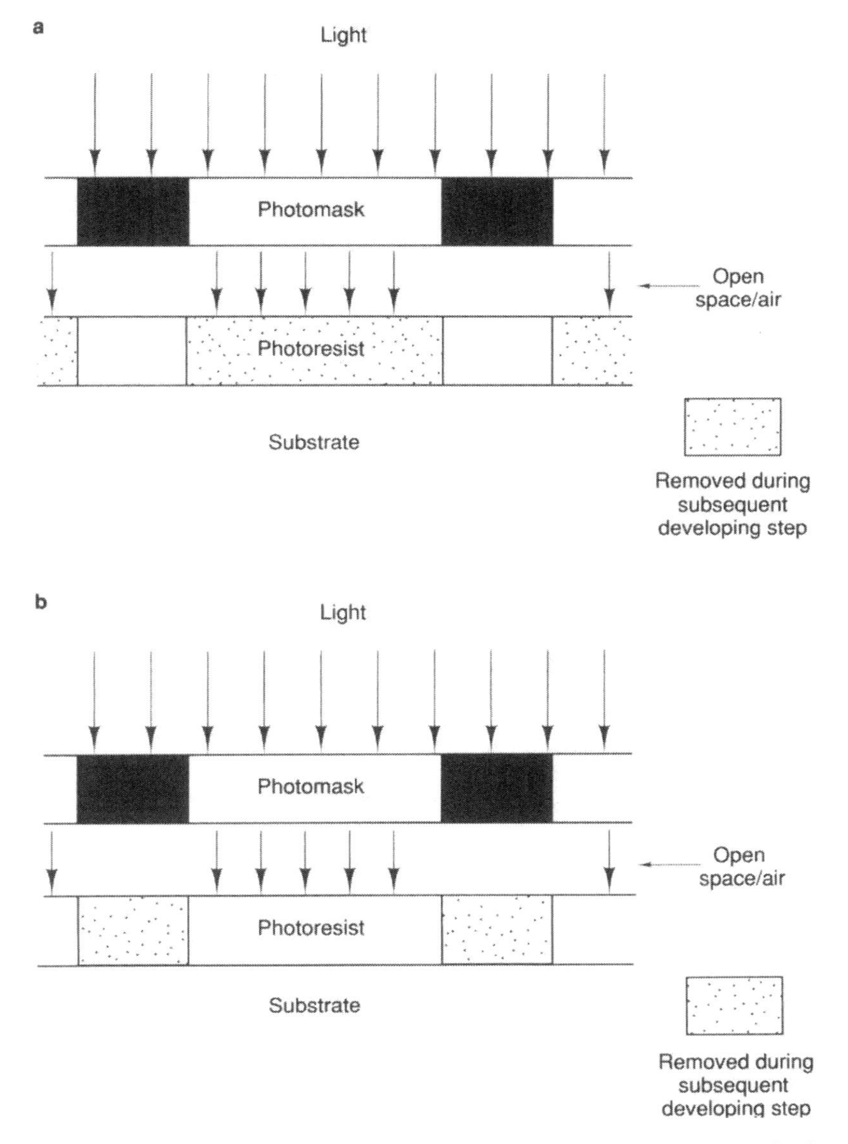

Figure 6-3. Resist Exposure (a) Positive resists permit exact duplication of a design from a photomask to a wafer. (b) Negative photoresists produce an inverted image to that seen on the photomask.

hardened is exposed to the light, meaning that the image on the photomask must be in reverse image in order to allow the photoresist to develop properly. For further description on some various types of photoresists, see Table 6-1.

Photoresist is typically delivered as a liquid of known viscosity. It is important that the viscosity be controlled exactly so that the deposition process can remain in control. Sometimes wafer fabs will mix various solvents with the photoresist in order to thin it. This is usually the case only if an unusually thin photoresist layer is required. There are so many types of photoresists available on the market that it usually is possible to find the exact formulation required without having to resort to imprecise mixing at the point of use.

The wafer surface is usually primed for the placement of photoresist before the resist deposition cycle. This is done by spraying a solvent, most commonly HMDS, onto the surface of the wafer to remove any moisture on the surface and to remove any other foreign matter. This ensures good, uniform adhesion of the photoresist to the wafer surface. If the priming is insufficient and moisture remains on the surface, the resist may lift or even peel back in places on the wafer. Even very slight lifting can cause severe etch discrepancies on the wafers, to say nothing of the difficulty of attempting to focus on and place an image on a film that is grossly uneven. In any event, the wafer is spun at high speed and the HMDS is sprayed onto the surface of the wafer, after which the wafer is dried either by heating it on a hot surface or spraying hot N_2 over the surface. After the priming step,

TABLE 6-1
Attributes of Photoresist Materials

Resist types	There are many photoresists manufactured for wafer fabrication. These types include positive, negative, X-ray, E-beam, and deep UV. They are often dyed to enhance contrast, reduce unwanted reflections, and control spectral exposure bandwidth. Multiple-layer resist technology is often used to extend capabilities in submicron devices. Refer to manufacturers' specs for specific details
Primary exposure wavelengths	
	g-line: 436 nm
	h-line: 405 nm
	i-line: 365 nm
	eximer laser: 248 nm, 305 nm, 193 nm
Exposure intensities required	
	g,i line: 175–250 mJ/cm^2
	eximer: 50–60 mJ/cm^2
Typical resolutions attainable (1989—commercial resists)	
	Positive resist: 0.35–0.4 μm
	Negative resist: 0.3–0.35 μm
	E-beam resist: 0.1 μm

there is usually a maximum fixed time that can lapse prior to the deposition of the photoresist. This time will vary greatly, depending on the type of photoresist and the conditions of the environment, especially the humidity. A fab or a laboratory that is not in tightly humidity-controlled environments, in New Mexico may be able to have a longer post prime sit time than a fab in Florida. After this time has been exceeded, the wafers must be reprimed. The time can be extended by placing the wafers into nitrogen-purged desiccator boxes immediately after the priming step.

After the wafers have been primed, they are ready for the photoresist deposition or spin process. As shown in Figure 6-4, the wafers are placed onto a chuck assembly capable of high speed rotation. Photoresist is then dropped onto the wafer. Sometimes the wafer will already be rotating at a slow speed; in other cases, the wafer will be stationary. The key aspects here are to make sure that the deposit is uniform, the

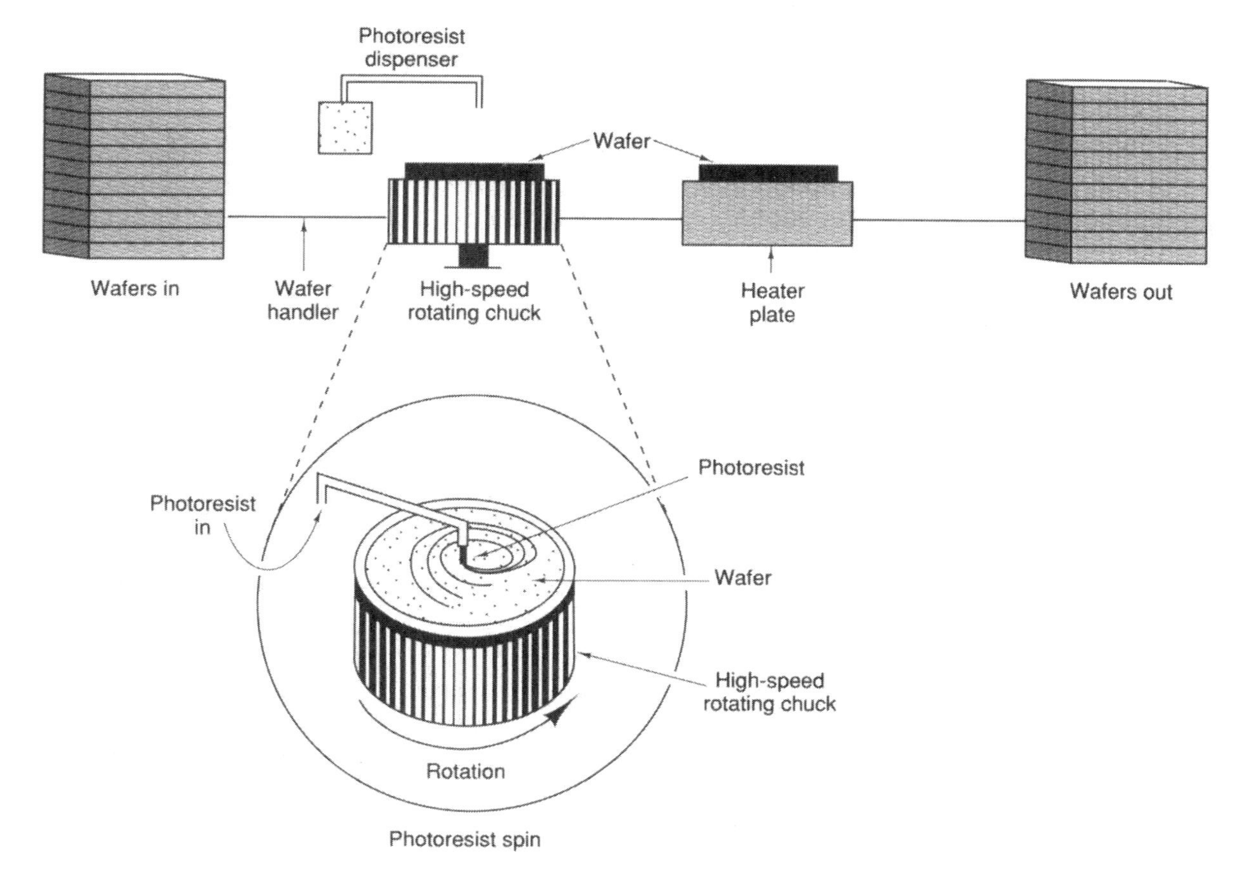

Figure 6-4. Block Diagram of the Photoresist Spin and Bake Operation Liquid photoresist is dispensed in a fixed quantity onto the wafer surface. The wafer is rotated on axis at 1000 to 4000 rpm. The speed of the rotation is determined by the viscosity of the resist and the thickness desired. The wafer is then transported to a heater plate where it is baked to harden it in preparation for exposure.

photoresist is clean, and there are no bubbles. Bubbles of all sizes can cause serious problems and can often be formed in the microbubble size range. Once microbubbles form, they are extremely difficult to remove from the resist. They are also very difficult to locate until after the resist has been baked. Microbubbles are usually identified quite easily by observing the shape of the defects found on the device. If they are nearly perfectly round, they are very likely to be microbubble defects. Other types of defects can form round patterns, but not with such regularity. In most cases, if one microbubble is found, others will also be found within a few minutes of inspection. They are rarely formed as individuals. In almost all cases, the resist will have to be removed and redeposited to allow further processing. There is a relationship between the amount of photoresist placed on the surface of the wafer and the overall thickness of the final film. The final film thickness is also a function of the spin speed of the process. The spin speed can range from 1000 to 4000 rpm, depending on the viscosity of the photoresist and the resist thickness that is desired. As seen in Figure 6-5, higher speeds will result in a thinner photoresist, but can result in bowl-shaped resist profiles if the viscosity of resist is too low. Conversely, lower spin speeds can allow for thicker resist films, but can result in resist films with humped appearances. In any event, the photoresist must now be of uniform and expected thickness.

The absolute thickness of the photoresist is an important parameter. It must not be thicker than the depth of focus of the lithographic

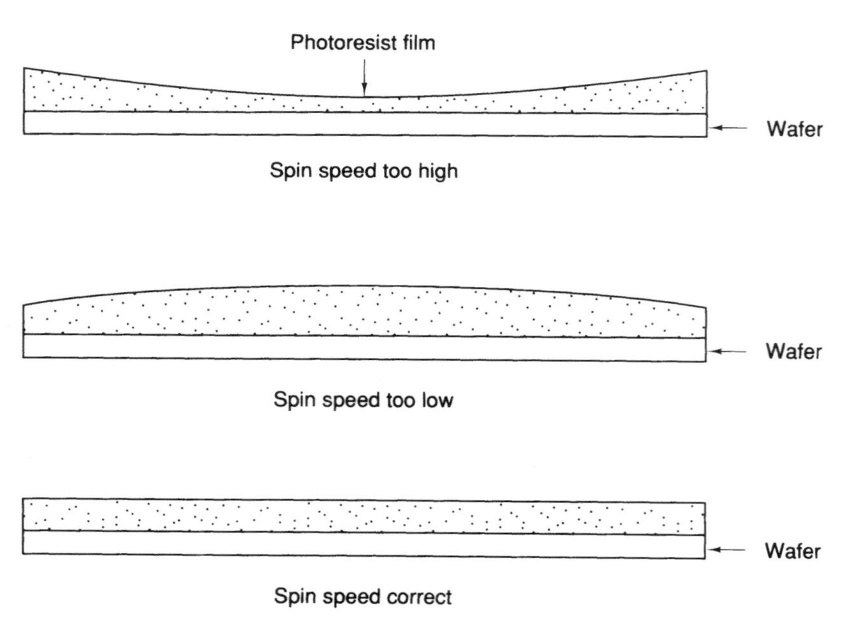

Figure 6-5. Nonuniformity of Photoresist Films Due to Spin Speed Variations Photoresist spinners must be characterized for each type of photoresist. Variations in viscosity or spin speed will cause variations of film uniformity.

imaging system or it will not be exposed properly. It also must be thick enough to prevent any problems from occurring during the other process steps. For instance, the etch steps usually attack the photoresist to some degree. If the film is too thin, there may be penetration of the photoresist with the etch plasma. Another possibility is that the substrate may receive an implant with an unpredictable amount of dopant if the photoresist is being used as an implant block and is too thin. Another implant concern is that a thin layer of resist is being used to absorb excess energy from the beam and is allowing a certain amount of the beam to pass through to the substrate. Clearly, a film thickness that is too great or too small will result in an improper dose or energy of the implant species reaching the substrate. Finally, an excessively thin film is more sensitive to pinholes, which will develop into point defects in the etch process. These pinholes can form due to film damage, particle contamination, or from microbubble formation.

Since the exposure of the photoresist is a function of the number of reactive photons impacting on a fixed number of molecules of photoresist in a certain time, it stands to reason that a thicker film will require more photons to equally expose than a thinner film. Thus, if a wafer has nonuniform photoresist on it, an even amount of light energy will produce a nonuniform level of exposure across the wafer. It is important to achieve uniform exposure levels or the linewidths of the devices will vary across the area of the wafer, which can result in die yield loss. Although there is a certain amount of tolerance allowed in the resist uniformity, it is usually expected that uniformities vary no more than 1 to 3%. As usual, these tolerances are continually being tightened.

Particles are the cause of a number of problems at the photoresist deposition step. The exact problem depends on the point in the process at which they are deposited. Large particles reaching the surface of the wafer during the high-speed spin will often leave characteristic "comet trail" patterns, as in Figure 6-6. They are so named due to the distinctive pattern in which the particle forms a "head" and a tail then follows where the spin flow was disrupted. In places, this tail may include tears or other damage to the photoresist embedded within it. In other cases, there may be only a localized thickness deviation, a "rippling" of the photoresist. Small particles deposited during the spin step in the process often display these types of rippling problems. The rippling effect may not be particularly noticeable unless it occurs on a test wafer that can later be examined on a surface-borne defect analyzer. Particles that are deposited during the spin step can also move on the surface, producing unusual patterns of errors and holes. Particles that are deposited after the photoresist spin is completed often appear as pinholes in the photoresist at later steps. The particles can also block areas of interest, such as implant areas, areas for etch, and so on. In any event, particles are extremely serious yield killers at this point. Care must be taken from this point forward that the

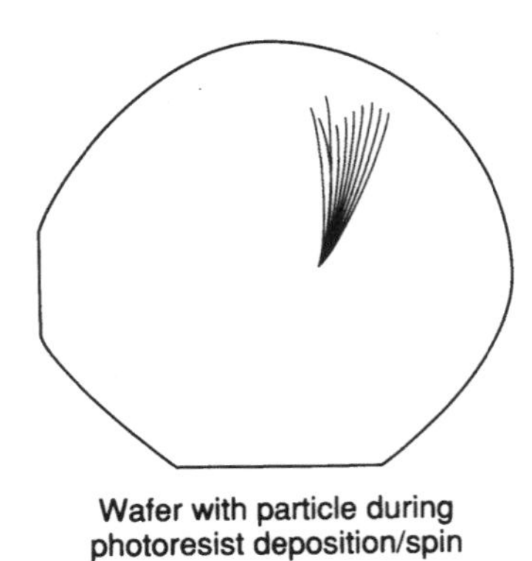

Wafer with particle during
photoresist deposition/spin

Figure 6-6. Comet Trail Defect When particles land on the wafer while the photoresist is deposited, the resist will deform as it is spun toward the outside of the wafer. As in this extreme example, the deformation will expand as it approaches the edge of the wafer. The size of the tail is largely determined by the size of the particle forming the "head."

wafers are not exposed to dangerous levels of light, especially in inspection stations, and that they are handled very carefully.

The photoresist is very soft, so contact with any surface will result in scratches that will cause failure of the IC at that location. In addition, large numbers of particles will be created which will cause defects elsewhere. Photoresist is so soft that a common test of verifying that an unknown film on a wafer is photoresist is by gently scratching the wafer in some unused area with something soft, preferably plastic. If the film visibly scratches, it is probably photoresist.

After the photoresist is spun onto the wafer to the correct thickness, the wafer is moved to a hot plate assembly, where it is heated to around 125 to 150°C for a few minutes. This is called a "soft bake" and it hardens and sets the photoresist film. This is done to prevent the photoresist from becoming damaged quite as easily by succeeding steps or by flowing once it is placed in a storage box. The soft-bake step is sometimes performed by placing a batch of wafers in an oven. These ovens are usually not very clean, and cannot guarantee the uniformity of heating and the time control of the in-situ hot plate bake step. As a result, bake ovens are declining in popularity in many modern IC fabs.

There are also a variety of photoresists that are used for electron beam lithography. They are typically only a few tens to a few hundred angstroms thick. They are not always spun on as a fluid, but can be deposited through a variety of other means, typically, CVD or epitaxial-like deposition processes. The resists must be extremely uniform in thickness to be effective. These resists are used with electron beam lithography processes, which can allow far finer resolution for the manufacture of submicron ICs. In general, the rules for handling E-beam photoresists are similar to those of standard photoresists.

6.1.3 Image Alignment

For the integrated circuit to operate properly, the various images that make up each layer of the chip must be placed carefully. Since each design is usually made up of between 8 and 15 separate photolithography steps, the precision of the alignment of each structure is extremely critical. In fact, the basic measure of the alignment and amount of etch for the photolithographic process is called the "critical dimension" or CD. There are a number of methods for placing patterns onto wafers. They include the use of projection alignment, where a clear patterned mask is placed over the entire wafer, and the entire image on the wafer is "shot" at once. This is the older technique and has limitations as far as minimum linewidths and uniformities are concerned. The most common method in use today is called the "step and repeat" or "stepper" methodology. In this technique the pattern of the chips is placed onto a glass plate called a reticle, which holds one to four chips per reticle. (Some very small circuit designs allow for a larger number of chips per reticle.) Each reticle is then aligned over a portion of the wafer and the image is exposed. After the image is produced the stepper moves the wafer to the next alignment site, and the pattern is exposed again. The advanced processes such as E-beam lithography allow ultrafine resolutions, down to 0.1 μm. These systems often suffer from reduced throughput time due to the method of raster scan exposure. Eximer lasers and X-ray devices are also being used to expose images on photoresists to produce submicron lithographs.

Projection Aligners

Figure 6-7 shows the design of a projection aligner. This machine consists of several parts: the wafer handler, the stage, the photomask, and the optics and light source. The photomask, which contains the patterns to be placed onto the wafer, is a glass plate slightly larger than the wafer. Each photomask contains one distinct pattern for one layer. The optical system usually consists of a series of lenses, which take the light from a light source, filters out unwanted wavelengths, expands the beam to the size of the photomask, and collimates the light beam, so that all of the light is hitting the surface of the wafer at exactly 90 degrees. Finally, the optics focus the light through the photomask and onto the wafer. Each of these lenses may be made up of a number of elements. The stage and the wafer handling system must be able to move the wafer from the cassette to a proper position under the photomask, and allow minor adjustments to be made in all axes.

The use of projection alignment equipment is usually limited to regions above 2.0 μm, although many manufacturers have perfected techniques that allow their use on structures as small as 1.25 μm or so. For the large structures, the projection aligner is ideally suited for mass production, since the entire wafer is exposed in a single pass.

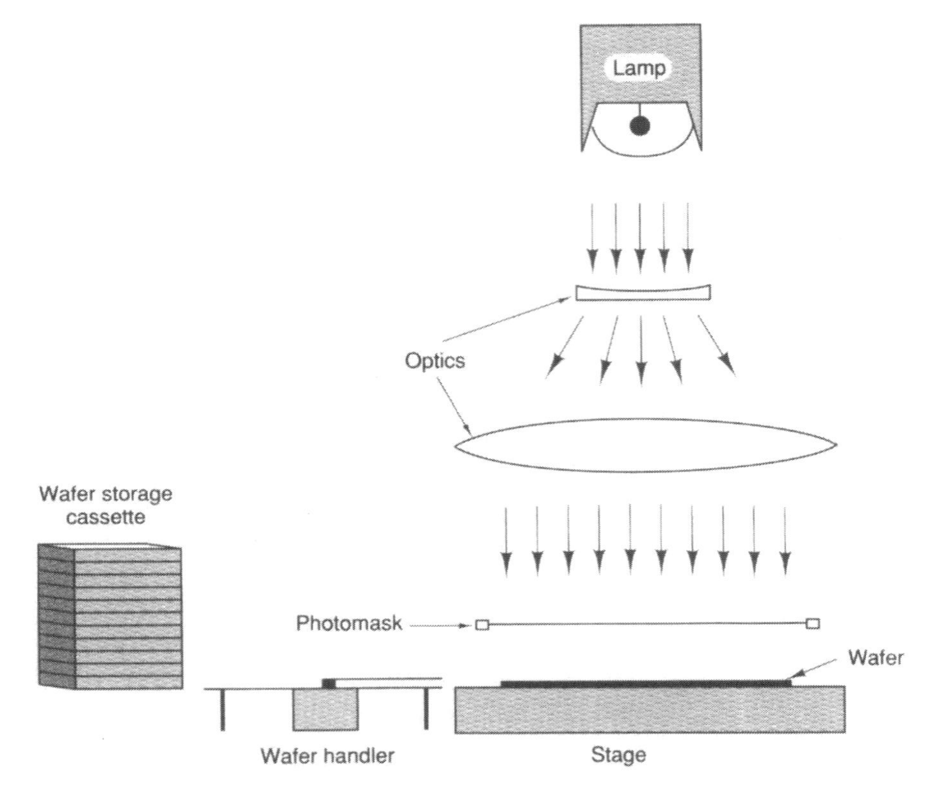

Figure 6-7. The Projection Aligner Shown is a block diagram of a projection aligner. This system provides a fast, inexpensive lithography process for small wafers (4 in. and below) and wider linewidths (≈1.5% minimum linewidth). Wafers are removed from the cassette, set on the stage, and aligned. Alignment is performed by adjusting the stage position while observing through a microscope. Light is produced by a lamp, expended, collimated, and then directed through a photomask onto the wafer. Each wafer is entirely exposed in one step. An alternate exposure method involves shining the light through a slit and then moving the slit across the wafer surface.

The projection aligner can produce many times more wafers per hour than a stepper, often 100 to 150 wafers per hour. Since the photomask contains the entire pattern of the particular layer in question, there are advantages other than speed. For instance, there is no setup time penalty for including test die on the wafer for allowing easy in-line parametric testing. This is a problem on steppers where special in-scribe test patterns must be devised. This often restricts effective parametric testing on devices in development. See Figure 6-8 for a simple description of the different types of test structures. In addition, the cost of projection alignment equipment is very low compared to stepper technology. As a result, for many custom and semicustom or ASIC applications, a 2.5 μm process might make the most economic sense.

Projection aligners can have problems that make the process engineer's life difficult. Misaligned optics, whether it is a lens that is slightly ajar, a mask that is placed incorrectly, or other anomaly can

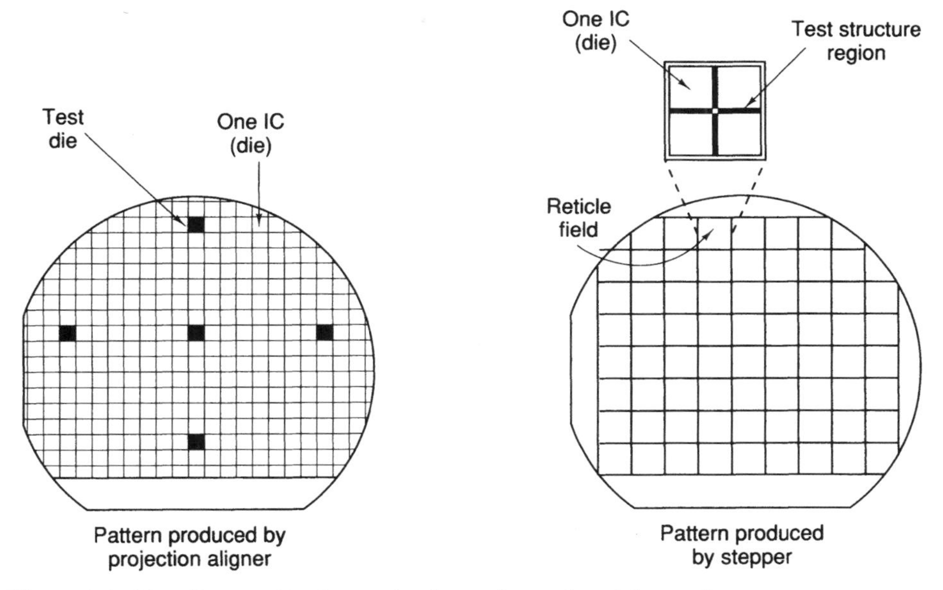

Figure 6-8. Test Pattern Regions As shown here, the ability to form test structures is limited with the use of a stepper. With projection aligners, areas on the wafer can be defined as test die, allowing plenty of surface area for these structures. Steppers cannot provide this without significant loss of time or loss of available die, so test structures must be placed into the scribelines between the chips.

cause the light that is impinging on the surface of the wafer to lose collimation. When this happens, the photoresist can be exposed in unpredictable ways. Clearly, if this effect is very pronounced it will be difficult to obtain a good focus over the entire surface of the image plane (in this case the entire wafer). This results in wafers having inconsistencies in linewidth or in other problems associated with poor focus. Only with the most careful optical alignment can the focus uniformity be controlled well enough to obtain $\approx 1.5\ \mu m$ linewidth resolution.

Another issue with the optical system is the removal of astigmatism and chromatic aberration from the lenses. The optical impact associated with these and related problems is depicted in Figure 6-9. Both of these issues are amplified because of the relatively large size of the lens systems required for the projection aligners. Lens edge effects can become quite pronounced. The problems associated with trying to remove these distortions tend to limit the maximum practical wafer size for use with projection aligners. As a result, projection aligners are typically used in wafer fabs using four-inch or smaller wafers.

A final issue when dealing with the optical system is that of light intensity uniformity. As the light beam is manipulated through the optical system, there is a significant loss in the amount of light delivered to the wafer. There is absorption by the optics, scattering and diffusion of the beam, as well as the reduction in light intensity that occurs as a result of the expansion of the beam. It is necessary to

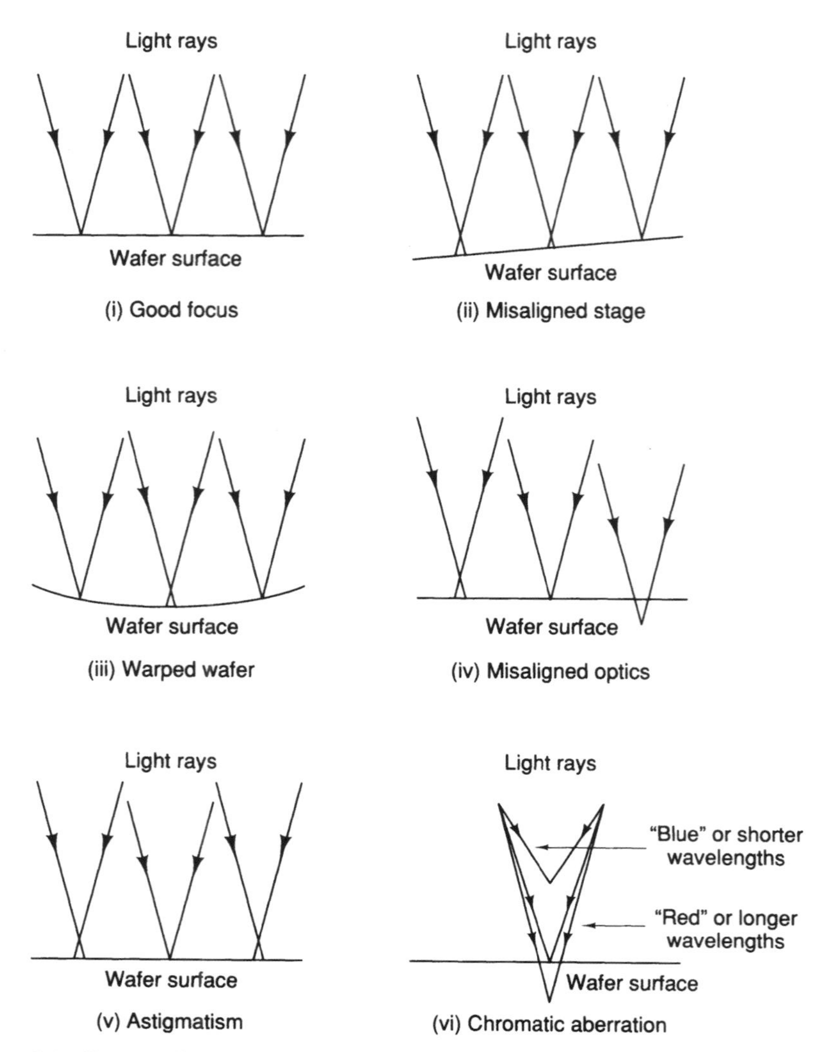

Figure 6-9. Potential Focus Problems at the Photolithography Steps Good focus uniformity is critical to proper device performance and high yield. A number of problems can affect focus uniformity. As shown, they include misaligned equipment, wafer warpage, and a variety of optical problems.

control these effects so that the proper intensity of light reaches every wafer, and that the uniformity of the light intensity does not vary over the wafer or the image. Fluctuations in light intensity will lead to areas of insufficiently exposed photoresist. If the exposure time is increased to improve the exposure characteristics of the areas with the dark spots, the photoresist in other areas of the wafers could be overexposed. Finally, lamp output should be monitored at all times so that the proper exposure times for the particular photoresist can be calculated. As light sources age, they will tend to lose their intensity, and this can be very significant as a result of other losses incurred in the optical system. This is particularly true of UV systems which will not only have short-lived light sources, but whose light is heavily absorbed by most glasses.

The stages that are used to handle the wafers must be held very securely and must be perfectly parallel to the photomask. Any vibration or motion will result in focus or misalignment problems, while a tilted stage will result in an unevenly focused image. The stage system must be able to accept minor adjustments from the operator to verify that the positions are correct in each of the axes, both the x and y directions, but also with regard to angle and sometimes tilt, as shown in Figure 6-10. Fortunately, the stages do not require the extreme position accuracy of the stages used on wafer steppers, as the pattern is aligned and exposed once per wafer. This helps to keep the price of the machines down.

Particle contamination issues are important when dealing with any photolithography system. However, the chances of an error affecting the fab output are minimized if projection aligner-based techniques are used. For example, if a particle lands on a photomask, or if some other defect appears on a mask, it will affect only the die on that particular spot, and it will occur on every wafer that is produced. While this clearly will affect the die yield of the device, it is relatively easy to trace the cause of a problem like this, and the impact to the fab is fairly low. There are ways of protecting photomasks so that even these defects do not occur, such as the use of *pellicles*, which are thin, clear membranes that are placed over the masks. Chemical

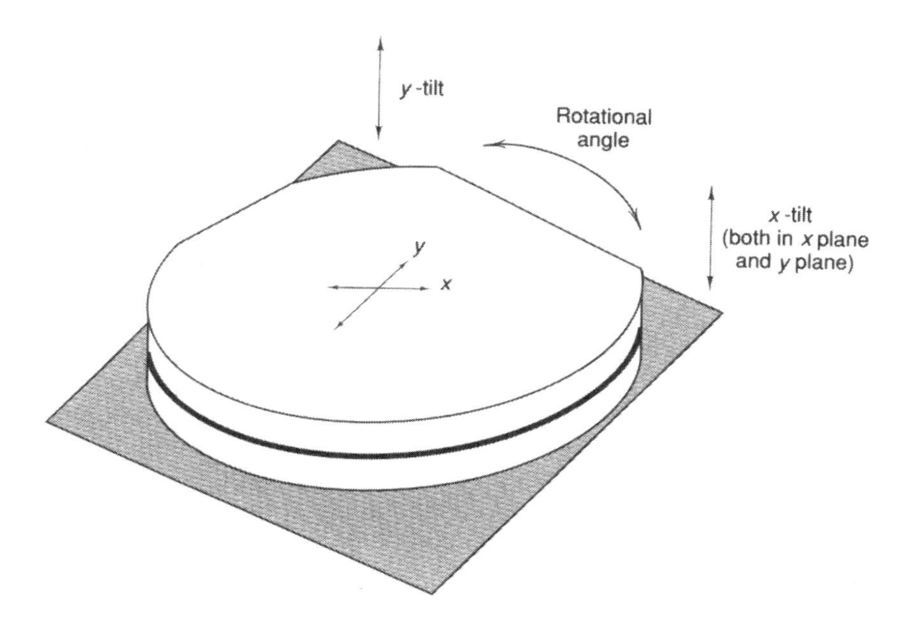

Figure 6-10. The High-Precision Lithography Stage Wafers are placed on high-precision stages for patterning. These stages must be controlled in both the x and y directions, as well as in the angle of rotation, so that the wafer can be precisely aligned in all dimensions. Some stages also include tilt capabilities.

contamination issues are not a serious problem with these systems as the temperatures that the wafers are exposed to are never very high. Mechanical damage is always a concern on systems that must handle the wafers frequently. Alignment tolerances and preventive maintenance issues must be checked regularly.

Wafer Steppers

Wafer pattern step and repeat instruments are today the most common systems for producing small geometries over large wafers. They are very expensive machines, and produce very few wafers per hour compared to the other equipment in the fab area. Typical throughput figures are around 20 to 30 wafers per hour. They are also prone to unexpected downtime, due to their complexity. As a result, there is a built-in conflict when making a decision to use steppers in manufacturing. On the one hand, to meet the output requirements for the manufacturing plan, a large number of systems must be purchased. On the other hand, the steppers are so expensive (often over $1 million each), and the facilities requirements so strict (read expensive), that the company typically must restrict the number of steppers that are actually purchased. Many fab areas turn to hybrid processing in which the steppers are used only for the most critical steps and projection aligners are used for less critical steps. Typically, steppers are used on layers where the geometries drop below 1.5 μm or so (depending on the ability of the company to maintain older technology projection aligners). There are a range of steppers available from lower end instruments which can easily produce imagery to about 1.0 to 1.25 μm and are smaller and less expensive, to high-end steppers which can produce designs to less than 0.5 μm and require specialized environments.

Figure 6-11 shows the major components of the stepper system. There are several similarities to the projection aligners described earlier, but in many cases the tolerances that are permitted are much tighter. Shown are the wafer handler and stage mechanisms, optical systems, reticle with attached pellicle, and the light source.

We will start with the heart of the optical system, which is the reticle. This is the equivalent of a photomask, in that it is a glass plate with a pattern described on it with opaque material (such as chrome). Each reticle has the pattern of a different layer on it. The reticle is placed over a section of the wafer to be patterned. This area is called the reticle field. Each reticle field may hold from one to eight integrated circuits, depending on the size of the circuits and on the available area of the reticle. The reticle may also contain test structures in the areas between the chips (the scribelines). This is different than photomasks where there may be one to five chip-sized areas containing test structures (so called "test die"). Test die are not used with stepper technologies for two reasons. First, the test die cannot be in every field, or you will be sacrificing many dozens of chips of potential

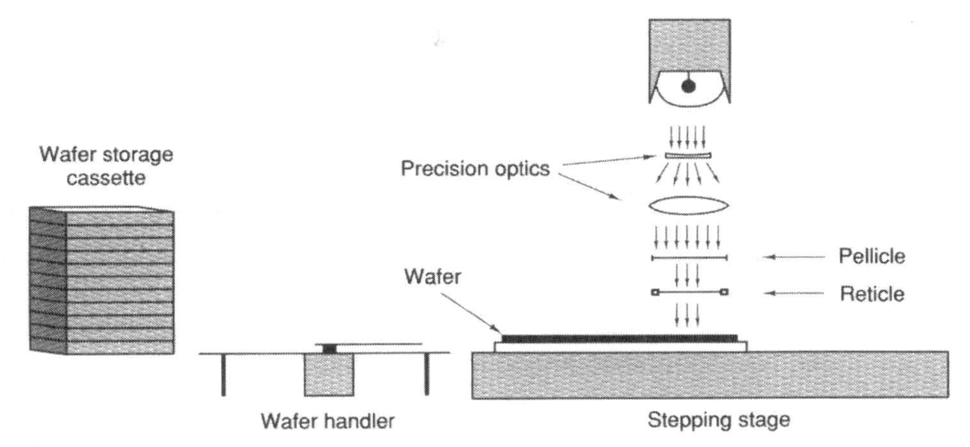

Figure 6-11. Block Diagram of a Stepper The block diagram of the stepper is very similar to that of the projection aligner. However, the stability of the system, coupled with the high-precision optics required for proper operation add to the cost and complexity of the stepper. The precision stages are accurate to tenths of microns. Optical systems utilize a number of wavelengths into the deep UV, depending on the manufacturer and the exact setup.

output. Second, it is very time-consuming to set up a stepper, so that the use of special reticles containing test structures becomes impractical.

Two types of reticles are typically used, the step-down reticle, and the one-to-one reticle. In the case of the step-down system, the image on the reticle is larger than the final image desired (usually about five times larger), and the optical system scales the image down to the right size. This has advantages in that it is cheaper and easier to manufacture the reticles, and defects that appear on the reticle are reduced by five times, as well as the pattern. Thus a 1.0 μm particle on a reticle will result in only a 0.2 μm anomaly on the finished product. However, these steppers require more complex optics, and are difficult to keep aligned and in tune. The one-to-one reticle is one that is the exact size of the final produced image. This system requires that the reticles are more carefully manufactured and are more expensive, and that they are more carefully maintained. Particles landing on the reticle will not be reduced in size and will appear as full-sized defects on the final image.

Defects falling directly onto a reticle can be disastrous. For example, suppose that the process uses a reticle field that contains four chips. At one of the critical layers, one of the chips has a defect in it that causes a short circuit or some other major failure to occur. If this defect is not caught immediately, as much as one-fourth of the production material processed from the time the defect was created until it was identified and removed, could fail. A loss of 25% of the output of a factory of any kind is clearly unacceptable. Therefore, reticles are tested prior to introduction into a fab area, and then are regularly inspected to verify that no defects have been introduced.

In addition to inspection techniques, the use of pellicles to control defect propagation has become widespread in the photolithography world. These are thin, clear plastic membranes that are placed over the reticle to protect it from contamination. Typically the reticle is placed in a frame, and then the pellicle is attached to the frame, as in Figure 6-12. The frame holds the pellicle about one to two centimeters away from the reticle, so that the pellicle is outside the focal plane of the reticle. When this is done, particles which land on the pellicle will be completely out of focus, and therefore invisible.

The optical system of a wafer stepper is similar in function to that of the projection aligner. However, the systems are much more precise, and are often much more complex. The ability to reach the submicron level is more a matter of precision in the system than any trick with the optics. Essentially, light is taken from the light source, filtered, expanded (although not to anywhere near the degree that projection aligners expand their beams), and then collimated and focused. With the step-down systems, the optics must include the ability to resize the

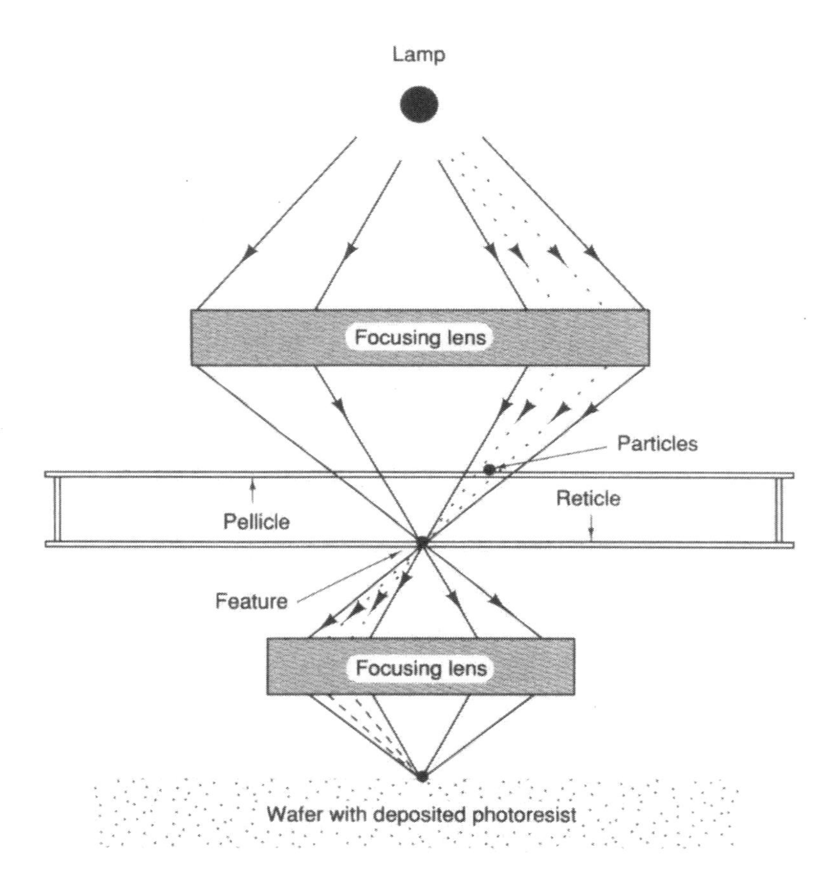

Figure 6-12. Using a Pellicle to Reduce Defects In this figure, we can see that a particle that is about the size of a feature will be trapped on the pellicle before entering the focal plane. This results in the prevention of a repeating defect, since the shadow of the particle will not have a significant effect on the resist on the wafer.

image, while keeping the beam collimated and the image undistorted. All of this must be done with regard to the problems of astigmatism, chromatic aberration, light intensity uniformity, and optical beam collimation. The ability to do all of this and still maintain resolution on images as small as 0.5 μm is actually quite incredible.

The wafer handling system is very critical on a wafer stepper. It is the stage that will make or break the consistent operation of the system. The stage must be able to be aligned in each axis for each field on the wafer. These include the x and y positioning to within 0.1 μm or tighter, theta positioning with a few seconds of arc, and the ability to tilt the wafer slightly to make up for any warpage or other anomaly of the wafer, stage, or optics. The stage must be able to closely reproduce the positions from wafer to wafer and from location to location. Usually, the system will have some sort of pattern recognition system that allows the system to align the pattern to the proper positions. In these cases, the computer will find certain patterns, called alignment keys, that are then used to adjust the stage position. These keys have shapes that allow the computer to easily analyze the orientation of the stage. For example, a plus-shaped key is used to determine the x and y alignment, as well as the rotational alignment, while a bull's eye-shaped key can be used to determine the wafer tilt. Both are used for adjustment of autofocus hardware. The system finds the key, adjusts the stage, and then checks the key again. Only after the system has homed in on the exact location for the image will the exposure be made.

The light sources for wafer steppers have been changing continuously over the years. For instance, there are a number of steppers that use high-intensity lamps and filter out any wavelengths that are not desired. However, this severely limits the amount of light available for photoresist exposure. In general, it is better to use a light source which can produce light at certain key wavelengths. To that end, halogen vapor lamps are used, with most of the light being produced in just a few wavelengths. Since the resolution of an optical system is dependent upon wavelength, stepper manufacturers have steadily changed their systems to shorter wavelengths, until now there is alot of emphasis on ultraviolet systems for attaining submicron linewidths. This can be an expensive proposition. Common types of glass absorb large amounts of ultraviolet light, limiting the number of optical elements. In addition, many materials exhibit unpredictable changes in refractive index at very short wavelengths. In most of these cases, the changes are not caused by the material itself, but by impurities in the materials. Suprasil® (a type of fused quartz) and magnesium fluoride are two of the more common materials used in the manufacture of ultraviolet optics. Reflective optical systems are sometimes employed to overcome problems with ultraviolet optics, but add immensely to the complexity of the optical system. Recently, more intense ultraviolet sources have been used in the manufacture of submicron technologies. They include deuterium light sources and eximer laser sources, which will be discussed in the next section.

E-Beam Lithography, Eximer Lasers, and Other Exotic Systems

In the search for the ultimate submicron lithography systems, a number of different avenues have been followed, including electron beam lithography, the use of deep-UV eximer lasers, X-ray sources, and hybrid step-and-scan mechanisms for image exposure. These systems perform their tasks in substantially different ways, although there are often similarities to more standard systems. Almost all of these methods are more in use in the laboratory than in production. However there have been production systems introduced using almost all of these technologies. Which one, if any of these, will prove to be more useful for the production of submicron technologies is a question still to be answered.

The method that is closest to the standard stepper methodologies in use now is called the step-and-scan method. This system employs a stepper system to move from field to field, but instead of shooting the entire image at once, the wafer is exposed slowly by scanning a slit over the surface of the reticle. This allows precise control of the exposure system, and permits finer resolutions to be obtained with optical systems. Systems such as these are expected to be able to produce the 0.35 to 0.25 μm linewidths to be required in the 1990s. The mechanics of these systems are expected to be quite complex, and they will not have a great throughput, but they also will not require the more exotic equipment that will be required with E-beam and laser systems. Some of the issues that could be expected to be troublesome would include variations in slit scanning speed, mechanical binding that could change the angle of the slit in relation to the pattern, changes in focus or alignment due to vibration of the mechanisms, or problems with diffraction edges in the event the slits are manufactured improperly.

Another system that is slowly gaining respect is that of the eximer laser. The system itself is not much different than a standard wafer stepper, with the exception of the laser light source and its required facilities. The laser systems are very powerful, allowing a far greater light intensity than any other standard light source. The short wavelengths, down to at least 193 nm, allow very fine resolutions to be obtained. Linewidths under 0.5 μm are easily obtained. There are obvious benefits in beam collimation also (since the laser beam is inherently collimated to a great degree). Ultraviolet lasers can run with various wavelengths, since various combinations of gases will give different particular wavelengths. The user is not limited to certain spectral lines or regimes. In fact, eximer lasers coupled with the use of tunable dye lasers can result in a nearly complete range of wavelengths at high intensity. These narrow wavelength ranges can also result in the use of photoresists that are sensitive only to the main wavelengths and are therefore easier to use and store.

There are some technical problems with eximer laser lithography, however. One is that the laser itself is pulsed, not continuous, and the output from pulse to pulse can vary by several percent on a well-tuned laser to 10 or more percent on a poorly tuned laser. As a result, the optics and computer systems must be prepared to adjust for these changes. Exposure times (as measured by exposure pulses) may need to be adjusted on the fly in order to guarantee uniformity. Another issue is that of beam uniformity. Eximer lasers do not have especially uniform beam densities. Optical systems have been designed to try to control the beam shape and intensity profiles (see Figure 6-13), but these systems cannot account for changes in beam density as a function of optical element degradation. Optics used in DUV laser systems degrade rapidly and in a nonuniform fashion, with crystal damage occurring in the optics wherever the light is absorbed. This formation of "color centers" limits the uniformity of the pulse intensity profile. Unfortunately, only specially coated optics can be used for mirrors, and the reflective coatings used can be damaged quite easily, and can become completely opaque to the UV light even when there has been little visible change in the mirror. The cost of these special optics is very high, often exceeding $500 for a single 2-inch round lens. Fortunately, eximer lasers do not exhibit the phenomenon usually seen with the common He-Ne laser that is called "speckle." This is primarily due to the fact that the eximer laser beam is not as coherent as the He-Ne beam. The existence of speckle would cause severe problems with resolution of the laser system, or would require more expensive optics to be installed in the system.

Until recently, the technology for producing eximer laser beams was not reliable enough to justify their use in a production environment. The inability to maintain a beam for a long time on a single charge of gas, short electrode lifetimes, and other objections have been brought up. However, most of these objections have been overcome due to the persistence and development expense incurred by the eximer laser industry. The newest nickel electrodes, with special reaction chambers, gas processing systems, and heavy-duty electronics have allowed the eximer laser to run unattended for many hours, with a reliability that is in line with the reliability of the other subcomponents of the lithography system.

The X-ray stepper has recently emerged from the labs into the production world. This device can produce wafers with linewidth resolutions down to 0.1 to 0.25 μm. these devices are very expensive and very slow, producing around five wafers per hour. The images are produced by exposing the wafers (covered with X-ray-sensitive photoresist) to a beam of X rays that are scanned across the surface of the pattern. A common method of producing these images is through direct writing, where the beam is moved in a raster scan pattern across

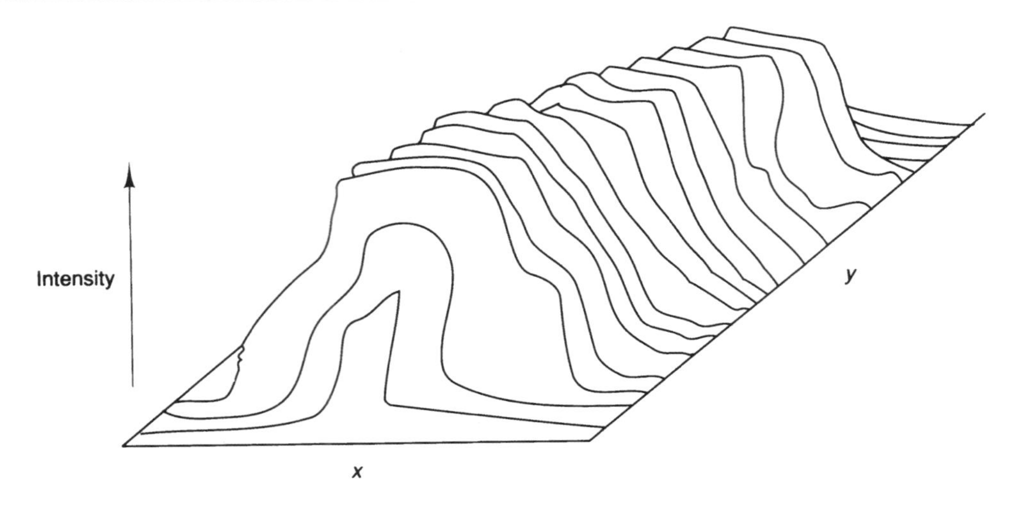

(i) Irregular, as-produced laser beam intensity profile

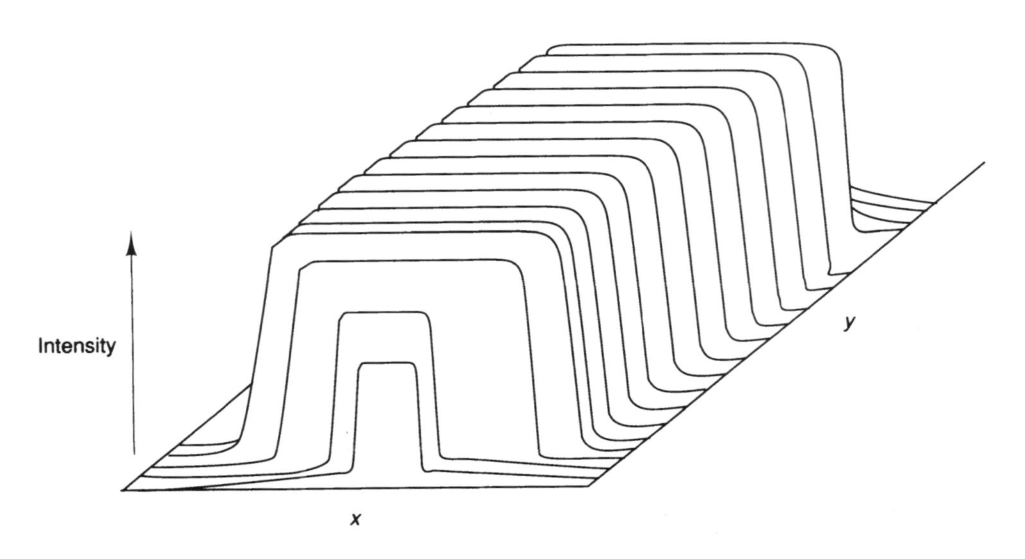

(ii) Filtered, regular, square laser beam intensity profile

Figure 6-13. Intensity Profile for Eximer Lasers While eximer lasers produce a large amount of UV light, the beam varies in nonuniform and often nonrepeatable patterns. This limits their effectiveness as a lithographic light source, since the light intensity must be very uniform for good imaging. Recent advances have allowed filters to be manufactured that significantly improve beam uniformity. While a significant loss of total power occurs, it is still sufficient to produce good results.

the wafer (or field if the system is stepper-based). Another technique is similar to the step-and-scan method described earlier where a slit is used to expose a portion of the pattern, and then is scanned down the length of the pattern. At this time, it is difficult to predict the problems that will exist as a result of the use of X-ray lithography. The resist will be very thin, in the order of hundreds of angstroms and will therefore have to be very uniform. The X-ray sources will have to be well shielded to prevent any safety hazard to the operator. Obviously, throughput time improvements will be necessary.

Electron beam lithography has been around for some time, as the resolution of the electron exceeds that of most photon-based lithography systems. It became obvious early on that the electron could be used in lithography as well as in analysis (e.g., SEMs). However, the costs of this equipment as well as the technical difficulties have prevented this technique from being competitive until recently. Even now, throughput time is slow and the instruments cost significantly more than optical wafer steppers. Current systems utilize direct write, raster scanning methods to expose the wafer to the electron beam. Beam density can be generally well controlled. The final production version of these machines will likely incorporate most of the features of the optical steppers, such as tightly controlled stages, step-and-repeat mechanisms, and so on.

The electron generation and beam control systems must be kept at vacuum at all times, which imposes a considerable number of constraints on the system, including requirements for ultraclean and vibration-free pumpdown steps, image alignment and focus steps, and exposure of the resist without saturating the sample. Overexposure of the substrate with electrons could result in circuit damage, therefore requiring tight control of all parameters before the actual E-beam exposure starts. However, since electron beam technologies can result in linewidth resolutions of under 0.1 μm, there will continue to be a significant amount of effort applied to find solutions to these problems.

We can see that there are a number of technologies that will be vying for leadership throughout the 1990s. It is difficult to predict which technology will ultimately prevail or if all of them can coexist in the wafer fabrication world. While the highest tech solutions can provide ultrafine resolutions, they are often saddled with excessive mechanical complexity and slow speed. However, the ever-increasing demand for high resolution will force the manufacturability issues to conform to the needs of the production fab area.

6.1.4 Image Exposure

We have discussed a variety of items surrounding the exposure of the photoresist, but have not discussed this item in detail. After the wafer positioning system has aligned the wafer under the pattern (we'll assume that we're using standard optical steppers, for ease of reference), the wafer with the photoresist is then exposed to the light until enough photons have reacted with the photoresist to clearly define the pattern and cause it to react completely. The exposure time of the process depends on the type and thickness of photoresist used, and the amount of light that is available. Overexposure and underexposure can both lead to linewidth variations. Figure 6-14 shows the changes in linewidth as a function of exposure. The image must be focused clearly all through the resist layer. This requires that the depth of focus of the optics used exceeds the depth of the resist, and is deep enough to allow proper focus on top of a variety of structures.

Photoresist is manufactured to react to a fairly specific wavelength to reduce the problems associated with chromatic aberration. This is shown in Figure 6-15, where an extreme example is shown. In this case, the incident light is white light, the photoresist is sensitive over the entire spectrum, and the optical system has enough chromatic aberration that the red light focuses below the image plane while the blue light focuses above it. As a result, there is poor focus and poor line resolution, as well as inconsistencies in linewidth from structures near the top of the step as compared to the structures near the bottom.

As mentioned, light intensity must be uniform across the surface of the wafer. Inconsistency will result in incomplete exposure of the photoresist. Incomplete exposures often result in a polymeric substance forming in the corners of the patterns. These substances will not be completely removed during the developing process, so that the substance is still on the device during etch. This can block the etch, leaving defects and other artifacts in critical areas. This partially exposed photoresist can be removed in a process called descum, but this

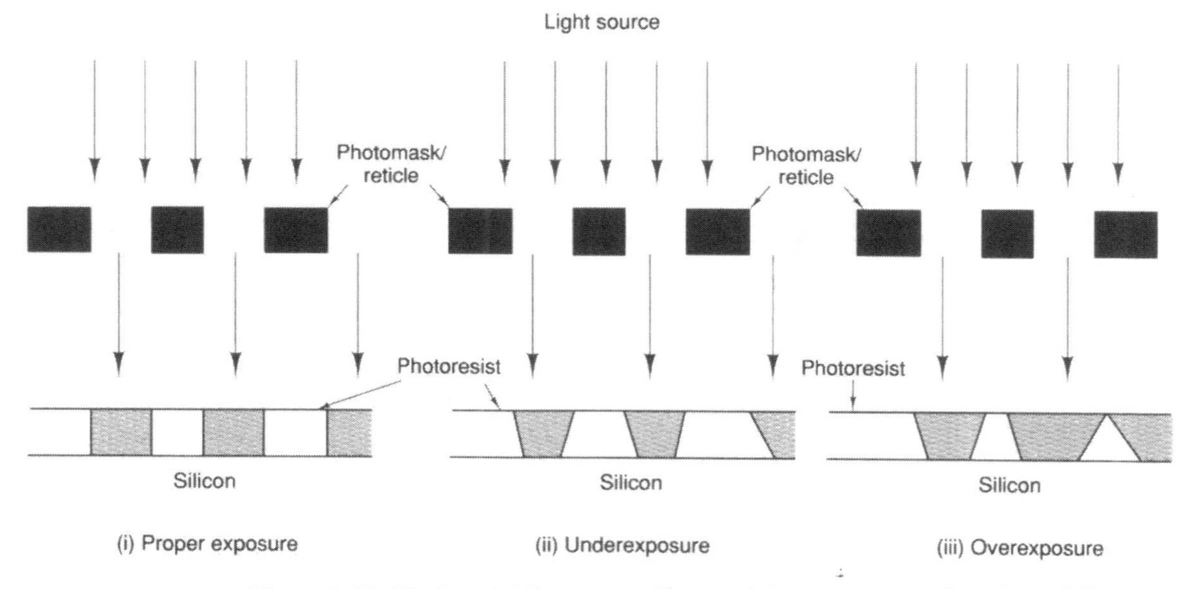

Figure 6-14. Photoresist Exposure Photoresist exposure must be set carefully, considering factors such as light intensity, photoresist type and thickness, photoresist exposure rates, and resist slope desired. As shown, a proper exposure will produce uniform blocks of resist where desired. An underexposed resist will be prone to residual resist formation and linewidth control problems. Overexposure of the photoresist will result in bridging and associated problems. Problems are similar but inverted when using negative photoresists.

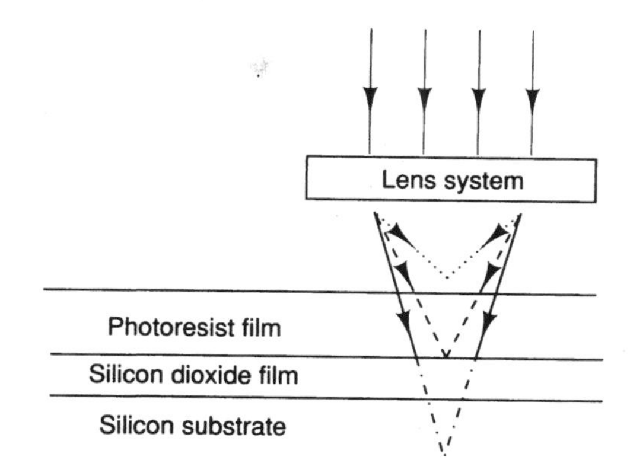

Figure 6-15. Imaging Problems with Severe Chromatic Aberration This not-to-scale figure demonstrates the problems with focus when severe chromatic aberration exists in an optical system. The ability to produce a finely defined feature will be limited be excessive chromatic aberration.

can add contamination and can cause other problems; therefore, the preferred method is to control the resist exposure properly.

It is also important to ensure that the light impinging on the surface of the wafer is perpendicular to the surface and is properly collimated. Otherwise, it may be possible to inadvertently construct unusual line profiles, including angled walls and influencing undercutting phenomena. Figure 6-16 shows the impact of exposure of the photoresist at a nonperpendicular angle.

Reflections in the optical systems, especially when trying to produce images on metal surfaces, can cause a phenomenon called notching. Notching is shown in Figure 6-17. This occurs when reflections expose areas of the photoresist that should not be exposed. This results in areas of the metal either protruding into areas where it should not be, or having small areas removed from the existing metal areas. Reflections can also cause lines to form across spaces, causing short circuits to form, or can cause lines to be separated, forming open circuits. These problems are usually corrected with careful alignment of the optical systems, and careful planning when creating the process. Recent developments include the deposition of films of differing index of refraction from the photoresist, with specific thicknesses that will reduce the reflections. These antireflective coatings are not 100% effective at removing these problems, but are very promising. These films must be transparent to the wavelength of light being used to expose the resist, and must be able to be removed with chemicals that do not attack photoresist prior to the developing process. Some of the films tested include deposited SiO_2 films, which have indices of refraction around 1.46, as opposed to the refractive indeces of photoresists, which are usually around 1.6 or so. In addition, the SiO_2 can be removed with an etch in HF, which will not affect the photoresist to any great degree.

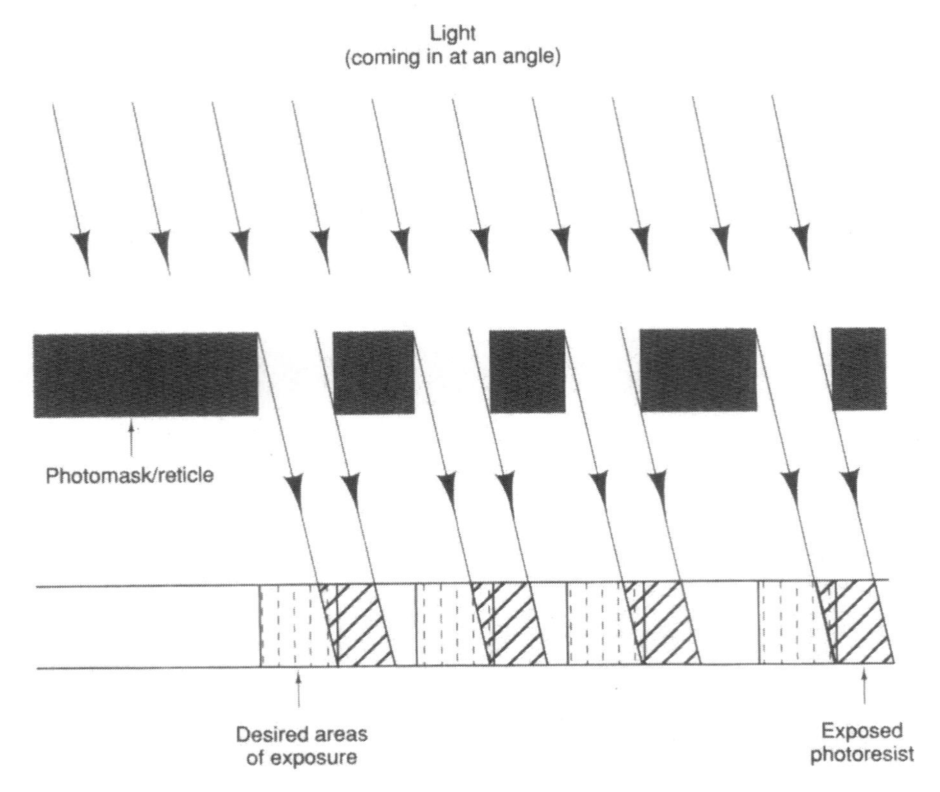

Figure 6-16. High-Angle Photoresist Exposure In this exaggerated case of light coming at an angle to the wafer, we can see that there will be a number of problems encountered. First, the lines will not be located in the proper position. Second, the linewidth will be reduced from that desired. Third, the exposed resist will not have vertical sidewalls.

6.1.5 Developing the Photoresist

This is a fairly straightforward operation, consisting of dipping the wafers in a bath of developer solution, or spraying the developer solution over the wafer surface as it spins on a chuck. Each of these methods is shown in Figure 6-18. The solutions used to develop the photoresist are picked specifically for the photoresists in use by the fab. Incredibly enough, some of the most common developer solutions use sodium hydroxide for a base material. Sometimes there are other metal ion-based solutions used, such as potassium hydroxide, although until the last few years they have not proved as successful as the sodium-based developer solutions.

The use of sodium hydroxide in the manufacturing seems to fly in the face of common sense when such chemicals are not allowed in any other area of the fab. However, in the earlier days of the semiconductor industry, the chips were not quite as sensitive as today's chips are and there were no other reasonable alternatives to the sodium-based developers. There are now many metal ion-free developers, some based

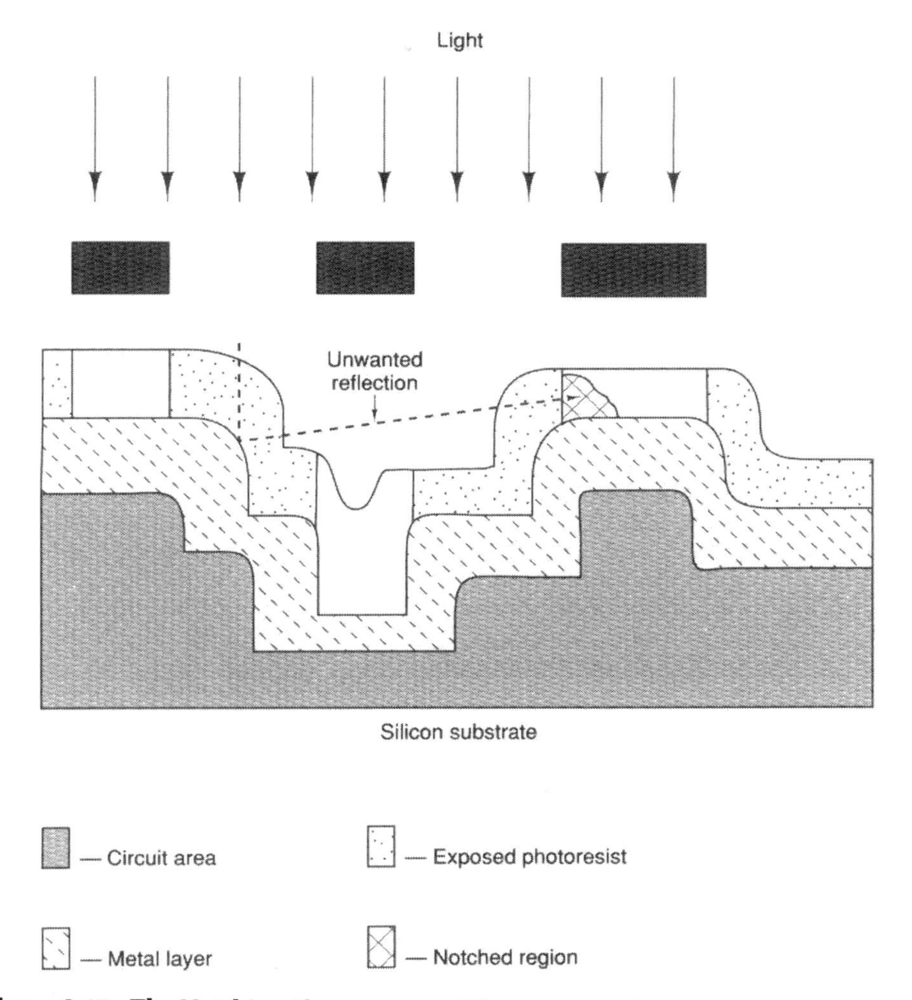

Figure 6-17. The Notching Phenomenon When unwanted reflections occur on the surface of the chip, undesirable photoresist can take place. As seen here, the reflection from one corner of the metal film is exposing the resist over another area of the metal. This region will then be attacked by the etch process, leaving a small region cut out of the metal line. This small region is called a notch, giving name to the phenomenon.

on ammonium hydroxide solutions, some based on choline solutions, others based on proprietary solvents. They are infinitely better to use than the metal ion based developer solutions.

The amount of time that the wafers are developed is fairly critical. Over- and underdeveloping the photoresist can result in linewidth changes and can assist in the formation of photoresist scumming. The length of the develop time can be adjusted by varying the concentrations of the developer solution. Most of the solutions can be diluted with deionized water. Another factor in the time required to develop photoresist is the temperature of the developer solution. Higher temperatures will cause the reactions to occur much faster, although uniformity may be sacrificed if the temperatures are too high. It is

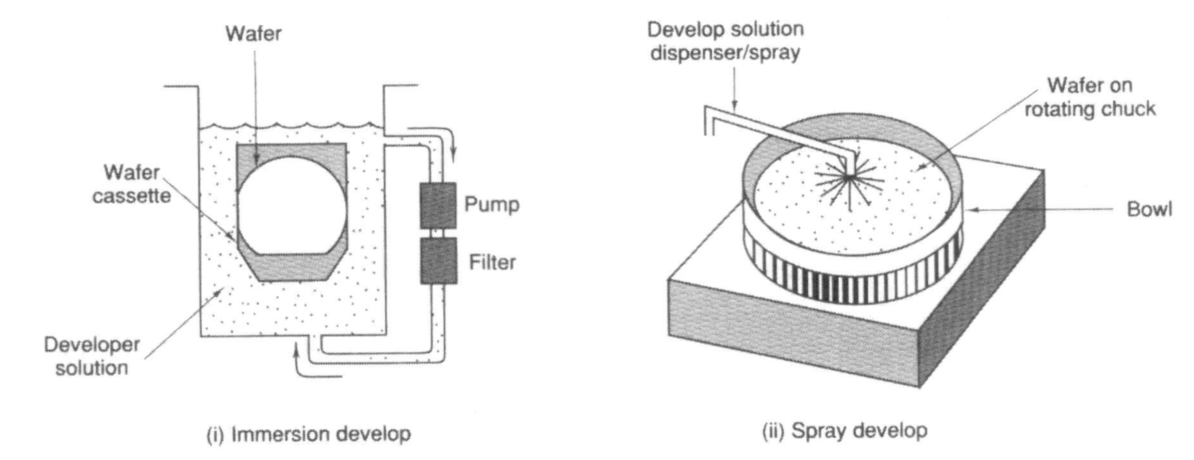

Figure 6-18. Photoresist Developing Like any photosensitive emulsion, photoresist requires developing. This is achieved in one of two very different ways. In the first case, a cassette is placed into a tank of continuously agitated developing solution. In the second case, the wafers are placed individually onto a spinning chuck and the developing solution is sprayed onto the wafers. Each of these options has advantages and disadvantages, and there is still no conclusive decision on which method will be preferred in the long run.

important to keep the developer solution circulating, for two reasons. The first is that the developer solution will weaken if allowed to stand, slightly changing the develop times in localized areas of the wafer. The second reason is that the solution becomes very dirty, very quickly. Particles may then be freed which will land in some particular area and prevent that area from becoming developed properly. In many fab areas, the batch develop processes use filtered recirculating baths, which must be cleaned frequently to prevent excess buildup in the tanks. Following batch develop processes, the wafers must be rinsed in a bath of deionized water and then dried in a spin rinse drier.

Single-wafer spray developers are being introduced to most new fab areas these days, as they permit a greater degree of control over the process. These machines spray controlled amounts of filtered developer solution over the surface of a spinning wafer for a fixed amount of time. Several potential problem areas must be controlled when using the spray developer. The first is that the machine must maintain proper spin speed or the photoresist may not be developed uniformly. Also, the developer solution must be applied consistently and must have consistent strength in order to guarantee uniformity. There has been on-going debate as to the most uniform method of developing photoresists. There are still many who prefer the use of batch (tank) developing.

The final issue is that the wafers must be properly rinsed on both the front and the back. There have been situations where the developer solutions have not been adequately removed from the back of the wafers. On these wafers, the developer solution was then driven into the back of the wafers in the etch and implant processes, and then later outgassed from the wafers while in the diffusion tube, causing total circuit failures on surrounding wafers. All of these problems are easily controlled as long as the systems are properly maintained and periodically tested.

6.1.6 Contamination in the Photolithography Area

Just as in the other areas of processing, contamination can be broken into three classes: chemical contamination, particulate contamination, and physical damage. The least serious of the contamination problems in the photolithography area is chemical contamination. This is primarily due to the fact that the temperatures reached in the photolithography area are modest compared to some of the other areas involved in the process. As a result, most chemical contamination remains on the surface of the wafers and can be cleaned off readily. One of the major chemical contamination problems that can occur at this stage of the process is the case of the improper removal of developer solution. In that case, the problem is that the sodium from the developer can move freely in the silicon substrate, and therefore can diffuse to a shallow level in the silicon, yet far enough in that the particular type of cleaning process in use cannot leach it from the surface. In virtually all other cases, the contamination that finds its way onto the wafer will be removed through subsequent clean steps.

Particulate contamination can be very serious in the photolithography area. Particles landing on the surface of a reticle can cause defects on a large number of die on every wafer. In the example used earlier, a reticle with four die on it will be the cause of a 25% die yield loss if any one die is rendered inoperable due to particles on a critical layer. Large enough particles landing on pellicles, or tears in the pellicles, can lead to a number of visual defects on the wafers. Particles can land on the wafers during develop, causing spots of undeveloped resist, or they can be deposited during the photoresist spin step, causing comet trails and other defects. The particles can even become entrapped in the photoresist, and later can be stuck in this surface of the wafer during an etch process or other subsequent process, and can sometimes remain trapped on the wafer permanently. Clearly, then, particles can cause a significant number of problems. They are usually controlled by filtering all of the various solutions used, including the photoresists, developer solutions, and water systems. Any edge bead removal process must be tested to verify that the edge beads that are removed do not end up as particle contamination on the wafers. (See Figure 6-19 for a description of the edge bead removal process.) All mechanical sources of particles, such as process chamber doors and other moving parts, areas of photoresist splatter, and other possibilities must be examined and cleaned up as required.

Several sources of mechanical damage can cause the image on the

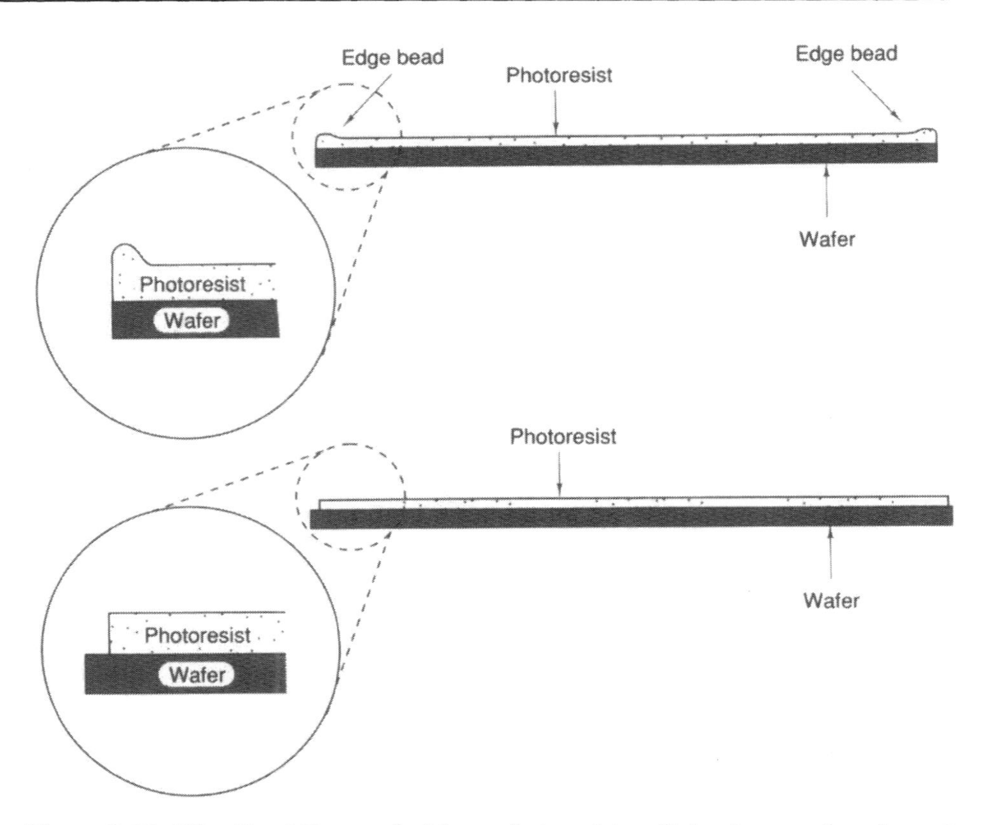

Figure 6-19. Edge Bead Removal Many photoresists will bead up at the edges of wafers during the spin operation. After baking, these small pieces will become brittle and break off, becoming particle contamination sources. This problem is eliminated by spraying the edge of the wafer with a solvent to remove the edge of the photoresist film.

wafer to be distorted or otherwise damaged. They include scratches, which can occur easily on photoresist-coated wafers. Photoresist is very soft, especially before it is baked. Any kind of scratching will completely destroy the wafer's image under the resist, and will also spread the broken-off pieces of photoresist all around the wafer as a form of particulate matter. These scratches can come from a number of places, but are often associated with wafer handler vibrations. These vibrations can cause problems with scratching if the vibration occurs when the wafer is under some object, for instance, a lens or a process door. Another type of mechanical damage that we have discussed in this context is that of microbubble formation. They are common in the relatively thick photoresists, and must be avoided at all costs. These bubbles, once included in the resist film after the initial bake, will not go away, and will lead to anomalies in the circuits after the etch process.

6.1.7 The Metrology of Photolithography

Before etch processing, it is important to check the wafers to make sure they have obtained the correct parametric values for the layer. This means that the linewidths, the alignment, and the overlay registration must be tested, as well as the physical condition of the wafers. Typically, before the etch step is performed, the wafers are run through a step called develop inspect. First, the wafers are inspected, under a lamp, for scratches or other gross defects, with care being taken not to expose any wafers. If no problems are seen, the wafers are next observed for particle count change. This involves a visual scan across the wafer. After this the linewidths and other parametric values are measured. The sequence of these steps may vary from fab to fab, as well as the quantity of wafers that are inspected through each step. Usually, only a sample of wafers is closely inspected, and from that it is assumed that other wafers in the batch are roughly equal in all of the various respects. Since we discuss defect and particle counting methods in other chapters, we will focus on the parametric issues in this section.

First, there is the issue of critical dimension, or CD, measurement. This measurement tells the operator how well the process has performed as compared to the average. When the reticle is designed, it will sometimes have a test area where a resist line of nominal width is placed in a somewhat isolated position. At other times, the design of the device will not permit this, and some particular structure on the device must be chosen for measurement. This parameter must be characterized by the process engineer and measurements for process control taken from these structures only.

The width of this line can then be measured by a variety of systems. The most common of these is the optical microdensitometer. This device measures an optical intensity profile across the line in question. See Figure 6-20 for a sample of an optical intensity profile. This device is limited in resolution to about 0.7- to 1.0-μm linewidths in the visible range, and to lower than that in the UV range. There are a number of technical questions relating to this technique, including problems with film contrasts, image interpretation, and optical alignment. The manufacturers involved in this field have spent a significant amount of effort guaranteeing that their systems can be shown to be reproducible over a long period of time.

The second major type of system used to measure linewidths is the low-voltage SEM. This device can resolve images down to around 0.1 μm. issues such as image interpretation and film contrast are also important when dealing with SEM linewidth measurements. The intensity profiles that are returned by the samples can sometimes appear distorted for unknown reasons. A significant cause of this is the fact that it is difficult to model exactly what an electron will do

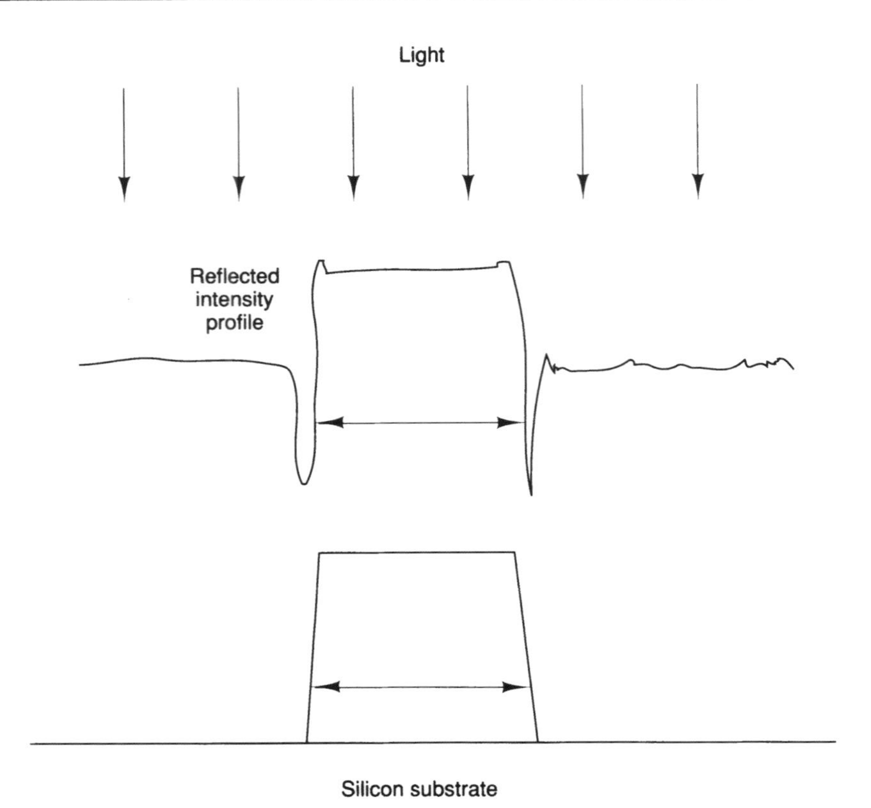

Figure 6-20. Linewidth Detection The linewidth of a structure can be measured by observing the optical reflection or electron emissivity changes of the structure. This is achieved (as shown for the optical method above) by shining a beam on the structure to be measured and then observing the reflected intensity profile. Careful calibration and edge selection are important for controlling the results of this procedure.

when it hits a material at a very high angle (when it hits the edge of a line).

There are other problems with the use of the SEM, for instance, the minute variations on the edge of the sample line may result in measurement errors. Issues such as these are resolved by using techniques such as spatial averaging to calculate the average CD, as shown in Figure 6-21. Another significant problem with scanning electron metrology is that the surface of the sample may become charged from the electrons impacting on the surface. Not only will this distort the images, it could cause damage to sensitive devices. Keep in mind that scanning electron metrology instruments are expensive, must be run with the wafer at vacuum, and are very slow when compared to optical techniques.

In almost all fabs, a set of standard linewidths is used to calibrate the linewidth measurement systems. These are often called the fab's "Golden Standards." Usually, these standards have no legal traceability to anyone else's standards. This is largely due to the difficulties

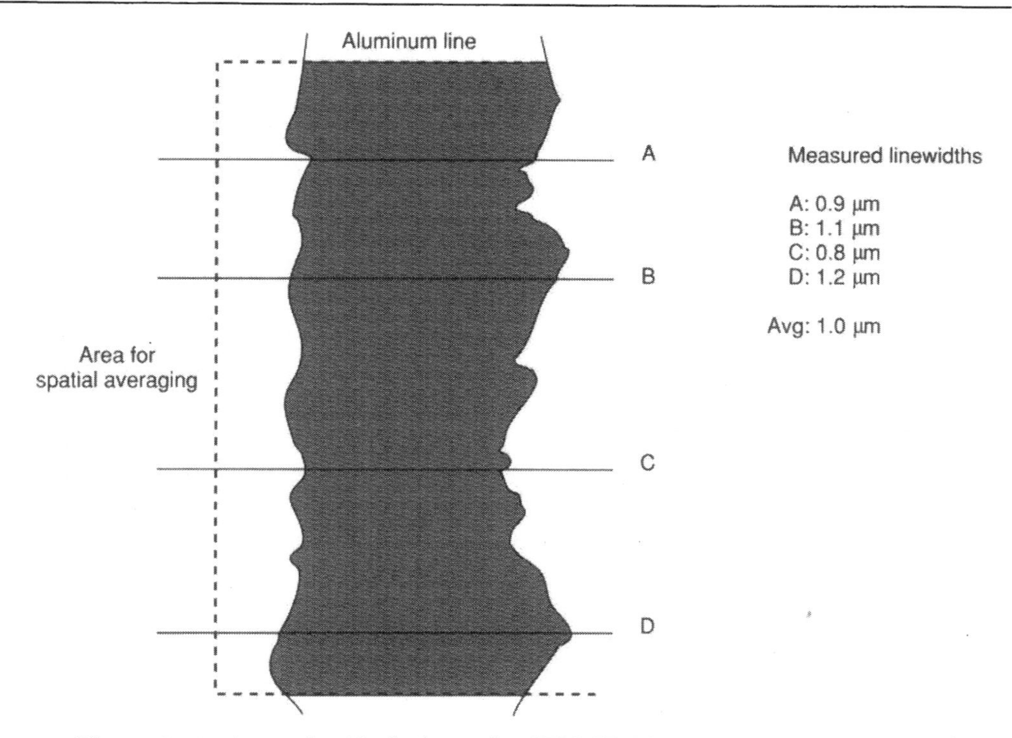

Figure 6-21. Averaging Techniques for SEM CD Measurement We can see that a single scan by an electron beam can result in a variety of anomalous readings, while a system which measures and averages a number of different locations produces a much more reasonable result, as well as providing information on the roughness of the line.

that the NIST has had producing standards for certain narrow linewidths and thin films. This has mostly been because of the technical difficulty in manufacturing and reproducing measurements on equal lines. However, since the production (golden) standards need to be only relatively correct, exact linewidth calibration is often only an academic issue. The various linewidth measurement systems should be tested against the golden standards on a regular basis to guarantee that the instruments' calibrations have not drifted. Then, when changes are seen in the linewidths of the production wafers, you can be assured that the errors are in the process and can be found.

Other parameters that must be tested are overlay registration and site alignment. These are measures of the ability of the stepper to place each field's image in the right location with the right orientation. A number of patterns are used for these measurements, displayed in Figure 6-22. These patterns can be used to indicate how far off the wafer is in the x and y directions for each field (measured by taking the distance being two otherwise known points). This is sometimes called run-out. Another indication can be inferred about the radial alignment or orthogonality of the patterns. This is done by comparing the alignment readings in a variety of locations around the wafer. Often, a structure called the "box in a box" is used to determine registration.

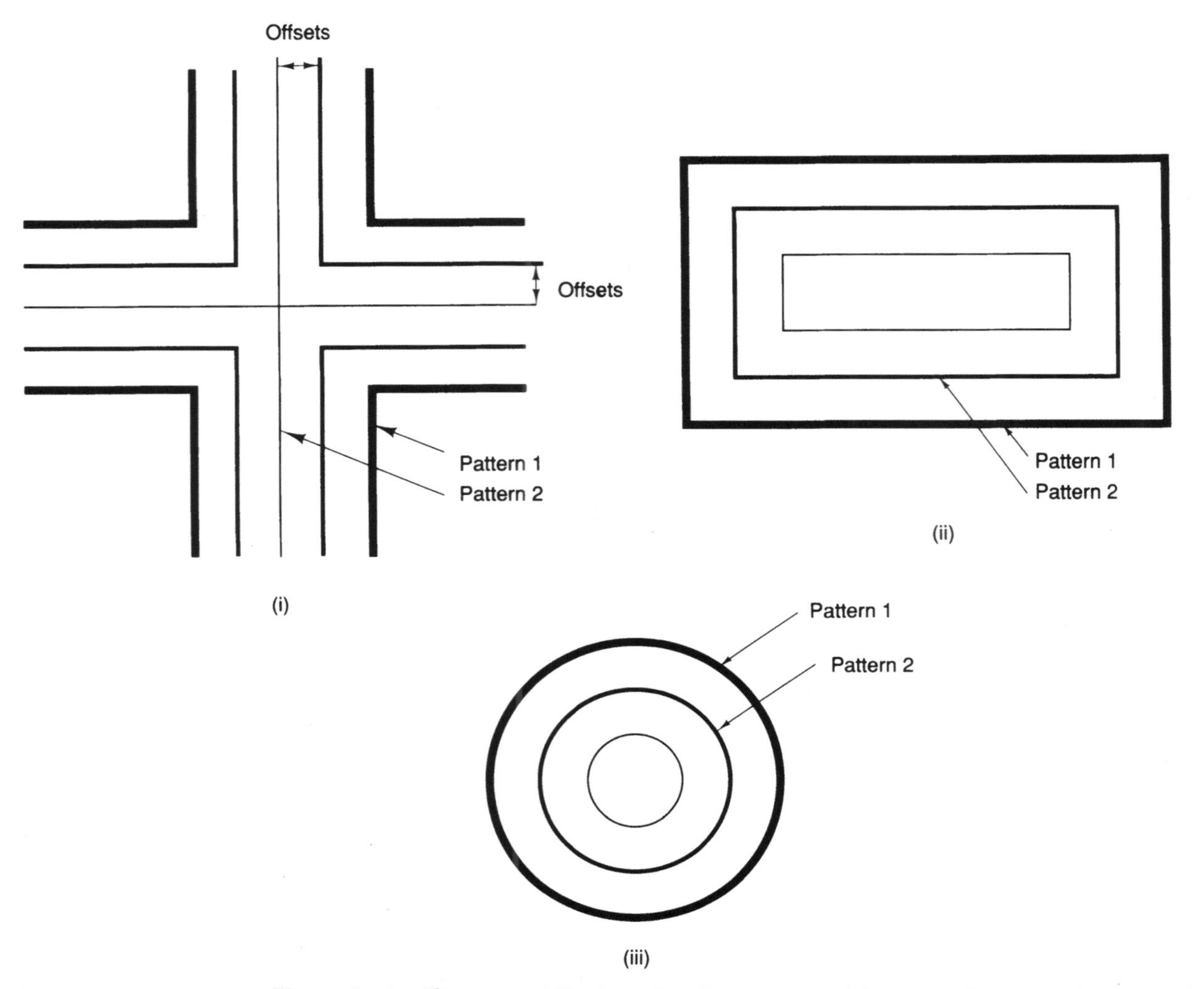

Figure 6-22. Alignment / Registration Targets In addition to the linewidth measurements already obtained, it is critical to have an understanding of how each pattern is aligned in relation to the underlying image. Thus, a number of different types of patterns are designed into the test structures to give these results. In each case, a previous layer has laid down a structure (here called "pattern 1"), while the new process defines the structure "pattern 2." These patterns are measured and the results compared against standards. If, for instance, the cross hair in target (i) is off to the right and up some, this distortion can be calculated from the offsets in the x and y directions.

This structure allows the stepper operator to determine most of the alignment issues in one step.

In many cases, errors will be found on wafers during the develop inspect step. They could include any of the problems listed above, from poor alignment to excessive numbers of particles. If defects occur, it is important to try to recover the wafers prior to the etch process. Any etching that occurs on these defects will permanently imprint the wafer with that defect. While wafer recovery or rework is often considered a simple matter, it usually involves a significant loss in die yield, and impacts manufacturability ("If there isn't time to do it right the first time, when will there be time to do it right the second?"). As a result, reworks are usually considered a last resort, performed only if the wafers are particularly valuable and cannot be easily replaced. A significant problem with reworking wafers is not a technical, but a psychological, one. A fab that does not allow reworks and throws away wafers that have been misprocessed will tend to produce fewer defective wafers overall than a fab that allows reworks, because a perfectionist attitude prevails in the fab that does not allow reworks. The fab that allows reworks will foster an attitude that poor quality is OK; if not, engineering will rework it. Thus, a reduction in reworks manifests itself in higher overall quality.

6.2 FILM ETCHING PROCESSES

The various dielectric and metallic films that have been placed on the wafers must be etched before they can be of use. The patterns placed on the photoresist layers above can serve no purpose without an adequate and well-controlled etch process. The etch process can be performed within a chemical bath or within a variety of machines that use gas plasmas to perform the etch. The type of etch that is selected depends on a number of factors, including the material to be etched, the chemistry to be used for the etch process, the linewidth requirements, maximum radiation exposure allowances, and the cleanliness requirements. It is important to verify that other materials that may be on the wafer are compatible with the etch process in use. In most cases, the chemistries offered by the etch equipment vendors are fairly specific to the films to be etched. However, there is almost always a certain amount of loss of all of the other materials on the wafer. The ability of the etch chemistry to attack the specific film desired is called the *selectivity* of the etch. A higher value of selectivity is typically preferred, with a good etch process having a selectivity better than 25 or 35 to one. As we can see in Figure 6-23, the ideal etch process will remove the film while retaining the original dimensions of the photoresist line, and while keeping the side walls of the feature nearly vertical and uniform. We will start the discussion with the various types of wet chemical etch processes.

6.2.1 Wet Chemical Etches

Wet chemical etches have been used for many years, and even with the availability of modern plasma etch systems are still in use in a variety of applications. There are a number of problems with wet chemical etches that limit their use with very fine geometries. They include excessive undercutting of the photoresist, difficulty in maintaining

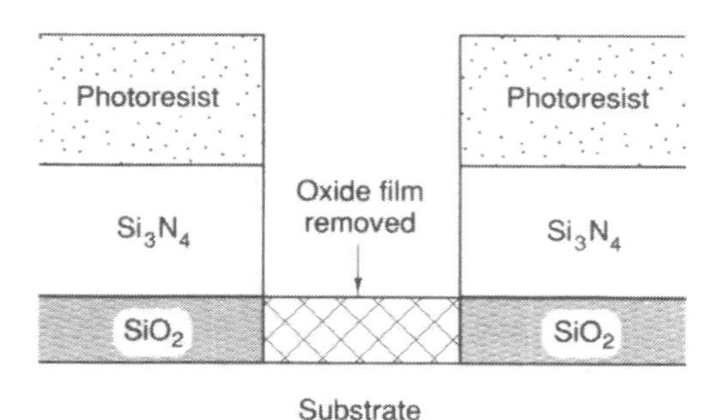

(i) Good (high) selectivity (oxide etch process) between oxide and nitride etch

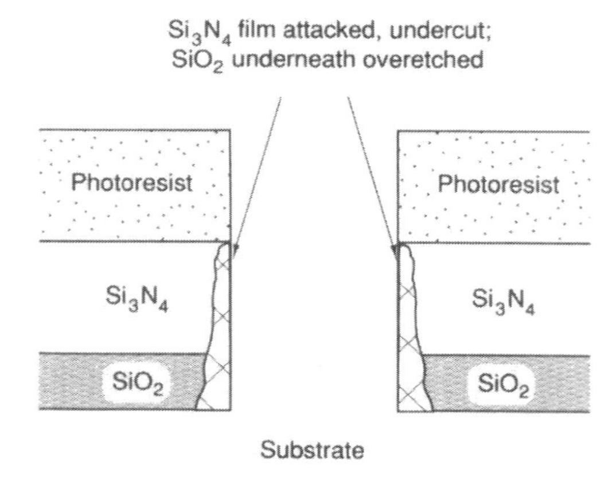

(ii) Poor (low) selectivity (oxide etch process) between oxide and nitride etch

Figure 6-23. Etch Selectivity The capability of an etch chemistry to attack only the film specified is called the selectivity of the etch. As shown, high selectivity will permit proper film removal. Poor selectivity causes the nitride film (in this case) to be attacked. This not only damages the integrity of that nitride film, but also permits the underlying oxide films to be overetched.

etch uniformity, excessive particle contamination, and inability to incorporate end-point detection. However, the processes are well characterized, can allow for batch processing, and are inexpensive when compared to plasma etching systems. They also produce no radiation damage, as can happen with plasma etch systems. Wet chemical processes can be very effective on broad area etches, such as nitride removal, as long as the chemicals are compatible with all of the films on the wafers. These processes can also include certain types of gate and source/drain oxide removal steps, as well as certain contact hole etch steps and the nitride removal steps. As long as they are used in appropriate situations, wet chemical etches can prove cost-effective.

We have already discussed one of the most common of the wet chemical etches in the previous chapter, that is, the HF etch. Many of the principles that hold true for that etch will hold true for the other wet chemical etch processes. The HF etch is typically used to etch SiO_2 films, but can be used for some silicon nitride and silicon oxynitride films. In some processes, there is a film of photoresist on the wafers so that only selected areas of oxide on the wafer are etched. In other cases, there is no photoresist on the wafers as the areas that are being cleared of oxide are usually very thin in comparison with the other nitride or oxide areas surrounding the region. However, caution should be exercised when using large-area HF dips on wafers that contain BPSG or other CVD SiO_2 films, as they etch much more rapidly than thermally grown SiO_2 films. These CVD films can be removed in a few seconds. The films also etch in a very nonuniform fashion, so that the wafers are fairly heavily damaged after immersion in HF. A common error in a wafer fab is to accidently dip wafers with CVD silicon dioxide films on them in HF solutions and strip them of the dielectric layers. This mistake is usually lethal to the wafers.

In reviewing the information on the HF solutions, we can see that they come in a variety of concentrations. The strongest solution used in the fab is called straight HF, although it is actually only 49% HF in water. (Pure HF is very unstable and produces large quantities of toxic gas.) This solution is very strong and can attack almost anything. The fumes are strong enough to etch SiO_2 (as was noted in the section on C–V plots, HF fumes can be used to clear the backside of wafers for a contact area), and can be toxic in very low quantities. The most common solution is 10:1 HF:H_2O solution (about 0.5% HF). This will etch thermally grown SiO_2 at a rate of about 5 Å/s, which is adequate for many needs. (This etch rate assumes that the acid temperature is around 25°C.) Other common solutions of HF include 5:1 (10 Å/s) or 50:1 ($\approx$1 Å/s) solutions. Slower etch rates are preferred when precise control over the etch is desired. If etch rates faster than 10 Å are required, a class of HF-based solutions are available. They are usually called BOE (buffered oxide etch) solutions. These solutions usually consist of some HF and water, with ammonium fluoride mixed in. This buffering tends to make the etch somewhat more uniform. These BOE solutions can be particulate generators, so they are usually used only with larger geometry devices. Diluted BOE solutions can also be used, with fewer problems with particles. The main problem with BOE is the formation of NH_4F crystals, which are water soluble. This can occur in filters that are not kept immersed or kept in tanks of chemicals.

Another of the wet chemical etches is the phosphoric acid etch, which is used to remove silicon nitride from the wafers. This etch solution is very slow, even at the elevated temperatures at which the phosphoric acid is kept. Often, the etch processes can take many minutes, depending on the thickness of the film. Since the phosphoric acid solution can separate easily over the span of the etch, it is necessary to have a magnetic stirring device in the solution. The phosphoric acid etch has a variety of disadvantages, such as its slow rate, the difficulty of producing clean wafers, and safety problems with fumes. As a result, it has fallen into disfavor in the past few years, with the plasma etch machines being used more often now for Si_3N_4 removal. Older fab areas, and fabs not willing to put the money into the required plasma etchers, will continue to use the phosphoric acid etch for some time to come.

Another one of the older etch solutions is the wet poly etch. This solution is a combination of water, HF, nitric acid, and acetic acid. There are various concentrations of these solutions, depending on whether the solution is to be used on doped or undoped polysilicon. The solution is designed so that the nitric acid will oxidize the polysilicon, which is then etched by the hydrofluoric acid. The acetic acid is added to act as a buffering agent to slow and control the reactions. The solution must not be so strong as to attack the photoresist films, but must be able to etch the polysilicon in a reasonable time. One of the problems with a wet polysilicon etch is that the etch is isotropic in nature, and will undercut the photoresist by some amount. This is undesirable, especially at very narrow linewidths. The wet polysilicon etch processes are limited to 2 to 3 μm linewidth.

The last of the wet etches that we will discuss is the aluminum etch. This is usually nitric acid-based, and is very messy. The wafers must be continually agitated to guarantee any sort of uniformity, and a number of devices have been built to control this agitation. This process is seldom used in modern wafer fabs due to the difficulties in controlling the uniformity of the etch, and due to cleanliness issues.

6.2.2 Plasma Etch Processes

The modern wafer fab typically uses a variety of plasma etch processes to remove films from wafers. These processes have been the focus of a significant amount of development in the last 10 years and are remarkably complex. The plasma processes can produce very uniform and isotropic etches. These are good conditions for producing fine geometries, and generally plasma technology can be used well into the submicron range. In addition, the etch process is considerably cleaner (if performed correctly) than the equivalent wet chemical etch. There are some disadvantages to the plasma processes. First, the equipment is expensive and complicated. A wide variety of maintenance issues must be controlled. For instance, mass flow controllers (MFCs) and RF power supplies must be kept calibrated or the process can produce undesirable side effects. The throughput times are limited by the more complex processing required. There can be a significant amount of radiation damage imparted by the glow discharge of the plasma. Finally, the by-products of plasma reactors are often very toxic.

Theoretical Aspects of Plasma Processing

First, we will discuss a few of the theoretical aspects of plasma processing. The basis of all of the plasma (or glow discharge) processes are the same. As shown in Figure 6-24, wafers are placed into a chamber which is then evacuated to a high vacuum. Then, gases particular to the process desired are introduced into the vacuum chamber and the pressure allowed to stabilize. At this point, a high-frequency electric field is placed across the gas, which ionizes, producing the plasma. The chemistry of this plasma stream allows the

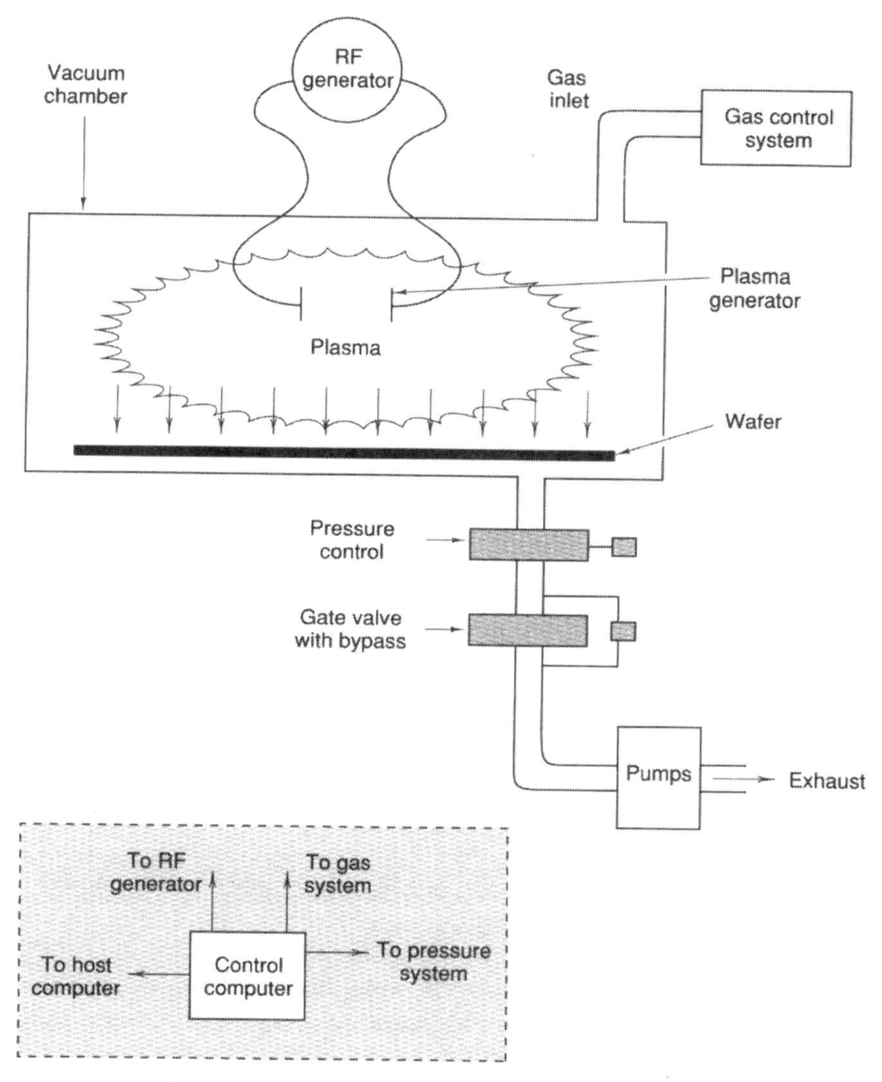

Figure 6-24. Block Layout of a Plasma Etch System Similar in layout to the plasma CVD reactor, the plasma etcher works by creating a high-energy plasma in highly reactive gases. These ionized molecules react on the wafer surfaces with the appropriate films and in a highly controlled and predictable fashion.

gases to react with the appropriate films on the surface of the wafer. After this reaction takes place, the by-products are swept away into the exhaust stream, and more plasma streams past the wafer. There is usually a detector which observes the reaction by-products, and when the detector shows that the by-products are no longer being produced, the etch stops. This is called automatic endpoint detection. After this, the wafer is brought back to atmospheric pressure and replaced in the cassette. Table 6-2 gives a variety of etch process parameters.

The plasma will interact with the surface of the substrate in a variety of ways, depending on the chemistry involved and the surface film type. Some of the reactions can be very complex, so further discussions of the specifics of this subject are left to the texts listed in the bibliography. For now, we can place the reactions into a few basic categories. One reaction that takes place is one in which the plasma reacts with the surface element to form a gaseous by-product directly, as demonstrated in Figure 6-25. For instance, silicon may react with chlorine-bearing plasmas to form $SiCl_4$, which can then be carried away in the exhaust stream. Another potential reaction involves a plasma gas reacting with a film to form a solid precursor product, which then allows another gas in the plasma to react with the solid to form a gaseous by-product. A third reaction occurs when the film on the surface reacts with the plasma to form a polymeric compound. They are usually difficult to remove, and in general are not desirable. There are cases where a slight amount of polymerization of the surface is desired in order to create certain effects in the etch process. Control of polymer formation and removal is delicate work and requires a

TABLE 6-2
Plasma Etch Process Parameters

Film	Gases	Etch rate (Å/min)	Selectivity
SiO_2	CF_4, CHF_3, NF_3	>5000	>15 : 1 over poly-Si
Si_3N_4	CHF_3, CO_2, NF_3	500–1000	1 0–50 : 1 over SiO_2
Poly-Si	CH_4, SF_6, NF_3	2500	>75 : 1 over SiO_2, 7 : 1 over resist
Aluminum	SF_6, CHF_3	5000	20 : 1 over SiO_2, 5 : 1 over resist
SiO_xN_y	SF_6, CF_4, CHF_3, O_2	600–900	~3 : 1 over SiO_2
Silicon	BCl_3, Cl_2, SF_6	>3000	10: 1 over SiO_2
Resist	Ozone	1000–1500	Near-infinite over SiO_2
GaAs	Cl_2, $SiCl_4$, CF_2Cl_2	5000	100 : 1 over SiO_2
Tungsten	Cl_2, BCl_3, CF_4, O_2	400	5 : 1 over SiO_2
WSi_2	Cl_2, BCl_3	3000	>10 : 1 over SiO_2

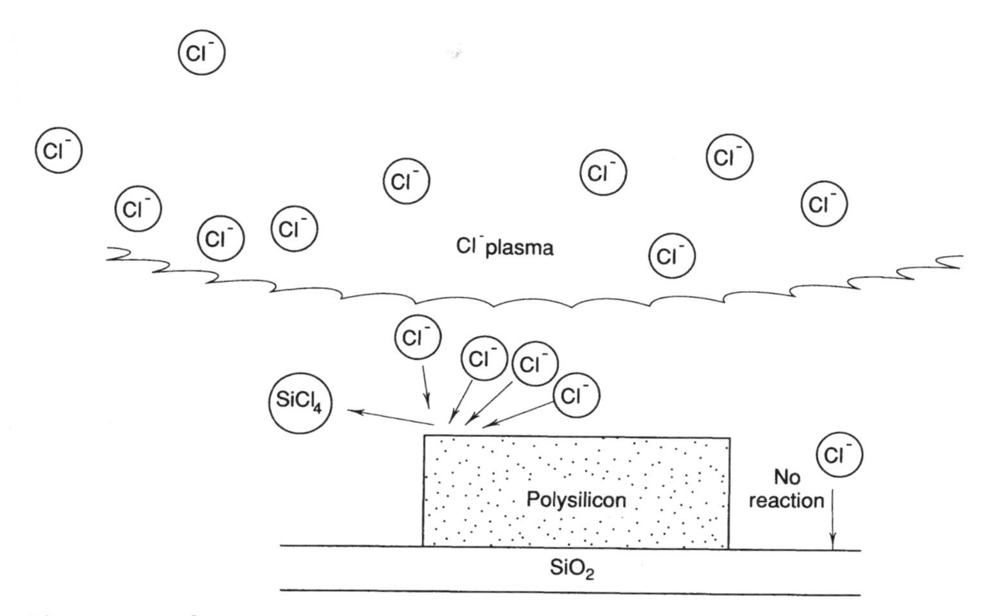

Figure 6-25. The Plasma Etch Process In this example, a chlorine plasma is used to etch polysilicon. The chlorine ions react readily with the silicon atoms, forming $SiCl_4$, which sublimates into the vacuum and is carried off. Since the Si–O bond is stronger than the Si–Cl bond, the chlorine will not attack the oxide. A fluorine-based chemistry could be used to etch the SiO_2 without significantly attacking the polysilicon.

somewhat complicated program for the etch reactor to execute. In a standard semiconductor process, the factors that control the type of reactions must be clearly characterized and optimized.

Just as the plasma can react with the surface in a number of ways, the gases that make up the plasma can recombine in a variety of ways while in the gas stream. This allows for some very unusual by-products. For example, a gas like SF_6 will break down to include small quantities of SF_4, SF_2, and so on. The films can sometimes react with these by-products, occasionally forming hard-to-remove compounds. In addition, these by-products are usually highly unstable and toxic and will be exhausted from the chamber along with the expected by-products. The exhaust scrubbing system must be constructed to handle all potential by-products.

The reaction rate of a plasma reactor is limited by a number of factors. Obviously, it is important to be able to control the reaction rate of the reactor in order to control the uniformity of the etch across each wafer and from wafer to wafer. These factors are essentially related to the reactive-ion bombardment rate and energy at the surface. The rate of the impact of the ions on the surface is limited by the total pressure (the total amount of gas available), and by the partial pressure of the reactive species (the percentage of ions in the gas stream that will react with the surface). As a result, two of the factors of etch rate control are control of gas mixtures and etch process pressure. The other major factor that affects the reaction rate of the films is that of ion energy. In general, higher energies will result in faster etches. The higher energies can be attained in a variety of ways, starting with the use of higher electrode potentials and higher frequencies of oscillation (although the frequencies at which plasmas form are sometimes very specific, in some cases it may be possible to form a plasma at a higher energy harmonic of the fundamental plasma frequency). Sometimes the reaction rate can be enhanced through heating the substrate. The major problem with increasing the reaction rates through increasing the ion bombardment energies is an increase in substrate damage due to the ion impact. In addition, boosting etch rates must be done with utmost concern that the etch process remains uniform.

One of the important features to bear in mind when designing the etch process is the selectivity of the particular chemistry to the various types of films that are exposed to the plasma. Incorrect choices could result in excessive amounts of undercut, improper edge slopes, or holes in films that should have remained unetched. A well-defined etch process might have selectivity between two films of 30 : 1, while a marginal process would have a selectivity of 10 : 1 or below. This number represents the ratio in etch rates between the two substances in question; for instance, the etch rate of polysilicon over SiO_2 in a polysilicon etch process. Selectivity can vary for a wide variety of reasons. Doped polysilicon will etch differently than undoped polysilicon. Bare aluminum films will etch at a different rate than aluminum films doped with silicon and copper. Thermally grown SiO_2 will etch much more slowly than deposited SiO_2, which will etch more slowly than BPSG films. All of these variables must be considered when designing the etch process.

Control of the slope of the walls of an etch trench is very critical. The slope required is usually as near to vertical as possible. This is in order to guarantee that each portion of the circuit can be manufactured as small as possible. If the slope is negative (undercutting, or creating an overhang), there will usually be a problems in subsequent steps. The problems develop when successive films are deposited or grown on the surface. In these cases, the space will allow holes to form under the films. They will become areas of inherent weakness and will eventually fail. Positive slopes are sometimes a desired condition. For instance, the sidewalls of trench capacitors in high-density DRAM devices usually have a slight positive slope so that the films can be grown or deposited with utmost care and so that the surface area of the capacitors used to store the memory charges can be maximized. This structure is shown in Figure 6-26.

Plasma Reactor Layouts

Several potential layouts can be used for the etch processes. The particular layout that is used depends on the type of process that the etcher will be used for. The older processes are still used for the less critical steps and for larger geometries. The equipment for these types

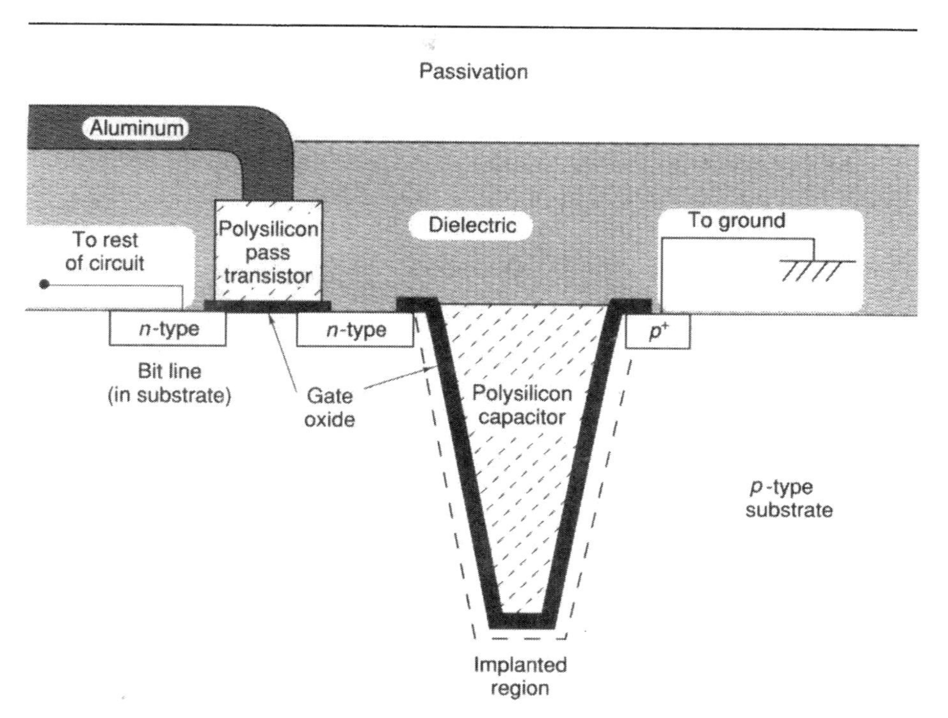

Figure 6-26. A Trench Capacitor DRAM Structure Comparing this to Figure 1-15, the trench structure reduces the surface area on the wafer required for each cell, while increasing the effective surface area of the capacitor. Advanced designs place the access/pass transistor above the cell in a double or triple poly layer structure for increased density.

of processes is also much less expensive than the newest versions of the plasma etchers.

The oldest version of the plasma etch system is called the barrel etcher, and is shown in Figure 6-27. The chamber and the boat (used to hold the wafers) are constructed entirely of metal, which unfortunately can add to contamination levels. A controlled stream of gases flows into the chamber and is exhausted from the other side. The plasma is formed or "struck" within the cylindrical chamber. The chamber is typically large enough to hold one boat of 25 wafers, which contributes to a high process throughput time. The typical process will take about 20 to 30 minutes to run, permitting a throughput of about 50 wafers per hour. The barrel etcher is excellent at high-speed wide-area etch and film removal processes, such as the photoresist removal and nitride removal steps. These types of processes do not require the precision of other etches in the photolithographic area. The barrel etchers are also able to etch lines for technologies down to about 2.0 to 2.5 μm with relatively good control. The disadvantage of this process is that the uniformity of the finished product will not be very good on linewidths smaller than this.

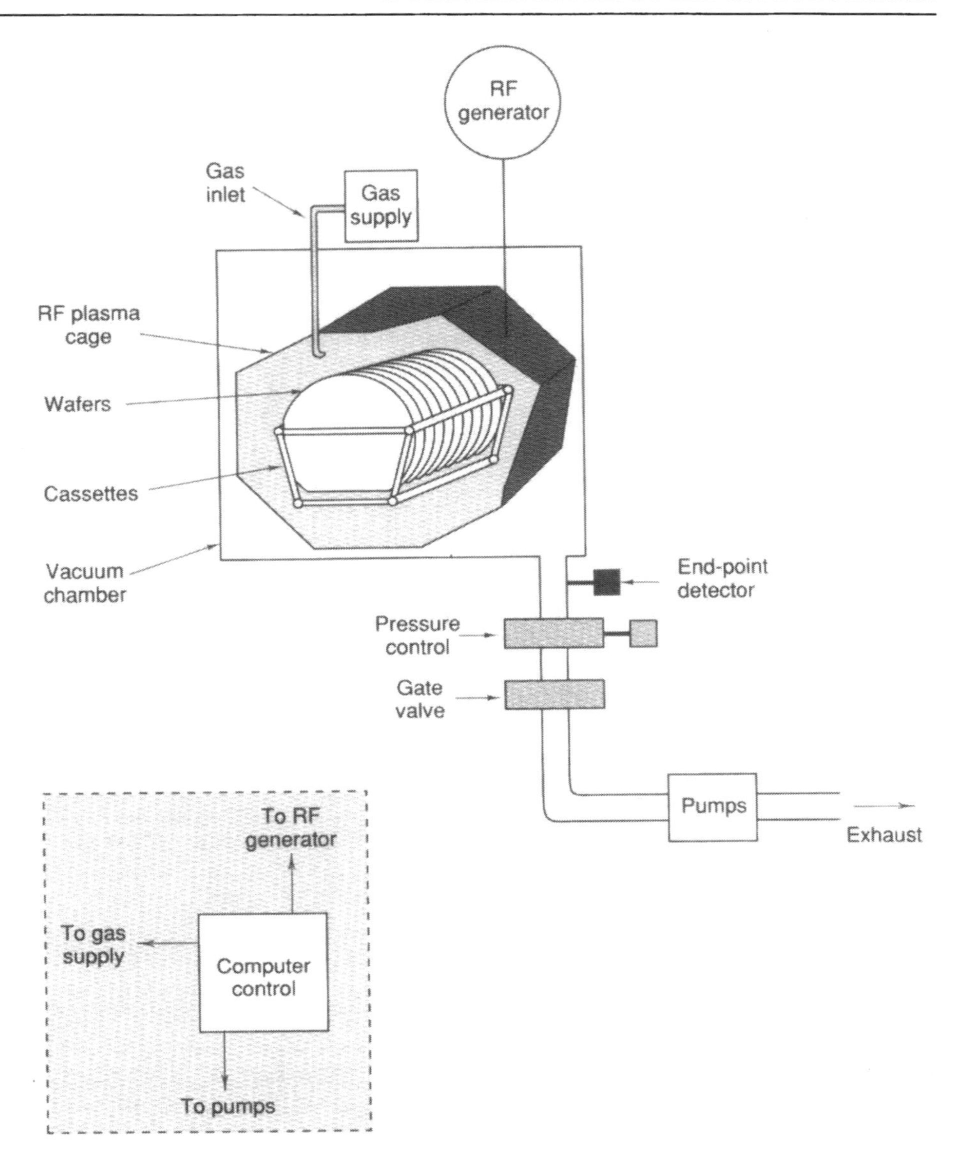

Figure 6-27. Barrel Plasma Etcher The earliest plasma reactors used in regular production areas were the "barrel" reactors, so-called because of the shape of the chamber. Twenty-five wafers are placed into the chamber, gases are sent into it, and a plasma is struck. The etch continues for a fixed time or until an end-point detector senses that all reactions in the chamber have ceased.

Another layout in common use in wafer fabs today is the parallel plate etch chamber, as in Figure 6-28. In this case the wafers are placed on the bottom plate of the chamber and the plasma is struck above the wafers. The gases flow into the chamber in a controlled fashion so that the pressures of the gases are constant at all times in the chamber. The electric field is generated across the plates with the wafers immersed within the field. This allows better uniformity and edge slope control since the active ions are flowing into the surface in a much more

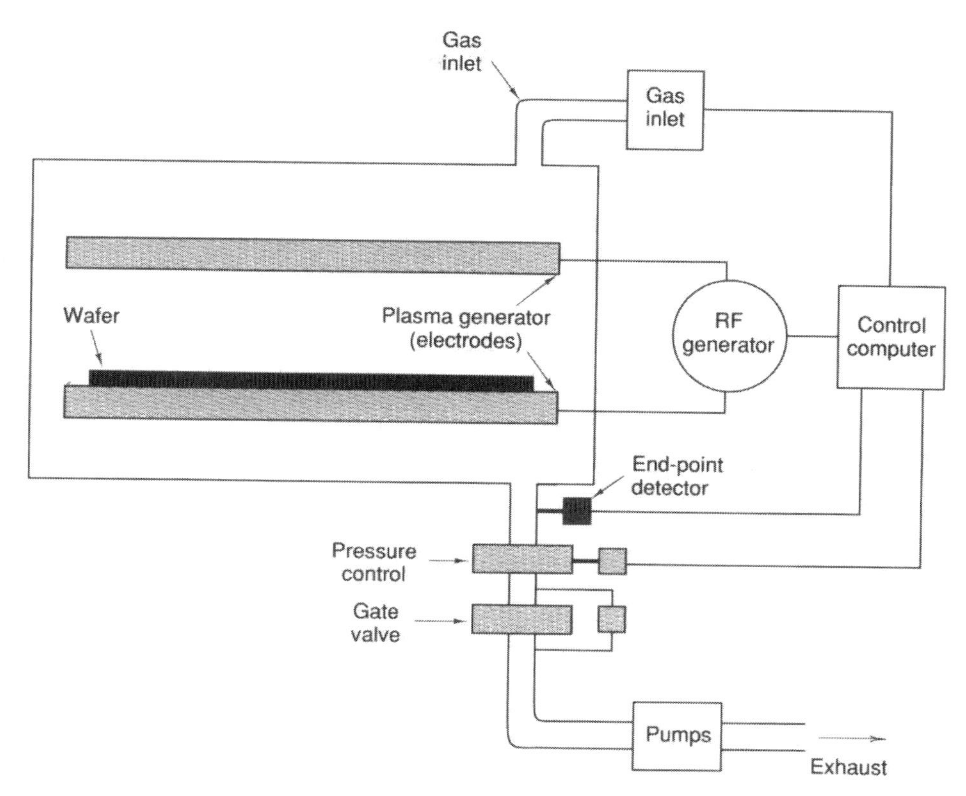

Figure 6-28. Parallel Plate Etcher The most common style of plasma etcher is the parallel plate etcher. In this system, the wafer is placed directly on an electrode which is biased with a positive charge to attract reactive species. While producing good uniformity, this system can produce a significant amount of radiation damage.

controlled fashion. In most cases, the ions are hitting the surface at a nearly perpendicular angle, as opposed to the barrel etcher, where the ions are coming in from all angles. Some etching machines allow batch processing, but an increasing number of etchers are single-wafer processors. This is due to the improved precision that is possible when processing one wafer at a time. A number of the newest machines use multiple chambers to allow for improved throughput. The parallel plate processor can produce lines in virtually any substance to well under one micron.

A third layout that has been developed is downstream plasma processing. This is also called an "afterglow" processing system. In this case, the plasma is produced in a separate chamber from the wafer and is then transported to the wafer chamber. This process has two advantages over parallel plate processing. The main advantage is that ion bombardment problems are minimized by keeping the wafers isolated from the electric fields required to generate the plasma. Another advantage is that the plasma can be delivered to the wafer in a more controlled fashion, thus leading to better overall process control.

These instruments work well to under 1 μm for most films. A device that works in a similar fashion is called a reactive ion etcher (among other names). In these devices, electromagnetic fields are used to cause reactive species to ionize and become much more active. In some cases, the results will not be a plasma as such but will work in a similar manner, by bombarding the surface with ions that will chemically react with it. The main goal of these more advanced processes is to reduce radiation damage to a device. The amount of radiation that can be absorbed by an integrated circuit is limited by the physics of the structure of the materials themselves and every dose of radiation introduced to the wafer will add to the total radiation dose and therefore slightly reduce the overall effectiveness and lifetime of the chip. The impact of damage caused by the radiation is covered in another section of the book.

6.2.3 Film Etch Process Types

This section gives a brief overview of some of the issues to be dealt with when etching a variety of different film types. It should not be assumed that this text covers all of the possibilities that are inherent within these processes. Etch is an ever-changing and dynamic field.

Photoresist Removal

This is probably the easiest of the plasma processes to discuss, although there are sometimes surprises even with these films. Photoresist is usually removed with an oxidizing type of plasma. Sometimes oxygen is used, other times pure or diluted ozone. The photoresist will then react to form CO_2 and water, along with various other by-products, depending on the exact makeup of the photoresist. Some of these other by-products are left as solid waste, thus leading to the term "ashing the photoresist." To remove this solid waste, photoresist removal processes are usually followed by sulfuric peroxide cleans to remove any excess photoresist ash.

Photoresist will become hardened to different degrees, depending on the processing it has received. If the photoresist has received only low-temperature bake processes and no ion implant steps, it can be removed from the wafers easily. Higher temperature bake steps will make the photoresist progressively more difficult to remove. The most drastic effect on photoresist hardening are the ion implantation steps. This is especially true of arsenic implant steps. Arsenic can harden some photoresists so much that 40 minutes of plasma oxygen ashing followed with 20 minutes of high-temperature sulfuric peroxide cleaning may not be able to remove all traces of the resist. Other elements harden the resist, but not as much as arsenic.

Since the photoresist removal process is a large surface removal process, is followed by sulfuric peroxide cleans, and is not usually critical to the yields on the wafers, particles are not usually a significant problem. However, the complete removal of the resist is of critical importance, since remaining resist could form nucleation sites for defects in CVD processes or pinholes or other defects in film growth processes.

Silicon Dioxide and Silicon Nitride Etching

Dielectric film etchers make up a significant percentage of the plasma etch processes. These films include silicon dioxide, the doped oxides, silicon nitride, and silicon oxynitride. While the chemistries for these etch processes are similar, the results can differ for various types of films. For example, the thermally grown SiO_2 films are much denser than the deposited SiO_2 films, which results in a slower plasma etch rate of the thermally grown films than that of the deposited films. In addition, deposited films are seldom as uniform as thermally grown films, which may result in uneven linewidths across the wafer on these wafers.

The deposited SiO_2 films can have a variety of dopants incorporated into the film. They include both boron and phosphorus, which are used to allow the film to flow more uniformly over metal steps. The existence of these dopants will change the etch characteristics of the film drastically. The nonuniformity of the dopant incorporation will add to the problems of etching the pattern into the films in a uniform fashion. Typically, the etch rates of the films will increase with increasing dopant content, and when the inconsistencies of film thickness, density, and dopant concentration are taken into account, an etch process for a doped SiO_2 film can be quite difficult to perform.

Silicon nitride also has several forms, which are all deposited, some using a high-temperature thermal method, others a plasma deposition method. Each of these forms differs in density and composition. The thermally deposited nitride films are denser and etch more slowly than the plasma deposited films. Usually, the thermally grown films don't have as wide a range in thickness or in uniformity as the plasma deposited films, and so have better etch results.

Silicon oxynitride films present a wide range of problems when an etch process is developed for them. First, the film itself can form a variety of combinations from mixtures of SiO_2 and Si_3N_4, to purer alloys of the film that can approach a uniform structure. Even in these forms, there can be slight variations in the exact quantities of the oxygen and nitrogen in the film. These variations can result in a variety of etch rates for various processes.

Plasma deposited films of all types tend to have a fairly high hydrogen content. On the Si_3N_4 and oxynitride films, the hydrogen can be attached to either the silicon atoms or onto the nitrogen atoms. This variation in hydrogen content can change the density and etch rate of the films in addition to the other effects already stated.

Most of the dielectric films involved in plasma etch processes are sensitive to particulate contamination. Particles can often lead to pinholes and the creation of preferential nucleation sites for future deposition processes, both of which can cause significant problems with yield consistency. Defects in dielectric film etches can often be related to reliability failures.

Uniform etching of dielectric films is a fairly complex science, and involves the balancing of a wide number of process parameters. Good control of the oxide etch process requires not only that the etch process itself is stable, but that the oxidation and nitridation steps are performed in a repeatable manner.

Polysilicon Etches

Polysilicon is one of most common materials for semiconductor devices, with uses as storage cells, resistors and conductors, and in other structures. In almost all of these cases, the polysilicon takes an active role in the operation of the integrated circuit. As a result, the etch processes that are used to define the structures in the polysilicon film must be very precise and well controlled. Defects in these processes can have severe consequences in overall yield.

Polysilicon is not a particularly easy film to etch. The film itself is made up of a multitude of miniature grains, which are of random orientation and size. While the orientation is not usually a factor in this sort of process, variations in grain size can make polysilicon etches quite challenging. The change in grain size can result in changes in etch rate, and sometimes even changes in uniformity, especially if the poly has been deposited in a poor, nonuniform manner. Sometimes there are very large grains and particles (which appear as "stars" under dark-field microscope inspection) embedded in the films. These spots can be very difficult, if not impossible, to remove. While attempting to remove them, there is often additional damage introduced into the film (if the etch time is extended to remove these particles, other parts of the film may become overetched).

Polysilicon is usually doped with phosphorus or boron to increase its conductivity. This procedure changes the grain size and the resistivity of the film. Both of these changes will result in changes in etch rate of the poly. Given the same etch process, doped polysilicon films will usually etch faster than undoped films. Polysilicon has a wide range of doping levels, which will change the etch rate accordingly. A film that has been uniformly saturated with phosphorus (a very common situation) will probably etch with good uniformity, since dopant levels and grain structures will be similar for all areas of the wafer.

Some polysilicon structures can influence the etch process itself. For instance, sometimes the etch rate of a polysilicon film will vary over a period of time in the etcher. At first, the etch rate will appear to be slow, since the etch process must first break through any native oxide that may have grown on the surface of the polysilicon. After that, the etch rate will reach a peak, and then start a very slow decline. This is explained by the fact that, for some structures of polysilicon, the film itself acts as one of the capacitor plates of the etcher. As large amounts of film are removed from the wafer, there will be a reduction in the ability of this film to continue to act in this fashion, which reduces the ion bombardment energy slightly, and therefore reduces the etch rate.

Clearly, any defects, such as film highlights (stars), particles, unwanted dopant species, and so on, can all affect the polysilicon film while it is in the etch process. Problems that result in incomplete film removal can result in the formation of short circuits. Problems resulting in overetch may reduce the polysilicon line's current-carrying capacity or reduce the ability of a storage cell to hold a charge. Contaminants in the gas stream can become embedded in the polysilicon and cause minor changes in the properties of the polysilicon structures, which can lead to eventual circuit failure or failure to reach operating specifications. Therefore, it is critical that the polysilicon etch machine is kept clean and under tight process control, and that the etch engineer understands the polysilicon deposition process as much as possible. Obtaining good results requires cooperation between the poly deposition processing personnel and the etch personnel trying to place patterns in the polysilicon.

Aluminum and Other Metal Etches

Metal etch can be a particular messy job, although good results can usually be obtained with modern etch processes. It is messy because the chemistries involved release a variety of toxic by-products, and there are sometimes problems associated with polymerization and other phenomena.

A significant issue when dealing with the etching of very thin metal lines is that the metals are by nature reactive and therefore will corrode rapidly in the atmosphere. There is usually a time limit of 8 to 24 hours that wafers can be left in the atmosphere after metal etch before the first passivation step. This corrosion occurs when the metal reacts with either water or chlorine residues (since many of the etching chemistries are chlorine-based). The corrosion properties are typically worse with multilayer metallization schemes, since there are often secondary reactions occurring already (e.g., titanium may be attacking aluminum), and is typically worse with Al-Si-Cu films than those of Al or Al-Si, since copper is highly reactive with chlorine.

Some solutions to these problems have been proposed. First is a relatively high-temperature bake to volatilize any remaining water or chlorine compounds. This procedure works well for extending the time span prior to corrosion, but will not prevent further corrosion as moisture slowly adsorbs back onto the wafer surface. Since water, and especially hydrogen peroxide, can initiate corrosion very easily, another method for controlling corrosion is to keep the wafers out of these solutions whenever the metal is exposed. This has the disadvantage of not permitting residual chlorine from being removed.

A third technique is passivation in a fluorine-containing plasma.

In many cases, the fluorine compounds will attack and volatilize the chlorine compounds, forming inert compounds. This passivation removes the corroding elements from the surface about as well as high-temperature bakes. Careful development of this reaction can permit a second beneficial effect, which is to permit the formation of a very thin fluorocarbon polymer, which protects the metal from the effects of moisture. Thus a careful postetch process can both remove residual chlorines and lay down a protective layer of polymer.

The metals come in a wide variety of types and combinations. I will not attempt to cover all of the combinations, although Table 6-3 does list some of the films and their various etch properties. Aluminum, for instance, can be used in its pure form, or mixed with silicon and/or copper. In each of the last two cases, the metal will become progressively harder to etch. In some cases, the metals themselves are tough and fairly impervious to common chemicals. Titanium films are examples of this case. There may also be effects due to the structure of the film itself. In a manner similar to that described for polysilicon, removal of the film from the surface of the wafer may cause changes in the etch rate of the process.

Some films and substrates may require that combinations of processes actually take place. For instance, it may be necessary to oxidize single-crystal silicon before removing the SiO_2 with the etch process, as opposed to attacking the silicon structure directly. This would be done to minimize potential crystal damage in the substrate. The exact technique to be chosen when developing an etch process must be determined by the requirements of the film and the effects of the chemistries involved.

A recent development has been the sputter-etch process. In this system, used primarily to create planarized films, the film is deposited and partially removed in a series of steps. The process has some promise in that the use of this technique significantly reduces the amount of effort and handling that would otherwise go into this process. The problem is that the technique is not especially clean,

TABLE 6-3
Metal Film Etch Process

Film	Gases	Etch rate (Å/min)	Selectivity
Al and alloys	BCl_3, CL_2, SF_6, $SiCl_4$, CHF_3	5000	20:1 over SiO_2, 5:1 over resist
Silicon	Cl_2, CCl_4, He	>3000	10:1 over SiO_2
Tungsten	SF_6, CHF_3, O_2	3000–8000	5:1 over resist
Ti-W	CF_4, O_2	700	3:1 over SiO_2
WSi_2	NF_3	2000–4000	20:1 over SiO_2
Poly-Si	SF_6/Cl_2, CF_4/Cl_2	2000–4000	>40:1 over SiO_2

with the etch portion of the process often leaving residual contamination on the wafer. If this contamination is coated with the film, it may build up into a serious particulate problem.

Also, as in the case of polysilicon, the metal films usually play a crucial role in the operation of the integrated circuit. Defects in these parts can result in improper current-carrying capacities, high contact resistance, aggravated problems with phenomena such as electromigration, and poor reliability. Particles and film defects often cause short circuits and other failures. Once again, there is a good deal of coordination required between the etch and deposition groups in order to guarantee the best results from any of the various metal etches.

6.3 PHOTOLITHOGRAPHY WRAP-UP

In this chapter, we have seen the various steps that are required to produce a pattern on a wafer. These steps are very critical and are always changing. The etch processes are in a particularly high state of flux. The ever-increasing demands for faster and more complex circuits reach their culmination in these steps, and it is the responsibility of the photolithography groups to manufacture the designs that are created. We have seen that the entire process is critical, from the deposition of the photoresist to the alignment and exposure of the wafer to the etch processes that actually define the patterns. The exact dimensions of all the structures must be maintained, and there must be no damage to other portions of the circuits while all of this activity is underway. This requires the utmost in care and precision.

Clearly, there have been great strides made in the field of manufacturing lithography equipment. Most of the problems that existed in the past few years have been resolved permanently through the use of more sophisticated equipment. However, due to the rapid change in the industry, many of these enhancements become outdated quickly, as new and more serious problems crop up. The next several years should continue to be very dynamic for the photolithography processes, as the push for submicron manufacturing intensifies. There will continue to be significant changes and improvements in the procedures and equipment used to develop these processes as the various companies and organizations (such as SEMATECH) set their goals on linewidth control to the 0.35-μm level and beyond.

CHAPTER 7

THE DIFFUSION PROCESSES

In order for an integrated circuit to exist at all, a series of films must be placed on the wafers. These are the films that are then patterned in the photolithography sequence. The films can be placed on the wafers or they can be grown from the wafers themselves. In the first case, the processes are usually called "LPCVD" or "thin-film" processes, while the growth processes are usually called diffusion processes. There is often a certain amount of overlap in responsibility in these two areas in a real manufacturing area. However, for the purposes of this book, we will discuss the two types of processes separately. Thus, the next two chapters cover each of these two aspects of processing, starting with the discussion of the diffusion processes.

The diffusion processes are best defined as those processes which permit or force the movement of atoms through the structure of the wafer itself. The term *diffusion* is derived from this activity. There are several processes that fall into this category, including the oxidation processes, the dopant deposition and redistribution processes, the anneal processes, and the alloy/sinter processes. In all of these cases, the wafers must be heated to a high enough temperature that the dopant atoms will redistribute themselves within the crystal structure. With silicon, the temperatures required are usually above 800°C, and often closer to 1000°C. Gallium arsenide diffusion processes must be carried out at a much lower temperature to ensure that the GaAs does not dissociate. The diffusion processes in a gallium arsenide fab range around 400°C.

Contamination in the diffusion process is a serious issue. The high temperatures of these processes can allow chemical contaminants to spread throughout the wafer, and can help to redistribute particles and mechanical damage as well as the chemical contamination. In addition, most diffusion processes are batch processes by nature, often allowing 100 to 200 wafers to be introduced to a furnace at any one time. Contamination can be spread from wafer to wafer quite easily in these environments. As a result, it is extremely important to maintain cleanliness at all times in the diffusion area to prevent damage to very significant numbers of wafers.

In this chapter, we will review the major subgroups of processes, the equipment that is used to run them, and techniques for keeping the processes clean and for controlling them. We will also discuss some of the film characteristics that are required, as well as some of the recent trends of diffusion processing.

7.1 DIFFUSION PROCESS EQUIPMENT

Most of the equipment for the processes to be described in this chapter is similar in most respects, varying primarily in the setup of the gas system and the temperature range. The main factors involved in the design of a successful diffusion furnace are the temperature control mechanism and the gas control system, along with the system process controller. Other items that must be considered include such system features as wafer handling techniques, cassette styles, and so on.

7.1.1 Diffusion Furnaces

A typical horizontal diffusion furnace is shown in Figure 7-1. The system consists of a number of subsystems, which are related to the gas, temperature, wafer handling, containment, or control aspects of the diffusion process. In this example, we are describing a horizontally mounted furnace, which is the most common type of furnace in use today. This preference is changing to that of the vertical reactor as space requirements become more critical and wafers become larger and heavier. Balancing a load of 100 eight-inch wafers on a horizontal loader is no simple task. Figure 7-2 shows a diagram of a vertical diffusion furnace.

We will start with the fused quartz tube, capable of handling temperatures of up to 1300°C. Everything in the process that comes into contact with the wafers should be made of quartz or other inert material. Fused quartz is used due to its material (SiO_2), the ability to manufacture it with required impurity levels, the wide temperature range that it can handle, and its relatively lower cost and ease of manufacture. The tube must have a diameter large enough to fit the wafers, the cassettes used to hold the wafers, and the wafer transporter and thermocouples and other instruments that may be required in the processing chamber. Usually, a tube has to be about 20 to 25% larger than the diameter of the wafers in use, perhaps more. It must be large enough so that no part of the system touches any other part of the

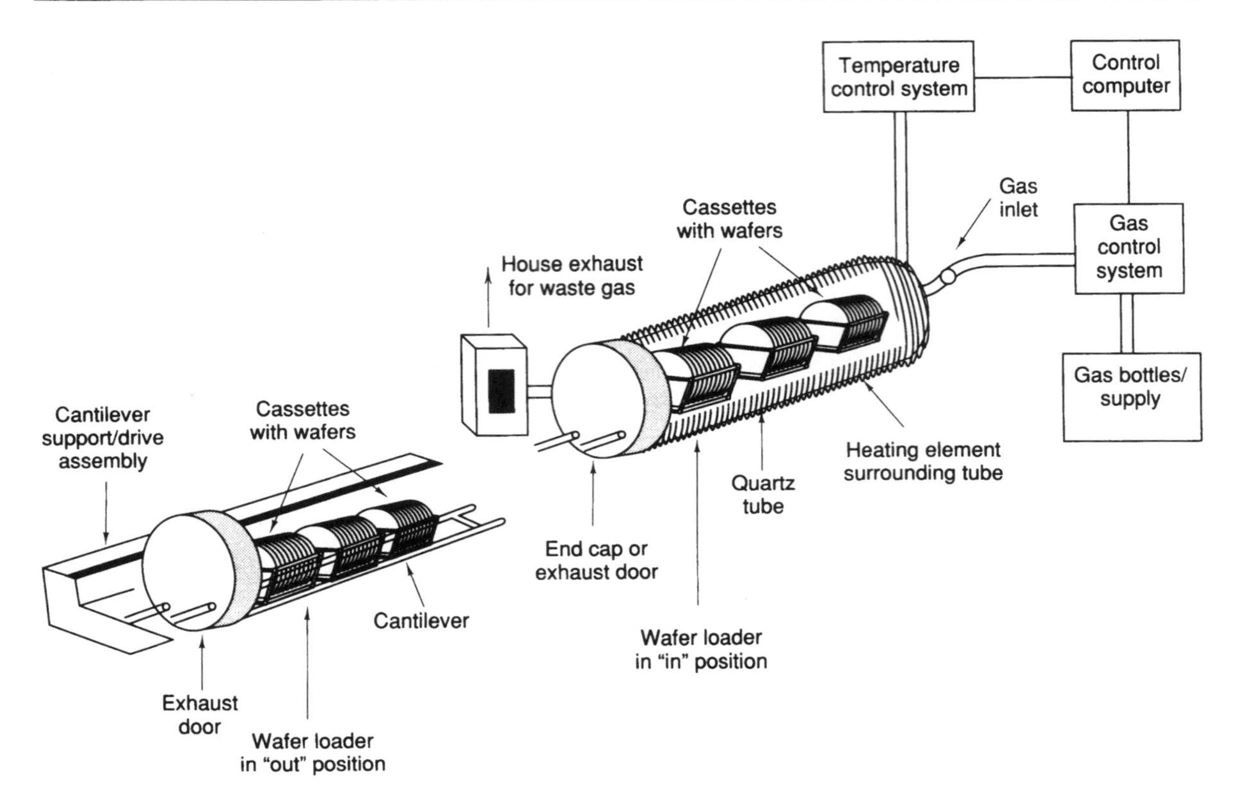

Figure 7-1. Horizontal Diffusion Furnace Layout This is a cutaway view of the horizontal diffusion furnace. The gas-control and temperature-control systems are shown, along with the wafer loader system, shown in the standby "out" position and in the active "in" position. Hot exhaust gases are trapped and vented away before reaching the cleanroom.

system. How well that can be accomplished depends on the type of wafer loader in use. It also must be long enough to allow a stable temperature-controlled zone in the center that is large enough to be useful (i.e., hold a lot of wafers). Diffusion furnaces are often eight feet or more in length.

The diffusion tube can be one of the most significant sources of contamination in the wafer fabrication area, as it can easily be contaminated with chemicals from a wide variety of sources. They can include the solutions used to clean the quartz (which should be as clean as the chemicals used to clean the wafers), hand marks and other handling problems, and contamination that is present in the heating elements of the furnace and even from the atmosphere. Chemical contamination can absorb into the quartz, where it resides until driven out at high temperature. Contamination can be introduced into the

→

Figure 7-2. Vertical Diffusion Furnace Layout The vertical diffusion furnace has the same components as the horizontal furnace layout, but has advantages with reduced leverage on the wafer loaders and reduced particle counts.

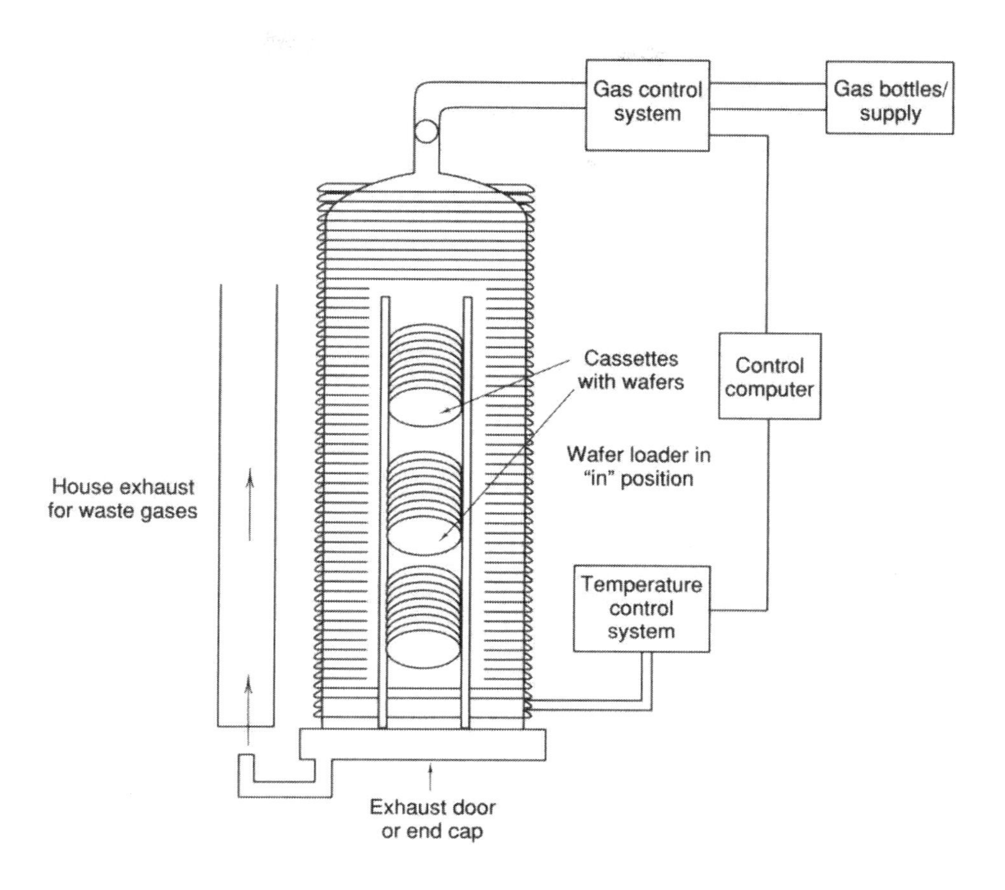
Gas control system
Gas bottles/ supply
Control computer
Cassettes with wafers
Wafer loader in "in" position
House exhaust for waste gases
Temperature control system
Exhaust door or end cap

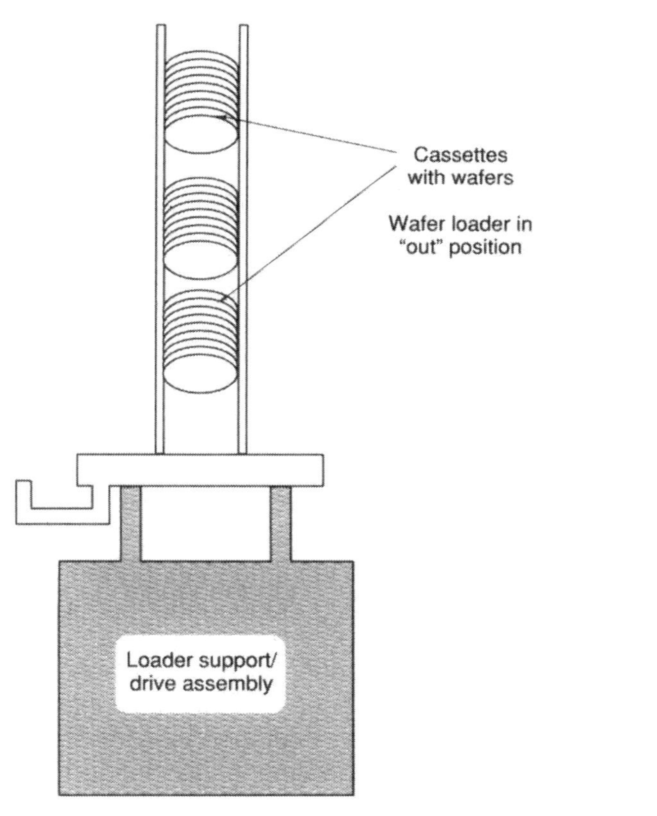
Cassettes with wafers
Wafer loader in "out" position
Loader support/ drive assembly

tube while at temperature (e.g., by melting a glove on the quartz), which can continue to be redistributed to the wafers for quite a while. Chemical contamination is often leached from the surface in special "steam" cycles, in which the temperature is raised and a steam and HCl mixture is run through the tube.

Another problem with quartz is that it deteriorates with time. When it is first manufactured it has a highly polished surface produced by running a special blowtorch over the surface of the part. The high-temperature cycling that the quartz experiences causes a significant amount of stress in the quartz structure. Impurities and contamination enhance the damage to the structure of the quartz. This process is called *devitrification* and will occur with all quartzware given enough time. The use of HF to clean the surface will increase the rate of devitrification. This is a particularly serious problem in thin-film deposition processes where films must actually be removed from the surface of the quartz. The phosphorus deposition tube also tends to have significant problems with rapid devitrification of the quartzware.

There are a number of issues that relate quartz devitrification to reductions in wafer quality. For example, the surface damage will allow chemical contamination to be absorbed more easily. It will also allow chemical contamination that is trapped in the quartz to be outgassed more easily. In addition, there will be an increase in particle generation as all of the damage surface edges break off. These particles can easily be transferred to the wafers. Finally, the damage in the quartz can weaken the integrity of the tube, which can lead to increases in cracks or breakage.

Cracks in the quartz will often lead to the influx of atmospheric gases into the quartz tube. This can cause several problems, depending on the exact nature of the process in question. When the process is an oxidizing one, in which the ambient is all steam or all oxygen, the nitrogen from the air will dilute the oxidizing species and slow the oxidation rate. In the event of an anneal process in an inert ambient, the oxygen and moisture from the atmosphere will cause some oxide growth. This could prove to be a serious problem in certain circumstances. In addition, it is also possible to get other contaminating species into the furnace through cracks. Clearly, major quartz damage will cause even more serious problems, from contamination of the gas stream to massive changes in the flow characteristics of the furnace.

The wafer loaders for horizontal furnaces come in two basic varieties, as shown in Figure 7-3. There is the older system, which relies on carriers with wheels (called boats) which could be rolled into the furnace. They produced huge amounts of contamination, could get stuck in furnaces, especially on phosphorus and other deposition processes, and sometimes broke and caused other problems inside the furnaces. These systems still exist, but are seldom used for anything but the oldest technologies or for basic research. The more modern method of handling wafers is in the form of a cantilevered loader. This

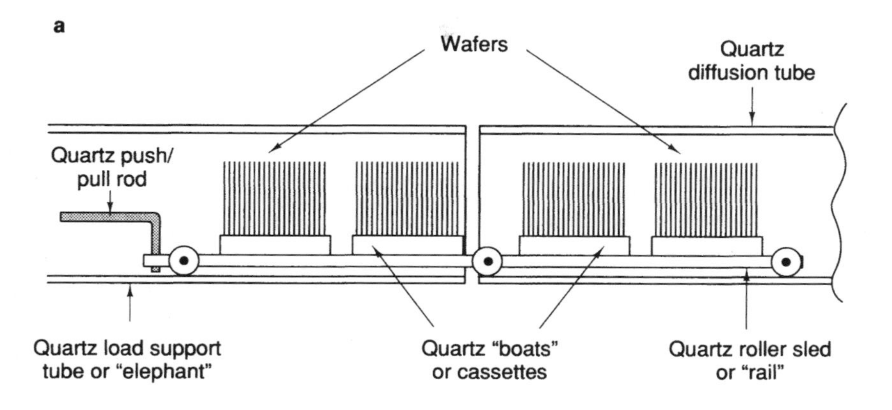

b

Quartz end cap/ door
Wafers
Quartz diffusion tube
Quartz securing bracket
Quartz "boats" or cassettes
Quartz securing bracket
Silicon carbide cantilever rod
Quartz cantilever sleeves

Figure 7-3. Quartzware for Diffusion Furnaces (a) An older, more manual type of quartzware loader for the diffusion furnace. Here the wafers are placed in quartz cassettes, which are placed on the roller sled. This is carried to the furnace in a short tube called an "elephant." The wafers are then pushed into the furnace for processing. An endcap is placed by hand over the end of the tube prior to processing. (b) The modern noncontact (except for the integral end-cap assembly) cantilever system. The wafers are loaded into the cassettes, which are then placed directly onto the cantilevers. The wafers are then pushed automatically into the furnace.

is a simple method to conceptualize, but a difficult one to implement. It took several years and several false starts before the industry received reliable cantilevers. The cantilevers help prevent contact between the various parts of the system, and can support thermocouples and other gear. The alignment of the cantilever to the tube is critical, as misalignment can cause the cantilever to bang into the side wall of the tube, creating particles. This is an even more serious problem if there is already a problem with devitrified quartzware. The typical cantilevered wafer loader must be made out of very expensive combinations of silicon carbide and alumina with quartz linings (see Figure 7-4). They are usually placed in some concentric fashion to provide mutual support. Sagging under the weight of the wafers at temperature is a common problem that can be solved only with the use of ceramic materials. The quartz lining is required because the ceramic materials, especially silicon carbide, are very rough and contain significant amounts of impurities which can contaminate wafers that come into direct contact with the ceramics. All of the materials used in the cantilever systems are brittle and can be broken quite easily. This is

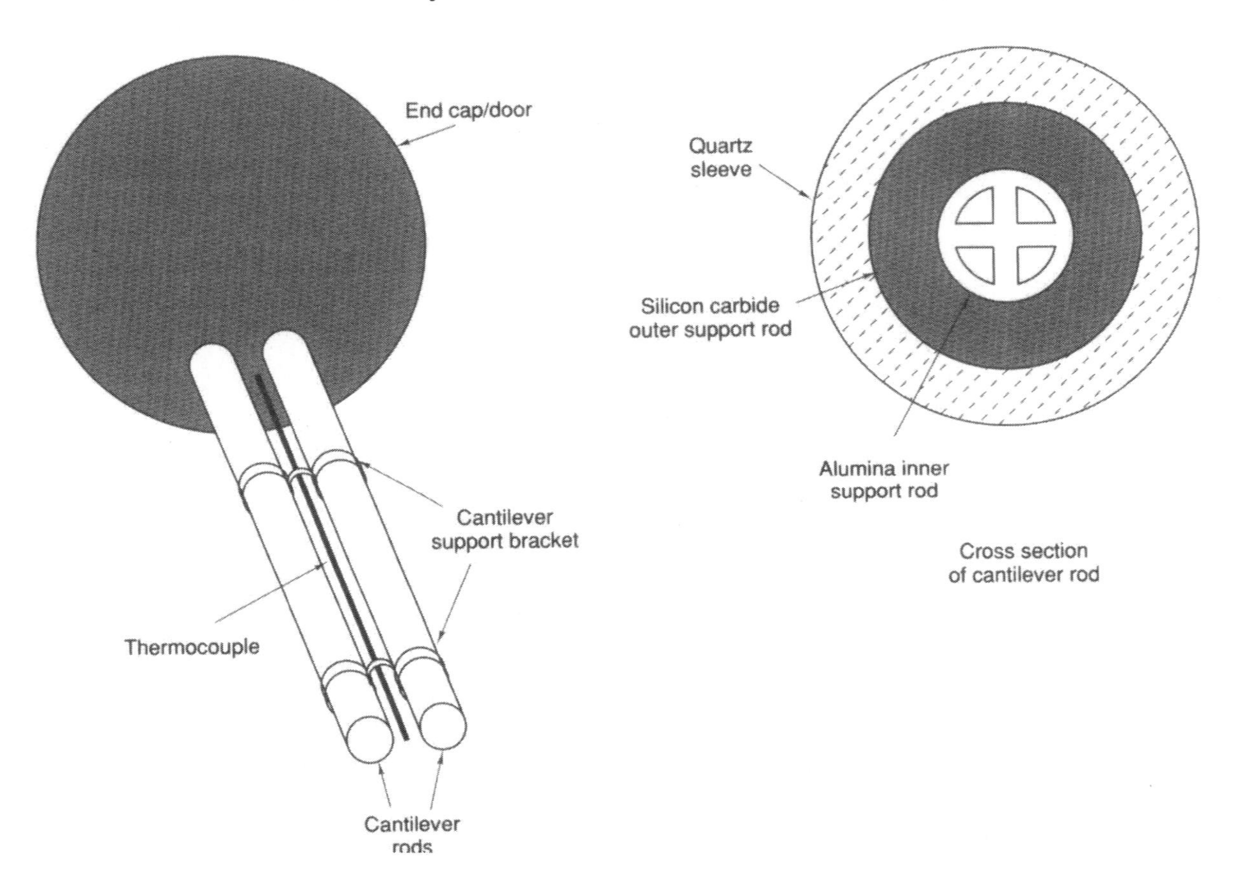

Figure 7-4. Cantilever Rods The cantilever system consists of the endcap or door, the cantilever support rods, two or more cantilever support brackets, and a multijunction thermocouple. The surfaces directly exposed to the high-temperature ambients are all made of quartz. The cantilever sheaths are made up of an alumina inner rod inserted into a silicon carbide cylinder. While single-material cantilever supports are often used, the combination of these materials, designed and optimized for strength, produces superior resistance to warpage, sag, or breakage. Both of these ceramic materials can outgas a variety of contaminants, so must be covered with a quartz cantilever sheath.

especially true if the parts are damaged by cleaning in strong HF or other improper chemical. Some cantilevers consist of only a silicon carbide paddle that must be prepared for use carefully.

All the parts of the system that are brought into the wafer chamber are typically made of quartz or enclosed in quartz sheaths. These parts include the wafer support cassettes, the wafer loader cantilevers, the thermocouples, and any gas injectors that may be required. All of these quartz parts will undergo the same stresses that the tubes will undergo, including devitrification. This is an especially acute problem when dealing with the wafer support cassettes, as the slots that the wafers are placed in can be damaged easily and can generate large numbers of particles. This is seen in Figure 7-5. Cassette slots can cause a number of problems when the boats are not maintained properly. The slots may become roughened, permitting particulate generation, or they may become too wide. This causes the wafers to tip against one another, ruining the local gas flow characteristics.

The next most significant subsystem of the diffusion system is the gas delivery system. This system consists of the main storage bottles, the gas delivery lines, line pressure regulation, mass flow control, and furnace injection. These systems are very critical, as leaks in them can easily cause contamination or dilution of the gas stream, and significant changes in line pressures. Many gases react violently with contact with air. For example, silane (SiH_4), a common gas used for the deposition of many types of films, ignites spontaneously in air to form SiO_2

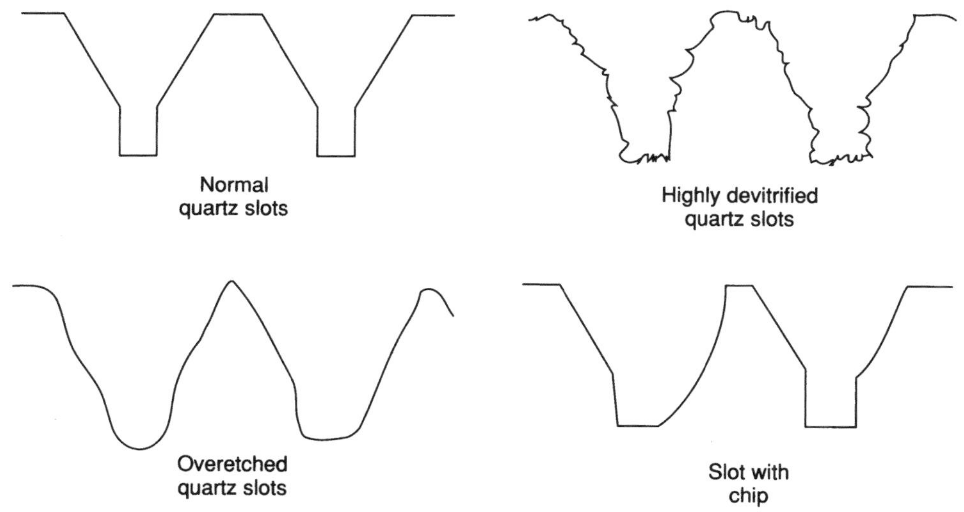

Figure 7-5. Deterioration of Quartz Cassette Slots Quartz cassettes deteriorate rapidly with use. Even careful handling will eventually result in the loss of quartzware. Normal slots have a specified shape, as shown. The quartz can be damaged in the high-temperature processes by warpage and contamination. Cleaning and etching quartzware causes both increased devitrification of the quartzware and the widening of the slots. The quartz can also be chipped or cracked. These areas cause increased particle contamination and/or increased wafer tipping, which can affect process gas flow and consequent wafer uniformity.

and H_2O. If there is a leak in a silane line, air may get into the line and react with the silane to form powder. This powder will plug filters, changing pressure levels, and can eventually permeate the entire system. Depending on the location of the leak, the wafers can become seriously contaminated in seconds. Tolerable leak rates are typically in the range of 1×10^{-9} cm^3/min, and must be tested periodically. Leaks tend to occur in fittings that are handled often or are in difficult areas to reach. Also, line pressure control is important in order to verify that the mass flow controllers (MFCs) operate properly. Most of the common MFCs on the market require a minimum pressure drop of 10 psi across the MFC to measure correctly. If the total gas system pressure drop is less than 10 psi, there will be improper control of the gas flows. Unfortunately, many MFCs will register that they are flowing correctly in this condition, making analysis of the gas system problems very difficult at times.

Gas flows must be optimized for the furnace size and exhaust requirements. While excess gas in the furnace does not usually cause a problem, insufficient quantities can lead to a number of uniformity problems. Improper flow can lead to areas of localized partial pressure imbalances which will give nonuniform readings both across any wafer and from wafer to wafer. It is also important to keep the overall gas flow rate in the tube high enough to prevent any potential backstreaming (atmospheric gases flowing back through openings in the endcap, diluting the gas stream). Gas flow rates will increase for larger furnace sizes. Typical total gas flow rates are around 8 to 12 SLM (standard liters/minute) for four-inch wafer processes (using five-inch tubes), up to 15 to 20 SLM for six- to eight-inch wafer processes.

Another critical subassembly of the diffusion system is the temperature control system. There are a number of issues to keep in mind here, including thermocouple (TC) selection for proper temperature detection. Table 7-1 gives a list of the proper types of thermocouples to use for the various ranges required. Use of the improper thermocouple type can result in processing temperatures considerably far from what is reported. For instance, in a specific case known to the author, a B-type TC was used in a furnace in lower temperature ranges then

TABLE 7-1
Thermocouple Requirements for Wafer Fab

Thermocouple type	Junction	Temperature range (°C)	Processes
B	Platinum/Pt–rhodium	800–1700	Diffusion, oxidation, RTP
R, S	Pt/Pt–Rh	0–1450	Alloy, CVD, PECVD, etch
J	Fe/Constantan	0–750	Alloy, CVD, PECVD, etch
C, G	W/W–rhenium	400–2300	Rapid thermal processes

specified. When the error was discovered and the proper type of TC inserted, it was discovered that the temperatures were over 30°C too low.

Another issue that must be controlled is the temperature response of the furnace. When the load of wafer is inserted into the furnace, the temperatures must be restabilized, and then the entire furnace taken to the processing temperature. The delay from the time of load insertion to stabilization should be characterized carefully, as well as the time it takes to stabilize the temperature after the wafers are taken from the standby temperature to the process temperature. Typically, there are a number of user-selectable values which will adjust the rate at which the furnace will apply power, and at what temperature it will start to respond. See Figure 7-6 for a sample temperature response curve. Very little advice can be given as to exact procedures to follow for this type of optimization, simply because every furnace and control system varies considerably. In general, the best type of test is to make the furnace take a step-function temperature change of a few degrees. The response values should be changed and the temperature response evaluated. The temperature response through time after that change should be recorded. This curve is used to analyze the response of the furnace and becomes the analog of the damped harmonic oscillator, with adjustments being made to adjust the damping forces. The goal of the testing sequence is to reach a stable equilibrium that allows long-term stability and minimizes the damping time in order to maximize throughput.

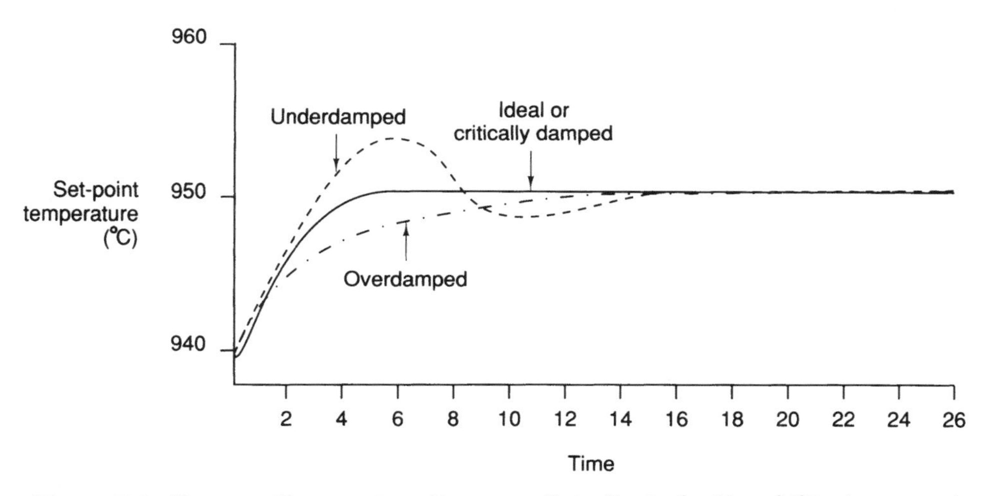

Figure 7-6. Furnace Temperature Response-Rate Control The ability to prevent temperature overshoot is important in controlling uniformity. Since processes usually run at temperatures higher than the standby temperature, a "ramp rate" in the range of 2 to 10°C per minute is applied. Control parameters can be adjusted to affect the ability of the system to home-in on the set-point temperature. These settings vary the amount of dampening of the control circuit/algorithm. Clearly, ideally damped systems will have no over- or undershoot, and will have reduced cycle times, and improved process repeatability.

There are a number of other components that constitute the diffusion system. For example, most systems now incorporate automatic systems to push the wafer load into and pull it out of the furnaces. This was not the case 5 to 10 years ago, when operators were required to push the wafers into the furnaces. This caused wafer warpage, contamination, excess breakage, and other related problems. Once the automatic push/pullers were developed, they rapidly replaced this operator task in most fabs. The devices brought in a series of problems of their own, however. For instance, the loader mechanism can fail with wafers in the furnace, trapping them there, and sometimes causing loss of the entire load. Misalignment of loader systems is often related to wafer damage or loss. Finally, the lead screws that are used to drive the cassettes into the furnace can produce huge amounts of particulate contamination. Early systems incorporated horizontal laminar flow systems to try to keep the air clean around the load area, without recognizing the effect of the lead screws, only to discover that the particle counts were terrible.

The final subsystem that we will discuss is the computer control system. There are a wide variety of types of these systems, running under a number of different operating systems, so only general statements can be made. As outlined in Figure 7-7, most diffusion systems come in groups of three or four tubes, called three-stacks and four-stacks. In most cases, there is a control computer for each furnace and a central computer that supervises the individual control computers. The control computer monitors the temperature and gas flows at all times and makes adjustments as required. It also will control the position of wafer handling systems and execute the appropriate sequence of events for the process. Usually, there is little one can do with the control computer directly, except in maintenance modes in which there is total control over every system, but at a very low and extremely manual level.

Programming and preparing the control computers is usually the job of the central computer. The central computer will normally have a disk drive for recipe storage, and will run an industry-standard operating system, like MS-DOS or UNIX. It usually has an editor, communications software, configuration software, and a variety of other tools that are required to make the job of recipe creation easier. Typically, the recipes should be created in a fashion similar to that used by programmers for writing professional programs. First, the basic functions are created. These are called the "recipe primitives." For instance, the primitive program "O2_ON" (Figure 7-8), turns on a number of gas valves, and sets the MFCs appropriately. The program "PUMP_DN" sets up a vacuum system for a pump-down cycle. There will be quite a number of these little modules, one for each function that is desired.

Above this level are the "recipe modules." They are portions of recipes that are common to a variety of recipes. For example, the

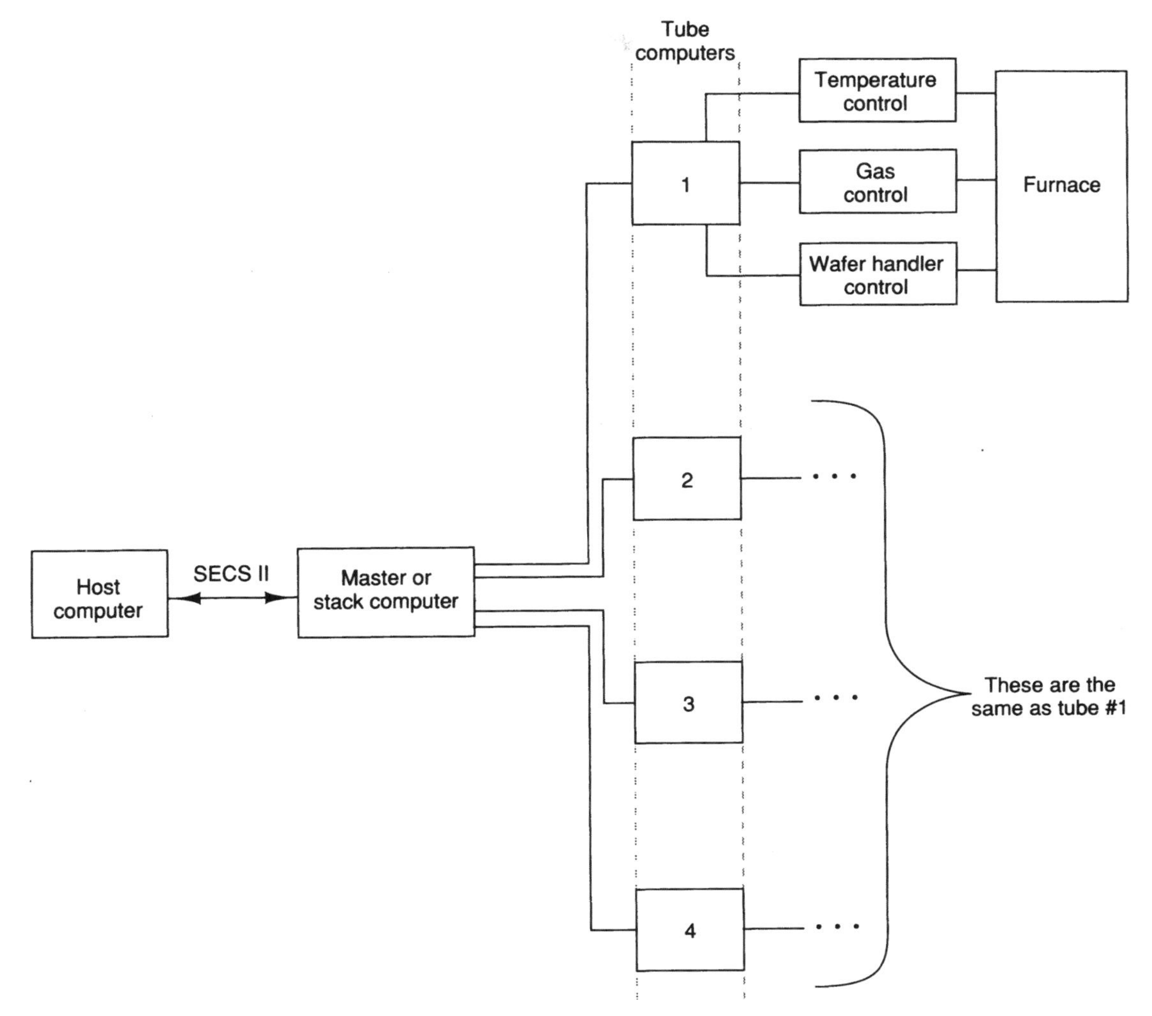

Figure 7-7. Block Diagram of Diffusion System "Four-Stack" This is a functional diagram of the typical diffusion system setup, called a "four-stack." The recipes or programs are usually stored in the master computer and are transferred to the individual furnace computers, which execute the programs and control the temperatures, gas flows, and other process variables. The stack computer sends data out to fab host computers using the SECS II protocol.

module "START_OX" (Figure 7-9), runs the sequence of the recipe primitives required to get the load of wafers into the furnace, stabilize the standby gases, and send the furnace to the correct temperature. There will be a variety of these modules which can then be used for the recipes.

The "recipe" level uses the various modules and an occasional primitive routine to build the various programs. These are then used to actually process wafers. "GATE_OX1" in Figure 7-10 is an example of a full-fledged recipe for a gate oxidation process. Actually, this programming technique will work for a variety of different processes

Primitive modules

O_2_ON	O_2_OFF	PUMP_DN
Open O_2 upstream valve Set O_2 MFC ramp rate Turn O_2 MFC on to set point Open O_2 downstream valve Monitor O_2 flow to stability (5–10s)	Set O_2 MFC ramp rate Turn O_2 MFC to 0 (zero) Close O_2 downstream valve Close O_2 upstream valve	Shut off standby gases (N_2) Verify chamber closure Open "slow-pump" valve Wait until pressure under 5–10 torr Open main gate valve Close "slow-pump" valve Turn on purge gases (low flow N_2)

Figure 7-8. Low-Level Primitive Modules This English pseudocode shows typical low-level (or "primitive") modules. Each low-level routine should consist of the series of steps required to achieve one particular goal. Each of these primitive modules is named with a mnemonic title that describes the function performed.

			Comments
START_OX:	N_2_O_2_ON	:	Turns on standby gases
	PUSH_WAFERS	:	Pushes load of wafers into the furnace
	SET_TEMP	:	Set temperature to process temperature
	STAB_TEMP	:	Stabilize to standby temperature

Figure 7-9. Recipe Modules Primitive modules are linked to form various small recipe modules, which perform larger tasks that are common to various recipes. For instance, most oxidation processes can be started using a recipe module similar to that shown here.

			Comments
GATE_OX1:	START_OX	:	Sets standby gases, temperatures, pushes wafers into tube
	SET_GATE_TEMP	:	Sets temperature to process temp, and stabilizes
	O_2_ON	:	Turns on oxygen and shuts off standby gases
	H_2_ON	:	Turns on hydrogen
	PROC_TIME	:	Holds for process time then shuts off O_2 and H_2
	ANNEAL	:	10–30 minute anneal at process temperatures
	RAMP_DOWN	:	Reduces temperature to standby temperature
	PULL_WAFERS	:	Pull wafers from tube, gives alarm when cool

Figure 7-10. Process Recipe As shown in this example for a gate oxidation process, the recipes are easily understood. The mnemonic name for each module identifies the action taking place, and the recipe sequence is obvious. Without this scheme, long programs become difficult to read and to change, or general tables of data must be used, which can restrict the flexibility of the systems.

and can be used in other processing environments, such as in the etch or implant processes.

7.2 DIFFUSION PROCESS SUBGROUPS

The diffusion processes are among the oldest and best understood of the semiconductor manufacturing processes. Many of the techniques for running the diffusion processes are descended from older metallurgical techniques. As a result, there is a tremendous amount of material describing the theory of the diffusion processes in detail. However, this literature does not often describe specific manufacturing problems or how the various techniques that are discussed can be implemented into a manufacturing environment. As a reference guide, a variety of various attributes of the diffusion processes are described in Table 7-2.

TABLE 7-2
Diffusion Processes

Process	Temperature (°C)	Time (hrs)	Gases	Purpose
Initial oxidation	925–950	2–2.5	N_2, H_2, O_2, TCA	Surface protection, first mask
Polysilicon oxidation	900–920	2–2.5	N_2, H_2, O_2, TCA	Insulates polysilicon layer
Field oxidation	950	6–9	N_2, H_2, O_2, TCA	Insulation and gate isolation
Gate oxidation	700–950	2–2.5	N_2, H_2, O_2, TCA	Transistor gate dielectric
Source/drain oxidation	700–950	2–2.5	N_2, H_2, O_2, TCA	Transistor source/drain (usually etched)
Sacrificial oxidation	900–925	2–2.5	N_2, H_2, O_2, TCA	Protection from implant damage
Tunnel oxidation	700–900	2–2.5	N_2, O_2	Programming EEPROM devices
Phosphorus deposition	800–820	1.5	N_2, O_2, $POCl_3$	Poly, substrate *n*-type doping
Phosphorus diffusion	900–950	2–5	N_2, O_2	Redistribute *n*-type doping
n-well drive	1050	4–12	N_2	Create *n*-type region for CMOS
p-well drive	1100	8–12	N_2	Create *p*-type region for CMOS
Other anneals	900–1000	2–5	N_2, O_2	Activate and drive implants
Alloy/sinter	400–420	0.5	H_2	Create good contacts
Contact diffusion	850–950	3	N_2, O_2, $POCl_3$	Saturate contacts with *n*-type dopant
Getter	850–950	3	N_2, O_2, $POCl_3$	Draw contaminants from substrates, films
High pressure	950–1100	0.5	N_2, O_2	High-pressure anneal, nitridation, and oxidation

Most of the processes that are encountered in the diffusion area can be simulated using a variety of computer analysis techniques. One of the oldest and best known of these is SUPREM, which models complex devices by calculating the dopant (or diffusing atom) concentrations, and determining the motion of the atoms through time at each of the temperatures encountered during the process.

In this section, we will break the diffusion processes down into their various subgroups, and discuss some of the characteristics of each.

7.2.1 The Thermal Oxidation of Silicon

There are almost as many theories and models relating to the thermal oxidation of silicon as there are theoreticians looking at the problem. Despite being one of the most widely studied of all processes, there is often little agreement on the exact methods by which the silicon oxidizes. Even the standard oxidation growth formula (the famous Deal–Grove equation) contains "fudge factors" to allow it to accurately predict the growth rates of SiO_2 over the entire range of thicknesses. We will attempt to reduce these many offerings into a single coherent package so that the individual involved in manufacturing can have a good understanding of the process.

Thermally grown silicon dioxide has three or four primary uses in a wafer fabrication area. It is used as the basic insulating dielectric to isolate all of the various areas of the chip, which is known as the field oxide. In this case, the oxide thickness is approximately 8000 to 12,000 Å and covers a significant portion of the area of the chip. This oxide layer is very strong and can usually handle several hundred volts or more prior to breakdown.

The next type of SiO_2 structure is the polysilicon oxide structure. This is used mostly to isolate the various polysilicon and metal layers from the substrate. It is usually grown from the polysilicon (thus resulting in the loss of a certain amount of the polysilicon), and is typically not of the same high quality and density of the field and gate oxide steps. The thickness of the polysilicon oxide/interlevel dielectric is typically 1500 to 4500 Å.

Another type of structure is the gate oxide. This film is usually from 200- to 500-Å thick and is used to isolate the polysilicon gate from the source and drain region in the substrate (see Figure 7-11). The main features required for this structure are consistency in thickness and in breakdown voltage capability so that the devices will operate with predictable speeds and for acceptable periods of time. Closely associated with gate oxides are the so-called source and drain oxides. These are either oxides grown to protect the contacts and are later removed, or they are used to protect the sidewalls of a structure from breaking down. These films are normally only a few hundred angstroms thick, and are generally not as critical as the gate oxides.

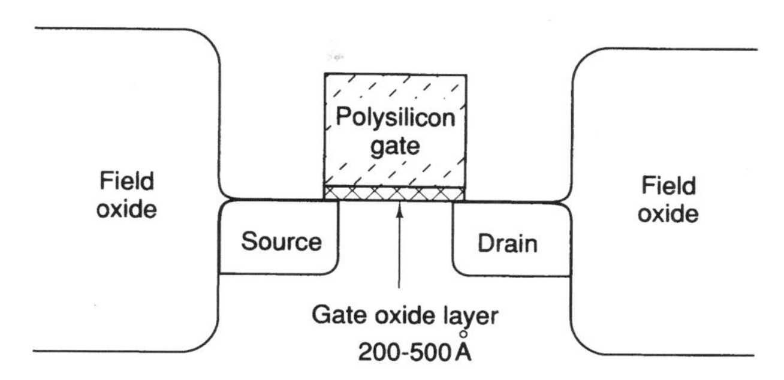

Figure 7-11. Location of Gate Oxide on Transistors The gate oxide layer is a thin silicon dioxide layer placed at the junction between the source and drain areas of a transistor.

The last type of oxide layer that is commonly used is the tunneling oxide. This is generally around 100 Å and is thin enough to allow Fowler–Nordheim tunneling to take place for charging and discharging floating storage cells on EEPROM devices. It is by far the most sensitive of the SiO_2 films. The main quality that is required for this process is that the oxide have very high purity and integrity, as any defect will reduce the lifetimes of the film.

There are other uses for SiO_2 that do not involve an active role in the circuit. For instance, silicon dioxide is also used as a base upon which patterns can be etched into the wafer. These patterns can be for the oxide or they can be etched into the silicon itself; for example, for trench capacitors on DRAM devices. Finally, an SiO_2 layer can be used to mask implants and to slow down the ions from an implant to prevent undue silicon damage.

In general, SiO_2 is amorphous for most integrated circuit applications. There are methods for forming more crystalline structures through the inclusion of various types of dopants but we will not consider them here. The molecular density of SiO_2 is about 2.2×10^{22} cm^{-3} as opposed to pure silicon, which has a molecular density of about 5.0×10^{22} cm^{-3}. From this we can see that a thermal oxide film grown from pure single-crystal silicon will actually take up 2.3 units for every unit of silicon consumed. Thus, a film of 2300 Å will consume approximately 1000 Å of silicon, as shown in Figure 7-12. This figure also shows an effect that occurs when oxide is grown near a nonoxidizing Si_3N_4 film. This is the formation of a structure called a bird's beak. SiO_2 has an energy gap (E_g) of about 9 eV, which places it well into the regime of an insulator. In addition, SiO_2 has an electric field breakdown level of $6–9 \times 10^6$ V/cm. At typical oxide thicknesses (say, 400 Å of gate oxidations) this results in a theoretical breakdown voltage of 24–25 volts. Of course, in reality, that number will be limited by a number of effects, including underlying silicon substrate defects, the method of oxidation, the existence of trapped charges in the bulk of the oxide, and the level of particulate or chemical contamination.

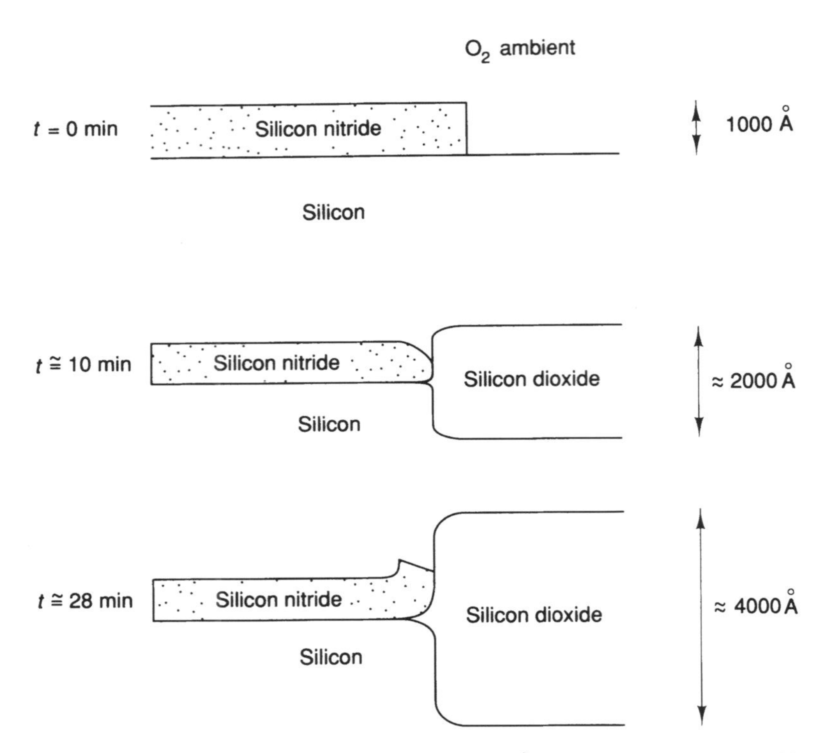

Figure 7-12. The Thermal Oxidation of Silicon As opposed to deposited films like silicon nitride, silicon dioxide is often grown directly from the silicon substrate. This reaction is diffusion-rate-limited at most thicknesses, so that we can see that it will take more than a doubling of time to double a thickness.

There are several methods for growing SiO_2 on silicon. The first uses pure oxygen, and is called a "dry" oxidation. In this case, the reaction is simply:

$$Si + O_2 \rightarrow SiO_2$$

In the second major reaction, the silicon is placed in a steam ambient, and the silicon then reacts with the steam. This reaction occurs at a rate nearly 10 times as great as dry oxidation processes. It appears that the actual molecule doing the oxidation is the hydroxyl radical as shown, although the basic equations for either potential reaction are simple:

$$Si + 2H_2O \rightarrow SiO_2 + 2\,H_2$$

or

$$Si + 2OH^- \rightarrow SiO_2 + H_2$$

At the temperatures that the wafers are being processed, a variety of other gas-phase reactions may occur within the processing chamber. This is especially true in light of the fact that most oxidation processes use HCl to help keep the integrity of the SiO_2 high. The reactions involved here are as follows:

$$O_2 + 4HCl \leftrightarrows 2H_2O + 2Cl_2$$

Thus, the small quantity of HCl that is used will also boost the oxidation rates slightly. The existence of these various gas phase relationships has caused theoreticians a good deal of trouble since the various reactions that can occur are somewhat unpredictable in any individual case. As usual, statistical techniques work better for developing models of oxide growth.

Oxidation takes place at the silicon/SiO_2 interface. Therefore, for oxidation to take place, the oxidizing species must diffuse through any preexisting SiO_2. For thin oxides, the limiting factor for growth rate is the actual surface reaction rate of the silicon and the oxidizing species. In most cases, this growth rate is assumed to be a linear function, although close examination shows that this is not exactly correct. The surface reaction rate is largely determined by the crystal orientation of the silicon crystal itself. For instance, <111> silicon will oxidize more quickly than <100> silicon. When examining the structure of the silicon (as in Figure 7-13), it is easy to see why. The oxidizing molecules have much clearer paths to travel through in the <111> crystal lattice. When thicker films of SiO_2 are required, the oxidation rate becomes limited by the diffusion rate of the oxidizing species

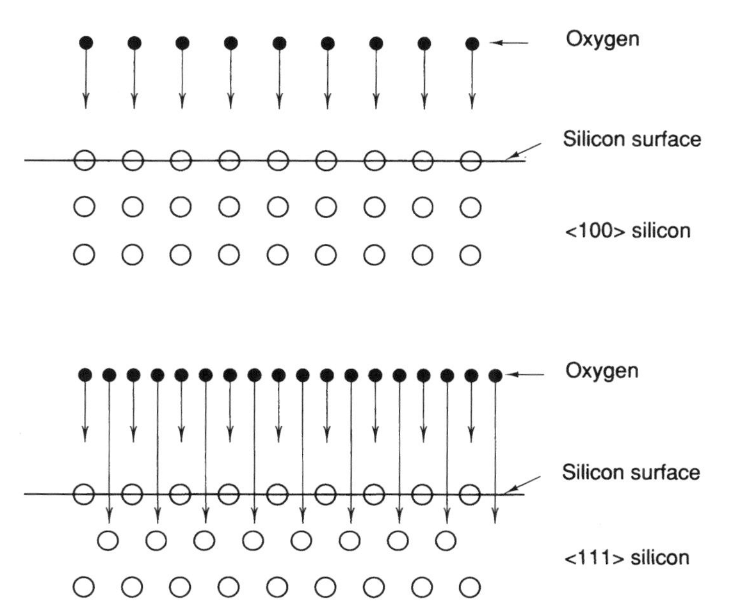

Figure 7-13. Oxidation of <100> and <111> Silicon Silicon with a crystal orientation of <111> will tend to oxidize faster than <100>. This can be understood by looking at the crystal orientation of the silicon at the surface of the wafer. The <111> silicon permits a greater number of atoms to be exposed to the diffusing oxygen molecules. Thus, an increase in oxidation rate is seen.

through the oxide. The oxide growth rate versus time for diffusion-limited processes is roughly a square root function. The classic formula for calculating oxide thicknesses was developed in 1965 by Grove and Deal and is calculated as follows:

$$T_{ox} = A/2\left[\sqrt{(1 + ((t + \tau)/(A^2/4B)))} - 1\right]$$

where

T_{ox} = oxide thickness
t = time
τ = "fudge factor" involving time required to grow an initial layer of SiO_2
A = $2D(1/k_s + 1/h)$
B = $2D(C^*/N)$
where D = diffusivity of the oxidizing species
k_s = surface reaction rate
h = mass transfer coefficient
C^* = coefficient including partial pressure
N = number of molecules of oxidizing species

There are a couple of problems with this theory which have led to a number of attempts at revisions of this equation. These problems include the use of the "fudge factor" for thin films, as many oxidation processes are actually working within 200-Å ranges. In addition, the model's accuracy decreases as film thicknesses reduce below 1000 Å. Alternative empirical models have been worked out by many researchers with results that are basically consistent with the Deal–Grove model.

As noted, the oxidation rate of a dry oxidation process is less than 10% of the rate of a steam oxidation. The exact rate of oxidation can be adjusted by varying the amount of the various oxidizing species in the ambient gas flow (partial pressures, which are the ratio of the active gases to the total flow of gas through the tube). Since most processes burn a mixture of hydrogen and oxygen to form the steam for the wet oxidation process, the growth rate of the oxide can be adjusted by varying the relative concentrations of these gases. Figure 7-14 shows the change in growth rate as a function of partial pressure of steam in an ambient that is composed entirely of steam and oxygen. Remember that additional HCl in the gas stream will also influence the growth rate of SiO_2, due to the creation of H_2O as a by-product. Total gas flows of the gas streams usually will have little effect on the overall process, as long as there is enough gas that the oxidants are replaced as fast as the wafers can consume them, and that the pressure remains at atmospheric pressure. If an insufficient amount of gas is introduced into the system, the diffusion system will go into a gas transport limited mode. In this case, the wafer uniformities will suffer and there will be an increased probability of atmospheric gases flowing into the tube (or backstreaming), causing even more serious uniformity and contamination problems.

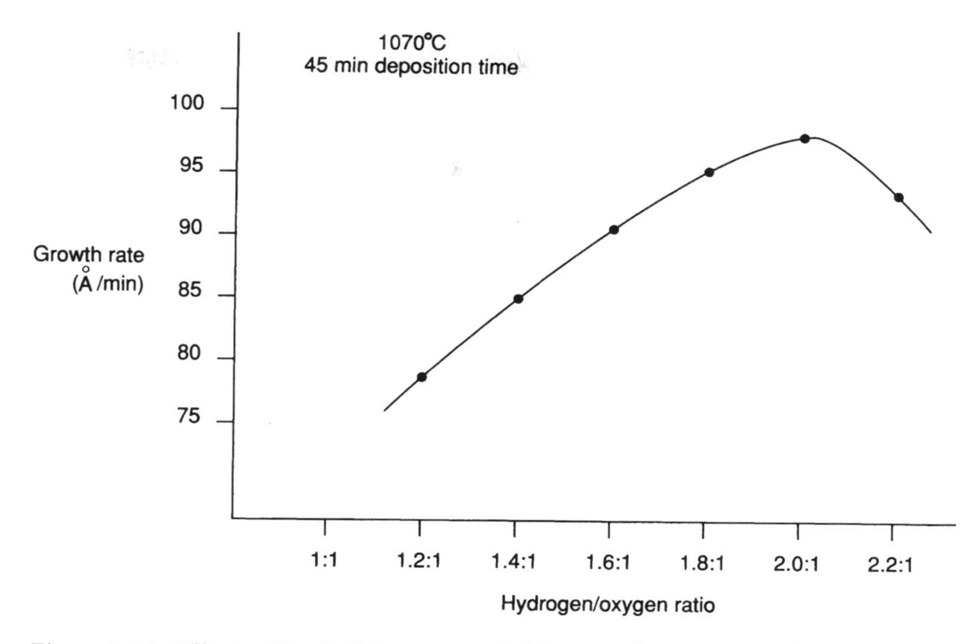

Figure 7-14. Effects of Partial Pressure on Oxide Growth Rate This chart shows that the partial pressure of steam in a pyrolytic oxidation furnace has a significant effect on the growth rate. In this example, the steam will be diluted by oxygen until the hydrogen/oxygen ratio equals two (i.e., $2H_2 + O_2 \rightarrow 2H_2O$). The decrease in growth rate at ratios greater than 2.0 indicates that the excess of hydrogen is diluting the overall gas stream. Too much free hydrogen can produce a fire hazard in the exhaust and should be avoided if at all possible.

Obviously, oxide growth is directly related to the temperature at which the reaction takes place. You will note however, that temperature does not come into direct account by the two equations specified above. That is because the effects of temperature are included as part of the terms for the temperature-dependent parts of the equations, which are the diffusivity rates, and the surface reaction rates, both of which are temperature-dependent. As can be seen in Figure 7-15, the growth rate is strongly affected by temperature. Typical fab area oxidation temperatures are around 920 to 950°C. Some oxidation processes are run as high as 1200°C. However, this is not practiced as often as expected. Interestingly enough, substrate damage can be removed with proper high-temperature cycling that can be designed right into the processing. In fact, we will see that the 950°C range is very nearly the worst range at which to oxidize wafers as far as substrate damage is concerned. Temperatures above 1200°C are avoided due to severe problems with quartzware warpage and other damage.

One of the reasons that temperatures are reduced in the early oxidation processes is a historical one, which makes little sense in light of today's automated equipment. It is that the wafers have a tendency to warp when elevated to temperatures above 1000 to 1050°C. This can happen even at 950°C if care is not taken in the

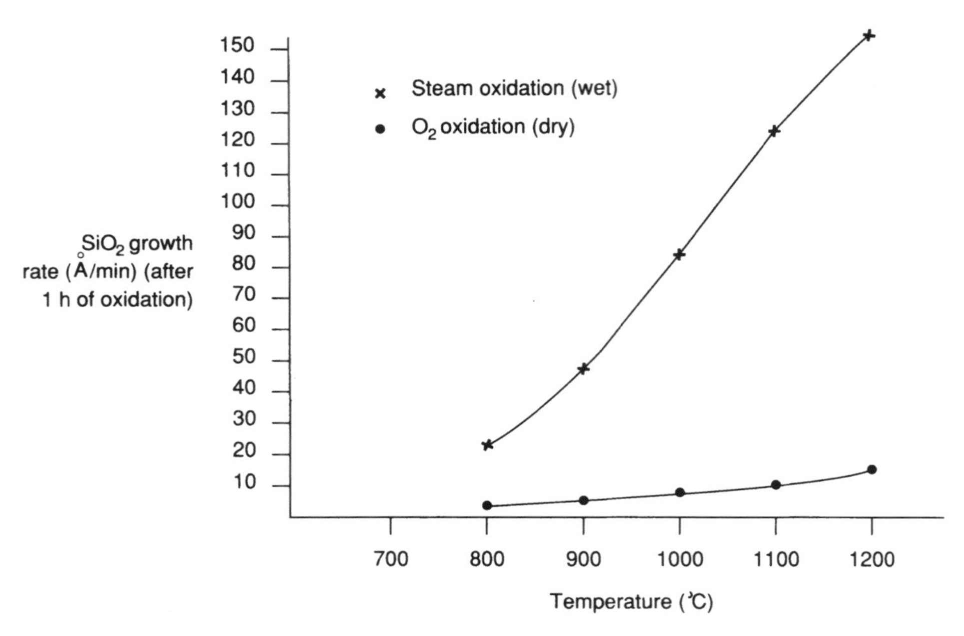

Figure 7-15. SiO_2 Growth as a Function of Temperature We can see that "wet" oxidation produces SiO_2 films at a much greater rate than pure O_2 ("dry") oxidation. This is an advantage when trying to construct manufacturable films that are both thick (around 1%) and very thin (under 100 Å). Appropriate combinations of gases permit careful control of the growth rate.

handling. This is the temperature range within which the silicon reaches its plastic point and becomes subject to plastic deformation. The more modern diffusion furnaces can control these problems by slowly and carefully ramping (changing) the temperature of the tube from the stand-by temperatures of around 800°C to the full processing temperature and back while controlling all conditions very carefully. This control enables the wafers to be oxidized safely at elevated temperatures.

Another reason that reduced temperatures are required for oxidation processes is the reduced thicknesses of the oxide layers required. In many cases, the oxides produced at higher temperatures will grow at too rapidly to allow for uniform film growth at very thin levels. Sometimes the oxidant species can be diluted with inert gases, but this does not usually prove to be cost-effective. Proper techniques for removing defects from the substrate at the earlier field oxidation steps will reduce the detrimental effects of lower temperature thin oxidation steps.

Oxidation can also be accomplished on polysilicon films. The growth mechanisms are nearly identical, but the difference in structure causes several changes in growth characteristics. To begin with, the polycrystalline silicon has a lower molecular density than single-crystal silicon. As a result of the lower packing density of the molecules and the granular structure of the polysilicon, the oxidizing agents can penetrate the poly faster, and will have more effective surface area to react with. This will result in a faster surface reaction rate. The SiO_2 produced by oxidizing the polysilicon may be of somewhat lower density than the oxide produced from single-crystal silicon, so that the diffusivity of the active species through the SiO_2 is somewhat higher. The end result of these two items is that the oxidation rate on polysilicon is higher than on single-crystal silicon, usually about 50% higher. Also, the ratio of silicon consumed to SiO_2 produced is somewhat lower on polysilicon (so that slightly thicker layers of the polysilicon are used to produce relatively thinner layers of oxide). It should be pointed out that when developing processes that oxidize polysilicon structures, care should be taken to make sure that appropriate quantities of polysilicon will remain on the structure after the oxidation has been completed.

Oxidation processes can be used to help in the dopant redistribution process. As seen in the example in Figure 7-16, which is the case of phosphorus doping of a silicon substrate, the oxide does not absorb the phosphorus to any great degree. However, since the oxidation rate of silicon in steam exceeds the diffusion rate of the phosphorus, the phosphorus will tend to pile up at the Si/SiO_2 interface. This phenomenon can be used to create shallow junctions and other structures. This effect does not work equally well with all dopant types. For example, the oxidation of arsenic-doped silicon will result in approximately 15% absorption of the arsenic by the oxide as it grows. This can affect device operation if not taken into account and if marginal amounts of arsenic are used. In the case of boron, the oxide will actually have an affinity for the boron as compared to the silicon. This will result in the absorption of up to 80% or more of the boron in a region of heavy oxidation. In the case of a field oxidation process, the oxide can absorb enough local boron from the substrate that there will

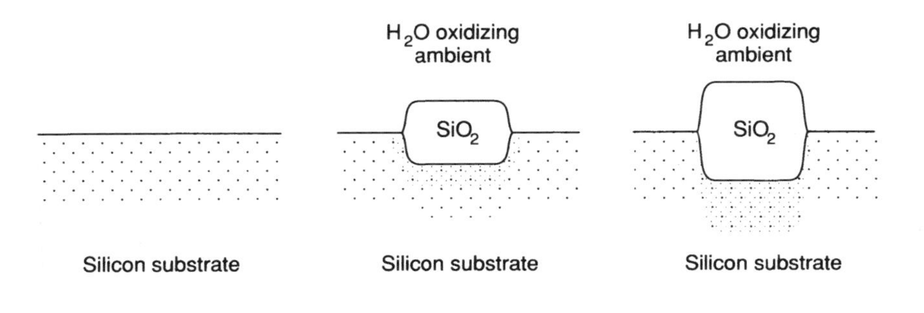

Figure 7-16. Enhanced Dopant Diffusion Using SiO_2 Growth Phosphorus does not diffuse into SiO_2 and also diffuses more slowly than the silicon oxidizes. As a result, the oxide layer is used to "push" the implant to predetermined depths, and can be used to affect the concentration of the phosphorus in the silicon.

be problems with device performance. As a result, many manufacturing operations implant the field area with boron prior to field oxidation in order to enhance the boron concentrations in the area.

During the oxidation of silicon a variety of types of defects can be generated, trapped, or otherwise incorporated into the film. As noted earlier, less than one spare charged atom in 1×10^6 up to 1×10^9 SiO_2 molecules will cause significant changes in the electrical properties of a film. Given that a gate oxide region is 350 Å and is perhaps 2×2 μm in size (or 0.14 μm^3, where 1 $\mu m^3 = 1 \times 10^{-9}$ cm^3), there need to be as few as 1500 to 1.5×10^6 or so ions trapped in the gate region to cause a problem. That is a very small number of impurity atoms, especially when you consider that even a 0.1-μm particle can contain somewhere around 2×10^6 atoms. A particle of that size can easily contaminate a large area of the chip if the particle is of a material lethal to the chips, such as NaCl.

The incorporation of contamination can affect the circuit in any number of ways. For example, the threshold voltage of the gate region may be altered or the breakdown voltage of the oxide may be reduced. The type and amount of damage depends partly upon the exact type of the contaminant and its location within the oxide film. As shown in the example in Figure 7-17, if a charge is trapped near the Si/SiO_2 interface, the effect on threshold voltage will be maximized, whereas the same charge trapped near the metal/oxide will have very little

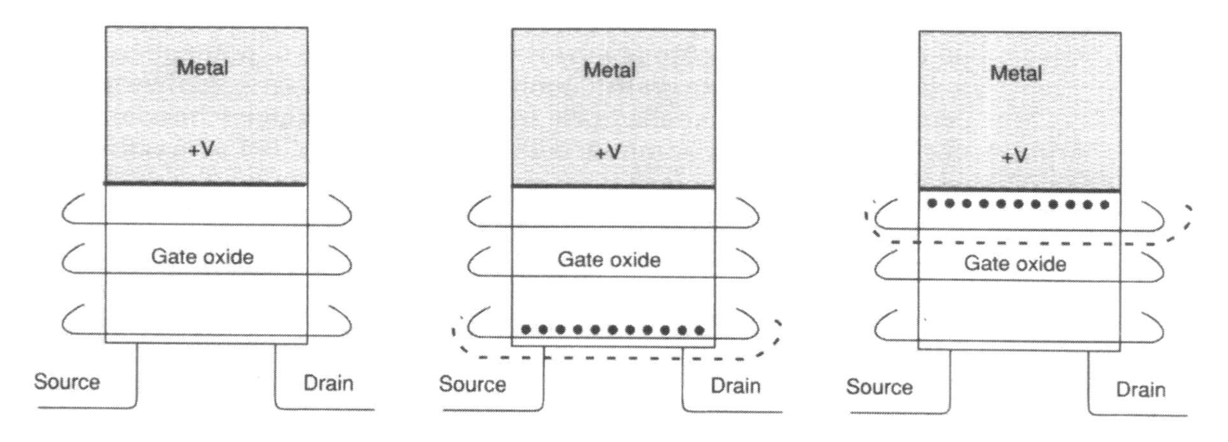

Figure 7-17. The Effect of Trapped Charges on Gate Thresholds Positive ions that have been trapped in the gate oxide will affect the gate thresholds slightly, depending on the concentration and location of the ions. The closer the ions are to the Si-SiO_2 interface, the larger the effect.

effect on threshold voltage. A charge found at the Si/SiO_2 interface is called a surface state charge and is identified as Q_{ss}. Charges that have become isolated in the oxide film are termed trapped charges, and are usually identified as Q_{ox}. Charges that do not vary as a function of changes in the gate potential will induce a constant offset to the threshold voltage. The voltage shift must be minimized and well characterized for any manufacturing process to be successful. Charges that appear and change in intensity during operation of the device can cause complete failure of a device due to unpredictable changes in gate thresholds. These fast surface states usually appear below the gate region in the substrate and are often related to crystal defects and other substrate problems. Crystal defects such as point defects can impart a fixed charge in an oxide when the structure is oxidized, especially in dry ambient. Another source of charge trapped in an oxide can be caused from radiation. If an X ray or a particle with an energy (either kinetic or intrinsic) of 9 eV or greater hits the oxide, there is a probability that an electron can be hit, and that it will attain enough momentum to escape the oxide. This results in an net charging of the oxide.

A similar effect is seen when electrons with enough energy to escape are released from aluminum or polysilicon. Some of these "hot electrons" can become trapped in the insulating SiO_2 film. If enough of them become trapped, they will cause a significant shift in the characteristics of the oxide. This problem is especially acute in the situation of EPROMs and EEPROMs, as in both of these cases electrons are forced to flow through the insulating regions in order to program them. In the case of electrons, the method used is hot electron injection, in which the electrons are given enough potential (voltage) to blast or avalanche through the oxide film. As a result, EPROMs can be programmed only a few dozen to hundred times. An EEPROM uses Fowler–Nordheim tunneling, which is a probabilistic effect that relies on the fact that, if a film is thin enough, the probability increases that electrons can jump across the forbidden gap of the oxide to the floating gate. This produces much less damage to the oxide even though some still occurs, and these devices are limited to a useful lifetime of a few ten to hundred thousand program/erase cycles.

We have already discussed the formation of the bird's beak in the region near an interface of a film that does not oxidize, such as silicon nitride. There are other physical side effects that must be accounted for or prevented while manufacturing wafers. One of the most significant effects is called the "KOOI" effect, or the "white ribbon" effect. This occurs near the edge of a nitride/oxide structure, and is a result of the high-temperature oxidizing ambient of the diffusion process (often field oxide is a major culprit). The oxidizing species diffuses into the thinnest regions of the nitride film and forms a narrow ribbon of silicon oxynitride. This can create all sorts of havoc later in the process. This effect can be limited through the use of dry O_2, although it does not eliminate the problem completely, and may result in excessive oxidation times. A more common approach is to use a sacrificial oxidation step prior to the growth of gate oxide to remove the ribbon. This is often done in conjunction with the use of a sacrificial oxide during an implant step.

7.2.2 Dopant Deposition and Diffusion Processes

These processes are at the heart of the diffusion area, as they involve the exact mechanisms that gives the process its name. This type of procedure is used to introduce various dopant species to the wafers. A typical application is to dope a layer of polysilicon with phosphorus. Phosphorus doping of the silicon substrate can also be accomplished through the use of these processes. Boron and other dopants can also be deposited in this way if required, although this is a less frequent process. This type of process is not generally acceptable for use with arsenic dopant sources due to the excessive toxicity of the fumes.

There are a variety of sequences used in this type of process. In some cases, the dopants are introduced to the wafer and are diffused to the proper depth in a separate step. In other cases, the deposition and diffusion operations take place in one furnace in one step. The choice of which process to use is largely based on yield effects of each type of process. Often, the one-step process produces the best results due to decreased handling and reduced chance of possible furnace problems. The biggest disadvantage is that the furnace walls may become a source for unpredictable amounts of excess phosphorus deposition due to the outgassing of excess phosphorus dopant from the quartz. Since this amount of doping will change as a function of furnace usage and of quartzware quality, this kind of problem can be very troublesome. These issues will be relatively less important in a procedure that is used to saturate a polysilicon film with dopant. When doping a region of the substrate, it is critical that the exact doping characteristics are reproduced on each wafer or there may be significant yield or performance problems. The use of a short phosphorus deposition step, with the diffusion step performed in a separate clean furnace, gives more precise control over the dopant redistribution process.

The sequence of events that occurs in the phosphorus deposition and diffusion steps is similar for both of the cases described above. In each case, the typical process calls for a nitrogen carrier gas to be bubbled through a $POCl_3$ (phosphorus oxychloride) container at a fixed rate and temperature. This allows a predictable amount of the chemical to absorb into the nitrogen. The flow rate of the nitrogen and the temperature of the $POCl_3$ bottle are critical to the final resistivity of the wafers. The bottle temperature should be checked prior to and during each run. It is also important to verify that the level is correct, as a too low $POCl_3$ level will lead to reduced amount of $POCl_3$ in the nitrogen stream. The bottles are usually replaced when there is less than one inch of $POCl_3$ in the bottle. This works out to many dozens to hundreds of runs, depending on the use rate and the bottle size.

The $POCl_3$ itself is a highly toxic acid that can seriously burn the skin and can cause liver and other internal organ damage. It should be handled in the most careful manner possible. Spills of $POCl_3$ are especially dangerous, as the chemical reacts violently with water to form toxic fumes. Inhalation of these fumes can result in a number of symptoms which can last for several days. The chemical also causes an annoying taste after inhalation that does not go away for several hours to several days. Any mixing of $POCl_3$ with water must be done with at least a 25 : 1 ratio of water to $POCl_3$. Fortunately, most $POCl_3$ is stored, delivered, and used in special shatterproof bottles that have special fittings preinstalled for ease of installation. Up to a few years ago, these precautions were not used and a large number of injuries, fab evacuations, and other safety-related issues were recorded as a result of $POCl_3$ spills.

As noted, a predictable amount of $POCl_3$ is absorbed into the carrier gas, where it is then sent to the furnace chamber. It is usually mixed with an additional amount of carrier gas at this point to allow the total flow of the gas through the tube to remain constant at all times. A small amount of oxygen is also introduced to the furnace with the $POCl_3$, although it is kept separate from the $POCl_3$ until the last minute. The furnace temperature is usually held to around 800°C for the deposition step. At this temperature, the oxygen readily reacts with the $POCl_3$ to form P_2O_5 (phosphorus pentoxide), which is the actual species that deposits the phosphorus on the wafers. The overall reaction can form a number of toxic by-products, such as P_2O_3. Clearly the phosphorus deposition step is not one of the safer diffusion processes.

$$2POCl_3 + 2O_2 \rightarrow P_2O_5 + 3Cl_2$$

The P_2O_5 reacts easily with the silicon to form silicon dioxide and free phosphorus, which absorbs into the silicon. This reaction occurs as follows:

$$2P_2O_5 + 5Si \rightarrow 5SiO_2 + 4P\downarrow$$

The silicon oxide that is formed will range from 300 to 600 Å in thickness after the phosphorus deposition step. The exact dopant concentration can be estimated from the oxide thickness and it is important to measure the thickness of the oxide as well as the resulting resistivity and to record these readings in the process control charts for the process. While the SiO_2 that has formed has a function in process control, it is usually of such poor quality that it must be removed (or deglazed) from the wafer surface immediately after either the deposition or diffusion steps, depending on the exact requirements for the ICs being manufactured. Obviously, test wafers must be deglazed after the oxide thickness has been evaluated in order to test the resistivity of the wafers. The exact location of the deglaze step in the process will sometimes be determined by whether or not the diffusion step will be doubling as an interlayer thermal oxidation step. Bear in mind that the phosphorus has only a small amount of absorbance in SiO_2 so that silicon oxide growth can act as a phosphorus diffusion rate and distribution profile modifier.

One of the processing side effects of this form of reaction is that there can be a significant amount of depletion of the P_2O_5 in the gas stream during the deposition cycle. If this occurs, the wafers closest to the door of the furnace (the "handle" end) will receive less dopant than required and will have high resistivity. This problem is especially acute with fresh quartzware in the furnace, which will absorb the phosphorus at an abnormally high rate. It is for this reason that most fab areas require the use of a predope cycle on each of the deposition process tubes when they have been freshly cleaned, or have sat idle for a long period of time. The opposite problem occurs when the tube becomes heavily saturated with the P_2O_5 after many uses. This can even result in the formation of puddles of phosphorus "goo" on the bottom of the furnace. This excess phosphorus can result in overdoping of the wafers, giving the wafers low resistivity readings.

Process engineers often make the mistake of trying to correct for gas flow problems by making temperature adjustments or other unusual procedures. When a phosphorus deposition furnace goes haywire, the first place to look is in the $POCl_3$ distribution system itself. This starts with an inspection of the tube. If no problem is seen, a quick excursion to the gas system may be required. A very common place for problems to occur is in the valve placed immediately downstream. This is often a three-way valve, called a source/vent valve or something similar. This valve often plugs with $POCl_3$ and can cause major problems in the event of failure.

An important point to remember when monitoring the phosphorus deposition process is that the resistivity readings move in opposite directions from the oxidation readings. In other words, a higher processing temperature or a longer processing time will result in a thicker oxide reading but in lower resistivity (V/I) readings. This makes sense when one observes the physical reactions that are taking place, as in Figure 7-18. Clearly, the same mechanism is at work, driving both the oxidant and the dopant deeper into the silicon (and sometimes adding extra quantities of dopant, although for this example we will assume that the amount of dopant reaching the surface is the same for all cases). As the dopant is driven deeper into the silicon, the channel becomes progressively larger, which provides reduced resistivity.

The phosphorus deposition process is an inherently dirty one and is therefore less sensitive to contamination than other processes. The process of depositing the phosphorus leaves a surface that consists of large quantities of condensed P_2O_5 and other phosphoric com-

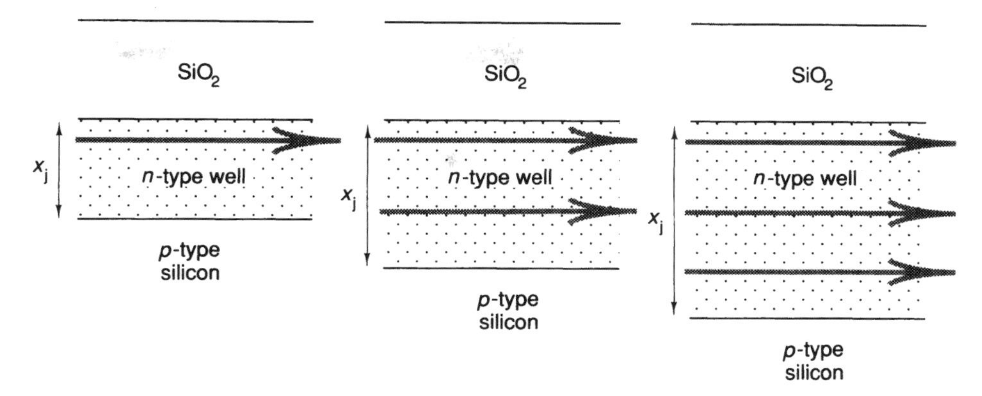

Figure 7-18. Resistivity versus Junction Depth, x_j Even though there is a reduced density of dopant ions in the same volume, the total depth of the channel can be increased through a sufficient amount of anneal time. This results in a net lower resistivity for the channel.

pounds, including phosphoric acid after the phosphorus comes into contact with moisture from the air. These particles are usually water-soluble, and will be lifted off with the oxide during the deglaze step. Particles that remain after the deglaze step are a more serious problem and must be removed if possible. They may become embedded in the silicon, and may be difficult to remove. One of the more serious problems in this point of the process is the potential for cross contamination. Metallic contamination or ionic contamination of the opposite polarity (i.e., a *p*-type dopant, such as boron) can drastically change the resistivity in localized areas of the wafers, which would have severe detrimental impact on the yields of the chips. The important area to watch out for is the preclean step, where a cross dopant could be transported to the wafers. Unfortunately, the phosphorus deposition process is so dirty that it is almost impossible to test for unexpected chemical contamination, as the traces of cross dopants will be masked with the phosphorus residues.

There are some other liquid-source dopant processes, although they are not in widespread use. Most of those processes use similar techniques to those presented here, and often have many of the same concerns and problems. The main purpose of these types of processes is to distribute a form of dopant to the surface in such a way that it absorbs into the substrate and leaves only easily removed by-products.

In the search for a clean method of doping, there have been a number of other methods invented for doping wafers within the diffusion furnace environment. The first is a spin-on method. In this case, a dopant source, for example, an arsenic solution, is deposited onto the wafer in a method similar to the deposition of photoresist. In this case, the fluid is poured over the spinning wafer so that the thickness is uniform (much as in the case of resist, the spin speed and viscosity must be matched to achieve optimum uniformity). The film is then dried in an oven, and the wafers are annealed in the diffusion furnace. This process has a number of problems. First is the control of the deposition process. In most cases, this does not provide the control required for modern IC manufacture, especially for modern shallow junctions. Another method, which is quite old, but which has experienced a resurgence in the last few years, is that of the dopant wafer. This is usually a slab 0.1 to 0.5 inches thick, approximately the same size as the wafers being manufactured, made out of boron nitride. The boron nitride can absorb a large quantity of chemical, which can be driven back out of the material simply by heating. This does not affect the structure of the boron nitride, preventing particle creation. While the slabs must be replaced periodically, the process is much cleaner and safer than any of the alternatives, which has aided in its resurgence. Another reason is that the techniques for uniformly impregnating the slab with dopant allow this method to be the potentially best method for depositing a uniform layer of dopant on the surface of wafers in the diffusion environment.

Many operations have switched to using implanted phosphorus in place of the phosphorus deposition process. This can add to the processing time of the wafers, and can be a factor in radiation-sensitive or implant damage-sensitive technologies. This process must be carried out carefully but has been shown to be effective in many cases.

After the deposition and deglaze sequence, the phosphorus will lie in a thin layer along the surface of the exposed silicon. It is necessary to drive the phosphorus in to the correct depth. This is done in the diffusion sequence. In the event that there is a one-step doping and diffusion process, the doping sequence will be followed with an elevation in temperature to the proper temperature for drive-in. If the process is a two-step operation, the wafers will be sent to a separate furnace for the drive-in process. That process will often involve further oxidation of the surface of the wafer. This is especially true in the event that the process is being used to grow an oxide layer in-between two polysilicon layers or between a polysilicon layer and a metal layer.

The diffusion step will usually take place at between 950 and 1000°C, and will usually take about two or more hours. The exact time required to reach a certain junction depth is given by the equation:

$$J_d = \sqrt{D \cdot t}$$

where D is the diffusivity of the dopant species through the silicon lattice at a given temperature and t is the time of the drive-in. J_d represents the depth of the junction.

Since the process used to drive the dopants into the wafer is typically a two-temperature process, there is sometimes some dispute over what temperatures to use to determine drive-in. Of course, computer analysis will give a precise answer to the question. However, a quick answer can be obtained by calculating the drive-in time from the point that the wafers reach about two-thirds of the processing temperature to the time that the temperature has ramped to below that same point. The processing temperature is determined from the top of the temperature curve. The dopants diffuse much faster in the early stages of a process than they do at later stages. It takes four or more times longer for the dopants to drive-in to a level twice as deep as another given level.

As a result, a device is designed with specific tolerances in mind, and with a total thermal budget as a constraint. This budget is the total amount of time that a wafer can be processed at certain temperatures throughout the entire process. The dopants will move the most within the first part of the redistribution process. Problems can come into play when misprocesses occur. For instance, an occasional error is to put wafers through the diffusion process twice. This can drastically alter the junctions and thoroughly damage the yield on the wafers. In the event that a diffusion process is aborted prior to completion (there are sometimes rework procedures that can be used, such as reprocessing for a fixed amount of time), that will allow the wafers to be "fixed." There is usually some yield impact, but if enough wafers are affected and the yield impact is minimal, these procedures may work quite well.

7.2.3 Anneal Processes

Associated with the diffusion processes are the anneal processes, which are used to drive-in dopants that were delivered to the wafers through implant processes. These processes are usually run in a non-oxidizing ambient, and may be run for many hours. Examples of these processes include *n*-well and *p*-well drive processes for CMOS processing. These processes usually run at high temperatures, sometimes as high as 1100°C or higher. Some of the drive processes can be as long as 20 hours. These extreme conditions are required so that the dopant can reach the required junctions for the deepest wells (see Figure 7-19).

Anneal processes are usually carried out in a nitrogen ambient, although sometimes argon or other inert gases are used instead of nitrogen. The use of the noble gases for the anneal ambient gas stream is due to effects that pure nitrogen can have on bare silicon substrates. In fact, it is important to make sure that wafers are coated with at least a minimal amount (75 to 150 Å) of SiO_2 or Si_3N_4, to prevent the nitrogen from coming into direct contact with the silicon. When nitrogen comes into contact with the bare silicon, it will react slightly to form a partial silicon nitride structure. This also results in severe damage to the surface of the silicon, in the form of surface pits and nodules of unremovable compounds, as well as contributing to the propagation of substrate crystal defects. This damage can be quite lethal to some devices, especially static RAM chips. There have been

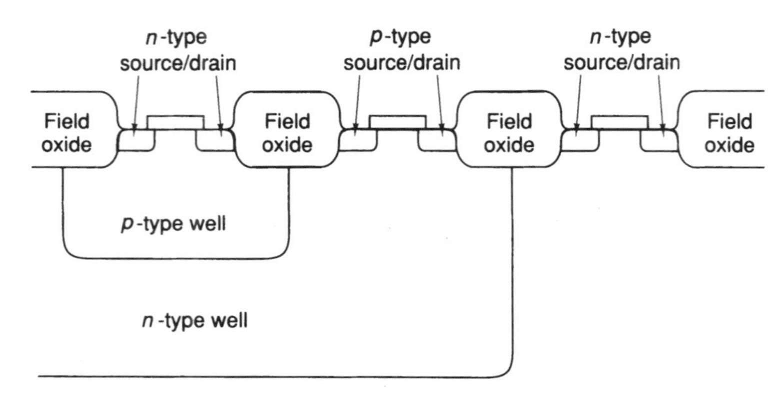

Figure 7-19. *n*-type and *p*-type Wells for Developing CMOS Technologies The transistors for a CMOS device are constructed in both p-type and n-type wells that have been diffused into the silicon substrate.

cases where allowing wafers to come into contact with N_2 at key steps has caused a greater than 60% loss in die yield. These types of problems can be avoided through the use of noble gas ambients, but the cost of manufacturing increases drastically (keep in mind that a diffusion furnace may use 10 liters a minute, which for a 20-hour process will add up to 12,000 liters of gas per process).

Anneal processes often are used to activate implants. When the ions from some species such as arsenic are implanted into the silicon, they do not always free all of their electrons. To force the atoms to reach the energy states necessary to release the electrons, a high temperature must be attained for a certain amount of time, usually half an hour or so. In most cases, the act of processing the wafers will produce enough activation that these effects are not noticed. A typical process is something like the source/drain oxidation process, sometimes called the source/drain reoxidation process (since the region has usually been oxidized previously at the gate oxidation step). As shown in Figure 7-20, the wafers are then implanted with arsenic in the source and drain regions. The arsenic needs to be redistributed and activated, so a drive step of about one-half hour to two hours will be required. After oxidation, the source and drain regions will act as appropriate conductors. In addition, the oxidation of the silicon is used to help push the implant and build a protective layer on the silicon interface. This contact area is then partially or completely cleared of the SiO_2. One of the purposes of the oxidation step is to process the implant-damaged silicon so that it can be removed. This will improve the quality of the contact region.

Another purpose of the anneal process is to cause deposited dielectric films such as BPSG to flow or densify over the surface of the wafer. These anneals smooth the rough, uneven surface of the BPSG, and cause it to flow over the edges of the steps in a consistent manner.

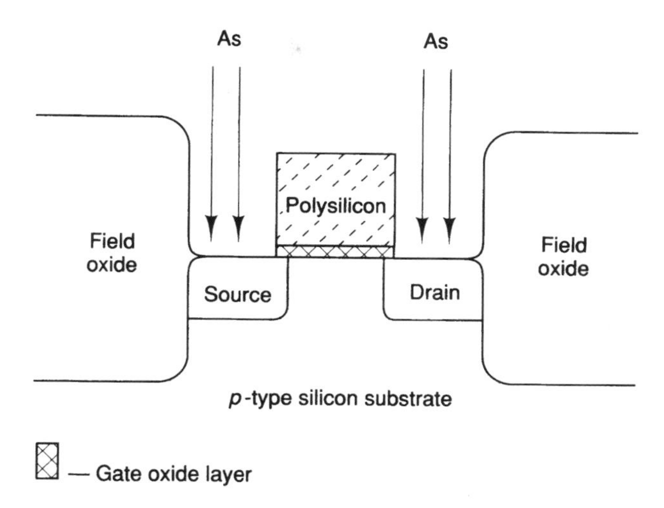

Figure 7-20. Arsenic Source / Drain Regions The source and drain regions of the transistor of an NMOS device are implanted with arsenic, followed by an oxidation/ drive-in/activation process. Remaining silicon dioxide on the source and drain regions is cleared so that good contact can be made later.

In the process of doing this, the film will become denser and stronger. This is done to permit a smooth deposition surface for subsequent metallization steps and to prevent cracking and other defects. These processes are carried out in a number of ambient conditions. The process temperature depends on the concentration of dopant in the film. Typically, lower dielectric film dopant levels require higher temperatures to flow uniformly. More dopant will generally provide a film that has better flow characteristics. However, there is a point at which these films become unstable.

We will discuss the dielectric films later. For now, we only need to know that the reflow processes range from 920°C to over 1000°C, although 950°C is the typical temperature. The densification itself will usually take only 30 minutes to an hour, again depending on the exact film used. The gas ambients will sometimes be inert, and other times oxidizing. The oxidizing ambients are useful in the case of these films as the reactions that formed the films may not have been entirely completed. The oxidizing conditions will complete the reactions and allow the film to be somewhat more stable. Finally, it is clear that, with the high temperatures required for the reflow process, it can only be accomplished before metals have been placed on the wafers.

7.2.4 Alloy, Contact Diffusion, and Getter Processes

These processes are all typically used at the end of the diffusion process. The diffusion processes are not normally used after metal deposition has occurred, but we will see that the exception to the rule is the alloy process. The other two processes (contact diffusion, getter) are used for preparing the substrate regions immediately before first metallization. The contact diffusion and getter processes are not used by all fabs, but nearly every fab has some sort of alloy process.

The alloy (or sinter) process is used to produce good contacts between the aluminum and silicon conductors. As we will see in the section on metallization, aluminum is a typical first metal layer, and is usually doped with small quantities of silicon and sometimes with copper. The silicon content of the aluminum is very important for the construction of good contacts. When the process is run, the aluminum and silicon at the interface will blend together much more easily if the aluminum is already mixed with silicon. If the process temperatures are unstable, or if improper gas conditions develop, serious contact problems can occur. They can include contact spiking, in which the aluminum penetrates the silicon contact deep into the substrate, or poor contacts leading to high contact resistance, both of which can become so severe that the chip will be prevented from operating. These conditions are outlined in Figure 7-21.

The alloy process is run in a relatively low-temperature furnace. Since this operation is performed immediately after metallization (typically aluminum deposition) the usual temperature is around 400°C. While some fab areas run their alloy processes closer to the maximum temperature of around 440°C, it is generally not recommended. Little can be gained by keeping the temperatures that high, and there is a good probability that deviations or instability in the alloy furnace could allow the aluminum to be damaged by overheating. One of the advantages of the low temperatures of the process is that the throughput time is minimized, with the typical process time being about one hour.

The gases used in the alloy process contain large quantities of hydrogen. Hydrogen is used here due to its ability to diffuse into and out of the aluminum films easily, in the process recrystallizing the aluminum film. In some cases, pure H_2 is used for this process. This brings with it a number of safety issues, as it can produce potentially explosive situations. Many fab areas have had hydrogen explosions of various types due to equipment or procedural failures. In a properly set-up hydrogen alloy process, the hydrogen is exhausted past an igniter that causes the hydrogen to burn off immediately. Even the best of these setups involve a significant possibility of fire. In other fab areas, a mixture of 5 to 15% hydrogen in nitrogen (called forming gas) is used. This mixture is typically not explosive or flammable, although it can become dangerous if improperly exhausted and mixed with air. Forming gas will usually work as well as pure hydrogen, especially with the thinner and narrower lines that are being produced today. The alloy process time is a function of both time at temperature and gas concentration. If the temperatures are reduced and the hydrogen levels are low, the process time will be increased slightly. In any event, the minimum time that the alloy process should be run is around 30

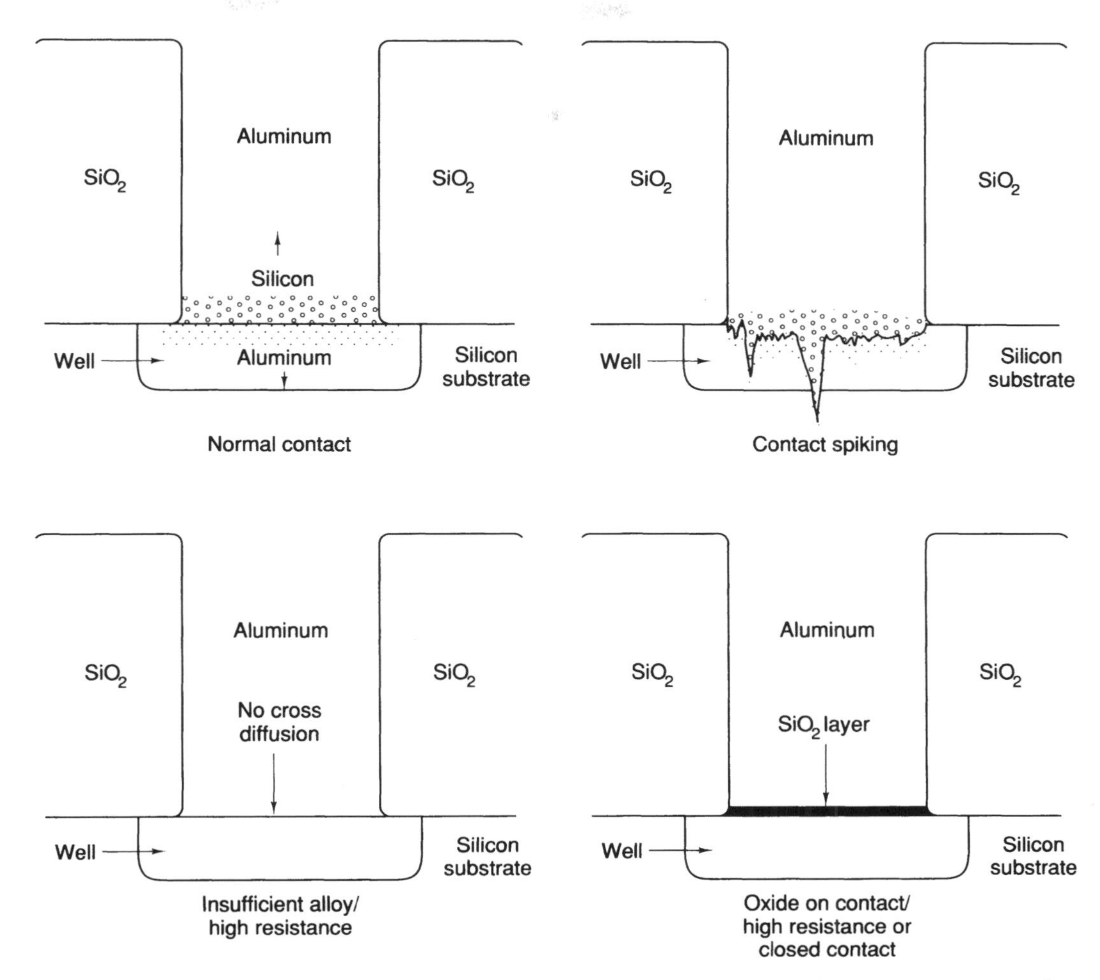

Figure 7-21. Aluminum-Silicon Contacts A normal contact between two materials, such as an aluminum film and a silicon substrate, is formed when atoms of each substance diffuse into the other substance. In the diagram shown above, some silicon will diffuse into the aluminum while much more aluminum diffuses into the silicon. Excess silicon depletion is prevented by using an aluminum/1–2% silicon alloy. Problems of excess silicon depletion, or with excess alloy temperatures include contact spiking, where aluminum crystal grains penetrate the well and make contact with the base substrate material. Contacts can present high contact resistance if the alloy temperatures or time were insufficient to produce good diffusion. They can also have high contact resistance or be completely closed with residual SiO_2 layers in the contacts.

minutes, while the maximum time of the process is around 60 minutes.

Associated with the construction of the contacts is a process called contact diffusion. There are often other names for this process, or it may be included in another process such as source drain reoxidation, but in all cases, there is some analog to contact diffusion. This procedure is performed prior to the deposition of the metal films.

Typically, there is a dopant implant or deposition step, over the cleared contact areas. In some cases, especially on larger geometries, the contacts are doped with a layer of phosphorus in the same manner as the phosphorus deposition and diffusion process described above. In other cases, the contacts are implanted with arsenic or phosphorus and the dopants diffused as described. The contacts must have the proper resistivity in order to act properly, and must be deep enough to thoroughly cover all of the conductive regions and prevent the spiking of the aluminum alloy through the contact and into the silicon.

An older process that has not been generally implemented in modern IC manufacturing (primarily due to the increasing use of epitaxial silicon with specially ground backsides) is the getter process. In this process, the wafers are processed through a phosphorus deposition process after the deposition of the PSG or BPSG step. First, all of the films will be removed from the back of the wafer. Then the heavy phosphorus deposition and diffusion will draw contaminants, especially metallic contaminants, toward the back side of the wafers where they will be lapped off later. In typical cases, the use of scribed or carefully damaged backsides on the wafers can cause stress in the crystal lattice on the backside of the wafer, which tends to also draw contamination toward the backside of the wafer.

7.2.5 Silicon Crystal Defect Control

The diffusion processes can be used to modify the silicon crystal defect levels. This involves a combination of high- and low-temperature oxidation steps, as well as controlled substrate impurity levels, particularly in regard to oxygen content. While most manufacturers have discovered the improvements that controlled oxygen concentrations can give, especially with static RAM performance, they have not typically intentionally incorporated the processing details that can optimize the yields of the devices. This type of processing is called crystal defect gettering or denudation and produces some startling results. It is based on the concept that, just as certain processing regimes can cause defects to propagate, other regimes allow the defects to shrink and eventually to disappear altogether. The removal of substrate defects can reduce problems such as parasitic transistor formation, latch-up, and other related phenomena.

The basic process starts with proper silicon selection. The substrate should contain about 30 to 40 ppm O_2. This is in contrast to other types of processing, where oxygen is usually limited to around 25 to 30 ppm or less. The reason for this high oxygen content is to force the creation of sites to which impurities and crystal defects can migrate. These techniques will work with silicon of lower content O_2 also, although the effects may vary from those described.

The denudation processes themselves actually consist of three separate subprocesses. In some cases, these processes can be designed to be part of the regular processing cycle, therefore minimizing the throughput time impact. For example, the denudation and nucleation cycle can be incorporated into the field oxidation step. Since the process works better in an oxidizing ambient, the only constraint on the process is the choice of the partial pressures of the oxidizing species.

The first step of the process is the actual denudation sequence. This is performed at a very high temperature, such as 1100 to 1150°C. Temperatures higher than this will cause the formation of surface pits and will have an unacceptably high probability of causing warpage. Temperatures below about 1075°C will not allow the formation of good, deep denuded zones. The depth and the rate of denudation are directly related to the temperature of the process. The process needs to be at the high temperature for about three hours, which means that a steam oxidation process can be used to grow the field isolation oxide. This film is typically about 8,000 to 12,000 Å thick and can be grown using 50% steam/50% O_2 mixtures. Usually a small amount of HCl (0.5 to 1%) will help the process along.

The physical processes that are underway during the denudation step are conceptually quite simple, as shown in Figure 7-22. The silicon lattice is taken to a temperature at which it reaches a very high vibrational state, just below the point where it starts to break down. This causes the atoms to vibrate around and settle back into the lattice when cooled. Since the energy pumped into the system (the crystal lattice) is so great, the defects will tend to "shake out" of the surface of the lattice. As the energy dissipates. the atoms will tend to settle back to the ground state, which will be its lowest possible energy state. This is the pure crystal state, that is, is with no dislocations or other defects. As a result, the surface of the wafer will start the realignment process and it will slowly progress through the wafer. The process that has been described will result in the formation of a denuded zone of from 15 to 20 μm. This is usually more than sufficient to provide the enhancements desired.

Since the silicon has a fairly high proportion of oxygen dissolved in it, the next step in the process must be to bring the oxygen out of solution to prevent it from migrating and reforming more crystal defects. This is done in the nucleation step (Figure 7-23), which takes place at about 750°C. During this time, the oxygen atoms are freed enough to migrate to nearby neighbors where they tend to combine or nucleate. These nuclei form the foundation for the steps that follow. The nucleation step takes about four hours and can occur as part of the denudation/oxidation process.

The last subprocess consists of the precipitation step. In this portion of the process the wafers are exposed to a temperature of around 1025°C for six to eight hours. During this time, the oxygen atoms are allowed to migrate to the nucleation sites provided in the step above. At the end of this step, there should be a large number of

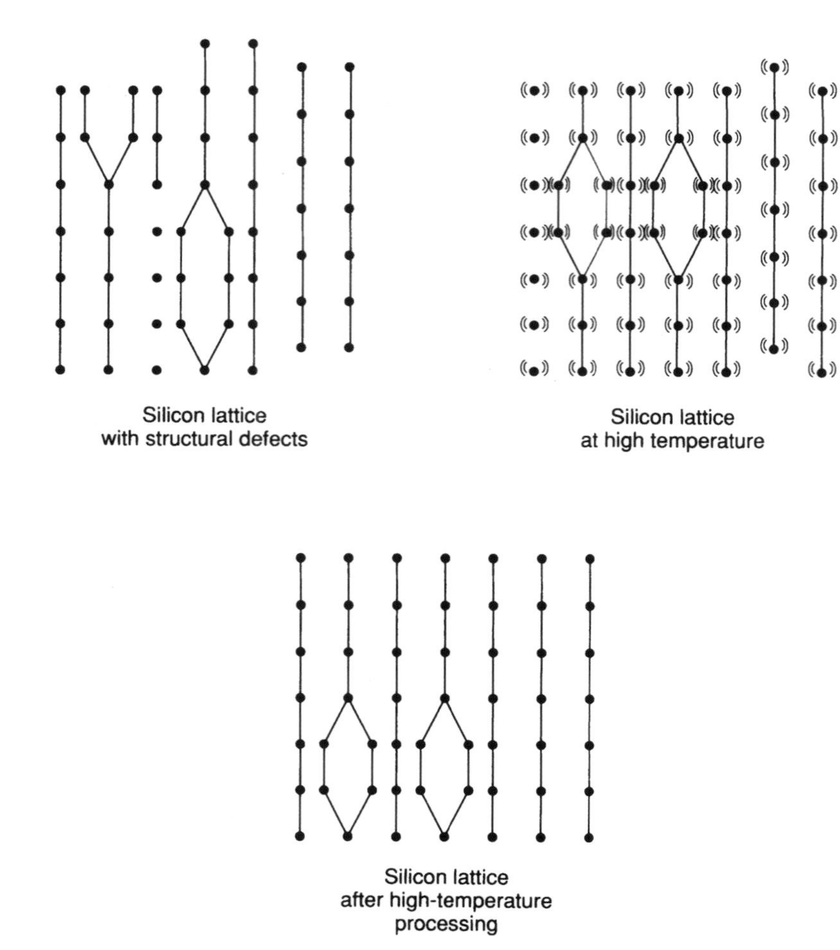

Figure 7-22. High-Temperature Defect "Gettering" in the Silicon Crystal When a silicon substrate with crystal defects is subjected to high temperatures, especially in an oxidizing ambient, the structure will begin to "rearrange" itself. As the silicon cools, the atoms will tend to come to rest in their lowest energy state, which is a single, defect-free crystal.

sites deep in the bulk of the silicon where the oxygen atoms have all clumped together. These sites will later act as gettering sites to trap various highly mobile ions. In the event that low-oxygen wafers are not used in the process this step may not be necessary, or may have a reduced throughput time.

It should be noted here that the standard oxidation temperatures of 920° to 950°C are in very nearly the worst range for silicon crystal defect propagation. Wafers processed in this way can have significant problems with things like high substrate leakage currents, low diode breakdowns, and so on. As seen in Figure 7-24, where a wafer was split in half and processed through a typical diffusion process on one half and a denudation process on the other, a wafer processed with the standard fab furnace cycle will have markedly greater defect levels

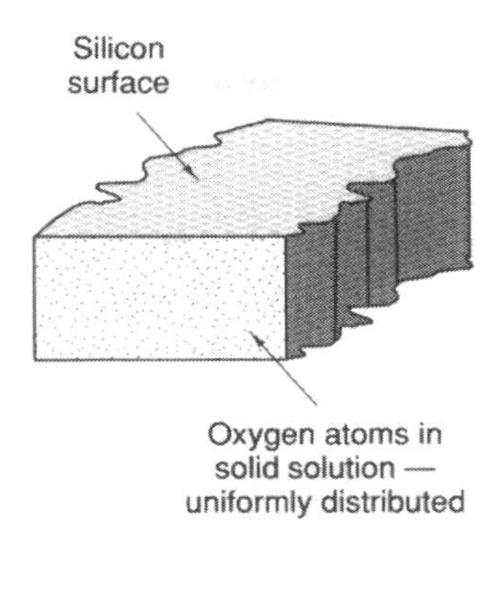

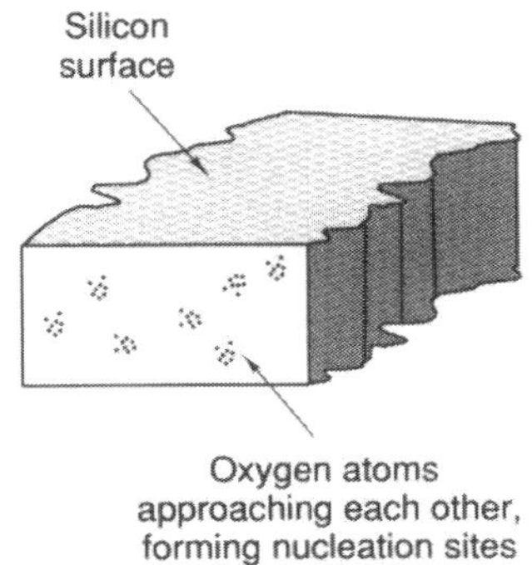

Figure 7-23. Nucleation of Oxygen Atoms in the Silicon Lattice Since oxygen is one of the most difficult elements to remove from silicon, and since it is a large, active atom, it is a significant source of lattice defects. By controlling its quantity, local concentration, and location, oxygen can be used to create sites to which defects may migrate. This is done during the nucleation step, which allows the oxygen atoms to clump together, helping to from a good denuded zone on the surface of the wafer.

than the wafer processed with the denudation process described above. In this case, this is the same wafer that was cleaved prior to processing so that all substrate variables would be reduced, and only the effects of the processing would be apparent.

7.3 PARTICLES AND OTHER CONTAMINANTS

Contaminants are a severe threat to all diffusion processes. The high temperatures associated with rapid diffusion rates, as well as the gas flow rates, unusual and corrosive gas combinations, and a large variety of moving parts in critical locations combine to become a hazard to the wafers at all times. As a result, the possibility of damaging wafers is fairly high. This requires an exceptional effort in cleanliness prevention to protect die and line yield. All of the three types of damage: particulate, chemical, and mechanical are present in the diffusion process area, and will be addressed in order.

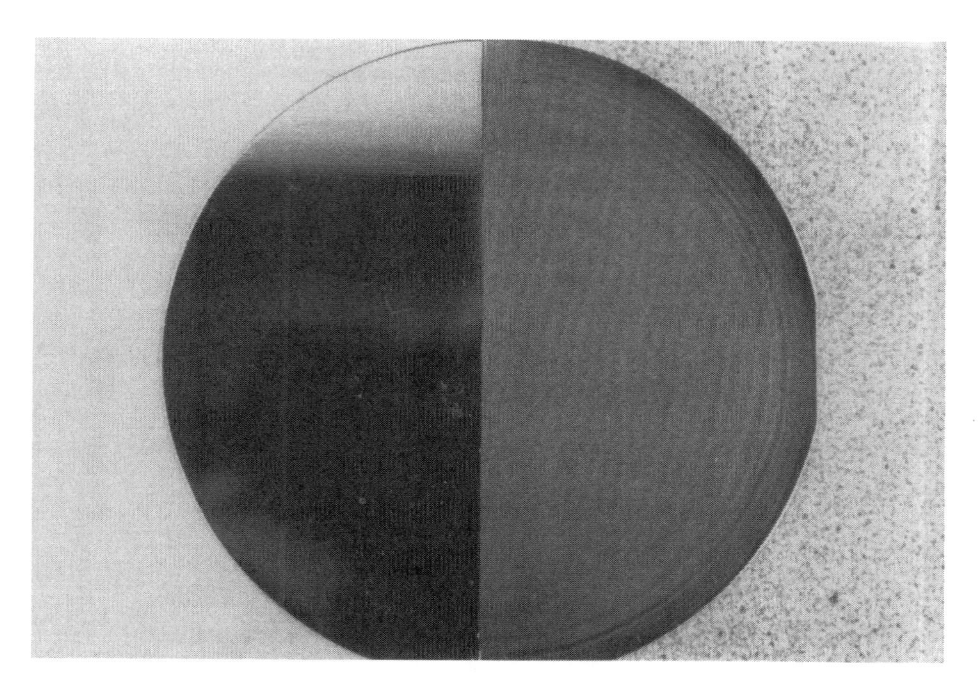

Figure 7-24. The Effects of Denuded Zone Processing In this photograph, one wafer is shown after having been split in two (to guarantee silicon consistency) and processed through the denuded zone processing on the left and a 950°C wet oxidation (i.e., a typical initial oxide process) on the right. After processing, the wafer halves were etched in the same chemical bath in order to highlight silicon crystal defects. The background shows photomicrographs of the surfaces of these wafer halves. As can be seen, the surface of the denuded wafer still reflects the image of the camera. The standard process half of the wafer shows heavy crystal damage. The rotational effect is due to minor variations in the crystal as it was formed in the ingot state.

7.3.1 Particulate Contamination

Particles can affect all of the diffusion processes. There are several effects that can occur as a result of particulate contamination. We will discuss the problems as they relate to the oxidation process first.

As shown in Figure 7-25, a particle landing on the surface of a wafer prior to an oxidation process can have a number of results, up to the inclusion of the particle in the silicon oxide film. Often, if the particle size is much larger than the film thickness, the particle may be jarred loose or cleaned off after the process, leaving an area of unoxidized silicon. That different particle sizes will result in somewhat different problems is not surprising when you consider how the size of the smallest detectable particles (0.25 μm or so) compare to the typical oxide film thicknesses (gate oxides are less than 0.04 μm, while the thicker field oxides are in the 1-μm range). Wherever a particle lands on the wafer during the growth of silicon dioxide will result in a short circuit between the conductive films that the oxide was insulating. In

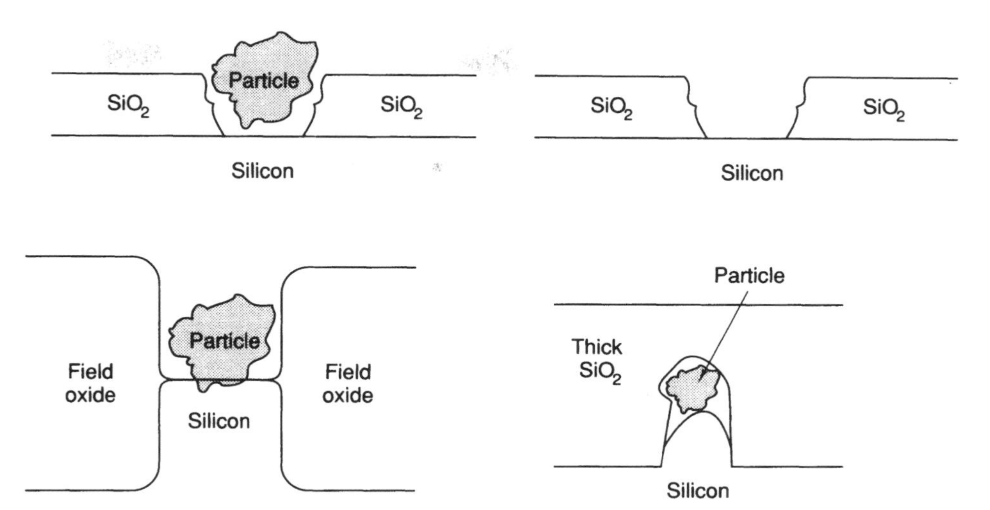

Figure 7-25. Side Effects of Particles in Oxidation Processes A variety of problems can occur as a result of particle contamination of a silicon wafer during oxidation. The particle can stay embedded, causing multiple-layer short circuits. It may be removed, causing a hole or gap in the film. Extreme cases can result in deep pinholes, or the particles can eventually become completely embedded within the film. Yield and/or reliability are both reduced with problems of this nature.

the case of a thin film and a large particle size, there will be no oxide protection at all. This results in the short circuit being created as soon the next layer of film is deposited. In the opposite case of a thick film and a small particle size, the oxide will have a chance to grow around the defects, embedding it in the film. This makes it impossible to remove and often very difficult to detect. These kinds of issues result in longer term reliability problems, reduced breakdown voltage problems, and so on.

The type of particles landing on the surface of the wafer during the oxidation process can also make a difference. This is a problem that is not as serious an issue in masking areas. For instance, a hydrocarbon will burn in the high-temperature oxidizing ambient, and giving off steam will locally enhance the oxidation rates. This gives the wafer a very spotty appearance. Carbon dioxide and other carbon compounds can become embedded in the wafer structure, causing reliability failures. These problems are often seen in an area where there is a "bump" of oxide with discoloration (therefore contamination) of the nearby films.

In addition, hydrocarbon-based particles introduced into the diffusion furnaces come from one of two sources, residual photoresist or human activity. In both of these instances, the hydrocarbon is contaminated with a variety of trace elements, including sodium, although the level of contaminants in human-generated particles is much higher. These elements can be driven off the particle and into the wafer.

In another case the particle may be metallic. A metallic particle residing in an insulating film is a sure killer, drastically changing the electric field profiles through the film, as shown in Figure 7-26. In addition, the metal atoms may be fast or slow diffusors. In the case of an element that does not readily diffuse through the wafer, the particle will probably remain intact until the testing steps, where the effects of the particle will be noted and the cell avoided or the chip discarded. In the case of a rapidly diffusing element, the particle itself may dissolve at high temperature and be absorbed into the surrounding substrate and films. In this case, a void will be left where the particle landed, and the atoms of the particle will have been dispersed to a large segment of the surrounding IC. In severe cases, where the particle has a low boiling point, the resulting phase change can result in a mini-explosion, more or less bursting a hole in the SiO_2. This can lead to the loss of the entire chip and in reliability failures of parts that pass sort as these metallic ions migrate to contacts and other crucial parts of the integrated circuit.

One of the most common types of particle is silicon dioxide. They may come from one of the deposition steps, or from scratches or other defects, but are usually contributed by the quartzware in the furnace. Devitrified quartz is a serious offender for this class of particle. It does not even need to be touched to generate particles which can be carried off by the gas stream and rain in on the wafers from all directions. The

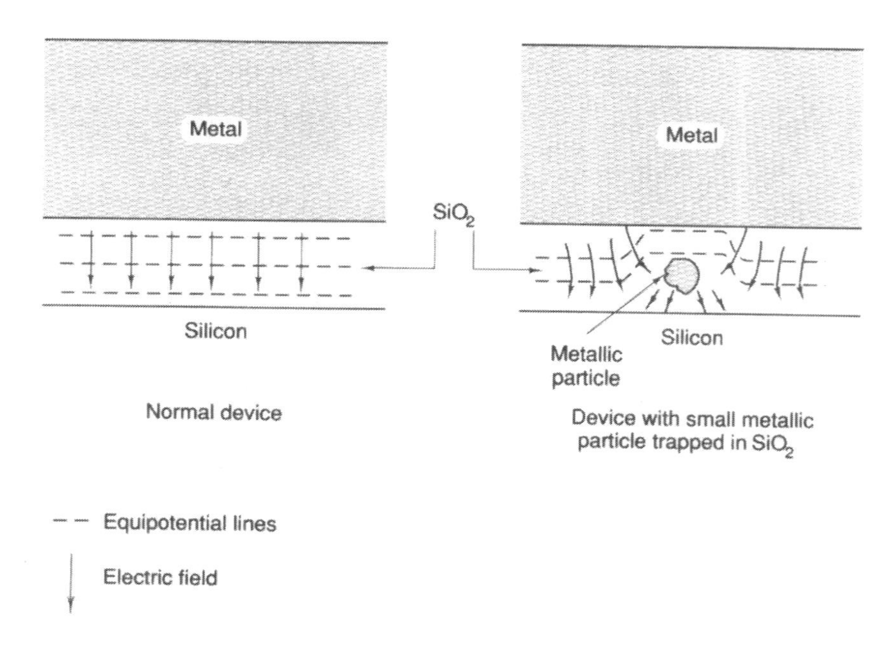

Figure 7-26. Metallic Contamination of SiO_2 Film Even tiny amounts of metallic contamination can significantly change the characteristics of a dielectric film. We see that the electric fields in a device distort and concentrate around the particles, which leads to early device failure.

silicon dioxide particles will often be contaminated with trace elements, will not be of the same composition or density as the growing silicon dioxide film, and will present an abrupt interface to the grown film. Leakage and breakdowns can occur along this interface, resulting in short circuits and related failure mechanisms. Silicon dioxide particles are usually quite large and do not melt or otherwise react unless scrubbed or etched off the surface. Unfortunately this cleaning will result in the removal of the protective oxide layer.

Particle problems in the nonoxidizing dopant redistribution (anneal) processes are similar in most respects to those described in the oxidation processes. These processes include the diffusion steps, reflow and densification steps, alloy, and all other nonoxidizing or nondepositing processes. In all of these cases we still have the same suite of contaminants: hydrocarbons which burn off, causing oxidation to occur (which is usually not at all desirable in these processes) and causing trace elements to be driven into the wafer; metallic particles that will melt on the surface of the film and absorb into it; and silicon dioxide particles that will usually not cause as severe direct damage to the IC, unless they are transporting trace elements into the process. Since these processes are usually used to drive controlled amounts of dopants into the wafer, the conditions are ripe for the redistribution of contaminant species also. Mixtures of types of dopants (i.e., mixing *n*-type and *p*-type dopants and contaminants) can result in changes to the electrical characteristics of the doped regions of the substrate or film. Some contaminants may be driven through the diode junctions and create parasitic transistors and increase diode leakage problems. In general, though, the anneal processes are generally the least sensitive to particle problems of those encountered in the diffusion area.

The last major process to cover while discussing particle contamination in the diffusion area is the phosphorus deposition process. During this process, the particles that land can create a number of problems in addition to the ones that can be described above. For instance, a large enough particle may block part of the deposition, causing a distortion in the deposition, as shown in Figure 7-27. This effect will be enhanced if the oxidation that occurs as a result of the contamination also reduces the diffusion of the dopant atoms and the concentration of the region. Metallic ions and other types of contaminants can also be dissolved into the doped region to cause reliability problems later on.

Clearly, since all of these problems are very small and very localized, occurring only in the region around the particle, the problems become catastrophic only in the event of a high density of particles. The definition of high density for particles varies from fab to fab and technology to technology. Some fabs tolerate defects as high as 3 to 5 per *field of view*, which is defined as the area viewed in an optical microscope at a fixed magnification. This area may only be a

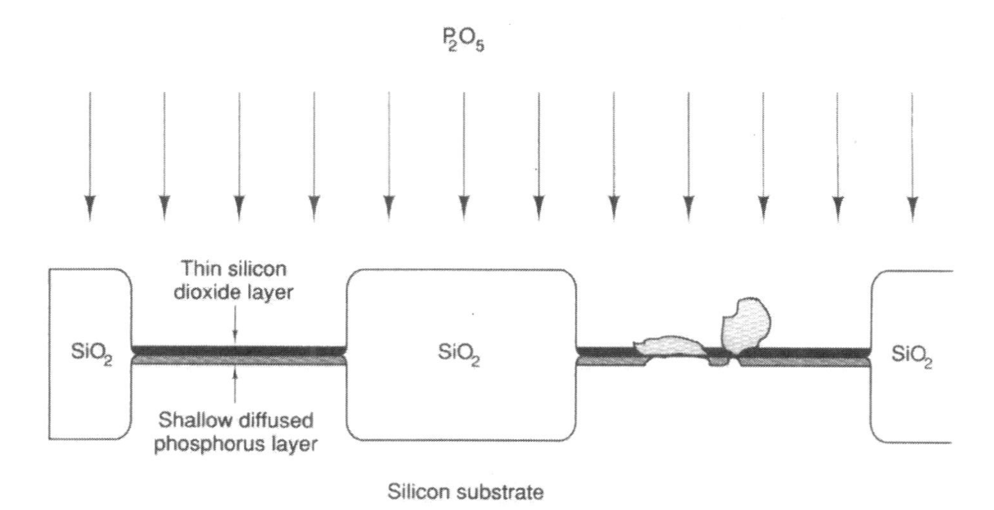

Figure 7-27. Effects of Phosphorus Doping When P_2O_5 is brought into contact with silicon, it reacts to form SiO_2, while leaving the phosphorus free to dissolve into the silicon. When particles land on the surfaces to be doped, they block the flow of P_2O_5 to the surface. This can reduce the concentration of phosphorus or eliminate the doping entirely in small, localized regions.

few square millimeters in area. Since the area of a six-inch wafer is 17,671 mm^2 (176.7 cm^2) this results in a huge number of defects, over 20 to 30 damaging defects per square centimeter. Most fab areas will not allow a defect density that high. A more typical value for a defect density in a process may be 0.5 to 2 defects per square centimeter. In the field-of-view terminology, that would be one defect per every two or three fields of view. This means the wafer would have a total of less than 100 defects on it. Many factories require an even cleaner process, and, in practice, the diffusion and oxidation steps produce only a few particles per wafer. Phosphorus deposition furnaces are often very dirty, however.

7.3.2 Chemical Contamination Problems

We have discussed a number of contamination types and methods of delivery to the wafer. Almost all of them ultimately are bound up with chemical contamination issues. There are a number of problems that occur as a result of chemical contamination. They differ depending on the location of the wafer in the process sequence, but they can be loosely classified. There are contaminants which reside in the surface films and cause changes in the performance of the film through time. An example of this might be trapped sodium atoms in the silicon dioxide film. There can be charges trapped at either the metal-oxide or the silicon-oxide interface which can induce changes in the device performance. There are contaminants that will weaken the structures of the various films and substrates and will diffuse readily, causing havoc everywhere. Gold is a good example of a highly destructive and very mobile atom. Damage such as silicon crystal defects, contact damage through junction spiking, increased diode leakage, decreased breakdown voltages, and related phenomena are possible outcomes from these contaminants.

Chemical contamination levels are generally tested using the C–V (capacitance-voltage) plot. This is a graph of the change in capacitance of a dielectric film as a function of applied voltage under high-frequency conditions. A more thorough description of C–V plotting is given in Section 3.2.1. The oxidation furnaces will usually be tested through a standardized oxidation process with all of the conditions held constant. While this is excellent at controlling the conditions of quartzware and so on, it does not reflect the conditions that the wafers see. As a result, some C–V testing is done by oxidizing the wafers in an ambient similar to that of the production process. Unfortunately, this somewhat reduces the ability of the fab to guarantee cleanliness of the systems. As always, there are trade-offs. The anneal processes typically use wafers of known-quality oxide grown on them prior to the test. In this manner it is expected that gross contamination will appear as a shift in the C–V curve. This is not always as effective as desired, but gives some level of control to the process. The phosphorus deposition furnaces are not generally tested using the C–V process due to the extreme dopant levels that exist within the furnace. Typically, contaminants cannot be discerned from the background noise.

7.3.3 Mechanical Damage in Diffusion

As always, there are sources of mechanical damage for the wafers during most of the phases of the diffusion process. Damage can occur in the wafer transfer sequence, during the push and pull cycles, during the operation of the process, and during the evaluation phase. As in most of the other operations, both machines and humans can create the damage. Machines can cause damage when parts fail and when misaligned. Humans are the cause with mishandling problems and with misaligning the equipment. In general, much of the damage that occurs in the diffusion area should be preventable, since the handling requirements are generally much less strict than in the masking area, where alignments are so critical.

The diffusion process effectively starts with the removal of the wafers from the box. From there, the wafers are transferred to the quartz boats from the Teflon® storage cassettes. This is done with vacuum wands or automatic transfer machines. In some cases, tweezers are allowed for handling wafers, but very seldom. Tweezers are generally banned from wafer fabrication areas, due to the difficulty of using them without scratching or otherwise contaminating the wafers. When wafers are handled with vacuum wands, the most common problem is bumping the wafers together while placing them in the quartz boats or replacing them in the Teflon® cassettes. This damage can be surprisingly severe. Wafer transfer systems must be kept aligned, or they may cross-slot the wafers or place the wafers in uneven slot numbers. When this happens, the next wafer in the boat or cassette will come into contact with the back of the adjacent wafer. This will scratch the surface, allow particles and other contamination to pass from one wafer to the next, and, in some cases, result in enough stress being placed on the wafer that it will break. This is especially true of substrates such as gallium arsenide or the various indium compounds, which are very fragile.

Wafer transfer systems should be tested for operation in the worst case conditions, such as power or vacuum loss, as these systems often contain several cassettes and there is a good chance of significant loss of wafers with insufficient safeguards. The quartz boats should be checked for alignment, slot size, and quartz quality prior to each run. Many wafer transfer systems can automatically choose and reject boats based on slot and size information. The choice of clean, vitrified quartz boats is still left up to the operator. It is imperative that the operators have the authority to reject any and all parts that do not meet quality requirements, and that they understand the importance of new, clean quartzware. Wafers that are placed in damaged boats can lean into one another, or can cause significant wafer damage when the quartz boat warps, bends, or otherwise stresses the wafers while at temperature.

The diffusion furnace will usually have an automatic push/pull assembly. These devices can fail for a number of reasons, with significant consequences to line yield in the event of failure. These problems can include warpage of the lead screw, damage to bearings and motors, improper lubrication on the lead screw, and other component failure that can lead to vibrations or oscillations of the load. This can cause the load of wafers to periodically contact the walls of the furnace tube, generating particles and causing wafer breakage. The push/pull mechanism can sometimes fail part of the way in the furnace, or can be running at a rate different than programmed. These types of problems can cause minor changes in dopant redistribution levels, and can increase the chances of warpage of some wafers. Other problems include the shedding of particles from the moving parts. This shedding can result in very high localized particle counts that can be blown all over the wafers by the laminar air flow system. This is especially important in the case of the bearings and lead screws of the automatic push/pull device. They should not be coated, and should not in general be lubricated. Only Teflon®-type lubricants in *very* small quantities should be used and only if absolutely necessary. Particle generation from these sources can contaminate entire loads of wafers at once. More advanced furnaces use belts or chains to drive the load into the tube.

Other furnace problems can result in damage to the wafers, and become safety hazards. For instance, an improperly set steam generation system, coupled with a failure of thermocouples, can get the furnace into an explosive condition. Most modern furnaces have redundant fail-safe systems built in, although hydrogen explosions have been known to occur on occasion. While, fortunately, they have seldom been injurious to the humans in the fab, these explosions usually turn all of the silicon in the tube into tiny fragments. Obviously, there is a problem with line yield in this situation, as losses in each explosion approach 100%. A more subtle problem occurs if the furnace temperature ramp rate is greater than or less than the programmed ramp rate. In this case, an inappropriate temperature change may cause the furnace to start the oxidation process at the wrong time, or cause the total time at temperature to exceed specifications. This last case is especially true in the event the furnace is allowed to ramp to a temperature without an adequate end-point time limit and has not been properly characterized to permit stabilization.

Even after the process has been completed, there are still a number of locations at which mechanical damage can be introduced. They are usually associated with test and inspection equipment. Misalignment or mechanical failures may cause microscope objectives or other parts to come into contact with the wafers. While this may sound like an obvious thing, it is important to keep these kinds of details in mind while evaluating equipment for purchase or use. Preventative measures should be in place for the machine, for instance, hard stops on microscope platforms that do not permit overextension of the focus. Preventive maintenance must be scheduled on all equipment, even if the work seems to be small. It is very often the support equipment that causes a significant amount of contamination and damage to the wafers while in the fab area.

7.4 METROLOGY REQUIREMENTS FOR THE DIFFUSION AREA

A number of attributes must be measured in order to control the diffusion processes. The exact procedure depends on the type of process involved. The main attributes that must be measured are film thickness, index of refraction, and film or substrate resistivity. Other attributes such as warpage, stress, pinhole content, and so on must also be performed on occasion.

Film thicknesses are measured by any one of a number of techniques, which include ellipsometry, spectroscopy, some variations on the basic spectroscopic methods, and a variety of optoacoustic techniques. The various techniques are shown in Table 7-3. All of them have their advantages and disadvantages. In almost all cases, the

TABLE 7-3
Film Thickness Measurement Techniques

Measurement technique	
Spectrophotometric	
Method:	Analyzes reflected interference sectrum
Advantages:	Fast, small spot, repeatable
Disadvantages:	Sensitive to changes in reference and in optical film characteristics
Ellipsometric	
Method:	Analyzes polarization angle of reflected beam
Advantages:	Accurate, good analysis of optical characteristics
Disadvantages:	Large spot, some thickness regions provide poor resolution
Modified spectrophotometric	
Method:	Analyzes reduced-range spectrum
Advantages:	Fast, inexpensive, repeatable
Disadvantages:	Large spot, sensitive to film characteristic changes, also prone to "order jump"
Prism coupling	
Method:	Measures changes in light deflection due to refractive index of thin film
Advantages:	Accurate, high precision
Disadvantages:	Prism must contact wafer. Film must be over 2500 Å (function of laser wavelength)
Profilometry	
Method:	Stylus surface contact
Advantages:	High accuracy, insensitive to film changes
Disadvantages:	Stylus must touch surface and must move over a step
Metal film measurements	
Method:	Analyze eddy currents with magnetic field
Advantages:	Repeatable, fast, measures opaque films
Disadvantages:	Affected by changes in film resistivity

methods have areas of weakness in which the repeatability or the accuracy of the readings may vary greatly. For example, there are a number of regimes where these methods have trouble resolving differences between a variety of potential readings. An example of this can be seen from a quick glance at the ellipsometric curves (see Chapter 10). In these regimes, the results can vary greatly as a result of minor discrepancies, such as signal noise and other problems. Sometimes these values can reach across several ranges. These abrupt changes are called order-jumps, and can be very annoying, as well as deceptive. Silicon oxide thickness ranges of around 5000 Å are notorious for being difficult to measure. Combinations of multiple films add even more uncertainty to the effort.

The manufacturers of these devices go to great lengths to improve the reliability of their instruments but the results of these efforts will go to waste if the machines are allowed to deteriorate or are not maintained. For instance, the aging of light sources and lasers or of photomultiplier detectors, as well as misalignment of the optical paths of these machines, can cause measurement problems to appear. It is often a requirement to have several film-thickness monitors in the fab area, using different measurement techniques to verify that all of the possible ranges are covered and that all of the machines remain in calibration. It is also important to keep the fab standards up to date. There are a variety of film-thickness ranges that can now be traced to NIST standards. However, they are limited in general to thermally grown silicon dioxide films of specific thicknesses. Unfortunately, they are in relatively easy ranges to measure, and do not reflect the performance of a film-thickness monitor on other films in any way (except to guarantee that the hardware is set up correctly).

Most of the instruments that can measure film thickness are capable of measuring index of refraction. This is not surprising since the data required to determine film thickness are the same as those required to measure index of refraction. In general, it is difficult to measure both index of refraction and thickness simultaneously. The data that are used to calculate this information can vary both by changes of index and thickness. As a result, it is possible to get a much more precise reading on thickness if the index of refraction is held constant or is determined through independent means. With oxidation processes, the changes of index of refraction will be minimal. We will see in the LPCVD chapter that, for deposited films, the index of refraction of any specific film can vary greatly. It should be pointed out that there are minimum thicknesses beyond which it is nearly impossible to measure index of refraction and film thickness simultaneously. This is usually below around 500 to 1000 Å, depending on the measurement technique.

Substrate and film resistivity are measured through two main techniques. The older and more common method is that of the four-point probe. This device contacts the surface of a test wafer after the phosphorus deposition or diffusion/redistribution process (and after the deglaze step). A voltage is placed across the probes and a known current is sent into the device. From this, the resistance of that film or substrate layer is determined. See Figure 7-28 for a visual explanation of this procedure. There are a number of disadvantages to this technique. The first, and most obvious, is that it cannot be used on product wafers. Since the wafer must come into direct contact with the probe, damage will result wherever the probe is placed. In addition, due to the complexity of the surface of the patterned wafer, it is doubtful that reasonable readings could be obtained from the instrument if it was somehow able to be used without damage. In emergency situations, it may be possible to read the resistivity of the product wafers or of the films by measuring the backside of the product wafer. While this will result in the loss of that wafer for any further use, at least one can obtain critical information on whether the process is in spec. It should

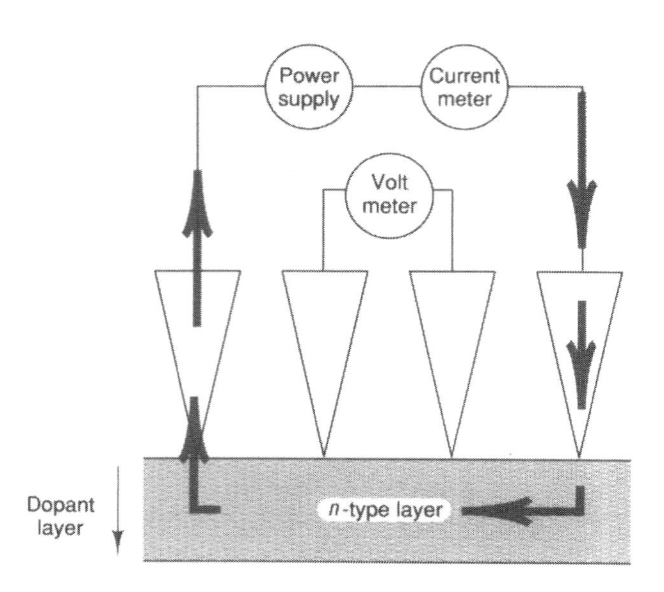

Figure 7-28. The Four-Point Probe The four-point probe is one of the primary methods for measuring sheet resistivity on test wafers.

be pointed out that polysilicon films will read much differently than single-crystal silicon, and there may be other changes between front and back readings, due to structural differences in the silicon. Thus, when product wafers are read, these differences should be considered. In most cases, backs of product wafers will have higher apparent readings than test wafers, although this should be characterized under known conditions before making any decisions or assumptions.

Some of the other devices used in the diffusion area include the surface-borne particle detectors and pinhole detectors. Since particles and pinholes are usually related in diffusion processes, the tests are sometimes associated with key processes. Most diffusion processes are tested for particle density and generation during each run. Pinhole detection is usually done on key test wafers only when a problem is suspected. Another common device used in the diffusion area is the stress gauge. This device measures the deflection of the wafer as a function of processing. Since films deposited or grown on the surface of the wafer will have different characteristics than the substrates, and differing thermal expansion coefficients, there will be a probability that the wafer will be stressed when the process is complete. One goal of the wafer process sequence is to minimize this stress. The stress tests have little impact on the quality of the wafer, so therefore can be run with product wafers if desired.

All of the various processes should have statistical quality control charts associated with them, regardless of the type of process, which should include thickness, resistivity, particulate levels, and other information as required. While we will get into the details of SQC in another chapter, we can see that a variety of different parameters must be regularly monitored. For instance, an anneal process may allow no more than a certain amount of growth, while gate oxidation processes will require growth within a narrow window. Field oxidation processes will require growth to within a fixed but larger window.

7.5 DIFFUSION PROCESS WRAP-UP

We have discussed the various processes that are available in the diffusion area. They are basically related to one of several activities: oxidation, dopant deposition, dopant redistribution, alloy/sinter, and anneal. These processes are usually critical to the operation of the devices, although there is a wide range of sensitivity with these processes. We have seen that chemical contamination is a serious issue with these processes, as contamination will be rapidly redistributed throughout the furnace tube, and can contaminate large numbers of wafers in a short period of time. Physical damage can occur in a number of ways, from the standard scratches and particle defects to the propagation of silicon crystal defects. The exact level of damage is dependent on the process sequence itself.

There are a wide variety of diffusion and diffusionlike processes that have not been discussed in depth here, but which will be becoming more important as time progresses. They include the development of films based on the high-temperature or high-pressure processing of wafers. High-pressure oxidation will be used to produce SiO_2 films at high rates and at very low temperatures. High pressure and temperature nitridation of SiO_2 can be performed, which allows for the creation of dense, stable films of silicon oxynitride for use as gate dielectric films and interpoly insulators. Many of the films in the refractory metal silicide group are produced by first depositing a film and then annealing the film in an appropriate ambient. As a result, we can expect to see novel techniques used for semiconductor wafer manufacturing as time progresses.

Finally, in general, there has been a continuing trend toward more precise and thinner films in all of the diffusion processes. Gate oxidation processes have been reduced from about 900 to 1000 Å in the early 1980s to about 300 Å or less today. The junction depths of the silicon diode have similarly been reduced to only a few tenths of a micron. The tolerances on these films have been reduced significantly, to the point where the industry has started to reach the level of intrinsic control available, and will need more precise instrumentation. For instance, many manufacturers are requiring oxidation thickness of 100 ± 5 Å. This can be very difficult if the measurement equipment has a repeatability of only ±2 to 5 Å, or an accuracy of ±5 Å. These types of problems must be resolved or the diffusion processes may become a limiting factor for quality in the fab area.

CHAPTER 8

LOW-PRESSURE CHEMICAL VAPOR DEPOSITION PROCESSES

Semiconductor wafers require the deposition of a wide variety of films on the their surfaces. The films may be metallic and therefore conductive, or dielectric and therefore insulating. The addition of various types of dopants to the films creates a wide variety of effects. Deposited films are differentiated from the films produced in the diffusion processes in that the films are not grown or diffused into the silicon substrate. The deposited films are produced from chemical or physical reactions in the environment immediately above the wafer surface. The reactions typically take place in a high-vacuum system, making the requirements for the process even more strict than for many of the other processes. An additional complexity for the deposition process is that the processes continually deposit films on all of the surfaces inside the chamber, which makes them self-contaminating. This limits their effectiveness to times when the system can be kept clean enough to produce films without excessive particle contamination.

There are two primary types of deposition processes used in a wafer fab. The first is the chemical vapor deposition, or CVD operations. The are a variety of subgroups of these, for example, LPCVD (for low-pressure CVD), APCVD (atmospheric-pressure CVD), PECVD (plasma-enhanced CVD), or LACVD (laser-activated CVD). These types of reactors are shown in Table 8-1. These types of reactors force gases to chemically react in the vacuum chamber, which results in solid materials precipitating out of the gas, which forms the deposit on the wafer surfaces. These processes will be covered in this chapter.

TABLE 8-1
LPCVD Reactors and Processes

Reactor	Process	Gases required
LPCVD	Si_3N_4	SiH_2Cl_2; NH_3
	Polysilicon	SiH_4
	SiO_2	SiH_4; O_2
	PSG	SiH_4; O_2; PH_3
	BPSG	SiH_4; O_2; PH_3; BCl_3 or B_2H_4
	W	WF_6, H_2
APCVD	SiO_2	SiH_4; O_2
	PSG	SiH_4; O_2; PH_3
	BPSG	SiH_4; O_2; PH_3; BCl_3 or B_2H_4
	Si_3N_4	SiH_2Cl_2; NH_3
	SiO_xN_y	SiH_2Cl_2, NH_3, O_2
PECVD[a]	SiO_2	N_2O; SiH_4; O_2; TEOS
	PSG	N_2O; SiH_4; TEOS; B_2H_6; PH_3; TMP; O_2
	BPSG	N_2O; SiH_4; TEOS; PH_3; TMB; TMP; O_2
	Si_3N_4	NH_3; SiH_4; TEOS
	SiO_xN_y	NH_3; SiH_4; TEOS; O_2

[a] LACVD, ECR, and other plasmalike technologies work with similar chemistries.

The second type of CVD process used in a wafer fab is that of physical vapor deposition. These are the processes, such as sputtering, where a material is removed from one surface (the target), such as a purified metal plate, and then is deposited on another surface placed opposite the target. The physical vapor deposition processes will be covered in the next chapter.

Temperatures in the CVD processes typically range from 300 to 800°C, and therefore are in temperature ranges that can permit redistribution of chemical contaminants that may have been introduced into the reactor. The temperatures in the metallization processes are typically lower than those of the dielectric deposition processes, although, after the metals have been deposited on the wafers, the succeeding dielectric films must be kept at relatively low temperatures, also. This is due to the lower melting temperatures for the metallic films. For example, aluminum films cannot be heated above about 450°C.

Clearly, particulate contamination and mechanical damage can be significant sources of problems in chemical vapor deposition processes, as the equipment has many inherent sources of contamination. Adequate preventive maintenance is very important with these types of process in order to correct problems quickly. In all of these processes, small problems are often known to grow into very large ones if not addressed immediately.

8.1 CHEMICAL VAPOR DEPOSITION PROCESS EQUIPMENT

Chemical vapor deposition processes are typically divided into two major categories: low-pressure and atmospheric-pressure CVD processes. There are a variety of techniques for depositing films within each of these categories, especially in the low-pressure regime. We will start with an examination of the equipment used to produce film using LPCVD processes, including the thermal CVD, plasma-enhanced CVD, and some of the more exotic processes. We will then proceed to a discussion of atmospheric CVD systems.

8.1.1 Standard Thermal LPCVD Furnaces

As shown in Figure 8-1, the classic LPCVD furnace looks very similar to the standard diffusion furnace discussed earlier. The gases are typically injected into the furnace in the front of the tube, or through specially constructed injectors along the bottom of the tube. The gases swirl through the tube and over the wafers, eventually exiting through the vacuum pump inlet at the back of the tube. This vacuum pump must be of sufficient strength to be able to pull the reactive gases out of the reactor quickly and in a smooth manner. The tube is contained within the furnace heating elements just as is the tube in a standard diffusion furnace. The energy from the heating elements is used to initiate the chemical reactions within the chamber. As a result, the standard LPCVD process is often called a thermal LPCVD process.

Other subsystems that the LPCVD system have in common with the diffusion furnaces are the gas handling and wafer handling systems, as well as the computer control systems. Since there is so much in common, a number of the equipment vendors that have manufactured diffusion furnaces have at some point manufactured LPCVD furnaces. In a similar manner to those discussed in the last chapter, the most common layout for the reactors has been horizontal. However, an increasing number of vertically mounted systems have been installed in wafer fabs in the last few years, due to floor-space and wafer handling requirements. A typical layout for a vertically mounted thermal deposition chamber is shown in Figure 8-2.

The various parts of the interior of the system are usually made of fused quartz. They include the tube, the cassettes or "boats," cantilevers, gas injectors, thermocouple sheaths, and other items as required.

→

Figure 8-1. LPCVD Furnace—Horizontal Layout Shown is a block diagram of a standard low-pressure CVD furnace. The furnace computer controls each of the various subsystem computers and controllers, such as the gas, temperature, and vacuum subsystems. In the figure, the wafer loaders are shown in both the "in" and "out" positions. The inset details the operation of the vacuum control system as described in the text.

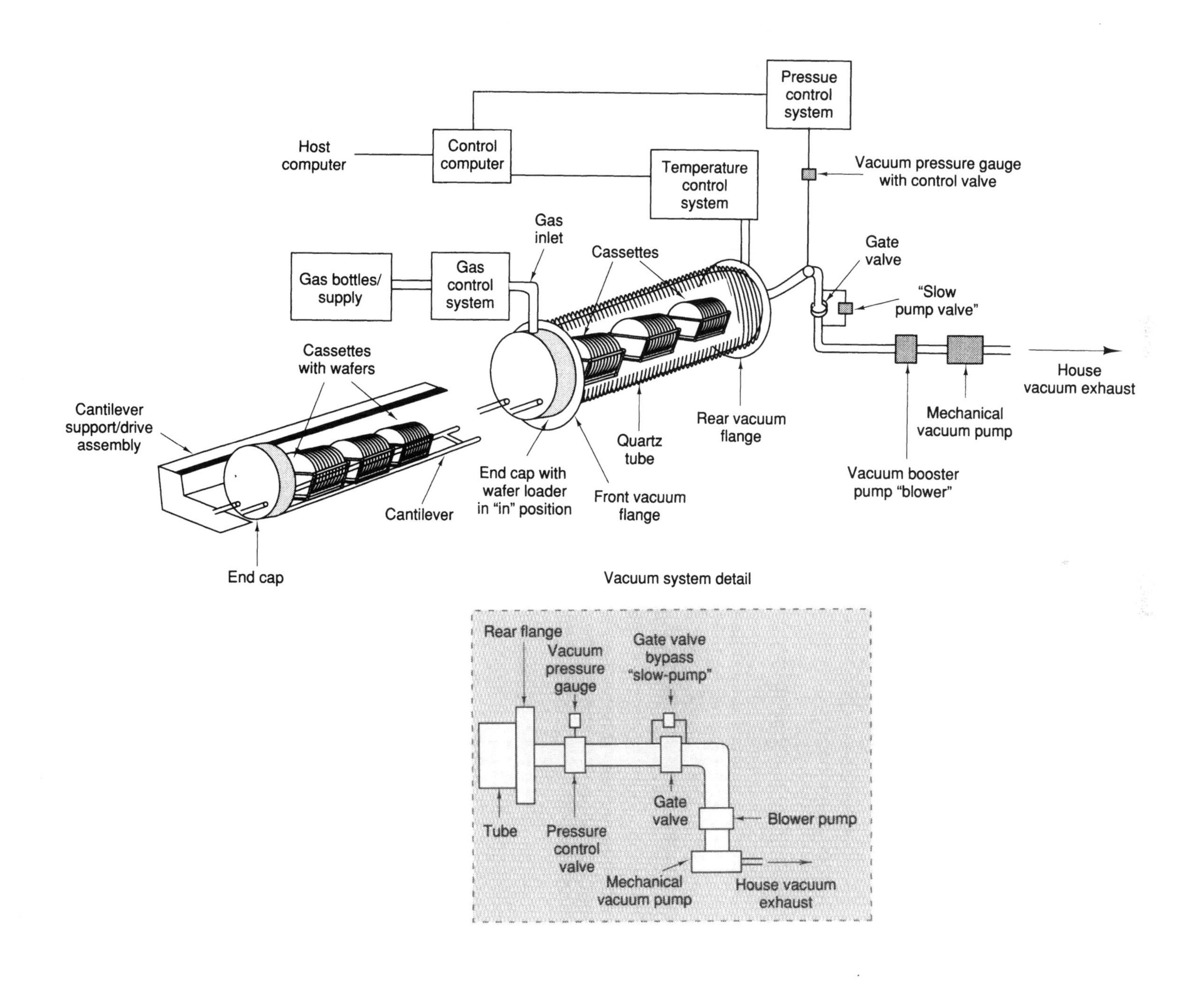
Pressue
control
system
Host
computer
Control
computer
Temperature
control
system
Vacuum pressure gauge
with control valve
Gas
inlet
Gate
valve
Cassettes
Gas bottles/
supply
Gas
control
system
"Slow
pump valve"
Cassettes
with wafers
House
vacuum exhaust
Cantilever
support/drive
assembly
Mechanical
vacuum pump
Rear vacuum
flange
Quartz
tube
End cap with
wafer loader
in "in" position
Vacuum booster
pump "blower"
Front vacuum
flange
Cantilever
End cap
Vacuum system detail
Rear flange
Vacuum
pressure
gauge
Gate valve
bypass
"slow-pump"
Gate
valve
Blower pump
Tube
Pressure
control
valve
Mechanical
vacuum pump
House vacuum
exhaust

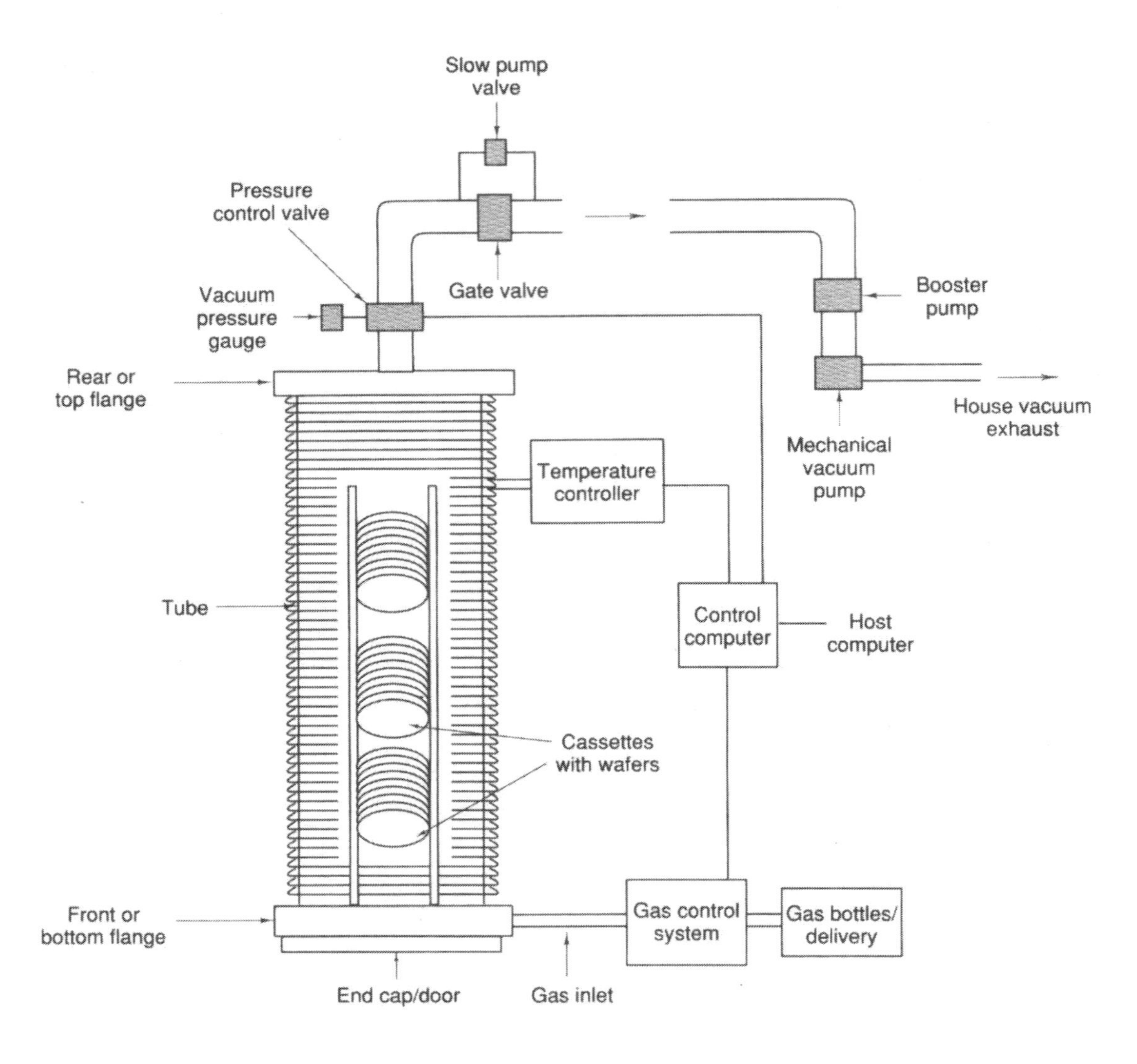
Slow pump
valve
Pressure
control valve
Vacuum
pressure
gauge
Gate valve
Booster
pump
Rear or
top flange
House vacuum
exhaust
Mechanical
vacuum
pump
Temperature
controller
Tube
Control
computer
Host
computer
Cassettes
with wafers
Front or
bottom flange
Gas control
system
Gas bottles/
delivery
End cap/door
Gas inlet

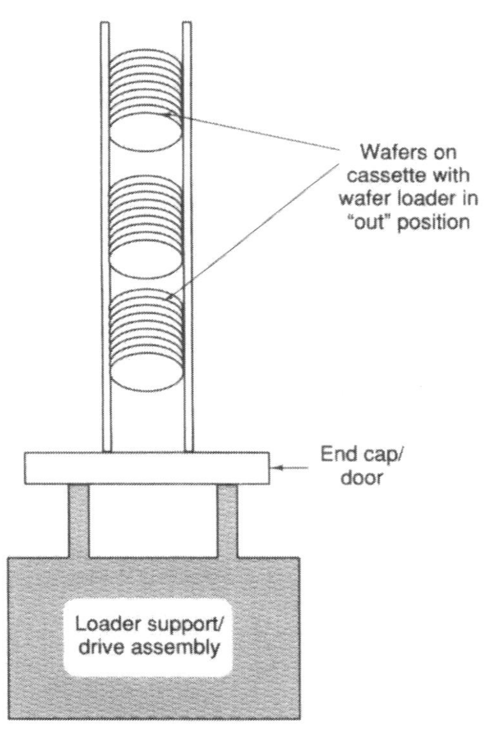
Wafers on
cassette with
wafer loader in
"out" position
End cap/
door
Loader support/
drive assembly

The issues that are involved with quartzware in the LPCVD processes are similar to those in the diffusion processes. Quartzware usage tends to be much greater in the LPCVD processes than with the diffusion processes. This is due to the fact that the films that are deposited on the quartzware will eventually crack and fall off, forming severe particle problems. The frequency of quartz changes varies greatly, depending on the process. A BPSG furnace tube may need to be changed every 15 to 20 runs, whereas a polysilicon tube typically does not need to be changed for well over 100 runs. The differential between the coefficients of expansion of the deposited films and the quartz, coupled with drastic differences in brittleness and adhesion, explains the differences in cleaning frequency for the various processes.

Quartzware that has had films deposited on it must be cleaned very carefully to prevent any damage to the polish of the quartz. Since films are actually placed on the quartz, they must be removed and the quartzware restored to as near to new condition as possible. The problem is that the films must be removed in HF acid, which attacks fused quartz also. The acid will etch into the surface of the quartzware, causing the slots of cassettes to widen, and the quartzware itself to devitrify. The loss of the surface polish on the quartz will ultimately contribute significantly to the particle problems of the furnace. As can be seen from Figure 8-3, a roughened surface can create particles in a number of ways. They include cracks from the attempted deposit over complex topology, from highly stressed multiple layer films, and from the rubbing of silicon or other quartzware on the roughened, loosened film pieces. In addition, the lack of a good surface polish will allow chemical contamination and moisture an easy avenue for penetration into the quartz.

Deposited silicon dioxide films, especially BPSG, are not especially stable and can be removed quickly even with weak HF solutions. One of the most significant problems with the cleaning procedures is overetching the quartz and damaging it, as complete removal of the film is not easily verified. Polysilicon films require a more complex process to remove, typically requiring the use of a mixture of nitric acid and HF. It is important to keep this combination weak (specifically, keeping low quantities of HF in the solution); otherwise severe damage can occur on the quartz before all of the polysilicon has been removed. Silicon nitride films are cleaned with a stronger solution of HF. The silicon nitride adheres strongly to the surface of the tube, creating a significant amount of overetching and devitrification. In all of these cases, it is important to quantify exactly how many depositions can be performed between quartz cleans, and to determine

Figure 8-2. LPCVD Furnace—Vertical Layout The vertical LPCVD furnace contains the same basic components as the horizontal furnace setup, but has advantages in reduced complexity, lower particle counts, and easier wafer handling.

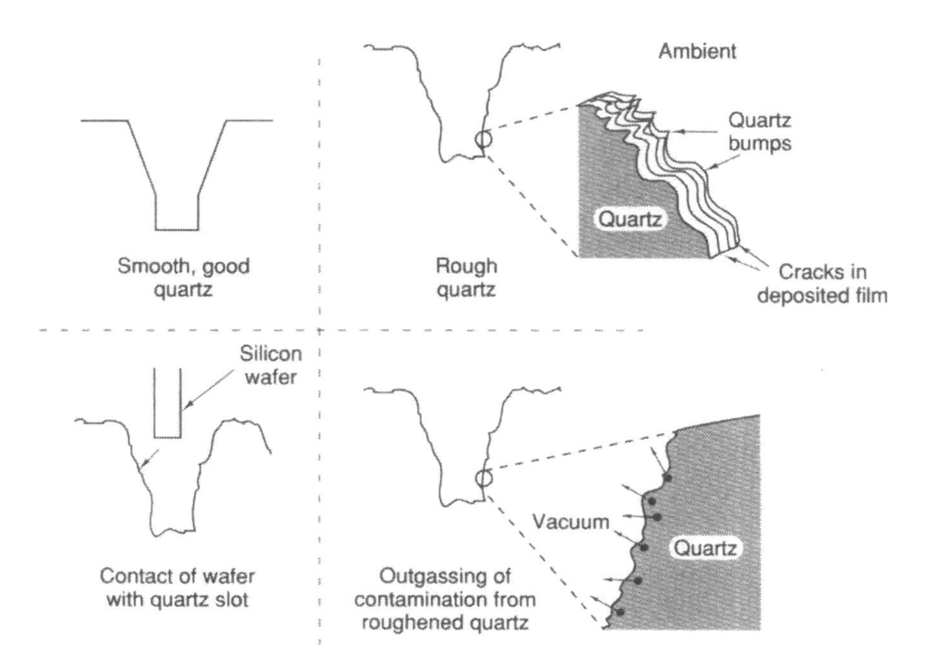

Figure 8-3. Effects of Poor Quartz Quality in LPCVD Furnaces The deposition processes are even more sensitive to poor quality, damaged or roughened quartzware than the diffusion processes. The continuing deposition of films increases the potential for particulate contamination, especially in the process of placing the wafers into the quartz cassette's slots. Chemical contamination and moisture and oxygen from the air can be introduced into the quartzware while it is out of the tube, or being cleaned. This contamination will be drawn out of the quartz when it is brought down to a vacuum.

the exact solutions to use for quartzware cleaning, to minimize quartzware loss through devitrification. Quartzware is very expensive, and should be kept in usable condition as long as possible.

Unfortunately, many LPCVD processes are too sensitive to contamination to allow for any chemical processing to take place on the quartzware. Sometimes the chemicals will not be completely removed, and will outgas in the vacuum environment, or the particles generated as a result of the devitrification of the quartz may be too severe. In some cases, the tubes are usable for a short period, but must be replaced very soon after cleaning, which impacts the downtime of the process. As a result, many fab areas do not allow LPCVD quartzware to be cleaned and reused after it has reached the end of its useful life. Instead, the old quartz is discarded and new quartz is installed. In this way, all quartzware used in the furnace is always smooth and film deposition on the quartz can be maximized. While this seems like a very expensive alternative, a close working relationship can develop between the fab and the quartz vendor, with the quartz manufacturers carefully cleaning and repolishing the tubes and other quartz pieces for use at a fraction of the cost of purchasing completely new quartzware. Unfortunately, once devitrification has proceeded too far, the quartz cannot be properly recrystallized and must be thrown out.

Many LPCVD processes require the use of special cassettes in order to guarantee gas flow uniformity, and therefore film deposition uniformity. These cassettes are shown in Figure 8-4. Since they are so large and must cover the wafers completely, they are particularly sensitive to damage and to chemical contamination. The furnace also requires a much larger tube size than the equivalent diffusion furnace in order to accommodate the extra diameter of the cassettes.

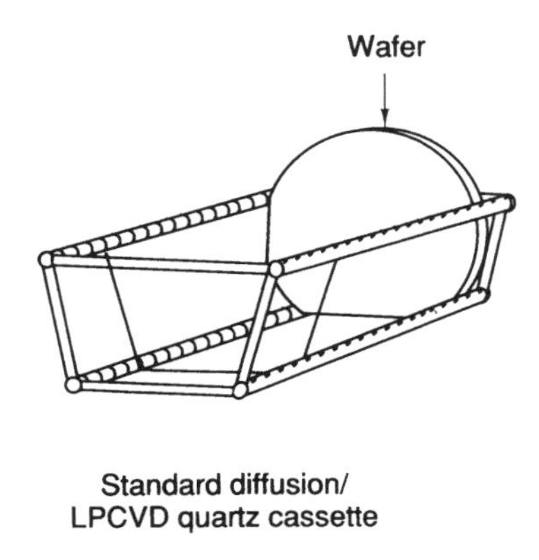

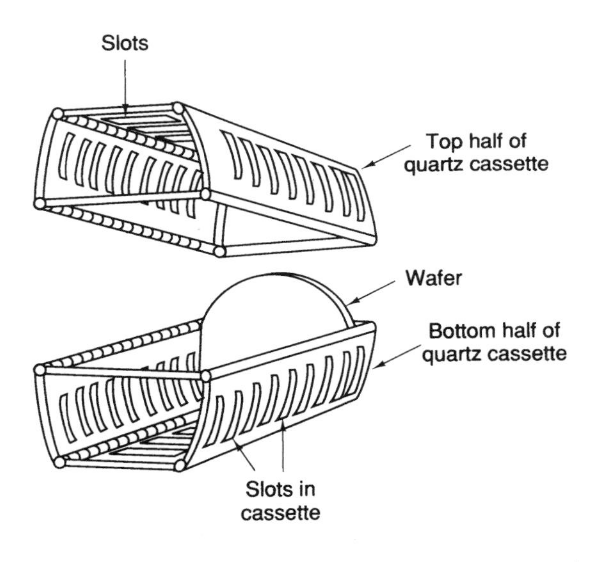

Figure 8-4. Quartz Cassette Types Shown are the two most common LPCVD cassette types. The one on the left is an open shell type cassette used most often in the nitride and polysilicon deposition processes. On the right is the closed-cage type of cassette used in SiO_2 and BPSG depositions. The type of cage is selected by comparing the residence time of the reactive gases with the reaction rates of these gases. The cage type of cassette will hold the gases in toward the wafers longer.

It is even more important in CVD processing to make sure that the cantilevers or other parts of the wafer handling systems do not make contact with the tube, or any other stationary pieces of quartz in the tube, than it is with the diffusion processes. Not only will particle problems be greatly amplified, but the possibility for quartzware cracking or chipping increases dramatically, due to the larger mass of quartzware moving and the greater stresses induced by the vacuum environment inside the tube.

Chemical contamination with LPCVD quartzware is nearly as significant a problem as it is with diffusion quartzware. Contaminants are easily redistributed through the systems and can contaminate a number of wafers at the same time, and must be kept out of the CVD furnaces at all times. It is critical to verify that the systems do not become cross contaminated, as this type of problem can create very subtle and difficult-to-trace problems.

The final issue when dealing with quartzware in the LPCVD processes is that of small cracks or chips in the quartz. These chips often lead to leakage problems, allowing atmospheric gases into the chamber. This almost always produces hazy or otherwise contaminated films. The leaks can develop in a number of locations, depending on the type of damage, as shown in Figure 8-5. Typical maximum leak rates for these types of furnaces are relatively high for a vacuum process, with ranges around 1 to 5 mT per minute allowable.

A significant area for leaks to occur in is the endcap/tube interface region. The endcap must be placed on the quartz tube in the correct manner in order to verify that it will not crack the tube or leak. Some of

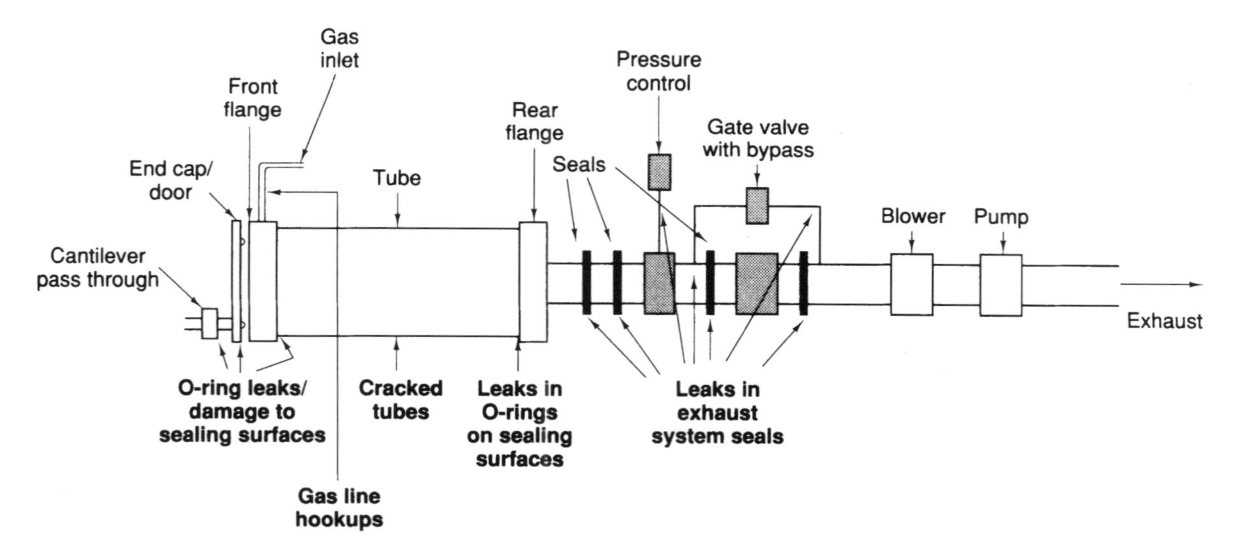

Figure 8-5. Potential Points of Leakage in an LPCVD Tube Potential areas for leaks to occur in an LPCVD furnace are pointed out in this figure. While the number of points is large, careful maintenance and layout can prevent most of them from becoming common areas of problems.

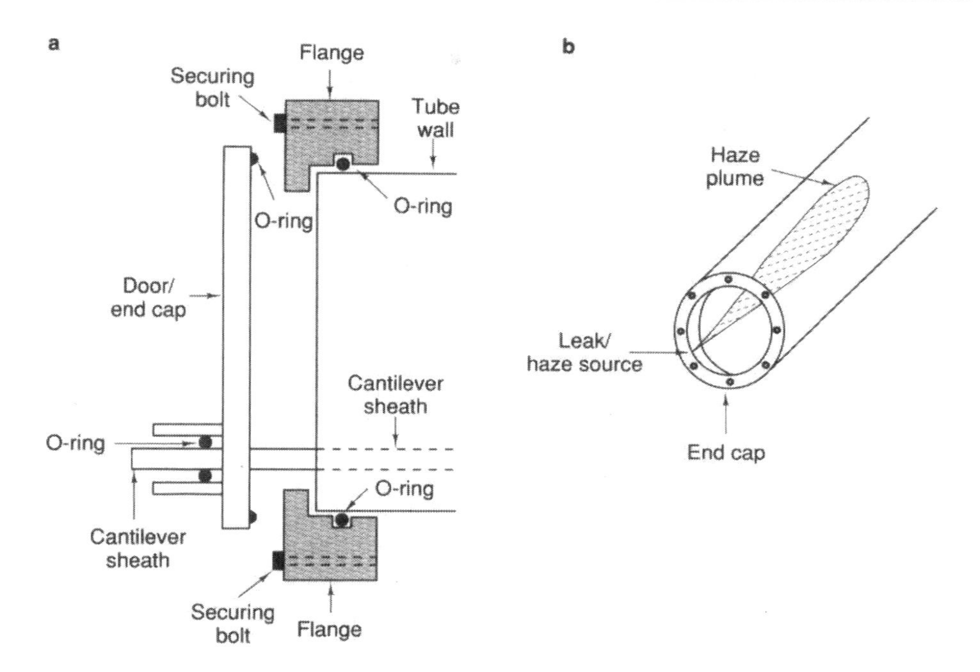

Figure 8-6. Details of LPCVD Endcaps (a) As we can see, there are a number of sources of potential leaks in the LPCVD tube endcap assembly. Each o-ring represents a potential area for leaks to occur around. Flat, cut, or frayed o-rings will develop leaks at the damage points. (b) Leaks in the endcap assembly will allow haze or smoke to form in the tube, starting at the leak point and spreading in a plume back toward the vacuum pump (and past the wafers, causing excess contamination).

these problems can be reduced by proper endcap design. As seen in Figure 8-6a, the endcap is placed over the end of the quartz tube, and is sealed through the use of o-rings. As was very dramatically shown in the Challenger space shuttle accident, a leak through the o-rings sealing two areas of vastly different pressures can cause significant problems. Several problems can occur in the use of o-rings. The first and most significant is that of scoring or cuts in the o-rings. A second problem is that of the o-rings overheating and melting. Any damage of the o-rings will cause leakage problems at that point. A third problem is that of inconsistent tightening of the o-ring. This can lead to bulges and thin areas in the o-rings, which also will allow leaks. A typical pattern of contamination that results from a leaky o-ring is shown in Figure 8-6b.

The prevention of damage to the o-rings is largely a matter of training and proper equipment. Screwdrivers and other metal tools should never be used to place o-rings into their slots, nor should these types of tools be used to remove them. There are special plastic tools that can be safely used on the o-rings. However, in most cases, there is nothing that can replace the (gloved) hand loading of an o-ring. Various organizations have different opinions of vacuum grease on o-rings.

In most cases, the vacuum grease is not desirable or necessary but, if required, the use of very small quantities of Fomblin-type low vapor pressure greases is recommended. The grease should be spread uniformly until it just coats the entire surface of the o-ring uniformly. If there is a visible sheen on the o-ring from the grease, you have used too much. Vacuum grease will sometimes help prevent or lessen the impact of imperfections in the surface of the o-ring and will ease its installation into certain types of o-ring grooves.

The use of stainless steel for the construction of the endcap results in a temperature control problem for o-rings used in high-temperature LPCVD processes such as silicon nitride deposition. At these temperatures (800°C and up), the stainless steel used with the standard endcap will not dissipate heat rapidly enough to keep the o-rings within specified temperature ranges and they melt. This causes severe problems, as the melting o-rings emit fumes into the chamber as well as atmosphere into the tube and thoroughly contaminate everything.

To alleviate this problem, there have been a number of solutions tried. Aluminum endcaps can handle the heat dissipation, but are soft and must be anodized (which cannot be done on the interior surface of the door or other critical parts), or otherwise treated or it will oxidize, forming particles. The use of dichlorosilane in the nitride systems means there is a significant chance of chemical reaction between the free chlorines and the aluminum in the door. Also, anodization layers can be physically damaged, and can form particles. Thus, the use of aluminum in the endcap is not recommended.

Another solution is to water-cool the endcap systems. This is a highly dangerous idea, considering the amount of current flowing in and around the LPCVD furnaces. This is especially true of the many setups that use the area under the wafer loaders for the main power supply location (which is where the leak would occur). The use of liquid-cooled endcap is strongly not recommended. There are also air-jet cooled designs. While not dangerous, these systems can leak also, creating a high volume of particles around the wafers as they are being pushed and pulled into the tube. Considering the earlier discussion of the effects of electrostatic forces on high-temperature devices, we can see that this could pose a significant risk to die yields, especially with manufacturing processes requiring a large number of CVD steps.

The best idea is to air-cool the endcap system using some design of radiation dispersion fins on the endcap, to aid in convective cooling, such as that shown in Figure 8-7. While this adds size and some cost to each endcap, and requires special cleaning procedures to make sure all of the parts are cleaned and dried properly, the ease of use when installed makes up the for extra effort. The fins may be made of a different material (such as copper or other thermally conductive material) than the endcap if required, and can be mounted on a removable

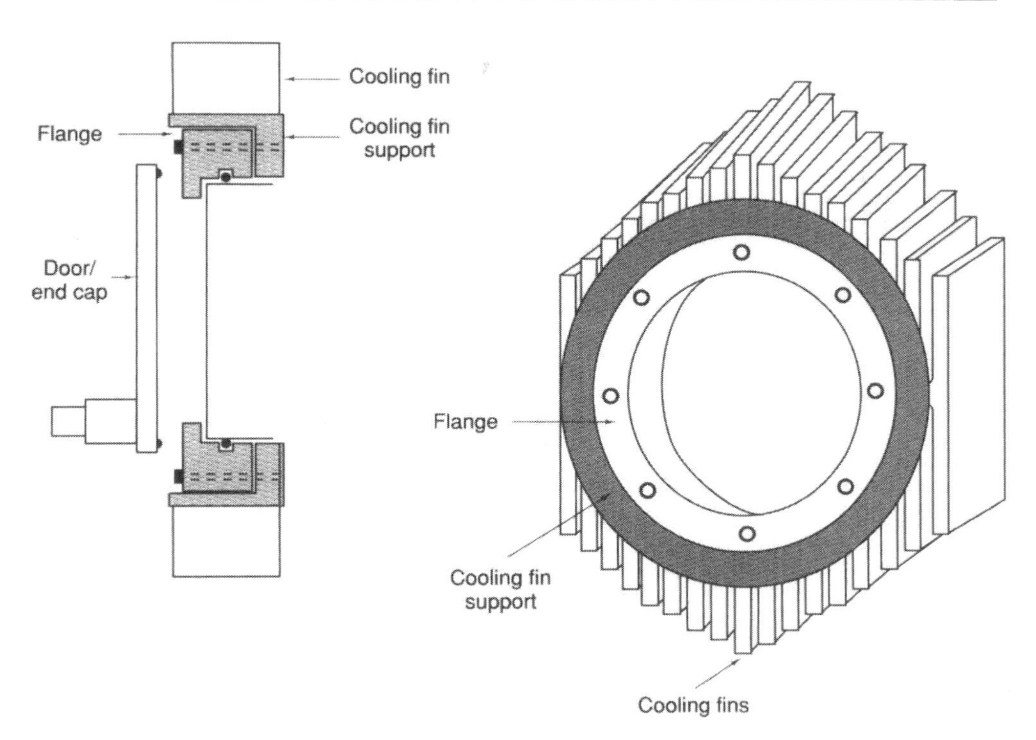

Figure 8-7. Air-Cooled LPCVD Vacuum Flange Assembly To dissipate heat away from the endcap and tube, a number of ideas have been tried, including water and high-pressure air cooling. Each has safety and contamination issues associated with its use. A passive, air-cooled convective/radiative heat dispersal system, such as shown here, can remove the heat without adding complexity to the equipment.

flange. This adds maximum flexibility with the various cooling requirements in the fab, and minimizes the downtime associated with cleaning it, while obtaining the maximum strength and stability on the endcap itself.

Clearly, control of the gas handling system of the LPCVD furnace is a significant issue when preventing contamination. Many of the gases used in these processes are pyrophoric, meaning they ignite in contact with air. In addition, significant portions of the gas system are at vacuum at all times, meaning that, if a leak occurs, atmospheric gases will be pulled directly into the gas system. This can cause severe particulate generation, gas flow nonuniformities, and other types of contamination problems as particles build up and plug orifices and filters as well as spread all over the wafers. These plugged locations in the gas system cause gas flow imbalances, causing improper gas mixtures to be present in the reaction zone above the wafers. This can lead to nonuniformity or chemical imbalances or other contamination of the wafers in that region. Leak rates in the gas control systems must be much lower than in the furnace tube. In most cases, the leak must be better than 1×10^{-9} cm^3/min. In a similar fashion to that noted in section 7.1.1, gas flow rates and exhaust flow rates, in addition to vacuum pressure, must be characterized for optimum conditions for each reactor type prior to use in the production process. Gas flow rates are typically much lower in the LPCVD processes.

In most cases, temperature control in the standard tube-type furnaces is more critical than the diffusion system requirements. Obviously, all of the discussion in chapter 7 about the critical nature of the thermocouples and temperature control systems is valid here, also. The reactions are highly temperature-dependent and will vary drastically by gas pressure and temperature changes. Most furnaces use gas injectors to send the gases down the tube in uniformly so that the temperatures of the tubes can be held constant. In the past, the gases were injected only into the front of the tubes and the temperatures of the tubes were "ramped" or "tilted" to compensate for the depletion of the gases as they react on the wafer surface and are no longer available for reaction. Figure 8-8 describes the effects of these changes. In this type of system, the center zone temperature is very critical and is never adjusted because the gas flow and reaction kinetics can be knocked out of balance with a change of as little as 1°C. The end-zone temperatures can usually be varied by a few degrees to provide fine-tuning of the tilt of process, and therefore the thickness of the film as deposited on the wafer. As we will see in our discussion of polysilicon, there can be

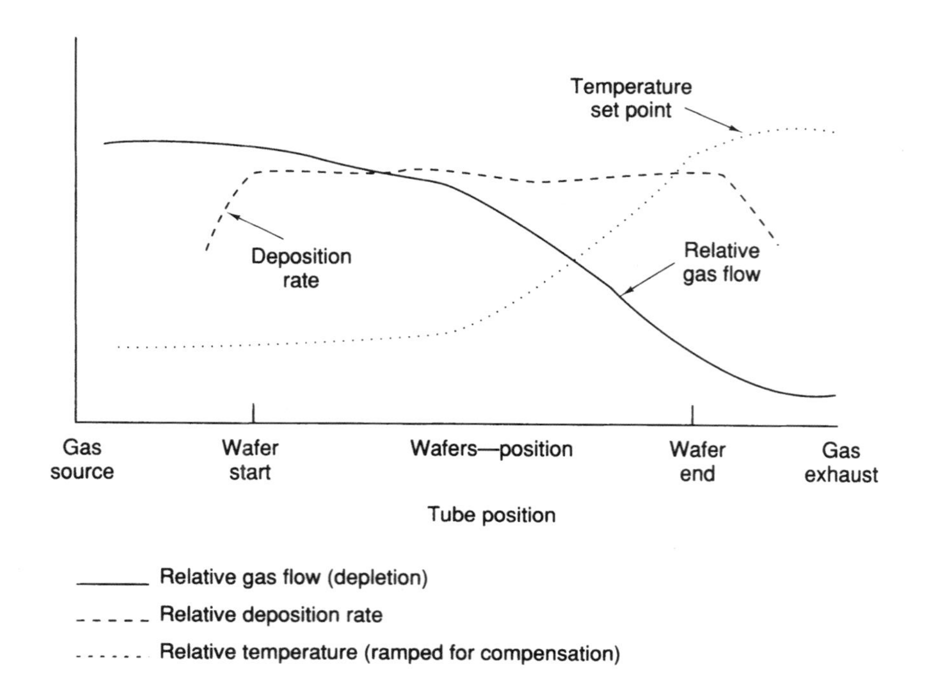

Figure 8-8. Control of Deposition Rate versus Gas and Temperature Ambient In a noninjector-based LPCVD furnace, a significant amount of gas depletion occurs. Normally, this would result in a significant reduction in deposition rate. This is compensated for by increasing the process temperature. The deposition rate stability is held up at a balance point with the temperature and gas depletion countering each other's effects.

other problems associated with this type of processing. Typical changes in reaction rate are highly dependent on the exact conditions of the process but changes in deposition rate of 2 to 3 Å/min per degree (1°C) of change are common.

The wafer handling and computer systems of the standard LPCVD furnaces are very similar to their diffusion system counterparts, although there will be differences due to the requirements of the vacuum systems, and extra safety requirements due to the use of more dangerous gases in the system. In addition, the cantilevers and other support structures must be able to handle the larger and heavier quartz cassettes.

A major structural difference compared to the diffusion furnace system is the use of the vacuum pump systems. Most of the LPCVD systems use a pump setup similar to that shown in Figure 8-9, with an oil-based "backing" pump and a Roots-type blower (used as a booster pump) providing system vacuum. They are capable of moving hundreds of cubic feet per minute (CFM) of gas at a time, and are used to increase the vacuum "well" available in the system in order to move the process gases through the system as rapidly as possible. This setup has a number of advantages, mostly pump capacity and low cost. The

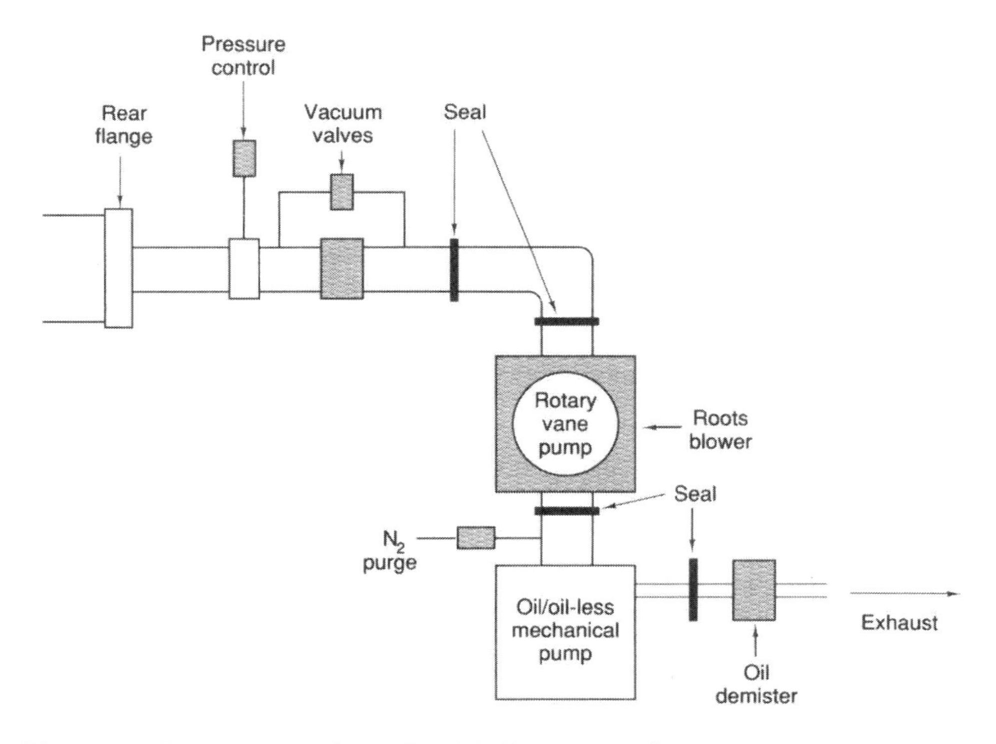

Figure 8-9. Vacuum Pump Setup for LPCVD Process The vacuum system consists of the pressure-control subsystems, the vacuum (gate) valves, and the pump assembly. The vacuum system consists of a Roots-type rotary vane pump (also called a "blower") and an oiled or oilless mechanical pump, which is also called the backing pump. An oil demister is used to remove oil mist from the exhaust stream.

disadvantages of this setup are its complexity and the use of oil-based backing pumps. There are a number of fittings that are used, all of which can leak. If they do leak, they can set up regions of pressure imbalances which can hinder the flow of exhaust gases through the system. Oil in the oil-based pumps must be changed periodically (at least once per month or more often if the process is very dirty), and can be toxic when removed.

A final problem related to oil-based pumps is the problem of oil backstreaming. It has been shown that, if pressures are allowed to sit below 50 mT for any length of time, oil mist can stream throughout the CVD system. This is usually countered by adding N_2 purges to various points in the gas systems. However, as shown in Figure 8-10, if these N_2 inlets are placed in the wrong locations, localized pressure imbalances can actually help push the oil mists throughout the system. There are oils such as Fomblin-type oils that are much better at preventing oil backstreaming due to ultralow vapor pressure, but they are very expensive, typically 5 to 10 times the cost of plain vacuum oil. The best method for preventing the backstreaming of oil into the process tubes is to use oil-free pumps for backing pumps. These pumps cost about three times as much as oil-based pumps, but do not need the maintenance and oil of the oil-based pumps and can pay for themselves in about three to six months, depending on the oil usage rates. Yield increases earned through preventing contamination related to oil backstreaming into the tube from the vacuum pumps can also be very significant.

It is important to maintain vacuum pressure stability in the chamber at all times. Pressure bursts, due either to MFC fluctuations, vacuum systems fluctuations, or other causes, can be the source of a significant number of problems in the deposition processes, from in-

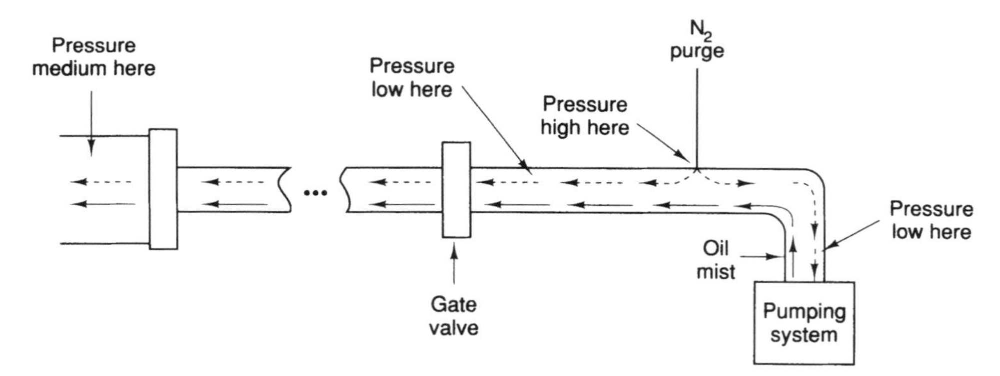

Figure 8-10. The Effects of an N_2 Purge Downstream of the Gate Valve Improper placement of the N_2 purge line can result in the forcing of oil mist into the furnace. If oil mist gets into the vacuum line (possible when the gate valve is off, and if the N_2 is off or at the improper flow rate), the mist can be forced into the furnace when the gate valve opens and the tube is at a fairly low pressure (after having been pumped down through the gate valve bypass).

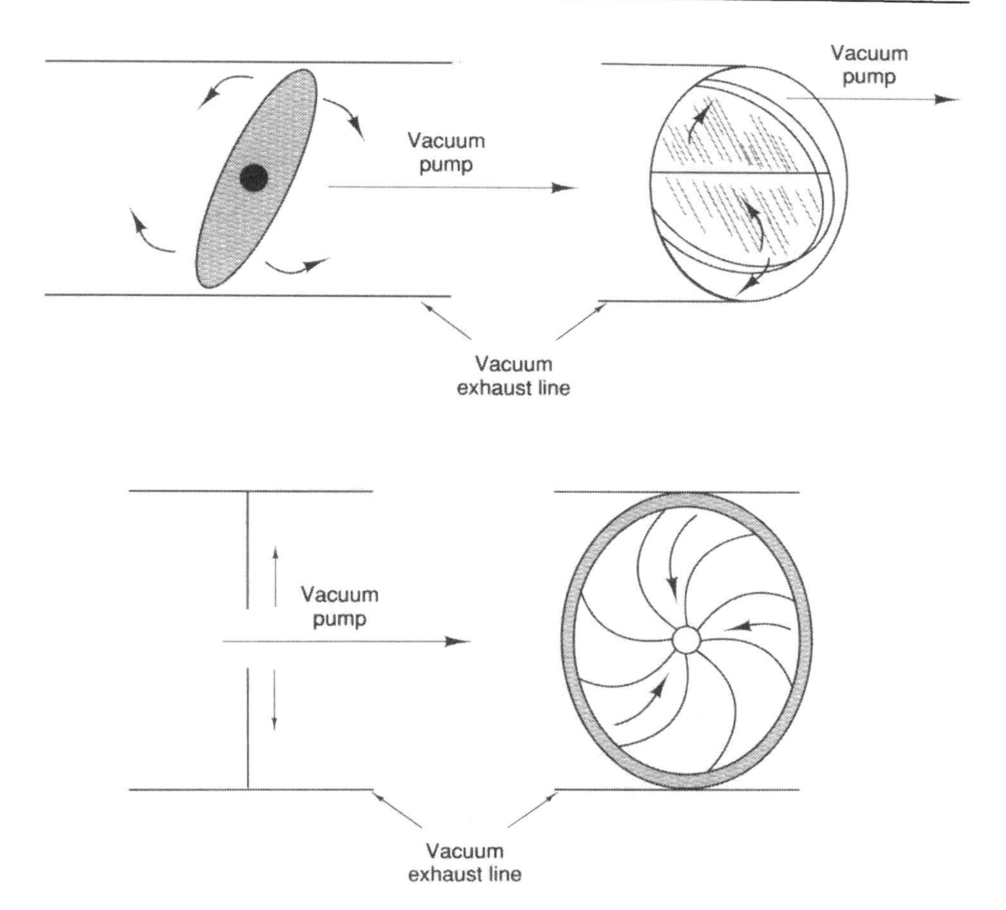

Figure 8-11. Vacuum Pressure Valves Vacuum pressure in CVD is controlled through the use of throttle valves. Newer designs use a solid disk to block the exhaust flow, since this type of valve is cleaner than the older butterfly valve.

creased particles to locally modified regions of crystal and haze. Pressure stability is usually maintained through the use of pressure control valves, as seen in Figure 8-11. Sometimes butterfly valves are used to maintain pressure control and sometimes N_2 purges are used. The butterfly valve, which works like a camera shutter, will have pieces of metal rubbing on one another, which will cause particles to be created. Using N_2 purges will usually set up gas flow streaming throughout the tube, bringing chemical and particle contaminants from the exhaust area directly into the process chamber. If used indiscriminately to adjust pressures, both can create more trouble than they solve in the long run. Pressure control subsystems must be monitored in order to look for causes of problems. For instance, if the control valve position is observed to run at a position of 67% open under normal conditions, and is then seen to vary so that the valve is open to only 30% of full, this may be an indication that an MFC has failed or that a gas line is plugged. In either case, the valve is opened due to an insufficient amount of gas being provided to the tube at any one time.

Pressure control valves cannot control the effects of mass flow controller "bursts." This must be accomplished through the use of good mass flow controllers and good control techniques. First, we will look at the design of the mass flow controller. It consists of a tube with a high precision, electrically operated valve, along with a sensor, as shown in Figure 8-12. This sensor consists of a tube connected in parallel to the main tube, which has a small heating element and a small temperature sensor in close contact. Since the amount of heat transmitted from the heater to the detector is related to the amount of gas flowing through the small tube, the gas flowing through the main tube can be calculated. Electronics attached to the sensors and to the valve controls how far the valve opens, which modulates the flow of gas through the MFC. Older MFC designs place the control valve downstream of the gas flow sensor. However, superior results are obtained using the more modern designs, which place the control valve upstream of the sensor.

Pressure bursts occur when the MFC is turned on suddenly, allowing the control valve to pop open. This causes the control valve to move well beyond the set point, therefore permitting an unusually large amount of gas to flow into the tube in an uncontrolled manner. This phenomenon can be very harmful for the wafers. Haze and excessive particle generation are common hazards with this kind of prob-

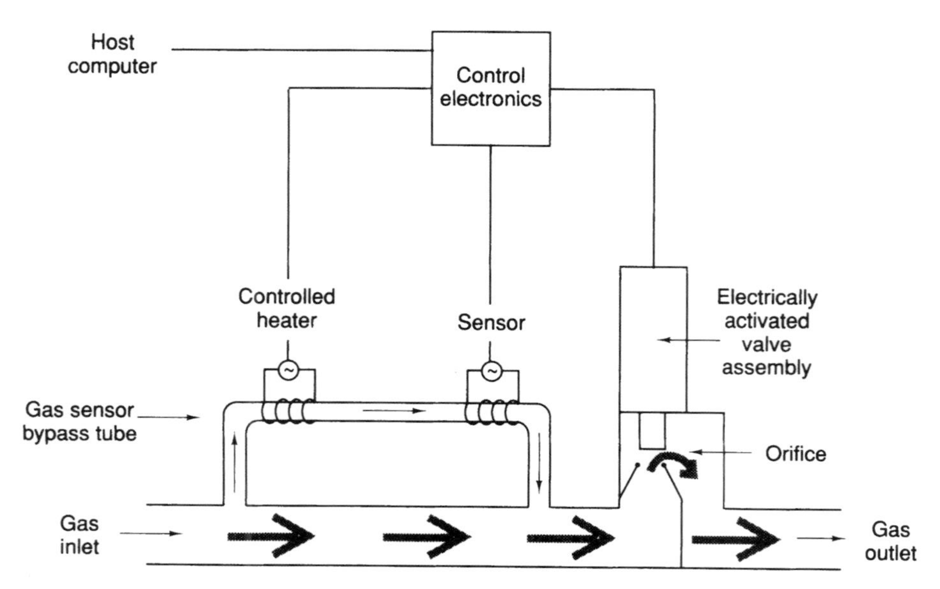

Figure 8-12. Mass Flow Control for Gas Systems Gases in the CVD, diffusion, and etch processes are controlled through the use of mass flow controllers (MFCs), which permit a controlled flow of gas through a small orifice. This flow is adjusted according to variations in the heat flow characteristics observed in the gas stream through the sensor bypass tube. If too much flow is occurring, the sensor will note a change in temperature from expected values and the control electronics will adjust the valve until the correct temperatures are obtained.

lem. This problem is controlled through careful, slow opening of the MFC to its set-point value under computer control. A 5- to 10-second period for opening the MFCs should be fine. The valve position should be monitored to verify that it is not sticking, which may be an indication of a contaminated or plugged MFC. Experience has shown that the MFCs should respond as rapidly as possible, and that the main system computer should monitor and control their operation; typically, slow start MFCs have not been as successful at controlling pressure bursts.

One final note concerns pressure detection systems. A significant amount of research has gone into the design of pressure transducers (manometers) that can maintain calibration while being immersed in an ambient that fluctuates from room pressure to high vacuum. While manometer manufacturers have attempted to remove this calibration problem, it can still occur occasionally. Some fabs have integrated special isolation valves that do not allow the manometer to come to atmospheric pressure at any time, or only during maintenance procedures. This practice has some drawbacks to it also, but these solutions have resulted in manufacturable processes. No perfect solution has yet been devised.

We have discussed the thermal deposition systems up to this point as if all of them were simply extensions of diffusion furnaces. While this is one of the most common types of systems, there are reactors that cause only one or a few wafers to be deposited on at once. These types of systems are more related to the single-wafer plasma reactors which we will discuss shortly.

8.1.2 Plasma-Enhanced CVD Processes

Plasma-enhanced CVD reactors come in a number of different styles, although in general two layouts are the most common. One is the parallel-plate single-wafer chamber, and the other is a tube-type reactor, similar to the diffusion furnace. With a few exceptions, we will not discuss the tube-type reactors here, as they are an older technology, and share most of the concerns of the standard LPCVD thermal reactors. Most of the systems on the market today are variations on the single-wafer–parallel-plate system. Some of the variations include downstream plasma generation (similar to that in the plasma etch processors), multiple chamber processors, and combination deposit/etch chambers for planarizing (smoothing) the films.

We will quickly discuss the differences between the tube-type plasma reactors and thermal CVD reactors. The systems look very similar. The biggest differences are that the quartzware is made of metal and an RF generator is hooked up to the boats. The boats are set up so that the wafers face each other and are hooked up to opposites poles of the generator. All other functions are done in a manner similar to that of the LPCVD processes. The temperatures in the furnace are reduced compared to those in the thermal reactors, with silicon nitride films being formed at as low as 350 to 400°C. The reactors are notoriously dirty, with large amounts of particles being formed in the chambers. Chemical contamination is also high in these chambers, as the metal plates absorb contamination while being cleaned and handled and then redistribute the contamination all over the wafers in the furnace. Of course, since the metal boats are not replaced, the contaminants can accumulate through time, making the chemical contamination problem more serious as the boats age.

The standard parallel-plate plasma reactor is shown in Figure 8-13. This configuration allows the film to be deposited in a controlled

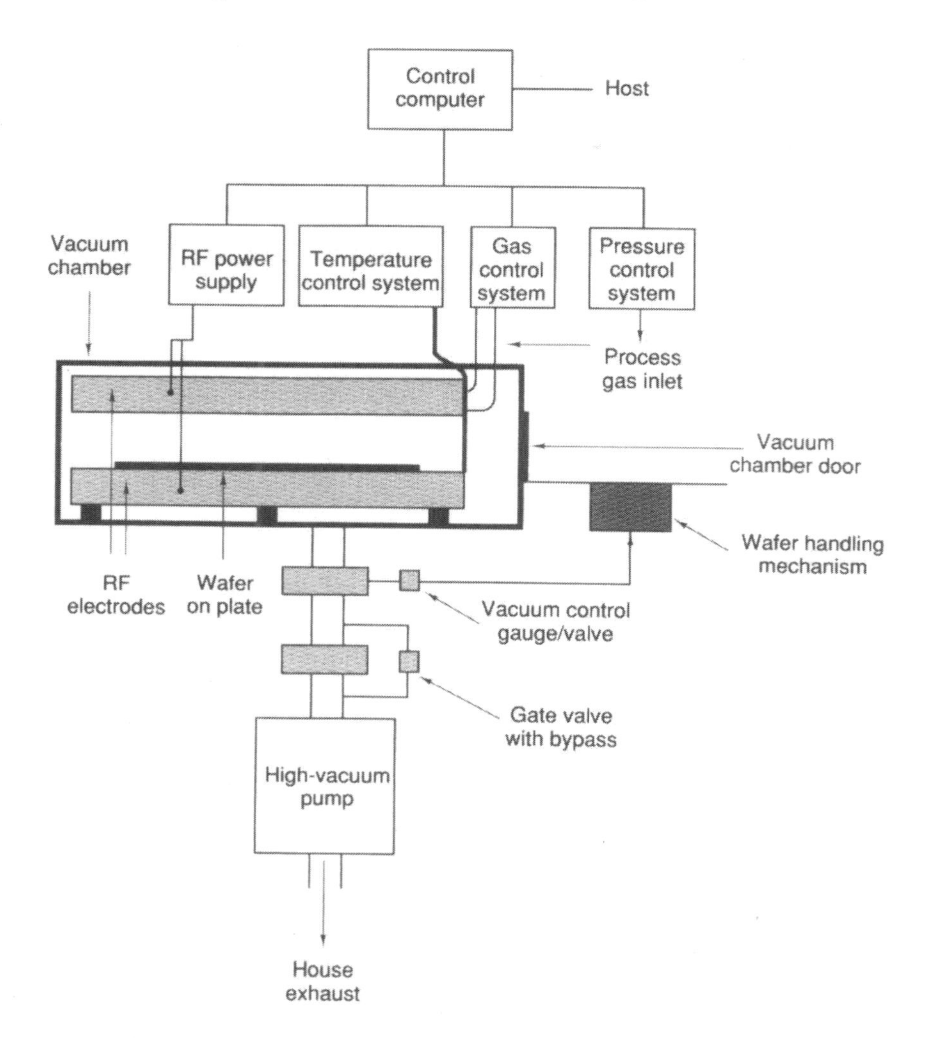

Figure 8-13. Block Diagram of a Plasma-Enhanced CVD System (Single-Wafer Layout) This is a schematic of a PECVD system. Many of the same features from the LPCVD system are used here. However, the wafers are processed individually and are placed on a plate in the vacuum chamber. This plate is used as one of the electrodes of the plasma generator. RF power is applied to the two plates while gases are sprayed down onto its surface. The wafer will be heated by elements located in the lower plate. This allows deposition of a wide variety of films, as noted in the text.

fashion. The gas stream is directed over the surface in a smooth flow that will permit the pressure to remain constant over the surface of the wafer. This helps maintain control on the uniformity of the film across the diameter of the wafer. In addition, this configuration helps control and contain the plasma that is formed over the wafer. By maintaining an electric field perpendicular to the wafer surface, good uniformity and good step coverage can be maintained.

Gas purity is even more important in plasma-assisted CVD systems than it is in thermal systems because the plasma systems can produce a large number of by-products in addition to the expected films if unknown concentrations of impurities are present in the system. The process can produce many particles if the gas stream is contaminated with oxygen, for instance. Typical gas system leak rates must be in the range of 1×10^{-10} to 1×10^{-9} cm^3/min. The biggest difference is that the chamber leak rates are also in that same range. Clearly, all of the other gas system attributes from the LPCVD and oxidation processes must hold true. Most of the time, the chambers on these systems are quite small; therefore, less-complex pumping systems are required for these chambers. Usually, only the smaller oil-less pumps are required to keep the process operating. In some cases, such as the multiple chamber processors, larger pump packages will be required. In all of these cases, an oil-free pump is recommended for vacuum pressure maintenance. Severe contamination of the wafers and processing chamber can result in the event that oil backstreams into the vacuum environment. Note that a process pressure of around 200 mT is typical for PECVD processes.

The interior of these chambers is typically made of stainless steel and is designed to remove corners and other locations where film stress can build up and crack, causing particle generation. The sides of the chambers are usually curved and smoothed to prevent undue gas stream turbulence. The small chambers can also be kept at the process temperature at all times, which reduces any cracking of the films due to the differences in thermal coefficients. Finally, the materials of the chamber will not be damaged in the way that quartz will be when cleaned. Obviously, there is the possibility of chemical contamination if the chamber is handled improperly or cleaned in the wrong solutions, but this problem is less likely to occur due to the smaller size and reduced cleaning requirements for the chamber parts. Many of the most modern systems incorporate automatic film-etch/chamber-cleaning steps that can be used to eliminate the buildup of film on the chamber surfaces after each set of deposition processes. This permits the system to be used for months with minimal cleaning requirements.

The most significant difference between the plasma systems and the thermal deposition systems is the energy source that causes the reactions to occur. Instead of high temperatures, the RF generator forces a tremendous amount of energy into the gases, forcing them to react. As a result, lower temperatures are typically required to produce the films, usually in the 350 to 380°C range. One of the main purposes of the temperature range used is to permit high surface mobility of the molecules landing on the wafer, so that good step coverage can be obtained. In theory, plasma CVD operations can take place at even lower temperatures.

Some of the biggest problems of the PECVD process are actually related to the use of the plasma as a reaction energy source. One problem is that of maintaining uniformity over large substrates, even though a significant amount of time has been invested in research to alleviate this problem, with some success. Another problem that is very significant is the effect of the plasma on the surface of the wafer. This radiation damage can be quite serious, especially in contact regions or other areas where the substrate, polysilicon, or metal layers are exposed. As a result, there have been some restrictions on the use of plasma processes for certain high-reliability and high-density devices.

8.1.3 Other Advanced Technology CVD Reactors

Various processes have been developed to try to obtain the advantages of the plasma-type processing, such as reducing the process temperature while simultaneously reducing the radiation damage on the surface of the wafers. These attempts fall into two broad categories: separation of the plasma reaction zone from the deposition chamber, and the use of alternative energy sources to initiate the required reactions.

The first type of chamber is often called downstream or afterglow processing, and is shown in Figure 8-14. The gases are sent to the reactor chamber, where the chemicals are given their reaction energy. The chemical precursors must then be delivered rapidly to the surface of the wafer in order to obtain a good deposition. These systems have more problems with deposition uniformity than the standard PECVD processes due to their sensitivity to minor gas flow perturbations. There are also potential problems with particulate matter if the reaction starts to take place on the walls of the chamber prior to deposition. These types of systems have been researched a great deal, but have not yet made a major market impact. The equipment for these systems is somewhat more complex than for the other plasma systems, and can add to the maintenance overhead of the manufacturing process.

Another type of chamber is called the electron cyclotron resonance (or ECR) CVD process chamber. This is shown in Figure 8-15. In this system, electrons circulate in a magnetic field, producing energy. This energy causes the reaction gases to activate and start the reaction. The gases react similarly to the PECVD processes and form the film on the wafer surface below. Problems that have been reported with this process have been related to uniformity and throughput issues. This process holds much promise for future development, as it offers a radiation-free, low-temperature, clean deposition process. The com-

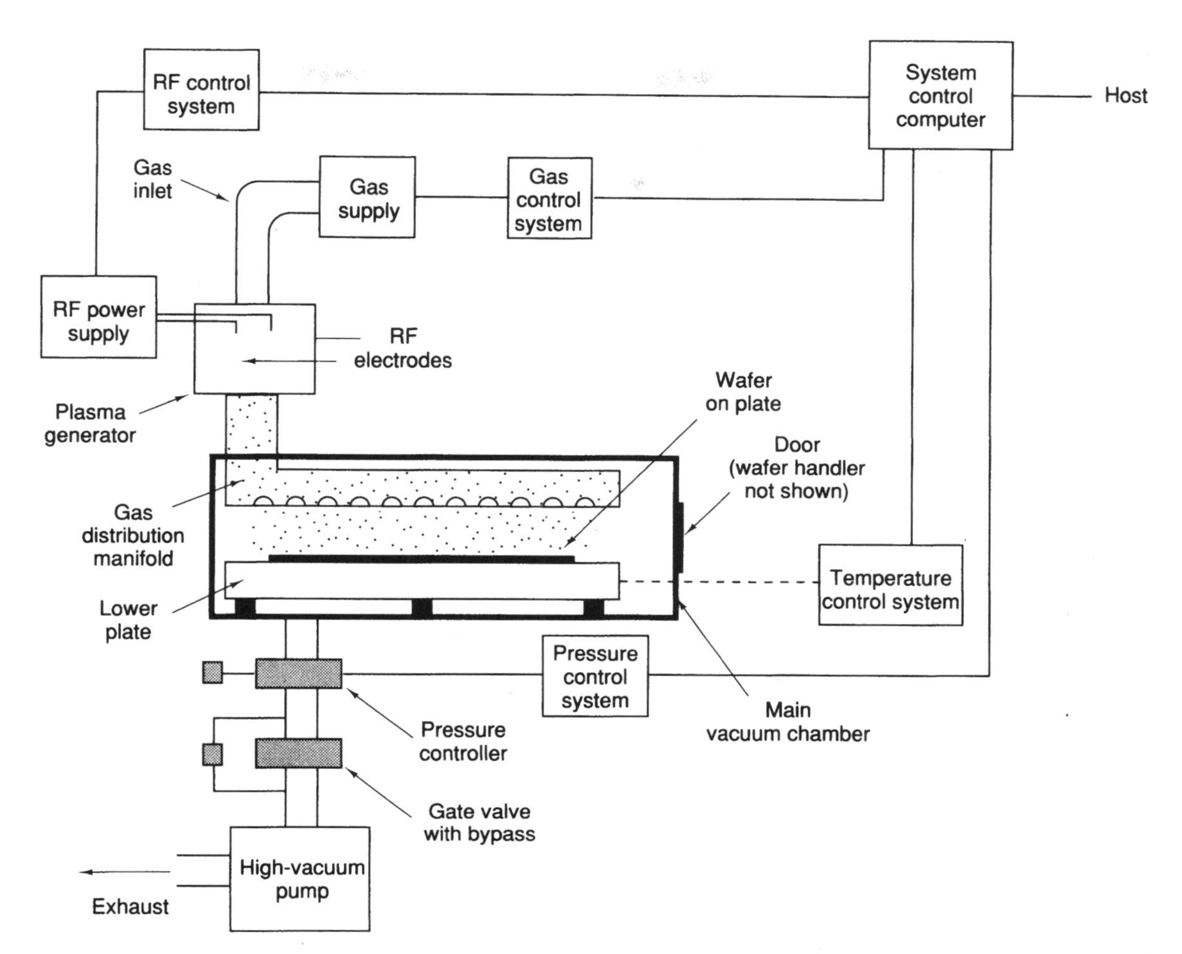

Figure 8-14. Downstream Plasma CVD Reactor This is a variation on the standard PECVD reactor layout. In this system, the plasma is not generated directly over the wafer. This reduces the amount of radiation damage to the wafer and removes the high electric field from around the wafer. This is accomplished by having a separate plasma generator to create the plasma, along with a gas delivery system that can deposit the gases into the reaction chamber quickly and uniformly.

plexity of the equipment and its associated problems is similar in many respects to that of the PECVD and afterglow processes.

The last type of deposition process is called the laser-activated CVD process, as seen in Figure 8-16. This is an outgrowth of other photochemical deposition processes. The laser provides much more intense radiation than the previous types of light sources used. In this photolytic system, ultraviolet light is used to dissociate molecules of the reactive gases to form the reaction precursors. They then land on the surface of the wafer to form the films. This process also provides for very low temperature processing, as well as a radiation-free environment for producing wafers. One of the convenient aspects of the process is that the laser beam attacks the bonds of the process

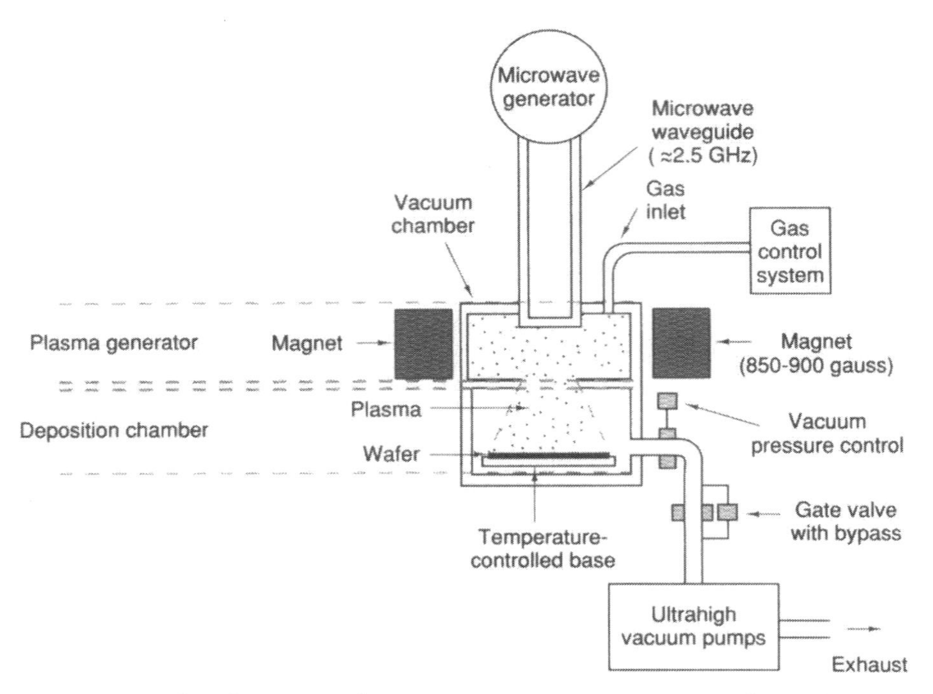

Figure 8-15. The Electron Cyclotron Resonance CVD Reactor The ECR reactor is similar to the downstream plasma reactor described earlier. The primary differences lie in the source of the plasma. In the ECR reactor, the plasma chamber is immersed in a magnetic field while a microwave beam is directed into the chamber. The high-energy resonance that develops forces a plasma to form. This plasma is then directed toward the wafer with a low accelerating voltage. This method allows excellent process results, due to the low ion energy (about 10% that of RF plasma reactors), low operating temperatures, and low operating pressures.

chemicals selectively, as opposed to plasma systems which generate reactions completely unrelated to the one that is desired. This also prevents any unwanted reactions from occurring. The specificity of the reaction with the laser beam allows the process to be used as an etch-and-deposition process. The chambers on these reactors tend to be quite complex, although the uniformities of these processes are be quite good and can accommodate any size substrate with only minor modifications. Deposition rates can be controlled very easily with these processing, permitting the deposition of very thin films.

Infrared and visible light lasers have been used to cause thermally induced CVD reactions. This is called a pyrolytic process and is quite effective although it does not share some of the attributes of photolytic reactions, such as low substrate heating. The equipment for laser CVD processing has not been used in mass production, although a significant amount of research has been performed in the field. There is a good chance that this technology will emerge as a significant contender for low-temperature, low-radiation, film-deposition processing by the mid-1990s.

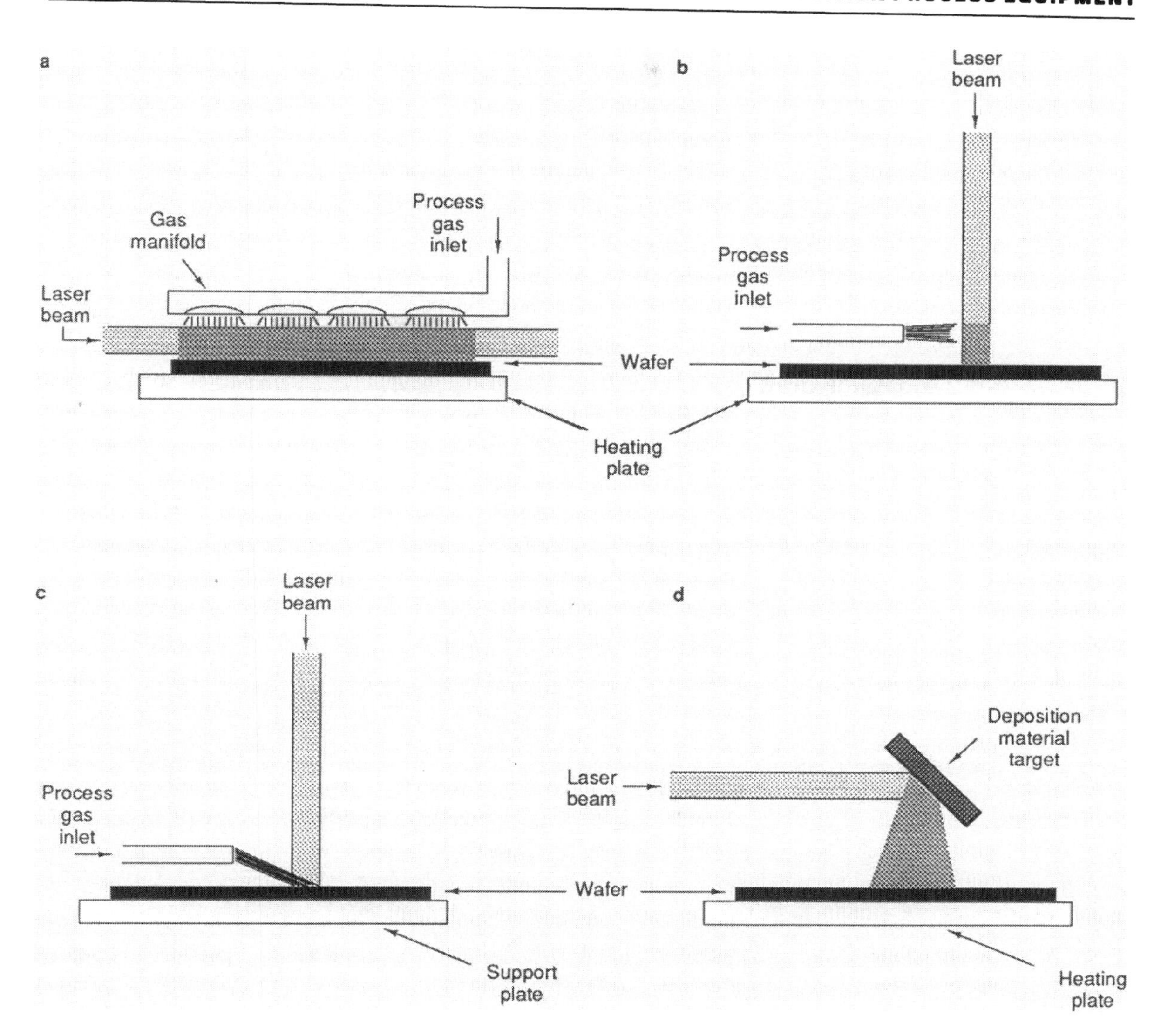

Figure 8-16. Laser-Activated CVD Processes Shown here are the four most common methods of laser-activated (or assisted) CVD processing. (a) The beam is directed over the surface of the wafer, parallel and slightly above it. Gas is then directed through the beam. The laser interacts with the gases which form a plasma to create the film. (b) The reaction is similar in that a wavelength that is not easily absorbed into silicon is directed at the wafer in the presence of reactive gases. They then form deposits under the beam, which can be masked for lithographic purposes. (c) The pyrolytic reaction in which the wafer, not the gas, is heated by the laser. This is similar to the standard LPCVD process. (d) A laser analog to sputtering in which the laser is directed at a target (say, tungsten) which ablates material off the surface and deposits it on the wafer. In all of the systems, either the laser or the wafer must be scanned in order to cover the entire wafer.

8.1.4 Atmospheric-Pressure CVD Processing

One of the older technologies used to deposit films on wafer is the atmospheric-pressure CVD systems. This technology is still efficient enough to permit the use of the process even with advanced silicon technologies. This system is outlined in Figure 8-17. There are a number of advantages to the use of this system, including the fact that it is well characterized, and can produce films of high quality, This process also provides a radiation-free deposition environment, although processing temperatures are somewhat high.

In the APCVD system, the wafers are moved through the reactor chamber on a belt. Gases are introduced directly above the wafers and are immediately exhausted. There are a number of sources of contamination and mechanical failures possible in this conveyer system. The belt can be damaged, can stop, or can move in unexpected ways; parts can scrape together; or chain drive mechanisms can be damaged. These situations can be reduced by careful preventive maintenance. The gases are blown over the surface of the wafers where they react to form the film, and then are exhausted into a manifold. Particles from this reaction can form on a number of surfaces, from the tops of the wafer support plates to the chain drive mechanisms to the inlet and exhaust manifolds. More gas is used in this process than in the LPCVD processes, and can therefore produce more contamination in a shorter time. Again, careful preventive maintenance can help prevent these problems.

The uniformity and deposition rates of the atmospheric deposition process are determined largely as a function of belt speed and

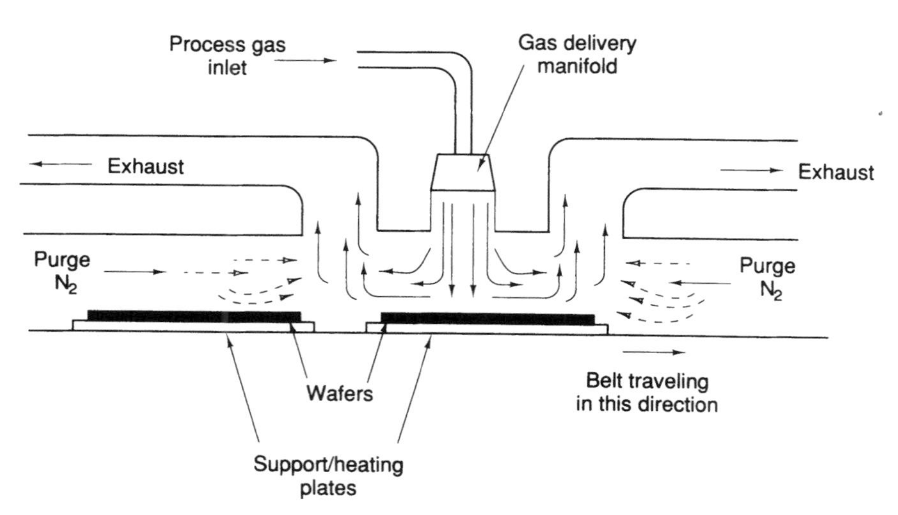

Figure 8-17. Atmospheric Pressure CVD Process In the APCVD process, the wafers are placed on belts or tracks, which are heated and moved under a gas delivery manifold. The process gases then react on the surface of the wafers to form the desired film.

exhaust/gas inlet flow rates. If these parameters are outside of the tolerance ranges, various problems will occur. For instance, a belt speed that is too high will result in a deposition that is too thin. Exhaust flows that are low on one side of the reaction chamber will result in a thicker film deposit on the wafers on that side of the chamber. These parameters must be checked continually throughout the process cycle to verify that the process is coming out as expected.

The processing temperature of the APCVD reactor is also critical. The temperature ranges on these reactors can vary widely depending on the exact process in use, but typical process temperatures are around 400°C. It is important that the wafers reach processing temperature quickly and maintain uniform temperatures, or the deposition will not take place in a consistent manner. Therefore, both the temperature detection and heater systems must be maintained, but on most of these systems are relatively easy to access for service.

Some of the more important advantages of the APCVD process include the following. First, high deposition rates are available. They can be several times that of the LPCVD processes, resulting in a high throughput for large-scale wafer production. In addition, the films tend to be quite dense and planar, and do not always require as much post-deposition processing as the CVD films. (This is highly dependent on the exact type of material being deposited.) In addition, the wafer has low stress compared to other processes, due to fewer environmental changes and shorter processing times. Damage to wafers due to machine-caused problems produces minimal impact to the line yields, since only a few wafers are actually in the machine at any one time. Finally, there is no radiation-induced damage, since there are no radiation sources.

Although the process is susceptible to high contamination, has high maintenance requirements, and high gas consumption rates, the advantages of atmospheric pressure CVD have permitted it to be used in many processing environments for SiO_2, PSG, BPSG, and Si_3N_4 films. It is likely that this type of coating equipment will be in use in many fab areas for a long time to come.

8.1.5 Wrap-Up of CVD Processing Equipment

As we have seen in these last several sections, there are many types of CVD equipment. The amount of effort, investment, and ingenuity that has gone into these systems is quite amazing. The reasons for this amount of effort are going to become much more apparent shortly, as the process problems that are related with the films that are deposited from each of these types of equipment is examined in detail. The field of dielectric deposition is still wide open, and any of a number of technologies or combinations of technologies must be developed to answer all of the needs of the semiconductor industry.

8.2 DEPOSITED FILM TYPES

While we have discussed the equipment in detail, we have not discussed the impact the various types of processes will have on the parameters of the deposited films. The impact on these films will vary greatly, depending on the process used and the film produced. Some of the films can be doped while being deposited, while others may be doped at a later stage in the process. The exact chemical makeup of the films is also quite critical to the functioning of the device, and this can vary within each process. As a result, we will need to discuss each film, and the changes that are effected as a result of the various types of processes. The films that we will discuss in this section of the chapter will almost all consist of dielectric films, with the exception of the polysilicon film. All of the metal films will be discussed in the next chapter.

We will start with a discussion of polysilicon films that are both doped and undoped. This will be followed with silicon dioxide films, which will include the doped PSG and BPSG films. Finally, we will talk about the varieties of silicon nitride and oxynitride that are available as final passivation layers. A number of common processing issues throughout this area cause a family of related problems with the wafers, many of which have been introduced in the sections on processing equipment. Clearly, the semiconductor manufacturing equipment industry has gone to great lengths to address the problems pointed out in this section.

8.2.1 Polysilicon Films

We will start our discussion with the properties of the polycrystalline silicon films. These are usually called polysilicon or poly processes. These films are some of the most important to the integrated circuit industry. They can be deposited on the wafers at a number of locations and are used for a variety of tasks. Typical uses include resistor and conductor lines, transistor gates, memory cell and other types of capacitors, and floating gate structures used on EPROMs and EEPROMs. Many integrated circuit processes have two polysilicon layers deposited on the wafers and may even have three layers if the required density of the circuit cannot be obtained in any other way. Most polysilicon films are deposited undoped, but in many cases there is no particular reason why these films could not be deposited with in-situ doping. Almost all polysilicon films will end being doped before the end of the process. Poly can doped with phosphorus or arsenic to make it *n*-type, or with boron to make it *p*-type, just as can single-crystal silicon.

The polysilicon process is a "thermal" reaction and is run at a temperature of around 600 to 630°C. By thermal, we mean that the energy is supplied to the reaction through direct heating and not through plasma or other means. PECVD systems are rarely used to deposit polysilicon. The deposition-rate and grain-growth mechanisms change quite drastically as a result of changes in temperature. Polysilicon films consist of small grains of single-crystal silicon that is oriented at random angles to one another, as shown in Figure 8-18a. The grains tend to get larger and less uniform in size as the temperatures are decreased and, as temperatures are increased, the grains will become smaller. When deposition temperatures become low enough, polycrystalline formation ceases and amorphous silicon is deposited. This film is a noncrystalline form of silicon, as seen in Figure 8-18b. The film forms structures that appear as thin slabs of material. This process requires temperatures in the range of 575°C. At the other end

a

Polysilicon film crystals — lines describe crystal orientations

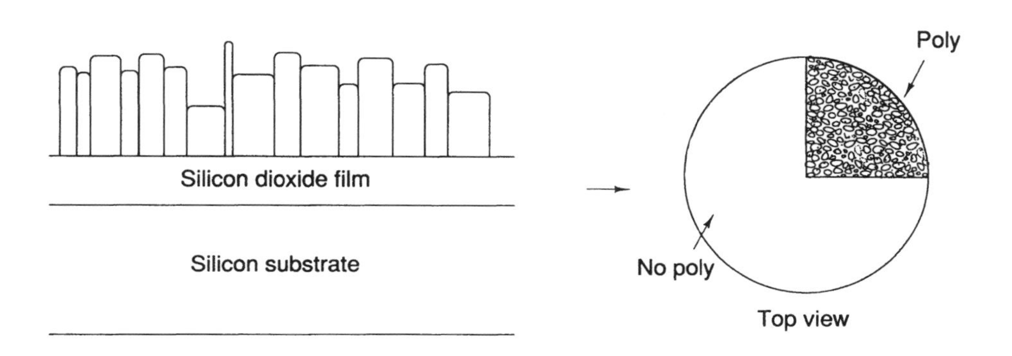

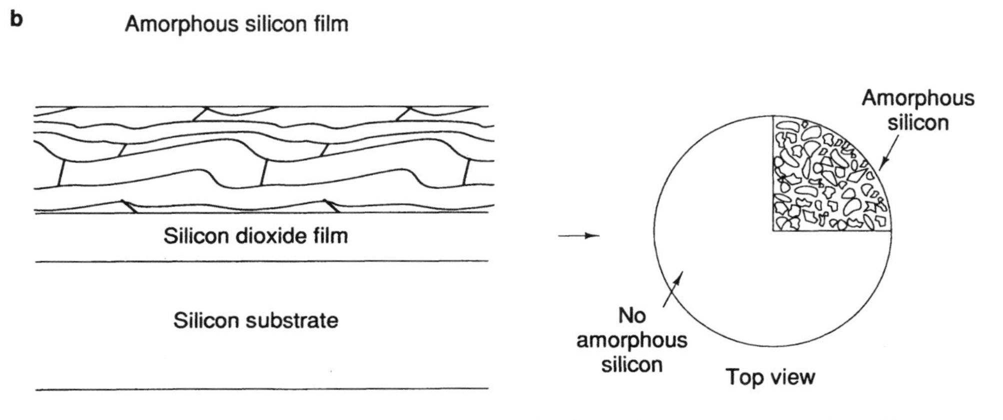

Figure 8-18. Deposited Silicon Films (a) Polysilicon forms as a series of small (0.1 to 0.5%) grains which appear quite uniform in nature. Often, they form in columnar structures. (b) Amorphous silicon has no distinct structure and appears similar to a pile of very small leaves.

of the scale, above 700°C, the polysilicon deposition process again ceases, and the films become single crystalline in nature. There are some applications where amorphous silicon can be doped and recrystallized to form the various structures usually occupied by polycrystalline silicon. Single-crystal silicon is usually grown in special systems called epi (for epitaxial) reactors. We will not cover amorphous or epi silicon processes in depth, but this should not be interpreted as reducing their importance. They are important films, but are not used in most of the standard production processes. Epitaxial films are often supplied on the silicon substrates, but this process is seldom performed by the wafer fab in MOS operations; it is usually performed by the silicon manufacturer. Bipolar processes commonly use epitaxial layers for the development of the buried-layer steps.

For a process to obtain maximum die and line yields, the polysilicon must be very consistent from wafer to wafer. This is quite difficult to achieve in the typical poly deposition process. Variations in gas flows, pressures, the constant depletion of the silane, and variations in temperature can all cause changes in grain structure. A very significant factor in the poly reaction is the depletion of the silane as the reaction proceeds. If the gas is injected in one port, which was the historic way of producing poly, the temperatures in the tube have to be modified to compensate for the reduced quantities of gas. As we can see, each molecule of SiH_4 reacts at the surface of the substrate and removes that molecule from the gas stream. The reaction is as follows:

$$SiH_4 + \text{heat} \rightarrow Si\downarrow + 2H_2$$

Thus, the gas stream becomes diluted with hydrogen and depleted of silicon. The process itself is stoichiometrically pure, in that there are no typical by-products that become trapped in the film, as occurs with many PECVD films.

To counterbalance the depletion effect, the temperatures are tilted in the tube, so that the reaction is slowed at the front end of the tube by running at a colder temperature than optimum. This temperature is usually in the 605 to 610°C range. The center zone of the furnace is set at the processing temperature desired, usually around 615 to 620°C. The zone farthest from the gas source typically has elevated temperatures, as high as 630 to 635°C which results in very inconsistent film quality. The dynamic nature of this setup also makes the process very sensitive to minor changes in any parameter. For instance, a regular occurrence with this process is the adjustment of temperatures to affect the overall thickness uniformity of the furnace. Often the temperatures in the end zones are adjusted, although a good rule of thumb is that more than a one or two degree change is excessive, and there may be other problems. The temperature changes usually cause a fairly predictable change in deposition rate, in the range of 3 to 4 Å/min per 1 degree of change. In general, the center zone is not adjusted, as modifications of that zone can drastically change the

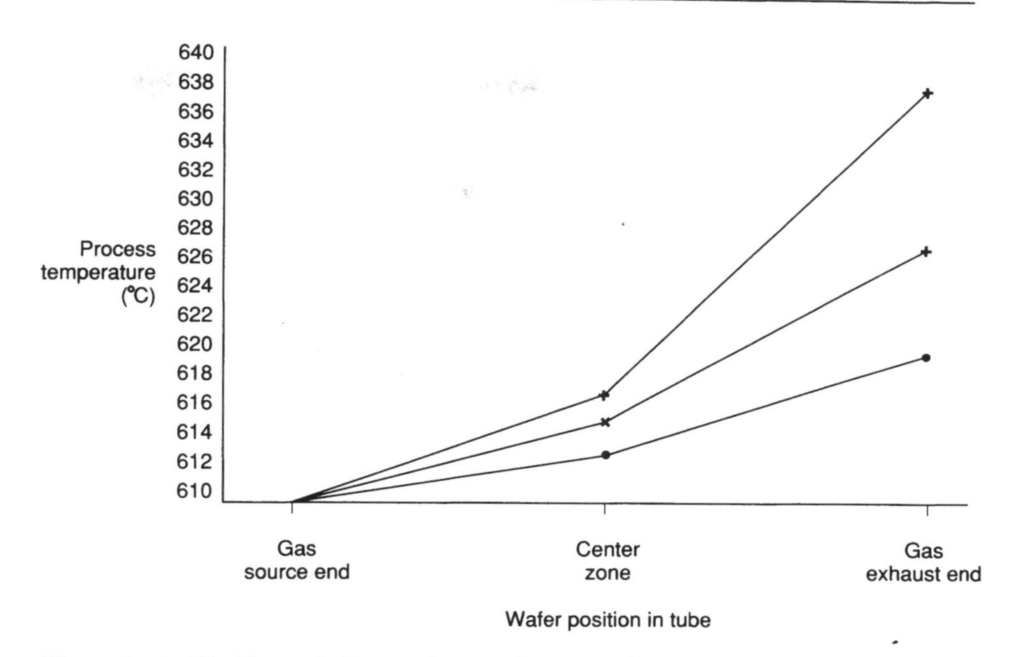

Figure 8-19. The Dynamic Processing Conditions in the LPCVD Furnace As shown in this hypothetical polysilicon deposition process, changes in temperature can drastically affect the dynamic conditions present in the tube. For instance, if the center zone temperature is lowered by one degree, there will be an excess of gas in the exhaust stream, so that downstream temperatures must be reduced to compensate and approach proper uniformity. Conversely, if the center zone is raised, excess gas depletion occurs and the downstream temperatures must be raised to increase the reaction rate and maintain uniformity.

dynamic condition of the system. This is graphically shown in Figure 8-19.

Modern polysilicon deposition systems use injector systems that allow the gases to be distributed throughout the tube in a more uniform manner than the single-inlet system. Using an injector, the depletion effects can be offset by adding gas, therefore permitting the furnace temperatures to be held constant. This results in a more consistent film quality. However, the addition of the injector adds additional maintenance issues, in that the injectors are usually made of quartz and are therefore quite fragile, and they can also become contaminated fairly easily. In general, temperatures are not adjusted on a run to run basis in these systems as they are in the single inlet systems.

Another feature of the LPCVD processes, in general, is the use of dummy wafers. As seen in Figure 8-20, these are wafers which are used to balance the load, so that every furnace run will have the same number of wafers in it. This is again related to the depletion reactions occurring in the system. If the total surface reaction area changes from run to run, there will be little control on the dynamic conditions of the process. For example, if there are fewer wafers in the tube the gases will not be consumed as rapidly, meaning that there will be more gas

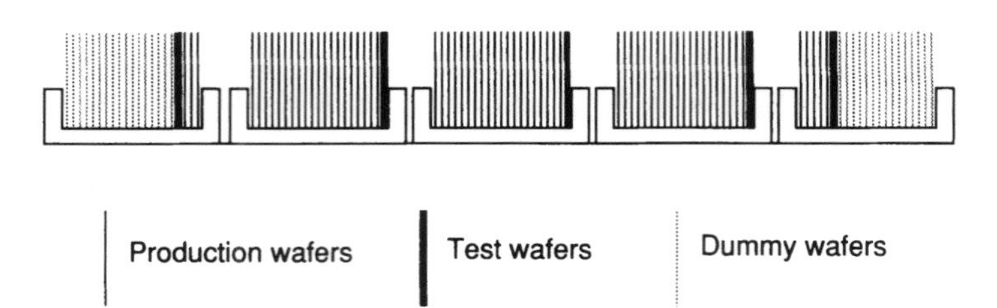

Figure 8-20. Typical LPCVD Wafer Layout The wafers that are processed in horizontal batch load LPCVD furnaces are set up in a specific way. There will be a fixed number of dummy wafers on either end of the load. The product wafers are loaded between these dummy wafers. Test wafers, if used, are interspersed at regular intervals throughout the load.

available to wafers closer to the exhaust. As a result, the films on these wafers will be relatively thicker than expected. If the load size is then increased unpredictably, the wafers toward the exhaust end of the tube will be starved for gas, and thinner films will be deposited on them. Dummy wafers solve these depletion problems at the expense of their own set of problems. For one, these wafers are not usually handled as carefully as the "regular" wafers (although they should be). The wafers are sometimes not cleaned for many dozens of runs or are handled improperly when moved around. As a result, they are often major sources of contamination in the deposition process. This is a shame as they are some of the easiest problems to solve for the long term.

To help reduce the difficulties with handling these materials, it is a good idea to place some inert film under the polysilicon so that cleaning processes can easily and quickly strip the polysilicon without destroying the silicon wafer. This is often done by growing a layer of SiO_2 on the wafer. A more effective method is to deposit a clean layer of silicon nitride on the wafer prior to the poly deposition. A final minor problem with dummy wafers is again related to improper handling, which is that dummy wafers that have had excessive deposits on them may not react at the same rate as wafers that have been recently cleaned and, therefore, can cause nonuniform deposition of the films, due to the varying concentrations of gas available to the other wafers.

Polysilicon is typically grown at a rate between 85 and 110 Å/min. Polysilicon grown at rates lower than this tends to be of poor quality, with inconsistent grains and a higher propensity for film anomalies. Growing polysilicon faster than this can result in low-quality films, poor uniformity, and high particulate counts (due to excessive gas-phase particle nucleation). Deposition rates can be adjusted through changes in three basic factors: temperature, pressure, and total gas flow. The change in temperature has already been discussed. Increasing pressure in order to increase deposition rates is fairly effective up to a certain point, but if the pressure rises above 200 to 250 mT, haze and surface anomalies can form. Typical deposition pressures are in the 150 to 250 mT range. Total gas flow can affect deposition rates, but this has a lower effect than the other factors. The more significant effect that occurs with changes in total gas flow are changes in wafer and run uniformity. When gas flows are too high, haze and other film problems can occur. To make sure that the poly process is in the "sweet spot," the pressure-control, gas-control, and temperature-control subsystems must be continually monitored for changes.

Polysilicon films do not tend to be particularly brittle and can conform to most shapes. This is convenient, as it allows the polysilicon to be placed on unusual structures and to be used to connect lines over steep terrain. Usually, a crack or flake is not apparent to the naked eye, either on the wafers or in the tube, unless some major problems have occurred. For example, sometimes a silvery threadlike material will form, which appears much like Christmas tree tinsel. This is associated with extremely thick films and with contaminated or oxidized films. This is especially common around the doors and the cantilevers in the systems. This problem is often exacerbated by the reaction of the atmosphere with the polysilicon on the quartz. Therefore, proper standby procedures should be followed, as will be discussed shortly.

Some devices can be sensitive to the subtle differences in films grown at the various deposition rates. Experiments should be run to determine the best combination of deposition rates, pressures, and temperatures for the specific devices being produced. Even details such as slot spacing on the boats should be considered if ideal processing conditions are desired.

As mentioned, polysilicon is almost always doped, either with boron (*p*-type) or phosphorus (*n*-type). Phosphorus is probably the most used dopant with the polysilicon films. These depositions are usually accomplished with the $POCl_3$ process described in Chapter 7. The polysilicon film is often doped until it approaches its solid solubility limit. This is 1.1×10^{21} atoms/cm^3, which results in a sheet resistivity of about 15 Ω/square. This results in low resistivity, but also results in the restructuring of the crystal structure. The crystals swell as they absorb the phosphorus, sometimes breaking into smaller crystals. Since the film must be doped uniformly, the polysilicon crystals must be uniform in size and texture before dopant deposition, or some crystals will have slightly different doping profiles. This can prove to be a problem later, if the differences in the crystals and their doping levels cause a change in the etch characteristics. This is not an uncommon situation.

It is important that the polysilicon deposition engineer become intimately involved with the polysilicon etch process, as the set of processes (polysilicon deposition, phosphorus doping, polysilicon etch) must be carried out properly. The existence of contamination or

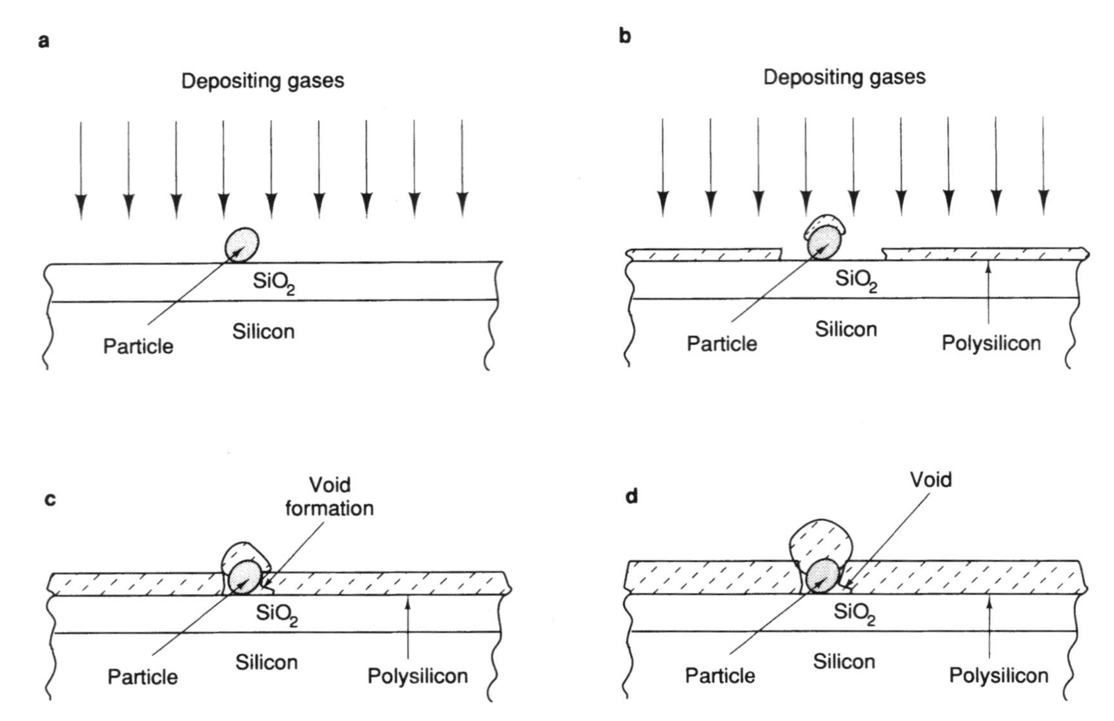

Figure 8-21. Particle under Deposited Film Particles can cause serious problems when films are deposited on top of them. The particles act as nucleation sites, causing enhanced deposition during the early stages of the process. As the deposition continues, microvoids can form as the films grow together. This intersecting region will be weaker than the surrounding films and can create etch anomalies and residual defects. In unetched regions, this sort of structure can cause electrical failures and can make succeeding layers difficult to deposit.

other defects produced in the polysilicon deposition process can create a significant problem during the etch process. As seen in Figure 8-21, a small particle placed on the wafer prior to polysilicon deposition can become a preferential nucleation site (this is an area which has a much higher affinity for the material in the gas phase, and thus becomes a seed for rapid local growth of the polysilicon crystal). This will permit the polysilicon to deposit there much more rapidly than in the surrounding area. This small particle thus becomes a bump on the surface of the film. This bump may absorb the dopant very differently than the surrounding film does. Then, when the wafer goes to the etch process, the etch gases may not etch the bump as rapidly as the rest of the film resulting in a residual defect that is much larger than the original seed particle. If there are further deposition processes, this operation will be repeated, with the defect slowly becoming larger and larger as it accretes material. However, these defects can appear very similar to other types of etch defects, and are therefore hard to distinguish from other etch problems, and may go undetected for a long time. It is important that both the LPCVD engineer and etch engineer communicate closely on all matters of importance, as changes in one area may drastically affect the results in the other area.

There are a number of things that can go wrong when polysilicon films are deposited. We have already discussed a number of things briefly, such as haze and surface "bumps." These surface anomalies can cause a lot of headaches. Defects of various types generally look like small highlights or stars in the field of view in a dark-field optical microscope, as illustrated in Figure 8-22. The defects may have a variety of appearances under SEM inspection, such as these seen in Figure 8-23a, b, and c. What is usually described as haze can actually be caused by a number of sources. The haze can be a result of a very high density of highlights, or can be the result of silicon oxide embedded in the film. Other contaminants can also become embedded in the film to cause haze. The sources of the contamination can be leaks in the tube, the gas system, or any other fitting; water remaining on the wafers after a cleaning step; moisture or contamination on the quartz; even contamination from other process steps can be transmitted through various subtle routes to the poly tubes to cause the development of defects and haze. Clearly, there are numerous avenues for the contamination of the polysilicon, and typically the poly deposition process must be controlled very tightly.

The various types of contamination can sometimes identify the source of the contamination. For instance, oxygen and moisture will usually react in the gas stream, forming particles that land on the surface of the wafer. After the particles land, the poly will build up until the process is complete, leaving the surface covered with a

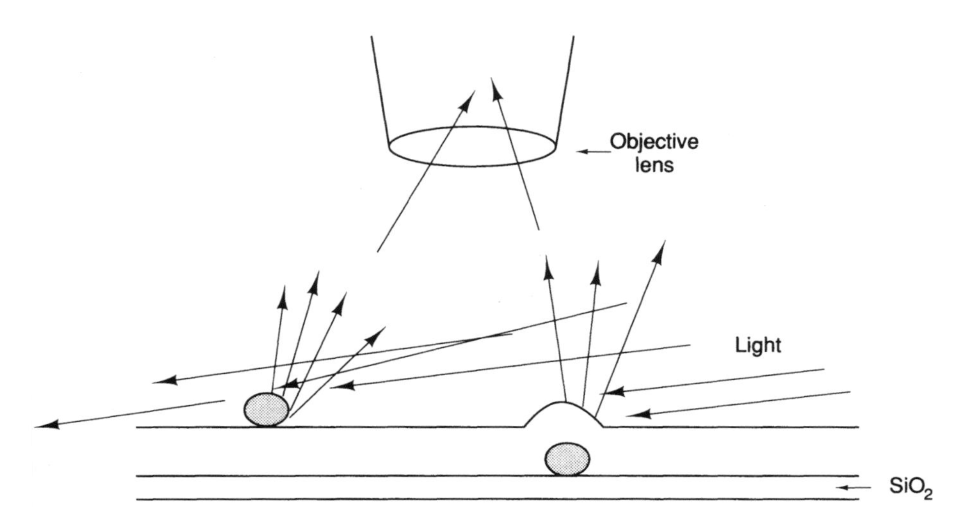

Figure 8-22. Particle Inspection in Dark-Field Microscope Particles and other film defects can be seen by inspection under the dark-field microscope. In this microscope, the light is shown at a high angle. When there are no features, little light will be reflected into the microscope. However, defects will produce bright "stars," which can be counted, and in some cases, identified.

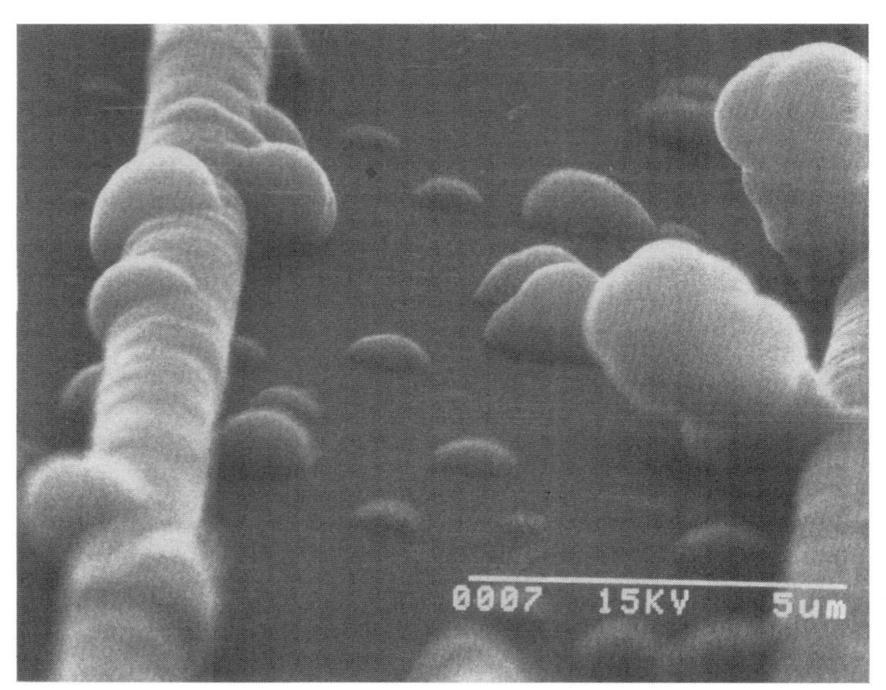

a

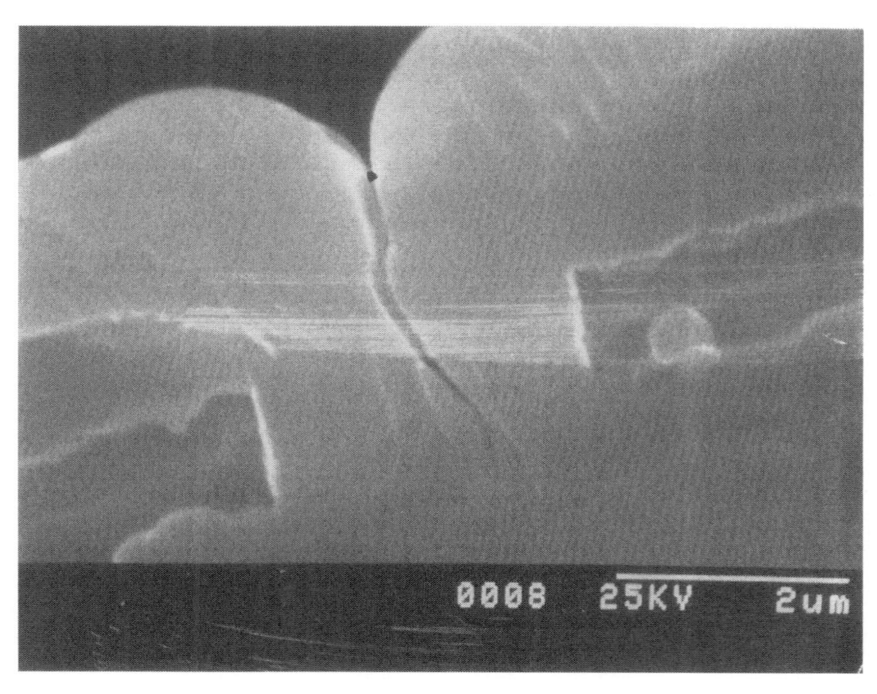

b

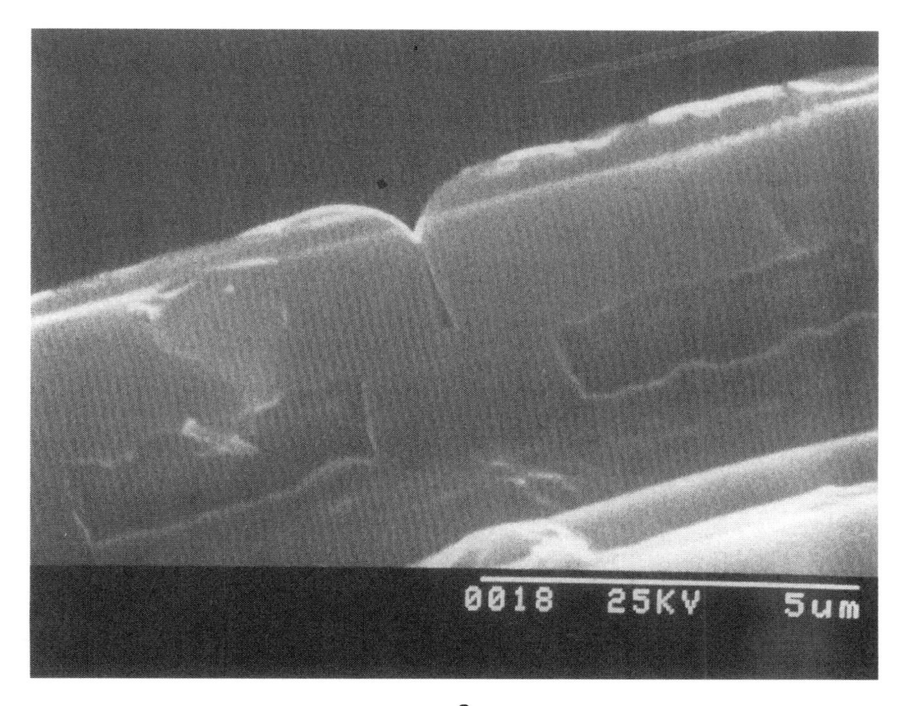

c

Figure 8-23. Defects Seen in CVD Films CVD films are prone to a number of different types of defects. They include particles generated during deposition, cracks due to excessive stress, contaminants or temperature change, and voids and other structural defects formed due to improper deposition characteristics. Photographs reproduced with the permission of Greg Roche.

variety of various-sized "spikes" or bumps. If the cause of this oxidant is a leak in a tube or an endcap or fitting, a characteristic haze pattern will be seen on the tube, quartzware, and other parts of the furnace. In fact, bad leaks can cause enough of a reaction to create smoke and actual ash deposits in the furnace. Even much smaller leaks will cause enough gas-phase reaction and nucleation to create potentially yield-killing defects. These types of problems often cause defects on the surface and within the film.

Haze can sometimes be generated from the quartzware in the furnace. This can occur if the quartzware has been improperly cleaned and dried, so that moisture trapped in the quartz can outgas and react with the silane. Another source of moisture and also oxygen is quartzware that is not kept in the proper standby conditions between runs. If the quartz stays in the atmosphere for very long, it will absorb the moisture and oxygen from the air. Generally, if the tube is be left idle for more than 15 to 30 minutes, the furnace should be placed into a vacuum standby condition. This moves the quartzware into the furnace, seals the door, and brings the tube down to vacuum. Note that devitrified quartzware absorbs moisture and gases much more quickly than fire-polished quartz.

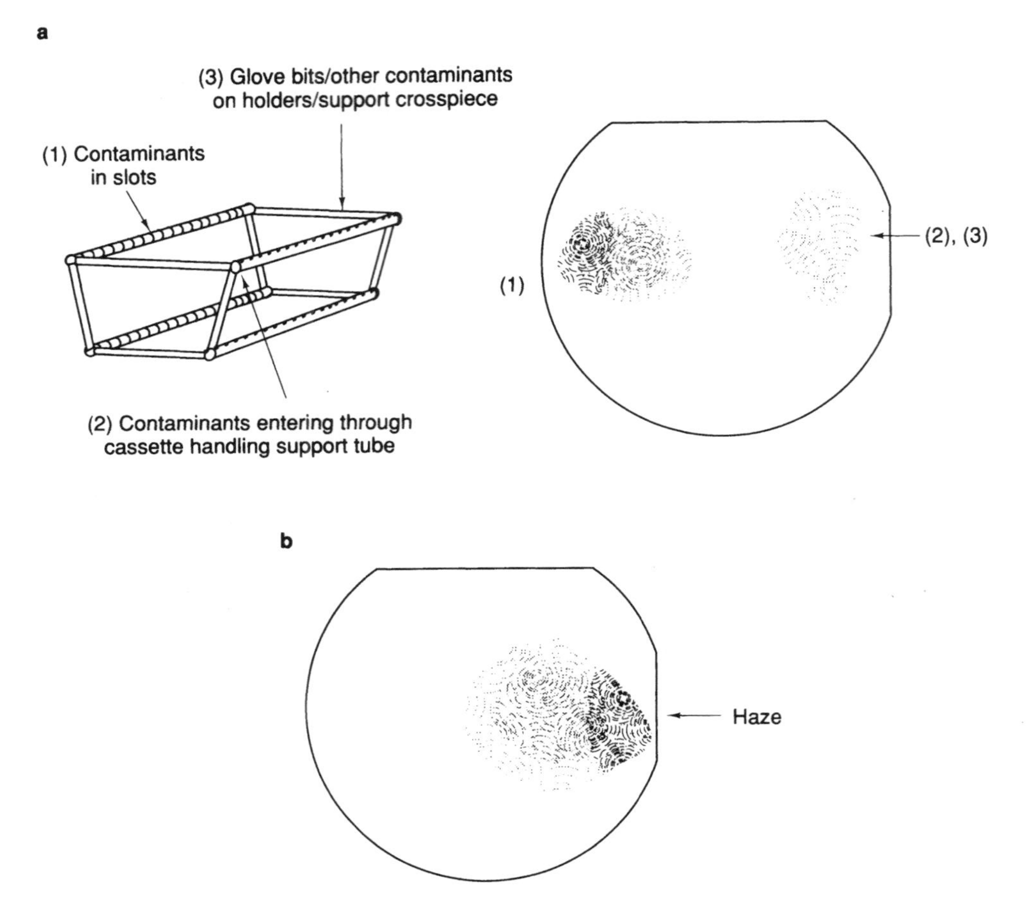

Figure 8-24. LPCVD Haze Formed by Contaminated Quartz Here we see the formation of haze due to contaminated quartzware. (a) Contamination is shown to have at least three sources, in the slots, on the surface of the support rods, or inside the cassette holder rods (which are usually just hollow tubes of quartz). The haze pattern will be more concentrated in the event of a contaminated slot, while other sorts of contamination will produce a more diffuse haze. (b) The slot contamination in more detail, where the edge of the slot (the "boat mark") is visible and the gradually decreasing density of the haze is prominent.

Other sources of contamination from the quartz includes things such as melted glove bits and fingerprints on the quartz which cause significant amounts of haze. Typical haze patterns are shown in Figure 8-24a. The haze will not be very obvious on the tube walls, and may be located only on specific boats. Often the wafers will have haze only on the regions near the boats. As seen in Figure 8-24b, if the wafers are loaded so that the flats face up, the haze will be prominent on the bottom of the wafers. To see haze in the tube more efficiently, a bright long-wave UV lamp can be directed down the length of the tube. Any haze on the tube walls will readily appear with this light.

A common practice with the deposition processes is to coat the all-new quartzware with a good, thick film prior to the introduction of any wafers to the tube, including all the boats and assorted parts. The film deposited is typically about 0.5 μm, or as thick as the process will deposit in a hour or so. The exact thickness is not critical. This coating reduces the possibility that moisture or other contaminants will get into the system from the quartz. The procedure also "glues down" any loose particles that may be residing in the tube. Problems with haze are sometimes reduced by running this coating cycle.

If the haze appears only on the wafers or in characteristic patterns on the wafers, such as a radial pattern (such as Figure 8-25), and does not appear on the quartzware to any great extent, the highest probability is that the contamination is being brought in on the wafers. The most common source of this kind of contamination is insufficient drying of the wafers. Residual moisture on the wafers is a common and very serious problem, and has led to the loss of a large number of wafers in almost every manufacturing area. A less frequent source of haze or surface contamination is absorbed oxygen and moisture from the atmosphere. This is usually a problem only in the event of a particularly sensitive product (the tunneling oxidation layer in EEPROM devices is an example of this). Other subtle, underlying problems on the wafers can be amplified by the deposition of the polysilicon on the wafers. For example, a slightly hazy SiO_2 film may make the wafers appear extremely hazy after polysilicon is deposited on it.

Particles that remain after previous film etches, such as silicon nitride etch, can often be highlighted and can appear as polysilicon defects. This occurs as a result of the accretion of material on small, nearly invisible particles. This can be an extremely complicated problem, as the source of these particles must be clearly identified before any progress can be made on resolving the effects of the problem. It is very easy to cause real damage to the deposition system when there is no real problem, by trying to solve problems that have subtle sources in preceding steps.

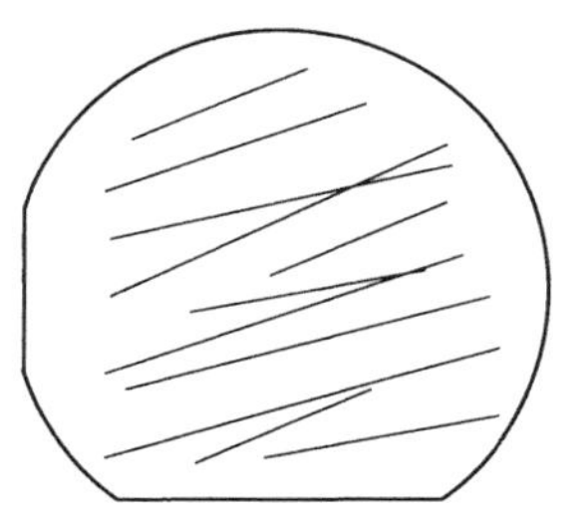

Figure 8-25. Characteristic Haze Patterns in CVD Processes Haze is often introduced to the CVD furnace from the previous process, especially if that process is a cleaning step. Moisture is considerably more difficult to remove than one would believe, so that improper drying often results in haze. In the figure to the left, the wafer appears to have been removed from a batch cleaning system with the center of rotation of the system to the lower left. The figure to the right has been processed in a spinner or an on-axis rinser-drier. In any event, insufficient time or conditions have existed to remove all of the moisture.

One of the problems with the grainy nature of the polysilicon films is that it is difficult to get a good count on the number of particles distributed over the surface of the wafer. The particle counters cannot distinguish easily between grains and particles, although this technology is progressing rapidly. At this time, the best method of counting particles is the tried and true method of visual examination. This is done by examining the wafer under a dark-field microscope and counting the visible highlights. However, surface-borne particle counters can readily gauge the haze content of the polysilicon film. This is a very convenient feature that can permit analyses on otherwise complex problems.

There are a number of tests that can be performed on doped and undoped polysilicon films. These tests include film resistivity, the recombination rates of the electron-hole pairs formed in the polysilicon, the capacitance obtained with the film, and the charge-retention characteristics of the film. These tests verify that the film will be able to carry its primary roles: memory (charge storage), capacitive and transistor devices (capacitor tests), and resistive elements (resistivity). The details of these critical tests will be discussed in chapter 10.

8.2.2 Deposited Silicon Dioxide

We have already discussed how silicon dioxide can be grown from silicon wafers or from polysilicon films by thermal oxidation. Another method for obtaining SiO_2 films is through chemical vapor deposition. Deposited silicon dioxide, also often called glass, can be produced using both thermal and plasma-enhanced methods. While there are slight structural and chemical differences between these films, they are all used for the same tasks, which are to form protective dielectric layers on top of (or between) metallic or nonoxidizing surfaces, and to "planarize" the surface topography so that further processing can occur.

Unlike thermally grown silicon dioxide, which is typically very pure and contains no dopants or contaminants, deposited silicon dioxide often contains a number of dopants, in order to alter its physical characteristics. The index of refraction is essentially the same at around 1.46; however, it can vary substantially, depending on the impurities or dopants embedded in the film. The film is less dense than thermally grown SiO_2, and is not generally acceptable for use in structures such as gate oxides or thin interpolysilicon-layer dielectric layers. The film can contain a significant number of defects, including particles, pinholes, and reduced structural strength. As a result, low breakdown voltage, high current leakage, and unpredictable electrical characteristics are typical of deposited silicon dioxide films. For the purposes of these films, these characteristics are not important.

Interestingly, a significant amount of research has been directed to the use of multiple layers of very thin (around 20 to 60 Å) films for use as gate and interpoly layer dielectrics. While very difficult to manufacture, these films have advantages such as the prevention of ions traveling through the film, changing threshold responses (the ions become trapped at film interfaces). Some of the combinations that have been successful include stacks of silicon oxide and silicon nitride, silicon oxide with nitridized silicon oxide (like silicon oxynitride), and thermally grown oxide capped with high-temperature, high-density deposited oxides. These techniques will become more critical in the 1990s.

The typical dopants that are used in these processes are boron and phosphorus. They are used primarily due to the type of changes they can induce in the step coverage stress, melting points, and other physical characteristics, and not for potential changes in electrical properties. The films that are produced by these dopants are called PSG (for phosphosilicate glass), BSG (for borosilicate glass), and BPSG (for borophosphosilicate glass). These dopants are typically mixed in the gas phase as the film is deposited. The exact mixture of dopants is very critical, with minor changes in dopant concentration causing a variety of problems, ranging from excessive defect generation to significant changes in the ability of the glass to flow over corners and steps.

The typical "thermal" (i.e., initiated by temperature) reaction for these various films takes place at around 400 to 430°C. The typical deposition pressures are in the range of 175 to 250 mT. The thermal reaction that forms the silicon dioxide is:

$$SiH_4 + O_2 \rightarrow SiO_2 + 2H_2$$

There are a variety of chemistries used in plasma deposition processing, with a broad variety of silicon sources in use. The most common reaction uses nitrous oxide as an oxygen source and silane, or disilane as a silicon source. In some cases, exotic chemicals with names such as TEOS (Tetraethylorthosilicate), or TMCTS are used as nonpyrophoric silicon sources. In most cases, these chemicals produce essentially the same end film as those produced by more standard means. For now, we will consider that most of the plasma reactions used are based on silane and nitrous oxide. The TEOS reactions are much more complex.

$$SiH_4 + 2N_2O + \text{plasma energy} \rightarrow SiO_2 + 2N_2 + 2H_2$$

The reactions for the dopant species are similar to that of the silane reaction. There are a variety of chemicals used to produce the reactions, depending on the type of process in use. We will discuss the pros and cons of these various processes shortly. For now, we can see that the dominant reactions used to form the dopant species in the thermal deposition process are:

$$2PH_3 + 5O_2 \rightarrow 2P_2O_5 + 3H_2$$

$$2B_2H_6 + 3O_2 \rightarrow 2B_2O_3 + 6H_2$$

$$4BCl_3 + 6O_2 \rightarrow 2B_2O_3 + 6Cl_2$$

As we can see, these reactions all compete for the available oxygen that is supplied to the furnace. This competition leads to some complicated reactions, and to a tight control requirement on the oxygen supply system. Variations in the oxygen content will change the exact balance of the reactions occurring in the process chamber, especially with the BPSG deposition process. Variations in the flow rates of the other gases will also alter the exact reaction taking place in the processing chamber.

Historically, the phosphine (PH_3) gas has been diluted with nitrogen and delivered to the system through independent gas lines. However, recent practice has shown that mixing the phosphine and silane together at the gas bottling facility allows the process to remain more consistent through time, and is safer. Phosphine is highly toxic, even when heavily diluted. Any gas leaks can cause serious illness and can contaminate a large area of a fab in a very short period. Phosphine has been recently linked to chromosomal damage and other serious long-term health risks. However, when mixed with silane, leaks in a gas line will cause the silane to react with the oxygen in the air. This will cause enough heat to ignite the phosphine gas also. As a result, instead of entering the environment as phosphine gas, the phosphorus is introduced to the outside world as a solid, which is simpler and safer to clean up. The major difficulty in developing a process using silane/phosphine blends, is that the gas mixture must be very carefully controlled and tested prior to delivery to the fab. Changes in the gas mixture in the bottle will result in changes to the dopant content of the films throughout the lifetime of that bottle (often three months or more). In general, if premixed gases are used, the PSG process is not much more difficult to run, and no dirtier than undoped silicon dioxide processes.

If the silane and diluted phosphine are mixed at the last second, then a number of subtle problems can arise. They include items such as pressure imbalances in the manifolds, which can lead to inconsistencies in the quantity of phosphine in the gas stream. This will lead to nonuniformity of the phosphorus in the film. Another subtle problem that occurs with having a separate gas system for the phosphine is the existence of extra lines and fittings that can leak, allowing additional sources of contamination.

The boron dopant sources have varied through time. Boron trichloride (BCl_3) has been used because its reactions were well understood and slow enough to produce reasonable films. However, it has the disadvantage of releasing quantities of chlorine into the processing ambient. While diborane (B_2H_6) produces a much cleaner reaction, it reacts very rapidly, and in the standard diffusion furnace-type reactor will not produce films of uniform boron content. As a result, there has been development work with both chemistries. The end-product films are similar in dopant content and other parameters.

The BCl_3 process is plagued with a wide variety of problems that do not exist with most of the other gases. For instance, the BCl_3 bottle must be chilled below 20°C, while the lines must be continuously heated from the bottle until the gas enters the furnace because the gas liquifies at about 25°C. If this happens inside the gas line, large quantities of particles can be generated. In addition, the excess liquid BCl_3 can get into the gas delivery system. Particles are created when the condensate escapes through the MFCs, or when it condenses abnormally after it passes the MFCs. These small particles are extremely lethal to die yields and can cause problems with increased downtime and other associated manufacturing problems, such as the need for increased inspections. One of the symptoms of this problem are BCl_3 lines that will not evacuate in a reasonable time. For instance, a six-inch leg of BCl_3 line that, when filled with N_2 gas, can be evacuated in a few seconds will take as much as 15 minutes to evacuate when filled with liquid BCl_3.

Another problem with the BCl_3 process is the chlorine by-product. The chlorine can react with several of the films that are used in the manufacturing process, such as aluminum and polysilicon layers. If this chlorine comes into contact with these films, the result is damage to the films, as well as chemical contamination of the circuit. This will result in long-term reliability problems if not immediate circuit failures. Thus, doped films are usually not placed directly on top of the metal or polysilicon layers. A layer of undoped silicon dioxide is placed on these films as protection. Placing metal films on top of the completed doped oxides is safe, since the chlorine has already been driven off. A very common problem is to have chlorine migration to the circuit through the "boat marks" left during previous undoped oxide deposition steps. This is seen in Figure 8-26. If

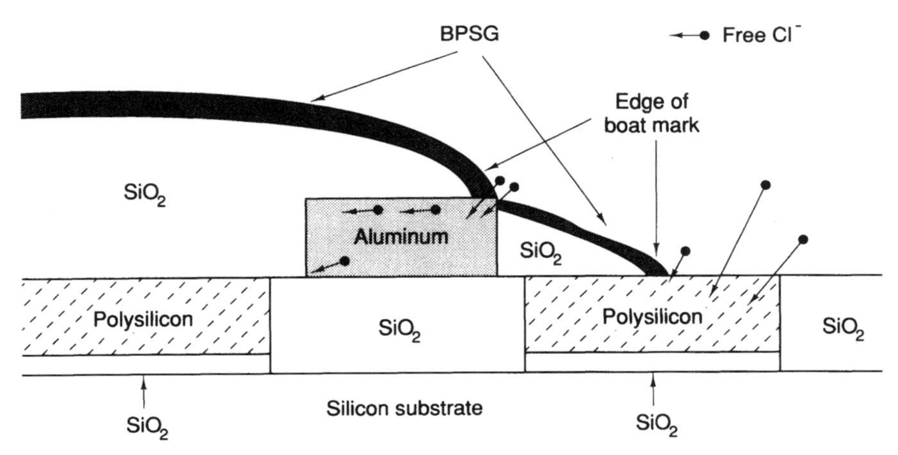

Figure 8-26. The Seepage of Chlorine under SiO_2 Dielectric Layers If the surface of the metallic portions of the circuit are exposed during processing (for example, around boat marks, large particles, etc.) and the subsequent process includes a BPSG deposition using BCl_3 as a boron source, this free chlorine can migrate under the undoped SiO_2 film. The best prevention for this is to use single-step, integrated passivation processes.

chlorine can migrate under the film being deposited, it can cause significant mechanical failures, for example, peeling off of the wafer surface. Another problem that is relatively rare is that of the effect of chlorine on other components of the circuit, for instance, thin oxide layers. While usually completely covered by the time BPSG is deposited, pinholes and other defects can allow the migration of some of these atoms into these very-sensitive regions. Finally, the chlorine also causes an increase in the corrosion and deterioration of the pumping and exhaust systems. Usage rates of pump oil will also be very high, as the chlorine will attack many types of pump oil. Fomblin oils are recommended for chlorine-producing processes, due to the minimal reaction between the oil and the exhaust effluent being pumped.

Diborane (B_2H_6) is much cleaner than BCl_3, even though it has a few problems of its own. For instance, it is very toxic, even more toxic than BCl_3, and must be carefully contained. It is also pyrophoric, although as delivered it is usually fairly heavily diluted in nitrogen, which helps to reduce its flammability and toxicity. It is also possible to mix the diborane with the silane, as with the PH_3. As previously mentioned, the chemical is very reactive with oxygen, reacting faster than both phosphine and silane. In a diffusion furnace layout, this means that the boron gas stream will be completely depleted before it reaches the entire wafer. This results in poor dopant uniformity. To make matters more complex, since the boron will be reacting abnormally fast near the injector side of the wafers, there will be a relative decrease in the amount of oxygen available to the phosphine. As a result, the phosphorus content of the films will be reduced. As the boron is depleted, the phosphorus content will increase until, at the far edge of the wafer, the phosphorus content will reach a maximum peak. In addition, all of this competition for available oxygen will cause changes in the reaction rate of the silane, thus changing the thickness in local regions, decreasing uniformity control.

Diborane processes are mainly used in two types of chambers. One of these is similar to the standard furnace process, but uses specially designed cassettes to contain the gases coming out of the injectors. This type of system has met with limited success. The other main type of system is the single-wafer chamber. This type of system permits gases to travel a short distance only and to react within a short period of time. They have had quite a bit of success in the market, with both thermal and plasma-enhanced versions of this type of chamber available.

Occasionally boron is mixed with the undoped SiO_2 to form BSG, although this is unusual. The most common films of choice are PSG and BPSG. The choice of which film to use is typically decided by determining how much annealing will be required later in the process. Almost all deposited oxide films are annealed to increase the density and to cause the film to flow smoothly around corners. The change in dopant levels will result in changes in the temperatures and time required to reach a certain level of "reflow" during the anneal process.

As can be seen in Figures 8-27a and b, the films will appear different after the anneal process, depending on the doping levels. Increases in the dopant levels typically reduce the temperatures and times required to achieve the required characteristics. The addition of boron greatly increases the amount of flow. The amount of dopant is quite crucial. If too much dopant is added to the film, it may leach out of the film and into metal lines or other crucial parts of the circuit. A similar problem will occur when the film is brought into the open air. Excess boron will crystallize as boric acid on the surface of the wafer as it cools. These crystals are not a serious problem in themselves as they are water-soluble and are easily removed by rinsing the wafers. However, they are symptomatic of an underlying process problem which should be evaluated.

Concentrations of dopants are usually expressed in mass percent. This makes the maximum phosphorus content around 6 to 8%, and maximum boron contents 4%, with total dopant content limits of around 6 to 8% (meaning less phosphorus will be needed when boron is introduced). Sometimes, the dopant levels are reported in other types of measures, such as mole percent or volume percent. These numbers appear somewhat different, usually in the low- to mid-teen percents. It is important to understand which set of units are being used to prevent confusion and potential misprocessing.

There are two basic structures of the deposited film, planar and conformal. The planar films are the more difficult to produce. These films must be created in such a way that the end product is flat. This is seen in Figure 8-28. The processes usually involve the deposition of a very thick film of oxide in order to overcome the effects of the underlying topography. This may lead to problems if the films have to be so thick that proper circuit operation cannot be obtained. In these cases, a new technique has been appearing with advanced plasma reactors, which involves the deposition of a film, followed with a surface etch on the top of the film, followed by another deposition cycle. As seen in Figure 8-29, this technique allows the slow buildup of film in the trenches of the wafer, while keeping the film on the top of the structures thin. This is a slow and detailed process, which many possibilities for errors, but has been gaining acceptance by the industry.

The deposition of conformal films is not terribly difficult, as most silicon dioxide films tend to be conformal by nature. One rule of thumb to remember for good control is that the best conformal silicon dioxide films tend to be formed at the lowest possible pressure and at the highest possible rate which implies the maximum reaction efficiency possible. The biggest problems with developing conformal films is preventing overhangs or "cusping" at the edge of the steps. The formation of this structure is shown in Figure 8-30. These regions of weakness will often become areas where shorts, breakdowns, and current leakage can occur. Other difficulties can occur in these regions during etch processes also, as the etch gases penetrate the cusps and cut even larger holes there.

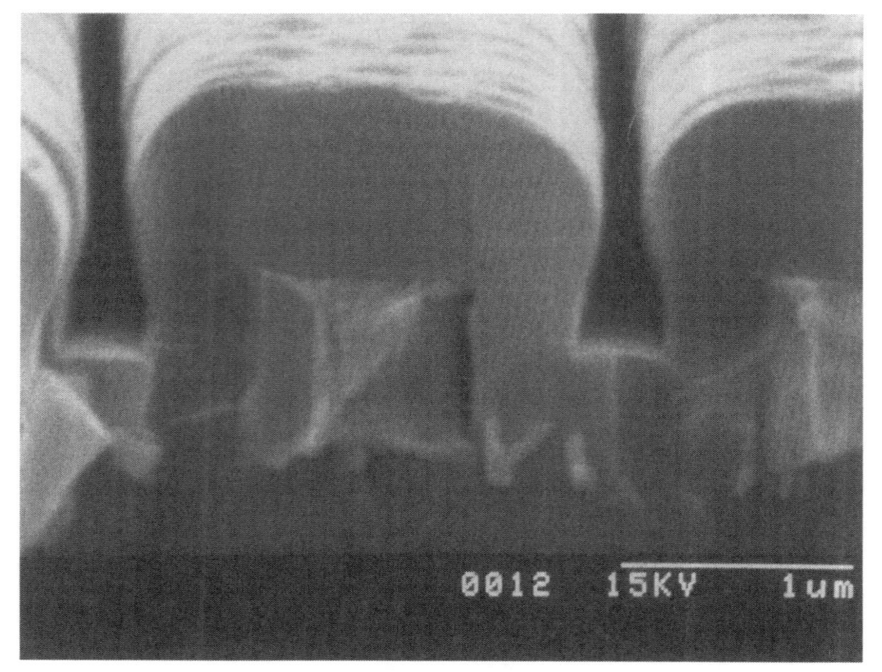

a

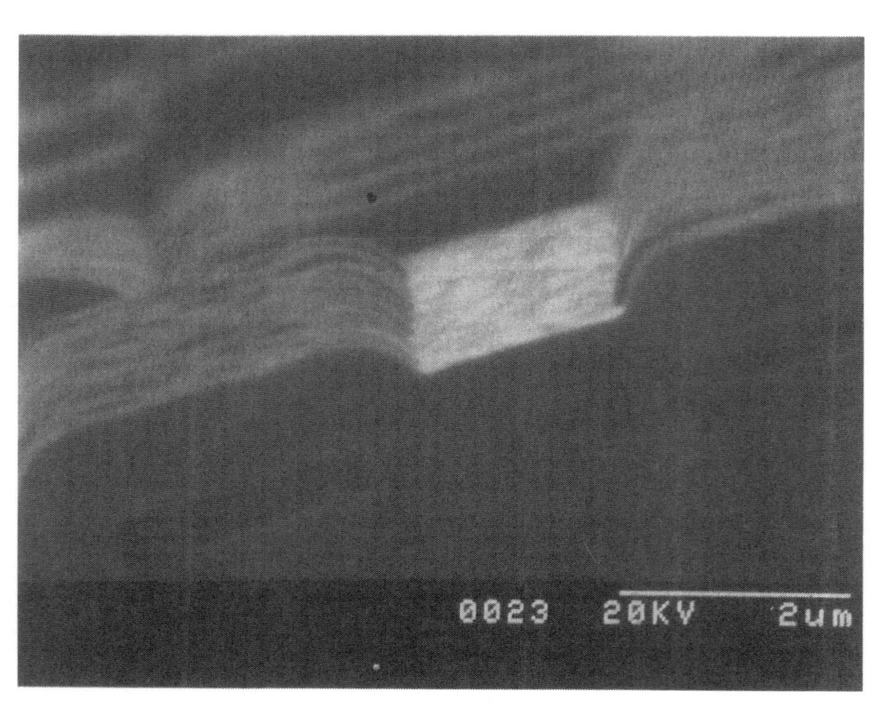

b

Figure 8-27. The Use of Planarized Films After many layers of films have been deposited and etched into multiple types of patterns, the surface topology becomes very complex. Deposition of subsequent films becomes difficult, as undue stress, cracks, and other related problems develop. The use of a dielectric film to smooth this rough surface is very desirable. After such a planarized film is deposited, subsequent films have nice, uniform surfaces to which to adhere. Photographs reproduced with the permission of Greg Roche.

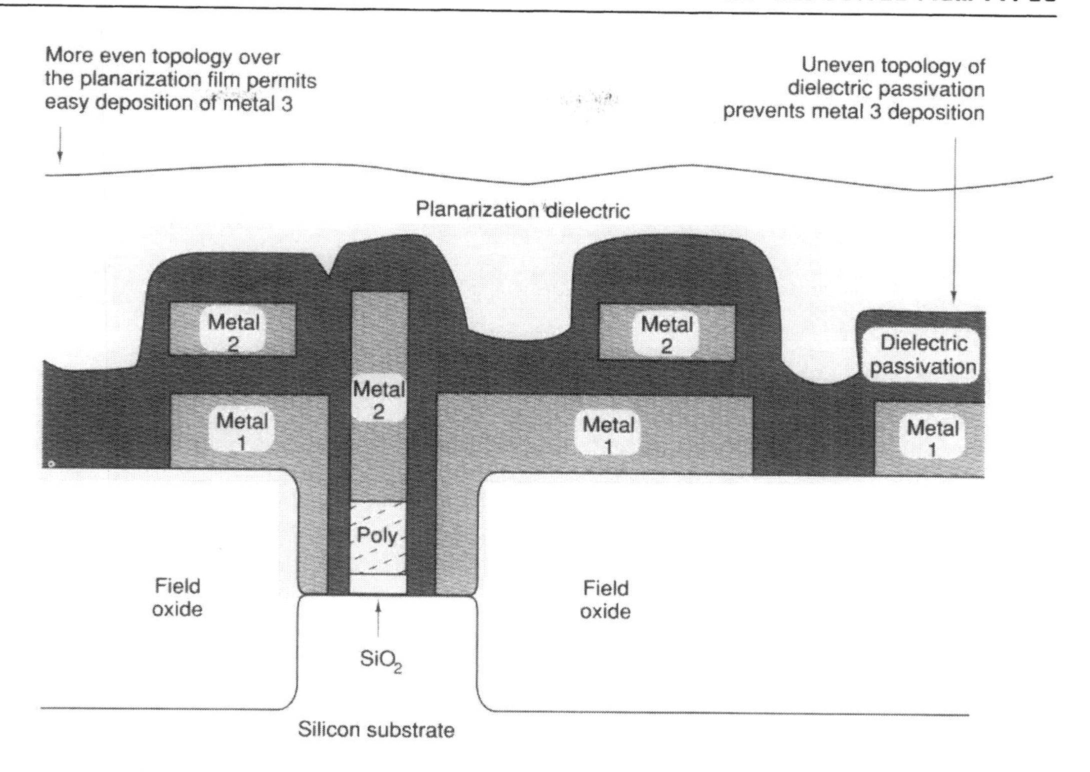

Figure 8-28. Annealing of Deposited Films As SiO_2 films are annealed, the films densify and re-flow over the steps to become smoother and more consistent.

It should be noted that almost all of the equipment and procedural problems and issues noted with the polysilicon processes can cause problems in the oxide deposition processes. This includes the requirement that dummy wafers are used to balance the load to prevent changing the depletion characteristics of the process. This also includes all of the statements about gas system leaks, especially in the various pyrophoric gas systems required with these processes. Leaks in a BPSG gas system can be very destructive. Moisture entering the system from improper dry cycles can cause a significant amount of haze.

A very common problem at the glass deposition steps is that they get blamed for a huge amount of contamination that they do not cause. Since the film is usually around a micron thick, any preferential deposition on a particle will result in a very large and visible defect. It is not uncommon for tiny polysilicon or nitride residual particles to be highlighted by the glass process. It is important to monitor the process very carefully, and independent of product wafer observations, to ascertain the source of particle problems early on. For instance, a bare silicon wafer should be tested for particulate addition to the next production run if high particles are seen on the product wafers during inspection and they appear to be coming from BPSG. If the test wafers show an insignificant increase in particle counts, it is very likely that

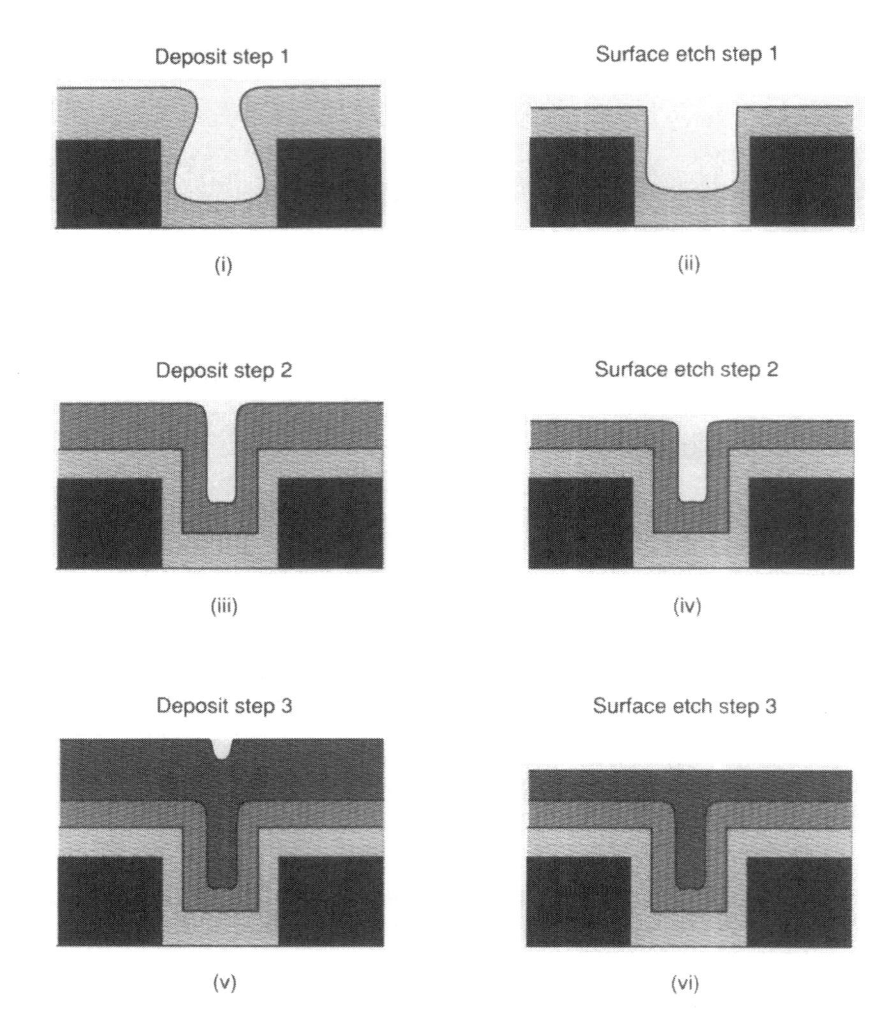

Figure 8-29. Planarization through Deposit and Etch Techniques One way to form a planarized film is to alternate the deposition of the film with an etch process that attacks the tops of the structures faster than the bottoms of the troughs. As shown, after several of these sequences a flat surface can be obtained.

small defects are being highlighted. This must be verified as soon as possible, both to remove the cause of the problem and to improve the visual quality of the material at the glass-deposition step.

Finally, we will discuss some of the more subtle problems that occur in and around the silicon dioxide deposition process. A very significant problem is the case where the topmost protective layer has flaws, pinholes, or design problems such that the atmosphere can leak in and reach the underlying doped oxide layers. When this occurs, the moisture in the air can slightly absorb into the film, reacting with and leaching out the dopants from the film. The dopants will react with the moisture to form acids (phosphoric and boric acids, primarily) which then attack the metal structures, leading to eventual circuit failure.

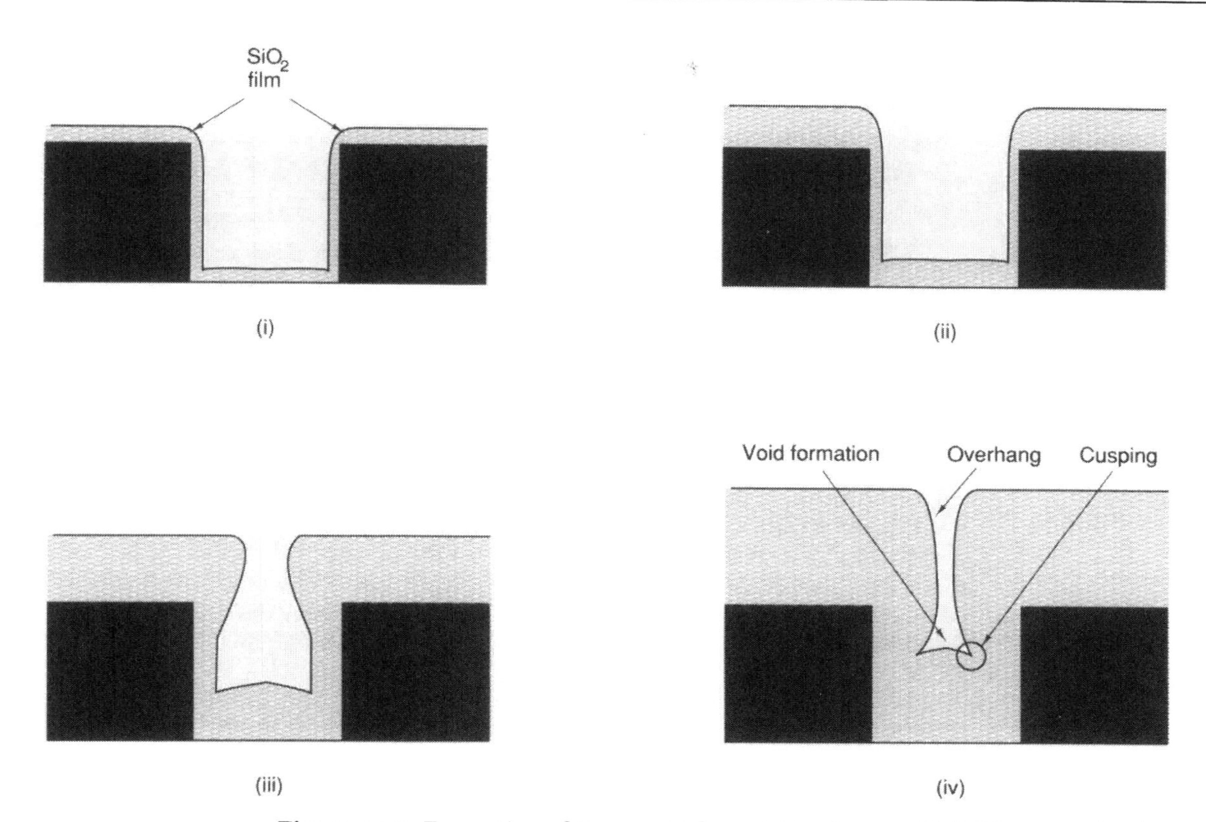

Figure 8-30. Formation of Cusps, Voids, and Overhangs CVD films must be characterized carefully to avoid the formation of overhangs and cusps which will ultimately result in the formation of voids, or in areas of reduced dielectric breakdown at the corners of the cusps. These phenomena occur because the deposition rates of the films vary on each of the sides of the steps. The deposition rates are highest on the tops of the steps, while the bottom of the step is next, until the gas channel is pinched off. Any remaining space here will remain as a void.

Under normal conditions, these reactions take place so slowly that it will be years, if not decades, before any significant changes are seen due to this problem. However, in humid, high-temperature conditions, the probability that this problem will have a reliability impact increases. Tests can be performed on the wafers to find out how serious the problems are. The primary test is the high-temperature "steam-pot" test. In this test, the packaged chips (and sometimes not packaged, if the wafers need to be tested directly) are placed in sockets and are operated while the chamber temperature is raised to around 140 to 150°C, with nearly 100% humidity. The wafers are then run for several hundred hours (usually around 500 hours, but this may vary depending on the severity of the test). Clearly, Mil-Spec parts must survive even the most severe tests.

Well-made chips should not fail for this problem, as the problem is preventable through proper passivation-layer sequencing. For example, you obviously should not use a doped glass film as the last step

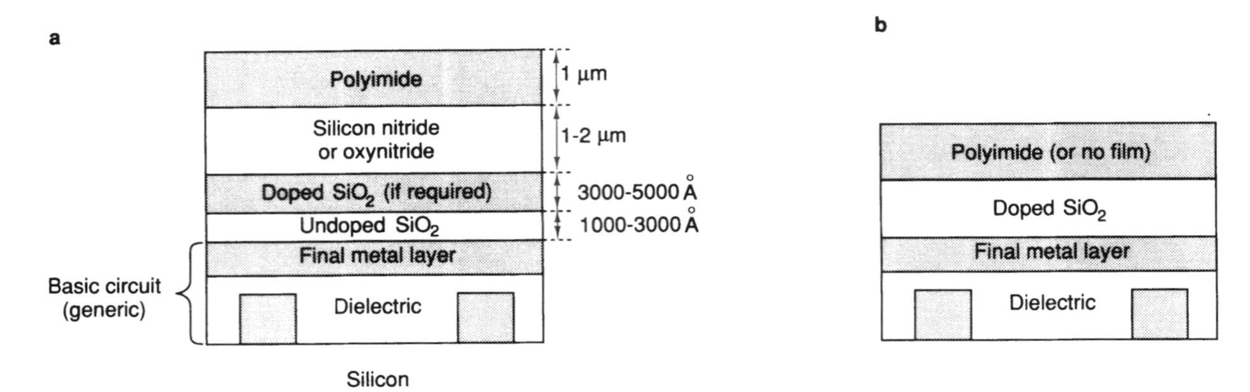

Figure 8-31. Passivation Layers for Silicon Chips (a) A reliable passivation sequence for a silicon IC. This sequence will prevent radiation, moisture, and atmospheric gases from penetrating into the inner workings of the IC. Some films such as the doped SiO_2 are optional, the doping providing a gettering location for contaminants to be drawn to (and away from) the main circuit. (b) This sequence is an unreliable step, since moisture can penetrate to the doped SiO_2 and form phosphoric acid, which then attacks and corrodes the topmost metal layer. The details of the final passivation sequence should be thought out carefully prior to implementation.

in the process, as moisture from the atmosphere will start to attack it, causing eventual damage to the aluminum bonding pads. A reasonable layering sequence is shown in Figure 8-31a, while an unreliable combination of films is depicted in 8-31b. As we can see, there is a thin undoped layer placed on top of the final metal layers. On top of this is a layer of doped glass, which has been annealed. This glass may allow the formation of corrosive acids and must be capped properly. As a result, the film must be covered completely, usually with a nitride or a oxynitride film. One of the crucial areas to cover are the edges of the films, where contacts, bonding pads, and other features may become exposed. Therefore, it is better to etch the doped film and then deposit the final protective coat (using silicon nitride) and etch it so that the edges of the doped glass are covered. Finally, the topmost layer may be this last thick nitride film or there may be an additional coating of polyimide placed on the wafer.

We have discussed the fact that deposited SiO_2 is often annealed at fairly high temperatures, typically above 900°C. These anneal processes are referred to by a variety of names: *Flow, Reflow*, and *Densification* are three of the more common names. Obviously, this cannot be performed on wafers with many metal films (such as aluminum) already deposited, but can easily be accomplished on top of polysilicon and many temperature-resistant metal silicide films. The ambient gas stream is sometimes oxidizing (usually, a steam ambient, as in a thermal oxidation process), and at other times nonoxidizing (nitrogen). The oxidizing ambient is used to verify that the various reactions that must take place are completed (the oxidant will attach to any loose silicon bonds). As a result, it is important to leave the wafers in the furnace long enough for the steam to diffuse completely through the glass. Since the film is so much less dense than thermally grown oxide, this diffusion will occur in 15 to 30 minutes (obviously varying depending on the total film thickness).

In both the oxidizing ambient and nitrogen ambient, the film will become much more dense as it is processed. While the density may increase by several times, it still may not match that of thermally grown silicon dioxide. The densified oxide film will also conform and seal to the underlying surface better than the as-deposited film. This has the benefit of producing a smooth surface to deposit subsequent films. The downside of this is that this densification usually places the wafer under a surprisingly high amount of stress. The oxide films that are deposited will usually place the wafers under compressive stress. The stress may range as high as 1×10^8 to 1×10^{10} dynes/cm^2, which is enough to bow the wafer by several microns, as shown in Figure 8-32. Highly stressed films are also much more prone to cracking. In

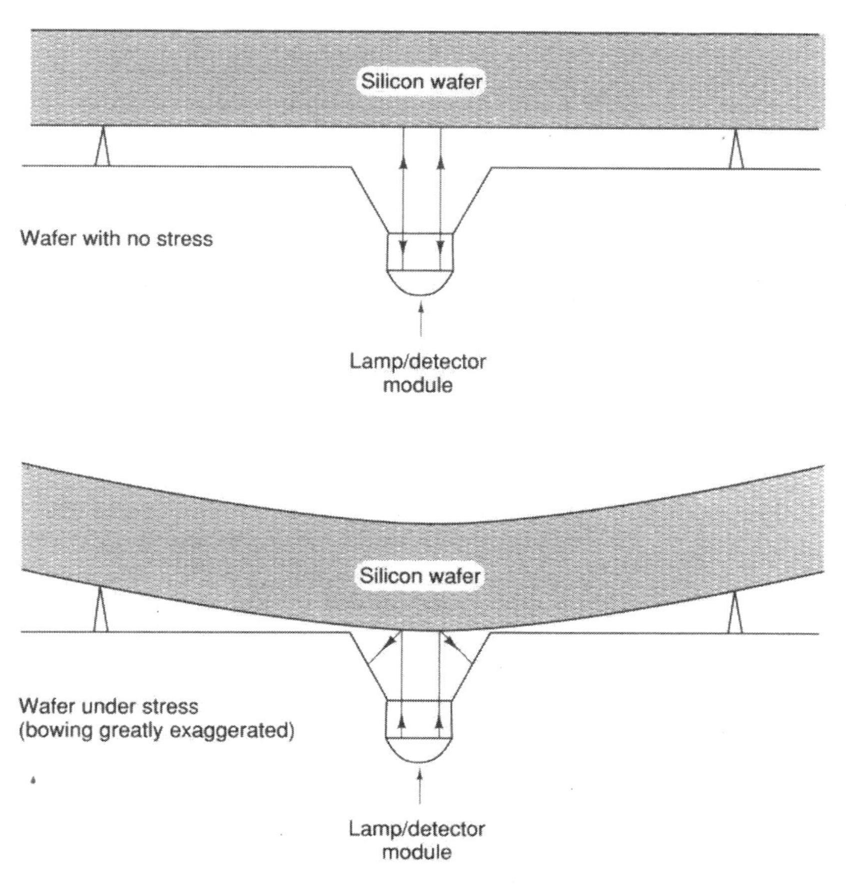

Figure 8-32. The Bowing of a Wafer under Stress When a film is deposited with high stress, the wafer will bow. This stress can cause the film to crack under subsequent processing. Here we see the bowing of the wafer being analyzed by a stress gauge.

addition, the high stress can place the underlying metal film under stress, thereby causing increases in the formation of hillocks and voids in the metal.

Since the oxide films that are placed on top of the metal layers usually cannot be annealed, it is important that these films have maximum density and are as smooth as possible in their as-deposited state. They are usually the films that require the specialized deposit/etch cycles. After these processes have been completed, further metallization can occur, or the chips can be finished and prepared for packaging.

There are some chemical differences between the thermally deposited and the plasma-deposited films. The plasma deposition processes tend to produce films that are high in hydrogen. This has some side effects, such as increasing film stress (counteracting the effects of lower processing temperatures), increasing the refractive index of the film above 1.46, and reducing the dielectric strength of the film. In theory, this hydrogen can be annealed out; however, since the plasma processes are usually performed on wafers after metal deposition, there is no practical anneal process that will produce reasonable results. The most common method to keep the hydrogen content down is to use silicon and dopant sources such as TEOS, TMB (Trimethyl borate), TMP (Trimethyl phosphate). They do not have the hydrogen by-products that exist in the other chemistries. However, hydrocarbons and other carbon by-products are often produced by these reactions. While most of them consist of CO_2, CH_4, and H_2O, some carbon can become trapped in the films. There have been cases where this trapped carbon has caused yield degradation.

8.2.3 Deposited Silicon Nitride

Silicon nitride is another dielectric film that is in common use in the IC industry. It has several properties that give it an advantage over silicon dioxide for a number of applications. The film itself is almost never doped, although the plasma processes often contain large quantities of hydrogen. Oxygen is often incorporated into the process to form a substance called silicon oxynitride, which we will discuss in more detail in the next section. It should be noted that many of the problems that occur with the oxide and polysilicon deposition steps occur in the nitride deposition process, so we will not repeat all of the details that have been covered that are common to the processes.

The nitride films differ considerably from the oxide films. The index of refraction of pure Si_3N_4 is 2.0, with the addition of hydrogen lowering the value of the refractive index. In addition, the dielectric constant of the nitride film is higher than that of silicon oxide. It is also much harder and almost impervious to chemical attack. The film is dense enough to prevent the diffusion of most of the contaminants that may enter the fab. While the film is harder than SiO_2, it can be produced in such a way that it is somewhat more flexible, reducing cracking problems. After reading all of these great features, one may wonder why it is not used for all of the various dielectric layers. The answers actually lie in the properties themselves. Its inability to accept dopants means it cannot be changed chemically, thus reducing its flexibility. The fact that it is chemically inert makes it more difficult to etch. In addition, no thermal oxidation will take place under the silicon nitride film, again due to the fact that little diffusion will take place through the nitride film.

Silicon nitride can be prepared from either thermal CVD or plasma-enhanced CVD methods. The thermal processes take place at 800 to 830°C, and at a typical pressure of 175 to 300 mT. A number of combinations of gases can be used to create silicon nitride films, but the most common are dichlorosilane and ammonia. The reaction that takes place is:

$$3SiH_2Cl_2 + 4NH_3 \rightarrow Si_3N_4 \downarrow + 6HCl \uparrow + 6H_2 \uparrow$$

Clearly, this reaction produces large quantities of by-products. The HCl can produce significant problems with both the wafers and the processing equipment, while the large quantities of hydrogen produced dilute the gas stream to amplify depletion effects, which are quite noticeable in this process. (Gas injectors are usually used in this process to prevent requiring a tilted temperature profile in the furnace). The HCl will be in a dissociated state in the vacuum environment, so there will be a certain quantity of Cl_2 in the exhaust stream of the furnace, as well as HCl. In an effort to reduce the effects of the reactivity of the chlorine, an excess amount of NH_3 is run into the tube. This causes a very large amount of ammonium chloride to precipitate from the gas stream as it cools in the exhaust lines:

$$HCl + NH_3 \rightarrow NH_4Cl \downarrow$$

This produces a continuous particle-generation problem, requiring frequent replacement of pump oil and cleaning of the exhaust system. The typical ratio of ammonia to dichlorosilane is a minimum of 3 : 1. Lower quantities of ammonia will start to give slightly silicon-rich and brittle films, with quality rapidly deteriorating below 2.5 : 1 $NH_3 : SiH_2Cl_2$. Symptoms of a low ammonia content include visual cracking of the nitride film, especially after a field oxidation or long anneal process, and also a slight increase in index of refraction. The use of more ammonia has no apparent deleterious effects, up to at least 11 : 1 to 12 : 1 ratios. In fact, the films produced under these conditions are often very pliable and have relatively low stress.

Due to the high index of refraction and high clarity of silicon nitride films, they produce some very striking and beautiful colors. This helps to highlight contamination, such as from haze, particles, or other sources. In fact, a surface-borne particle counter can pick up much higher apparent counts of particles if not calibrated properly as a result of this phenomenon. There will be such a contrast between the clarity of the films and the contaminant that even very small reflections from the particles will appear very bright (interpreted as large) to the instrument's detectors. Haze is very easy to see visually on most nitride films, although visual particle detection can sometimes be difficult for the same reasons that it is difficult with particle counting machines.

The thermally deposited silicon nitride films are typically used in the early steps of the process, where they are used to mask areas of the wafers, preventing certain reactions from occurring. For instance, during the field-oxidation process, silicon nitride is used to prevent the oxidation of the silicon in the gate and other areas that will later become active regions of the circuit. Silicon nitride is also used to block implants from reaching areas of the wafers in which they do not belong (e.g., separating the *p*-well and *n*-well regions of the substrate in a CMOS technology). These films are typically about 1000 Å thick, although there can be a fairly large range of thicknesses, depending on the manufacturing area and process. The nitride films are usually deposited on top of a thin (500 to 1000 Å) layer of SiO_2, since the deposition of nitride directly on the silicon substrate can increase the incidence of substrate crystal defects.

Thermally deposited silicon nitride films are often placed on top of polysilicon films as interlevel insulating films. In many cases, silicon dioxide is grown from the polysilicon, and the nitride is deposited on top of the oxide film. Since the dielectric characteristics of nitride are intrinsically better then those of oxide, the material works well in this application, as long as it is a defect-free film. This makes a manufacturing process using an interpoly nitride layer even more sensitive to particles and other film defects than one using standard interpoly oxides. Some of the advantages of the process include higher breakdown voltages, higher circuit noise immunity, less current leakage (if the film has no pinholes), and increased capacitance than SiO_2 films. The disadvantages are almost always manufacturability-related. For instance, there is the cleanliness issue. There is also the problem that nitride is hard to etch, and most nitride etchants will also attack SiO_2. Finally, the nitride will not allow further growth of oxide, and may form structure such as the bird's beak structure, seen in Figure 8-33. All of these issues make the ICs difficult to manufacture, especially on processes using narrower linewidths.

The plasma-deposited films exhibit similar characteristics as the thermally created nitride, with a few exceptions. First, plasma-deposited nitrides often contain significant amounts of hydrogen. This hydrogen can be bound to either the silicon or the nitrogen atoms. The nitrogen–hydrogen bonds are typically more stable, and may allow somewhat better results. The amount of hydrogen bonded to each species can be varied to some degree by alteration of process gas flows. Logically, an increase in NH_3 will intensify the production of the

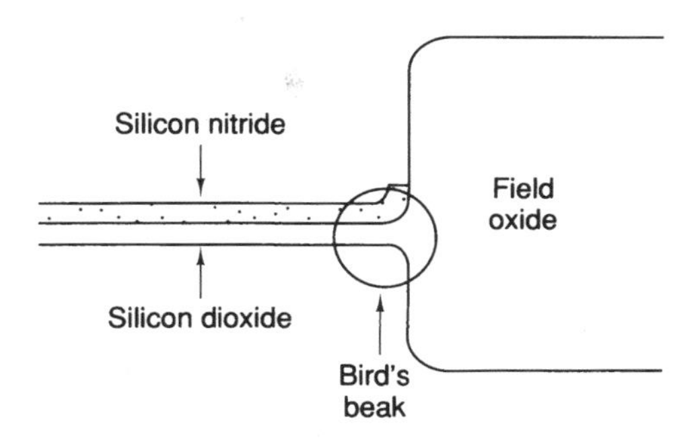

Figure 8-33. Bird's Beak Formation When the field oxide is grown, the oxidizing species can diffuse sideways as well as down through the silicon. When this sideways (lateral) diffusion occurs under a nitride film, the film bends upward, and the structure called the bird's beak is formed.

nitrogen–hydrogen bonds, while an excess of silane (or other silicon-source gas) will intensify the production of silicon-hydrogen bonds. The incorporation of the hydrogen into the film causes a number of side effects including a reduced index of refraction (which can vary to as low as 1.7 to 1.75), increased film stress, and decreases in film quality, such as reduced dielectric breakdown voltage, increased leakage currents, and increased permeability of ambient conditions (primarily moisture and oxygen from the air). Some of the hydrogen in the film can be removed by annealing, although this is rarely an option, since the wafer is often near completion when plasma nitride films are deposited, and cannot receive further high-temperature processing.

Nitride films are often under tensile stress after deposition, often causing them to crack over steep steps, especially if the deposition process produces a brittle film. Stress can range as high as 1 to 5 $\times$ 10^9 dynes/cm^2, although more acceptable values are typically in the 10^7 to 10^8 dynes/cm^2 range. Unfortunately, once introduced to the film, there is no efficient means of reducing film stress. As a result, significant efforts have been underway to identify specific process conditions to minimize this stress. Typically, film stress can be reduced by adjusting the gas concentrations. Gas streams that are rich in ammonia tend to produce more flexible nitride films that will not crack as easily and, in general, careful control of the gas mixture is critical. For instance, a nitride film that is produced in an ammonia-poor gas stream will tend be crack very easily, getting so bad that visual peeling can start to occur. Another method of keeping stress under control is to keep the temperature changes to the wafer slow and constant. Sudden temperature changes can induce subtle substrate warpage which can increase film stress levels drastically. The increase in film tension will typically force more warpage, resulting in more stress or film cracking. Another method for reducing stress is to deposit film of similar coefficients of thermal expansion on the wafer. While difficult in some cases, there can sometimes be intermediary steps to achieve the same results, for example, placing an oxynitride between a nitride and oxide layer. There are a number of other "tricks" used by the various equipment vendors to reduce film stress, most of which are proprietary in nature (and usually quite specific to the reactor producing the film).

Nitride films are often tested using the steam pot test described earlier. A good nitride film should allow no penetration of the ambient into the inner workings of the chips. The most significant source of this kind of leakage is particle contamination of the wafers before or (more likely) during the deposition process. When significant losses occur during this phase of reliability testing, immediate attention should be paid to the sources of any particles. A second common source for leakage of this kind are cracks in the film at the edges of steps. This problem can be very subtle and very random, since it may occur only in one area of a reactor where the gases may not be in the correct proportion (for instance), and may not occur with every batch of wafers. The only sure way to identify the exact cause of the leakage of these structures is through SEM analysis, if the exact location of the fault can located.

8.2.4 Deposited Silicon Oxynitride Films

This film is a combination of silicon dioxide and silicon nitride, just as the name implies. It has just been developed to the point of widespread use in the wafer fabrication environment. The film is produced only in plasma or other "enhanced energy" reactors, and is difficult to produce even in those conditions. The film has some inherent advantages to it, such as lower and adjustable index of refraction, while retaining the strength of the nitride films.

The silicon oxynitride chemistry is rather complex and cannot be described easily. In addition to the reactions for SiO_2 and Si_3N_4, there are a number of possibilities for oxynitride reactions to take place:

$$SiH_4 + N_2O + NH_3 + (\text{small amount})O_2 \rightarrow SiO_xN_y \downarrow + \text{several by-products}$$

Usually, about 1 to 3% oxygen is introduced to the reaction chamber. This acts almost as a catalyst, forcing the reaction to take place (after all, oxygen does not need the plasma to react with the silane). Sometimes, disilane and other compounds are used as silicon sources to form silicon oxynitride films. The pure silicon oxynitride film will usually have an index of refraction of about 1.65 to 1.70. The various combinations of films that can be produced in the oxynitride reactor can range from nearly pure SiO_2 (1.46) to that of nearly pure Si_3N_4 (2.00).

The biggest problem with the silicon oxynitride films is the production of single-phase compounds. Silicon oxynitride-like films can be easily produced that maintain both silicon dioxide and silicon nitride molecules. This alloy of the two materials is sometimes useful, but does not attain the strength of pure silicon oxynitride. In some instances, if the film is insufficiently dense, low dielectric breakdown voltages and other related problems can occur. Observing FTIR spectra of the various films shows the convergence of silicon oxide and silicon nitride peaks to form one single peak (see Figure 8-34). The narrow peak is indicative of a tight crystal structure, and shows the formation of true silicon oxynitride.

Like the other plasma-deposited films, the oxynitride films contain a significant amount of hydrogen. Again, the control of the exact mixture of gases is critical to the performance of the process. Stress is always a problem with deposited films and oxynitride is no exception. However, the mixture of the usually tensile nitride and the compressive oxide films can help in the overall reduction of the stress on the wafers after deposition.

Silicon oxynitride films are typically used for final passivation and protective layers. They are usually about 1.0 to 1.5 μm thick. The film is usually deposited directly onto the final doped oxide layer on the wafer, and not onto the metal lines directly. Sometimes the oxynitride will be coated with polyimide films as a further protection from the elements, and especially from radiation damage. The film is not typically used in the active portions of the device at this time.

The oxynitride films should also be able to pass the steam pot reliability test with virtually no failures. Again, the main culprits for the failures are particles and film cracks in steep topography. Scanning electron microscopy analysis must usually be performed to identify the problem and then the process must be adjusted or cleaned up as required.

8.3 CHEMICAL VAPOR DEPOSITION WRAP-UP

We have covered a lot of ground in this chapter. CVD processing is one of the most complex and lively of the processes. Keeping these systems operating is quite a challenge in itself, much less finding ways to improve the process. There are a number of subcomponents of all of these processes, with a variety of failure modes. The number of possible problems adds a tremendous amount of uncertainty to the processes. Fortunately, there are a number of common threads that make the job of diagnosing and controlling the CVD processes easier.

The common problems of the CVD processes start with the equipment. All of the processes are sensitive to internal chamber contamination and can create hazy or otherwise poor-quality films with even small amounts of moisture or other reactive materials getting into the chamber environment. As a result, all equipment must be cleaned and dried thoroughly. This concept is very basic, but it is surprising how often these procedures are completed incorrectly.

The CVD gas systems, both gas inlet and exhaust, must be leak-free and must allow for maximum conductance, except in the regions

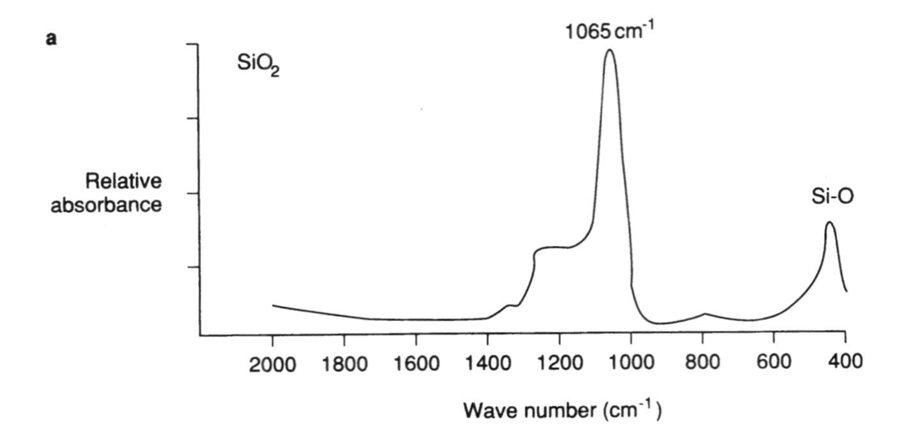

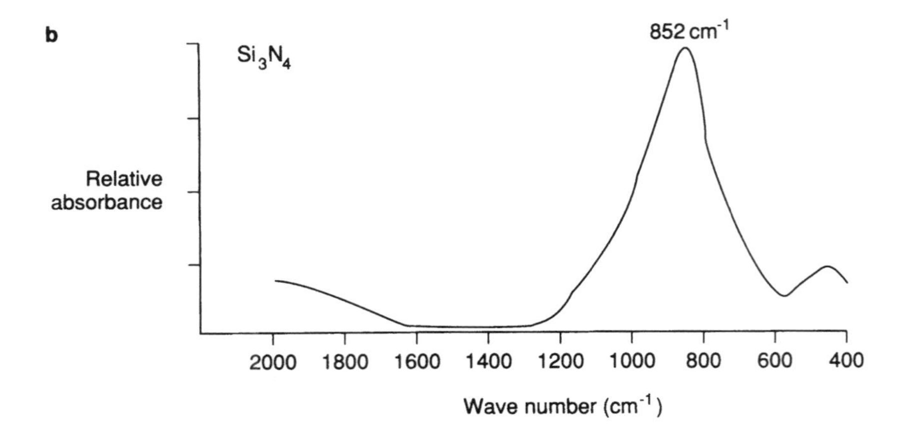

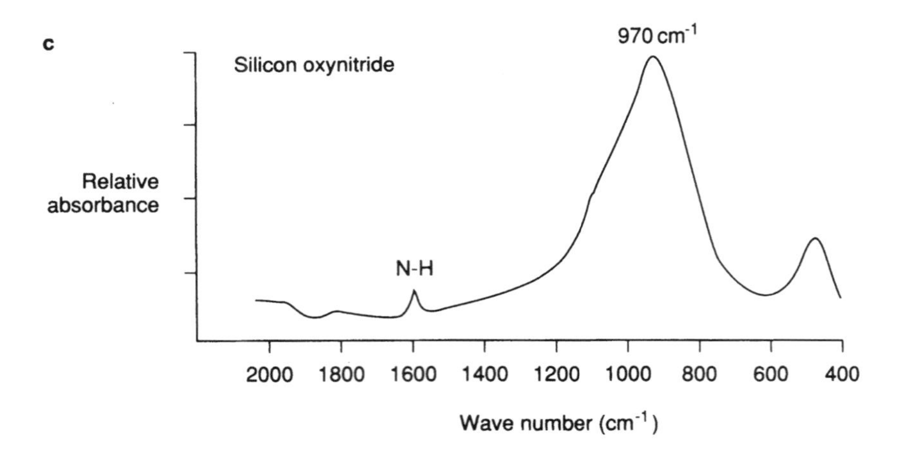

Figure 8-34. FTIR Spectra of Some Dielectric Films Silicon dioxide and silicon nitride have distinct FTIR spectra, with peak positions that can be used to monitor and control the stoichiometry of the films. The oxide and nitride films can be alloyed together to form silicon oxynitride. The exact composition of the film can vary significantly, with peak positions corresponding to the particular combination.

where the gas streams are intentionally constricted. Leakage in the pyrophoric portions of the gas system can result in a large number of ailments, from miscalibration or incorrect flow from MFCs, to improperly closing vales, to completely plugged lines. These problems are usually avoidable if the system is properly checked and purged prior to use. A typical silane gas line pump out will take anywhere from 12 to 24 hours, with nitrogen purges every few hours. If these purges are done properly and are not rushed, no problems should be seen.

A common problem is that of the loss of integrity of the lines around the seal of the gas bottles. These areas undergo a lot of stress during the lifespan of the gas system. If these areas become contaminated or damaged, the gas can mix with the air through the leaks and start to cause problems. A more typical issue is the lack of proper purging of the gas system when a gas bottle is changed. Even though the gas lines being fed by the bottle are vacuum-purged, it is crucial to leave the gas lines filled with nitrogen with enough positive pressure that air cannot get into the gas line easily. It is a mistake to leave the gas lines at vacuum prior to breaking the vacuum seal, because the air then rushes into the pipe, filling every part, reacting with any residues, and leaving behind gas lines permeated with water and oxygen.

Contamination problems can be introduced or magnified by contamination in the fixtures inside the processing chamber. Since most of the parts for standard furnaces are made of quartz, there is a significant possibility for contamination being transported on the quartz to the processed wafers. While the parts for other chambers are typically made of high-purity stainless steel, it is still relatively easy to improperly clean, dry, or handle these parts, so that contamination can be delivered to the chamber from these other parts.

Gases for all of the CVD processes must be extremely pure, with the plasma and polysilicon deposition processes requiring the highest purity. In most cases, 99.9999% pure gas is the minimum acceptable level. If contaminants are present in the gas, they will surely be incorporated into the films. Polysilicon is especially sensitive to gas-borne contamination. Associated with the purity of the gas is the uniformity of its distribution to the processing chamber. Mass flow controllers must be kept in calibration, and in-line manifold pressures must be remain properly balanced or the gases that reach the chamber may be improperly distributed and cause nonuniform dopant distribution and thicknesses.

Good pressure control is also a significant factor in the development of a good CVD process. In all of the processes described, the deposition rate is greatly influenced by changes in pressure. Defect generation, gas-phase nucleation, and particulate precipitation are all influenced by process pressure. In general, higher quality films are created when the deposition rate is kept at a stable point (particular to each process) at the minimum pressure and gas flow that will allow that deposition rate to be maintained. Unstable or incorrect pressure can lead to numerous problems and should be well controlled with throttle valves. Pressure bursts associated with the opening of valves and mass flow controllers can cause problems in all of the LPCVD processes.

Other equipment-related issues include furnace temperature control and stability. All of the LPCVD reactions are temperature-sensitive, even the plasma-enhanced reactions. The thermally induced batch processes, such as polysilicon deposition, are more sensitive to temperature than the single-wafer processes, due to the depletion effects involved. In all of the LPCVD processes, process temperature is often related to film density, as well as crystal structure and grain growth characteristics. Temperature control can usually be maintained in modern process environments in most circumstances.

Particle contamination of the wafers prior to deposition of the CVD films causes a number of ailments common to all of these processes. Even very small particles (0.1 μm or so) can create problems that can lead to circuit failures later because the gases will often precipitate preferentially on the particles, causing the particle to grow into a bump far larger than its original size. Particles can come into the process from many sources and can be generated within the process chamber itself, both as a result of processing and of mechanical actions. These sources must be minimized to maintain contamination free processing.

Film electrical and mechanical properties have a number of elements in common also, with the notable exception of polysilicon. In these films, high dielectric strength, low stress, and smooth contours (whether conformal or planarized films are required) are all positive attributes. For the polysilicon film, good grain and dopant (thus resistivity) uniformity, as well as good thickness control, are additional attributes.

Overall, the process engineer will find that these processes are ultimately much more similar in nature than a simple look at their chemistries would indicate. Many of the problems that are seen in one type of CVD process are often seen in slightly different manifestations in the other CVD processes. It is often useful for the CVD engineers to become familiar with the associated etch process for the film being deposited, since many of the problems of a CVD film do not become obvious until after the etch process.

Thus, we can see that the proper engineering of the various CVD processes is very important to the operation of the fab area and its output. Not only do these operations cause the most havoc with die and line yields when problems occur, but they also can have serious impacts on the output of the fab area when contamination or other issues cause excessive downtime. It is up to all of the personnel that run the CVD operations to make sure that the many variables are all balanced properly to allow the processes to run within tolerance and produce high-quality films.

CHAPTER 9

ION IMPLANTATION PROCESSES

The next major process in the "front-end" part of the wafer manufacturing process is the ion-implantation process. In this process, ions of particular elements are accelerated toward the wafer surface and impacted onto the surface of the wafer. These implants are used for a wide variety of tasks, one of the most critical being the formation of conductive regions, either in the substrate or in films, such as polysilicon. Often, these regions are going to be used as resistors with particular values, so that the complete implant and anneal sequence must be carefully controlled. Implants are also used to produce other effects, such as the creation of buried SiO_2 layers.

The implant processes place a number of stringent requirements on the manufacturing equipment due to the placement of the ionized substances into the wafers through physical, as opposed to chemical, means. Contamination must be tightly controlled, stray magnetic fields must be controlled, and extremely high vacuum environments are requirements in order for ion implantation to be carried out effectively. Implant processes are very sensitive to particulate and chemical contamination and are prone to easy contaminant redistribution due to the high accelerating voltages present in the systems. Contaminants that are present in the chamber have a relatively high probability of reaching the wafers, where they are likely to cause electrical parameter changes. There are a number of potential sources of contamination in the implant equipment, including the vacuum system, the wafer handling systems, and air leaks and other problems.

In addition, since these are high-vacuum processes and require high-voltage power supplies, the systems are by nature complex to operate and maintain, which results in higher production costs and more meticulous maintenance requirements. We will discuss the details of these process requirements in the sections that follow.

As mentioned, ion implantation steps are typically used to dope, or change the resistivity or type of particles regions of the substrate, while precision doping of polysilicon layers is being performed with ion implantation. The implant concentrations, or doses, are carefully controlled and are set in conjunction with subsequent anneal steps to ensure that the implanted regions of the wafer are of the correct junction depths and the correct resistivity. In the case of polysilicon, the film is usually implanted almost to the solid solubility limits (similar to the case of $POCl_3$ doping of polysilicon discussed earlier). Otherwise, the implanted areas of the substrate are doped to a depth, with a concentration that can be described by a Gaussian curve. This leaves the area with a heavily doped, or even supersaturated, layer at the predicted depth in the wafer. The subsequent anneal step allows the dopant atoms to migrate rapidly into the crystal structure of the substrate. After the anneal step has been completed, the resulting dopant profile in the silicon will be described by a slightly modified Gaussian curve. An example of this distinction is shown in Figure 9-1.

9.1 ION IMPLANTATION PROCESSES

A number of elements and their isotopes (also called "implant species") can be implanted into the surface of the wafers by accelerating them in a vacuum and directing them onto the surface of the wafer.

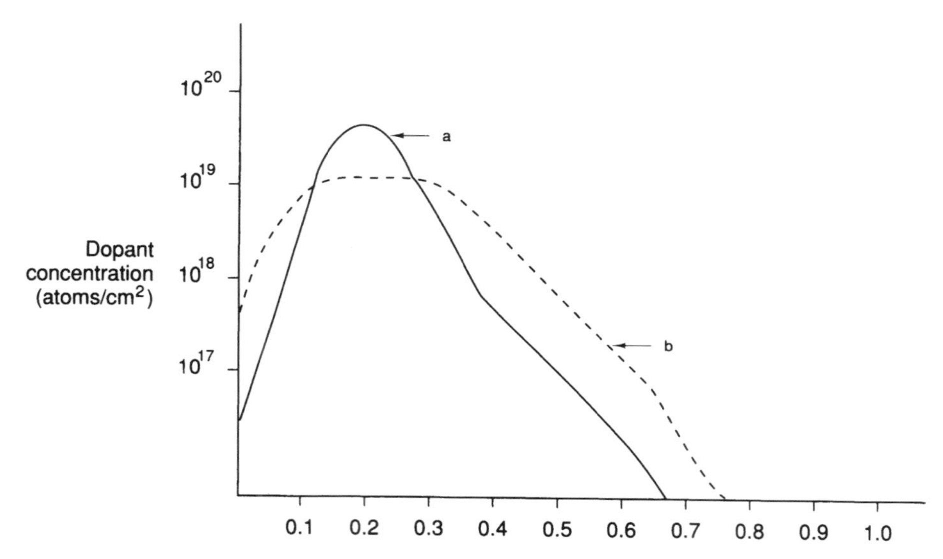

Figure 9-1. Typical Dopant Concentrations before/after Anneal As can be seen by this generalized example, an anneal step for about 45 minutes at a sufficiently high temperature will result in a different, less peaked profile. With a sufficient amount of time at temperature, very deep junctions can be achieved. Curve a represents as-deposited implants while curve b represents post-anneal concentration profiles.

The techniques for almost all of the implant processes are similar. The exceptions are usually due to technical differences in systems producing different ion energies. Elements that are usually implanted in silicon to produce electrical doping effects include boron, arsenic, and phosphorus. A list of the various ion implantation processes is given in Table 9-1.

The species used for a particular implant depends on a number of factors, including the dopant type required (*p*- or *n*-type), the ability of the species to give up electrons within a known energy band, the substrate type being implanted, and the depth of implant required. Associated with this is a decision on how much silicon crystal damage is acceptable, since the implantation of these ions does cause damage as the ions slow to a halt within the crystal structure. Clearly, a larger or more massive ion traveling at higher speeds will cause significantly greater amounts of crystal damage than a smaller ion at lower speeds. Other processes are usually implemented in conjunction with the higher energy implants to help reduce and repair this damage. This includes the succeeding anneal step which will help reduce some of the worst surface damage. Another common implant damage-protection mechanism is the sacrificial oxide, which absorbs most of the punishment, allows for very shallow implant depths, then is removed, permitting further processing.

There are also processes requiring the implantation of other elements, such as oxygen, for a wide variety of purposes. They often require very specialized, and very high energy implanters. We will

TABLE 9-1
Typical Implant Processes

Process	Dopant types	Purpose
Field enhancement	B	Form heavy p^+ region under field oxide; counter boron consumption by oxidation.
n-well	P	Form deep n^+ region for CMOS structures (PMOS devices)
p-well	B	Form deep p^+ region for CMOS structure (NMOS devices)
Threshold adjust	B, P	Stabilize and enhance gate threshold voltages
Source/drain	As	Form source and drain regions for the transistor
Film/substrate doping	P, B, As	Dope substrate or polysilicon film structure, such as resistors, capacitors, memory cells, etc.
SIMOX	O	Form buried layer of SiO_2 in order to isolate surface silicon for shallow junction devices

cover these other processes only as required and will spend most of the discussion covering only the usual species. Most of the issues that occur during the implantation of these species are common to the implantation of other species.

As mentioned, the implant processes will deposit the ions into a layer with a concentration profile through the depth of the silicon that can be described by a Gaussian function. This implant step is then associated with an anneal step, either in a furnace or in a rapid thermal anneal processor. This anneal process is set to a particular temperature for a specific amount of time, so that the implanted species can be driven to the desired depth in the silicon. An additional by-product of the anneal process is that some implant species require activation (freeing of extra electrons from the valence shells) as a result of neutral ion formation during the impact. This high-temperature process will allow that activation to occur. After the anneal step has been completed, the dopant profile will also be described by a Gaussian function, and will have an effective current-carrying depth (called the junction depth) that is determined by the diffusion rate of the dopant and the time at high temperature. This junction depth is shown in Figure 9-2, and is described by the following equation:

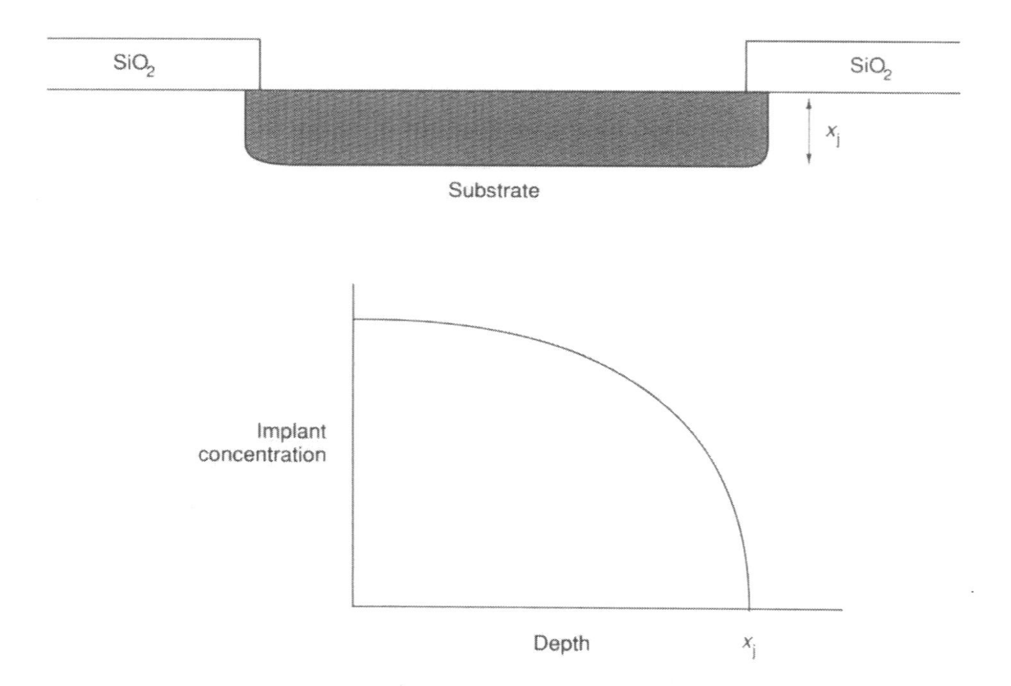

Figure 9-2. Definition of Junction Depth The junction depth, x_j, is defined as the depth at which the concentration of dopant goes to zero. At this point, the type of silicon will change and the diode is formed. The depth of the doped region is determined by the diffusivity of the particular species, the temperature of the process, and the total time at temperature.

$$x_j = 2 \cdot \sqrt{(D \cdot t)}$$

where x_j = junction depth of the implant
D = diffusivity of the dopant species
t = time

From the viewpoint of wafer manufacturing, successful ion implantation is usually more of a function of the maintenance of the systems than it is of the complexities of the implant process itself. While research into the field is continuing, the majority of the processes used in the wafer fabs are quite standard and quite well characterized. The physics of the implant process is well understood, and usually is sufficiently simple enough in concept that the process can be very easily implemented. This is in contrast to many of the other process, particularly the plasma CVD and plasma etch processes, in which many of the mechanisms of the reactions are not exactly known.

9.1.1 Ion Implanters

The basic structure of the ion implanter is shown in the block diagram in Figure 9-3. In this structure we can see the source area which contains the material to be implanted. The ions are extracted from the source material, and enter the main ion extraction and acceleration area. In this region, the ionized atoms of the source are accelerated to a

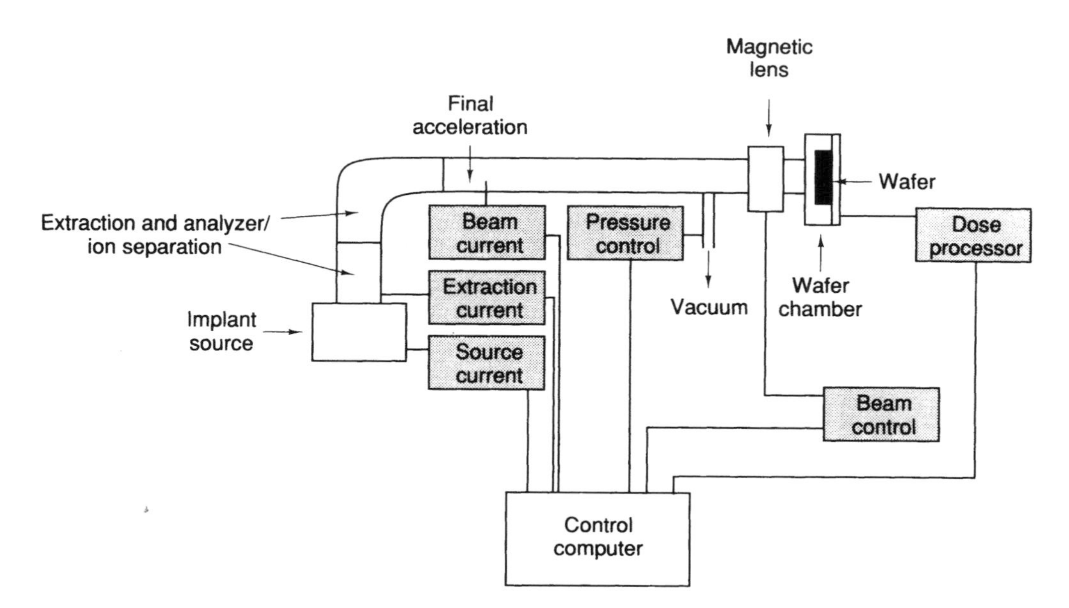

Figure 9-3. Block Diagram of Ion Implanter The ion implanter consists of a number of major subsystems, including the vacuum/pressure control system, dose monitoring system, magnetic field control systems, and the beam direction control system (especially critical with raster scan implanters).

fairly low velocity and then shot through a magnetic field that further separates contamination that may be present in the source. In addition, the extraction current, which controls the magnetic field density, is set to permit the exact isotope of the source element to be extracted. The ions then enter the acceleration region of the implanter, in which they enter a very high magnetic field. In most implanters, this section is merely a type of linear accelerator, since the ions can be accelerated to reasonable energies within a reasonable distance. In some cases, the accelerator may use a curved path in order to allow higher implant energies with heavier ions. Finally, we have the target area where the wafers are located. The system also includes high-vacuum equipment and a number of high-voltage power supplies.

The flight path of the ions through the implanter in shown in Figure 9-4. The source material for the implant species may be either solid or gaseous in form, each having various advantages and disadvantages, mostly related to containment of the by-products and wastes from the source chamber. Many of the implant sources used are poisonous (e.g., the use of arsine, AsH_3, as an arsenic source). The material is ionized by passing a strong current through it and stripping the ions from the source with a strong magnetic field.

Depending on the type of material, there are a number of ways of ionizing the component elements. The most common methods are shown in Figures 9-5a and 9-5b. In the first case, a filament heats a

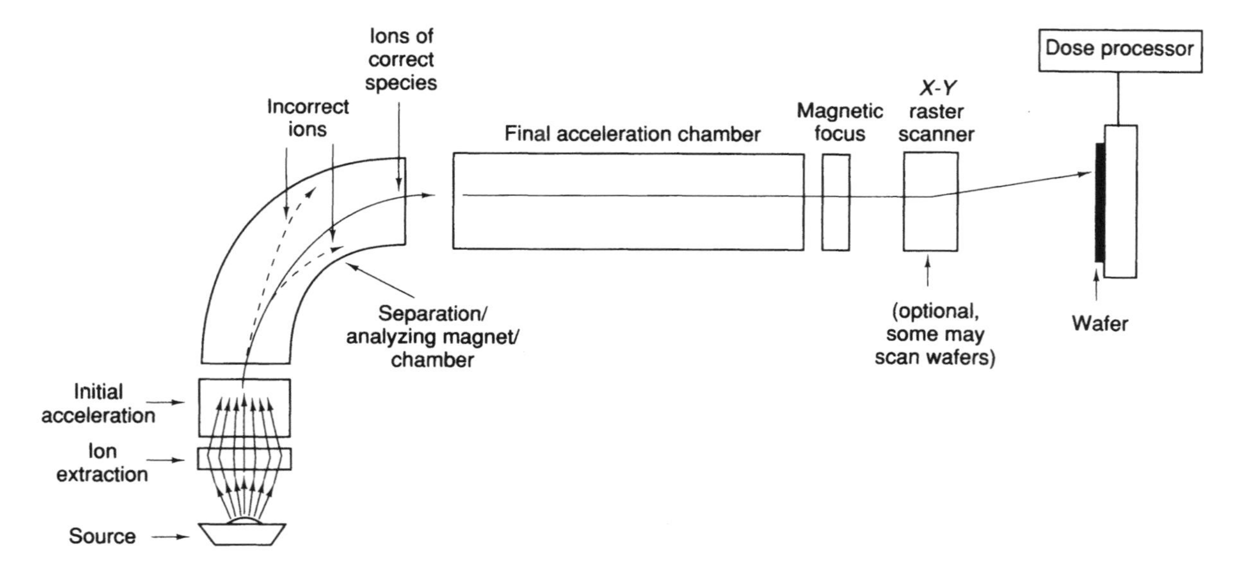

Figure 9-4. Path of Ion through Ion Implanter An ion must travel through a number of areas prior to impacting the wafer. First, a source is ionized and the ions pulled away with the ion extraction magnets. The analyzer magnet/chamber is used to strip all undesirable ions and isotopes from the source ion beam. The beam is then accelerated to the desired energy level, focused, and directed onto the wafer. Different types of implanters have different methods of achieving uniform scans, some moving the beam, others the wafers.

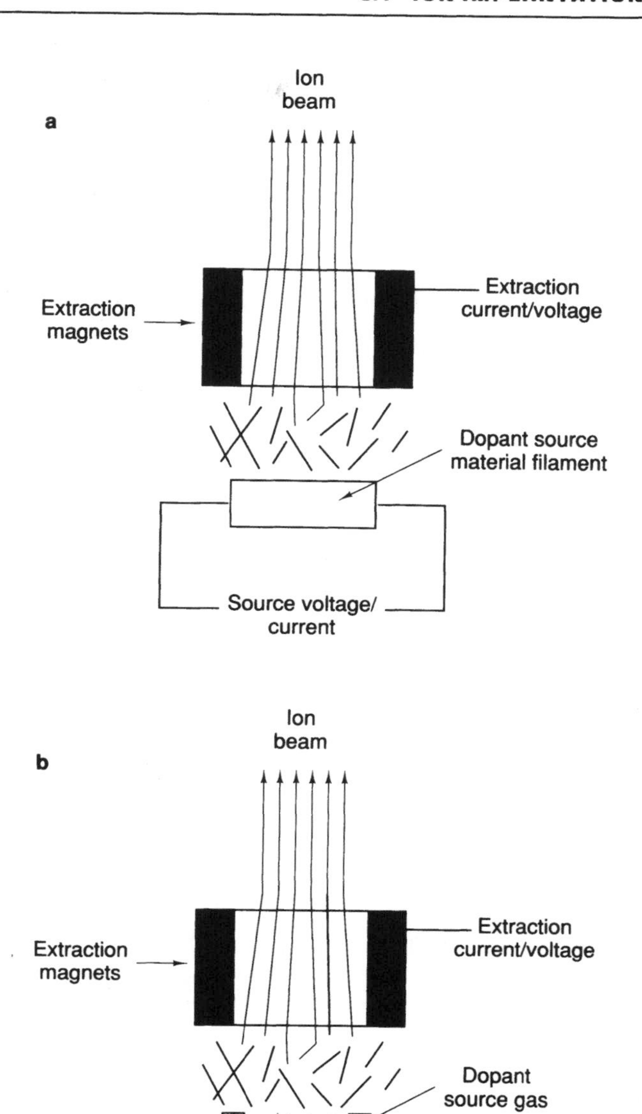

Figure 9-5. Dopant Sources for Ion Implantation Ion implant dopant sources are used in either gas phase or solid phase. In both cases, an electric field is used to ionize the source material, which is then drawn out with the extraction magnets. (a) Dopant sources in filament form; (b) dopant sources in gaseous form.

solid source element and, in the second case, a very strong electric field is placed across a gas source. In both of these cases, the current flowing through the source (the "source" current) determines the number of dopant ions produced in a given amount of time. As noted above, once produced, the ions are extracted from the source by a magnet, the power of which is determined by the setting of the extraction current.

Problems that can occur in this portion of the implanter include power-supply fluctuations resulting in changes in both the ionizing source currents and in the electromagnets used for extracting the ions. These fluctuations will impact the consistency in the number of ions produced. Changes in the pressure of the source chamber can also cause changes in the number of ions produced. Changes in the temperature of the ionization filaments (due to damage, oxidation, or other contamination) also cause variations in the number of ions produced. Since the final resistivity of the wafers is determined by the total number of ions driven into them, changes or instabilities in the number of ions available for acceleration can result in a reduction of the quality of the implant process as a whole.

In general, most of the power supplies used in modern implanters work well when new but, as the implanter ages, the various power-supply parameters will start to vary. This can cause subtle changes in the implant process. All power supplies should be checked periodically (every three to six months) to verify that the supplies are stable and holding to the correct specs. Elemental sources and source chambers can also become contaminated, which can reduce the number of dopant ions extracted from the source. This reduction in dopant-ion quantity can be caused by the changes in the ionization rate of the source mentioned above, or by the replacement of the normally extracted ions of the correct species with those of the contaminating species. Since few, if any, of the contaminant atoms that are emitted from the source will make it through the extraction system to reach the wafer surface, a common result of source contamination is reduction of the total available ions in the beam. This leaves the wafers with insufficient implant dosage or other changes in the implanted regions. Dosage variations due to these types of problems have been reduced through the use of dose processors, which estimate the total number of ions that have struck the surface of each wafer.

Now, the ions that have been extracted from the source are accelerated to a fixed velocity and shot through a magnetic field that is placed across the path of the electron beam. Since the radius of curvature of the ion beam path is determined by the ion mass, velocity, and magnetic field intensity, proper selection of the magnetic field will result in the dispersal of the various ions throughout the chamber. Since each of the elements will end up in specific characteristic locations in the chamber, it is easy to isolate the primary species that is required for implantation. Small adjustments in the magnetic field strength permit the isolation of separate isotopes of the source element. The dispersal of the elements in the chamber is shown in Figure 9-6. The magnetic field strength required to bend the ion beam to the correct orientation is given by:

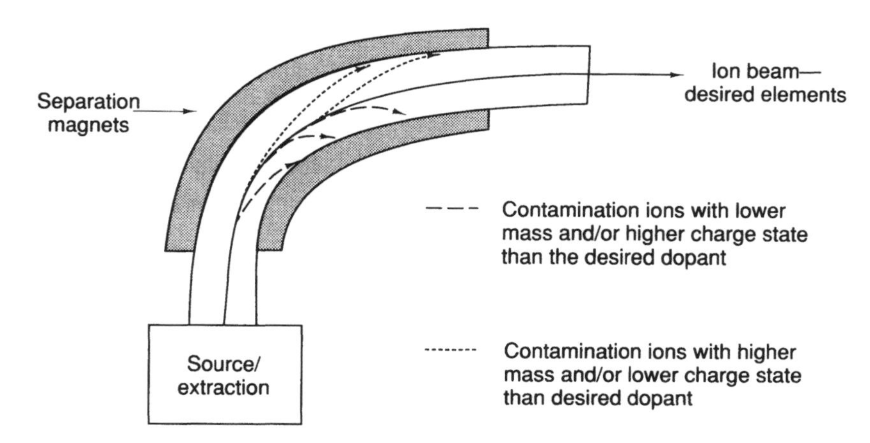

Figure 9-6. Purifier of Ion Beams As the dopant ions are accelerated through the implanter, the ion beam is stripped of contaminants by streaming it through a curved chamber immersed in a magnetic field. Ions of undesirable elements will not traverse the chamber, but will become embedded in the chamber walls.

$$R = (m \cdot v) / (q \cdot B)$$

where R = the radius of travel of the ion beam
m = the ion mass
v = the ion velocity through the magnetic field
B = the magnetic field strength
q = the charge on the ion

or:

$$B = (m \cdot v) / (q \cdot R)$$

By placing a slit in the correct location at the exit of the extraction area, only the proper species will emerge into the acceleration chamber. This slit area, as well as regions within the implanter that are in areas characteristic of key contaminant species (such as carbon, oxygen, or sodium), should have special cleaning cycles prepared in order to verify that no contaminants can escape from the walls of the chamber and become redistributed onto the wafers.

It is possible that contaminants can be present in the implanter due to poorly cleaned parts, leaks, fingerprints, or from excessive buildup of source contaminants, as mentioned above. This will allow undesired elements to get into the ion-acceleration and separation regions of the implanter at random speeds in random directions, and then be accelerated toward the wafer along with the desired species (as seen in Figure 9-7). While the velocity of the contaminant ions may be much lower than that of the primary species, the existence of the contaminant at the very surface of the wafer or film will cause many electrical problems. It is nearly impossible to prevent this type of

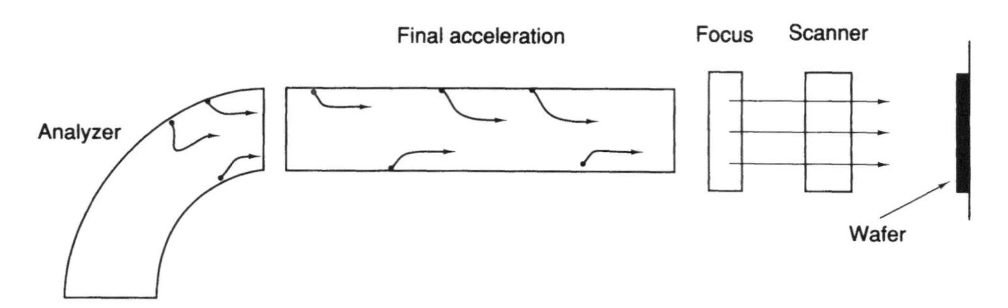

Figure 9-7. The Introduction of Contaminant Species into the Implanted Wafer Contamination outgassing from the walls of the acceleration chamber can become a problem if it is charged and is accelerated by the magnetic fields. A final beam separation "filter" is sometimes used to bend the main beam one more time to remove this source of contamination. Careful cleaning and elimination of residual material on chamber walls is a better long-term solution.

contamination from reaching the wafer surface after the species separation and extraction. Problems in controlling the process and in predicting the location and magnitude of the contamination are magnified by the fact that the contaminants are moving in random ways and do not always move in predictable ways within the chamber. As a result, it is very important to verify that all of the parts in the sections of the implanter after the extraction and ion-separation sections are kept contamination-free.

In some low-energy implanters, enough acceleration is gained in the separation area to permit the implant. In most cases, however, there will be a final acceleration stage after the species extraction area. This will occur particularly when the initial separation and acceleration fields are very low. With many modern implanters, the initial acceleration is kept low so that the species extraction zone can be kept small for minimal floor-space requirements. The implant species must then be brought up to the correct beam energy. This, of course, is required in order to verify that predictable implant parameters are obtained. This last stage is accomplished through the final acceleration magnets, which are controlled by a parameter called the beam current. As we have discussed, any contaminants (of the appropriate charge type) that enter this region will be accelerated toward the wafers as well as the implant species desired. As noted, it is important to verify that the chambers are kept clean to keep contaminants from getting into the final accelerating region. Both chemical and particle contaminants can accelerate during the final acceleration stage.

Particles will not attain the velocity of the ions and will not usually penetrate the surface of the wafer (which would cause a considerable amount of damage to circuits). Nevertheless, particles in the chamber gain a static charge, which results in the particles becoming secured to the wafer surface where a number of side effects can occur.

We will discuss the impact of particulate and chemical contamination of wafers during the implant operation in a moment.

During this final phase of beam acceleration, the beam is often focused through a magnetic lens to reduce the random scatter of the beam so that the uniformity of the implant can be improved. The ion beam is aligned and focused here and is critical to successful operation of the process. In the case of the raster scan method described in the next few paragraphs, it is the only way the beam can be controlled enough to be used. The attempt is to create a nice, uniform, Gaussian-shaped ion beam. This will permit the best uniformity control possible. The ability to focus the beam can depend on the energies (and thus the velocities) required for the process and the mass of the dopant species.

At the target area, there are two typical methods for getting the beam to cover the surface of the wafers. The first is the raster-scan method, as shown in Figure 9-8. In this case, the beam that emerges from the magnetic lens is directed into a set of magnets that are located

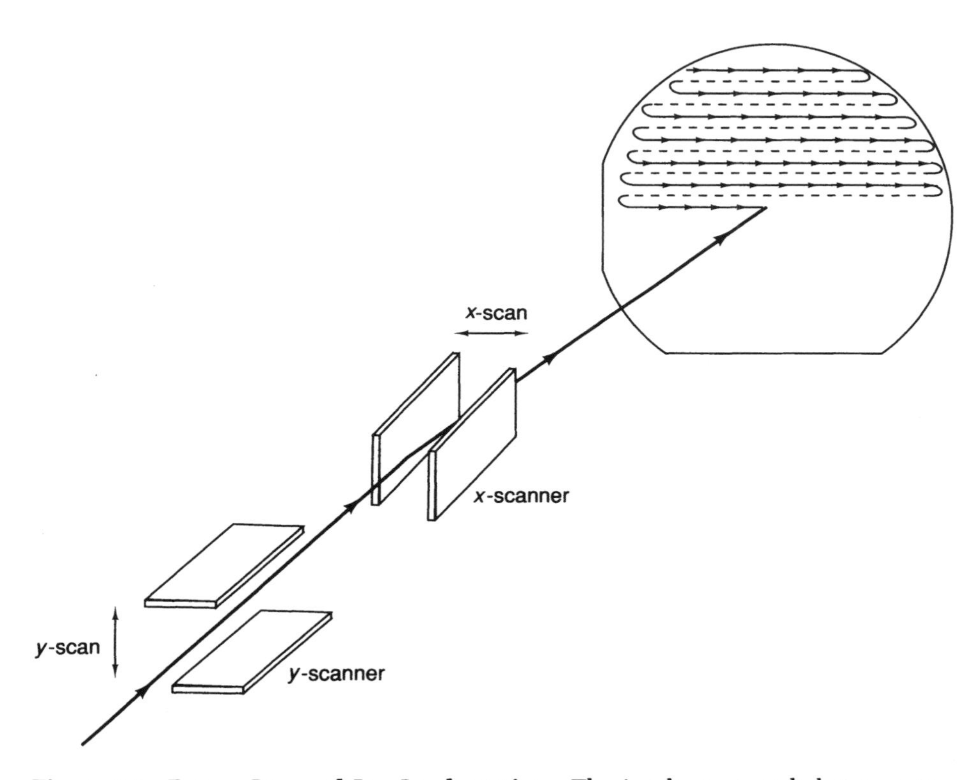

Figure 9-8. Raster-Scanned Ion Implantation The ion beam travels between two sets of magnets to direct the beam in the x and y directions. The beam is moved across the wafer in the x direction until the edge of the wafer is reached. The beam then quickly moves back across the wafer and steps down the width of the beam in the y direction. The beam is then scanned across the wafer again. The wafer is repeatedly scanned in this fashion until the appropriate dose is reached.

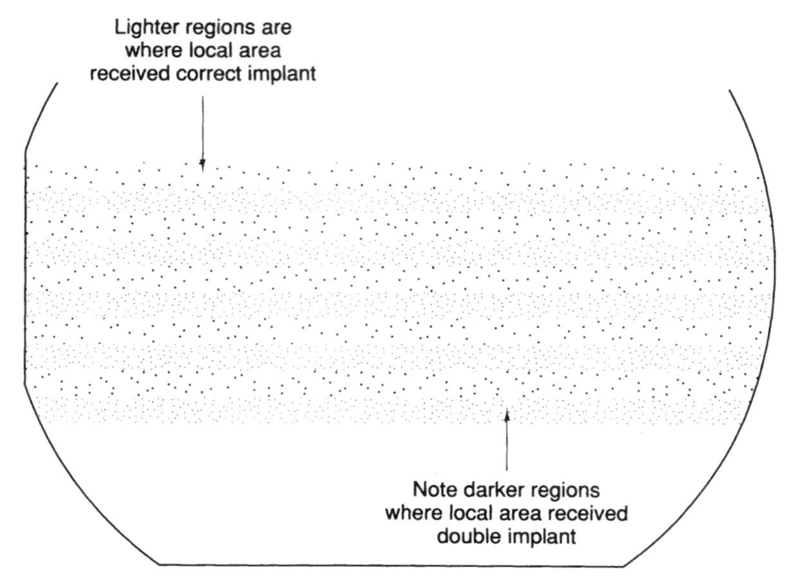

Figure 9-9. Striped Implants Due to Miscalibration of *y* Scanner One of the problems of a raster-scanned implanter is that improper tuning or misalignment of the scanners can cause regions of excess implant doses. As seen here, a slight overlap has resulted in striped areas of nearly doubled implant dose.

in an x–y plane perpendicular to the direction of the ion beam. This procedure is very similar to the raster scanning process used to produce a television image. The beam sweeps across the target area in one direction, for example from left to right in the x plane. At the end of the scan, it may be "blanked" with an electronic shutter, or prevented from reaching the surface of the wafer, and then the beam is moved back along the original track in the x plane. When it gets to the start location, it then moves the beam one step in the y direction.

One of the main issues with the raster scan method of implantation is the difficulty in maintaining uniformity of the beam across the entire wafer. The beams must not overlap or there will be stripes of excessive dopant on the wafers. Similarly, since the beams are typically not uniformly dense and square, if the raster spacing is slightly too wide there may be stripes of insufficient dopant. Both can be serious problems to the yields of the devices. Inconsistencies in the magnetic fields of the scan devices, especially in conjunction with inconsistencies in the beam (possibly due to source or chamber contamination), can lead to localized areas of incorrect implant dosage that will reduce quality. This is demonstrated in Figure 9-9.

Another problem with the use of the raster scan method is that of maintaining a perpendicular (to the wafer surface) ion beam. If an incident angle of 90° is not maintained, implanted regions of the wafer can be shadowed by nearby structures, as shown in Figure 9-10a. The wafers may become misimplanted when the implant species comes in

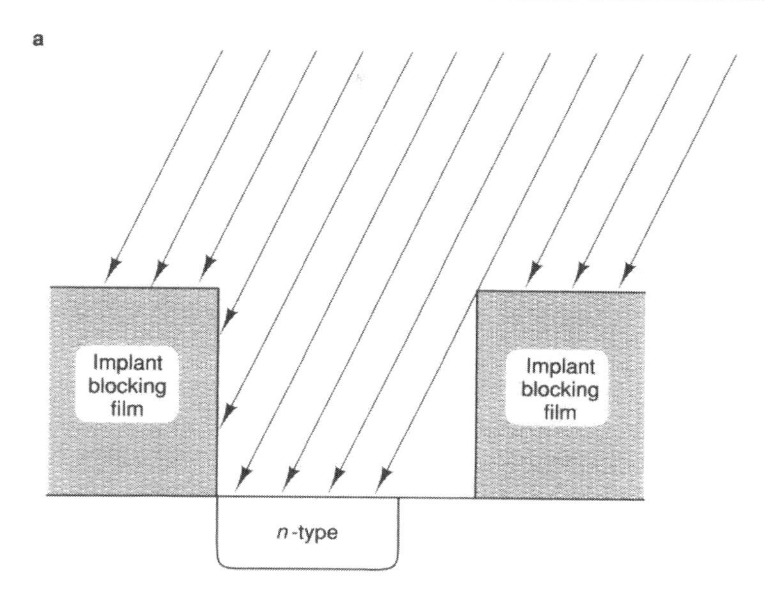

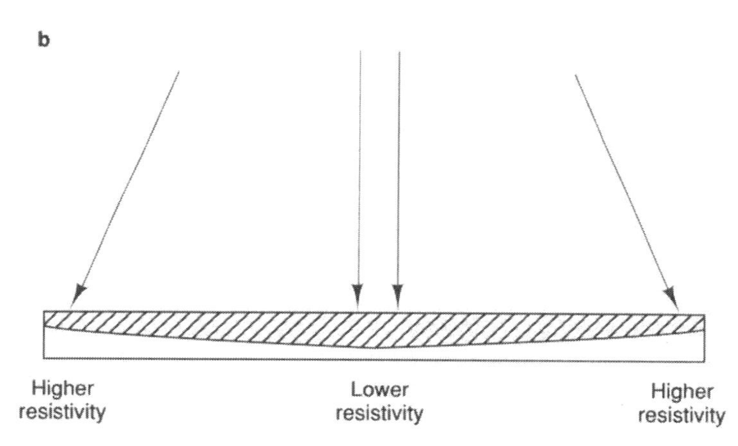

Figure 9-10. Side Effects of Angled Implants Here we see the effects of ion implant beams approaching the wafer at an angle. This can occur in older raster scan implanters or misaligned targets (wafers) in the implanter. While these drawings are grossly exaggerated, the shadowing and variation in implant depth can cause die yield degradation. (a) Implant shadowing due to angled ion beam; (b) nonuniformity of resistivity due to nonparallelism of ion beam.

at an angle in comparison to the crystal orientation of the wafer. If the implant beam is too distorted at the edge of the wafer, the dopant atoms will reside at shallower levels at the edge of wafers rather than the center of the wafers. This becomes a serious problem for larger wafer sizes. The results of this possibility are shown in Figure 9-10b. Many older (>2 μm) processes use implanters intentionally angled at about 6°. This is done to prevent channeling effects in the substrate. For narrower linewidths, the shadowing effects usually overpower the advantages from reduced channeling, and yields end up being reduced.

The second type of targeting mechanism involves holding the beam in a fixed position and moving the wafers in front of the beam. This has the process advantage that the wafer can be held perpendicular to the beam at all times. In addition, the focusing requirements are reduced, and heavier elements with higher energy requirements can be more efficiently implanted than with the raster scan method. These processes include many of the high-energy arsenic implant steps. The disadvantage of the system is that, by definition, the wafers must be moved in relation to the rest of the system at all times. Of course, all extra wafer motion is viewed as a potential particle and contamination source. Typically, as shown in Figure 9-11, several wafers are placed on a large platter or platen, which is raised to a point perpendicular to the beam direction. The platter is then rotated at high speed and scanned across the ion beam. Since the scan rate is slower than the rotation speed of the wafers, the beam crosses the wafer in an overlapping fashion over the entire surface. Since the beam is somewhat more diffuse than the raster-scan implanter, the overlap effects are reduced if (and this is a big if) the scanning mechanism is very stable.

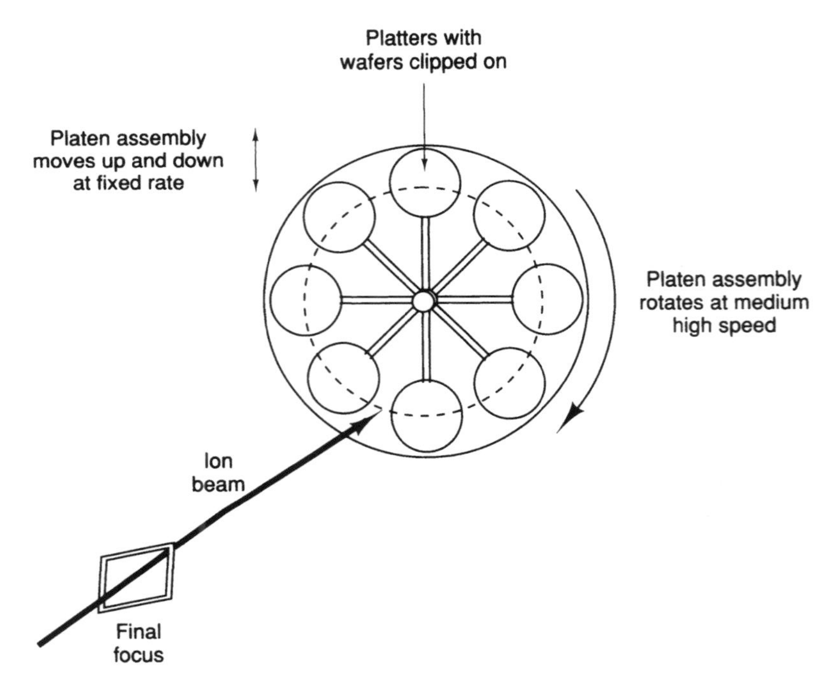

Figure 9-11. Rotating-Platen Ion Implanter This type of implanter uses mechanical motion to produce the scanning required to cover the surface of the wafers. In these systems, the platen assembly simultaneously rotates and scans vertically. After a given number of full scans and rotations, the wafers will have uniform, predictable implants. This type of implanter is well suited for high-energy implants of heavy ions.

Obviously, mechanical considerations become very important when dealing with this type of machine. It is very large and complex, with many mechanisms that can fail. The rotation of the platen must be very uniform, with no periodic or random vibration modes. In many respects, a regular vibration pattern can have a more significant yield impact than one that is more random in nature, since the first may set up periodic regions on the wafer that are not correctly implanted. This is similar to the effects seen in Figure 9-9. Also, the scan mechanisms must run in a very stable manner and at consistent speeds. Binding, sagging, or other changes in the scan mechanism as the machine ages can create significant uniformity problems. The effects of a scanning mechanism that binds, then unsticks and jumps forward can include a number of bizarre phenomena, such as implant "stripes" that indicate regions of excessive implant surrounded by areas of insufficient implant.

Additional issues with the mechanical parts of the implanter include the wafer handling devices that are used to move the wafers from their cassettes to the platen, and the mechanisms that move the platen to its proper position. These subsystems can create particles and can be the source of serious mechanical difficulties. Finally, much of this activity must be done in a high-vacuum environment, which adds a wide variety of constraints to the type and selection of robotics gear. For instance, heat is not dissipated quickly in a vacuum. Thus, proper heat sinks must be installed to allow the excess energy (from the decelerating ions) to dissipate into the chamber walls.

The safety systems of the implanters are very important. There are a number of areas of major concern when dealing with these systems. The most visible aspect of the system is the power distribution network. The various electromagnets must all use high voltages and high currents to achieve the magnetic fields required to force the ions to the chosen energy levels. For heavy ions, such as arsenic, the fields required are very high, requiring very large systems. As a result, the power supplies must be carefully controlled and contained, with good short circuit protection to prevent electrocution of personnel, especially maintenance personnel.

Another serious area for concern is the source materials and the waste materials from the source chamber and other parts of the system. Many of the chemicals used for the implants are dangerous, such as arsine (AsH_3), phosphine (PH_3), and diborane (B_2H_6), as well as a number of solid element sources that are used. These chemicals are very active and dangerous in their own right, and, in the case of arsenic, there are additional long-term health consequences that result from its ingestion. The potential for environmental contamination occurs during the source change and replenishment cycles, and during equipment cleaning cycles. Implanter parts are often cleaned in sealed, environmentally safe glove boxes, using solvents such as TCA and TCE, which are (unfortunately) some of the best cleaning solvents available for removing the contaminant glaze that form on the stainless steel parts. Of course, the use of these solvents adds to the toxic waste problem. Chemicals such as alcohol and choline are also being used to clean the implanter parts in an effort to at least reduce the toxic by-products.

A final area of concern is that of radiation output. The impingement of the various high-velocity ions in the walls of the implanter can produce a lot of high-energy radiation. In some cases, it will be in the form of alpha or beta radiation, but this will usually be trapped within the walls of the chamber (which will mean that an old chamber may have a certain amount of radioactivity in addition to contamination with toxic chemicals). The most dangerous radiation will be emitted in the form of high-energy photons. Usually, this radiation is effectively shielded from the environment. However, if the shielding material is improperly secured, radiation can escape into the environment. Fortunately, the types of radiation produced in an ion implanter are short-lived and at very low levels, keeping the risk of serious contamination low.

The various types of implanters come in a wide variety of configurations, but the most common are the raster-scanned medium current implanters, which are commonly used to implant boron and phosphorus atoms, in other words, the lighter, lower energy implants. The scanning-plate implanter is typically used for high-current implanting, which is used for the heavy ions like arsenic and for accelerating lighter elements such as oxygen to extremely high energies for deep penetration. Modern electromagnets have permitted the development of raster scanning high-energy implanters, and the development of high-T_c superconductors will eventually help increase the ion-beam energies, and reduce the system size required to produce these energies.

Thus, we can see that the systems that are used to implant the various types of ions into a wafer can contribute significantly to the results of the implant processes. While the procedure for creating and distributing the atoms sounds quite simple, the mechanics are not that straightforward. There are wafer handlers, vacuum systems, gas handling systems, and several large electromagnets with associated high-voltage power supplies. Any or all of these subsystems have enough of an effect on the process so that even small variations in these systems can result in significant variations in wafer yields.

9.1.2 Ion Implantation Processes

The most common elements that are implanted into silicon wafers are boron, arsenic, and phosphorus. When implanted into silicon, boron is a *p*-type dopant, while phosphorus and arsenic are *n*-type dopants. Each of these elements also imparts certain mechanical and chemical attributes to the wafers, as well as the more obvious electrical ones. In addition, all of the processes are sensitive to chemical and particle contamination. These types of contamination can be the cause of various types of yield loss. We've already discussed some of the main sources of the contamination, which can include mechanical failures, leaks, and improperly cleaned materials. Other sources of contamination can include impurities in the implant source, contamination introduced to the implant chamber with the wafers, and so on.

From the point of view of actual wafer processing, the implant process is relatively straightforward. The wafers will usually be covered with a layer of photoresist which is then exposed so that the regions where the implant is desired are left clear of resist. For most purposes, a photoresist layer of 0.75 to 1.25 μm is adequate to prevent the ions from reaching the surface of the wafer. The exact thickness required is a function of the total velocity and mass of the ions impacting the surface, and will sometimes need to be thicker, possibly up to 3 to 5 μm thick. In general, the film thickness is chosen to be substantially more than adequate to block the implant entirely. Marginal resist thicknesses could result in problems if the implant beam energy is off calibration and is unusually high. In these cases, the implant will penetrate through the photoresist and will change the electrical parameters of the underlying substrate.

Sometimes, silicon oxide and nitride layers are used as implant blocking films. Since these films are much denser than photoresist films, thinner films in the range of a few thousand angstroms can effectively block the implants. Of course, the oxide, and particularly the nitride, films are much more difficult to remove than the photoresist, resulting in more potential circuit damage from undercutting or other problems with film-removal procedures. For example, nitride films can be used early in the process for protection from the field implant/field oxidation. At this stage in the manufacturing process, the nitride film can be deposited, etched, used as an implant block and as a field oxidation growth block, and then removed with little or no damage to the active regions of the silicon. The effects of implant blocking films are shown in Figure 9-12.

There are additional uses for implant block films. For instance, very thin silicon oxide films, around 150 to 500 Å thick (called "sacrificial oxides"), are used to reduce the total energy of the ion beam, primarily to reduce the amount of damage inflicted on the silicon substrate by the impact of the ions into the crystal structure and to reduce channeling effects. These types of thin-film processes are often used with arsenic implants due to the large size and mass of the ions. In addition, the various active regions of the chips (such as the source and drain regions of transistors that are often doped with arsenic) are more likely to use sacrificial oxides, due to the detrimental consequences of damage to the substrate at this point. These kinds of damage can result in increased latch-up and creation of various types of parasitic transistors and current paths through the silicon crystal structure. The thin oxide layer can reduce this damage by up to 90%.

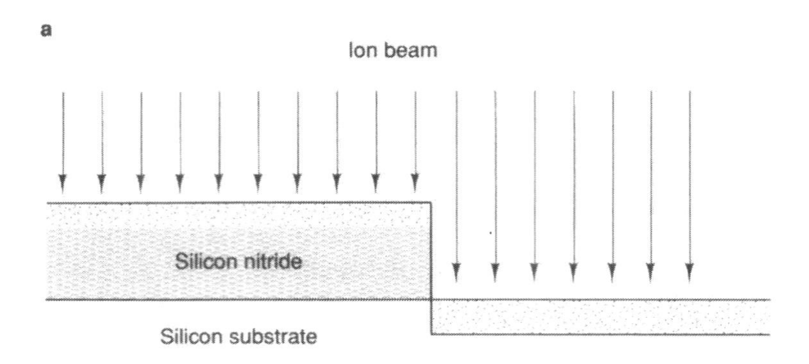

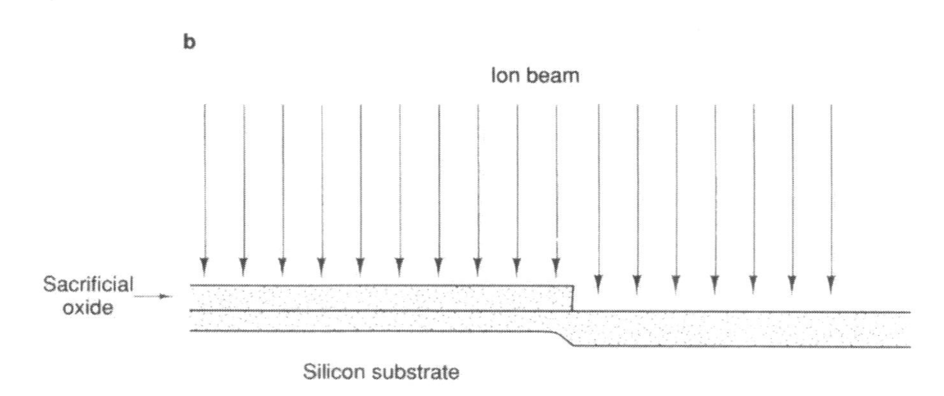

Figure 9-12. The Uses of Dielectric Blocking Films for Ion Implant Dielectric films are used in implant operations for several reasons. One is to block the ion beam from penetrating to the silicon surfaces, as seen in (a). In (b) the sacrificial oxide layer is used to reduce damage to the silicon crystal by allowing some of the energy from the ion beam to be absorbed into the oxide. This has the side effect of reducing the implant depth. Sacrificial oxides are removed immediately after the implant or subsequent anneal processes.

It is important that the thickness of this oxide layer is carefully controlled, since variations in the thickness will result in variations of ion implant depth and dose. Usually, a tolerance of no more than 5% is permissible, that is, a control requirement of ±8.75 Å at an oxide thickness of 175 Å. This thin oxide film is of very poor quality after the implant process is complete, filled with some of the implant species ions and contamination, as well being heavily damaged. As a result, it is removed immediately after the implant process has been completed (although in some cases, the oxide may be left on the wafer during the anneal process, to prevent any further oxidation or crystal damage from occurring during the anneal process).

The wafers are implanted with the ions at a fixed energy level and with the ion beam as purified of unwanted elements and isotopes as possible. When these parameters are known, the average implant depth can be calculated. Since the energy of the individual ions will vary in a random fashion and since the collisions that occur within the silicon lattice after the ions impact the surface are also fairly random (not completely random, however, due to the regular crystal orientation of the silicon), the ion concentration will vary around the average implant depth in a modified Gaussian distribution. The depth of maximum concentration is modified by adjusting the implant energy. For instance, shallow high-speed channels can be created by using low-energy beams, whereas applications such as SIMOX and SOI (silicon on oxide or insulator) technologies have been developed where high-energy oxygen implants place the atoms at a specific depth in the silicon and subsequent anneal steps allow the silicon and oxygen to interact to form SiO_2 within the silicon, while the surface atoms of silicon reform to make single-crystal silicon. These applications are shown in Figure 9-13.

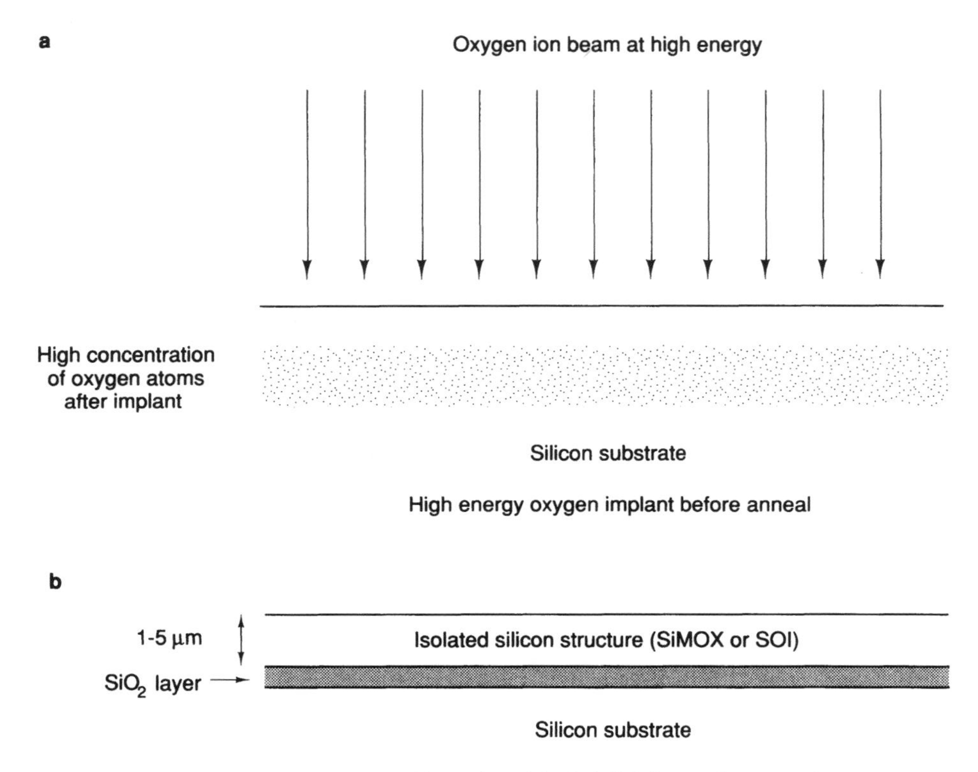

Figure 9-13. The Development of Implanted SIMOX Structures To create very shallow junctions, it is sometimes beneficial to construct circuitry on silicon that is electrically isolated from the substrate. This is sometimes done by depositing thin layers of epitaxial silicon on SiO_2, but has been more recently manufactured using a high-energy implant technique. Here, the oxygen is implanted deep into the silicon. The wafer is then annealed at a high temperature, which causes the silicon and oxygen to combine, forming a layer of SiO_2 buried within the silicon. Proper temperature cycles can cause the surface silicon layer to recrystallize into a uniform, defect-free layer.

The point where the n-type doped region of the silicon substrate effectively stops and the p-type starts is called the junction depth. This was shown in Figure 9-3, and is described by this equation:

$$x_j = 2 \cdot \sqrt{(D \cdot t)}$$

where x_j = junction depth of the implant
D = diffusivity of the implant species
t = time

This depth is usually modified through the anneal steps and subsequent high-temperature processing steps. In many cases, this additional anneal time is significant enough that it becomes critical to track the entire "thermal history" of each implanted region throughout the process to prevent unexpected dopant atom redistribution. In most cases, though, most of the redistribution of the dopant atoms takes place in the early stages of the anneal cycle and subsequent dopant drive-in is minimal. The exact junction depth of a particular section of the device is usually determined during the implant process and its associated total time at temperature. There are a number of computer programs that can model the effects of the total thermal history of the manufacturing process on each of the implanted regions.

The junction depth is not the only important parameter of the implant step. The total ion dose being placed into the wafer is also critical. The concentration of the ions on the surface will affect both the total resistivity of the device and the way that the dopant atoms are redistributed during the anneal step. If the concentration is too high, resistivity readings will be low, and the dopant will move farther, possibly hampering device performance, for example, by reducing gate threshold parameters. When the concentration is too low, the converse is true: The resistivity will be high, junction depths will be shallower, and gate thresholds will be increased, slowing the devices and increasing power consumption requirements.

The total ion dose is monitored through devices that are attached to the targets (wafer holders) of the implanter called "dose processors." These devices measure the total amount of charge that has accumulated on the plates during the implant cycle. While they are not always exact measures of the number of ions actually settling on the surface of the wafers, they are repeatable enough that the processes can be characterized, allowing the process to be halted if a deviation has occurred within the dose processor, and the wafers in question to be examined in more depth before a large number of wafers are misprocessed.

At this time, the average implant depth cannot be precisely monitored while the system is in operation, as can be done with dose measurement control. Due to the physics of the process, it is usually assumed that, if the electromagnets have well-controlled power supplies, there should be a minimum of implant depth variation. This usually proves to be true. If a question comes up as to the average depth of implant that a certain implanter is producing at any given time, it may be possible to measure the wafer temperature for given dose levels. If the average implant depth is varying, then the beam energy must vary, and since this adds energy to the overall system, and there are no other outlets for this energy, it will appear as excess wafer temperature if the energy is too high or reduced wafer temperature if the energy is too low. This is not a standardized procedure, however.

After the implant has been performed, the ions will all reside in a thin, relatively dense layer at the surface of the silicon, with a significant amount of crystal damage already inflicted. In addition, the collisions of the ions with the silicon substrate will also have caused some of the ions to return to the ground state. This has the effect of preventing the dopant from performing its task, and changing the resistivity of the circuit element. As a result of these effects, the wafers must be sent through an anneal process after the implant step. In some cases, there is no separate process step identified, but the anneal is performed in any event. This is true in the case of the boron gate threshold adjustment implants that occur immediately prior to many gate oxidation steps. Since the gate oxidation temperature is near the range that the anneals will take place, these two steps are combined into one.

In some cases, the anneal steps can be quite long. For instance, the *n*-well and *p*-well drives used to create the various regions of *p*- and *n*-type material (see Figure 9-14a) for CMOS devices are long (12 to 24 hours), and run with temperatures as great as 1,100°C. The drive time is a function of the process temperature, the substrate and dopant species being used, and the designed depth of each of these regions. In most processes, however, only a much shorter time will be necessary to achieve the ends desired. For instance, in the case of a source/drain implant (Figure 9-14b), the junction depths will be less than 1 μm and can be obtained with an anneal time in the one- to two-hour time frame that is required for a standard diffusion process (considering ramp up, stabilization, and ramp down times, as well as the oxidation, and standard 15-minute anneal steps). The reionization or activation of the implanted elements takes place very quickly after the process temperature has been increased beyond the activation temperature, usually around 950 to 1000°C.

Since a typical enhancement transistor works by allowing the voltage on the gate to open a small channel for electrons to flow through (as seen in Figure 9-15a), it follows that reductions in the channel length will permit faster devices that require less power to operate. However, it is not always practical to develop a totally new design technology in order to shorten channel lengths. Usually, drastic reduction in the channel lengths must be associated with reduction in all other parameters, making it prohibitively expensive for a company to build a cleanroom that can permit the construction of the

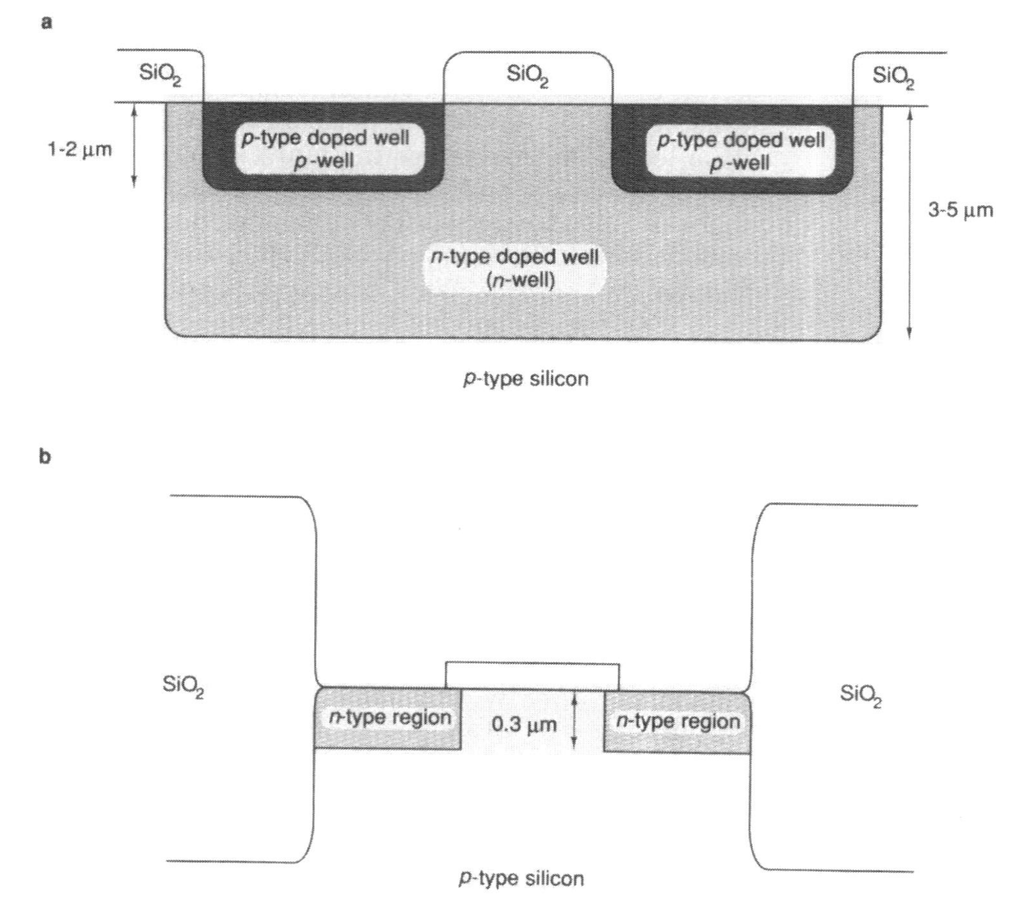

Figure 9-14. Implanted Wells in Silicon Implants can be used to develop a number of different structures in silicon substrates. With high doses and long anneals, dopants can be driven-in several microns to form regions of varying types for use in CMOS technology as shown in (a). Shallower implants of heavier ions, such as arsenic, are used to form source and drain conductive regions as shown in (b).

small geometries required for high-performance devices. As a result, there are design tricks that can be used to enhance the device performance. For instance, the drain side of the gate region can be lightly implanted so that a shallow partial channel already exists, as shown in Figure 9-15b. As a result, when the gate is activated, the effective channel is reduced in length. Usually, this drain extension is limited to about one-third or so of the true channel length to prevent unwanted channel leakage.

There are a number of difficulties associated with the implant steps. They can usually be related to the setup of the implanter or to some minor malfunction, but nevertheless can be quite damaging to the wafer. One of the most serious of these problems is charging. This problem is often traced to an improper ground or a ground loop in the system, or to improper processing (the wafers may need to have any

a

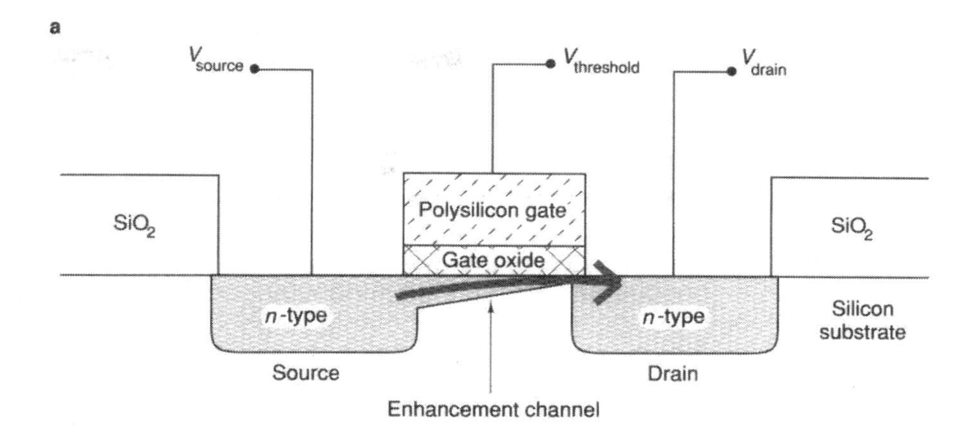

b

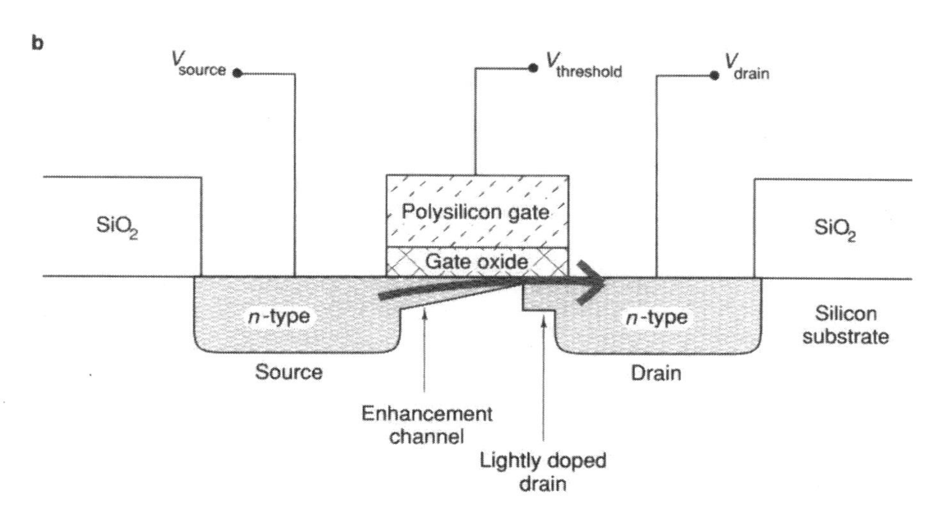

Figure 9-15. Lightly Doped Drain Structures (a) As $V_{threshold}$ increases, the enhancement channel enlarges, until contact is made between the source and drain regions and current can flow through the gate. (b) With a structure called the lightly doped drain (LDD), the enhancement channel length required for gate operation can be reduced, lowering power requirements and enhancing speed.

oxide or nitride films removed from the back of the wafer to provide good contact to the implanter plate). However, charging is a problem that can occur with normal operation of the implanter, especially if a large area of the wafer consists of dielectric films, which trap the charged atoms and do not permit discharge causing the surface of the wafer to repel the incoming atoms.

At first, the repulsion will cause a generalized reduction in the depth and/or total dosage of the implant, but increasing trapped charge intensity will deflect the incoming beam, causing distortions and uniformity problems. Finally, if the charging becomes too strong, the beam can be completely deflected away from the wafer surface, resulting in little or no implant reaching the surface of the wafers. Obviously, all of these problems can severely impact the die yields of the wafers affected.

We will now discuss the impact of particulate contamination on the implant process. This impact will vary, depending on the technology of the devices. Small linewidths with CMOS devices are far more sensitive to implant problems than wider lines and older types of MOS and bipolar technology. In general, particles can act as localized implant blocks, preventing the proper implantation of the substrate, as seen in Figure 9-16. The amount of damage this will cause to the yield of the wafers is determined largely by the size of the particle. A very large, thick particle can completely trap the ions before they reach the surface of the wafer. Smaller particles will allow some of the ions through, so that the implant depth will be nonuniform throughout the region of the implant, as also shown in Figure 9-16.

The particles that are created in the chamber will often receive strong static charges as a result of coming into contact with so many charged atoms. This problem will become more noticeable as the implant energy is increased. As the particles become more highly charged, they will adhere more strongly to the surface of the wafer. The extra adhesion of these particles means that they will be much more difficult to remove at the clean steps after the implant, which can cause short circuits and other problems if the particles are permanently trapped in the active regions of the circuit. Often, even neutral particles floating around in the vacuum of the implant chamber will capture the ions in the ion beam and become charged. This will tend to draw the particles toward the wafer surface within the implanter, which further increases the overall total defect density of the process.

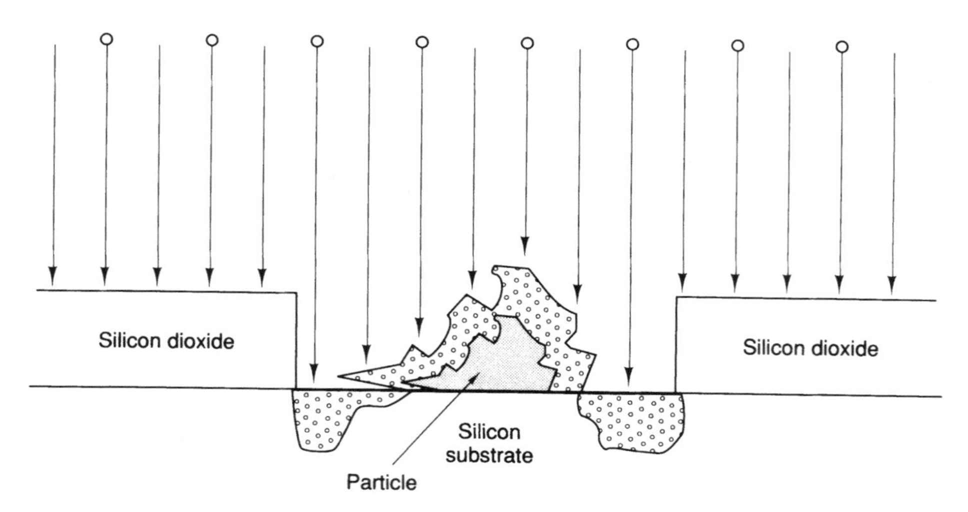

Figure 9-16. Implant Blocked by Particle Particles landing in an area being implanted can block the incoming ions partially or completely, hampering device performance.

Particles can come from wafer handler parts that are rubbing together, not properly designed or exhausted, or not kept clean. They can also be delivered from contamination that outgases from the walls of the acceleration columns. Improperly baked or rinsed photoresist films can also outgas any remaining solvent or developer that is in the film. This can end up as chemical or particle contamination, which depends on whether the outgasing is of a high-enough quantity to allow the formation of conglomerates to form particles. Any scratches on the surfaces of the wafers, especially in photoresist films, will cause a huge number of particles to be created. Scratches will have the additional problem that the ions will more easily penetrate the photoresist or other ion blockage film, and will be implanted in regions uncovered by the scratch.

Another topic that must be considered is that of implant process control. There are a number of issues related to process control that make the procedure difficult. The main problem is that of the requirement that the implants must be annealed before any electrical testing can be performed, to guarantee implant quality. Therefore, if only diffusion processes are performed on the wafers, an implanter will be shut down for the length of time that the diffusion process runs, plus test wafer preparation time (removal of oxide layers, for instance). Then, finally, the wafer can be tested using a standard four-point probe. When the anneal process is performed in a rapid thermal anneal system (or when a rapid thermal anneal process has been calibrated to simulate the longer diffusion process), test results can be obtained much more rapidly. A recent development in ion implantation testing has occurred with a device that uses an infrared laser (to which silicon is transparent) to force dopant atoms to either fluoresce or absorb the beam (depending on the exact wavelengths used). By studying the intensity of this fluorescence or absorption, the approximate quantity of ions of the dopant species can be determined.

The results of the resistivity or dopant concentration tests (or both, if available) should be recorded on the process control charts. This can be tricky, since some implanters run as single-wafer implanters, others as multiple-wafer processors. This must be compared to the rapid thermal anneal systems, which are typically single-wafer processors, or diffusion furnaces which typically process 100 or more wafers at a time. From a process control and fab manufacturing viewpoint, the simplest implant setup has a single-wafer implanter connected directly to a rapid thermal anneal system and from there to the measurement system by robotics, so that a wafer can be implanted, annealed, and tested in a single manufacturing step. These systems have not yet been sold as units; however, it is likely that combinations like this will appear in the marketplace before much more time passes.

As we have seen, the ion implant process is not terribly complex, and is limited mostly by the physics of the materials and the mechanics of the implanter. There are many process parameters that must be controlled, including the amount of dopant extracted from the source, the isolation of the exact species required, and the final acceleration and focus of the ion beam. These parameters are typically modified through the use of variable current levels through the various components. In older systems, they are set by the operators, but in the newer systems, the system computer can handle many of these tasks. They are required in order to maintain a consistent implant dose and depth of implant across the entire wafer surface. As in all other processes, yield optimization requires the use of process control charts that monitor all of these parameters, in addition to the resistivity results from test wafers that have received the implant and anneal processes.

9.2 WRAP-UP OF THE ION IMPLANT PROCESSES

We can see from looking at the material presented in this chapter that ion implantation occurs more through purely physical than chemical means. In most cases of ion implantation, a chemical source of dopant is vaporized, stripped from the source by a strong electromagnetic field, and then flung into the wafer surface. Problems crop up when contaminants are introduced into the ion beam or are accelerated by the internal magnetic field toward the wafer surface. Also, distortions in the magnetic and electric fields within the system, whether from wafer charging or from contaminants on chamber walls, will impact the yields.

Clearly, the quality of manufacture in this area is critical. This includes superior maintenance on the systems, since the implant processing equipment depends more on the mechanics of isolating and accelerating the correct atoms than with the chemistry of the processes. Defects, scratches, particles, and contamination of all sources can result in significant yield degradation in the implant area. It should be remembered that the current flow of many of the parts of the circuits travel through the regions of the semiconductors that have been implanted with a dopant. As a result, these steps are directly in the middle of the critical path for manufacturing high-quality wafers. It is important to the success of the manufacturing process that all involved personnel be made aware of the significance of these processes.

CHAPTER 10

METAL FILM DEPOSITION PROCESSES

The last of the major wafer manufacturing processes that we will discuss are the metallization processes. These processes are used to deposit films that are then used to provide the conductive paths and interconnections for the various parts of the integrated circuit. These conductive paths must be constructed in three dimensions and in some cases require specialized selective metal deposition processes that permit deposition only at the interconnection points between the various layers of metal films (called "contacts"). This is in addition to the more typical processing by broad-area deposition methods. In all cases, the various conductive lines and contacts must be defined by the photolithography and etch processes.

A variety of metal films are placed onto the wafers late in the processing sequence, after all of the higher temperature oxidation and anneal steps have been completed. The types of film used in the manufacture of ICs range from the most common metal, which is an alloy of aluminum and silicon, to other metals, such as tungsten and metal silicides that can be used for broad-area deposition. Metal deposition processes, like selective tungsten deposition, are used to fill the contact holes (also called "vias") without depositing any metal on the dielectric layers. The metal films exhibit a wide range of different characteristics, necessitating some amount of compromise and innovation in developing the correct type of effects and resistivities required. These characteristics include varying amounts of resistivity, different degrees of immunity from electromigration and hillock or void formation, different amounts of flexibility (for deposition on steep terrains), and varying resistance to corrosion and oxidation. In some cases, there are advantages to being able to oxidize the metal so that insulating dielectrics can be created directly from the film, reducing the need for another dielectric deposition step.

This wide variety of films results in the development of a fairly large number of different processes. Each process has its own set of characteristics. However, many of the major aspects of these processes are common to all. A list of various metal films and some typical process and test parameters are shown in Table 10-1. As we can see, there are both physical deposition techniques, such as sputtering and evaporation, and chemical vapor deposition techniques that can be used to create metal films. Selection of the process is largely determined by the availability of appropriate chemical reactions for the CVD process and the price/performance of each particular option for producing a film.

In addition to being used to produce the metal lines used for conductors, vias, and so on, some metal substances are used as "barrier" metals in contacts. These barrier metals will prevent some of the problems discussed in the chapter, such as contact spiking, as well as providing benefits such as reduced contact resistance.

In general, the most common methods of metal film deposition include the physical vapor deposition processes, including sputtering and evaporating. These are the processes that will be covered in this chapter. Chemical vapor deposition processes are also used for a number of metal film deposition steps. Since we already covered the CVD processes in some detail, we will cover only the items of interest that are specific to the metal deposition processes. In general, many of the process requirements for the other CVD processes will be requirements for the metal deposition processes.

We will see that in many respects the metal film deposition steps are controlled and run much like a CVD process, although the techniques for analyzing the data are quite different. For example, the metallization processes utilize high-vacuum environments, often use a plasma generator to increase the total energy of the reaction, and are prone to highlight underlying particle contamination problems. The films, however, are opaque and cannot be measured by optical means except when very thin (in the hundreds of angstroms). As a result, electromagnetic means are usually employed to measure the thicknesses of these films. This type of measurement introduces a different set of concerns when film parametric data are determined.

The metallization processes are very sensitive to particulate contamination and prone to easy redistribution of contaminating atoms due to the high vacuums and high voltages that are present in the chambers. Particles that are present in the chamber can cause a number of problems, including short circuits, open circuits (depending on the types and locations of the particles), poor contacts, and so on. Chemical contamination can become trapped within the metal films, causing changes in film resistivity and enhancing the probability of

TABLE 10-1
Metal Film Processes

Film	Process	Deposition rate	Temperature (°C)	Purpose	Advantage	Disadvantage
Al-Si-Cu	Sputter	10,000	400	Conductors	Well known, low resistivity, easy to process	Hillocks and high electromigration, near-current-carrying capacity limit
W	CVD	75	300	Vias, barrier metal	Selective deposition on Si, low electromigration	Poor adhesion on SiO_2, two-step process
Ti	Sputter	4000	400	Barrier metal	Reduces electromigration and hillocks, better contact stability	Enhances Al corrosion, oxidizes rapidly
Mo	CVD	600	600–1300	Vias, barrier metal	Very selective deposition, low electromigration, one-step process	Poor adhesion on dielectric, MoF_6-etches SiO_2
$TiSi_2$	Sputter/ Anneal	100	700	Barrier metal	Reduces contact resistance, good barrier metal	Lateral growth in source/drain regions
TiW	Sputter	1000	1000	Conductors	Can handle high-temperature, strong material	Requires further development
Superconductor	Various	—	—	Conductors	High speed, low power, huge current-carrying capacity	Requires LN_2 to operate, high-temperature anneals during process
Copper	Sputter	—	—	Conductors	Low resistivity, good stability, low electromigration	Hard to plasma-etch, easy corrosion
WSi_2	CVD	500	700	Interconnects, gate metal	Low resistivity, good stability	Difficult to control process
$TaSi_2$	CVD	300	700	Interconnects, gate metal	Good adhesion	Difficult to control process

hillock formation, electromigration, and other detrimental effects. There are a number of sources for this contamination, including the vacuum system, the wafer handling systems, air leaks, gas contamination, and associated problems.

10.1 METALLIZATION PROCESS EQUIPMENT

We will start our discussion of the metallization process equipment with the most common system in use today, that is, the sputtering process. In this system, shown in Figure 10-1, the wafers are placed on a flat plate that serves as both a support and an electrode. Another plate, consisting of the material to be deposited on the wafers, is placed a few centimeters away. This second plate is usually called the sputtering target. The space between the plates is evacuated to an extremely high vacuum level. Inert gases are then introduced into the chamber and a plasma is struck in the inert gas. The ions in the plasma then strike the surface of the sputtering target (which has a "biasing" voltage applied to it which draws the plasma toward it and away from the wafers) and propel small quantities of metal away from the target

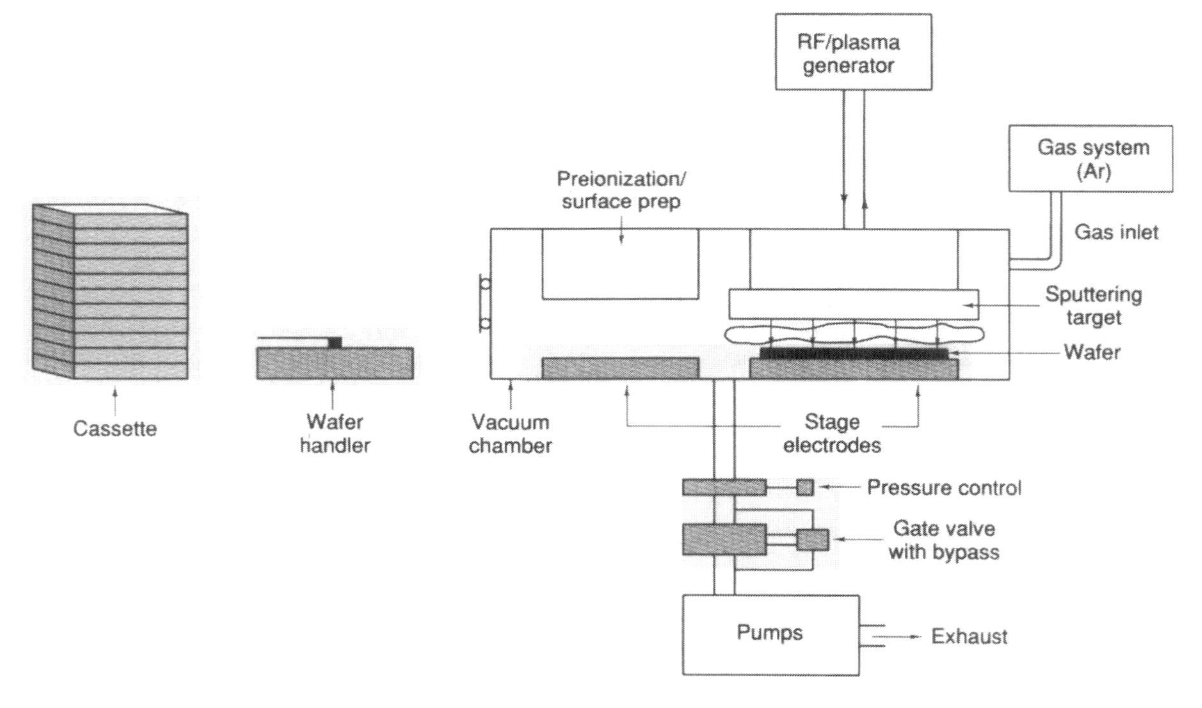

Figure 10-1. Aluminum Sputtering System The sputtering system is quite simple in principle. The chamber is evacuated and then filled with argon. A plasma is struck and the ions are then accelerated with a biasing voltage into the sputtering target. Metal from the target sprays off of the target and lands on the wafer. Deposition rate is affected by pressure, bias voltage, RF power, and the condition of the sputtering target.

and toward the wafers. This process continues until the desired film thickness on the wafers is obtained.

Since the types of materials that can be sputtered can vary greatly, there are a significant number of variations on the basic process. The parameters that are most important are typically the ones that will cause a significant variation in the total energy imparted to the inert gas molecules. These factors include the process temperature, frequency of the plasma generator, and electromagnetic potential across the plates. In addition, parameters such as the plasma/chamber pressure, chamber gas type, and target material type can greatly affect the deposition of the material. For instance, the use of heavier atoms in the plasma gas (e.g., xenon instead of argon) will permit higher sputtering energies, which is useful for targets of material harder than aluminum.

First, we will discuss the wafer handling mechanism for the metal deposition system. Obviously, there is a need for a wafer handler, since the metallization process is performed at vacuum and since the plates are kept very close together. It would be very difficult to reach into a sputtering chamber to retrieve wafers without damaging them. A robot has been developed with an appropriate arm that reaches into the cassette, and retrieves a wafer. This arm will then retract until the wafer is centered over the robot arm axis. The robot then turns around and extends the arm until the wafer is properly positioned in the chamber. It then sets the wafer down and retracts out of position. There are many variants on this basic design. The problems related to these wafer handlers are similar to those found in other wafer handling systems. They revolve mostly around the reliability of the subsystems and the quality of the maintenance that they have received. For example, a wafer handling robot that is vibrating as it moves through the air lock door may permit the wafer to come into contact with some part of the system, scratching its surface. It may also shake contamination loose from other parts inside the system.

Another major subsystem for the sputterer is the high-vacuum system control. The sputtering process requires a very high baseline vacuum (exceeding 1×10^{-8} mT), and is very susceptible to leaks. Problems with leaks will result in a number of film problems, which will be reflected in deviations in die yield. These problems range from entrapment of foreign atoms in the films to oxidation of the metal films (which can lead to the creation of insulating regimes within the film itself). Entrapment of the molecules from the air will, in most cases, increase the resistivity of the film and, in some cases, the molecules will even react chemically with the metal. Vacuum oil fumes can be very hazardous to the metal film deposition processes. As a result of this potential for backstreaming and of the low vacuum pressures required, sputtering chambers tend to use very high-powered dry pumps, cryogenic, or turbomolecular pumps. Entrapment of metallic atoms from other contaminant sources is not usually as critical, although migration of atoms within the metal films can be a problem.

Particles can also be created quite easily in a chamber that contains leaks, and they can severely affect the yield of this step. In many cases, the particles will be aluminum oxide or silicon oxide, due to reactions between air and the sputtered atoms in flight. If the particles land in the proper locations, they can cause current leakage paths or shorts between adjacent lines (as in Figure 10-2a), shorts between layers (as in Figure 10-2b), open contacts (in other words, the contact has not been made, as shown in Figure 10-2c), or a broken line (Figure 10-2d). Any of these problems will be likely to cause chip failure, especially if the line that has the particle on it is one that carries high currents.

Finally, changes in deposition rates, haze, and reflectivity parameters can occur as a result of leaks within the system due to the chemical reactions taking place between the atmospheric gases leaking into the system and the metal film as it is being deposited. The changes described here are similar to those described above, except that, in these cases, the leakage problem has reached a fairly gross level, so that changes are visible on the macro (human) scale.

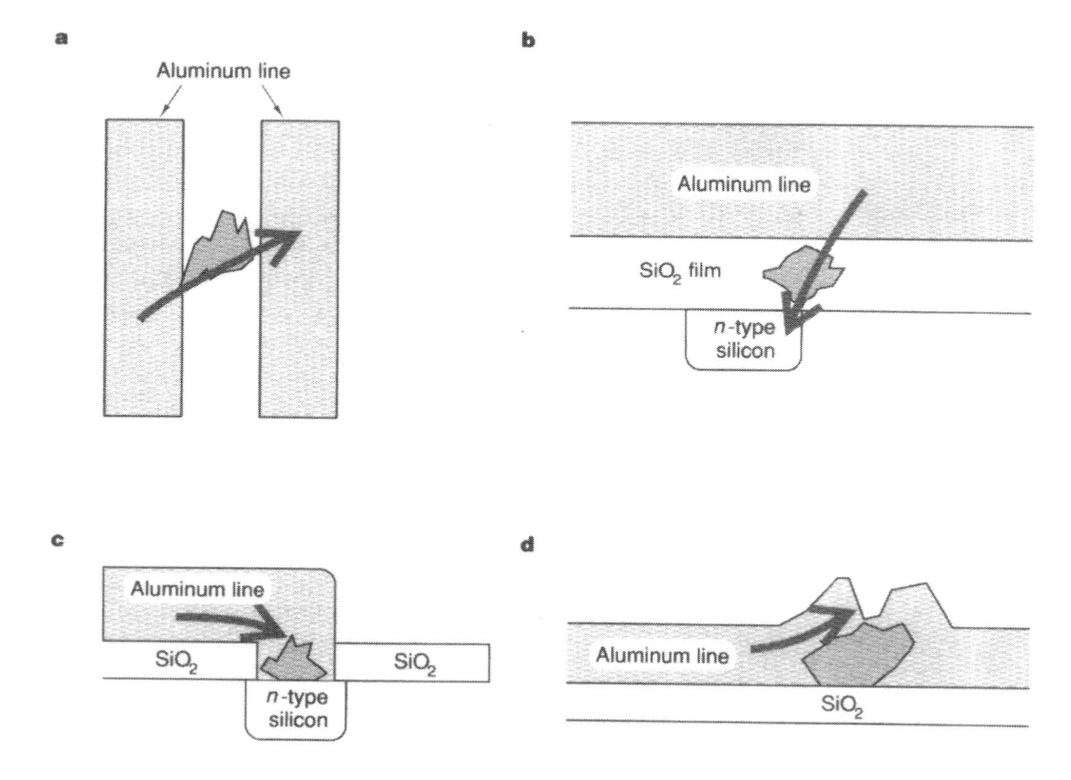

Figure 10-2. Effects of Particles at the Metallization Step As can be seen here, particles can cause several types of failures, resulting in both open and short circuits, depending on the location and type of the particle. (a) Particles shorting aluminum lines; (b) particle shorting conductor line to substrate; (c) particle blocking contact; (d) particle damaging deposited metal lines.

Obviously, since the process is so sensitive to leaks, it will also be sensitive to other possible sources of contamination. They include impure processing gas (delivering noninert molecules in the gas stream), impure or dirty metal sputtering targets, other contaminants such as residual chemicals or water remaining on the wafer after a cleaning operation, and fingerprints and other similar contaminants. All material that can outgas in the high-vacuum environment should be removed from the wafers before they enter the deposition chamber.

To maintain uniformity, it is important that the plasma and electric field densities remain consistent throughout the sputtering region. A number of parameters must be kept in control to provide this consistency and therefore provide acceptable uniformity. For instance, the power supplies that drive the plasma must stay calibrated in order to verify that the frequency, potential, and current set points are being obtained. All connectors and so on must be secured and checked, especially after maintenance procedures have been completed. Instabilities in the fields inside the chamber can create a number of problems, from changes in deposition rate and uniformity to crystal growth pattern changes.

The purity of the sputtering target must be maintained throughout the depth as well as across the width of the target. The target is typically manufactured somewhat larger than the deposition zone it is intended to cover. The sputtering target must be changed on a regular basis. There are two reasons for this: The main one is that the metal will eventually be entirely sputtered off of the target. The other reason (which is the main reason that the targets are changed) is that there will be a significant impact on the uniformity of the film. The target will deteriorate as it is used, leaving it with a deeply etched region, as shown in Figure 10-3. After this deterioration has reached a

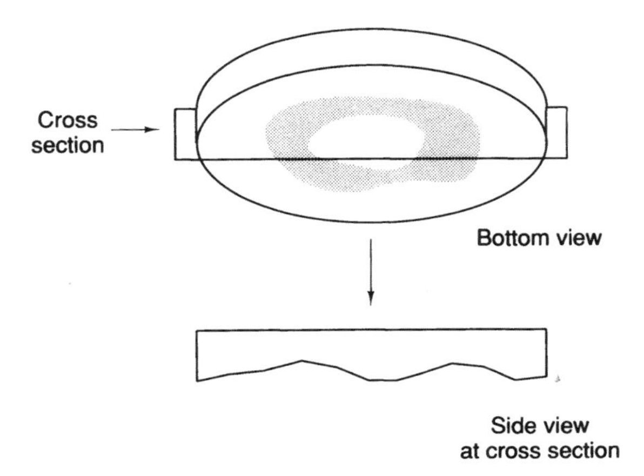

Figure 10-3. The Deterioration of Sputtering Targets As the sputtering target ages, deep grooves and pits will be etched into its surface. When this damage is sufficiently deep, the deposition uniformity will change. In addition, other subtle changes such as in grain structure uniformity can occur with aged and deteriorated sputtering targets.

certain level, the metal atoms coming off the target will vary in number and possibly in composition (if the sputtering target has any variations in silicon content as a function of depth).

In any event, as we can see in Figure 10-4, a change in the uniformity of the electric fields caused by the deterioration of the target will result in changes in the uniformity of the film thicknesses. Also, drastic changes in these parameters can cause changes in other parameters, such as crystal structure and growth mechanisms. Therefore, the sputtering target should be examined periodically for wear, or should be well characterized so that it is replaced after a fixed number of deposition cycles. In most cases, the remaining piece of the target can be returned to the manufacturer for refurbishment. Various types of materials can be used for sputtering targets, from a number of metals and metal silicide combinations to quartz (also called bias sputtered quartz processes).

Clearly, the mechanics of the process are somewhat simpler than some of the other processes requiring high vacuum. For instance, there are no special magnetic field systems for bending ion beams, as in the implant process, nor are there toxic or pyrophoric gas handling systems as in the CVD systems. These features make the systems relatively easy to maintain. Like any deposition system, it is very important to adhere closely to the preventive maintenance schedule in order to keep excess deposition and contamination from causing unnecessary downtime and other, more subtle, problems.

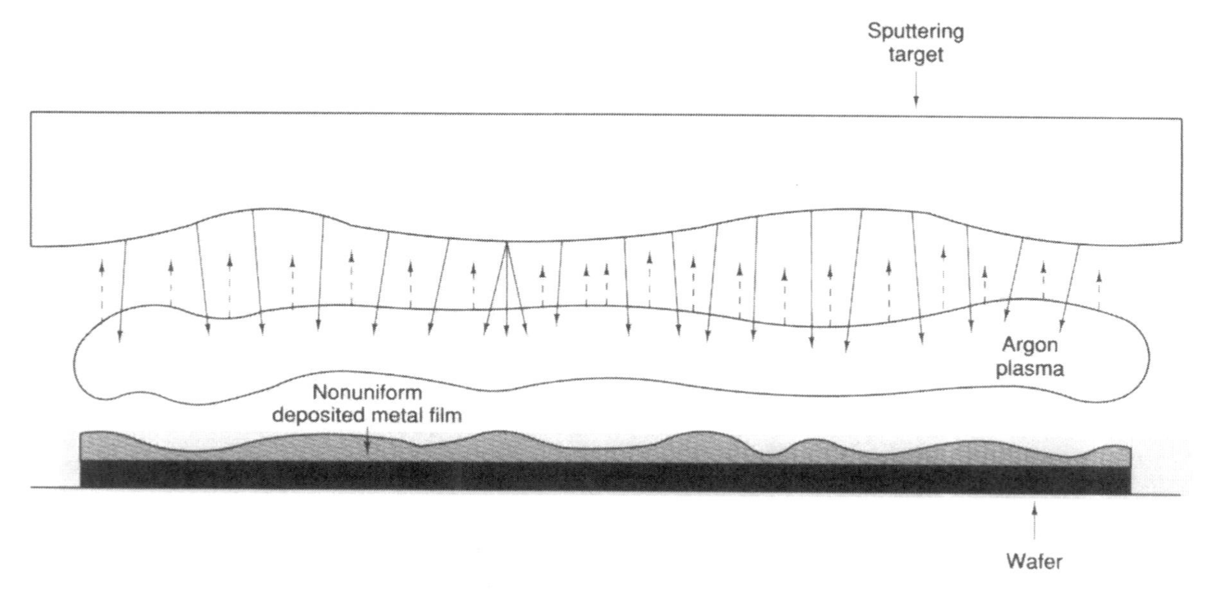

Figure 10-4. Nonuniform Metal Deposition Due to Worn Target A severely worn sputtering target will affect the electric field density throughout the sputtering chamber. This, in turn, affects the speed and direction of impact of the argon ions, producing small, but significant, variations in the local density of metal ions. These variations affect the uniformity of films after deposition.

Note that, unlike implanters and CVD operations, sputterers can be constructed in very small spaces. For example, desktop gold sputterers are often used for scanning electron microscopy sample preparation. These small systems can produce films only on substrates of limited sizes and at slow rates, so are not up to the demands of heavy-duty manufacturing, but can be used for some lab applications. Even the full-scale manufacturing systems can be reduced to very small systems in comparison to other equivalent processing systems.

Another type of manufacturing system used for metallization is the evaporator. A block diagram of this system is shown in Figure 10-5. In this system, a fixed quantity of aluminum or aluminum/silicon alloy pellets are placed in a small crucible, where they are heated or bombarded with electrons until they melt. The wafers are placed into

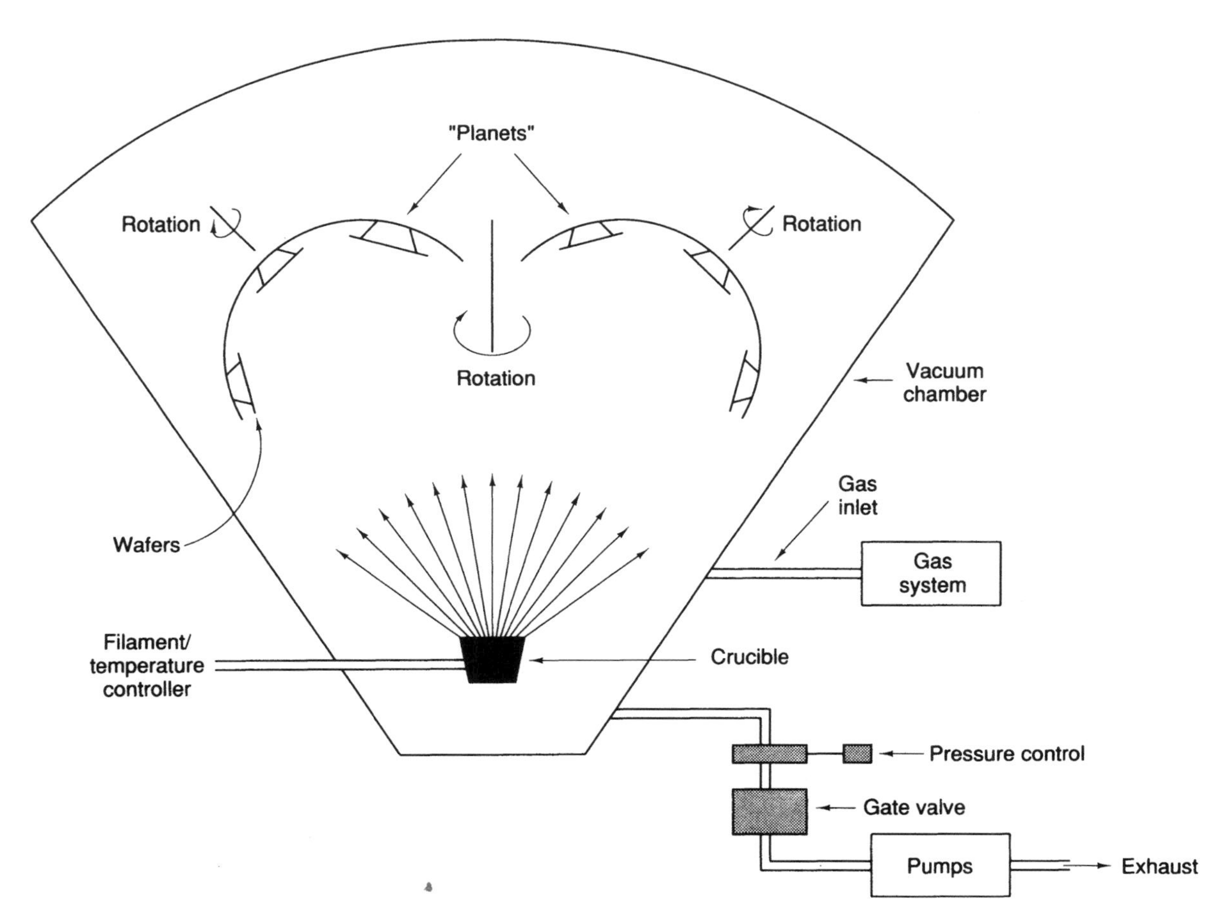

Figure 10-5. The Metal Evaporator The evaporator is an old method for depositing aluminum. In this system, metal is heated and then vaporized by a filament or an electron beam. This metal sprays up through the system. The wafers must be rotated on special structures called "planets." Each planet rotates about its axis, while the entire planet is rotated around a central axis. This mechanical set up does not produce exceptional uniformity and is prone to particle contamination.

large structures (called "planets" due to their motions) that rotate inside the evaporator so that all points on the wafers are exposed to the metals being deposited. The system is brought down to a vacuum condition and the wafers are put into motion. At this point, the pellets are melted and, at the proper point, the melt is vaporized, by either bombardment by electrons or superheating by filament. The vaporized aluminum then sprays down onto any of the surfaces of the chamber that are in the paths of travel, including the wafers. Rotation of the planets is required to even out nonuniform metallization. A variety of other evaporation methods have been developed, but are not in common use.

In general, evaporation has been replaced by sputtering, due to the increased film quality control that is available with the sputterer. Evaporators produce films with poor thickness and uniformity control, along with poor film composition control. There is also much less control on the crystal growth mechanisms, so that the crystals form in random orientations. Also, contamination from the melt crucible can become trapped in the film. One final point that limits the use of evaporators is the motion of the wafers on the planets. This is a significant source of particle contamination and permits the possibility of wafer breakage, with the consequent contamination of all the wafers in the evaporator.

Other methods that are used to produce metal films on wafers are the CVD methods (called MOCVD or metal-organic CVD). Typically, the chemical source gases used are of the metal-organic type. A list of the various source gases possible, as well as the end result films is given in Table 10-2. Most of these chemicals are very volatile, available only as liquid sources, and poisonous, so extra safety precautions must be taken. These processes are otherwise very similar to the other low-pressure thermal and plasma-enhanced CVD processes that were discussed earlier. These processes allow some degree of process control with metal films that cannot be sputtered easily. However, for most films, sputtering is still the preferred technique for metallization.

One of more common uses of metal CVD processes is development of selective metal deposition processes. In these techniques, gases are reacted together over the surface of the wafer. The chemistry of the process is selected so that the metals will crystallize on metal surfaces much more readily than on dielectric surfaces. Then the metals will deposit inside contact holes (vias), and not on the surrounding dielectric.

The final method of metal deposition we will discuss here is molecular beam epitaxy. This technique can be used for more than metal film deposition, and is commonly used in research labs for the deposition of very thin layers of compound semiconductors. In this process, the process source gases are introduced into a reactor chamber, where the chemicals are ionized and delivered to the wafers electromagnetically (thus the term molecular beam) in very tiny quan-

TABLE 10-2
Epitaxial/Metal CVD Processes

Film	Gases used
Mo	MoF_6
Ti	$TiCl_4$
W	WCl_6, WF_6
$TiSi_2$	$TiCl_4$, SiH_4
$TaSi_2$	$TaCl_4$, SiH_4, H_2
WSi_2	WF_6, SiH_4, SiH_2Cl_2
Al	TMAl[a]
GaAs	TMGa, AsH_3
AlGaAs	TMGa, TMAl, AsH_3
Polysilicon	SiH_4, PH_3 for in-situ doped

[a] TMxx stands for the "trimethyl" group of the element required. For instance, TMAl is trimethylaluminum, TMGa is trimethylgallium.

tities. These chemicals react on the surface to form very thin layers of the desired compounds, sometimes only a few monolayers thick. These types of devices can be used to form ultrasmall electronic and optoelectronic devices. Devices such as the quantum transistor are more easily manufactured using these very small structures. By the end of the next decade, these processes will be in common use for the highest performance devices. At this time, the processing throughput times are too long to permit molecular beam epitaxy processes to be of practical use in the current generation of integrated circuits.

We can see then that there are a number of techniques for depositing metal films on semiconductor wafers. They have a variety of advantages and disadvantages, but the industry has found some stability in the now-standard sputtering systems. The MOCVD process and molecular beam epitaxy processes promise to continue to bear promise for even more significant advantages in the future.

10.2 METALLIZATION PROCESSING

The metals that can be deposited on the wafer surfaces vary significantly in a number of characteristics. There are differences in these films for such parameters as current-carrying capacity, heat dissipation, and the ability of the film to withstand higher processing temperatures. There are mechanical differences in these films, too, such as their relative ability to be deposited on steep terrain without cracking, or their immunity to problems such as electromigration. In addition, the differences in the deposition processes and film characteristics result in varying degrees of stress placed on the wafers during the process. The various films and some of their more typical parameters are shown in Table 10-3.

The most common element used in the deposition of metallic films on the wafers is aluminum. This metal is soft, has a low melting point, and is easily deposited onto the wafer surfaces. The purification technology for aluminum is also well characterized, so that ultrapure targets and sources can be manufactured readily. The aluminum film is usually deposited using an evaporative or sputtering technique. As a result, this film is a broad-area deposition process, permitting the entire wafer to be metallized quickly. The aluminum films are then given a pattern in the photolithography area, and the films are etched to form the structures required.

A typical aluminum or aluminum alloy film is about 5000 to 7000 Å thick, with linewidths of 0.5 to 1.0 μm. The ratio of the height to the width of the aluminum line is called the aspect ratio, and is determined by the following relationship.

$$\text{Aspect Ratio} = \text{Step Height/Width}$$

The step height is the height from the base of the line to the top edge, while the width can defined by either the linewidth or the pitch of the lines (which is the sum of the line and spacing widths). Lines with high aspect ratios are more difficult to produce and to coat with dielectric films, and are much more prone to problems like void formation in deep troughs (see Figure 10-6). As a result, progressively thinner metal films are required to allow manufacturability. Of course, any reduction in film thickness results in loss of current-carrying capacity,

TABLE 10-3
Metal Film Parameters[a]

Film	Typical maximum temperature (°C)	Resistivity ($\mu\Omega \cdot$ cm)
Al/Al–Si	420	2.7
W	700	5.6
Ti	>1100	41
Cu	>800	1.7
WSi_2	>1000	70
$TiSi_2$	>900	13–25
$Mo/MoSi_2$	>1100	90–100
TiW	450	65–75
$TaSi_2$	>1000	40–50
n^+-poly	>900	500

[a] Metal film parameters can vary greatly. Two of the key parameters required in the selection of film are the typical maximum temperature that the film can handle and the resistivity of the metal film.

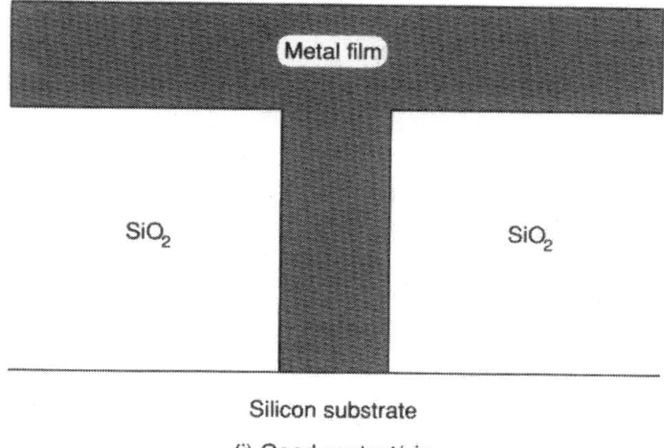

(i) Good contact/via

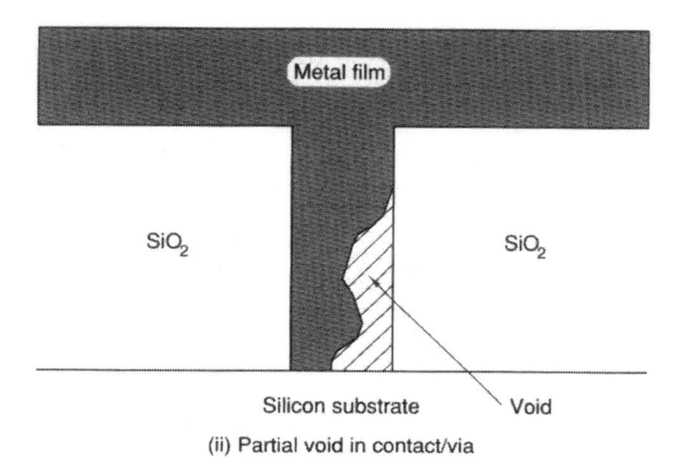

(ii) Partial void in contact/via

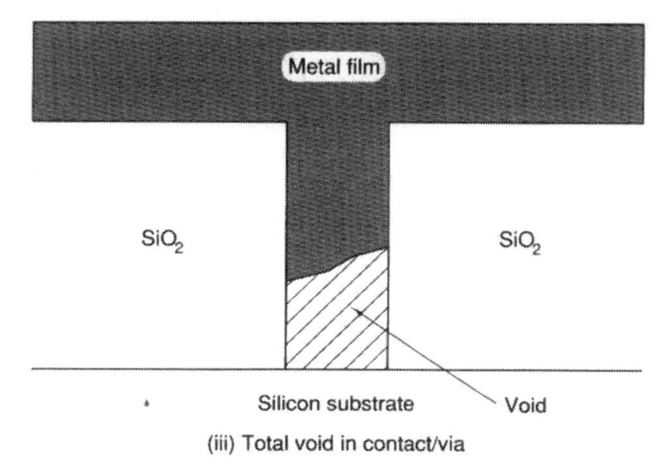

(iii) Total void in contact/via

Figure 10-6. Void Formation in Metal Contacts and Vias Voids can form during metal deposition in deep troughs, contact through-holes, and vias. They can end up as partial voids which affect device parametrics, or as complete voids, preventing the structure from operating.

so that only noncritical lines can have arbitrary film-thickness reductions. We will see that there are other tricks to permit production of chips within these manufacturing and technological constraints.

Aluminum is usually mixed with a small quantity of silicon and/or copper. The content of the silicon is about 1 to 1.5% by weight, which is near the solid solubility limit of the aluminum. Without this concentration of silicon dissolved in the aluminum, the contacts between the aluminum film and the silicon substrate or polysilicon layers would be very poor. When there is not enough silicon in the aluminum, the contacts will form interface junctions which will have a relatively high contact resistance. This problem can slow device operation or prevent the device from operating entirely. With the addition of the silicon, the contacts will form consistent junctions with a continuous change in alloy type, allowing current to flow from one substance to another with a minimum of resistance. The contact alloy is depicted in Figure 10-7.

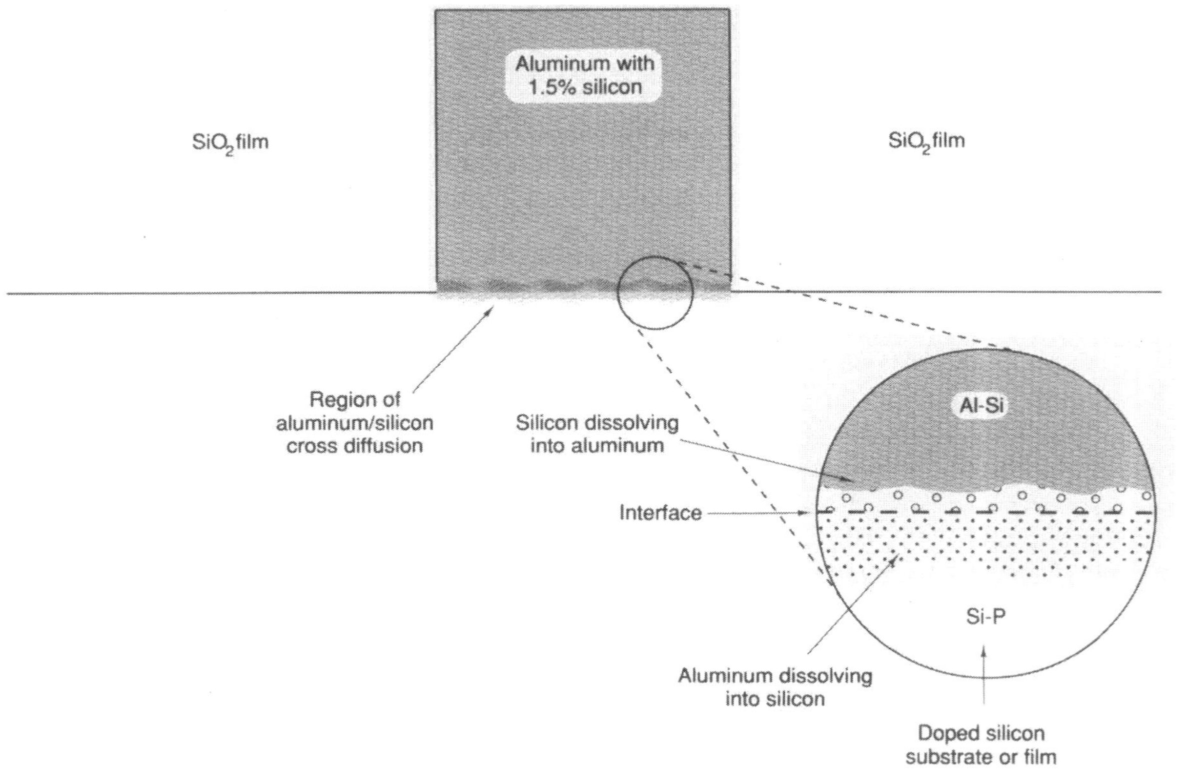

Figure 10-7. Metal-to-Substrate Contacts The contact is the region on the semiconductor device where metals of differing types or on different layers connect to complete the circuits. In the case of the aluminum–silicon-to-substrate contact shown here, the metal film is deposited and then alloyed to allow a smooth continuum of resistivity from one material to the other. The aluminum has about 1.0 to 1.5% silicon dissolved in it, which is near the solid solubility limit, to help control the cross diffusion at the contact interface. Contacts are formed reliably when the aluminum can dissolve into the silicon freely, and when silicon only dissolves into the aluminum film to replace the atoms that were dissolved into the silicon.

The process for alloying the metals together is also important. For a good contact to be formed, both metal layers must be heated enough that the atoms of each metal (silicon and aluminum) will diffuse into the other. This usually occurs somewhat below the eutectic point, at about 385 to 420°C. If the temperature is too low, a good alloy will not be formed and the contact may not operate correctly. This is similar in principle to a cold solder joint on a circuit board. However, if the metals are overheated, or if the incorrect alloys of aluminum are deposited, an effect called spiking can occur (see Figure 10-8). In this case, the aluminum atoms force their way through the surface structure of the silicon contact region and form crystals, which usually penetrate through the contact region and into the substrate. At the same time, silicon atoms diffuse up into the aluminum layer in unpredictable ways if the metal has insufficient silicon content. Obviously, these effects are not good for the device, causing a variety of yield-damaging effects.

Control over contact resistance is important, since instabilities here can cause significant and sometimes hard to find circuit problems. A number of elements can affect contact resistance. Spiking can

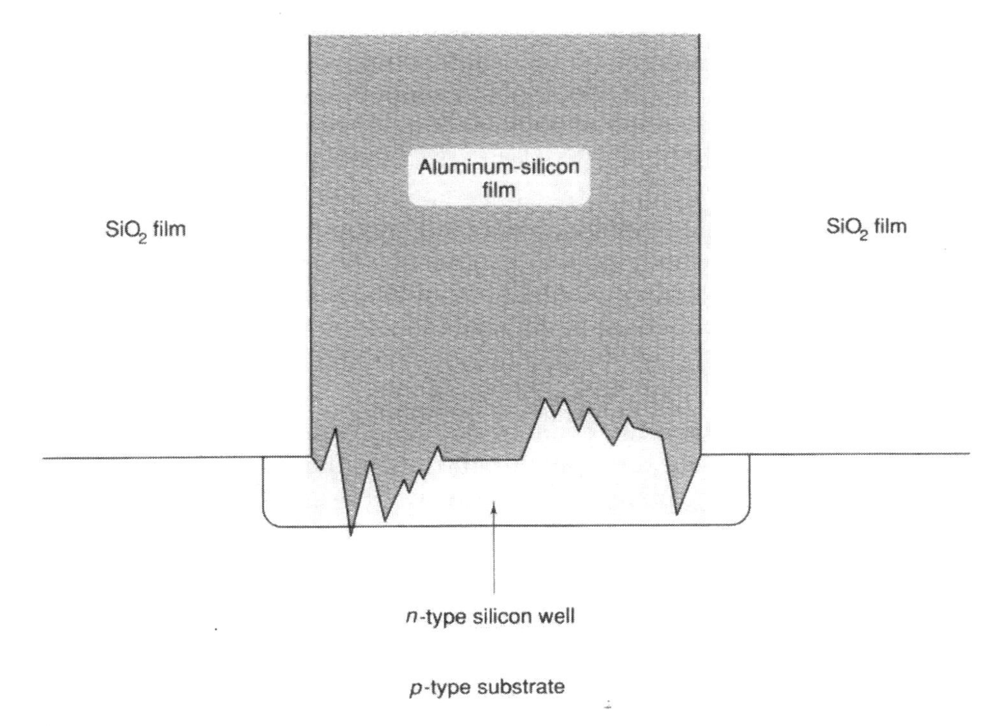

Figure 10-8. Contact Spiking Contact spiking occurs when the aluminum film penetrates the "well" region and connects to the substrate material, There are a number of deleterious side effects to this, including increased substrate currents, parasitic transistor development, short circuits, and poor long-term reliability. The spiking is caused by the formation of aluminum crystals which almost literally push the silicon aside as they are produced. This effect is most serious when alloy temperatures are too high, but is also present when the silicon content of the aluminum is too low, so that excessive cross diffusion occurs.

affect the contact resistance, usually increasing it. The junctions between the spiked aluminum crystals and the silicon crystals do not form a good joint in this case. Another source of problems with contact resistance is oxide in the contacts, which can either increase the contact resistance or close the contact off completely, depending on the amount of oxide remaining. Insufficient alloying can result in poor contacts, also resulting in high contact resistance. Both metal cracking at the top of a contact and void formation at the bottom of the contacts result in open circuits and consequent circuit failure. The contacts can be tested thoroughly only by electrical means. However, careful visual and in-line SEM inspection of side effects of these problems can often permit some evaluation of the problems.

To reduce the effects of contact resistance variations and to prevent spiking and other problems associated with the joining of aluminum and silicon, many new IC designs, especially high-speed devices, have gone to more complex metallization schemes. For example, a fairly common sequence of metallization is to form a metal silicide layer on the top surface of the silicon, for example, WSi_2, followed with one of a number of potential barrier metals, including titanium–tungsten, molybdenum, or titanium nitride films. These films are then connected to the via fill metal, often tungsten. The next conductive metal (aluminum–silicon) will be placed on top.

Obviously, a number of complications can occur when developing a process with as many components as this. For example, particles trapped in the contact may result in pinholes in the very thin barrier metals. Insufficient anneal times can result in a poor formation of metal silicide, resulting in high contact resistance. Etch and photoresist removal processes can result in damage to the contact region causing corrosion or mouse bites. Other etch processes cause damage to barrier metals. All of these issues should be observed closely during processing to prevent yield loss.

Another problem that occurs with overheating the wafers during the alloy cycle is that the aluminum will melt, destroying the crystal structure developed during deposition and also causing other severe problems such as the melting and breaking of conductor lines and the formation of small nodules of aluminum called hillocks (shown in Figure 10-9). The formation of aluminum hillocks is therefore also exacerbated by excessive alloy temperatures. The hillocks can later lead to increased electromigration failures.

The alloy process is usually carried out in an atmosphere containing hydrogen, as this gas easily diffuses through the aluminum, facilitating the alloy process. It also diffuses back out of the film relatively quickly after the hydrogen source is removed. Fortunately, the amount of hydrogen that is required in the ambient gas stream is relatively low (5 to 15%), so that nonexplosive combinations of nitrogen and hydrogen (called "forming gas") can be used for the alloy processes. The typical alloying time is about 30 minutes, although times from about

(i)

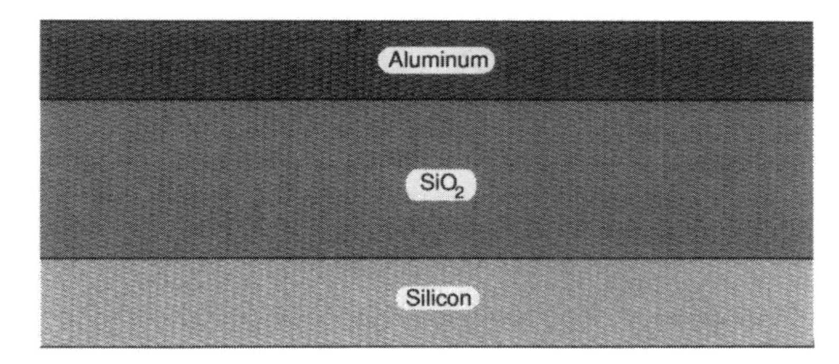

(ii)

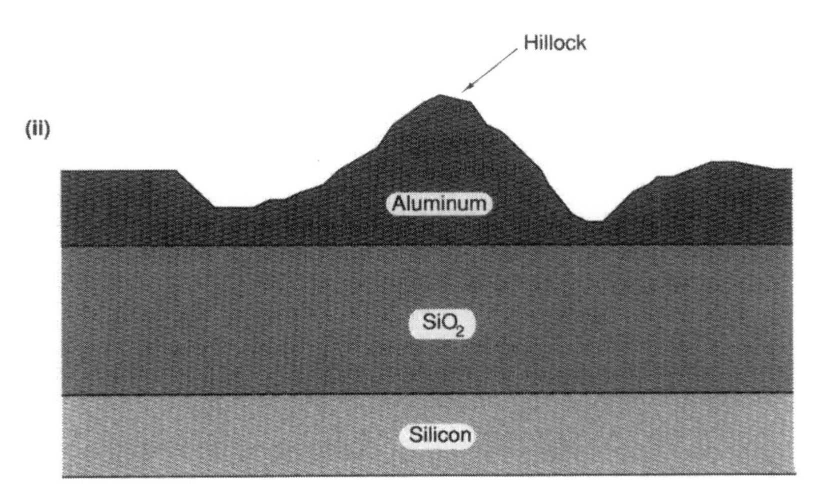

Figure 10-9. Hillock Formation Aluminum films that are overheated, stressed, or otherwise misprocessed can form small nodules, called hillocks, in the lines. The hillock is formed by material from the adjoining portion of the line, causing troughs, breaks, and voids in the aluminum. Long-term problems include reduced reliability and increased problems with electromigration.

20 minutes to over an hour may be acceptable, depending on the type of devices and contacts used.

There are other consequences that can occur as a result of high-temperature (over 300°C) processing after the aluminum film deposition process. One is the formation of hillocks, which are small nodules of aluminum that form as the film remains heated for a length of time. They become much more prominent as the time at a given temperature is increased, and with higher temperatures. Their formation rate and size increases when the aluminum film is deposited under stress, or when consequent dielectric films are deposited under stress. To minimize this problem, the lowest possible deposition temperatures and films or processes exhibiting the lowest amount of stress are used. The formation of hillocks has been associated with defects such as film cracking, electromigration, and other phenomena that occur with metal degradation. In addition, hillocks are almost always associated with the formation of voids, since the material to form the hillock must come from somewhere. Typically, the void or reduced thickness region is immediately adjacent to the hillock, which further aggravates device reliability problems.

Another element that is mixed with the aluminum film is copper. This is usually added in only a small quantity, 1% or less. The addition of this copper to the alloy helps prevent an effect called electromigration. Electromigration occurs when the device has been in operation for a long time. As current flows through the circuit element, the electrons continually collide with the atoms of the metal lines, slowly moving them around, until cracks or other defects occur that cause device failure. The addition of the copper makes the aluminum film harder, somewhat more stable, and less prone to this phenomenon. Addition of copper to the aluminum film also makes the film much more difficult to etch, as the etch chemistries for removing the three-substance alloy, without attacking other films nearby, are very complex.

In addition to the use of copper dopants in aluminum films for prevention of electromigration, there is an association between electromigration and the crystal structure of the aluminum film. Aluminum appears to be less likely to form hillocks or to be as prone to electromigration when it has formed regular crystals that are aligned vertically, perpendicular to the line of electron flow (the crystals form structures that look like columns, as shown in Figure 10-10). While the crystal formation can be adjusted through variations of some process parameters, the formation of vertically columnar crystals also appears to be facilitated by the addition of copper to the alloy.

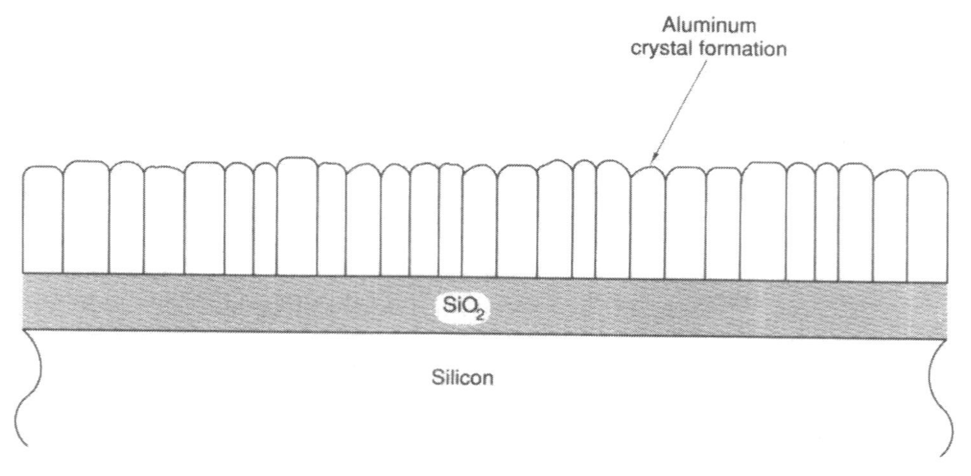

Figure 10-10. Development of Aluminum Crystals The most reliable aluminum films are formed into rows of virtually aligned crystals. This uniform structure reduces hillock and electromigration problems and provides stability for high reliability and low resistivity.

There are often two or three layers of aluminum placed on the surface of the wafer during processing. Each of these layers will provide quite different functions for the devices on the chips. In some cases, the lines will be used for interconnections between transistors, for example in ASIC or microprocessor devices. Two of the layers may be used as X and Y addressing decoders for memory arrays. A third layer may be used to provide power to the components, or may be used as ground conductors. It is critical that the design of the chip take into account the current-carrying capacity of the aluminum lines when setting up linewidth specification windows.

Each of the layers of the metal must be separated by a dielectric layer for insulation. Therefore, after each metallization step and its associated lithography and etch processes, the wafer will be coated with a film, usually silicon dioxide. Sometimes the oxide will be doped (i.e., PSG or BPSG films are used), although there it is generally preferable to use undoped glass since reactions can occur between the phosphorus or boron and the aluminum films. They are greatly amplified in the presence of moisture, for example, if the packaging of the chip is leaky, cracks have appeared in the passivation layer or the die is otherwise defective or improperly handled. Usually, the films separating the metal lines are quite thick, often around 0.75 to 1.0 μm. The primary reason is that the films must not break down, even if the metal lines are carrying very large current loads or unusually high voltages. There are other reasons for using thicker films for dielectrics, as we will now see.

Clearly, multiple layers of films piled on top of one another can make chip surface topology very complex, as described in Figure 10-11. Series of criss-crossing metal lines on top of stacked polysilicon gate transistors, with deep contacts cut down to the surface of the silicon, make the surface very rough and very difficult to work with. It is especially hard to deposit metal or dielectric film on these wild topographies without the films cracking or forming voids or other undesirable features. As a result, a substantial amount of effort has gone into a technique called planarization. There are a number of ways to create a planarized surface, which we will consider in a moment. In general, planarization is any process or series of processes that can be used to level the rough topology for a smooth, planar surface for the next level of deposition. While simple in concept, the ability to remove the effects of topology is quite difficult to achieve in practice.

A number of techniques are used for achieving planarized surfaces. We start off with a wafer that has a layer of annealed BPSG and its first metal layer already deposited and etched. At this point, another dielectric film must be deposited. One type of film that is used is the spin-on glass. This technique allows a liquid chemical source of SiO_2 to spread out and flatten on the surface like photoresist. One of the problems is making sure there are no bubbles or voids formed at the base of contact holes.

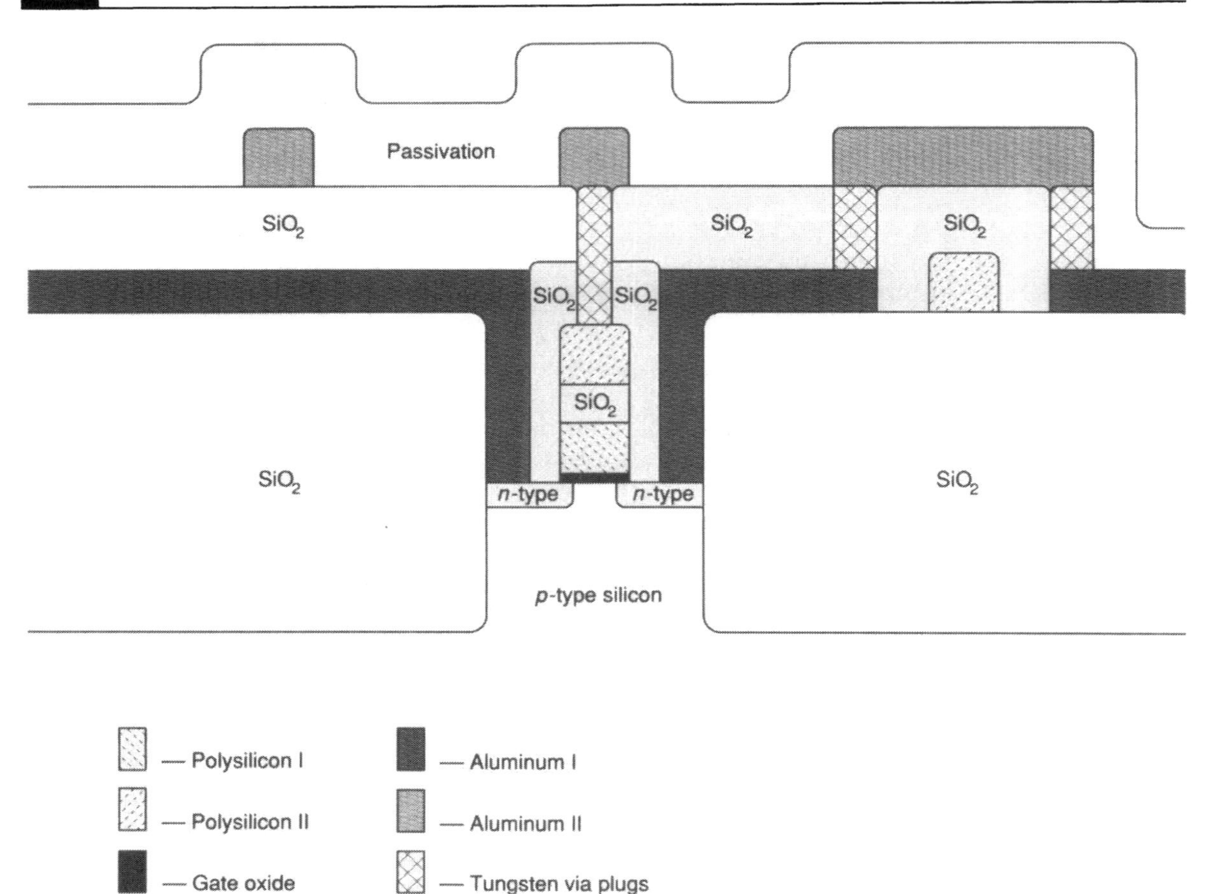

Figure 10-11. The Complex Topography on an IC We can see from this hypothetical dual-poly/dual-metal process that chip topology is quite complex. This uneven top surface is difficult to use as a base for any further device development, meaning that, with standard design considerations, there will be a maximum limit on the number of layers that could be used in the chip. To extend this limit, planarization techniques must be used.

Another technique is to use a plasma or laser-enhanced processor to deposit and etch the dielectric films. As discussed earlier (Figure 8-29), this procedure deposits a thin layer of film over the entire surface, and then removes a small amount of that film from the very top of the highest peaks with a short plasma etch process. This allows the film to build up more rapidly in the trough areas than on the peaks, which eventually allows the film to be made roughly level. After an acceptable certain smoothness is obtained, the dielectric film can be deposited to a reasonable thickness. Finally, before the next layer of metal is deposited the contact holes, or vias, must be etched into the dielectric film. The use of these deposit-etch and sputter-etch processes can leave residual contamination in the form of particulate contamination and metallic ion contamination.

Once the vias have been etched, the actual contacts must be made. It is possible to perform a blanket deposition of aluminum, but the film will then be conformal to the new topography and will still be prone to cracking at the corners of the contacts. While the problems of cracking, void formation, and film stress are reduced in comparison to the more complex topography, the poor reliability implications of the blanket deposition are still significant.

As a result, another process is often interjected here which allows for the vias to be filled without deposition on other surfaces. This process is called a selective metallization process, and is a complicated process. It is important in this process to be able to fill the vias completely with no void formation. It must be grown to a point where it is close to level (planar) with the surface of the dielectric film. However, the film must not deposit onto the surface of the dielectric. This can result in short circuits, current leakage paths, and other unusual operating problems. One advantage to the addition of this step is that, while it is difficult to control and characterize, the manufacturer now has the ability to deposit metal in the contacts without having to perform additional lithography or etch processes.

One of the most common metals used for via filling is tungsten since the chemical reactions that allow the tungsten deposition do not involve the dielectric molecules, and the dielectric films do not provide the nucleation sites needed for the film to adhere that exist with metal or silicon films. In addition, there are other benefits to the use of selective tungsten processes. For example, the tungsten can act as a barrier metal between other disparate substances (such as silicon and Al/Si alloys), thus reducing the contact resistance. It also acts as a physical barrier, reducing the probability that contact spiking will occur. These effects help reduce the electrical effect of excess junction leakage current.

There are two basic reactions that are used to provide tungsten films. The first is used for the first few hundred angstroms of tungsten film, and then will stop, preventing further deposition. This reaction is based on the reduction of WF_6 with silicon as shown:

$$2WF_6 + 3Si \rightarrow 2W\downarrow + 3SiF_4\uparrow$$

The silicon in this reaction can come from a number of sources, including the silicon substrate, polysilicon layers, aluminum/silicon alloys, and the various refractory metal silicides. The reaction rate and thickness of the tungsten layer are largely dependent on the amount of silicon available to the reaction before the diffusivity and reaction rate balance out and no further reaction occurs. For instance, tungsten forms layers up to 5 to 10 times as thick on pure silicon than on aluminum silicon alloys. However, the bonds of silicon dioxide, silicon nitride, and silicon oxynitride molecules are too strong for the WF_6 to steal the silicon away, resulting in a reaction that is fast, selective, but self-limiting in that, after the initial coating of tungsten has been completed, no further deposition will occur. The process works as selective metallization because the reaction can occur only on silicon metal (or alloy) surfaces, and will not occur on adjacent dielectric regions.

Several phenomena occur at the interfaces to dielectric layers. In the cases we have been discussing, where the contact holes and vias have been etched into the dielectric films, a phenomenon known as encroachment can occur as the deposition time increases. As seen in Figure 10-12, the volume of the tungsten layer is less than the volume of the silicon it replaces, allowing the WF_6 to slip in under the SiO_2, and forming a thin conductive layer that slowly advances through the device at that interface. This problem can severely impact the processing of VLSI wafers, especially in the submicron regions and must be controlled.

Another phenomenon that occurs is called creep, and is seen at interfaces between thermally grown oxides and silicon regions, in the bird's beak area. This is shown in Figure 10-13. In this case, there is enough tungsten diffusivity through the very thin silicon oxide layers at the edge of the bird's beak that the tungsten can react with the underlying silicon to form a very thin film of tungsten on the top of the dielectric film.

As a result, the amount of time for the silicon-reduction reaction is minimized to that required to adequately form the barrier layer. Some manufacturing processes that do not require specialized techniques to plug the vias can stop at this point and take advantage of the benefits of the barrier metal, which, as mentioned, are reduced contact resistance, reduced spiking, and reduced current leakage. For those processes that require that the vias be completely filled prior to the broad-area deposition step, a second deposition process is required. This technique is accomplished quite simply by adding hydrogen to the reaction chamber during the process. This changes the tungsten reaction to one of the reduction of tungsten hexafluoride by hydrogen as shown:

$$WF_6 + 3H_2 \rightarrow W\downarrow + 6\,HF\uparrow$$

In this reaction, the tungsten metal will deposit on anything that provides a good nucleation site on which the crystals can form. The selectivity between the deposition of the tungsten on metal films and dielectric films is very high, but tungsten can be deposited on dielectrics if the surface has received proper conditioning. Some process

→

Figure 10-12. Tungsten Encroachment During the first few hundred angstroms of tungsten deposition, the process is one of silicon reduction by WF_6. However, since the tungsten metal is denser than the silicon, the film that is deposited will be thinner than the layer of silicon consumed. This results in a slow formation of voids under the edges of contact walls. This problem is called tungsten encroachment, and will cease to occur once the silicon reduction reaction stops (after 200 to 500 Å of tungsten deposition).

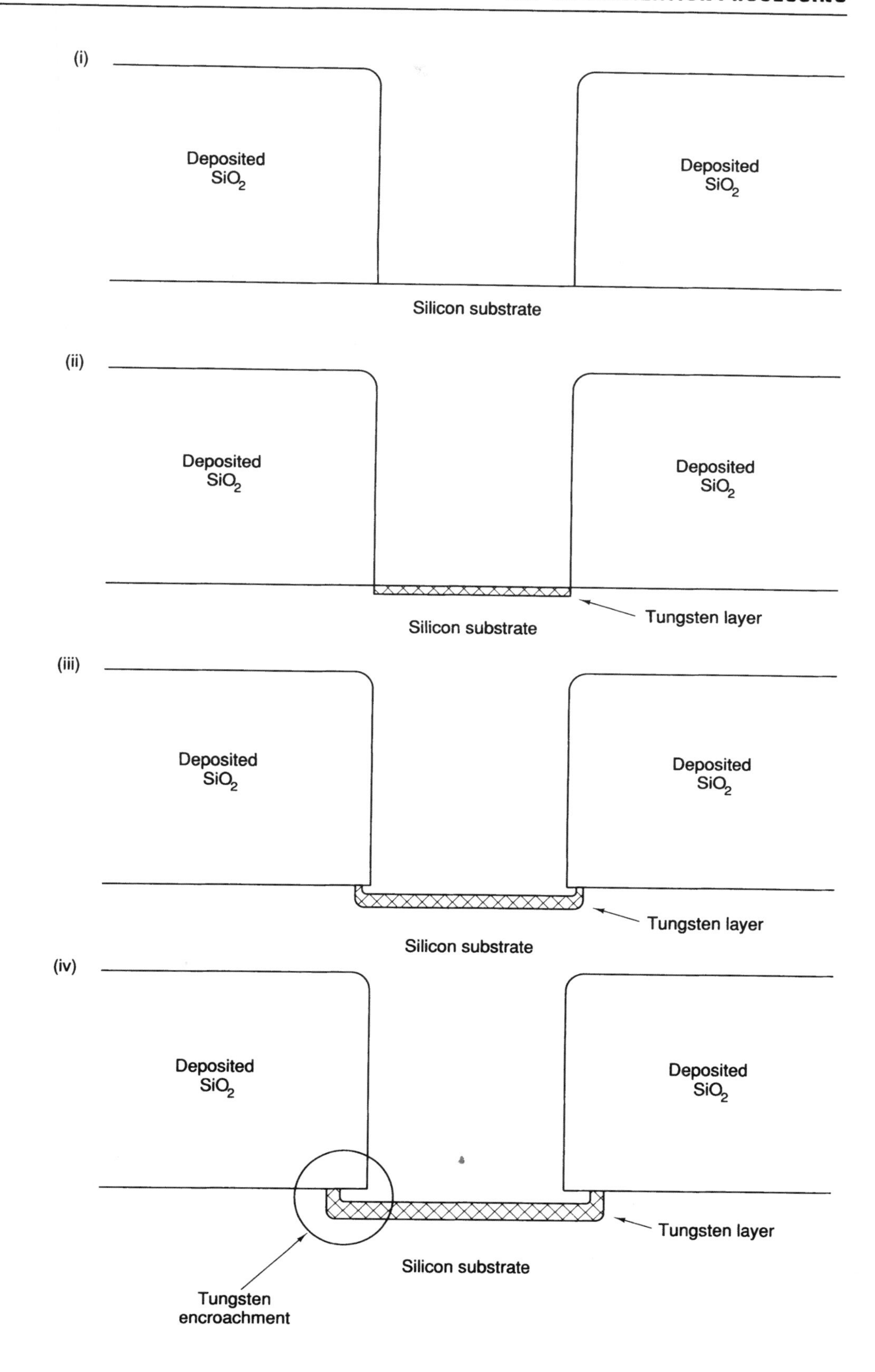
(i)
Deposited SiO_2
Deposited SiO_2
Silicon substrate
(ii)
Deposited SiO_2
Deposited SiO_2
Tungsten layer
Silicon substrate
(iii)
Deposited SiO_2
Deposited SiO_2
Tungsten layer
Silicon substrate
(iv)
Deposited SiO_2
Deposited SiO_2
Tungsten layer
Silicon substrate
Tungsten encroachment

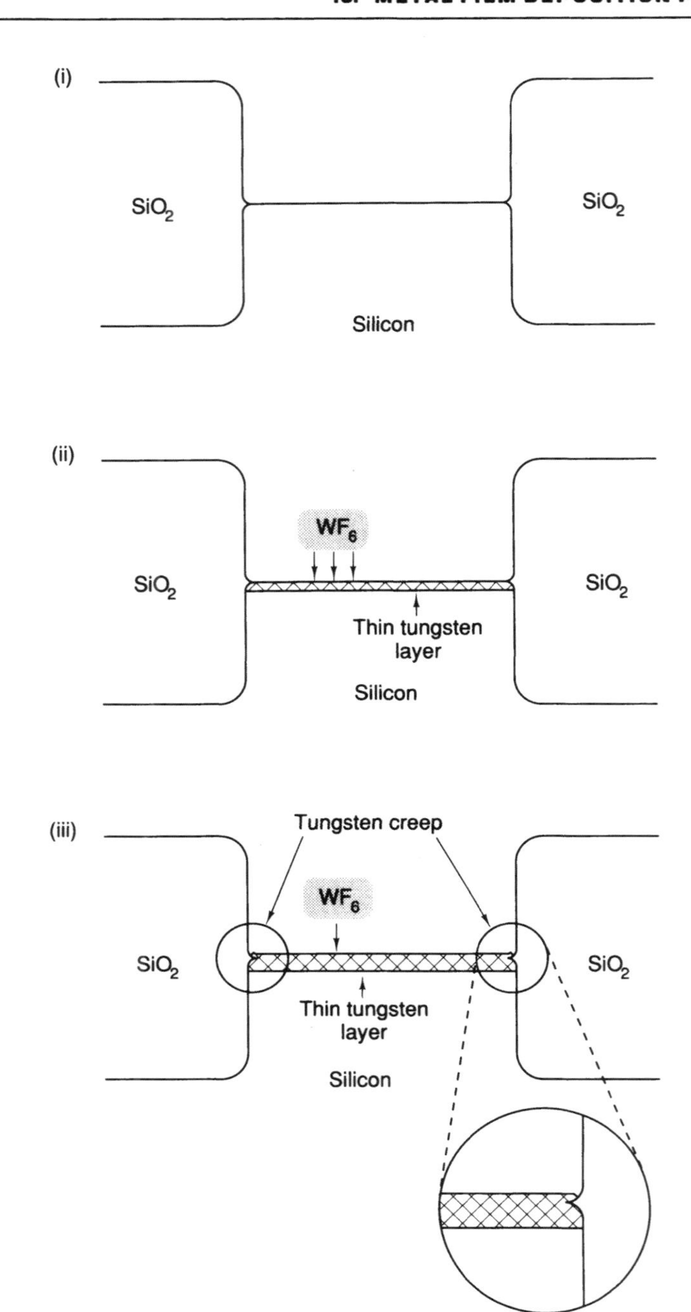

Figure 10-13. Tungsten Creep over Dielectric Film During the early stages of tungsten deposition, the tungsten can be deposited over the top of some dielectric structures. For example, a close look at tungsten deposited in a contact bounded by field oxide (with bird's beak structure) shows that the tungsten has deposited over the bird's beak. This is defined as tungsten creep.

variables can be adjusted to affect the selectivity of the process. For instance, selectivity improves as the process temperature is decreased; the total pressure of the system, as well as the partial pressure of the WF_6, is reduced; and as the total time in the reaction chamber is reduced. Metal films, films of polysilicon, silicon substrates, metal silicides, and other tungsten films provide excellent nucleation sites, whereas the surfaces of pure silicon dioxide and silicon nitride provide very few opportunities for nucleation. However, nucleation sites can occur on dielectrics, especially if the film is contaminated. Moisture, chemical residue, particles, and other defects can create nucleation sites upon which the metal can start to grow. It will then spread over the surface of the die until serious damage has occurred to the structures affected. As a result, the selective tungsten process area must be kept absolutely clean. Other factors which can affect this selectivity breakdown include the surface structure of the film. For instance, the roughened surface of a CVD oxide provides more sites for nucleation than the smooth surface of a thermally grown oxide layer.

The tungsten is then quickly deposited, which can take a few minutes, based on a typical deposition rates in the few hundreds of angstroms per minute range. The tungsten via plug will then be used as part of the planarized base for the next level of metallization, as shown in Figure 10-14. After these various films have been deposited, the whole structure is alloyed (sintered) for about 30 minutes at around 400°C to ensure excellent conductivity between the metal films. Clearly, this same pattern of planarization can be repeated for the next layer of metal, so that, in theory, many metal layers could be produced.

One of the reasons it is necessary to go through all this effort is that aluminum is used for most of the blanket metal deposition processes. The film itself exhibits poor step coverage on steep terrain, depositing in such a way that voids can occur easily, and in such ways that the film stress can be excessive and can contribute to film cracking. When considering the problems involved with electromigration and low current-carrying capacity, it surprising that aluminum is still in such widespread use. The reason for this is actually quite simple. All of the other films exhibit technical problems which limit their usefulness and the reliability of the chips.

For instance, the hydrogen reduction of tungsten can be used to create a broad-area blanket deposition over the entire surface of the wafer. In fact, that process results in films of excellent conformality, with good step coverage, low stress, no hillock formation, electromigration some two orders of magnitude lower than aluminum, and very low likelihood of film cracking. This makes the film sound like a good substitute for aluminum. However, the feature that makes tungsten so practical for selective tungsten deposition works against chip manufacturers; the tungsten does not adhere to the dielectric films and thus cannot be patterned easily. The adhesion of this film to the dielectric

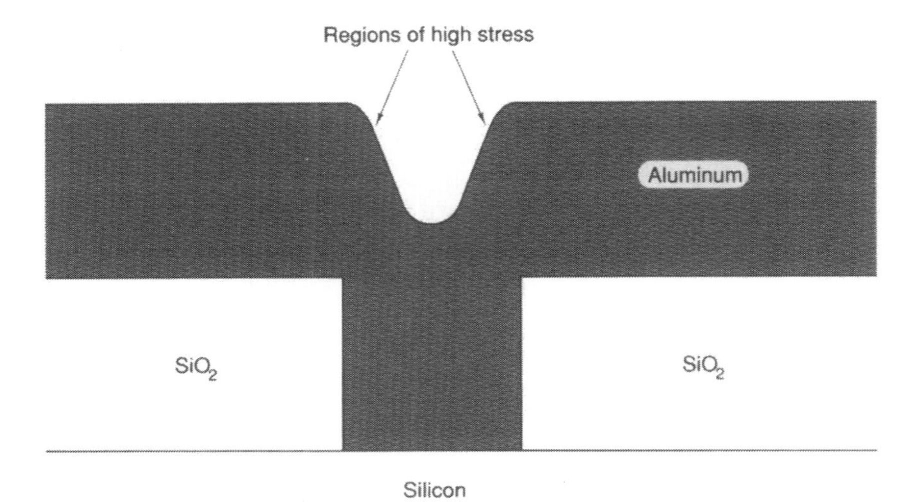

(i) Nonplanarized metal via fill

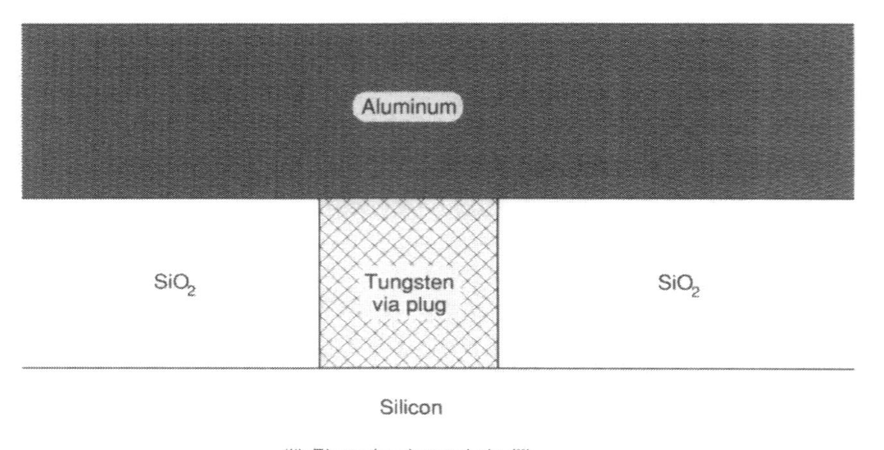

(ii) Planarized metal via fill

Figure 10-14. Planarizing Metal Layers When films are deposited over steps, stress points are developed at the edge of the steps. When metal lines are involved, these stress points lead to cracking, increased electromigration, and other problems. To eliminate this stress, the contact holes and vias can be filled with other metals (tungsten is often selected). By carefully processing the wafer in this way, the aluminum film is deposited on a nearly flat surface and has significant reductions in stress/step-related failures.

surface can be improved with the deposition of a thin layer of molybdenum on the dielectric films, but this adds a significant amount of complexity to the process. Additional problems that can occur in tungsten films include degradation through time due to hot carrier injection, mobile charge entrapment, and so on. Tungsten has not been viewed as a reliable alternative to aluminum.

Similar problems exist on other films, although in some cases the parameters of the films appear better suited for semiconductor applications. In many cases, the processes are either too new or too expensive to have been used in the wafer fab. In many cases, there are still technical problems such as the adhesion problems that limit the implementation of various films. A promising alternative metal film is titanium or titanium silicide. This film has higher current-carrying densities, can handle a high processing temperature, and is manufactured with enough silicon that it can be "oxidized" to form a film of Ti-SiO_2, which is dielectric and quite strong. Some of the problems involved with the use of titanium include its extreme hardness, difficulty in etching the films, and reduced conductivity.

There are a number of parameters that must be studied and recorded on process control charts as a result of metallization processes. These parameters will help the process engineer determine what kind of problems related to metallization processes are occurring when chips fail. For instance, a contact resistance problem may be the result of an insufficient amount of silicon in the alloy, or could be due to oxide in the contact area immediately prior to the metal deposition step.

One of the main parameters that is checked is the reflectivity of the aluminum. This is used as a rough measure of the purity of the film. A pure aluminum film will be extremely shiny, while the addition of silicon will make the film somewhat less shiny, although it is not really "cloudy" in appearance. An excess amount of silicon or contamination from a leak or other source can result in a truly cloudy film. These issues show up readily in any type of contamination problem. There is no accurate standard for the "correct" reflectivity for a process. Usually, the reflectivity is characterized by comparison of these readings against the yields of wafers of known reflectivity readings. Reflectivity at certain wavelengths can sometimes be used to measure the approximate thickness of a metal film, although this measure is not very accurate.

Another test that is performed is a thickness/total resistivity test. This is generally performed by placing the wafer into a magnetic field and observing the eddy currents that occur as the magnetic field is adjusted. The intensity of the eddy currents will be a function of the total resistance of the film in the local region, which is associated with total film thickness, based on the expectation that the metal film will have a constant resistivity for all types. This may not always be exactly true, but is usually close enough for most applications.

A final test may be performed to guarantee the uniformity of the sputtering targets. The aluminum film is deposited onto the surface of a test wafer with oxide on it. The wafer is then placed in sulfuric acid, which attacks the aluminum but has no effect on silicon. As a result, the silicon will precipitate out of solution with the aluminum into small nodules that can then be observed on the wafer surface. If the target is in good condition, the silicon nodule distribution should be fairly uniform on the wafer. If it is not, the test should be repeated and, if the results remain, the target should be studied further prior to use on production wafers. X-ray fluorescence tests can also be performed to obtain silicon concentration values for maintaining process control.

Clearly, metallization processes are critical to the operation of the integrated circuits, since most of the current moving through the chips passes through the metal layers. Changes in the films can result in chips that run too slowly, or too hot, or do not operate at all. The problems associated with the development of submicron geometries has made the metal line issue even more critical than previously assumed. Reduction of the line sizes has pushed the current-handling capabilities of aluminum lines to their limits. This opens the door for development of other processes, including both pure metals, and for metal silicides. We can be assured that within a few years aluminum will not be the metal of choice for most high-density IC applications, and that one of the many alternative metals or metal silicide films will be used instead. At this time, however, there is no clear leader in this competition.

10.3 WRAP-UP OF THE METALLIZATION PROCESSES

In this chapter, we have covered a variety of the metallization processes that are available for the manufacture of integrated circuits. Typically, the films with the lowest resistivity, highest strength and temperature handling characteristics, and the most stability and immunity from problems such as electromigration are the most desirable for IC devices. Unfortunately, with most of the films available today there are trade-offs that must be evaluated when selecting a particular film or film combination.

While most devices in production today use aluminum and its alloys as the primary conductive film, we are seeing a more widespread acceptance of other films in the industry. The issue of contact reliability has led to a number of complex metallization schemes, requiring the use of multiple thin layers of metal on each contact, to provide a continuous and high-speed joint through which electrons can flow easily.

The importance of attention to detail during manufacture is as critical in the metallization processes as in any of the other processes. Contamination can lead to many significant problems, from shorting the devices out, to reducing the reliability of the chips. Contacts can be closed, poor contacts can be formed, film resistivity may not be uniform across the circuit, or hillocks may have formed, with associated voids, increasing the probabilities of electromigration. These factors must all be monitored and controlled in order to produce metal films of exceptionally high quality.

CHAPTER 11

INTEGRATED CIRCUIT TESTING AND ANALYSIS

As we have gone through the various processes that make up the integrated circuit manufacturing cycle, we have referred many times to the testing of the wafers. Clearly, since the manufacturing process is so complex and defects can form at so many points in it, the wafer manufacturer must test the parts thoroughly and often so that problems can be identified quickly. It is also of paramount importance that the personnel who are making the decisions required in manufacturing wafers have at least a rudimentary understanding of the test and sort aspects of the process. If this part of the puzzle is not clearly understood, the decision-making process will be hampered and the possibility of yield-reducing failures increases. In fact, since there are so many processes performed on the wafers, the probability of making good wafers decreases rapidly if the quality of the product dips even slightly. Looking at Table 11-1, we can see that, if it takes 140 process steps to manufacture an integrated circuit, even very small levels of defects uniformly distributed through the process can result in very drastic losses in fab output.

Due to the additional requirements placed on the manufacturing process, and as integrated circuits become more complex, circuit testing costs have risen to the point where they are as high as the actual manufacturing costs. The testing problems have been reduced somewhat by building some self-diagnosing and self-correcting features into the chips, but the part must nevertheless be tested properly in order to verify all results.

There are a variety of tests that are performed on the wafers at nearly every step, from mechanical tests, such as film stress or wafer

TABLE 11-1
Impact of Defects on Die Yields[a]

Step	Cumulative defects A	Cumulative defects B	Cumulative yield loss A	Cumulative yield loss B	Die left A	Die left B
0	0	0	0	0	10000	10000
10	270	100	54	20	9946	9980
20	513	200	103	40	9897	9960
30	810	300	162	60	9838	9940
40	1053	400	211	80	9789	9920
50	1350	500	265	100	9735	9900
60	1593	600	319	120	9681	9880
70	1863	700	373	140	9627	9860
80	2133	800	427	160	9573	9840
90	2403	900	481	180	9519	9820
100	2673	1000	535	200	9465	9800
110	2943	1100	589	220	9411	9780
120	3213	1200	643	240	9357	9760
130	3483	1300	697	260	9303	9740
140	3753	1400	751	280	9249	9720[b]

[a] Column A assumes 27 defects per process (3σ for 10,000 ICs); assumes one die lost per every five defects. Column B assumes 10 defects per process; assumes one die lost per five defects.
[b] Net die gained with improvement in yield: 471 or 4.71%.

bowing, to parametric tests such as film thickness, resistivity, and particle levels. These tests can describe the quality of the manufacturing process, but cannot really tell the quality of the integrated circuits themselves. In some cases, the tests will be very revealing about the quality of the films produced in a certain reactor or furnace, such as the C–V tests; however, in most cases, this "luxury" does not exist. For example, there is little point in running test pattern wafers in a photolithography room, since the wafer's real topography will usually be much more complex. As a result, the process parametric results are only indicators that the process itself has not changed or drifted. There is no indication as to whether yield or reliability of any specific wafer has been affected.

Thus, a wide variety of electrical tests are performed that can tell the yield enhancement and product engineers what kind of problems are being seen on the wafers. These electrical tests are performed at a number of locations, both during the processing cycle and after the cycle is complete and the chips are being sorted. There are also tests that can be performed on low-yielding chips that can provide insight into the failure modes of those specific parts. The results of these electrical tests must be correlated to the results of the in-line fab tests to permit the manufacture of high-quality integrated circuits. This chapter, then, will deal with the variety of tests that are performed on the wafers during and after the manufacturing cycle to ensure that yield and reliability are maintained.

11.1 IN-LINE NONELECTRICAL TESTS

We will start by covering the in-line fab quality tests. In some cases, we may have covered the tests in other chapters, but we will review a variety of techniques here. The main parameters we will cover are film thickness, linewidth, particle contamination level, film and substrate resistivity, and film stress tests. We will also briefly cover the optical inspection techniques that are often used in controlling wafer quality. The tests that are used in the fab area are listed in Table 11-2.

11.1.1 Film Thickness Measurements

We will start by discussing the measurement of the thicknesses of the films that are created through growth or deposition. The first consideration is whether the film is transparent or opaque. If it is transparent,

TABLE 11-2
In-line Process Metrology Equipment

Equipment	Technique	Parameter monitored
Particle Counter	Laser/white light	Particles, haze, scratches, stress
Film thickness analyzers		
	Spectrophotometer Ellipsometer Prism coupler Acoustooptic Profilometer	Film thickness, refractive index, optical constants
	Eddy current (for metal films only)	
Linewidth analyzers		
	Optical SEM	Linewidth, pitch, registration
Film stress reflectance test	Interferometry	Wafer bowing translated to stress
Inspection	Optical	Visual inspection for defects
	Electronic/optical	Automated defect analysis
	SEM	High-resolution defect inspections
Resistivity	4-point probe IR spectral analysis	Dopant concentration, resistivity

one of a number of optical techniques can be used. If the film is opaque (which is usually the case with metal films), resistive, magnetic, X-ray fluorescence, or infrared absorption techniques must be used. Usually, the thickness of a film must be controlled within 3 to 5%, although some structures, particularly electron tunneling devices (such as EEPROMs), may require tighter tolerances. This control must extend to all points on each wafer, and to all of the wafers that are in a lot.

Most transparent films in the range of 500 to about 50,000 Å can be measured without much difficulty using spectrophotometric methods. As long as the various optical parameters of the film are known or can be calculated (such as refractive index, absorption coefficient, Cauchy coefficients), the interference pattern of reflected light can be used to accurately measure the thickness. This technique will also work on a film that has low levels of absorption, up to a certain level also. Figure 11-1 shows a typical spectral curve for a few-thousand angstrom SiO_2 film. Films that are under 500 Å in thickness become nearly completely reflective and do not return these "nice" curves. As a result, the ability to measure the thinnest films used in the IC industry (80 to 100 Å) is severely limited using these techniques, although recent advances have improved the ability of these systems to accurately and repeatably measure these thin films. Spectrophotometry is the most common method of film thickness measurement for the dielectric, photoresist, and other clear films.

Another technique used to measure thin films is called ellipsometry, and is based on the fact that the polarization angle of light will change as it deflects through a film. This is shown in Figure 11-2. In the ellipsometer, a laser beam is focused onto the surface of the wafer at an angle of around 70°. It is shone through a polarizing filter that produces elliptically polarized light. As this polarizer is rotated, a maximum in reflectivity will be observed. At this point another filter, this time a plane polarizing filter, is rotated in front of a photodetector. When the maximum intensity of this light is found, the polarization angle can be determined and a calculation made which accurately describes the thickness and index of refraction of the film. This device is often used to measure the index of refraction, except for very thin films, which are very difficult to determine. The ellipsometer has extremely good precision to a thickness of under 20 Å, if the index of refraction is known.

Another method for measuring the film thickness is similar to the spectrophotometric technique, in that white light is shone through several narrow bandpass filters and the amount of light reflected from the surface measured (see Figure 11-3). This can produce very repeatable results, but suffers from its inability to discern between different film thicknesses that have similar optical properties at the wavelengths chosen. This results in multiple solutions and difficulty in result interpretation.

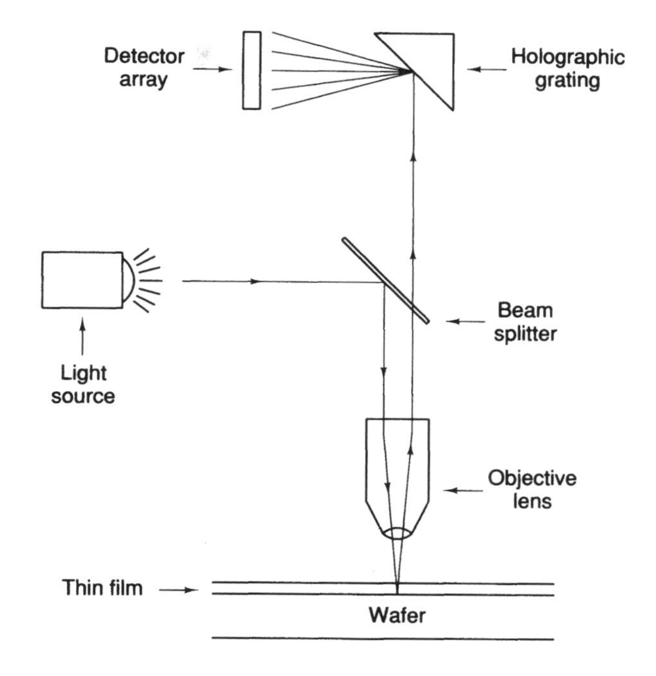

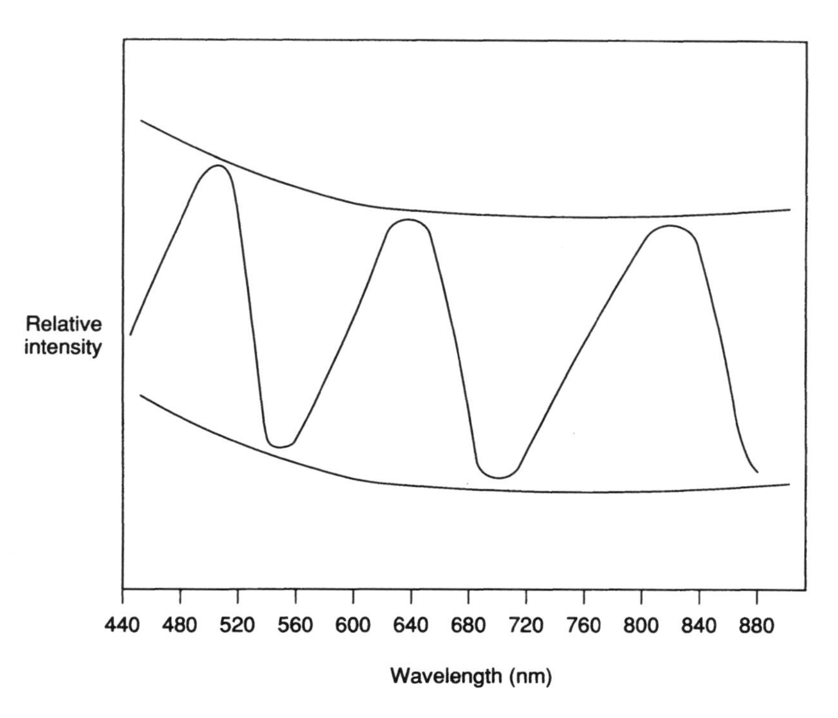

Figure 11-1. Spectrophotometric Technique for Measuring Film Thickness In this system, white light is focused onto the wafer surface, where it is reflected back in a characteristic manner. This reflected light strikes a grating which breaks the light up into its spectrum. A typical spectrum is shown for a film of several thousand angstroms. The skew of the spectrum at shorter wavelengths is determined by the index of refraction of the film.

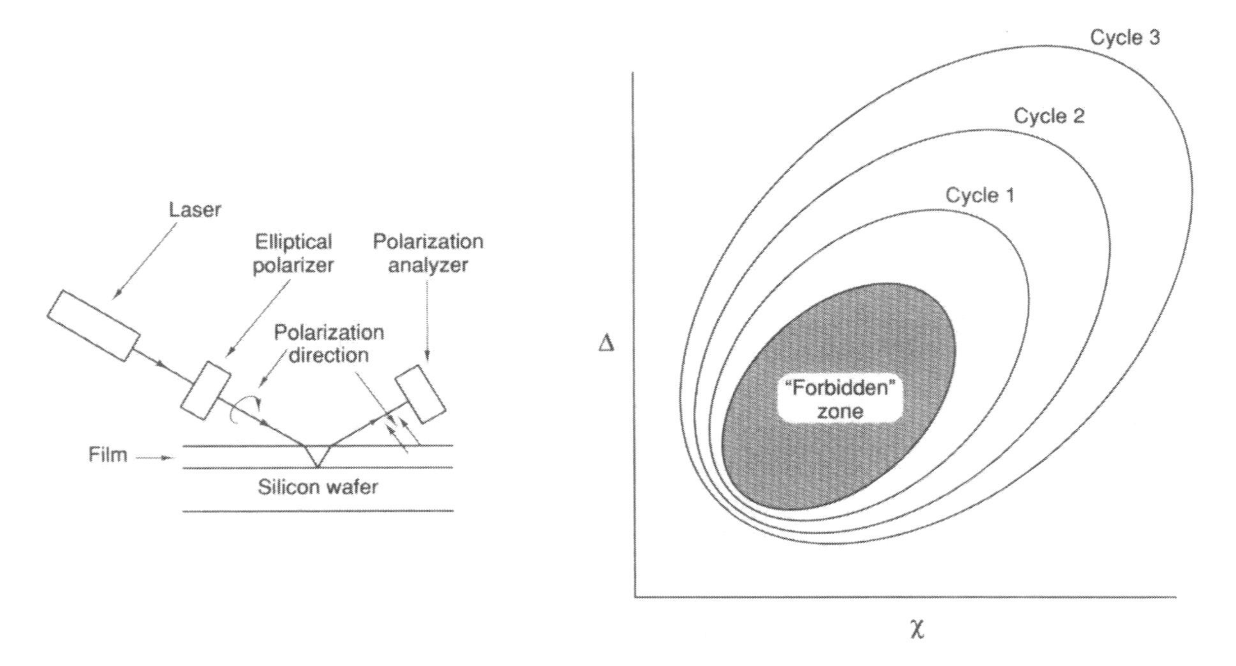

Figure 11-2. Ellipsometric Technique for Measuring Film Thickness The ellipsometer uses the change in polarization angle of an incident elliptically polarized laser beam. The elliptically polarized light will change to plane-polarized light at an angle that is determined by thickness and refractive index. The thickness is determined by analysis of the two constants delta and phi. These values travel in cycles around a "forbidden" zone, so that every 2000 in. or so in increasing thickness, the cycles approach one another closely, and the accuracy of the measurement is reduced.

A final method for measuring thicker films (over 2500 Å) is called the prism coupler. In this device, a prism is mechanically held to the surface of the wafer and a laser is shone through the prism, as shown in Figure 11-4. The amount of deflection of the laser gives the instrument the ability to measure the film parameters very accurately, especially index of refraction. Since these devices come into direct contact with the wafer surface, they are limited in application to test wafers. In addition, this technique is limited to films above 0.25 μm (2500 Å) thick.

When the films to be measured are opaque, the main methods of measurement are electrical, primarily because the opaque films currently used in the semiconductor industry are usually conductive and easily tested. In some of these devices, an oscillating electromagnetic field is held near the wafer and eddy currents within the film, which can indicate the thickness, are measured. Probing carefully designed structures can also indicate metal thickness, although these techniques may be too complex or time-consuming for manufacturing work.

Another method of measuring the opaque wafers utilizes the amount of infrared absorption or X-ray fluorescence. In both cases, a

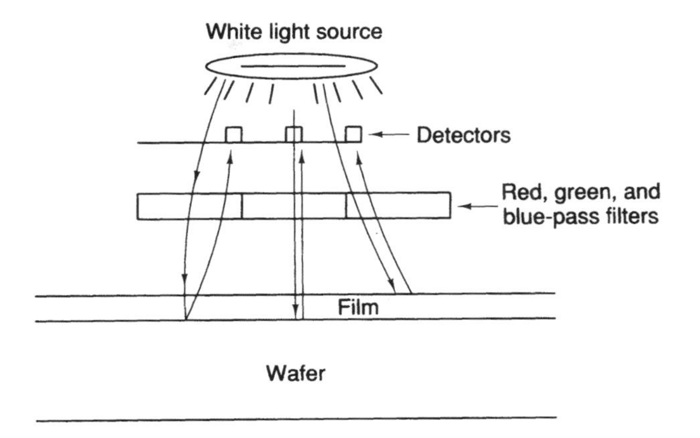

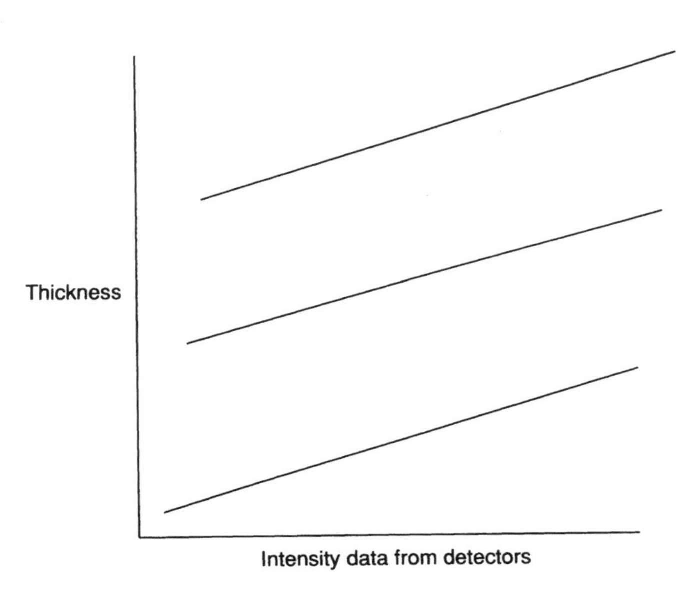

Figure 11-3. Modified Spectral Technique for Measuring Films In this rapid, large-area measurement technique, films are analyzed by observing the results of calculations performed on the reflected data. The technique can result in some problems with order determination, since spectral response for many is nearly duplicated at different thicknesses.

source is placed in front of the detector and energy of a known wavelength is directed onto the wafer. Any absorption or retransmission is picked up by the detectors, which are calibrated for the film. These devices work well and produce repeatable readings, but do not have especially good resolution (in other words, they may not be able to detect a change of a few angstroms).

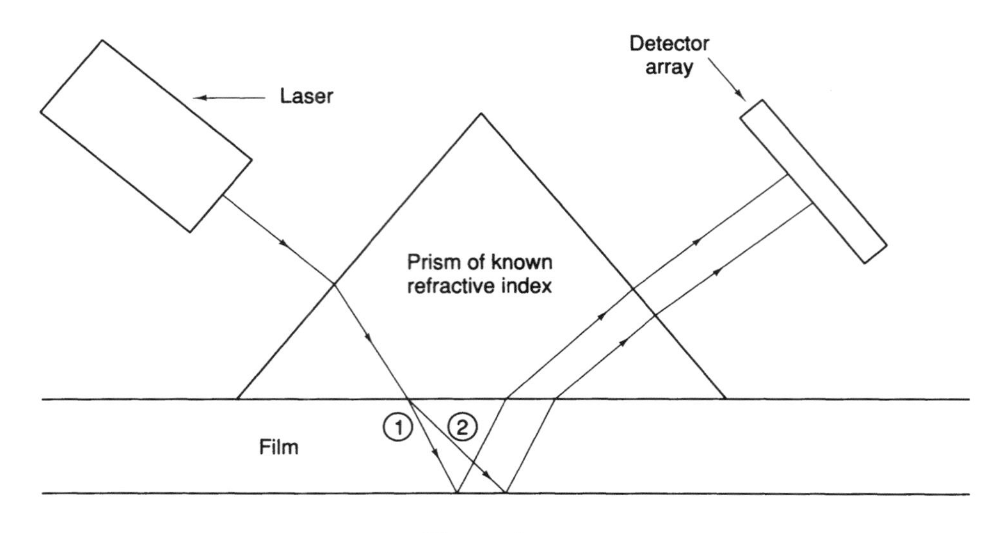

Figure 11-4. Prism Coupling for Refractive Index Measurement The prism coupler is an excellent method for determining the index of refraction of films over 2500 Å. It works on the simple principle that changes in refractive index cause light to bend in varying, predictable ways. The index of refraction can then be measured through simple, geometrical techniques. The disadvantage of this technique is that the current generation of tools require direct contact between the prism and the wafer surface.

11.1.2 Linewidth Measurement

One of the most important parameters to measure on a wafer is the size of the various lines, spaces, contact holes, and other component parts of the circuit. These parameters are called critical dimensions and are set for each layer of each device. The success of a photolithography group is often a measure of the ability to produce and measure the very narrow lines in the design. There are two basic techniques, one optical and the other a scanning electron microscopy technique. The techniques are somewhat similar in concept, and have a number of implementations. In both cases, the system shines a source (photons or electrons) onto the surface of the wafer and measures the reflectivity (or re-emissivity) of the surface at a number of points across the line (see Figure 11-5). The reflected intensity is then plotted as a function of position. This data is then analyzed for calculation of the linewidth.

There are three primary methods of optical linewidth measurement. In the first case, a white light is shone onto the surface of the wafer and the image is projected onto a slit that is scanned across the image, as depicted in Figure 11-6. A detector is held behind the slit, and the intensity of the light is measured at regular intervals as the slit is scanned across the image. This data is then converted into the intensity profile. It is up to the engineer to decide where the line should be measured on the slope of the intensity curve. A threshold of

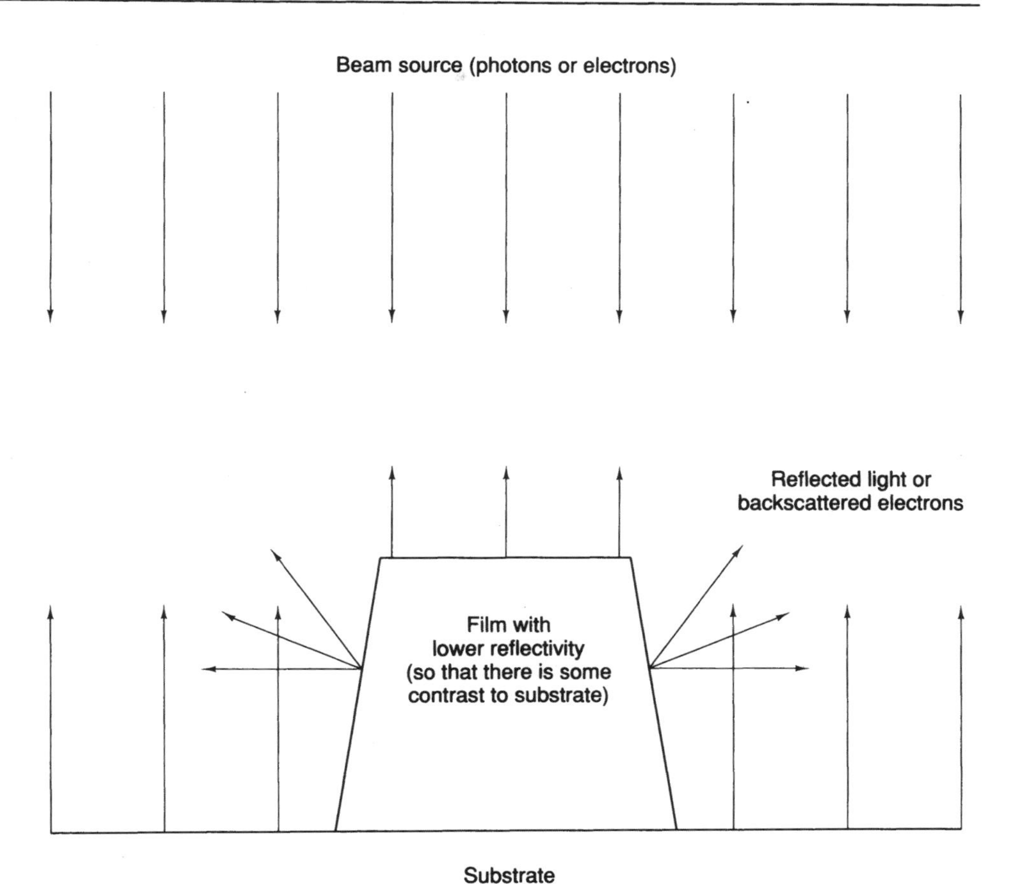

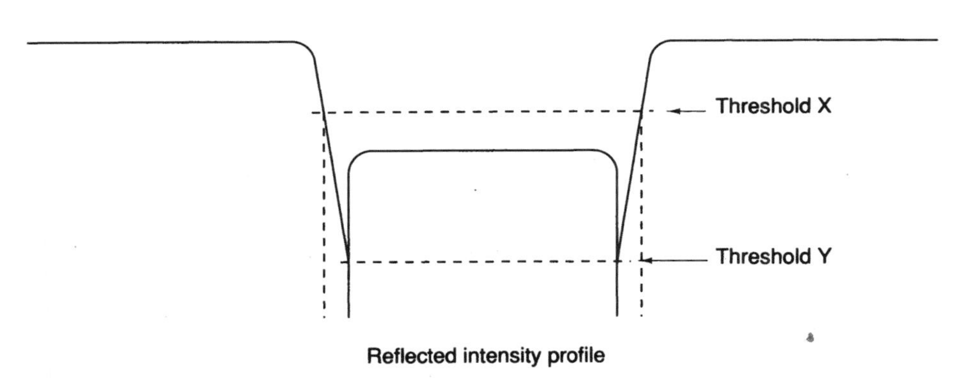

Figure 11-5. Intensity Profile Used for Calculating Linewidths When a line of some material is produced on a substrate, light will reflect off of the material and off of its edges in predictable ways. The intensity of this reflection can be observed to detect and intensity profile. The linewidth is then determined by selecting two points on the profile (called the threshold). As we can see, different threshold selections can produce slightly different results.

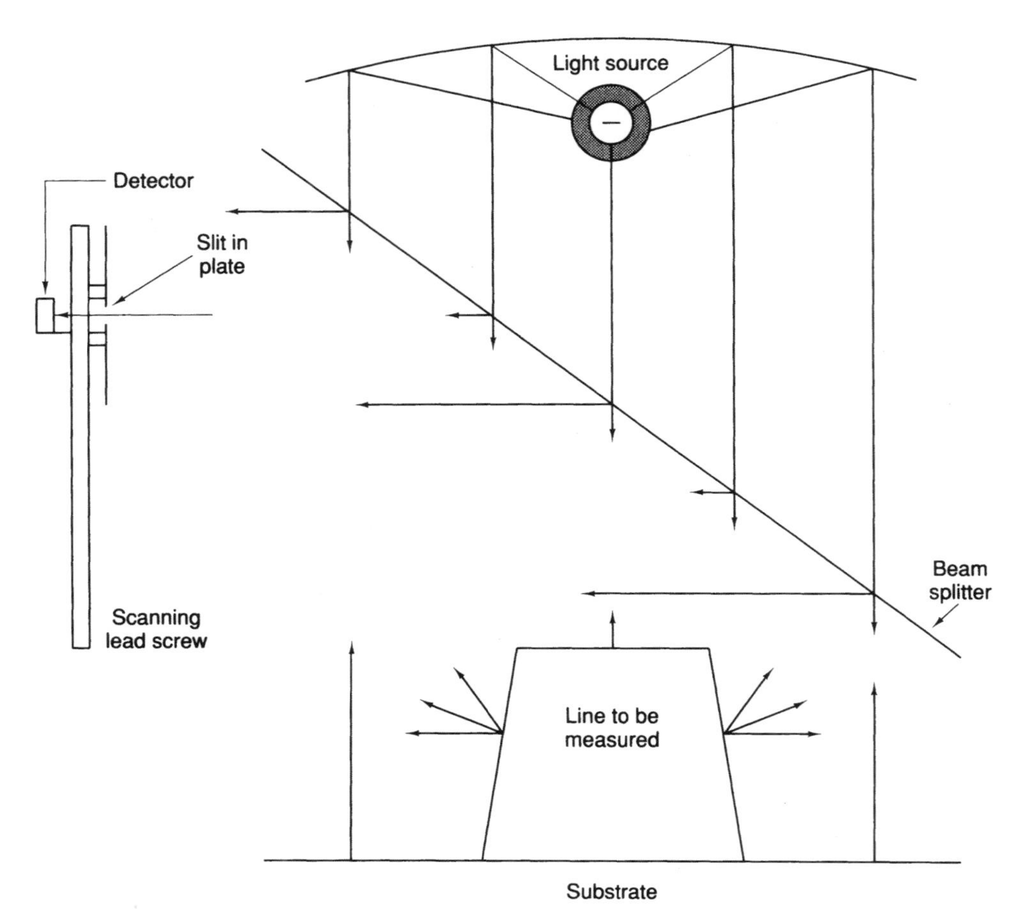

Figure 11-6. The Scanning Slit Profilometer In this device, the line intensity profile is measured by observing the intensity of the reflected light behind a slit that is positioned and moved so that the full image is recorded. The profile is analyzed in the same manner as in 11-5.

50% will measure farther up the slope than will a threshold of 35%. Whatever value is used, it must be adhered to after the process has been characterized.

The second method is similar to the first, except that a scanning slit is not used. Instead, a CCD array or camera is placed at the focal point of the image and each element of the array is examined for intensity. From this an intensity profile is generated and the linewidth calculated using the same threshold requirements as in the first optical linewidth system. To reduce the effects of linewidth variation on measurement precision, the intensity profile is observed for a number of rows on the CCD array. This is described in Figure 11-7, and will be discussed in more detail when we discuss SEM linewidth measurement.

The third method of linewidth measurement is the use of a laser in place of a white light to obtain the optical intensity profile. Other-

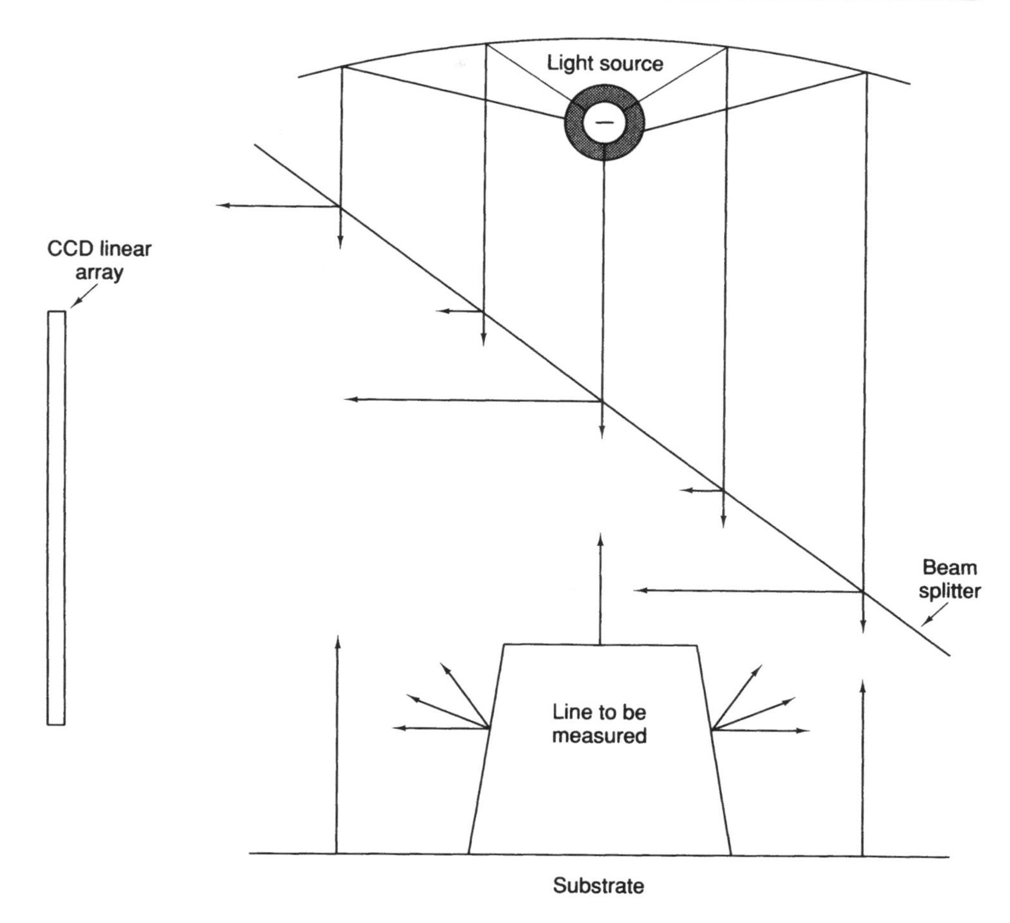

Figure 11-7. The Linear Array Profilometer The advent of modern CCD image detection has permitted the construction of a solid state profilometer. There is no significant difference between the types of profilometers, as long as all other factors, such as focus control, lamp intensity, and so on, are kept constant.

wise, the techniques for measuring the lines are similar to those used in the other optical systems. The laser systems have better submicron feature measurement capability than the white light, but are not truly competitive to the SEM systems that are currently in fashion for this measurement.

The optical linewidth systems work well for linewidths from about 0.75 μm and up. Below this width, the wavelengths of the light used becomes comparable to that of the linewidths and interference patterns become apparent that make actual measurement quite difficult. The laser-based systems are accurate to approximately 0.4 μm, due mostly to the monochromaticity of the laser light.

Typically, narrow linewidths (under 1.0 μm) are measured with SEM techniques. The wafer is placed in the SEM chamber, aligned to the correct measurement location, and the surface is exposed to a stream of electrons. Typically, a low-voltage source is used so that total electron acceleration is around 1000 volts. This low voltage allows the use of secondary electrons (as opposed to the backscattered electrons, which are the electrons observed in most high-voltage general purpose SEMs). These electrons are produced from a much shallower region of the surface, therefore producing an excellent image of the material's top surface. This image is then examined and the image intensity profiles are displayed and the linewidths calculated.

Since the lines on a wafer will not be perfectly smooth, there will be a problem measuring the width of the line if it is very rough. This is a more significant problem with SEM and CCD array technologies than with the scanning slit method, due to resolution of the array or electron beam. This phenomenon is shown in Figure 11-8, where A is the measure of linewidth at a wide point in the line and B is the measure of the linewidth at the narrow point of the line. As a result, a number of linewidth test instrument manufacturers have developed algorithms that average the readings across a larger part of the line. This average will be a good measure of the effective width of the line, while the standard deviation can be used to gauge the uniformity of the etch process.

11.1.3 Resistivity Tests

The films deposited onto the wafers that will be used as resistors or conductors must be tested for film resistivity. This is also true for any substrate doping that may occur. If the films or substrate regions are improperly doped, they will not operate correctly. It is important to determine this problem immediately in order to fix it as soon as possible. There are currently two main tests in use for measuring resistivity.

The first of these tests is called the four-point probe, and is shown in Figure 11-9. In this device, four needlelike probes are placed onto the wafer surface (since this procedure requires contact, it is performed on test wafers only, not on product wafers). Known quantities of current are pushed with known voltages through the outermost probes. Observing the effects on the inner two probes allows the system to determine the resistivity of the layer. Readings on this device can be easily skewed due to damaged probe tips, penetration of the film layer, and so on. These devices are used mostly to test $POCl_3$ doping of silicon and polysilicon, and can also be used to test implant doping levels. It is also possible to measure metallic films with this technique, but it is discouraged.

Some of the metal film thickness devices mentioned in the previous section can be used to measure resistivity of metal films. Metal films will usually exhibit consistent resistivity, as long as proper process conditions are maintained, so that this measurement may not be as important as the measurement of doping levels in the silicon resistors, and so on.

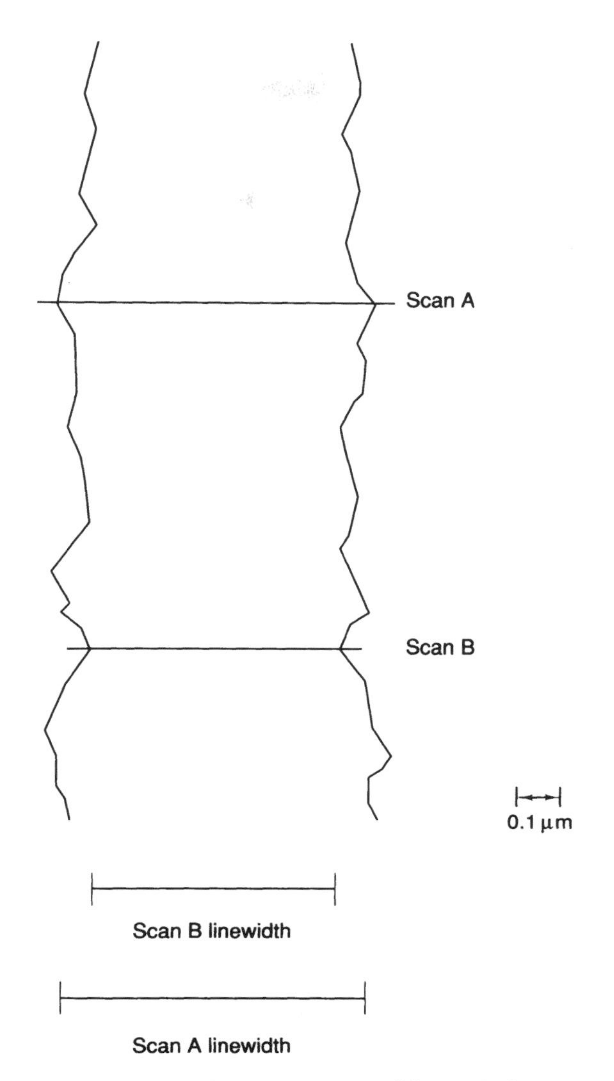

Figure 11-8. Electron Beam Width versus Linewidth Variation As seen here, the high resolution of the electron beam can result in variations in readings. Scan A has a width of 0.68% while scan B is 0.56%, a 0.12% difference. Scan-averaging techniques can be used to produce a more accurate 0.60%.

One of the newer methods for measuring the dopant levels of a wafer after implantation uses the absorption and re-emission characteristics of the dopants, shining an IR laser into the silicon and observing the returned spectra. They give excellent results, and can be used with a wafer immediately after implantation. (Use of the four-point probe at the implant operation requires that the test wafers be annealed prior to measurement. Often a small furnace or rapid thermal anneal system must be supplied for to anneal test wafers.)

It should be noted that there a number of techniques for measuring resistivity of product wafers. For instance, it may be necessary to

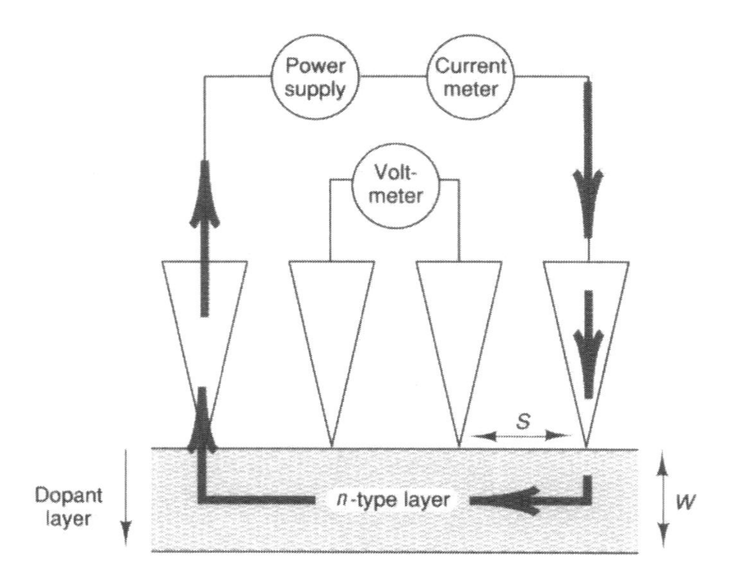

Figure 11-9. Sheet Resistance Measurement with Four-Point Probe The four-point probe is a simple device that measures the voltage drop across a fixed distance. By simultaneously measuring the current across the circuit, the resistivity can be determined by the equation: $R = 2\pi S(V/I)$.

measure a product wafer for verification of a test wafer result. Although a wafer is lost, there are often test wafer problems (most often mixing test wafers, and using them twice) that mask the true nature of the product wafers. You never want to scrap wafers that are good, so it always pays to experiment with a few wafers if a lot is suspected of being so far out of spec that it will be scrapped. In any event, it is possible to get product wafer readings by measuring the backs of the wafers. It is important to note that the readings will probably not match those of the test wafers. As a result, it is recommended that each type of wafer be tested to verify what readings to expect when the product is within the specification limits. Usually, polysilicon films will read higher in resistivity than single-crystal silicon, and will often be the film exposed on the back of the wafer.

11.1.4 Particle Counts

We will not spend much time on particle counters here, since we have covered that in detail already. In most cases, test wafers are the only items tested directly for particles, although a number of devices have recently appeared on the market that can detect and report particle counts even on patterned production wafers. This is done using holographic techniques, and appears to work in at least most of the key applications (after CVD processes), although these devices are relatively slow. Most of the particle counters for wafers are fashioned after

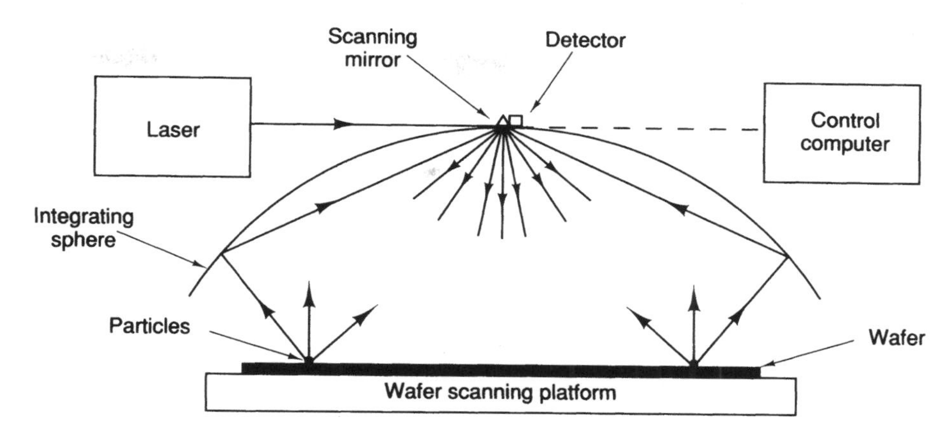

Figure 11-10. Wafer Surface-Particle Counter In this device, a laser beam scans the wafer surface. Particles on the surface will create specular reflections which are detected and analyzed. The size of the particles and the extent of haze on the wafers can be determined by this method.

Figure 11-10. In these systems, a laser is reflected off the surface of the wafer and any unidentified specular reflection is identified as a defect.

Particle counts on wafers are performed after the following steps: all LPCVD operations (nitride, polysilicon, oxides, etc.), metallization processes, photoresist deposition processes, diffusion processes, and etch processes. Particles are sometimes counted after cleans and other steps, also. Obviously, since particle defects are critical to the yield of the devices, the ability to accurately discern the smallest particles is extremely important.

11.1.5 Film Stress Testing

We have touched on this subject a number of times, so we will limit this discussion to the basics. The test itself is important since films that are overstressed can crack, and can contribute to the degeneration of other films on the semiconductor wafer. Stress testing is performed in a number of ways. However, all of the methods are essentially the same, that is, to measure the deflection or warpage of the wafers. This is performed optically, with two primary methods of implementation in production lines.

The first device, as shown in Figure 11-11, measures the deflection by placing a wafer on a platform. A bright, white light shines on the back of the wafer, and the amount of light reflected gauges the distance to the wafer. The wafer is then processed and the test repeated. If the wafer has been deflected, the amount of reflectivity will change. Obviously, if a film is deposited on the back of the wafer, or there are other changes to the reflectivity of the back, this device can give results that are skewed.

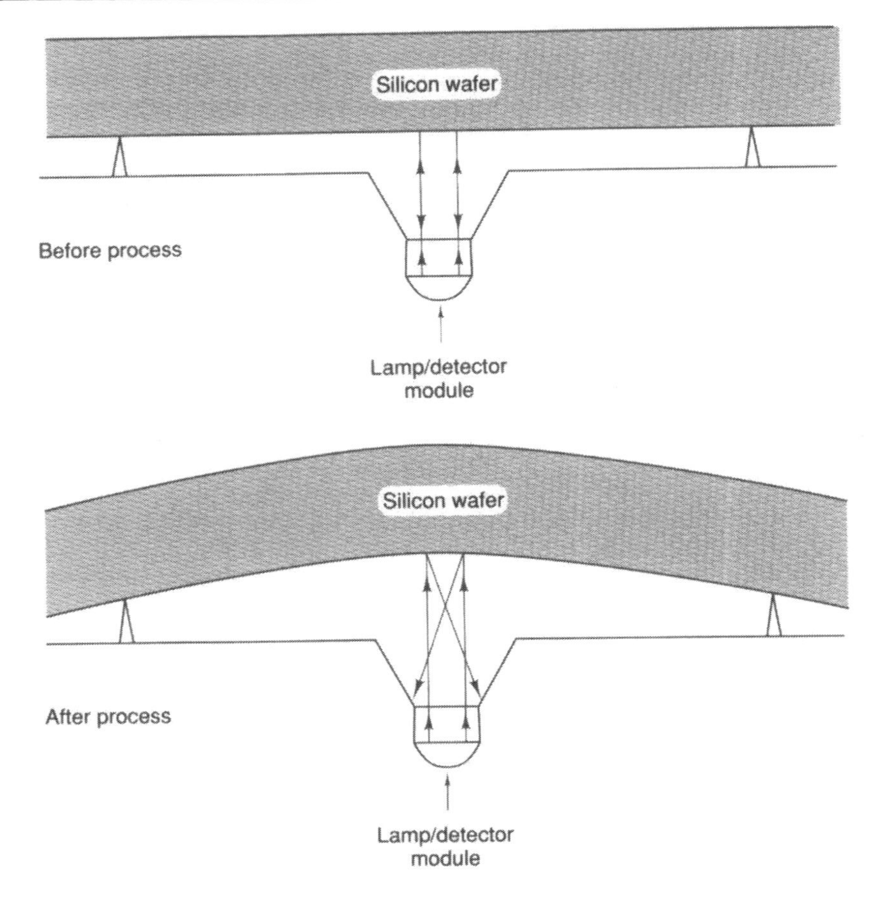

Figure 11-11. Reflective Methods for Determining Film Stress As can be seen here, a deflection of the wafer due to stresses induced in a wafer during processing cause a change in the amount of reflected light returned to the detector, as compared to the amount measured prior to processing.

Another instrument used to measure the deflection of the wafer is the interferometer. The interferometric technique is shown in Figure 11-12. When the interferometer is used, a warp on the wafer will be noted by interference fringes appearing on the wafer (or on its reflected image in a detector). The amount of warpage of the wafer is determined by counting the number of fringes. To actually run the test, the wafer must be examined before the process to determine the number of fringes already on it. After the process has been completed, the fringes are counted again, and the difference between these two values is used to determine the amount of change in the flatness of the wafer.

11.1.6 Optical Inspection Techniques

The last of the nonelectrical tests that we will discuss is the optical inspection. This is performed at a number of steps in the process in order to ascertain the number of defects on the actual production

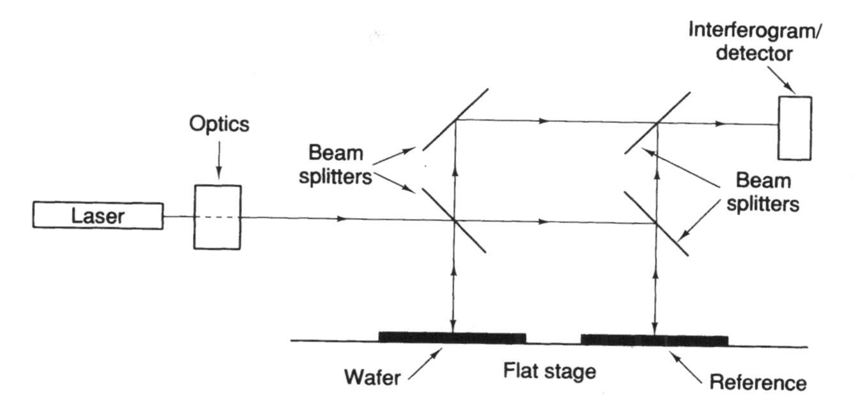

Figure 11-12. Block Diagram of Wafer Stress Interferometer In this simple block diagram of a wafer stress interferometer, we see that the laser light is shone on the wafer while being compared to a reference (which may be a stored computer image of the wafer prior to processing). When aligned, the interference pattern between the wafer and reference is eliminated. After processing, any wafer bowing will show up as a definite interference pattern. The amount of deflection can be determined by counting the number of fringes in the interferogram.

wafers. Variations of this technique are the oldest methods used to detect contamination on the wafers, so are the most characterized. While optical inspection gives the manufacturing area a certain real-time defect analysis and rejection capability that is difficult to program into a machine, there are still some drawbacks to the procedure including the extra operators required to perform the inspection, and the increase in the amount of handling that the wafers will receive. As a result, optical inspections are usually limited to the most critical operations, such as immediately following CVD processes or after etches. "Operator calibration", getting everyone to agree on what they are looking at, can be a significant issue.

One of the requirements for the optical inspection program is that a large enough area of the wafers is examined that a reasonable estimate can be made on the defect density of the wafers. This can make the inspection procedure difficult if the process produces defects in many relatively small areas of the wafer. A typical method is to scan the wafer continuously from top to bottom and then from side to side, as defined in Figure 11-13. Knowing the field of view of the microscope objective used and the size of the wafer make it easy to calculate the area of the wafer observed: Area = 2 × FOV (in mm) × wafer diameter (in mm). A maximum number of defects will be permitted before the wafers are held for further rework or disposal. Since the allowable particle sizes and wafer quality can vary greatly from fab to fab, there is no generally accepted standard for the number of particles allowed. Some fabs get nervous if two particles show up in this small area, while others are comfortable with 5 to 10 or more particles. This number can also vary from process to process. For instance, BPSG

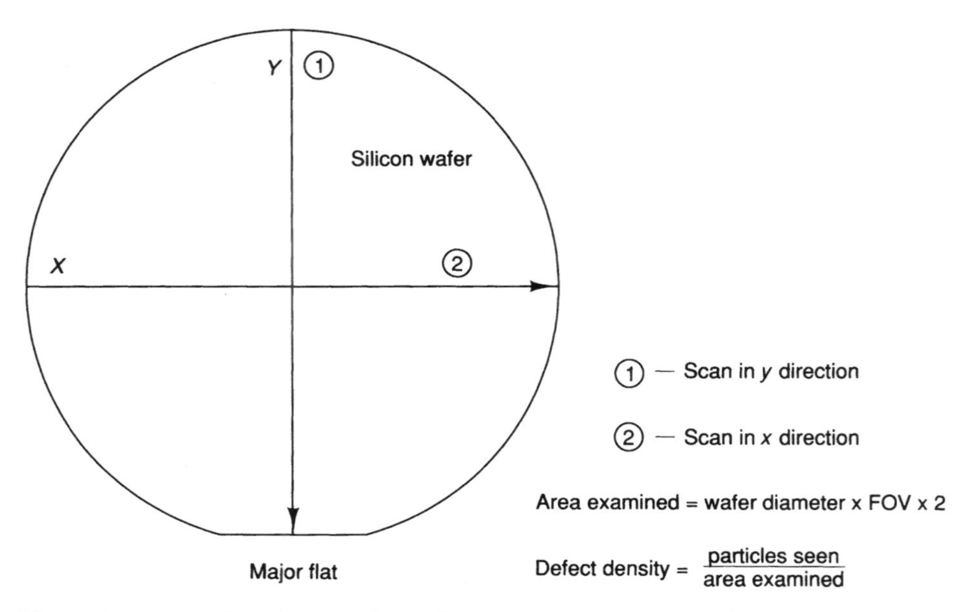

Figure 11-13. *x-y* Scan Pattern for Defect Density Estimation This standard pattern is often used for visual examination of wafers. Very few particles should be observed in this test, since the area observed is so small. For example, on a 150 mm wafer with a magnified field of view of 1 mm, only 3 cm^2 are seen out of a total area of 176 cm^2. Thus, each particle observed represents 0.33 defects/cm^2.

films are typically dirtier than polysilicon films, but at the same time do not need the high quality of the poly films.

Typically, inspections are performed on a microscope at a magnification of around 100× (total: 10× objective, 10× eyepiece). Lower powers will not give the magnification required to see the very small particles, while higher powers have restrictions in field of view, depth of focus, light-gathering ability, and so on. Magnifications beyond 1000× are not useful in an optical microscope. Inspections can be performed on SEM instruments also, although the speed of the inspection is reduced significantly. Electron microscopy defect scanning will not become useful until its throughput can at least approach that of an optical system, which permits experienced operators the ability to inspect a wafer in much less than a minute.

11.1.7 Nonelectrical Parametrics

We can see from the preceding discussion that there are numerous tests performed on the wafers during the manufacturing process. In most cases, the tests are performed on test wafers, although an increasingly large percentage of measurements are now taken directly from production wafers if at all possible. As a result, the test instrument vendors have attempted to come up with optical noncontact methods for obtaining test results. They have also attempted to automate their equipment to a much greater degree so that the operator interaction (and consequent wafer contamination) is minimized. The ability to produce wafers of the density and quality required is controlled to a great degree by the test instrumentation used, so that a significant amount of development will continue to occur in this field.

11.2 IN-LINE ELECTRICAL TESTS

After the wafers have had metallization layers deposited on them, it is possible to get the first true electrical results from the wafers themselves. Up to this point, the devices are too incomplete to test properly. There are a lot of tests that can be performed on the chips, unfortunately more than there is space for. As a result, there has been a significant of amount of clever work done to minimize the space requirements for testing, while increasing the amount of information that can be obtained from each individual test. As an example of the difficulties involved, we will first look at the test-pattern requirements. We will then examine a number of the tests that are most commonly performed in a wafer manufacturing operation. It would be nearly impossible to include every test that every manufacturer performs, even if it were legal to do so, since we can produce so many variations of these tests. For instance, any one particular type of device will have as many as 70 to 100 individual electrical tests that can be performed. A list of common electrical tests is given in Table 11-3.

TABLE 11-3
Representative Electrical Tests

Resistor tests	Diode tests
Poly resistors	Reverse bias breakdown
Substrate resistors	Forward bias breakdown
Aluminum/metal contacts	Diode leakage
Current-carrying capacity	Latch-up
Contact test	Transistor tests
Contact resistance	Speed
Poly to substrate	Gate threshold voltage
Poly to poly	Maximum current flow
Metal to substrate	Power consumption
Metal to poly	Channel resistance
Contact leakage	Assorted tests
Capacitance tests	Floating gate capacity
Poly cell capacitance	Program/erase time
Charge retention time	Short-cycle test
Interlayer capacitance	
Dielectric breakdown	
Field oxide	
Gate oxide	
Poly top and sidewall oxides	
Tunnel oxide	
Interlevel dielectric	
Passivation dielectrics	

11.2.1 Test Patterns

The vast majority of the electrical tests can be provided for off-chip, so that precious design space in the circuit area does not have to used. However, the structures that are developed to perform these tests should closely approximate the structures that are in use in the actual integrated circuit, so that results are comparable to the expected device performance. Almost all of the tests involve one or more contact points, so that full input and output provisions must be accounted for, including the design of a special probe assembly and support circuitry as required to operate and manipulate the test structures and link them to the outside world.

In the days before steppers, the electrical test patterns were simple to develop. Since the designer placed several hundred chips onto a full mask, the placement of two to five test chips was relatively simple, as shown in Figure 11-14a. The designer would lay out the test samples in the same space as a normal chip would occupy, and then have the test pattern placed into the proper location on each mask. However, with the development of steppers, this has become extremely impractical. This sort of scheme requires that two sets of reticles be made for each layer, one with test pattern, one without. Not only is this expensive, but the manufacturing time would be excessive. As a result, the test patterns must now all be crammed into the scribelines between the chips, as shown in Figure 11-14b. Since there are only four to eight chips in a field on a reticle, there is a very limited amount of space on which to place test structures. It takes clever design work to get all of the functionality required into these areas. Most of these test devices must double as test structures for some of the nonelectrical tests, such as film thickness measurements.

The probe cards must have exceptionally finely tuned controls and probes in order to hit the small contact pads that are permitted. It is very easy for a misaligned prober to miss the scribeline test pattern and enter into the region of the chip, often causing damage. There have been some innovative solutions to these problems also. For instance, there is an electrical test prober that uses E-beams rather than electrical connections to perform certain tests. While the designs of the test structures may have to be changed to optimize for this method, it is otherwise promising for the nondestructive testing of ICs.

11.2.2 Resistor Tests

The first of the electrical tests we will discuss is the resistor test. This is a very simple test, no more complex in principle than checking the value of a resistor by placing it in circuit with an ohmmeter. The difference is that there are a number of different resistors and conductors to test, and unexpected changes in resistivity must be interpreted properly. For example, an increase in resistivity could mean that

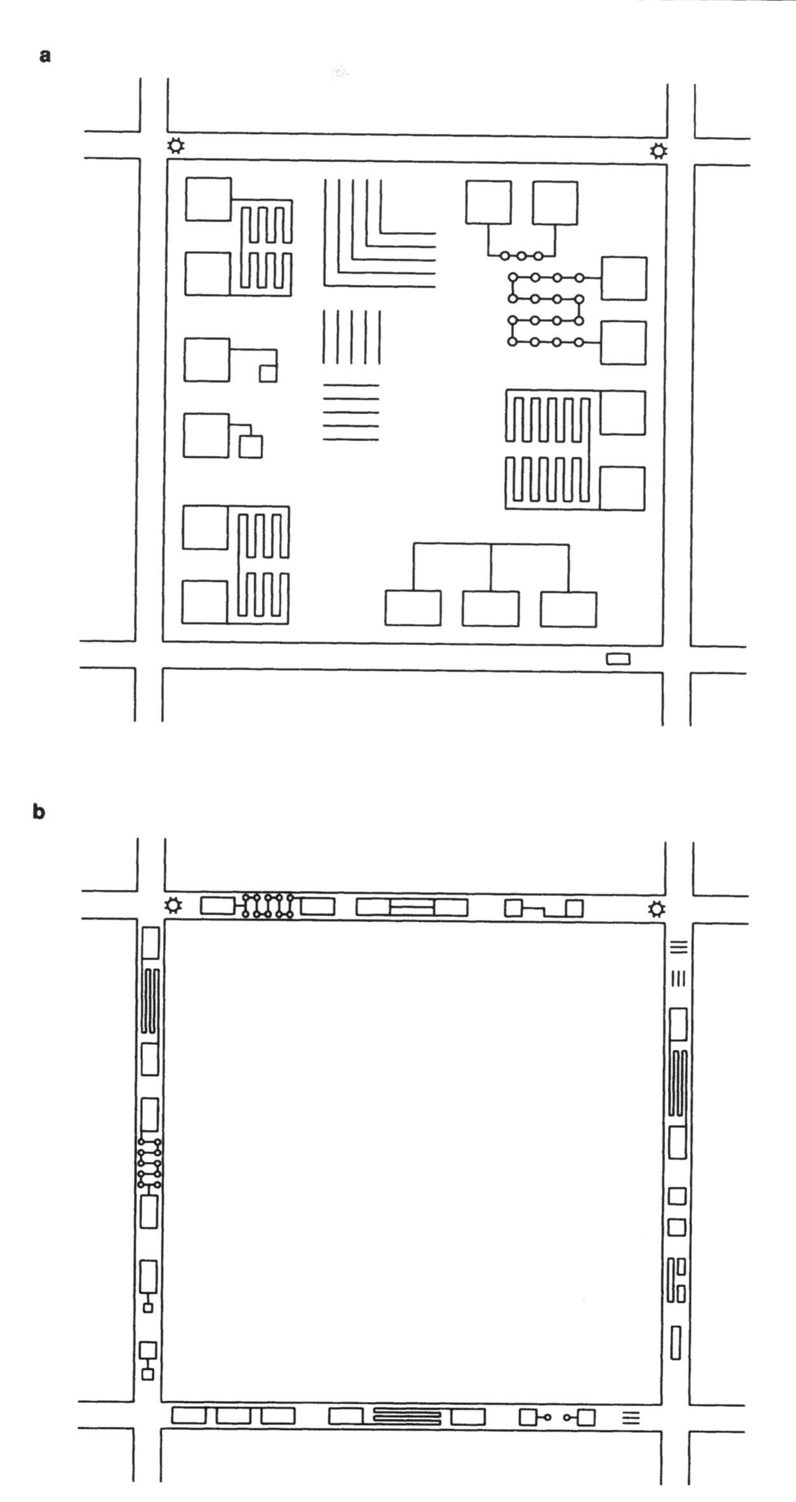

Figure 11-14. Test Structure Locations As we can see in this model, there is much more space available for testing, but with careful layout, nearly all of the tests that can be accomplished on the test die (a) can be carried out using the smaller space in the scribeline (b).

insufficient dopant reached the wafer, the anneal cycle was too short, or the film was of the incorrect thickness. It is important to isolate these variables as much as possible during the tests.

These tests start with a series of resistor networks, which are just resistors of doped polysilicon or other conductive materials, with several linewidths of each material. Clearly, the resistor must be long enough to give a good reading on the resistance, so serpentine designs are sometimes used. The resistor tests are shown in Figure 11-15. In general, the voltages used to test the resistors are the same used in the devices, typically 5 volts, with some technologies requiring other power-supply voltages (+/−12 volts and about 3.5 volts are other typical on-chip voltages used).

There are similar devices constructed for each of the various materials as required. For instance, metal silicide resistor films can be tested using this method. Since metal films are typically very conductive, these short resistors are not good measures of the film's conduc-

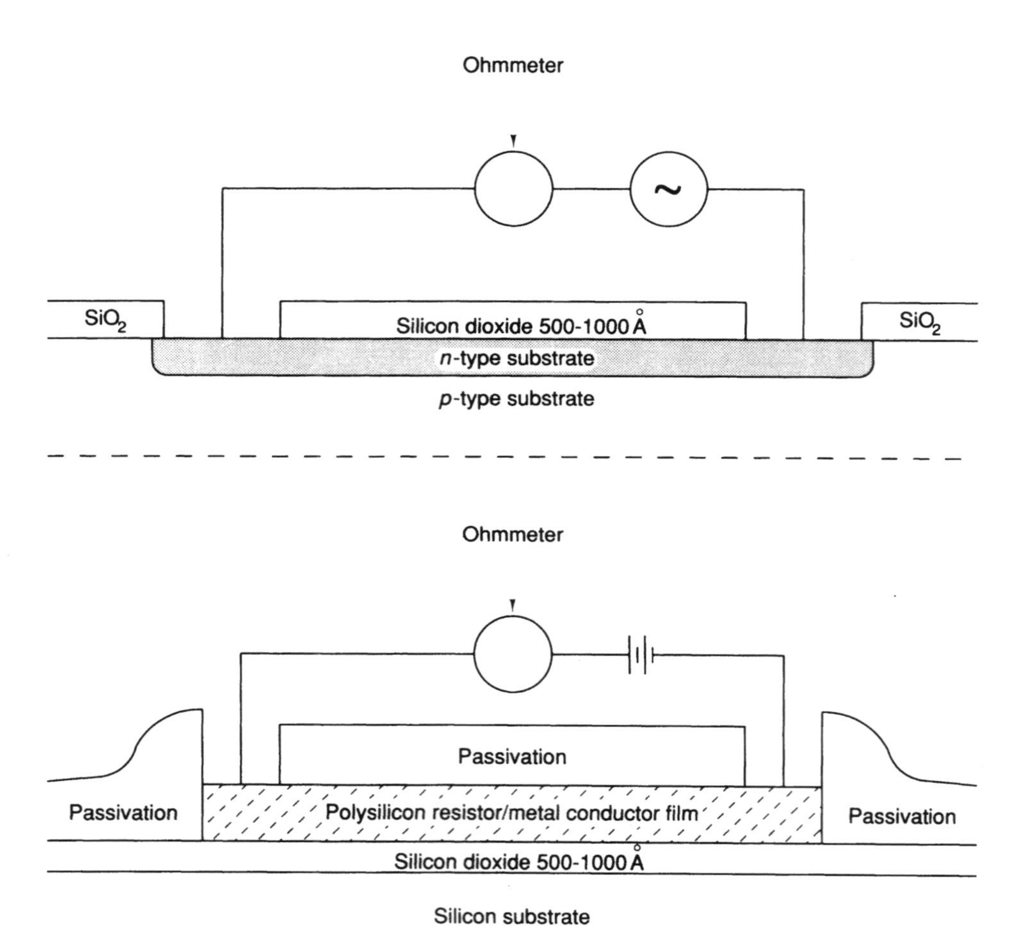

Figure 11-15. Resistor Test Layouts The two basic test structures for resistor and conductor lines are shown here. The top structure shows a substrate resistor structure while the lower picture shows a thin-film resistor structure.

tivity. Metal tests can usually be conducted on chips with ease, however, by testing for speed or for other side effects of poor conductivity. For instance, measuring the amount of power required to operate certain sections of the circuit, and measuring how much heat is produced can provide a significant amount of information on the conductivity and current-carrying capacity of a metal line. If either of these values is too high, it could indicate high resistivity in the metal film.

The final resistance/conductance test that we will consider is the contact chain test. In this test, as shown in Figure 11-16, a series of metal lines and conductive substrate regions are linked together in close proximity. Since the resistivity of the doped substrate region can be determined through testing another resistor on the test pattern, and the metal resistance can be assumed to be very low, excessive circuit resistance will indicate a potential circuit failure (due to oxide remaining in the contacts, spiking aluminum, or other problems). Typically, these contact chains should have between 25 and 100 contacts on them for the best control possible. However, space restrictions usually limit the number of contacts available to 11 to 25.

11.2.3 Capacitance Tests

Capacitors are produced for a wide variety of reasons on a semiconductor wafer. They are used in circuit networks for timing devices, they are used to store charge for memory cells, and for many other purposes. The capacitors are usually made from the polysilicon films, due to the ability of the polysilicon to effectively store charge.

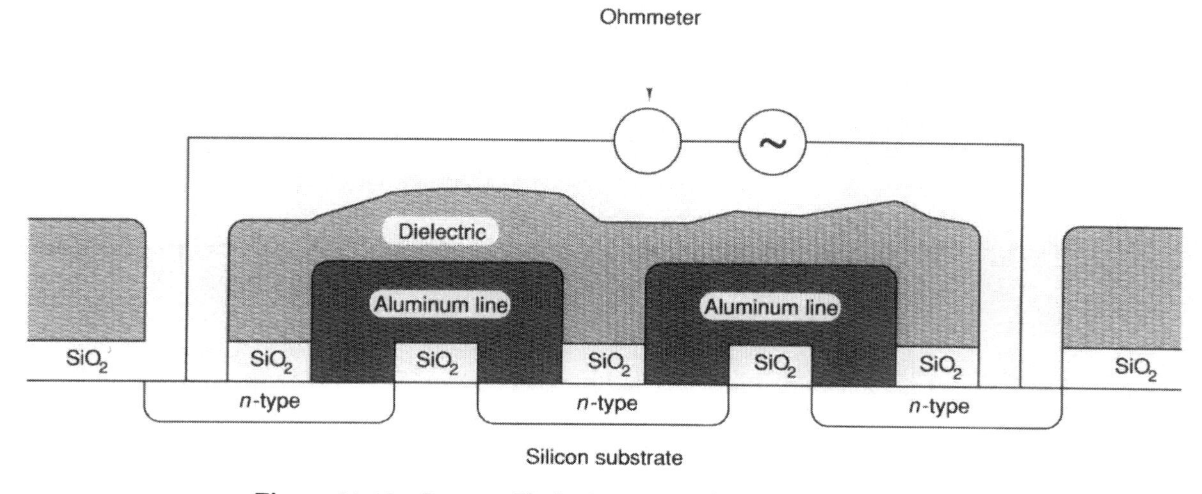

Figure 11-16. Contact Chain Layouts The contact chain test consists of up to 100 contacts. Structures will be constructed to test both film to substrate and film to film contacts.

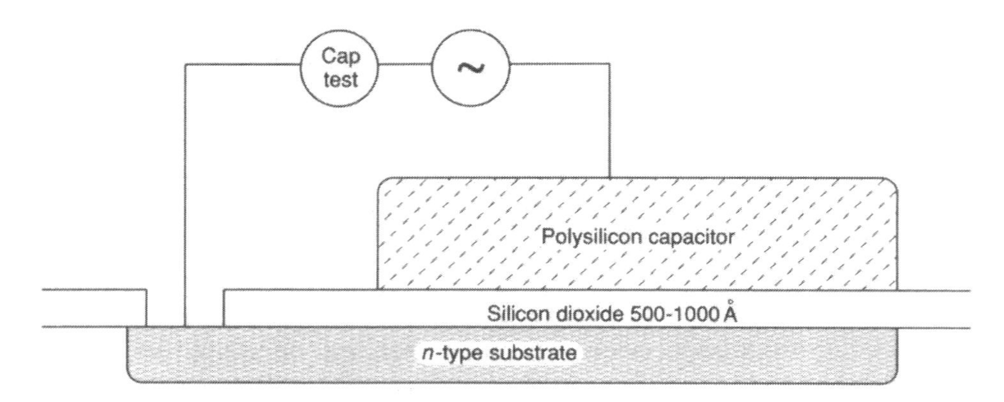

Figure 11-17. Capacitance Test Structure Capacitors are tested by placing the top polysilicon film on a surface of oxidized, n-type diffused silicon. A charge is placed on the capacitor plate, and its characteristics (such as charge retention, discharge time, and charge to breakdown) are measured.

The test procedures can be quite simple, as in Figure 11-17, consisting merely of two contacts, one of which is directed to the substrate and the other to the film. A charge is then placed on the film to measure the capacitance of the dielectric film and the leakage (discharge) rate. Typical capacitive results are in the picofarad range, and are a function of the surface area of the capacitor. Both of these parameters must remain in specified ranges or circuit timing parameters may change drastically. The ability to hold charge for at least the minimum specified time is extremely critical in DRAM devices, which must be able to retain the charge on the cell long enough for the memory recharge circuitry to refresh all of the other cells on the chip, before being recharged.

In many cases, the tests are performed in a somewhat more complex manner than that described above. In most of the tests that are performed on wafers, there are a number of different-sized capacitors, usually with the same sizes as the various device capacitors along with a few that may be somewhat smaller or larger in total area. They are hooked into the input/output contacts with a circuit that allows the engineer to test each of the capacitors separately. If various types of polysilicon layers are used for capacitors, or if other combinations of metals and dielectric films are used for device capacitors, there are usually test capacitors representing those devices, also.

11.2.4 Dielectric Breakdown Testing

The dielectric films that insulate the various conductive layers of the integrated circuits must be of the highest quality. If the films allow current to pass through them at relatively low voltages, the devices will eventually fail. Depending on the severity of the processing problem that is causing the high current leakage, the device may exhibit immediate failures during the wafer sort operation or may fail later in the field or during long-term reliability testing. This type of failure can occur within any dielectric film, so test structures must be manufactured for each type of film on the wafer.

Often, the dielectric breakdown tests are performed on the capacitor test structures after the capacitance has been tested. This not only saves space, but can also save testing time. These structures are typically available only for the key film types that are used for device capacitors. As a result, there will often be additional testing to study the breakdown characteristics of other films. For example, a floating polysilicon structure, such as in an EPROM or EEPROM device, will have a probability of current leakage or dielectric breakdown across the edges of the structure as well as across the broad area of the device. As shown in Figure 11-18, test structures and sequences must be developed to evaluate these combinations.

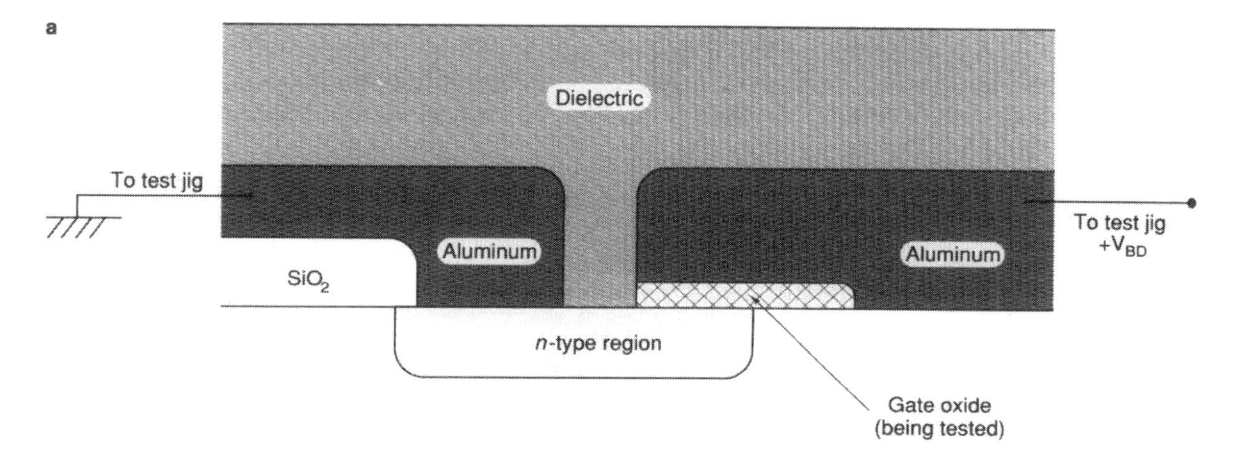

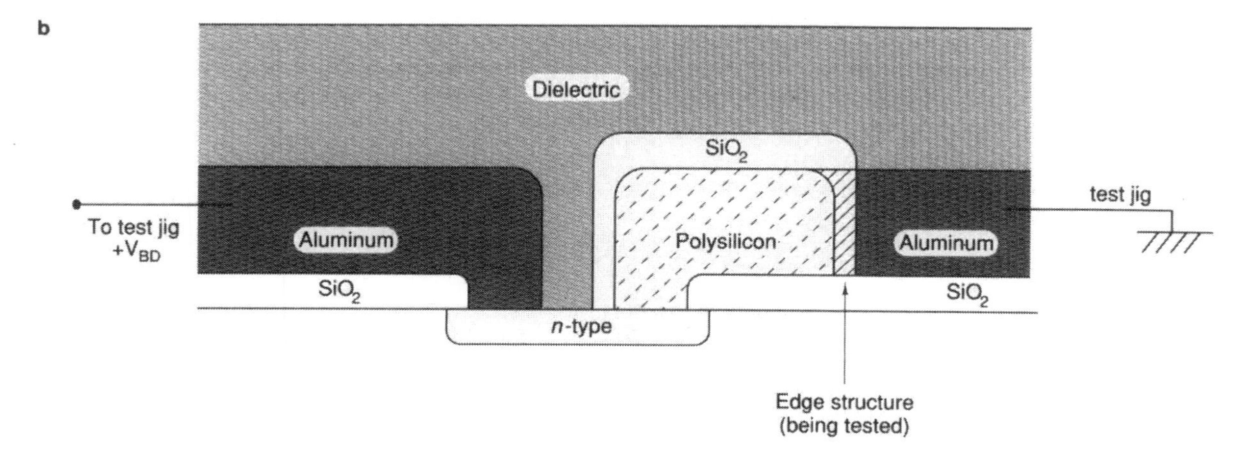

Figure 11-18. Dielectric Breakdown Tests Two of the most common dielectric breakdown tests involve the breakdown of films surrounding polysilicon structures. This information is required to verify film quality and expected long-term reliability. (a) Surface/area breakdown test; (b) Edge breakdown test.

As we can see from Figure 11-19, the dielectric breakdown tests are performed by placing a steadily increasing voltage across a capacitor of known thickness. An ammeter is placed in circuit to register any current flow. As the voltage increases, current will eventually begin to flow. At first, the increase will be gradual but, at a certain point, the current will avalanche across the dielectric layer, destroying it, as shown in Figure 11-19b. Thus, the dielectric breakdown test, properly performed, doubles as a test of variations in leakage current as well as dielectric breakdown and, consequently, provides information on film parameters such as density, pinholes, and contamination levels. Most

a

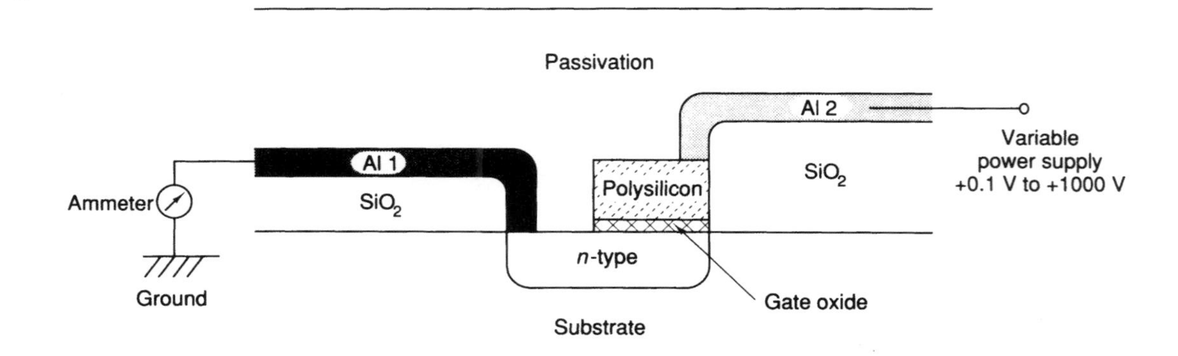

b

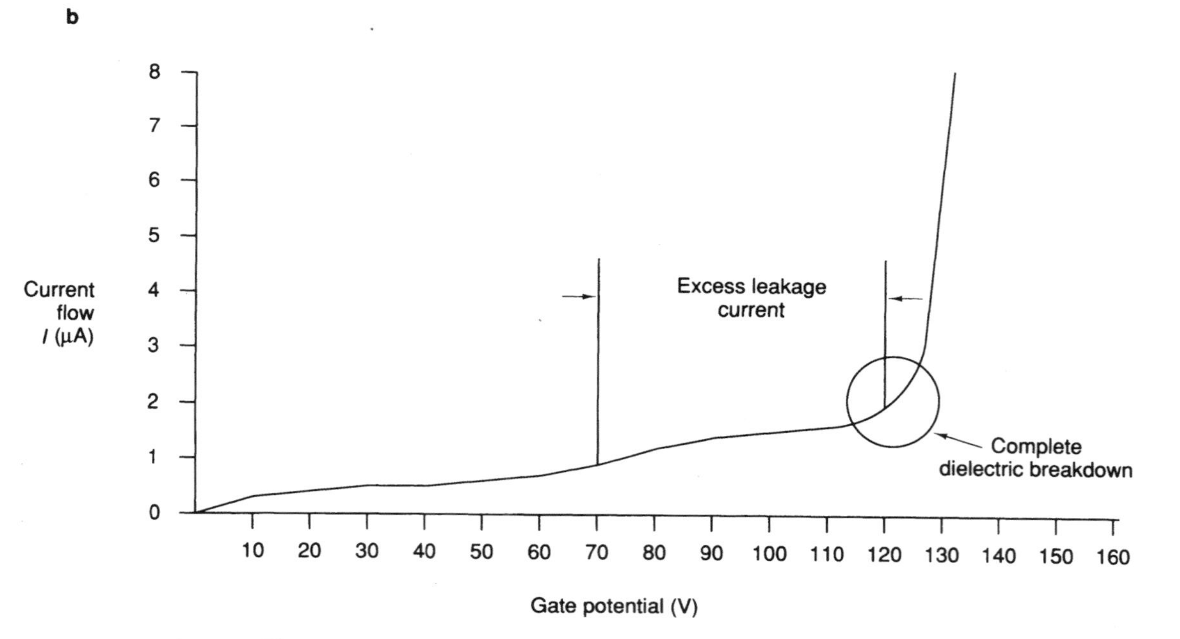

Figure 11-19. Dielectric Breakdown Curve Voltage can be increased across a dielectric with little change in leakage current until a point of catastrophic failure is reached. Generally, there is a region preceding the breakdown voltage where current leakage exceeds acceptable limits. (a) Dielectric breakdown is found when current flows to the ammeter at known voltages; (b) dielectric breakdown curve.

manufacturers stop the test and call it a failure if the leakage current exceeds some arbitrary value, often 1 to 3 pA. The voltages required will typically range from 0 to 100 or more volts, depending on the technology of the devices. For instance, a power IC must be tested at much higher voltages than an MOS digital IC, due to the much higher operating voltages encountered there.

11.2.5 Diode Testing

Diodes are used throughout the design of the integrated circuit. In many ways, every junction between dopant types becomes, in effect, a diode. Since these diodes are used to channel the current in the active region of the circuit to the proper locations, they must not leak and must perform consistently at all times. Problems with diodes can result in the formation of parasitic transistors, and can influence latch-up problems and a number of other substrate-related problems. Causes for leakage of diodes can include aluminum spiking, poor anneal cycles, poor doping or implant cycles, or other process problems. A well-tuned process should not produce leaking diode junctions.

Diodes can be tested in two ways: through front contact or through front-to-back contact. They are shown in Figure 11-20. In the first case, probes are placed into two adjacent regions of opposite dopant type and charge is placed into the well. If any charge is picked up by the second probe, there is a significant amount of diode leakage. In the second instance, the second probe is attached to back of the wafer. Other than that, the test is performed in a similar manner. The primary difference is in the location of the area being tested. In other

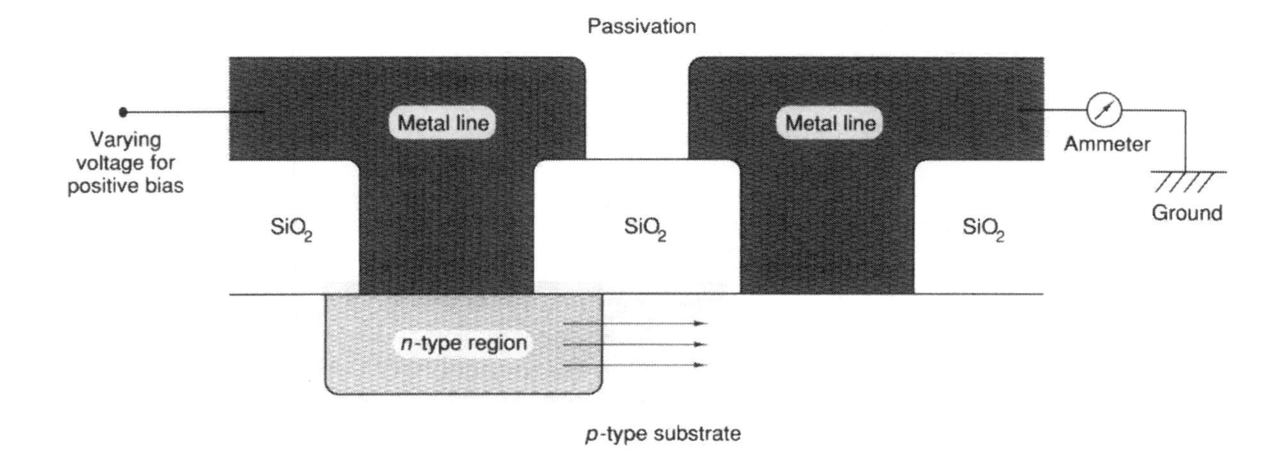

Figure 11-20. The Diode Test Structure The diode test structure is simple, requiring a high-voltage power supply and an ammeter. Excess current leakage or reduced diode breakdown voltages are undesirable effects. The power supply can be connected to the p-type contact to test the diode's reverse biasing effects.

words, the front contact probes will test the edge of a well, while a front/back probe provides more information on the bottom of the well. In many cases, whether a device leaks across the edge or along the base of the junctions is irrelevant, and the type of diode test is chosen for convenience.

If sufficient voltage is applied to a doped well region, localized electric fields can be produced that can cause a breakdown in the substrate material and allow current to leak directly through a device (e.g., a transistor). In normal operating ranges, there is little likelihood of attaining diode breakdown. As a result, the devices are usually tested by changing the diode voltage steadily from a positive bias of some reasonable voltage (say 11 to 20 volts) to a negative bias of as much as 100 volts. Leakage will again be in the picoamp range over most of this range until the breakdown voltage is attained, at which time there will be complete failure, and a large surge in current flow is obtained.

Integrated circuits are often constructed so that a voltage can be applied to the back of the chip during operation. The primary purpose of this "back-biasing" is to enhance the operation of diodes and to reduce the effects of diode leakage, latch-up mechanisms, and parasitic transistors.

11.2.6 Transistor Operation Tests

The transistor is a slightly more difficult device to test, mostly due to the fact that the transistor is an active device, not a passive device like a resistor. It has a number of parameters which must be tested to verify that it will operate within specified ranges. They include the threshold or turn-on voltage, the maximum current flow through the device, the switching speed of the transistor, and the resistance of the channel (to determine total power consumption). These parameters are usually checked for each of the various types of transistors constructed. With most of the standard technologies, most of the transistor lies in the substrate, while the gate material may vary. Polysilicon is the most common gate material, although aluminum and metal silicides are often used to obtain the maximum speed possible from the design.

First, we will look at the threshold voltage. This is the voltage that causes the substrate material to create an inversion layer at the surface of the silicon (allowing free electron travel). Threshold voltage tests are often carried out in conjunction with the C–V tests, using similar techniques, in which a high-frequency signal is placed across the device, as in Figure 11-21. The voltage then changes from −5 V to +5 V (some technologies may require different ranges). The gate threshold voltage relates closely to the voltage at which the transistor turns on. The channel forms at a shallow angle, and becomes connected when the threshold voltage is reached. This is shown in Figure 11-22. It reaches its full-on condition at a slightly higher gate voltage. Often,

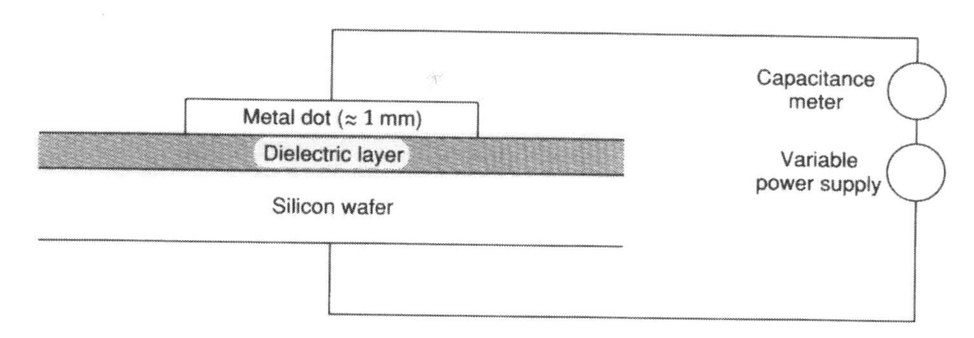

Figure 11-21. The Basic C–V Structure The C–V test structure consists of a round metal dot, approximately 1 mm in diameter, deposited on top of the dielectric to be tested. The dielectric is usually in a gate oxidation thickness range (400–1000 Å). The dot will be from 5000 Å to 10,000 Å in. thick and will consist of the standard metal used in the fab (Al, Al-Si, Al-Si-Cu) film. The probe is placed on the metal dot while the back side of the wafer is used for the second contact.

these two are equated for convenience since the voltages used in real devices far exceed the requirements of the transistor device itself.

The voltage at which the transistor turns on is specified very tightly, and is usually around 2.75 V. When voltage higher than the threshold voltage is placed on the gate, the electrons will flow freely to

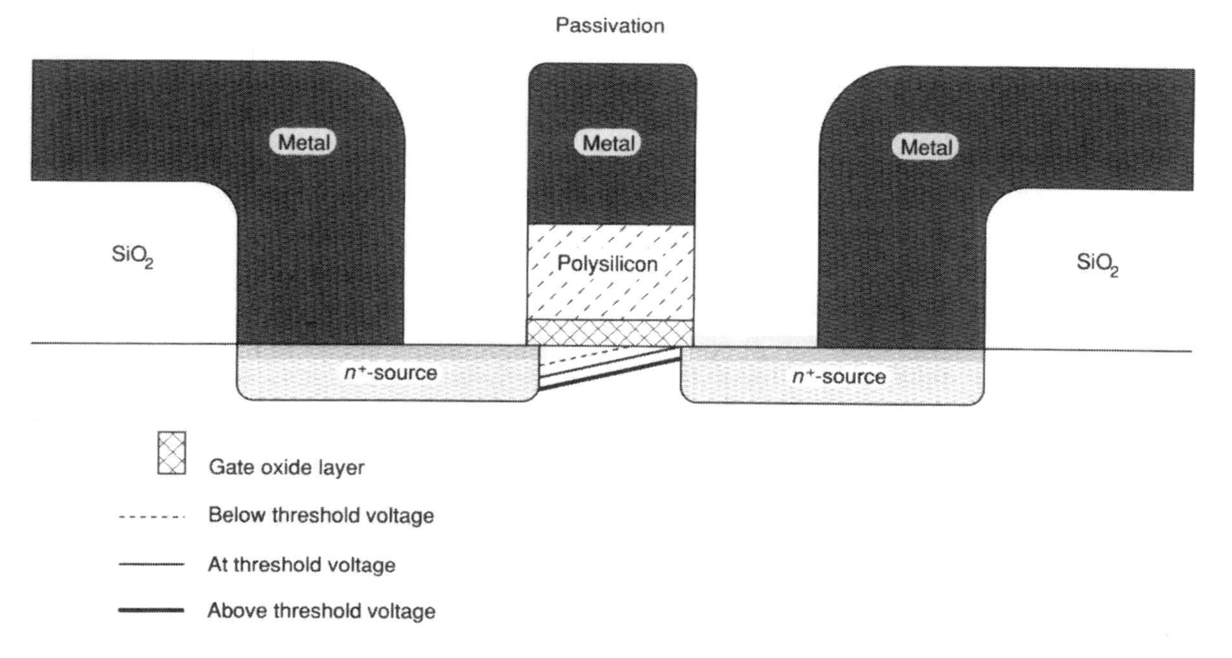

Figure 11-22. Channel Turning at Threshold Voltage As the gate voltage applied to the polysilicon layer is increased, the inversion layer under the gate oxide deepens. Below the threshold voltage, the channel between the source and drain is not open. At the threshold voltages, the channel just opens. Typical gate operating voltages are higher than the threshold voltage to allow a sufficient large channel.

the interface region, where they form the channel, allowing the gate to operate. Similar principles apply for the operation of the depletion transistors with the voltage required to stop the free flow of electrons called the turn-off or pinch-off voltage.

Threshold voltage tests are often applied to structures other than transistors, for example, broad field areas. The concern here is that the flow of current through the upper metal layers, possibly at high voltage (12 V or more) will induce changes in the device structures below them, thereby causing changes in other device parameters.

In most of the threshold voltage tests, the capacitor test pads can be used with ease (as long as these tests precede the dielectric breakdown tests). This preserves both space and speed of testing and does not preclude the use of the transistors themselves to test threshold voltages.

A low value for threshold voltage may result in occasional accidental device operation. This will often lead to the corruption of some data in the machine at some point. This corruption can occur when electrons have enough energy to start the traverse to the other side of the gate. Of course, as the electrons approach the drain of the transistor the potential increases, propelling them into the drain. If this leakage continues for a long time, it will cause small but critical damage. Eventually, these transistors will burn out, allowing current to flow continuously, and eliminating their ability to act as a switch.

Another important parameter is the resistance of the wells and the channel to current flow. An excess amount of resistance (due to insufficient doping of the regions, insufficient size of channel due to high threshold voltage, or severe silicon-crystal damage in the channel region) will result in the device having a number of problems. For instance, the maximum electron flow (or gate current) will be limited to a certain value. In addition, the device can heat up or consume excess power if the resistance in the channels and other parts of the transistor is too high. These are all serious problems when you consider the number of transistors on each chip. Excess channel resistance will also cause the operation of the device to slow, possibly enough to change device parameters or affect the internal timing sequences of the chip.

Another electrical parameter that is tested on transistors is their switching speeds. It is critical that the device turn off and on very quickly, and in a very stable fashion. Transient effects must be minimized for proper device operation (see Figure 11-23). In most MOS devices, transistor switching speeds are usually in the 11 to 20 ns range, although many slower, and some faster, devices exist. CMOS devices have been reported in the 5 ns range. Many bipolar, ECL, and other devices have faster than 1 ns switching speeds, while GaAs transistor have switching speeds even faster than those of silicon.

Obviously, if key transistors are operating at different rates, the integrated circuit will have an increased probability of failure. Usu-

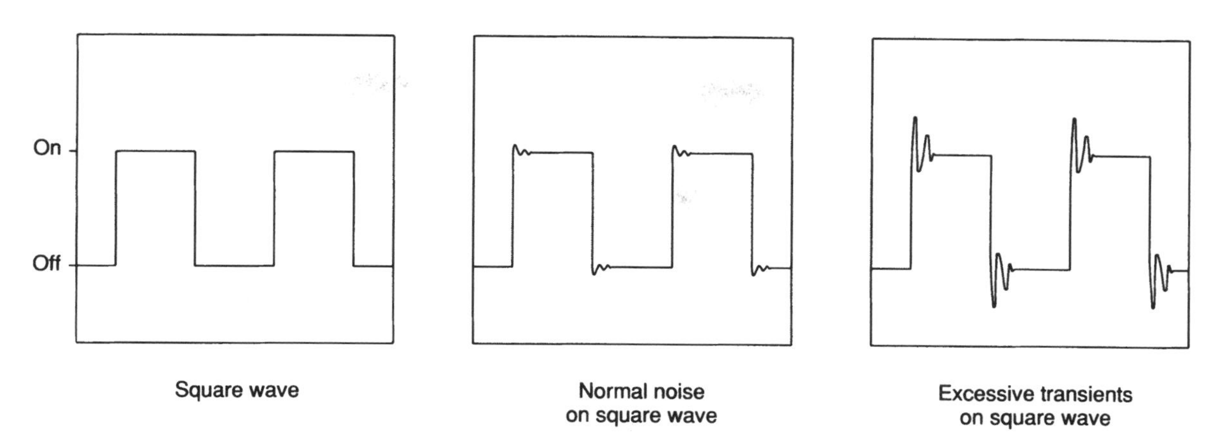

Figure 11-23. Transistor Transients and Voltages Spikes While real devices always have some switching noise, the magnitude of this noise must be kept as low as possible. Excessive noise or transient spikes can cause premature device failure or other unusual effects.

ally, a variance of even a few percent will drastically alter device performance. If all transistors are slowed equally, the chip will probably operate, but at a reduced rate. As a result, these chips are often downgraded and sold as lower speed chips at lower prices. While not as serious as complete yield crashes, the impact of this type of problem on fab profitability can be enormous.

There are several reasons why the switching speed of the transistor can vary. We have already spoken of some of them, which can include high channel resistance and incorrect threshold voltages. Other causes for incorrect transistor switching speeds can include improper oxide thickness (thick gate oxides will often result in a slowdown of the device) and improper voltage levels on the chip.

11.2.7 Electrical Test Wrap-Up

While there are usually technology-specific tests that are run on integrated circuits, most IC technologies will test for these basic items at the in-line electrical test operation. Occasionally, there will be more intensive tests covering small sections of circuitry. However, in most cases, the in-line electrical tests are used to test parametric values and not circuit parts. As a result, the main structures tested are those most commonly used: resistors, conductors, diodes, capacitors, and transistors. The test structures that are designed for these circuit elements must cover the entire range of device and film possibilities. Usually, the values of these parameters must vary by only a few percent in order to maintain device functionality. Higher speeds, smaller linewidths, thinner films, and more complex devices can increase the requirements of these parameters and make the electrical test procedures even more crucial to the success of an integrated circuit.

11.3 INTEGRATED CIRCUIT SORT PROCEDURES

After the chips have been completed, the manufacturer must test all of them to determine which are good and which are useless. This is done prior to the chips being diced and packaged separately. These tests must run the chip through all of its paces, for instance, testing all the instructions of a microprocessor, or checking all the memory cells of a memory chip. Every chip must be tested and the results stored for analysis. In addition, the chips that have failed must be identified so that they do not accidently end up in a package after the wafers have been diced. These issues will be discussed in this section.

11.3.1 Testing Procedures

There may be some preparatory procedures performed, such as UV erasure of EEPROM and EPROM devices. Then each wafer is placed on a test jig, where probes are placed into contact with the bonding pads of each chip. Obviously, these procedures must be done very gently and carefully, or damage may occur to the chips that can prevent device operation after packaging. After good contact is made, a test computer runs the tests to be performed through the probe and then checks the responses. If all of the responses are correct, and no problem is found, the probe is moved to the next chip. If the chip is found to have a problem, it is first analyzed to determine whether it is a lethal error or a recoverable one.

A recoverable error usually involves programming the device to use on-board back-up circuitry. For example, memory devices often contain a number of additional cells that are redundant and are not used unless memory-cell errors are found. Usually, there will be a enough redundancy to "repair" up to about five defects. If a memory cell is found to be damaged, the test computer will program the chip to reroute the bit to the redundant data area. This is a convenient way of reducing the number of chips rejected and is practiced by nearly all memory manufacturers.

Another type of recoverable error is one in which the device is found to be operating at low speeds, excessive power consumption amounts, or has other minor defects that prevent its sale as a prime part. They are often sold at a reduced price and with reduced specifications. Usually, these chips will be identified at sort, but cannot be explicitly marked to prevent complete device failure. There may be other recoverable errors that are specific to a particular technology or device that will prevent the loss of all chips that have marginal functionality or are slightly out of specification tolerances.

If a chip is found with a nonrecoverable error, it is considered a failure and the chip is marked with a red dot of indelible ink. Chips with red dots are destroyed after wafer dicing. The data from the chip are stored, and an entry is made in the "sort bin" record. This can be thought of as a computer storehouse where chips are thrown into boxes or bins, depending on the test results. Each "bin" represents the existence of a failure for each of the tests performed. Typically, a good die will be considered to be in sort bin 1, while marginal or repaired chips will go in bin 2, and so on. A sample bin setup is shown in Table 11-4. We can see that each test is associated with a bin number.

The sort procedure may be broken up into a number of steps. This will often involve a presort, where devices have basic operations tested. This allows the final sort procedure to be somewhat faster and more efficient, and can allow for certain preliminary tests to take place. For example, nonvolatile RAM devices will often be programmed and erased up to several hundred times to verify cell operation and charge-retention capability. Other chips may be placed in a standby condition (with special test cards) at an elevated temperature for several days prior to final sort. After the presort tests are completed, the chips that have passed to that point are subjected to the final sort tests.

TABLE 11-4
Example Sort-Bin Setup

Bin	Test performed	Disposal of chip
1	All tests passed	Ship it
2	Device passes tests, but is slow	Ship at discount[a]
3	Device passes tests, power use high	Ship at discount[a]
5	Infant mortality	Throw away
10–20	Electrical tests, 10 shown here, but this can vary significantly	Disposition or throw away
22	Short circuits on chip	Throw away or LYA
23	Open circuits on chip	Throw away or LYA
25	Stuck bit	Use redundant cells
26	Rapid discharge rate	Use redundant cells
27–30	Current leakage through films, junctions	Throw away or LYA
35	Total chip operation (works through tests, fails on full operation)	Throw away or LYA
37	Cycle test to write/read cells	Throw away or LYA
38	Timing error	Throw away or LYA
40	Other assorted errors	Disposition each

[a] Whether these can sell for a discount is largely a matter of market acceptance and need for such a product.

11.3.2 Post-Test Operations

After the chips have been tested, all data stored, and failed chips marked, the wafers are sent to a number of further operations. They are first coated with a protective layer, and then the wafers are ground to about one-fourth to one-third of the original thickness. The wafers are then cleaned (the grinding is a very dirty operation). After the wafers have been ground, they become extremely fragile and must be handled very carefully.

The wafers are often coated on the back with a micron or so of gold to allow a backside ground point or backside bias voltages. This structure allows the device to be more stable and reduces the effects of problems in the substrate. The chips are then sawed apart on the scribelines using a high-precision diamond saw. The protective films are removed from the working chips, which are then sent to the packaging operation. The failed chips are thrown into a silicon reclaim bin.

At the packaging operation, the chips are glued into the package, and wires are soldered delicately between the bonding pads on the chips and the leads from the pins on the package. Lids are then glued on and the chips are stamped with a log and a manufacture date or other code. You will find that most chips are dated with a work week and year code. For example, a code of 4489 means the chip was packaged and stamped during the 44th week of 1989. This is convenient since some chips have had long lifetimes (e.g., the Z-80 microprocessor) and may be available in several forms and quality levels, depending on the age of the chip.

After packaging, the chips will be tested again by being placed into a final test machine. In this device, the operation of the part is assured, and the continuity of the bonding wires is verified. If burn-in testing or high-temperature cycle testing is required, it is performed at this time. Special test machines have been constructed that allows several hundred chips to be placed in these machines and operated, while the entire system is heated to around 150°C. This kind of high-temperature testing is commonly performed on military-quality chips, as well as selected chips of standard commercial-grade integrated circuits. Finally, chips that pass these tests will be labeled with the company logo and the date.

Assuming that a chip can survive the rigors of the processing environments, all of the tests that are performed, and still looks good and performs within specified ranges, it can now be sold. At this point, there are no further tests to be performed on working circuits, unless the manufacturing area decides to perform high-yield analysis on working chips (see the next section on yield analysis). Further testing will usually be performed on failed chips to determine the cause, and to provide feedback to the fab area as to what problems have occurred, so that the fab can fix the cause and improve yields.

11.3.3 Sort Binning

As we have discussed, we can see that there are a number of tests that can result in the rejection of the chips. Each result is then registered as a failure in a specific bin. We will first discuss the various tests involved in the sort operation. There are many tests that can be performed on a chip; thus the exact suite of tests depends on the particular device. We will devise an example test bin suite in this section.

The first bin is used for the chips that fail instantly on start-up, and do not work long enough to indicate the mode of failure. Failures that fall into this bin are usually defined as "infant mortality." Sometimes the causes of these failures cannot be identified, since they are usually gross failures. Other times the problems can be identified under visual inspection, as noted in the yield analysis section. Problems that cause infant mortality often include physical problems such as scratches and other major defects.

In most cases, the next set of tests will involve testing for parametric values. This is similar to the electrical test procedure, except that the information must be deduced from the operation of the device, as there are not usually specific test structures set aside for these types of parameters. If parametric values are found that are far out of the specified range, the device can be considered a failure and will fall into an appropriate bin. Parametric values can be out of specification ranges for a variety of processing problems, for example, high resistivity due to low dopant quantity.

The devices are then tested for conductor continuity. Any short circuits or open circuits that are detected will be grounds for failure of the chip at that point unless the defect is in a portion of the circuit that is protected through redundancy. Short circuits and open circuits are most often caused by particle contamination. Sometimes the particles can actually be embedded in some film and are acting as the conductor for the short, or they may have existed on top of a layer of resist or film during an etch process, preventing proper pattern generation. Therefore, particles are seen as a common source of these kinds of failures.

Associated with short circuits is the stuck bit, or stuck transistor, failure. This occurs when a bit in a memory cell will not change its charge (therefore its value). This occurs most often because the charge leaks off the cell almost immediately. This will result in a charge value of zero. Whether this results in a logical "1" or a logical "0" depends on the type of circuit into which the bit is designed. (Some designs use inverted logic for certain technical reasons, so that a cell with no charge will be equal to a "1.") Stuck cells can occur as a result of particles on top of the gates of transistors or on top of memory cells. They usually cause the gates or cells to fail immediately, resulting in stuck bits early in the test sequence. If bits start to become stuck after the device has been tested for awhile, the problem may be caused by chemical contamination or pinholes or other leakage in the gate dielectrics. There are a number of symptoms for different devices that can be traced to these types of failures.

As noted, if the device is a DRAM, the storage capacitors will discharge rapidly, before the refresh circuitry can get around to the cell to recharge it. If the device is of a nonvolatile nature, the charge will leak off of the cell, and the data will change after awhile. Since nonvolatile memory devices are generally installed at a very deep level of a computer system and then forgotten about, the consequences of failures of these parts could be very serious. Problems are not restricted to memory chips, as even microprocessor chips can exhibit these failures, which result in registers losing data, and related problems.

Device operating speeds are also tested, and the chips sorted by speed. There may be two or bins for various speeds, depending on the marketability of the slower chips. If it costs more to manufacture the slow chips than they can be sold for, the lower speed bins will be eliminated and the chips will be considered low-speed failures. Device speed is often affected through process problems, such as thick gate dielectric films, or excessively high resistivity of the various conductive films.

Current leakage through dielectric films or from various well regions into the substrate is another serious problem that will result in a failure of the device at sort. Serious problems of this type can be produced through the actual breakdown in dielectric films, breakdowns in diode junctions, or through short circuits from a metal conductive line to some key region on the substrate or other film layer. Smaller amounts of leakage can also cause device operation failures. A variety of chemical and physical effects can cause the current leakage problems, including film contamination, contact spiking, and so on.

Ultimately, a number of small problems or marginal parametric values can add up to circuit functionality problems. The symptoms often start with timing problems which occur when a portion of a circuit is operating at a rate below that of other components of the chip. As a result, the total operation of the chip must be slowed to the rate of the slowest component or it must be thrown away, depending on the severity of the problem. If the clock mismatches are too far off and are based on the same basic clock pulse rate, the results will occasionally fall out of sync, causing data to be lost (see Figure 11-24). This type of problem is often found by repeated execution of complex instructions to see if data are ever lost. Usually, the designer can tell the test engineer which circuits will have timing problems in the event of process variations. Sometimes, new errors pop up in unexpected areas. In all of the cases, it must be determined which portion of the circuit element is out of phase with the majority of the circuit, and then analyze what process or design problem has caused the timing problem.

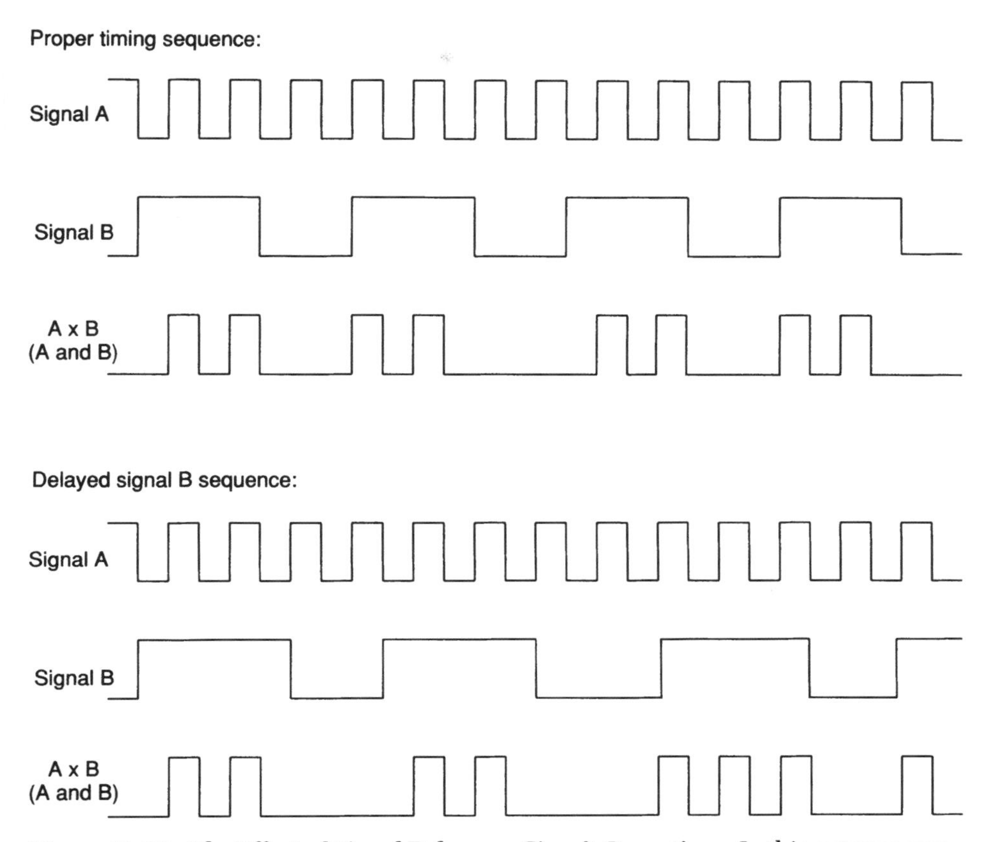

Figure 11-24. The Effect of Signal Delays on Circuit Operation In this extreme case, signal B is delayed by 25%. This is sufficiently out of sync to completely change the signal pattern of the output signal (A and B). This would create some very serious problems in a computer system.

Timing problems can be especially complex on microprocessors and on application-specific devices, primarily due to the relative complexity of the circuits. Memory devices can be as prone to timing problems as other chips but, since the chips are mostly large arrays of identical circuitry, analysis becomes relatively simple. Microprocessors must execute anywhere from a few dozen to several hundred operations, with each instruction coded into the circuitry of the CPU, along with all internal communication and subfunctions, such as floating-point calculation and memory management. The timing requirements of a one million transistor CPU are very specific and can be very complex. Minor inconsistencies or processing problems (especially minor reticle defects) can cause particular instructions to return incorrect results or not operate at all. Depending on the nature of the instruction, this could be catastrophic (preventing the chip from executing a normal operating system, an instruction like a MOVE or COMPARE instruction), or it could be fairly benign, occurring only occasionally (as in a floating-point math operation).

Defects other than timing errors can cause instruction set errors. These errors occur when data input to a chip are processed and come back with an answer different than one would expect. They can be the result of processing errors or design flaws. In some cases, there is some unknown limit or "cliff," which causes yields to fall off drastically when certain types of processing problems occur. A typical example of this would be that of EEPROM devices. The best yield usually occurs when the tunnel oxides of the devices are at exactly the design thickness. Thicknesses lower than this value are often associated with rapid oxide deterioration and/or rapid leakage; in either case the device fails. On the other hand, if the tunnel oxide is a little on the thick side, the yield is not affected as greatly. This phenomenon is described in Figure 11-25.

Problems like these can be very confusing and can result in a significant amount of finger pointing between the design and manufacturing groups. This is due to the very difficult test procedures involved and the amount of time required to perform the tests. The amount of time required to redesign problem areas is very significant, often months, and is usually a method of last resort. Even if the decision is made to redesign the problem area, the company may need to produce the chips. If it can be determined that the chips can run under certain manufacturing constraints (e.g., additional testing, such as 100% sorting of wafers at the key operations), they must be applied. These constraints can be extremely disruptive to wafer fab flow if not handled correctly. In these conditions, the designers must work as quickly as possible.

Memory devices have their own set of problems. Since the designs call for extremely dense arrays of very tiny transistors, almost any minor deviation can cause a chip failure. In addition, since the memory device is accessed so often (every direct CPU access, plus every refresh cycle in the event of a DRAM), the probability of eventual chip failure increases drastically. As a result, certain specific types of tests are created to cover all of the various memory configurations. This is accomplished by storing a series of values that exercise all the cells of the array at once, then reading those values back and comparing the exact value returned. For instance, if an array of cells is eight-bits wide (this arrangement is only for convenience of the example), and the data stored are AA in hexadecimal (1010-1010 in binary), any other data that are returned will indicate which bit has failed. For example, if the hexadecimal value A3 (1010-0011) is returned, we can say that two bits of that row have failed bits in them. Since the value of AA exercises certain bits only, the test will usually alternate the date stored to something like 55 in hexadecimal (0101-0101 in binary). This allows all bits to be operated in an off and on condition.

The array tests can take a number of different forms, depending on the results required. In some cases, it is more important to see if the cells can handle repeated programming and erasing, in which case

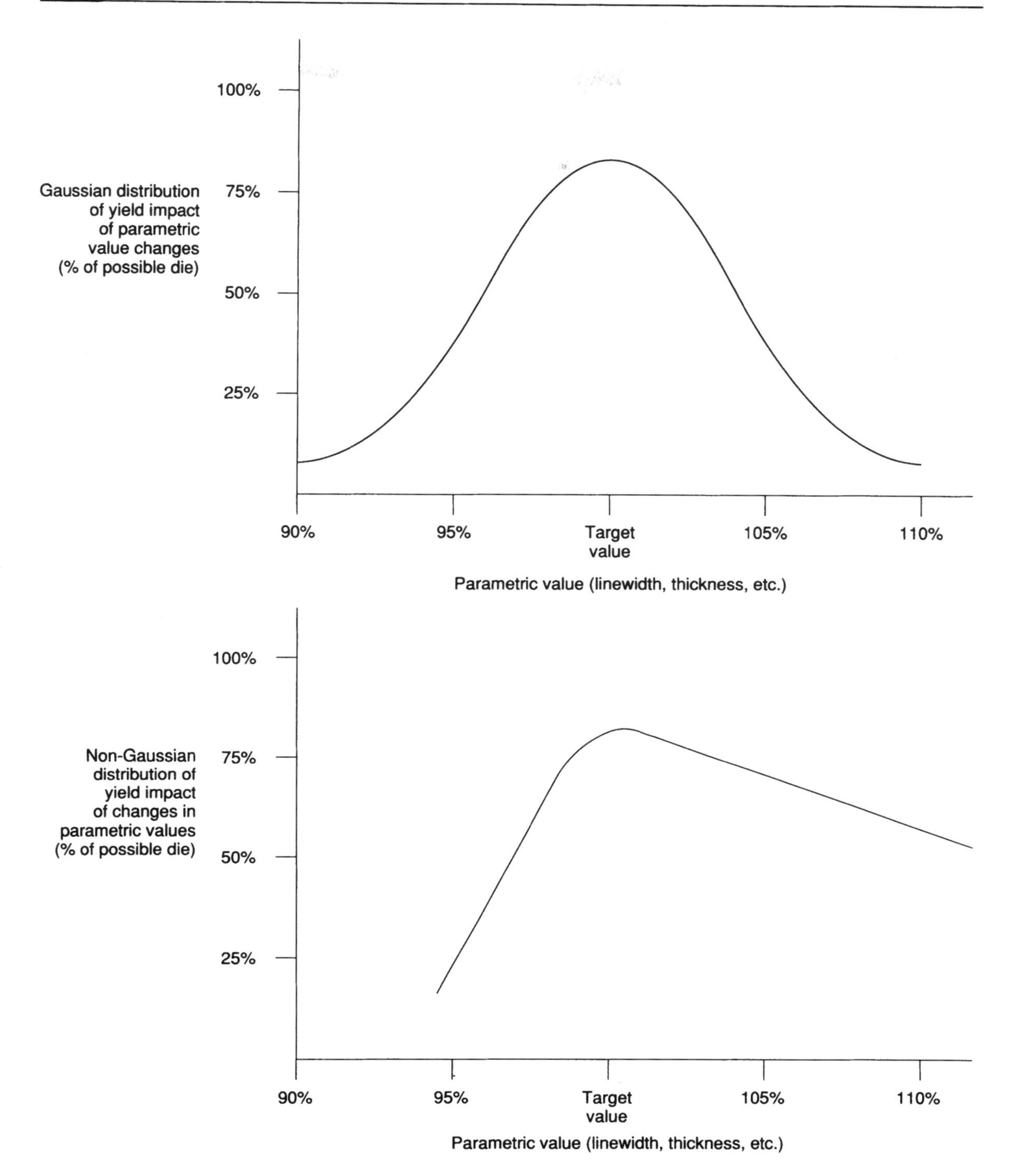

Figure 11-25. Possible Yield versus Parametric Distribution Curves When a large number of wafers are analyzed, a number of trends can be discerned when observing die yields and key parametrics. In some cases, yield loss may be symmetric about the target value. In other processes, such as thin oxidation processes, parametric variation in one direction (thinner oxides, in this case) is more deleterious than variations in the opposite direction.

the memory cells are rapidly cycled. In other cases, it may be more important to see how well the memory cells hold their charges, in which case the program/erase cycles may be much longer. Typically, a compromise must be reached, in which a chip may undergo a large number of cycles, then be studied for charge retention, and then cycle-tested again, with this cycle being repeated until specification targets are met.

The sort tests that are used on real devices are usually a combination of all of the tests mentioned above, plus any others that are useful for the specific device in manufacture. Results are obtained in a tabular format, and the problems identified if at all possible. It is sometimes useful to construct a table similar to the one shown in Table 11-5

TABLE 11-5
Sample Sort-Bin Prioritization

Bin ID[a]	No. (in bin)	Priority[b]	Notes
1	1285		
2	12	8	
3	3		
5	21	4	
10	4		
11	6		
12	12	6	
13	2		
14	1		
15	18	3	
22	9	10[c]	
23	4		
25	3		
26	32	1	Specific location known
27	33	2	This and No. 26 may be related
28	12	5	
35	4		
37	9	9[c]	
38	1		
40	42	7	Catch-all category
		Total loss	: 228 Die
		% yield loss	: 15%
		Total yield	: 85%

[a] Uses bin IDs from Table 11-4.
[b] After all of the die have been sorted into bins, the problem areas must be prioritized. This is done by analyzing the extent and impact of the yield loss and by determining how complex the solution may be. It is best to attack the "easiest" problem with the largest yield impact first, with lower-impact, complex problems addressed later.
[c] These may rate higher if LYA shows an obvious solution.

to analyze and prioritize the problems. The bin with the highest number of failures is usually the one that should have first priority, but you must be careful to make sure that only a small variety of failures may all cause one particular failure mechanism, or that the bin is not some sort of "catch-all" bin for otherwise unpredictable failures. If this is the case, determining which problem has the most significant impact may be very time-consuming. If, instead, two or three bins can be selected that have the same number of chips lost, but have specific causes that can be acted upon, it is better to prioritize these problems to the top of the list.

The data that are obtained at wafer sort can be analyzed in a number of interesting ways. This is especially true if there are a significant number of wafers analyzed at the electrical test area. The most immediately useful method is to plot the die yield of each wafer with the values of various parameters on the opposite axis. This is shown in Figure 11-26a. In many cases, it may be interesting to compare the yield to a ratio of two parameters, such as polysilicon oxide breakdown voltages on the sides and the top or bottom of the structure, as shown in Figure 11-26b. Clearly, the more wafers that are available for this type of analysis the better.

11.3.4 Wafer Sort Wrap-Up

We have seen throughout this section that it takes a combination of process engineers, design engineers, and yield/product engineers to fully analyze, interpret, and act on a serious process problem. Often, a problem may be caused by a certain interaction of factors that was not apparent before the chip was manufactured (this is very common with new, high-tech devices). Only through cooperative teamwork can yield problems be resolved. It is critical that the design engineer understand the expected operation of the device, that the product or yield engineer be able to interpret electrical and sort data well enough to determine the underlying process problem, and that the process engineer modify the correct set of procedures or processes to prevent the problem from recurring. The desire to place "band-aids" on problems must be overcome at all costs at this critical point in the operation, as they can easily become dogma. With the long lag time between the discovery of a problem at sort and its final resolution, the need to find long-term solutions becomes more critical.

We have seen that there are a wide variety of failure mechanisms for integrated circuits. While some of these failures may have obvious causes, a significant number of them occur without any indication of why they failed. The purpose of the sort test suite is to categorize these failures to prevent their occurrence in the future. As a result, the test sequence should be fluid so that new bins (or tests) can be added as the need arises. This aids in the long-term understanding of the process. Careful inspection of the test data, in association with electrical test

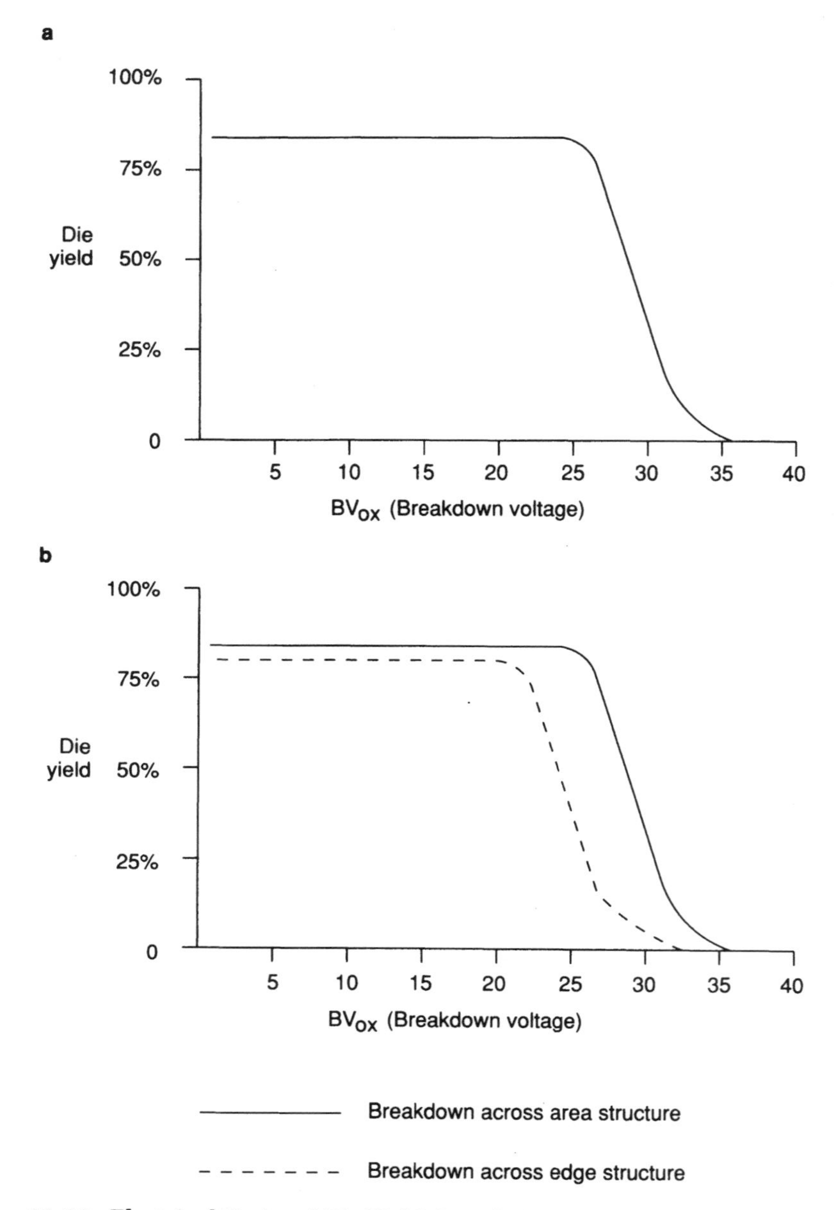

Figure 11-26. Electrical Test and Die Yield Correlations Die yield can be compared to electrical test results for clues on how to improve yields. For instance, if it is found that some sort of breakdown is reducing yields, the breakdown voltage tests would be examined. As in (a) the standard BV_{ox} may not show any abnormality. However, examining the BV_{ox} of the edge breakdown structure as in (b) may show a critical breakdown that may be associated with the yield problem.

data and process control data, can usually determine the cause of a problem and point to an answer. If none of these techniques succeeds, the wafers and die that fail must be sent through low-yield analysis procedures, which permit the wafer to undergo a postmortem, and determine failure mechanisms.

11.4 LOW-YIELD ANALYSIS

Wafers or chips that have failed at the final sort operation must be analyzed to prevent further loss of material from that problem. Usually, this information can be obtained through the electrical test, process control, or die sort data. However, a certain number of failures will remain unidentified. If the number of chips falling into this category is very large, more than a few percent, the fab area is in a very unstable state indeed. Sometimes, though, new problems come up and must be understood and the effects eliminated. The job of finding otherwise-unidentified failures falls to the low-yield analysis (or LYA) group. This group may also take exceptionally high-yielding wafers and study them in detail (in effect, high-yield analysis), to determine optimum conditions for high yield. This can allow the fab to maximize its process control system's efficiency and maximize fab output in the long run. However, this is usually possible only at companies that can afford the loss of a few of their highest yielding wafers, which may be very painful indeed (it's easy to destroy chips that don't already work, it's a different story when the devices have revenue value).

Low-yield analysis techniques are destructive to the chips, but this is of little consequence since the chips have already failed the test procedures. The procedures start off in a benign enough way, by studying the run card and the process control logs for the dates that the lot in question was processed through each of the various steps. One of the key items to look for is any item that appears to be common to other runs with similar failures. For a simple example, perhaps the wafers run in a certain furnace on a certain day of the week come out with reduced yields. If it is found that the furnace has its quartz cleaned each week on that day, a simple experiment may prove that this is the source of the yield loss. Another example may be that the wafers fail for excess leakage only at certain times, and then for full lots at a time. Further examination of records may point to a common problem with one vendor of silicon.

Another, somewhat more subtle form of this kind of analysis occurs when wafer parametrics reach extremes. For instance, a wafer that has a gate oxide that is thin (but within acceptable specification ranges), but has thick polysilicon (also with specification limits), may be prone to excessive gate leakage. The combinations of critical factors are almost endless when the total number of process steps are considered, but, with some study, a few key parameters can be combined to form even more subtle analysis. To take an example, it is possible to determine the yield impact of three or more variables. For instance, the yield effect of thin gate oxide, thick polysilicon, excess phosphorus dopant, and thin polysilicon oxide layers can be determined by sorting the data in a spreadsheet or database program. Obviously, computer support is required for this kind of analysis, although many results can be obtained through these simple sorting procedures.

It is amazing to see the results of this relatively simple sort of low-yield analysis. Very often, a few specifications can be slightly changed to optimize the key process for a particular parameter, and the process boundaries more tightly defined so that the manufacturing area can have solid guidelines for wafer dispositioning. It is not uncommon to see a steady progression of yield improvements due to this type of analysis, with a final outcome of a 10 to 30% yield improvements.

The next procedure involves performing electrical tests on the failed parts. Obviously, if only a few chips have failed for a particular cause, more care must be taken during analysis. The electrical tests are similar to those performed in the other sections of this chapter; for instance, breakdown voltage, diode leakage, and circuit timing tests are all possible tests that may be performed. The main difference in these tests is that they are performed on the circuit at almost any cost. People have been known to wire connectors directly to portions of the circuit for test, or to use lasers to isolate circuit sections for analysis, although it is more likely that the majority of the analysis can be performed using standard bonding pads and probe cards. Special tests may be performed to induce certain other types of failures to determine how quickly related problems may appear. The chips may also be operated to device wear-out to determine all of the major wear-out mechanisms at work in the chip. Unfortunately, these types of tests often will often supply only a little more data than those obtained through the standard electrical test procedures.

At this point, visual examinations are started. First, each chip that is to be tested is examined in intimate detail. Sometimes enlarged photographs of the chip are used (with plastic overlays to permit marking and labeling) to specify defect locations and track other visual anomalies. All defects must be located and their location within the circuit noted. The location information is critically important when determining a problem. For example, if a problem exists within a memory array, and no particles or other defects were seen but one particle was seen in an isolated field area, it would be very difficult to maintain that an array failure was the result of particulate contamination.

When a defect is found in an appropriate area of the circuit to cause a failure of the type noted, the inspector must then examine similar regions on other chips that exhibit the same type of failure. If the defect is found on all of them, it is safe to say that removing that defect will eliminate a portion of that problem. The exact amount of improvement to expect from the removal of the defect will be based upon experimentation. (After all, you have no idea how many working chips have that very same defect—it may only be coincidental or an indication of some other marginal circuit component.) If a number of defects are found that can be traced to a specific problem, they must be cataloged and the effects of their removal noted.

If no defects are found on a surface examination, it becomes necessary to remove some of the layers of the wafers. First, the upper passivation layer is removed, which exposes the topmost metal layer. This is examined for any defects that are attributable to the film removal process. Problems include hillock formation, voids or cracks in the metal lines, and other related problems. Embedded particles can also be a severe problem. The topmost metal layer is removed next, and the interlayer dielectric observed for cracks or leakage, and so on. This continues until all of the films have been removed or until the wafer has become so damaged that further inspection is impossible.

Even after all of this analysis, there will be a certain low percentage of chips that will have failed for no known reason. Sometimes a reason will be known, but there will be no technology available that can prevent that particular set of circumstances from occurring. While it is important to keep track of all of these failures, and to keep good records of test procedures used, only a certain amount of time should be spent on completely unknown failures. While it is possible to discover some otherwise untapped potential for yield, the existence and maintenance of a good data control system, and a well-planned strategy of yield improvement will result in an ever-decreasing number of unknown device failures and in a steadily increasing yield of working chips.

11.5 GENERAL DISCUSSION OF WAFER TEST

Every semiconductor chip must go through some sort of test procedure. In the simplest integrated circuits, the tests are fairly straightforward. The test procedures and the associated time requirements and costs escalate steadily as chip complexity increases. For the most powerful microprocessors and memory devices, the test procedures can add up to more than a third of the manufacturing cost of the chip. Therefore, it is important to control the devices in the test area most carefully, as they have been essentially completed with the manufacturing costs almost fully incurred.

Since the wafers must be tested completely and then diced into separate chips and glued and bonded into their packages, the possibility of losing good, working chips is high. The amount of handling each chip receives individually is higher during these steps than at any other time during the manufacturing process. For instance, the test probes must be placed onto the chips without damaging the bonding pads. This is more difficult than it sounds. If the probe wires are too sharp, or have too much pressure applied, or are improperly bent, they can easily penetrate the bonding pad and cause the device to fail.

Some of the tests may themselves cause device failures if the test system sends improper signals to the chip, such as applying excess voltage to the circuits. These types of problems can occur as a result of mistuned or failed test equipment (for example, a miscalibrated dial that reads 5 volts, when the actual voltage is 12 volts), or as a result of operator error (sending the wrong program to the prober, for instance). These types of failures can sometimes have subtle causes. As an example, noise from an unknown RF source in the test area may be picked up by the tester and transmitted to sensitive areas of the chip's circuitry.

Particle contamination is not terribly significant at this point in the process, although chemical contamination can create a number of problems, and physical damage is real possibility. As a result, the cleanliness of the test and assembly areas is far below that of the wafer fab itself, but the wafers still must be handled with vacuum wands, and cannot be touched on the front surface. Scratches in the protective layers on the chip can allow the penetration of chemical contamination into the chip. Moisture is a serious offender here, seeping in through cracks to attack the underlying metal layers. Chemicals that can induce device failure include sodium and other metallic salts delivered by skin oils.

Many of the problems that occur during these operations result in long-term failure modes. For example, an improperly bonded wire may fall off after the device has been operating for a while. The stress of the continual heating and cooling of the chip (while turning the computer off and on, for instance) will eventually cause sufficient metal fatigue to cause the chip to fail. The chemical contamination problem spoken of in the preceding paragraph also results in a long-term reliability problem. Therefore, a full-chip reliability test must take place after the chips have finished assembly.

The methods essential to the testing and analysis of integrated circuits are, in concept, quite simple (as usual, the simplest idea is usually the most elegant). They start with finding a commonality between several parts that have failed. The clues that are gathered there allow you to determine the next logical course of action. This decision must be based upon acute observation of the significant data trends. For instance, if the chips are failing due to a low gate breakdown voltage, further study might show that the gate oxide thicknesses were low for those lots in question. Process control logs may then show that the oxide thickness is low for some reason on one of the gate oxidation furnaces. The true test for these analysis techniques is to predict a failure rate on lots in progress based on their parametric values. Only a very stable fab area will be able to reach this level of control.

Another use of data trend analysis was described earlier, when we displayed die yield versus parameter, and parameter versus parameter plots. In these cases, die yield improvements can be made just by slowly and methodically "tweaking" each process to its optimum point.

The association of the visual defect analysis at LYA with the on-line particle control data is critical. Particles that can cause chip failures can occur in any number of steps, and the step associated with failure (e.g., a short between two polysilicon layers) may be caused by particles deposited at unusual locations (in this example, the particle may well have been added at the cleaning steps prior to the interlevel polysilicon oxidation step). Therefore, high particle-count data that show consistent trends should be analyzed. When a lot is found with a particular type of failure, the process control logs should be studied to see if there is a day where high particle counts occurred at any step. All reasonable possibilities should be checked. Then the records and trend charts should be carefully compared to see if lots that had run through the operation on previous days when the operation also showed high particle counts, also exhibited evidence of the same failure mode.

Finally, if it is determined that a particular problem has a source at an operation, an experiment should be run, splitting a lot or two and processing the halves through various arrangements of the process in question. These "split lots" will permit you to get an idea of the amount of damage that is being caused due to a certain problem, or to get an idea of the improvement possible with a process modification. Split lots were discussed in a previous chapter, but remember that the samples must be large enough to have statistical significance. Due to the cost of manufacturing wafers, the number of wafers permitted for engineering experiments are limited, and operational chips will be taken away and sold, often removing part of the control portion of the test. These are political issues that should be resolved prior to initiation of the split lot, so that there is no conflict when the wafers are completed.

We also have the matter of reworks. Most fabs do not allow rework procedures to be performed on standard production material. Some, however, do allow the reworked material to be sent on under engineering control so that the effects of the rework procedures can be studied. Even though there is usually only yield loss associated with reworks, some surprising things can happen. For instance, some of the improvement seen from the use of sacrificial oxides to remove substrate crystal damage was largely discovered through rework procedures, as were some of the unusual effects of oxygen dissolved in the silicon substrate. Therefore, not all material should be thrown away. Each set of material should be examined and allowed to continue if the procedure has not been attempted before. Obviously, excellent documentation should exist on attempted rework procedures and their results, so that the work can be easily referenced in the future.

Finally, there is high-yield analysis. These procedures are similar in most respects to those of low-yield analysis. In these cases, an exceptionally high-yielding lot is taken from the fab line for analysis.

Some of the testing will be destructive, resulting in the certain loss of some good chips. This is done to find out why they work so well—what makes this lot better. First, the lot's history will be examined closely to find out exactly what each process parameter was. Other lots that had high-yielding averages will be compared and common trends analyzed. Finally, the wafers themselves will be analyzed. A few will be taken for chemical and surface analysis, while others will be subjected to SEM surface and cross-section analysis, and the visual inspection and film removal procedure described earlier. While high-yield analysis is somewhat expensive in the short run, it can be very lucrative for the long run, in that you may stumble upon otherwise unnoticed details that can allow process changes that will improve yields dramatically.

11.6 WAFER TEST WRAP-UP

We have seen in this chapter that testing must occur throughout the wafer manufacturing process. In many cases, the testing procedures are the source of handling and contamination problems. Nevertheless, the wafers must be analyzed at every step to verify that correct operation will result. Historically, test wafers have been used for most process monitoring, and special test structures on production wafers have been developed for electrical parametric testing. However, the steadily climbing costs of wafers, along with the cost of real estate on those wafers combined with the historical problem of preventing the test wafers from contaminating the process, have caused IC manufacturers to use the production wafers and chips for these tests directly.

Parametric tests are performed in the fab and cover a variety of areas. They include film thickness, resistivity, dopant content, and so on. These tests are performed in real time, so that the manufacturer will know at any time whether the process is running in control or not. It is critical that process control be utilized in association with parametric testing, since variations often occur with the process.

Electrical tests are performed after the metallization layers have been placed on the wafers, sometimes after the protective layers have been deposited. The data obtained include diode leakage, resistor value tests, contact resistance, capacitance, and other basic test values. These tests are performed some time after most of the critical processing has taken place, so that problems found here will have a lag time before any fixes put in place will have effect. This lag time will be directly proportional to the throughput time of the wafer fab.

The sort and reliability tests are performed at the end of the manufacturing process. The chips are separated from each other and the good ones sent to customers. The bad ones are thrown away or tested further, depending on the problems that have been found. Problems discovered at this late stage may not be resolved for many months, depending on the severity of the problem.

Due to the complexity of modern integrated circuits, it is becoming increasingly difficult to test them properly. As a result, a number of manufacturers have instituted self-test programs within their chips. These test modes, sometimes documented, sometimes not, permit the user to apply certain voltages or issue certain instructions and the internal circuitry will execute a test program that will return a code indicating any areas of trouble. While these test programs cannot cover all possible failure modes, they can make the job of the test engineer much easier.

Through the 1990s, the importance of integrated circuit testing to the overall success of the fab will increase. The complexity of the devices means that the test procedures are complex and prone to misinterpretation. As a result, manufacturing requirements must become tighter in order to keep the technology well within its boundaries. The costs will also continue to soar, causing many more manufacturers to build self-test circuitry into complex devices, or otherwise develop new, more cost-effective methods of chip testing. Some excellent ideas are already appearing, such as the use of E-beam probes rather than electrical wire probes. They allow sufficient current flow for device operation, but cause little or no damage to the semiconductor device. Many more of these types of systems will exist by the end of the decade.

We can see that, as a result of all of the data generated, a very organized system is required to control the information. This is provided by the statistical process control system, which is the subject of the next chapter.

CHAPTER 12

STATISTICAL QUALITY CONTROL

The term Statistical Quality Control has been used in a wide variety of ways in the semiconductor manufacturing industry. The term is typically synonymous with the term Statistical Process Control, and they are both usually referred to by the acronyms SPC or SQC. Although there are various methods of implementation, I will be describing the ones that have seen the most service in the semiconductor industry. It should be noted that there are a number of differing viewpoints on the exact makeup of SQC, which is a topic covered quite well in the published literature. Technically speaking, *Statistical Quality Control* is the act of observation of a process and is more of a reactive or passive term. The term *Statistical Process Control* implies that an active role is being taken in the control of the process. This action is taken in accordance within the guidelines laid out by the Statistical Quality Control procedure. However, there is little strict adherence to these terms so they are often used interchangeably. There is probably no "right" method for implementation of the SQC or SPC methods; instead, it is probably more important to be consistent in the methodology used so that the repeatability of the process is maintained.

The effectiveness of the program has varied drastically from company to company and from fab to fab. There are many reasons for this inconsistency, but the main reason, I believe, that there has been a problem is the lack of consistency in methodology and lack of standardization in the area. A significant amount of work has been performed by AT&T and Grant and Leavenworth. This work forms the basis of many of the SQC programs that have been implemented.

Note that there are terminology differences from place to place in the literature and in the real world, especially with regard to limit names, and the phrases "defect" and "defective." We will not get into these issues in depth, but instead will just define our terminology. The terms engineering control limit and specification limits are defined below, but are usually equivalent to the 2σ and 3σ limits, respectively. In some cases, these are set at 4σ and 6σ. The term *defect*, in this chapter and throughout the book, has meant that a chip has a particle or some other single occurrence of something not specified in the process. The device may or may not operate with a defect. A *defective chip* is one which fails to operate within specification ranges. For a chip to be defective, one or more defects will be present.

The process of implementing SQC must include the development of data reporting formats, graphical reporting formats, and the ability to include trend analysis, as well as specify response procedures. The trend analysis may be in the form of a specification stating procedures to follow in event of a problem, or it may be a computer algorithm that generates a reasonable answer based on evidence available at the time (this could include the use of an expert system, for instance). The response procedures need to include actions to be taken and responsible parties. We will attempt to develop part of these requirements in this chapter, and give direction for implementation of SQC procedures.

We have discussed the improvements that are possible through the use of control charts in many places of the manufacturing process. However, these improvements are possible only if the program is used properly and has the required dedication from all of the groups involved. Half-hearted efforts may even cause more harm than good in many cases. When fully and properly implemented, the SQC program becomes more than just a "team project" and becomes the basis for an overall high-quality manufacturing environment. Ultimately, the SQC procedures themselves become tools for a more global method of fab management known as Management by Exception. In this system, which assumes very tightly controlled conditions, new decisions are based only on exceptions to specifications, with the specifications covering all details of the processing sequence or maintenance work required.

The SQC program (especially when related to the control of particles) is probably the most complex and interdisciplinary program to implement in the fab. This is mostly due to the tremendous flow of information and the requirements that process problems be solved immediately to prevent undue downtime. Critical information must be filtered out of the overall flow of information (often by nontechnical people) and developed into a form that is useful and significant. Data from different groups must be correlated to see if any trends are visible and to identify problem areas. Action must then be taken to prevent the problem from recurring, and the process tested to verify that the problem is resolved prior to starting production in that area. Finally, it is important that the Quality Assurance department audit records at periodic intervals to verify that all record keeping is being performed correctly. When manual SQC systems are used, errors can easily crop up, corrupting the data that the engineer must use to analyze a problem. While errors can be reduced by implementing computerized systems, it is still critical that the QA audit be performed on a periodic basis to ensure the best process control possible.

Thus, we will develop guidelines and procedures for the gathering, analysis, interpretation, and distribution of SQC data to the various groups. We will also combine them with potential response scenarios to show how a management by exception system might work. We will start with the implementation of data-gathering methods.

12.1 DATA COLLECTION METHODS

Since the semiconductor manufacturing process consists of so many steps that must be performed to exacting specifications, it is critical that information is kept for all parameters that can cause deviation from these specifications. While each process step will have details that are specific to it, the basic formats for the log sheets can have a number of features in common. In an effort to reduce paperwork, a computer network should be hooked up which allows the equipment data logging features to download all of the processing information available for any one run. In general, though, even if direct downloading of the data is not possible, it is not a good idea to skimp on data collection. Even though many manufacturing personnel believe that the data collection is not required, the three minutes or so required to go through a "preflight" checklist is more than justified by the several hundred thousands of dollars worth of material that could be lost in each lot. If a process develops problems, it is critical that the source of the problem is identified as rapidly as possible, before more material can be processed through the equipment or before some production causes unacceptable downtime and production line backup.

12.1.1 Electronic Data Collection

First, we will discuss the data that must be obtained from the wafer manufacturing equipment. Almost all of the wafer manufacturing hardware today is built with a special communications standard built in. This standard is called SECS (Semiconductor Equipment Communications Standard), and is in the second generation, so is usually called SECS II. This communications standard consists of a set (or "stream") of data that contains functions that can operate the systems and parametric information. Some types of processing equipment can be operated entirely from a remote host computer. This allows for a complete data transfer of all parameters. The biggest problem with this type of mass data transfer is that the storage space required for all of the equipment in the fab would quickly overwhelm most mass storage devices. As a result, it is usually necessary to extract only the key information from the data stream and discard the rest. That means that critical data should be stored for each run, and less critical data thrown away at the end of the run unless it is outside of the predetermined values.

Key parameters that should be obtained will depend on the type of processing equipment involved. These will include items such as processing or exposure time, furnace temperature, process gas flows and pressures, RF power levels, and so on, as shown in Table 12-1. If there is enough processing power available, the manufacturer may want the host computer to sample each piece of equipment several times during the process, and the host computer should be able to immediately analyze any alarm condition. The critical pieces of data that are obtained should be stored in a database that will allow random access based on a wide variety of input, such as requests for all process pressures in a certain furnace on a certain day.

Semiconductor test equipment can also use the SECS II interface to allow a host computer to specify particular wafers to measure or to download test data. This test data could include things like raw thickness or resistivity data, and electrical test or sort data. The data may also be transferred as analyzed data—that is, the data that are transmitted to the host have already been evaluated, so that the data will include wafer averages and standard deviations for the parameter in question. Some equipment requires that data be downloaded through a floppy disk to a remote PC. While this works for a wide variety of situations, there are problems associated with the management of data in distributed databases that go far beyond the scope of this book.

TABLE 12-1
Some Process Parameters Requiring Real-Time Monitoring

Diffusion/LPCVD	*Ion implant*
Temperature	Pressure
Gas flows	Control currents
Pressure	Dose monitors
Process time	Source monitors
Photolithography	Implant time
Resist thickness/uniformity	*Etch*
Develop solution temperature	Pressure
Exposure time	RF power
Lamp intensity	Temperature
Alignment/registration	End-point detection
Plasma Deposition/Sputtering	Gas flows
Temperature	Bias voltage
Pressure	*Metrology*
RF power	Wafer/lot IDs
Gas Flows	Cassette positions
Bias voltages	Metrology info, as needed

Suffice it to say that the best solution is to have all of the available wafer data stored in one master file or system, and to maintain tight control on the ability of the user to change the data in that database. Due to the amount of information in the database, it will be virtually impossible to keep the data from being corrupted if they are not controlled in the strictest fashion. The danger of database corruption, of course, is unreliability in results. Serious cases of this could lead to yield crashes and other unpleasant side effects.

Finally, a number of manufacturing systems have started to include self-diagnostics software, along with the ability to communicate directly with the vendor's service department over modem to determine the source of problems quickly and effortlessly. In the future we will see this concept extended into the area of *artificial intelligence*, in which expert systems will be used to diagnose and resolve most processing problems.

12.1.2 Manual Data Entry Logsheets

At this time, the number of fab areas that are hooked up in the network described above is very small, although growing rapidly. A number of fabs have pieces of the system in place, usually requiring manual data entry of key parametric values by the operators. If the procedures for this data entry are not set up for maximum efficiency, this inherently time-consuming task can significantly reduce the level of productivity in the fab. In order to reach maximum effectiveness, it is important to design logical logsheets and data entry screens.

There is no perfect layout for a logsheet, and virtually everyone who tries to design one will come up with something different. In this book, we will try to cover the major styles of logsheet and I will give my opinion as to the most effective. However, the logsheet that is most effective in any fab is the one that contains all of the significant data required and is easy to read and use. If the operators who must use it actually like it, it is likely to be a success. However, do not compromise on data collection, since one of the complaints will always be "There's too much writing." As mentioned before, it makes little sense to worry about two or three minutes of process verification and data collection, when hours worth of process time and several hundred thousand dollars worth of semiconductors are at stake. It seems a small price to pay for process control. Many fabs are trying to go to "paper-less" environments without logsheets and run cards, although this is very space- and computer-time intensive.

There are three basic formats for logsheets. The selection of the basic format is made by the complexity of the process and the number of variables involved. Each piece of equipment and process should have its own logbook, including all clean and inspection stations. When multiple processes are run in the same chamber, for example, in the diffusion processes, the logsheet should be produced in a format that is flexible enough to allow it to be used for all of the various processes used in the chamber. The only drawback to this approach is that there will be some loss of logsheet space due to the need to have all of the various data.

We will start with the simplest logsheet, the single-line log. In this type of log (shown in Figures 12-1a and 12-1b), each process run is described in one line. The log can be arranged in a vertical or horizontal layout, but we will use the horizontal layout example. This type of log assumes that each run consists of only one lot, as in the event of an implanter or single-wafer etcher. It also assumes that the amount of information required is limited. The example describes a log for a oxygen plasma etch process. In it, we see that there are a number of columns for each of the primary parameters. We will go into each of these in some detail. The other types of logsheets will use much of this information in them so we will not duplicate the description there.

a

Run #	Date	Time	Lot #	Number of wafers	N_2	O_2	SiH_4	Pressure	Process time	Temper-ature	T_{ox}	Notes
1	10/9	14:30	4429	22	6.6	4.2	0.12	180	18:00	420	3018	
2	10/9	18:30	4422	18	6.5	4.2	0.12	181	18:00	420	3029	
3	10/9	22:00	4336	24	6.6	4.2	0.13	180	17:50	421	3005	

b

Run #	1	2	3						
Date	10/9	10/9	10/9						
Time	14:30	18:30	22:00						
Lot #	4429	4422	4336						
Number of wafers	22	18	24						
N_2	6.6	6.5	6.6						
O_2	4.2	4.2	4.2						
SiH_4	0.12	0.12	0.13						
Pressure	180	181	180						
Process time	18:00	18:00	17:50						
Temperature	420	420	421						
T_{ox}	3018	3029	3005						
Notes									

Figure 12-1. Single-Line Logsheets Shown here are two possible layouts for single-line logsheets. They are ideal for single-wafer or single-lot processors, and are easily modified from these examples to incorporate key parameters for other processes. (a) Horizontal layout; (b) vertical layout.

First, the run number is recorded. This is typically defined as the number of runs since the last cleaning cycle or preventive maintenance cycle. Usually, the run number is allowed to increment up to the point where a clean must be performed, and then it is reset to one. This run number is usually used to correlate with the process control charts, which will be discussed later. The run number is used in conjunction with the date and the time to uniquely identify all runs and the time frame of the process. This may be required in the event that wafers at sort or electrical test are discovered, with a problem traced to the particular process.

The next two columns contain the lot numbers and number of wafers in each lot. As mentioned, this type of logsheet is conducive to single-lot processing, since each line has space for only one lot. It is possible, of course, to use multiple lines for each process, although this is not recommended (one of the other formats will work better). Of course, this data is required so that the lots can be traced through the various steps.

The next several columns contain the actual processing parameters that were obtained during the process. For example, the first two columns contain the gas flows for this particular reactor. This data should be obtained from the equipment itself during appropriate steps in the process. Typically, these gas flows must fall within certain predefined limits. For instance, in the O_2 etch process the exact flow of oxygen must be regulated within a few percent. The flow rate should be recorded so that changes in yields at that step may be able to be associated with changes in gas flows. Another parameter that is associated with gas flow control is that of process pressure control. Changes in process pressure can cause significant deviations in device performance and yield. As a result, this is a value that must also be observed and controlled at all times, with all variations recorded for future analysis.

Another set of columns in this example are those relating to the power input to the system. In this case, if we are describing a plasma etch process, the key parameters would be RF energy and frequency, as well as chamber temperature. We will record this data for future analysis, also.

Of course, we now have several parameters that have one place to enter data, yet have effects that last for several minutes to several hours. For example, there is only one line to enter a gas flow. If the gas flow is not stable and changes after the first check, how does one record this information? Also, if the process is very long (as in an anneal process), the gas flow should be checked several times for verification of correct flow (note that most systems have alarm conditions that can be defined by the engineer to trigger an alarm on discrepant parameters). As a result, the amount of data entry required should be considered when making the decision of which logsheet to use. If variations in parameters are possible or the process is long enough to warrant multiple checks, a more comprehensive format should be chosen.

Next, the actual processing time is entered. This is defined as the amount of time that active processing is going on. Not included are times such as vacuum pump downtimes, temperature stabilization cycles, post-processing anneals, and so on. In our example, this time is the amount of time that oxygen is flowing in the system and RF power is applied to the gas, forming the plasma. This is obviously a critical parameter, and is often one of the few parameters that can be adjusted. (Temperatures and gas flows are rarely adjusted in any process; the only variable then becomes time, so that minor variations in thickness can be accounted for.)

There is then a column to record the particle counts for the process, This data will usually be the results for a single test wafer, although more data may be made available, and should be recorded. Finally, there is a column to record notes. Changes in processing parameters (such as the fluctuations noted above) should be noted here. Other unusual occurrences, both bad and good, should be recorded. The space to record notes on the logsheet should be large enough to allow legible notes.

With this type of logsheet, the amount of process measurement data that can be recorded is quite limited. In our example, we are recording the thickness of a structure on the wafers before and after the etch process. This is done to calculate the dep or etch rates of the process. If space permits, a column could be added for recording the calculated etch rate. In most cases, the manufacturing personnel need only a "go–no go" decision, and will only care that both of the values are within specified ranges. As usual, these data can be of excellent benefit when characterizing and optimizing process yields.

The biggest advantage of this type of logsheet is that a large number of process runs can be placed on one sheet of paper. This not only saves paper, but permits somewhat easier viewing of all the trends. This is especially true if a rudimentary process control chart is placed at the edge of the logsheet. Part of the problem with this setup is that the data is presented in a dense format so that a full logsheet of data can be almost overwhelming in its complexity.

The second type of layout for a logsheet is a partial page layout, as seen in Figure 12-2. In this setup, each page of the logbook is made up of 6 to 10 smaller data entry regions. They will include much of the same data as described in the single-line logsheet, but will permit more data to be stored. For instance, the gas flows and temperatures can be monitored at a number of different points during the process. This allows the process engineer to have more information available to analyze process control trends. For example, analysis of this data may show that variations in oxide thicknesses is due to an insufficient temperature stabilization cycle prior to the oxidation cycle.

Run #		Lot #	# of wafers	T_{ox}				Standby	Process	Anneal
Date					Avg.		N_2			
Start time					Std. dev.		O_2			
End time					Range		SiH_4			
Operator							Pressure			
							Temp.			

Process time

Notes:

Run #		Lot #	# of wafers	T_{ox}				Standby	Process	Anneal
Date					Avg.		N_2			
Start time					Std. dev.		O_2			
End time					Range		SiH_4			
Operator							Pressure			
							Temp.			

Process time

Notes:

Figure 12-2. Partial Page Layout for Logsheet The logsheet layout allows the entry of about six runs' worth of data per page. This layout permits significantly more data to be stored for each run, as well as permitting batch processes and more complex processes to be more closely monitored.

In addition to being able to monitor changes in process parameters at a number of different points, this type of setup is well suited for operations that permit multiple lots to be processed at one time, such as diffusion and some thin-film deposition operations. The extra room allows ample space for the entry of lot numbers and sizes. By controlling the location of the log entry on the logsheet, some estimates can be made later as to the position of the wafers within the tube. The measurement data can also be entered in appropriate locations, and from multiple points within the furnace. This data can be derived from test wafers placed in three to five locations within the furnace or from the product wafers themselves. Analyzing the position-to-position measurement data with process control techniques can point out problems such as unbalanced temperature profiles, unstable gas flows, and so on.

The complexity and density of each data collection region is a key factor in the success or failure of this type of logsheet. Trying to put too much data into each area can result in the logsheet becoming cumbersome to use and therefore not filled-in properly by the manufacturing personnel. Thus, while it may be possible to incorporate this type of logsheet with a process control chart, it is not recommended. Once again, however, if the information is important enough to cause an impact on the process, the data should be recorded.

The final kind of logsheet layout is the full-page layout, as described in Figure 12-3. In some cases, it may be possible to break the sheet up into two parts, if there is enough room to prevent crowding. This type of logsheet is used when all possible information is desired. In this case, the same type of information is recorded as in the partial page layout, only in this case more extensive data analysis can be performed.

One of these additional features is the provision for a process "preflight" checklist. This is a convenient device to verify that all systems have been checked and that every possible precaution has been taken prior to starting a lot of wafers in a process. This procedure should be followed in any event, whether the checklist is formalized or not. The benefit, of course, is the verification.

Another potential feature for the full-page layout is the ability to calculate and record a larger amount of statistical data than was possible with the other layouts. This statistical data can be used in conjunction with the process control charts to control and predict the process parameters with a very high degree of precision. This data can involve more complex data analysis of the individual wafers (by analyzing the data from each wafer), or of the entire run of wafers.

A final possibility for a feature for the full-page logsheet is a worksheet for a control algorithm. This is a feature that is only occasionally used in wafer fabrication. This could be handy in certain LPCVD operations where temperature adjustments of a degree or two are routine. Normally, these changes are performed by engineering, but are certainly not so complex that they could not be performed by manufacturing, given the proper guidelines. Thus, a worksheet that prompts the operator as to what changes to make can free time for the engineer to work on improving and updating the process and not making minor adjustments.

Run # ______ Date ______ Start time ______ End time ______ Process time ______

Lot #	Device	# of wafers	T_{ox}			Preflight checklist		Initials
______	______	______	______	Average	______	Gas setpoint	______	______
______	______	______	______	Max.	______	Temp. setpoint	______	______
______	______	______	______	Min.	______	Program installed	______	______
______	______	______	______	Std. dev.	______	Wafers cleaned	______	______
______	______	______	______	Range	______	Wafers loaded	______	______
______	______	______	______	$\frac{\text{Max. - min.}}{2 \times \text{mean}}$	______	Program started	______	______

Step	Push		Ramp up		Oxidize		Anneal		Pull	
Parameter	Set	Actual	Set	Actual	Set	Actual	Set	Actual	Set	Actual
N_2										
O_2										
SiH_4										
Pressure										
Temperature										
Operator initials										

•
•
•

Figure 12-3. Full Page Layout for Logsheet This layout can take from a half page to a full page, depending on the amount and types of data required. In this case, the logsheet is used as a real-time record and checklist of the process, not just a record of the runs manufactured.

The full-page logsheet layout is for the very committed only. The logsheet takes up paper, space, and even more time to fill out. However, if the optimum in process control is desired, complete data logging is required. It is for this reason that direct communications links should be set up. This same type of logging system could be set up from a host computer to terminals at each operation, permitting the computer to verify that all of the human interactions have been accomplished. All of the equipment parameters can then be checked directly from the host computer.

If data must be stored on the logsheet and then later hand-entered into a computer database (a very common situation), the format of the data on the logsheet should be in the same order as that on the computer data entry screen. This prevents some of the data entry errors that will crop up.

We have seen that there are a number of logsheet styles available. No one logsheet will work for all processes, nor can any generic logsheet (other than the most generalized full-page layout) cover all possibilities. Therefore, the final decision on the logsheet style should be made after careful consideration by all of the groups involved.

12.2 PROCESS CONTROL CHARTS

Process control charts are typically generated from the results of the measurements from each process run. These data may be film thickness, resistivity, linewidth, or particle count values, and in most cases the process control charts can be handled in a similar fashion. In most cases, the process control chart is generated from the process data in real time, that is, after each run is completed. This is done to allow all process trends to be analyzed prior to the start of a subsequent process. If a problem is seen, changes should be implemented prior to the start of the next run to prevent further loss of material.

The process control chart is typically filled out by the production operator. As a result, the methods that are required to produce the process control data should be easy and fast to calculate and to record. There are often calculators distributed throughout the fab to assist the operators while calculating these data. Of course, in an environment where the data are automatically uploaded to a host computer, the operator interaction will not be required, saving production time and effort.

The process control charts should be associated tightly with the logsheets of the process under control. In other words, all of the data required on the process control charts should be available in raw form on the logsheet. In addition, the proper design of a logsheet and process control chart has the data entry in logical, consistent locations throughout each document. The proper way to set these up are shown in Figure 12-4, which also shows a setup that is improper. Notice in the incorrect setup that you are forced to jump around from place to place on the logsheet, which wastes times and adds confusion. If the data can be transferred easily from the logsheet to the process control chart, there is a much better chance that all of the data will be transferred correctly.

One final point before starting the discussion of the process control charts themselves. While the engineering staff usually is charged with the responsibility to make process adjustments, the manufacturing personnel should understand what adjustments are being made

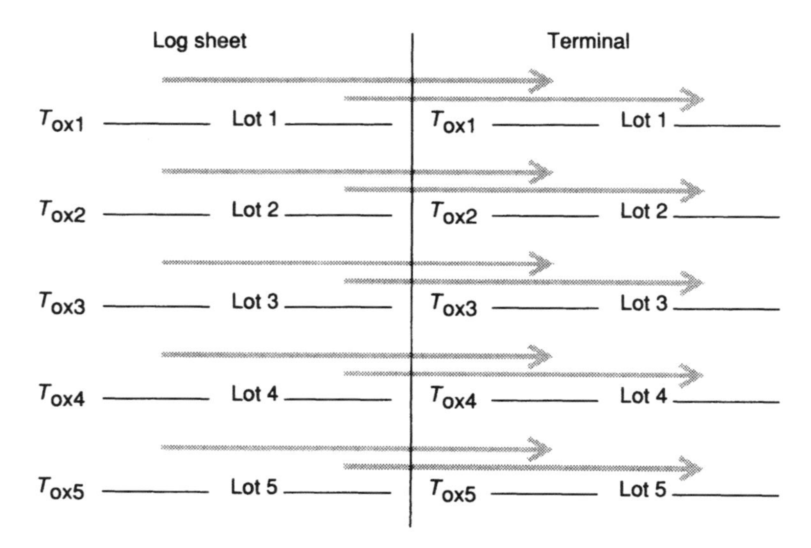

Proper data transfer scheme

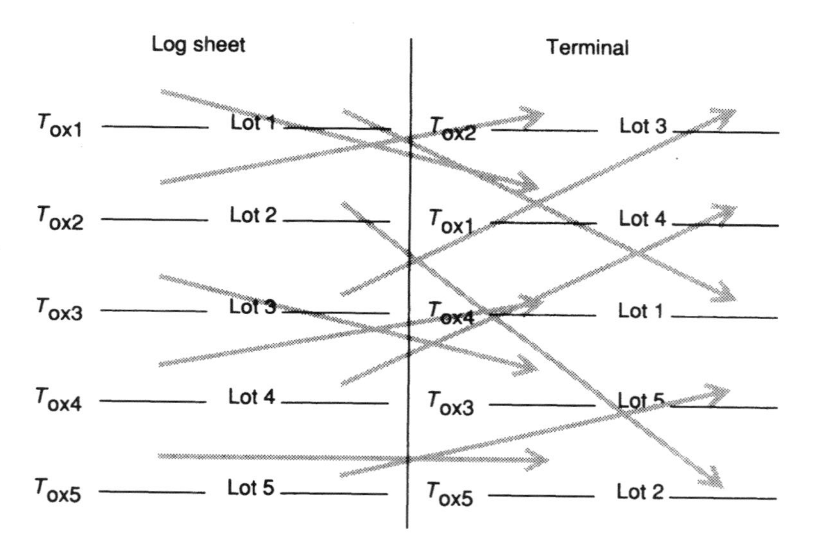

Improper data transfer scheme

Figure 12-4. Efficient Data Entry To effectively transfer data from the logsheet to a data entry screen on the computer-based wafer tracking system, it is critical to use a consistent format. As shown in the fairly obvious example here, it is much more confusing to replace data if the sequences are inconsistent. Not only is the probability of error increased, but the time required to enter the data is increased.

and why. This entails understanding the operation and interpretation of the control charts. This can result in a number of positive side effects. First, it can help to speed the process correction cycle when a run is discrepant, as the operator will be able to report the problem accurately, possibly proposing a solution if it is apparent. Next, the operator can eventually learn to predict the operation of machines so that through subtle clues, he or she may be able to prevent discrepant runs from occurring in the first place. Finally, with sufficient experience, the operator will be able to tell if the engineer or maintenance person has indeed repaired the process. If this repair work is insufficient for the purpose, the engineer or maintenance staff can be called back until the problem is resolved.

Now we can start to discuss the process control charts. Actually, the standard process control chart in a wafer fab is a fairly simple thing. As shown in Figure 12-5, it consists primarily of two types of line graphs, along with associated run and lot identification data. Although process control charts can be defined in either a vertical or horizontal layout, we will use the horizontal layout for our examples. Typically, the results of 10 to 25 runs are placed on any one process control chart. The number of runs recorded is largely a matter of individual preference, as long as enough data are available to show process trends.

The first line graph shows the run parametric averages. For instance, it might contain the average thickness from an oxidation run,

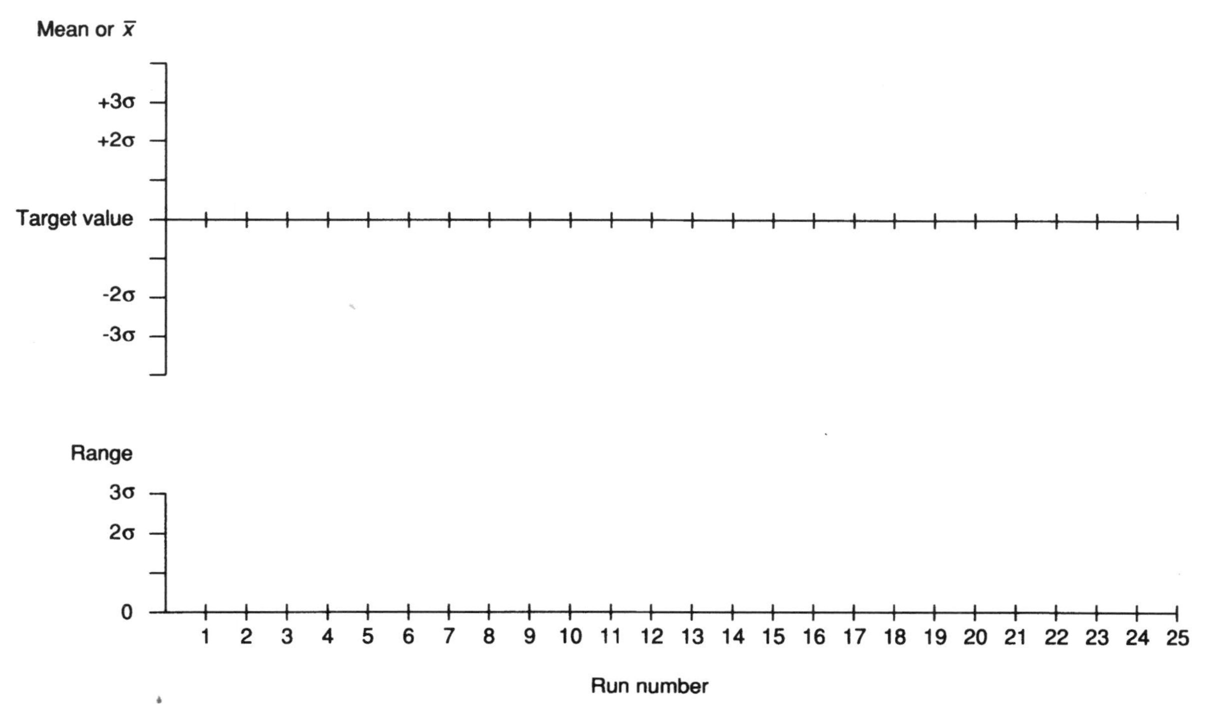

Figure 12-5. The Basic Mean and Range Charts for Statistical Quality Control Statistical quality control charts are used to gauge the control and repeatability of the processes in the fab. The vertical axes of the charts are identified with the process parameter desired. The horizontal axes are aligned with the run number of the process. This is an arbitrary number usually associated with maintenance cycles. Record the average of the test wafer readings on the top chart, and the range (maximum value − minimum value) of the readings on the lower chart.

or it might contain the average resistivity of a lot of wafers after an implant operation. It may also show the average of the etch rate of a plasma etcher. This chart is called the X or X-bar chart, or the "Mean" chart. Each process run is assigned one column and the run number that is used to identify the particular run on the logsheet is used to identify the run on the control chart. This is recorded in the space provided. The average thickness for the run should be calculated from the parametric values entered on the logsheet (we will discuss the methods of calculating various statistical data in the next section). The average value that is used should be entered onto the logsheet, and the value located on the graph in the appropriate column and a hatch mark entered. The hatch mark should then be connected by a line to the previous data point. A typical control chart is shown in Figure 12-6.

The average process data is used to determine the overall process response. For instance, you will be able to tell if the process time needs adjustment or if the process pressure is unstable. When a specification quotes a value, for instance an oxide thickness specification of

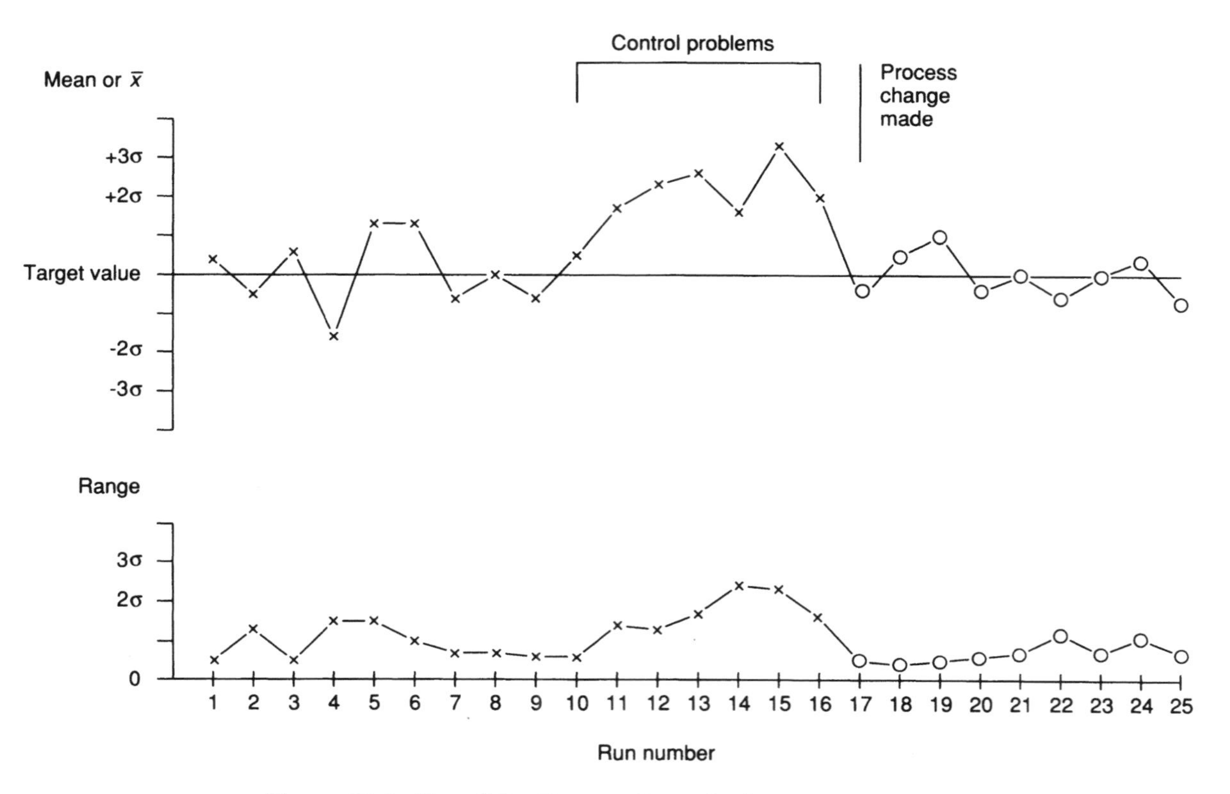

Figure 12-6. Use of the Process Control Charts Here the process control chart is analyzed. In this case, the process was running in control up to about run 9. At that time, the process trends moved up until at run 15 the specification limits are exceeded. A process change is implemented at run 17 to put the process back into control. In actual use, the process adjustment would have been made earlier, so that process specifications would not have been exceeded.

1000 ± 50 Å, that means that the average thickness of the run (or each of the lots in a run) is between 950 and 1050 Å.

The vertical axis on the process control chart is labeled with the possible values that the data should fall within. The central value of the chart will be the specification target. The optimum process will run very near this axis at all times. Some variation is to be expected, so as long as the variation consists of a periodic oscillation around the average value, the process is considered to be under some control. The amount of this control is gauged by the amplitude of this oscillation, which is measured by calculating the long-term standard deviation of a number of runs.

There are two sets of lines placed on the mean control chart. The inside set refers to the Engineering Control Limits. These limits are used by the engineering staff to prevent lots from going outside of the specification limits. If the process starts to drift to any great degree, the average values will exceed the control limits. Procedures are then initiated to correct any potential problem. The Engineering Control Limits can be set in a number of ways. The most common way is to arbitrarily set the limits at about two-thirds of the specification limits (i.e., if the specification range is ±30 Å, the control limits are set at ±20 Å). Another common method for setting the control limits involves running a series of tests (or production runs), and determining the standard deviation for these runs. The control limit (and sometimes the specification limit) is then set as a multiple of the standard deviation from these tests. Usually, the values of $\pm 2\sigma$ are used for the upper and lower Engineering Control Limits, while the values of $\pm 3\sigma$ are used for determining equipment specification limits. For instance, if the furnace can repeat its performance to within ±5 Å to one standard deviation over 40 to 100 runs, an engineering control limit could be set at two standard deviations of ±10 Å, and a furnace specification limit of ±15 Å.

The equipment specification limits must always be within the specification limits imposed by yield loss mechanisms within the circuit. If the system just described is being used to manufacture chips with a yield tolerance at that step of ±10 Å, more than 5% of the wafers run through the system will probably fail during the sort operation for incorrect film thickness. In fact, the integrated circuits should be able to handle a process average drift of over $\pm 1.5\sigma$ without going outside of the yield imposed limits. As shown in more graphic detail in Figure 12-7, this means that the worst case data points would still permit high yields to be obtained. Thus, for a process requiring ±10 Å control, the system would have to maintain control of about ±6 Å at 3σ, if an additional 1.5σ is allowed for process drift. This leads to a worst case variation of ±9 Å, which is within yield tolerance ranges. The engineering control limits of $\pm 2\sigma$ would lead to a variation of ±4 Å in this case.

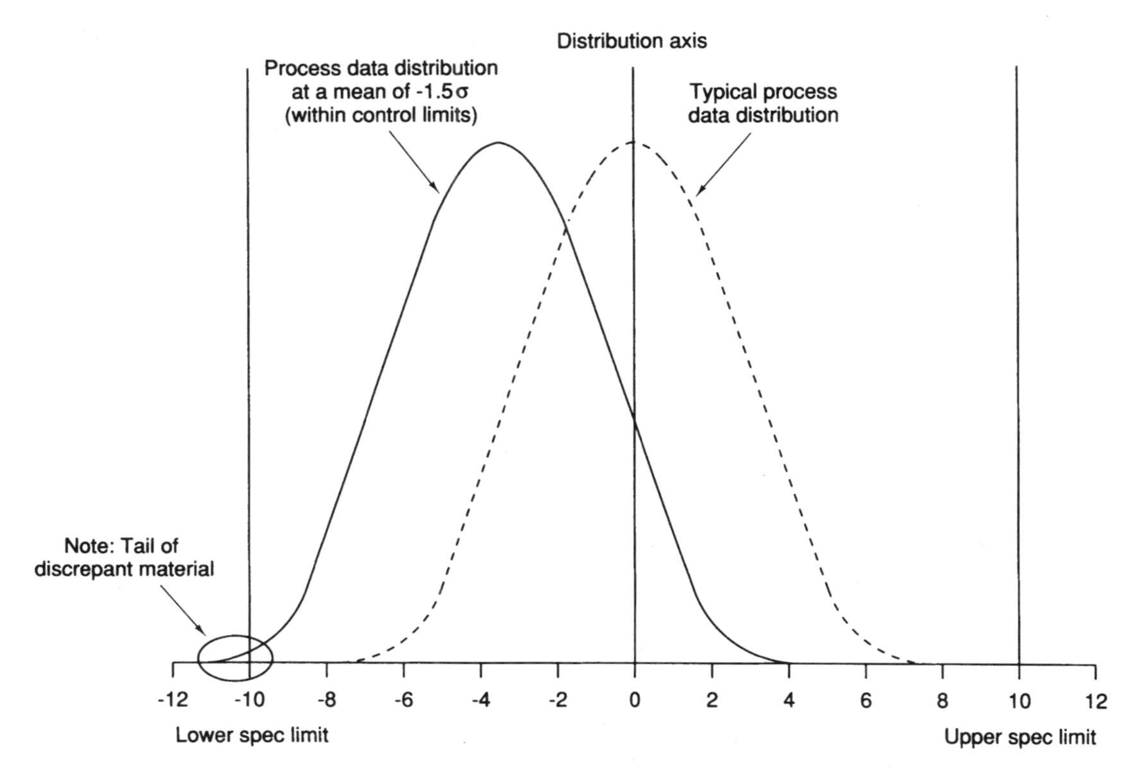

Figure 12-7. Variation in Distribution of Data with Process Drift Here we see the effects of process drift on the distribution of data points resulting from measurements of a parameter of the process. For example, a gate oxidation process may use 10 Å specification limits. If the furnace produces ranges of more than 7 Å in a run, there will be a significant probability that some amount of material will fall out of spec if the process average (or mean) moves by 1.5σ (3 Å). Since this value is within the 2σ control limit, no adjustments are made, and material is processed with discrepant readings. Without tighter control on the process, then, these losses will slowly add up to become quite costly.

The next set of lines on the control chart are the specification limits. They are usually determined by design rule criteria, by the impact of the process on die or line yields, or by the ability of the operation to remain in control, whichever is the tightest. When the process drifts outside of these limits, it is usually shut down, and no further material is run through the system until it is proved that it is back in control. Usually, the control limits will have been exceeded before the specification limits are exceeded, so that there should be some warning that the furnace is having problems. Sometimes, however, the furnace will move directly from a within-spec to an out-of-spec condition. This can be a serious problem if it happens very often, as it implies a process running at the edge of some drastic process change that is otherwise uncontrollable.

Of course, the average thickness cannot contain all of the information about a run. As a result, the parameter called range is used.

This is the second line graph used on the process control chart. The range is defined as the highest reading minus the lowest reading. There are a few different parameters that could be used in this chart, including standard deviations, which would be more correct statistically. However, these other parameters require somewhat more complex calculations, so are not typically used in a manufacturing environment.

Since the range figure consists of the difference between two values, the end result will always be positive. The ideal process range approaches zero as closely as possible. The specification limits are usually set to about the same value as the specification limits for the mean control chart. While the actual lots may be within specification, the probability in increased variation in electrical parameters will be increased if the range is very high, as will the probability of yield loss if the run average also varies. The specification limits of the range values are thus set by the process tolerances allowable to control the process, and not directly by the impact on yield.

The control limit is set in a method similar to that of the mean control chart control limits. Depending on the amount of characterization desired, the control limit can be set arbitrarily to about two-thirds of the specification limit, or it can be set at a lot level through analysis of process trends. In either event, the purpose of the control limit for the range parameter is the same as that for the mean parameter: to allow the personnel running the operation the ability to predict the performance of the processing systems.

We can see, then, a process control chart serves two purposes: to predict the performance of a system, and to monitor the quality of the product. Problems can be seen in a process, whether they are global in nature (which will show up in changes in the average values), or are localized within the furnace (which will show up as changes in the range parameter). After this, further analysis can be performed to permit a more detailed understanding of the process variables. However, to make sense of the data that are presented in the control charts, it is necessary to investigate more closely the interpretation of the statistics surrounding the control charts.

12.3 INTERPRETING STATISTICAL QUALITY CONTROL INFORMATION

There are actually a number of different uses to which SQC data can be put, as shown in Table 12-2. The specific use depends largely on the type of organization using the technique. For instance, the most common use of SQC is in the manufacturing area. This makes sense, because the primary function of the SQC is to verify that the vast majority of wafers produced are within specified ranges. However, SQC methods can be used by equipment manufacturers to guarantee

TABLE 12-2
Uses for Statistical Quality Control Data[a]

	User	Need	Method
I	Fab manufacturing	Process repeatability	Fixed limits for best control
II	Equipment vendor	Equipment reliability	Calculate limits after processing tests
III	IC design/engineer	Yield enhancement	SQC data compared to yield data
IV	Process engineer	Metrology control	Calculate limits on known sample

[a] There are a variety of uses for SQC, including production process control, equipment reliability evaluations, and yield control. While the techniques for analyzing the data are slightly different, the mechanisms for producing the SQC data are quite consistent.

that their processes remain in control and to determine where improvements can be made. Thorough analysis of equipment in idealized situations can reduce the number of uncertainties when a problem is found in the wafer fab.

A third area where SQC analysis can make an impact is during the process development stage. Careful review of process control charts, especially when compared to electrical test and sort data, can give the manufacturer solid information on the tolerances of the various processes before they go into production. This will prevent the creation of specification limits that are too tight or too loose. Associated with this is the ability of the process engineer to further refine the process as it gets into full-scale manufacturing. The data obtained from high-volume manufacturing can give much-needed insights into areas of yield improvement.

The last area in which SQC methodology can be used with effect is in the gauging of metrology systems. It is critical that the measurement systems read the same from day to day and from operator to operator. Through running controlled tests over a long time and studying the results, the repeatability and control of the metrology systems can be observed and the systems properly maintained. All types of equipment can be tested in this way, although the key to the procedure is to make sure that the test samples remain absolutely unchanged from day to day.

The most important value that is determined is the mean value of the process. This is calculated simply by adding together all of the data points, and dividing by the number of points. This is performed as follows:

$$\text{Mean} = (\text{Data}_1 + \text{Data}_2 + \text{Data}_3 + \ldots) / (\text{Number of points})$$

Another common calculation is for determining the range of the process. This is defined as the difference between the maximum and minimum values of the function, as shown in this equation:

$$\text{Range} = X_{max} - X_{min}$$

SQC methods work much better when proper statistical principles are adhered to. In many cases, approximations are used for simplicity in calculation. With the number of modern calculators available, this does not seem to be an issue. One of the most common approximations is to use the formula:

$$\begin{aligned}\text{Approximate Deviation} &= (X_{max} - X_{min}) / 2 * \text{Mean} \\ &= \text{Range} / 2 * \text{Mean}\end{aligned}$$

This value approximates one standard deviation, and for some distributions may be valid, but it will not give a true estimation of the process if there is one point that is for some reason way out of range. (This may be due to a point of contamination or to a problem with the measurement system.)

The most common value calculated in SQC procedures is the standard deviation value. This statistical construct works well for determining the distribution of data when the process produces a Gaussian (or bell-shaped) distribution of points. While many processes will behave in this manner, there is no guarantee that they all will. It is therefore important to verify the distribution of the data before ascribing too much significance to the standard deviation value. A classic example is that of particle count data, seen in Figure 12-8, which shows a distinctly non-Gaussian appearance. If the standard deviation is calculated using the Gaussian distribution formula, it is possible to get a value such as "average particle size = 1.1 ± 1.3 μm." This is quite interesting as it implies particles of a negative size. Since there is obviously no physical significance to this, then a proper distribution formula must be used. Table 12-3 demonstrates a variety of different data distributions, along with the type of curves, and the formulas for estimating the standard deviations. Some distributions can be quite complex, forcing us to use more difficult techniques for analysis. They are well beyond the scope of this book, but can be interesting topics.

There is seldom much interest in the so-called one sigma (or one times the standard deviation value) other than as a baseline value, simply because the process control at that level is too crude to be useful. However, there is a tremendous amount of interest in the control of the process at three times the standard deviation value. This level of control means that over 99.3% of the parts (whether wafers or chips, and depending on the test type and the distribution type) will be within specification ranges at any one process step. This is generally considered acceptable although, after some hundred steps, the actual loss in die yield is considerable. The number of die lost after a certain number of process steps is shown in Table 12-4. This table also shows the impact of slightly higher and lower per-step defect rates, at

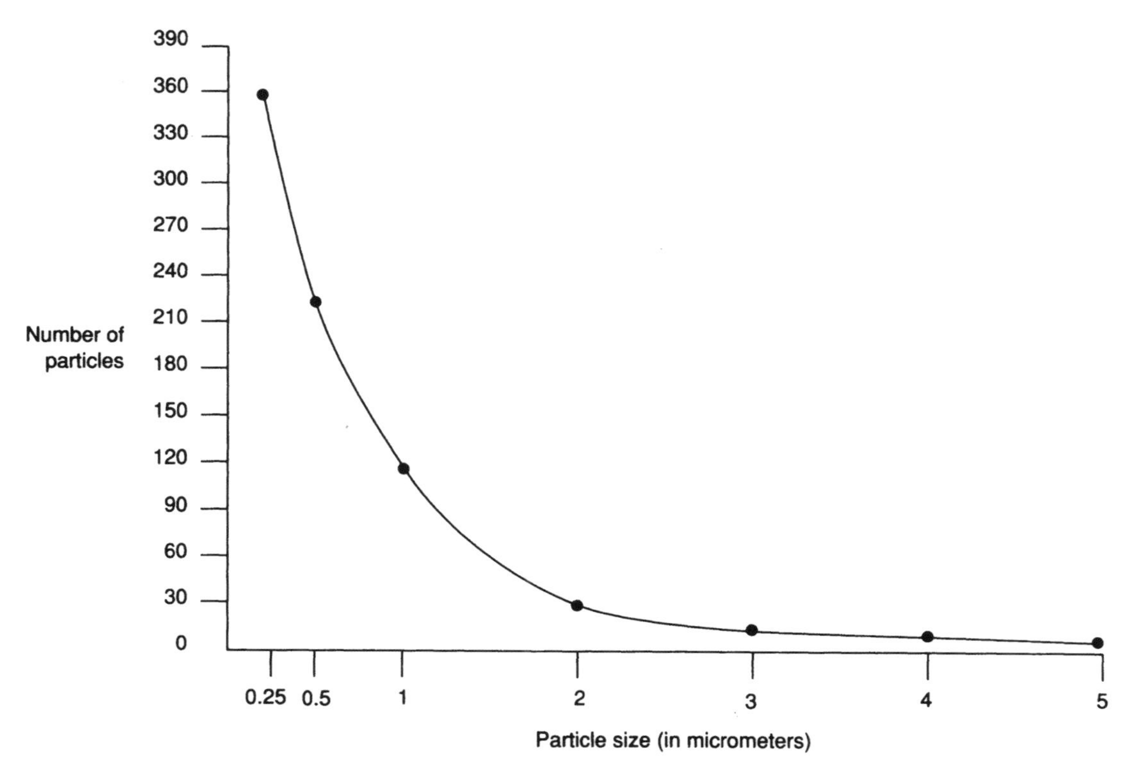

Figure 12-8. Distribution of Particles Particle counts by wafer size is one type of distribution which must be handled carefully statistically. Clearly, while an average particle size is reasonable, the standard distribution must be asymmetric, since a particle cannot have a negative value for size.

one, three, four, and six sigma. The six-sigma values are being used to an increasing degree in wafer fabrication due to the tighter control available.

The long-term averages of the parameters are often of interest. The average values for each of the runs should themselves be averaged together at regular intervals. This may occur every 5 to 10 runs, or every time a new process control logsheet is started. This "universal average" or "average of the averages" permits the long-term trend of the process to be estimated. If this value changes through time, adjustments or long-term fixes can be put into place to maintain tighter control of the process.

The reproducibility of the process can be determined by examining the standard deviation of the run averages. If they vary greatly from run to run, it will be much more difficult to predict the process performance and prevent yield loss in the event of process average drift. In addition, the probability of process drift is much greater if the repeatability of the process is poor.

Determining the control of a process through the use of SQC techniques is called trend analysis. These techniques are used to

TABLE 12-3
Different Possible Data Distributions

Distribution type	Description	Deviations	Curve shape
Gaussian	Uniform distribution around a central mean. Most processes follow this distribution.	Std. dev. = $\sqrt{\frac{1}{n-1}\sum(x_c - \text{mean})^2}$	Mean (most probable value); σ (symmetric)
Poisson	Nonuniform distribution. Mean does not equal most probable value. Typically seen in particle count distributions, other defect distributions.	Std. dev. = $\sqrt{\text{mean}}$	Most probable value; Mean; σ

TABLE 12-4
Defect-Free Die at Various Quality Levels

Number of process steps[a]	Die remaining after steps[b]		
	93%(~2σ)	3σ	6σ[c]
10	4839	9733	9999
25	1629	9346	9999
50	265	8735	9998
75	43	8164	9997
100	7	7631	9996
200	0	5823	9992
500	0	2587	9980

[a] We see here that there is a rapid decline in yield if process control parameters are not controlled tightly enough. Yield loss is greatly amplified by the number of steps involved in manufacturing wafers. If the processes can be controlled so that the 6σ limits are within the yield loss limits, then very high productivity will be seen.
[b] Started with 10,000.
[c] Notice that if a process can be controlled to ±6σ, within the yield limits of all processes, very significant improvements in yield occur.

identify true process drift from statistical fluctuation. The problems of process control can be made even more complicated when multiple processes are run in one furnace or piece of equipment. As noted, when the process is in control, the long-term average will be very near the center line. As the process drifts around, we see that there will be times when the process average ranges on one side or the other of the center line. In addition, the process may occasionally vary outside of the $\pm 2\sigma$ limits. If these excursions are monitored carefully, they can be used for analysis of process trends.

First, we will discuss the grounds that are used for suspecting that a process has a constant offset or deviation. A process should be under suspicion when there are a consistent number of points on one side or the other of the center line. The number of points that defines a problem can be set in a number of ways, with the definition being almost entirely dependent on the discretion of the manufacturing organization. The process should always be checked under the following conditions:

1. The data was on the same side of the process average seven times in a row.
2. The data was on the same side in 10 out of 11 process runs.
3. The data was on the same side in 12 out of 14 process runs.
4. The data was on the same side in 14 out of 17 process runs.
5. The data was on the same side in 16 out of 20 process runs.

There may be times when more stringent requirements are necessary, especially with very advanced technologies. For example, it may be necessary to start observing the process very closely if the process runs on one side or the other of the process for as few as 9 times out of 12.

Another trigger mechanism from the process control charts occurs when one of the Engineering Control Limits is exceeded. If the limit is exceeded only once in a while, there may not be a serious indication of trouble. This is due to the fact that the Control Limits are usually set at 2σ, which leaves room for approximately 5% of the readings to fall outside of that limit. However, if any of the following circumstances occur, there is a reasonable amount of certainty that the process is going out of control:

1. Engineering Control Limits exceeded on 2 out of 3 process runs.
2. Control Limits exceeded on 3 out of 7 process runs.
3. Control Limits exceeded on 4 out of 10 process runs.

Usually, if the process has these symptoms, it will also be seen to have symptoms such as the data points running on side of the central line or the other. Obviously, a furnace problem can be identified if the data fall outside of the specification limits. However, again, there is a small probability that a reading will exceed these limits. For most types of 3σ Statistical Quality Control implementation, a failure in 1 out of 35 or 2 out of 100 is considered acceptable. However, for the precision required in wafer manufacturing, this is very high. A more reasonable expectation is that, if a process is in control, less than 0.5 to 1% of the data points should fall outside of the specification limits (e.g., if a polysilicon process is being monitored, the overall process average should fall out of the process specification ranges less once out of some 150 runs).

Since control of the process actually depends on controlling the long-term average, as well as the long-term deviation of the process, it is more useful to use a combination of methods to trigger the process response teams into action. For instance, the process control flag might be triggered when four data points out of six are on one side of the central line and, simultaneously, two points during those six runs exceeded the control limits. This type of control is much more complex to analyze, but you can be more assured that the process is running under control at all times.

A number of factories use moving averages or control limits in an effort to optimize their processes. This is a dangerous practice unless controlled very tightly, since even small deviations can cause major problems. For example, if a furnace slowly deteriorates through time, and is scheduled for maintenance only once per month, by the end of that month the control on the process may be significantly worse than at the beginning of the month, if moving averages are used. For an example of this, see the illustration in Figure 12-9. In this instance, the system variation is always within the specified ranges, but since it is barely within these ranges, the next set of specification limits will be wider allowing the process to run even more out of control, so that the next time the ranges are adjusted, they are made even wider. Soon, the spec ranges will have exceeded the point at which serious yield problems will start to occur.

12.4 CONTROLLING MULTIPLE PROCESSES IN A SYSTEM

We will now cover the situation that occurs when there are multiple processes that are being used in one piece of equipment. There are a couple of solutions here, some more obvious than others. The most obvious solution is simply to include a process control chart and logsheet for each individual process in the system. This can get very messy in a hurry, since some systems are used for five or six processes or more. Obviously, this technique can work with a computerized system, but will be much more difficult in manual data entry and analysis systems. Another drawback to the technique is that it will be much more difficult to discern overall equipment trends, especially if the several processes can be run interchangeably.

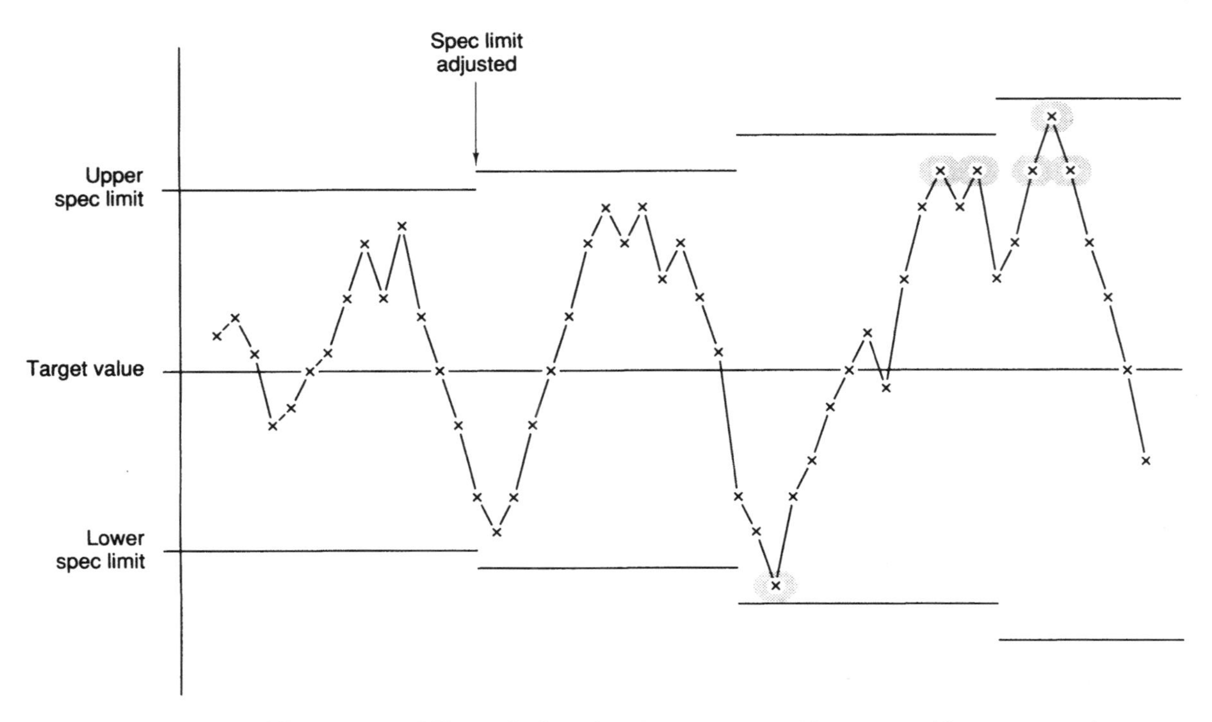

Figure 12-9. Effects of Changing Spec Limits With an Unstable Process If processing data are used to calculate new spec or control limits, precise guidelines must be followed to prevent an unstable process from drifting and pushing the calculated limits out of control. In the case shown, the process eventually becomes so unstable that runs are able to be processed outside of the original processing limits.

Thus, there is another technique that can be used to chart data for multiple processes in a single furnace. First, there must be some consistency to the specifications of the processes scheduled for a system. This technique will not work with widely varying processes. For example, there may be 180, 220, and 250 Å gate oxidation processes that can be run in the same furnace. These processes may have specification limits of ±5% or ±9, 11, and 12.5 Å each. With some testing to verify yield impact, it should be possible to round these spec limits to ±10 Å with little impact on the processes. Clearly, the 220 and 250 Å processes will not be subject to yield loss. If the spec windows are too tight or if the process specifications vary too far from one another, there may be excessive downtime incurred due to "false" reasons.

Now that these specification limits have been determined, we must enter the data onto the control chart. Obviously, if we place just the absolute values of the thickness on the control chart, the graph will become useless very quickly. Instead, subtract the specified target value (the center line value) from the real value of the run (from the test system), which leaves a relative offset from the specified value:

$$\text{Offset} = \text{Actual value} - \text{Specification target}$$

This offset is then plotted on a control chart that has been modified so the mean chart looks like Figure 12-10, with a central value of zero. As can be seen, the central value becomes zero (since the actual value is equal to the specified value) and deviations from that point are measured in relative terms greater than or less than zero. For example, if the specification target is 1.2 μm and the value obtained is 1.15 μm, the point on the control chart that would be plotted is −0.05 μm. on the other hand, if the real reading was 1.32 μm, the value plotted on the control chart would be 0.12 μm.

Looking at the previous example of three gate oxidation processes in a single furnace, we can see that, as long as the specification and control limits are equal, the relative difference from the specified target can be plotted consistently on one process control chart, as shown in Figure 12-11. Obviously, for this to work, all processes run in the system and monitored by this control chart must have some common structure. For instance, you would not want to place gate oxidation and field oxidation data on the same control charts (typical targets for these two processes are 350 and 8,500 Å, respectively). However, they could be run in the same furnace if two control charts are kept (this is not recommended).

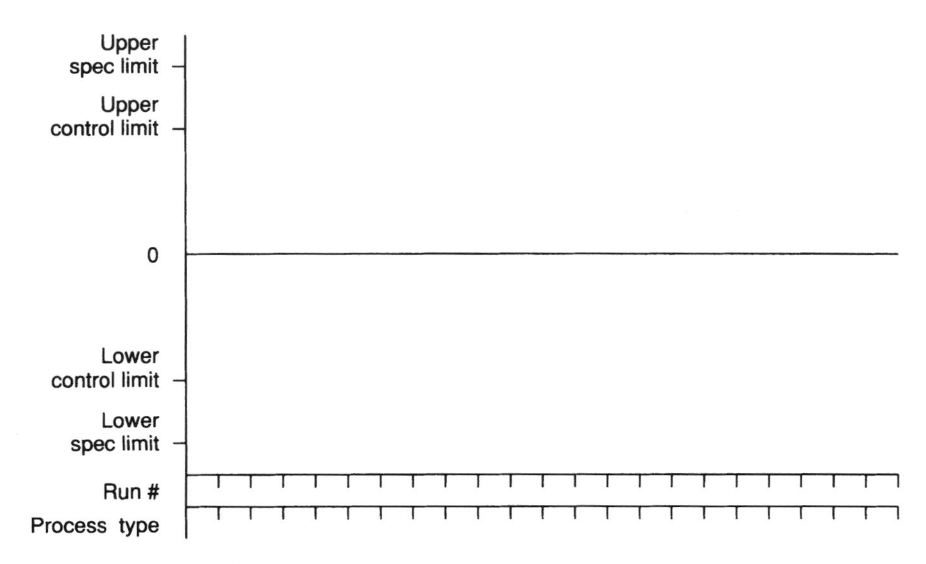

Figure 12-10. Modified Mean Data Chart This chart can be used when similar processes are run in a single furnace, for instance, 400 Å and 300 Å gate oxide processes are each run through a diffusion furnace at random intervals. This chart can reduce the number of individual charts required, by combining the data onto a single page, and can provide insight on the operation of the equipment that may not be visible while observing the individual process control charts.

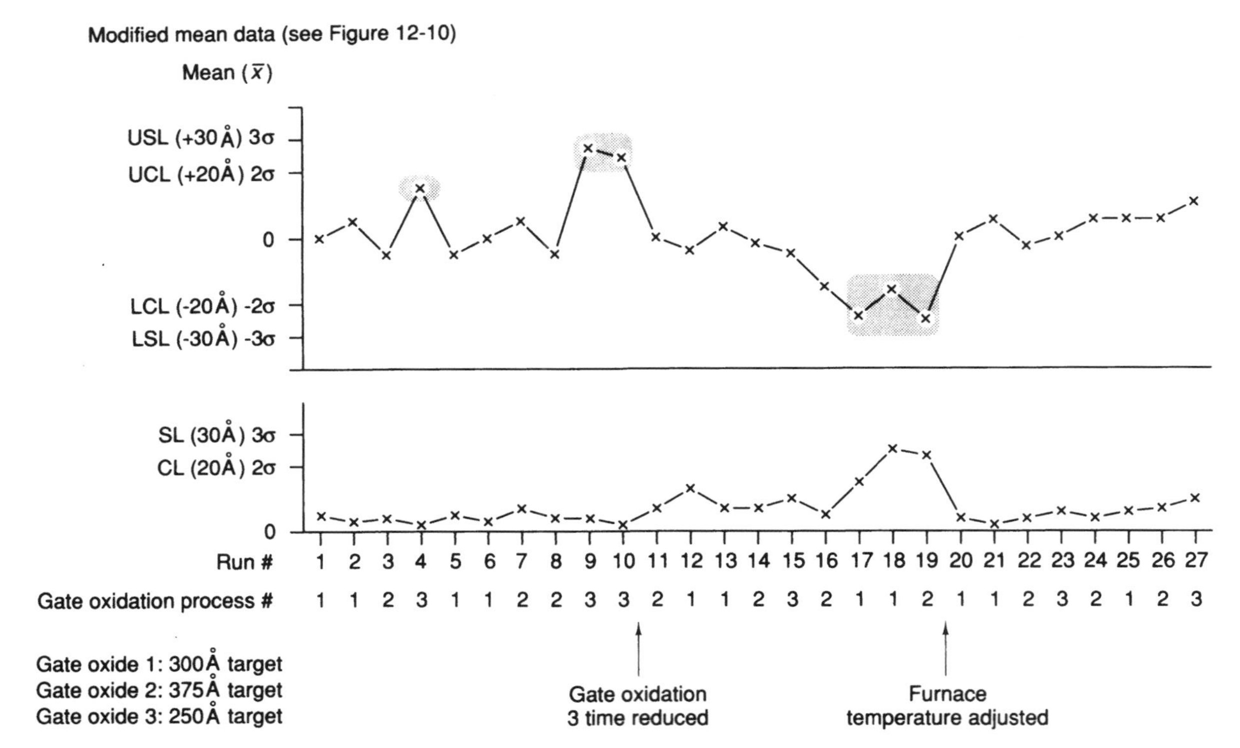

Figure 12-11. Process Control Chart for Multiple Processes in a Single Furnace The multiple-process SQC chart is used in equipment that runs a variety of processes. As an example, a diffusion process may have three gate oxidation processes (labeled as Gox1, Gox2, Gox3). Both individual process and furnace process problems can be diagnosed using this chart. The gate oxidation 3 process shows unusually high thicknesses, so requires a time reduction, which brings consequent control to that process. However, by run 16, the furnace demonstrates poor performance for all processes, so a temperature correction is required. Note that the modified mean data chart uses 30 Å away from a target value [which is defined as (actual value − target value), with zero being perfect] and 20 Å for the control limits.

12.5 DEFINING AND CHANGING SPECIFICATION LIMITS

A main point of the process control effort is to obtain some degree of process predictability. This predictability concerns both the quality of the material being processed through the system and also the availability of the system for production use. You do not want to place material in a system that is going to cause yield loss, nor do you want to shut down the production equipment unnecessarily. The extremes here are that, without any process control, this visibility is completely lost and yields will vary uncontrollably from day to day, whereas with process control limits that are set too tight, the system will be down too often, sometimes for no reason.

As a result, the system should be run with reasonable process control specifications that are a careful balance between these two extremes. This is possible only through thorough analysis of the yield effects of each process, in conjunction with thorough analysis of the process control data. This is very time-consuming work at the start if done properly, but can be very rewarding in the long run and can reduce the amount of effort required to run the fab. There are any number of methods to obtain these goals, but the following is suggested as a blueprint for the actions required to set specification limits. I am assuming that the design is new since we will discuss changing specification limits in a moment.

As an integrated circuit is designed, the chip designers will be able to identify critical parts of the circuit, and may be able to predict where process specifications must be tighter than usual or where they can be relaxed. If the process itself is new (not just a new design using a existing process technology), the designer may not have this knowledge. In any event, the design will usually have a nominal tolerance of no more than 5% in its various electrical characteristics. This translates to a process control limit of $\pm 5\%$ for the individual processes. It is usually important to get the process engineers and designers together at this point, to prevent problems due to limitations on one group that are not known to another. For example, the designer may want to use a $0.9 \pm 0.1\ \mu m$ line on the design, but the process engineer knows that the fab cannot reliably produce lines at $0.9\ \mu m$. Before the design is specified, then, some compromise will have to be made.

Now the process engineer must run statistical tests on all of the critical pieces of equipment. For instance, can a sputterer produce high enough quality aluminum with the required repeatability for the new design? The system should then be tested for its stability and repeatability. This data can be obtained from standard production runs if the proper data collection systems are in place. All equipment that meets the requirements should be identified. If no equipment can meet the need for the new process, a decision must be made to purchase a newer piece of equipment or modify the design specifications.

When the system is tested, it should be observed for as many runs as possible, preferably somewhere near 100 runs. From this, several pieces of data may be obtained. One is the long-term control of the process, which is determined by calculating the average and standard deviation of all 100 runs. If the long-term average exceeds the designer's specification limits, further work must be done to isolate the cause. At this point, unless the failure was very bad, the equipment would not necessarily be out of the running.

The next step is to take the 100 runs and break them into 10 groups of 10 runs each, in sequence. Each of these groups should be averaged, and the standard deviations determined. If trends are seen here which take the process out of spec, but are controllable, the

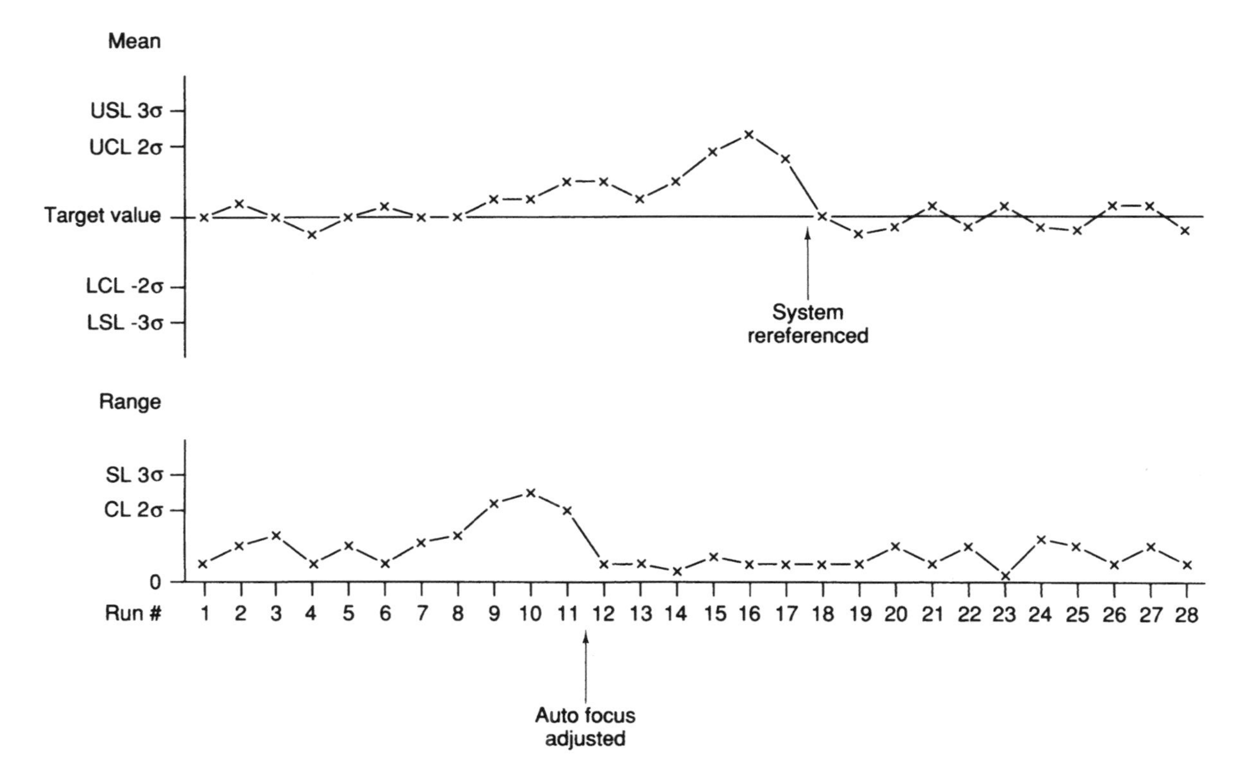

Figure 12-12. Use of the SQC Charts for Equipment Evaluation The statistical quality control technique can be used to gauge processing and metrology equipment. Here, a film thickness analyzer is tested by reading a single test wafer of known thickness every few hours. This gives the engineer an idea of the long-term repeatability of the system. The specification and control limits will usually be very narrow for this test, perhaps 1.5 and 1 Å, respectively. The range chart is used as a measure of short-term repeatability, as each data point should consist of a significant number of readings, possibly 25 to 100. For instance, auto focus variations can cause a significant effect on short-term precision, while rereferencing is used to eliminate long-term drift of the system.

equipment may be suited for the new process. However, if the standard deviations from these 10 groups are very high (comparable to the design limits), then the equipment will not be as suitable for the process. A new piece of gear may be required. This procedure is shown in Figure 12-12. In any event, use the largest standard deviation value that is found for the process while it was running in control (don't take a period of time where problems were known to have existed) to set the Specification Limits and Engineering Control Limits for the first pass. As usual, the Specification limits are set at a 3σ level, while the Engineering Control Limits are set at two standard deviations.

After the new process control procedure is defined, the first 100 runs should be observed closely. Again, any unusual trends should be noted. After the 100 runs are complete, the data should again be broken down into 10 run groups, and analyzed for short-term deviations and trends. The performance of the process should be compared to the expected values and the specification and control limits adjusted if required. System downtime should also be examined at this point to see if there has been unnecessary downtime due to anomalies in the process control chart system. If this all checks out, the process control system should be operational.

While the process control system is being developed and analyzed, the die yields for lots coming out of the critical processes should be examined very closely and plotted against the average parameter. The best way to do this is to sort the lots that have gone through the equipment into bins by parametric value. For instance, 100 lots of polysilicon deposition may result in thicknesses ranging from 3300 to 3500 Å. The lots can then be sorted into 25-Å wide groups (3300 to 3325 Å, 3325 to 3350 Å, etc.). When the lots are processed through sort, record the average die per wafer and look for anomalies in the electrical test values that are associated with that process. If there are problems seen near any edges of the process control window, the specification limits should be tightened. Otherwise, if all yields are stable across the range of thicknesses, the process control system is probably in good shape. At this point, it will take several weeks to months before any further opinions can be obtained, since the operators will have to get used to the system and work it into their job flow, and process bugs must be worked out. Later, if problems are seen, or if process optimization is desired, the process control charts can studied in depth again.

Once the SQC system is in place there should not be many changes needed, unless process optimization is desired. Any changes that are implemented should first be classified as major or minor, based on whether it is a paperwork change or whether the wafer data sampling or wafer processing will be impacted. If there is a change required in the wafer processing, studies should be performed ahead of time to find potential pitfalls in the proposed change. These studies may include split lot experiments, theoretical research, or even short test periods (where the idea is implemented in a controlled fashion for a short time—two to four weeks). If no problem is found, changes can be made; otherwise more study is required.

12.6 RESPONSE TO PROCESS CONTROL PROBLEMS

The second main function of SQC methodology is not predictive in nature, but instead is reactive or responsive. In other words, the SQC methods will help identify and resolve problems that come up on a regular basis, also preventing yield loss or unnecessary downtime. Since problems occur continually in the wafer fab, it is important to obtain as much information as possible and the perfect forum is the process control charts and logsheets. A number of issues can be resolved very quickly once close observation of the control charts is established.

Through observation of the trends, it may be possible to use past performance as indicator of immediate future performance. For instance, in a BPSG system, if the operator notices that the particles start rising above a certain point two to three runs before a scheduled clean, which is sometimes followed with a discrepant run (due to high particles), the operator can act on that information and have the quartzware pulled earlier than scheduled to prevent any loss of quality. Certain types of problems will have characteristic types of failures which appear in the process control charts. Some of them are shown in Table 12-5.

Usually, if the problem is identified clearly enough, it can be submitted directly to the correct group for analysis. If simple preventive maintenance is required, the production technicians usually are assigned the task. If the problem appears to be within the more complex mechanical or electrical portions of the systems, the maintenance groups can be called in, and, if the problem is unknown, engineering will be called.

SQC charts can also be used to establish which type of problem a system is having. For instance, a uniform rise in thickness may indicate a temperature-control problem, whereas a nonuniform etch rate may indicate a problem in the gas delivery or exhaust system. Careful control should be maintained of all records associating process control failures to certain equipment failure modes. Quick reference to this information can significantly reduce the amount of time needed to respond to and resolve a problem.

Finally, the SQC system can help the operator check whether or not the problem has been resolved. The use of one or two test runs is usually enough to identify any additional problems. In almost all cases, a manufacturing system should be tested after maintenance work has been completed, and the results compared to SQC data already obtained for the process.

TABLE 12-5
Some Common Problems Discerned through SQC Analysis

Problem	SQC symptom	Area
MFC failure	Sudden change in X-bar, no alarms	Diffusion/CVD
MFC drift	X-bar trends one way, range degrades	Diffusion/CVD
Temperature off	X-bar or range changes after PM	All heated
TC failure	Range degrades, cannot be adjusted	All heated
Nearing PM	Parameters varying, range increases	All
Material variations	Parametrics have periodic spikes	All
Lamp/RF drift	Changes in CDs at FI or DI	Photolith

12.7 OTHER SQC CHART TYPES

Probably 95% of the SQC charts used in the semiconductor industry use the first process control chart method that was defined in this chapter, that is, using the target value as the central value of the chart. While it is the most common style, it is not the only control chart style possible.

We have already discussed another method of control chart that can be used when multiple processes are run in a single system. As shown in Figure 12-11, this system is basically the same as the basic SQC system, except that it uses a relative value for the central line, that is based on the difference between the specified target and the actual parameter.

Other variations of process control charts are possible, such as the composite control chart shown in Figure 12-13, or the multiple or

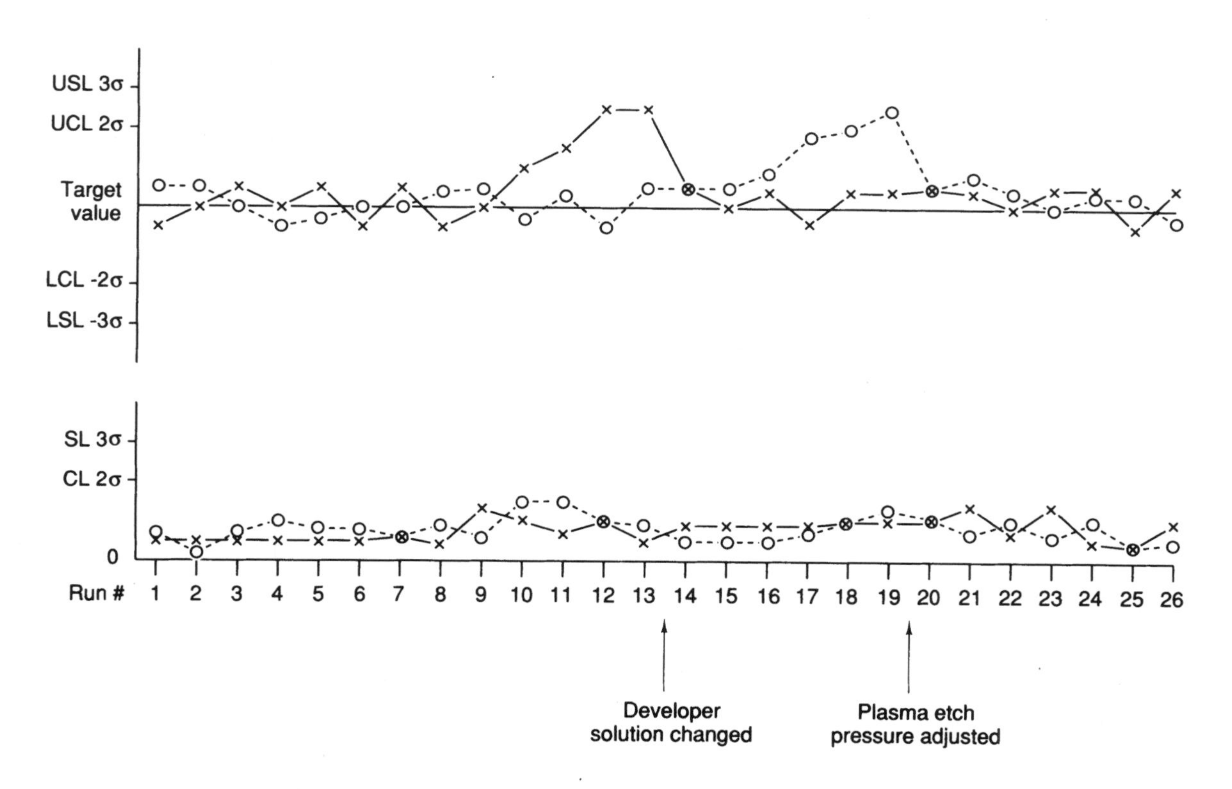

Figure 12-13. Composite Statistical Quality Control Chart The composite SQC chart can be used to identify trends in two closely related processes. While busy, this chart is useful in comparing the results of these processes. For instance, pre-etch (develop inspect) and post-etch (final inspect) results can be plotted and analyzed with this system.

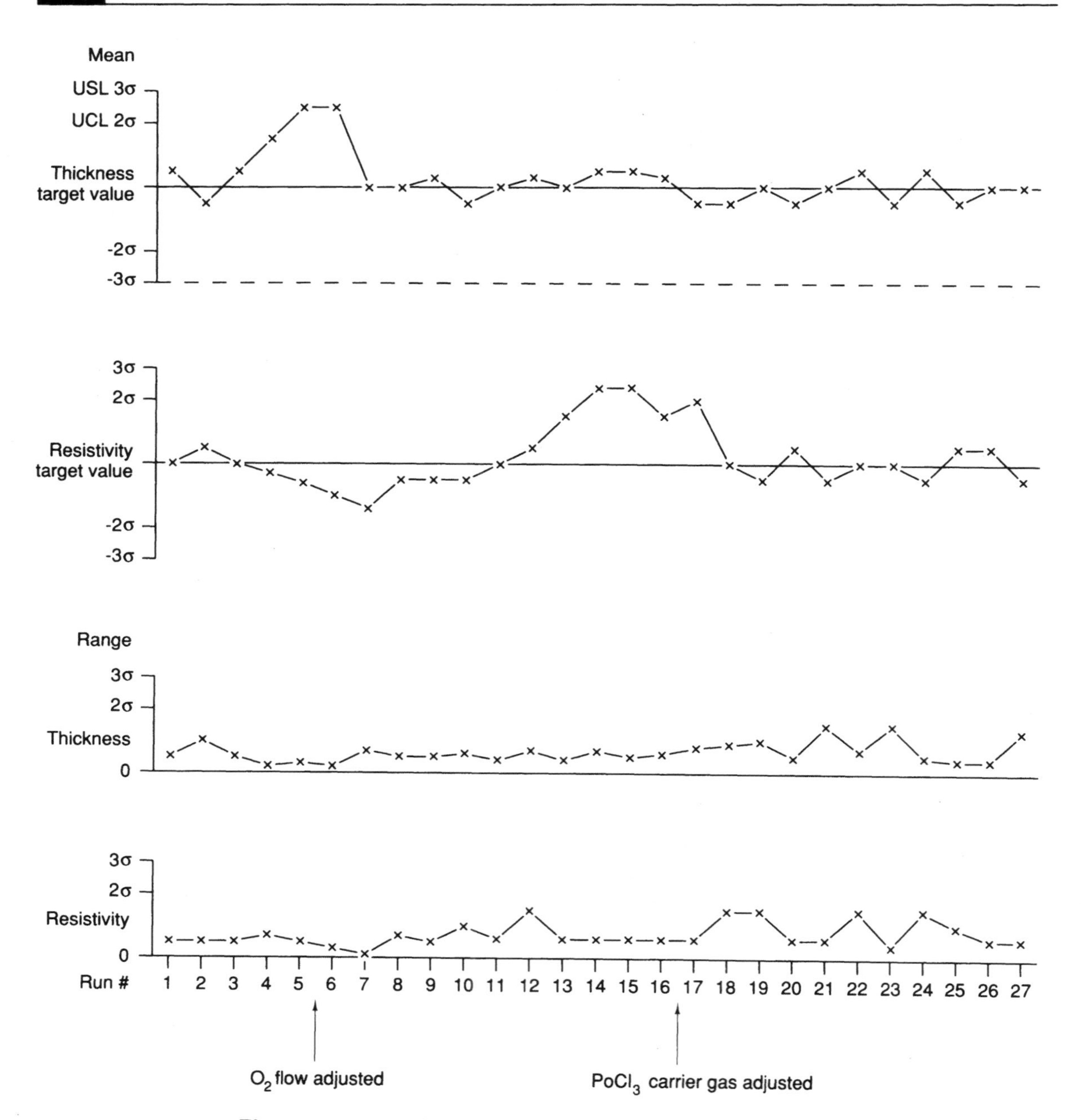

Figure 12-14. Combination Process Control Chart The combination SQC chart permits analysis of a process in which two variables are measured but are dissimilar. An example is the $POCl_3$ doping process, which has two parametric variables, oxide thickness and resistivity. As shown here, excess O_2 flow to the system can enhance oxide growth and retard dopant diffusion into the silicon layer. Variations in N_2 carrier gas flows can affect the amount of $POCl_3$ in the furnace, and therefore change the resistivity of the doped layer.

combination control chart shown in Figure 12-14. In the first example, two related parameters are plotted on the same graph, for example, Develop Inspect and Final Inspect critical dimension values for a masking operation. In the second example, two processes that have no relationship in parametric values but have a substantial physical relationship to one other are plotted on two separate, parallel control charts. An example of a process where this type of control chart would be useful is a phosphorus-doping operation where the oxide thickness and resistivity of the substrate are both critical parameters.

12.8 STATISTICAL QUALITY CONTROL WRAP-UP

We have seen in this chapter that Statistical Quality Control is an important consideration in the wafer manufacturing environment. Due to the vast number of steps involved in manufacturing integrated circuits, even small levels of imperfection can add up to serious problems with yields and device reliability. One of the most effective tools to use in monitoring the process is the SQC system, with appropriate logsheets and process control charts available and kept accurate at all times. The control charts themselves are not too complex, consisting in most cases of a chart of process run averages (Mean or X), and a chart of ranges (taken from the maximum and minimum values of the data). However, to make SQC systems work, understanding and cooperation are required from all of the groups that work within the wafer fab.

The SQC system is virtually universal in its functionality. Almost every process can utilize some technique similar to those described in this chapter to maintain control of the process. Even less critical industries could use a dose of SQC, not to the extent that wafer fabs need, but this system does help to maintain the quality that is required by today's consumers.

Process control is used to help in predicting system and process performance, and can be used to prevent system deterioration and line or die yield loss due to predictable system failure modes. It can also help pinpoint the source of a problem and can speed the repair process by enabling the manufacturing personnel to call the correct groups into action immediately.

While it is true that the extra logsheets and control charts add complexity to an already complex job, as well as adding paperwork and some time to the fab, the implementation of a process control system is well worth it in the long run. The spread of computerized wafer tracking systems, and the consequent process control capabilities, permits the complex fab control problem to be resolved into a problem of data entry and database control.

In fact, to some degree, a lot of the information provided in this chapter will be of interest only to certain smaller organizations in the future as more and more fabs go to automated quality control schemes.

These large systems can store, generate, and plot more data than could ever be recorded in a fab area, and can allow incredible insight into the workings of a process through time. Large host computers are clearly more efficient, drastically reducing the amount of time and effort required to perform and analyze SQC procedures, and are highly recommended. Most of the computerized systems that are on the market follow an SQC methodology similar to that described in this chapter. Essentially all that is required is to call up the chart and analyze it.

The analysis of SQC data is actually the most complex part of the process. One must look for trends in the data and must always be exploring the edges of the process windows in order to learn more about the process itself. While clues are given throughout the book on how to respond to the problems that will be found through statistical quality control, it is only through experience that one can judge all of the aspects of the SQC data and make correct interpretations. This is why it is so critical to record the experiences found in a logbook or specification so that when future occurrences of the problem appear, the solution can be readily found without retracing all of the steps.

Interpretation of the SQC data is probably the most critical aspect of the analysis procedure. Determining what a process is doing is far less threatening than trying to make it do something else. Improper interpretation of the process data can result in completely improper solutions being implemented, thereby creating more problems. As a result, it is recommended that anyone in charge of maintaining a process using SQC techniques become fully versed in its details. The success of his or her fab may depend on it.

CHAPTER 13

THE PEOPLE WHO MAKE THE CHIPS

Throughout the book we have referred to the various groups that perform tasks in the process of manufacturing integrated circuits. These groups include the manufacturing group, the maintenance group, and the engineering groups, as well as the fab management team. All of these individuals must work together in order for the complex product to work, and to be manufactured profitably. Another very serious concern when dealing with all of these people is their safety. The IC industry uses many toxic chemicals, unusual radiation sources, and high-power electrical equipment. The possibility of danger has been reduced as much as possible, and the semiconductor industry is known as one of the safest in which to work. Keeping a good safety record also requires teamwork between the various groups.

Each of the different groups has a number of subgroups associated with it. In some cases, they report directly through the chain of command, at other times they are related more by function than by structure. There are many informal "dotted line" structures in the IC industry, where a member of one group works for another group full time for extended periods. We will not deal with these types of structures, but instead will attempt to present what is the "usual" fab organization. There are wide variations in these organizations, but the organization charts that are shown here will describe most of the functions that must be performed.

13.1 THE MANUFACTURING ORGANIZATION

The manufacturing group is really the group everybody else functions around. They are the only ones who actually make money (everyone else just spends it as fast as they can), and will make or break the wafer fabrication process. Since they are the primary group in the fab, all of the other groups' responsibilities will revolve around production first. If a process goes down, all other work stops or slows to a crawl until the problem is resolved. This is because the costs of operating the fab are very high indeed, often in excess of $15,000 an hour.

To meet the high costs of the wafer fab, the manufacturing group is chartered with the responsibility of making as many chips as possible, but in no case producing fewer than the predicted output. The predicted output is usually already sold, and budgeted for. If targets are missed, severe cost cutbacks must occur almost immediately to make up for the shortfall. Chips produced in excess of the targeted goals are usually almost all profit and are therefore very valuable. Thus, the number one priority for the manufacturing group is the mass production of the highest quality wafers possible.

Clearly, it is critical that the highest quality standards be maintained consistently throughout the production process. There are numerous reasons for this, from the obvious, such as keeping die yields from crashing (or maximizing die yields to maximize profits), to the less obvious, such as the reduction in fab morale if quality problems are allowed to become serious. Morale and attitude problems are very serious issues in the perfectionist fab environment, and situations such as these can lead to the complete failure of the wafer fab and its eventual closing.

The manufacturing group also has the ultimate responsibility to make sure that all procedures are followed and that all specifications are met. They essentially "own" the fab area; therefore they are the ones who maintain the cleanroom discipline, verify that all personnel (including vendors or service people) are properly attired, and guarantee the quality of the process. The manufacturing group will almost always be monitoring some part of the fab or another, testing for airborne particulate, water purity, gas pressures, air flow through the exhausts, and so on. While parts of these tasks may be performed by other groups, it is still the responsibility of the manufacturing group to make sure that all of the variables are in order, and that a quality product can be manufactured. As a result, in the ideal situation the operator should have the authority to refuse to operate a piece of equipment that he or she knows is defective until it is proven to be operating correctly. Similarly, he or she should have the authority to reject any and all parts that are not in all respects up to specification levels. After all, the operator is expected to perform nearly perfectly on the job. This is impossible unless the proper support is there.

In this same vein, the manufacturing group is also in charge of running all test runs. It is assumed that they know the most about the operation of the systems they run from day to day, so they should be able to tell if problems have been resolved or if the system is otherwise ready to run. Test runs are performed by the manufacturing group after almost all maintenance work (unless it was very minor). Most fab areas also have time windows for the processes, such that if the furnace has not been tested or run in the past 24 hours, a test run is performed. These test runs are processed to the same specifications as the production runs. If the results of a test run are within the specified ranges, the process can be turned on, and production wafers processed through it. If the results are outside of the specified ranges, the reason must be discerned, fixed, and another test run processed before production is started. Running test runs involves a certain amount of logistics, such as purchase, storage, and reclaim of test wafers. These logistics are also the responsibility of the manufacturing organization.

The manufacturing group must coordinate the movement of the wafers through the fab, and must also orchestrate the required maintenance and production schedules. Clearly, incorrect timing can cause severe line balance problems if a process is shut down just before a large number of production wafer get to it. This kind of incorrect timing is one of the points of trying to implement a Just-in-Time system for inventory management. With the JIT structure in place, these imbalances can be minimized. It definitely is in the interest of the manufacturing group to do a good job on the scheduling, for if they do not they will not meet their own production forecasts.

Manufacturing is also responsible for performing some of the more routine preventive maintenance tasks. These types of jobs include cleaning quartzware, changing process chemicals, "profiling" equipment temperatures, and calibrating and cleaning various pieces of equipment. They must schedule the maintenance accordingly, just as with the standard maintenance procedures. Typically, these tasks take only a short time to perform individually (at least in terms of impact to the fab area), as opposed to maintenance work, which may take from hours to days to complete.

Thus, the manufacturing group is large, and maintains many responsibilities. They effectively own the fab and the equipment and therefore get both the credit for its performance and brunt of its failures. To make sure that these responsibilities are handled as expected, the manufacturing group is usually broken up into several functional areas. As usual, there is no universal rule for setting up the group, or universal agreement on the exact titles. In most cases, the positions described will exist in some form in almost every manufacturing area. Two possible setups are shown in the organization charts in Figure 13-1.

In both cases, the group is led by the manufacturing manager, who oversees the supervisors (or general foremen, who manage the

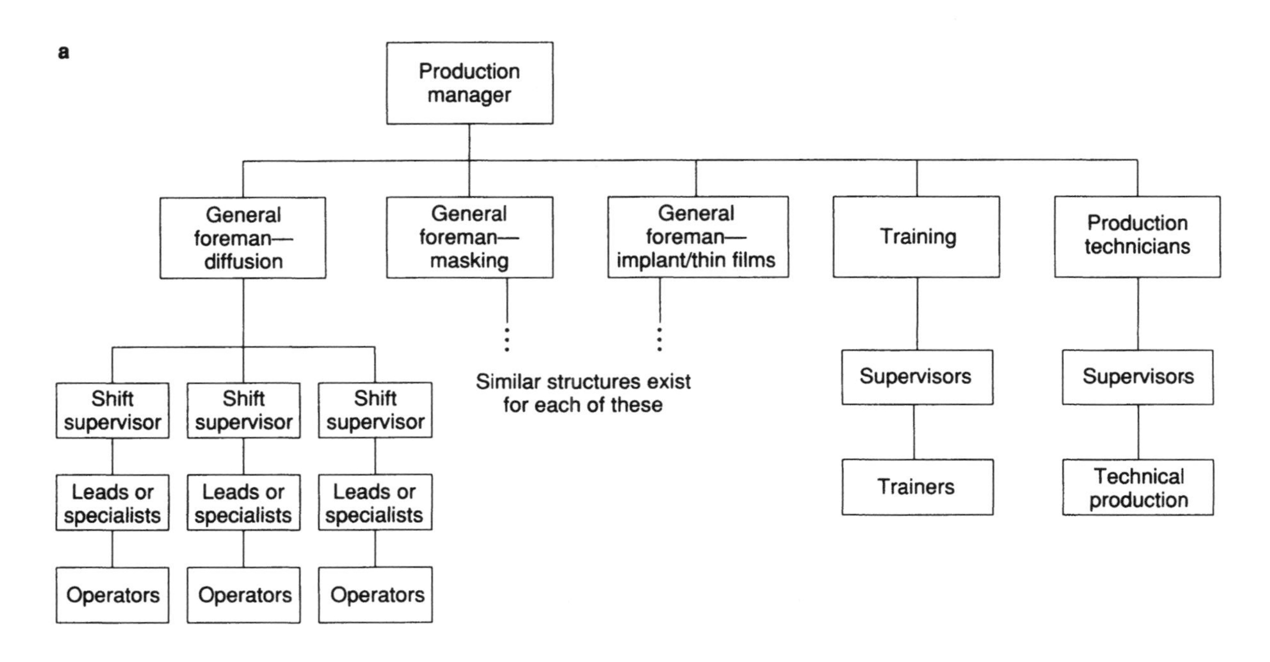

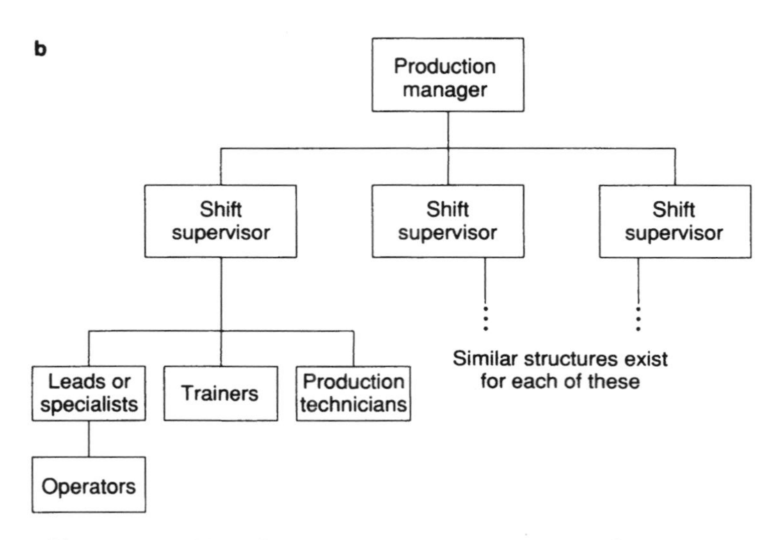

Figure 13-1. Manufacturing Group Organization Charts Here we see two potential organizations for manufacturing groups (a) large group; (b) small group. Many such structures can exist and the exact structure must be determined by analysis of the output expectations and degree of automation of the wafer fab.

supervisors in large organizations) of the various functional subgroups. These groups include production, production technicians, and training groups, as well as some of the process control functions. There may be a few engineers working within the manufacturing organization to provide coordination and technical support for (and manufacturing input into) fab area upgrades and other cleanroom projects. Each of these groups will typically be broken down in their various functions as described in the following paragraphs.

The production group starts with the supervisors of the various areas within the fab. Usually, at least two supervisors are required, one for diffusion, thin films, and implant, and one for photolithography and etch. In large operations, where each area may have many operators, more supervisors are needed. If there are more than two production supervisors, a general foreman will sometimes be brought in to coordinate the activities of the supervisors. The supervisors play a large role within the organization of the manufacturing group. They are not only directly responsible for getting the wafers manufactured, they must oversee the fab for conformance to specifications and maintain order and morale among the operators and other personnel.

Reporting to the supervisors are either the operators or "lead" or line specialists. The lead person usually works with the operators to clarify specifications and special instructions, coordinate operations, and verify the equipment setups and test results. Finally, the operators run the equipment and actually manufacture the wafers. Their importance should not be underestimated. They are, in many ways, the most important individuals in the company. If the operators cannot handle their jobs effectively, or are hampered from performing their tasks by inappropriate specifications or poor management style, the entire company suffers. It is important that good communication links be maintained between the operators, the supervisors, and the management team through regular meetings and get-togethers, so that each group can express their needs clearly to the others.

The production technicians are the individuals who perform the routine maintenance tasks in the wafer fab. These tasks include changing and cleaning quartzware and other system components; replenishment of chemical solutions, including photoresist, developer solutions, and acids; checking and recalibration of process equipment temperatures; and maintaining an on-hand supply of spare parts and consumables. Usually, there will be two or more production technicians in the diffusion/thin films area, and two for the photolithography area, although in smaller organizations only one will be assigned to each area. For safety reasons, there should never be less than two production technicians on shift at any one time, and they should maintain contact with one another throughout the shift. The production technician's job is one of the most dangerous in the wafer fab, since they deal with hazardous chemicals, hot quartzware, and high-voltage equipment every day. Most of the people performing these jobs are well aware (and well informed, in most cases) of the hazards, so that the accident rate has remained low.

The training department exists for three main reasons. The first purpose is to train newly hired employees on the mechanics of fab operations. In the case of operators new to the wafer fabs, this means training them on all aspects of wafer handling, cleanroom procedures, documentation procedures, and so on. This can be very extensive and requires that the new operator pass a number of proficiency tests before becoming a certified operator. This training period is critical, since new personnel typically cause a significant amount of damage to wafers (inadvertently, of course, there is a lot to learn, much of which does not make sense to a new person). Training procedures vary from company to company, but most involve both classroom training and in-fab training for all new personnel that lasts from two weeks up to two months, with follow-ups and cross training (training the individual on other operations) going on after that for some time. Experienced individuals are usually given less formal training, consisting mostly of specification and safety training. The experienced person will need to pass the same set of qualification tests as the inexperienced person (perhaps even more extensive ones).

A second function of the training group is to train the operators who have become proficient at one step so that they may do another process step. This makes the operators more valuable by increasing their overall knowledge, reduces monotony, and allows more flexibility in the operation of the fab by having the ability to bring in extra trained people to run key operations in a crunch. This cross training is of even more crucial importance in small wafer fabs, where the number of personnel is limited.

The third main task of the training group is to keep the operators current on all specifications. This means periodically recertifying the operators to make sure they are aware of all process details at all times. It also means that they must verify that all operators know about specification changes. These occur on a regular basis in the wafer fab, so a mechanism must be provided for the information to be passed on reliably to all personnel who need to know.

In case the issue of operator certification seems trivial, rest assured it is not. There are, however, a number of interpretations for certification. The legal definitions require that the individual be able to perform all specifications properly. It is critical for all personnel to be properly certified to run each operation in order to obtain military and other high-reliability certification of the fab's product. Unfortunately, the reality of the situation is that it is much easier to be certified on paper to operate a piece of gear than it is to be actually qualified to operate it. Often, training departments have been put in the position of having to train an excessive number of people in too short a time. They must then get the new personnel to pass the tests for certification even though they may not always have an adequate understanding of the process. This can result in a number of serious side effects, sometimes resulting in reduced die or line yields. Nevertheless, having the proper training and certification for all operators can significantly reduce the possibility for error.

The final major task of the manufacturing group is that of process or production control. These activities are usually coordinated with those of the quality assurance organization. Typically, the manufacturing group will monitor many of the aspects of the cleanliness of the fab, and will verify procedures. For instance, the production group may perform the wafer layout tests and air- and water-borne particle tests. If any of the test results are out of the specification range, that area will be cleaned and retested. Areas of chronic problems or critical to device performance are checked by the Q.A. department. Another area in which the production department assists the Q.A. department is in the inspection of wafers. Usually, most inspections of in-line wafers are carried out by production operators. These inspections are often performed immediately after deposition steps and after etch steps. Other key processes may also be followed with inspection steps. In most of these cases, these inspections are carried out by the production operators.

13.2 Maintenance Department

The maintenance department is responsible for the proper operation of all facilities and equipment. Their duties include both preventive maintenance and repair work. It is also common for the maintenance department to coordinate evacuation and recovery procedures in the event of an emergency. They are also usually responsible for the implementation of safety procedures and specifications, in the absence of a separate safety department. Finally, they are usually given the charge to schedule the workload as much as possible, with minimal impact to wafer production, and to maintain adequate supplies of spare parts for most of the possible contingencies. Ultimately, the primary goal of the maintenance department must be to supply the highest possible quality levels in the manufacturing equipment at all times, and to prevent all unnecessary downtime. This is done in order to maximize the output of the fab area. Figure 13-2 shows two typical organization charts for the maintenance organization. The significant difference in these organizational setups is that in the one situation, the individual departments report to the plant manager as opposed to reporting to a single maintenance manager or director.

The first—and most visible—of the maintenance organizations is "line maintenance." This organization consists of a number of subgroups that perform tasks in each of the primary areas of the factory. For instance, the steppers will be maintained by one group of personnel, while the etchers will be handled by another group. These two groups may have overlapping responsibilities, especially when considering support of the peripheral equipment (cleaning systems, driers, spinners, etc.).

Another area within line maintenance includes the diffusion and thin-film support personnel, who maintain the furnaces and the related gas-handling and vacuum systems. Sometimes, the personnel who support the vacuum systems (such as the LPCVD and PECVD

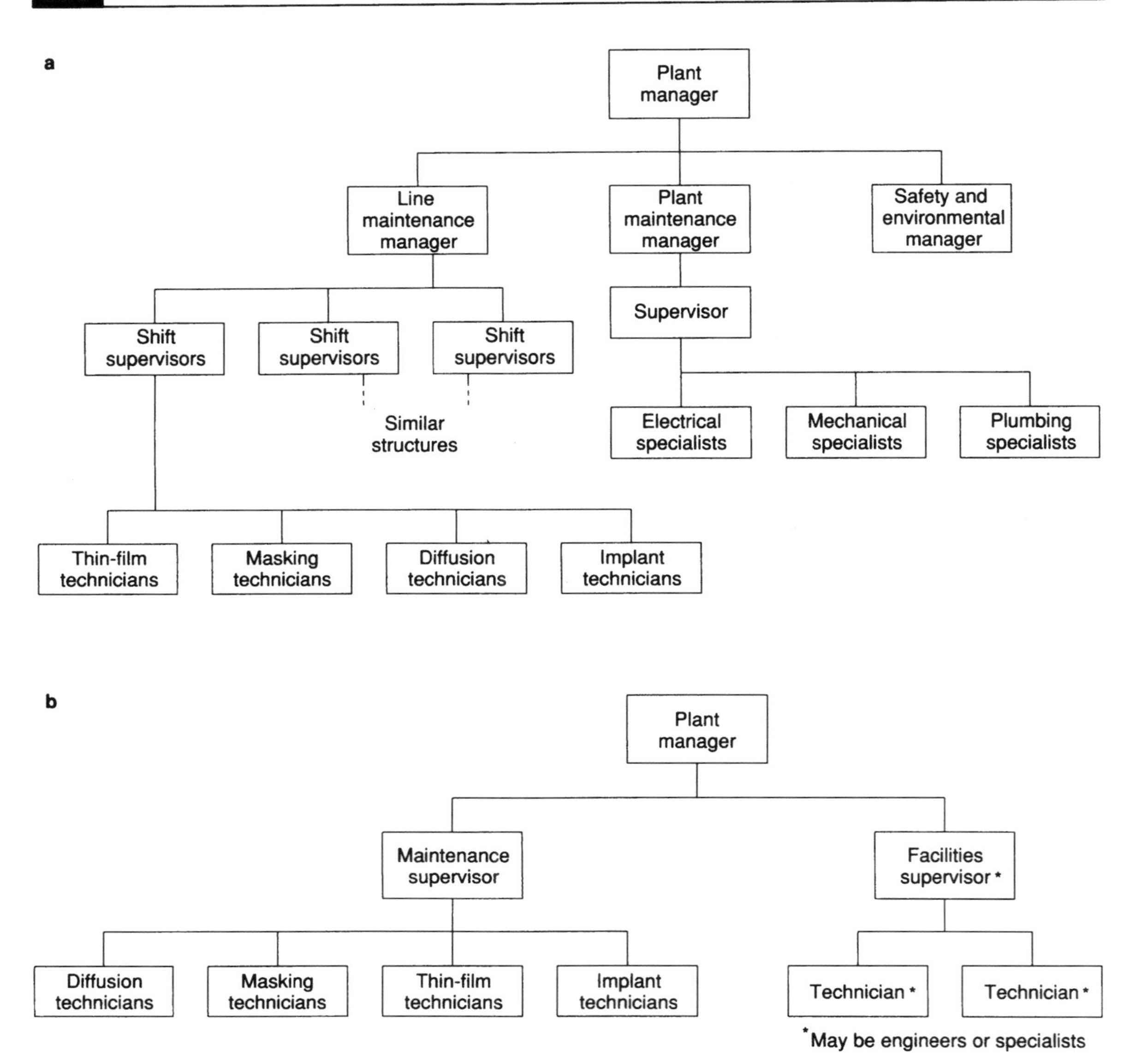

Figure 13-2. Maintenance Group Organization Charts The maintenance organization can be arranged in a number of ways, depending on the size of the organization and the types of equipment in the fab. (a) Large group; (b) small group.

systems) are organized in separate groups, since their tasks are more specialized. Typically, though, most of the tasks, such as handling the electronic controls and gas control systems, can be shared by all of the diffusion and thin-film deposition personnel. Again, there will be some overlap when dealing with the various support equipment in the diffusion area.

The ion implanters usually have maintenance personnel assigned directly to them, due to the complexities of the machine. These people must be able to safely and effectively handle toxic gases, ultrahigh-vacuum equipment, and extremely high voltage and radiation levels, while keeping the systems operational.

The number of maintenance people needed to keep the fab area operational depends largely on the amount of equipment that is used, and the uptime requirements of the fab. The higher the uptime requirements (which results from restricted fab capacity due to strong market demand or low-yielding material), the more maintenance people required. The level of training and quality of workmanship are also key factors. If the manufacturing equipment can operate nearly full time without unscheduled maintenance, and preventive maintenance truly prevents most problems, maintenance personnel requirements will be reduced.

A minimum number of personnel are available on the off-shifts. These individuals must be well versed in a number of areas to be able to cope with problems that come up during the shift. In many cases, the system problems cannot be resolved immediately due to insufficient resources, but the assistance of the maintenance personnel will ensure that every possible wafer is saved in the problem process. It is especially critical that the off-shift people be aware of all safety and emergency procedures so that there is full-time coverage for all contingencies.

The line maintenance crew is usually assigned the task of coordinating a schedule for preventive maintenance (PM) procedures, which usually involves analyzing how long each particular task takes and blocking out an amount of time appropriate for the job, perhaps 25 to 50% more than the expected time of the maintenance work. (In other words, if a task is expected to take about 30 minutes to complete, one hour of shutdown should be allotted to the procedure to allow for unexpected eventualities.) The procedures should be scheduled at regular intervals, and then the plan should be cleared and coordinated with the manufacturing group. It is important to be flexible, and not overload the schedule on one day so that, if something comes up that allows an opportunity to perform preventive maintenance at a slightly different time than scheduled, it can be performed then. For example, if a vacuum system is scheduled to have a biweekly oil change on a Tuesday, and the system goes down to the production technicians for a clean on Monday, the oil change schedule should be flexible enough to allow that vacuum system to get the oil change while the quartzware is being cleaned. In this case, it is better to schedule a few oil changes every day than to schedule all of them on one day. This way, the oil change dates can be modified as the need arises.

The ability of the line maintenance department to provide manufacturing equipment of high quality can have a significant impact on the success of the wafer fab. This quality is provided only through discipline by the maintenance organization. They must always follow cleanroom rules, even though these rules can make their jobs more difficult. For instance, it can be difficult to perform many delicate tasks while wearing gloves. Also, no paper may be brought into the fab (a major offender here is the vacuum sealing or VCR gasket, which comes individually wrapped in paper containers that must torn open). All tools should be degreased and cleaned before being used in the fab for the first time and should be cleaned regularly thereafter. Spare parts should be handled, cleaned, and prepared for use with extreme care. The parts can become contaminant sources or can cause further problems with downtime if they are damaged even slightly. In addition, spare parts for the systems are very expensive. The VCR gaskets cost several dollars apiece, and are used in large quantities, while a vacuum pump can cost $25,000 to $30,000. A common gas control valve can range from $75 to $450. These high prices can add up in an incredible hurry if the parts are mishandled.

Quality of workmanship includes making sure all appropriate procedures are followed and documented. All dates, repairs effected, and other key information should be recorded for future reference. Details of all problems and their solutions should be recorded, so that repeated occurrences of a problem will be resolved with reduced downtime and/or reduced wafer loss. All procedures should be clearly and unambiguously specified and the specifications followed to the letter. Some jobs, such as gas system modifications, require a long time to carry out correctly. This time must be fully allotted, and all other groups (such as manufacturing) must understand the need for this time. Short cuts should never be taken, because they will almost invariably lead to bigger problems later on.

The second major group that works within the maintenance organization is the plant maintenance group. This group keeps the facilities of the building maintained and in top condition. The line of responsibility is usually drawn at the equipment interfaces themselves. For instance, the line maintenance organization will work on the gas systems of the equipment, replacing mass flow controllers, filters, and so on, up to the point where the main gas lines are connected. From there to the gas bottle outlet, plant maintenance will have responsibility, while the gas bottle changes are line maintenance issues. There are, of course, exceptions to this general rule, depending largely on how far it is between the gas bottle and the manufacturing system, and how complex the piping issues are. Similarly, the house exhaust and chemically scrubbed exhaust systems are usually handled by the plant maintenance organization, while the responsibility of properly venting equipment to the exhaust is left up to the line maintenance organization.

As opposed to the specification prevalent in the line maintenance organization, only one group of plant maintenance personnel takes care of all of the various tasks in the facility. The personnel include individuals who can maintain the substantial electrical systems, water handling systems (both deionized and city water), the various chemical and water drainage systems, the gaseous and fluid filtration and purification systems, vacuum exhaust systems, and all other building and design issues. The group maintains trained support personnel to assist the specialists in the event of emergencies or other serious trouble.

The plant maintenance organization is usually in charge of the implementation of environmental protection procedures. Wafer fabs produce a wide variety of extremely toxic effluent in all forms. The liquid effluent, while largely water, contains large quantities of chemicals that can be neutralized chemically, such as hydrochloric acid, ammonia, and choline. Some chemicals, such as hydrofluoric acid, which are toxic in the environment in any form and cannot be neutralized, must be removed separately and treated as toxic waste disposal. Other chemicals in this category include chromic acid and certain other heavy-metal solutions. The liquid effluent is cleaned by running it through tanks containing chemicals that will react with the effluent to form salts that will precipitate out of the solution to leave pure water. Usually, additional water is supplied to the system to further dilute the effluent. As shown in Figure 13-3, this usually requires several stages to remove all of the toxins. The separate toxic waste removal system does not permit any of the effluent to escape into the environment, but merely places the waste in storage for later disposal.

Gaseous exhaust is usually scrubbed by bubbling it through large columns of water and neutralization chemicals. There may be a number of stages of this cleaning also, as shown in Figure 13-3. The types of

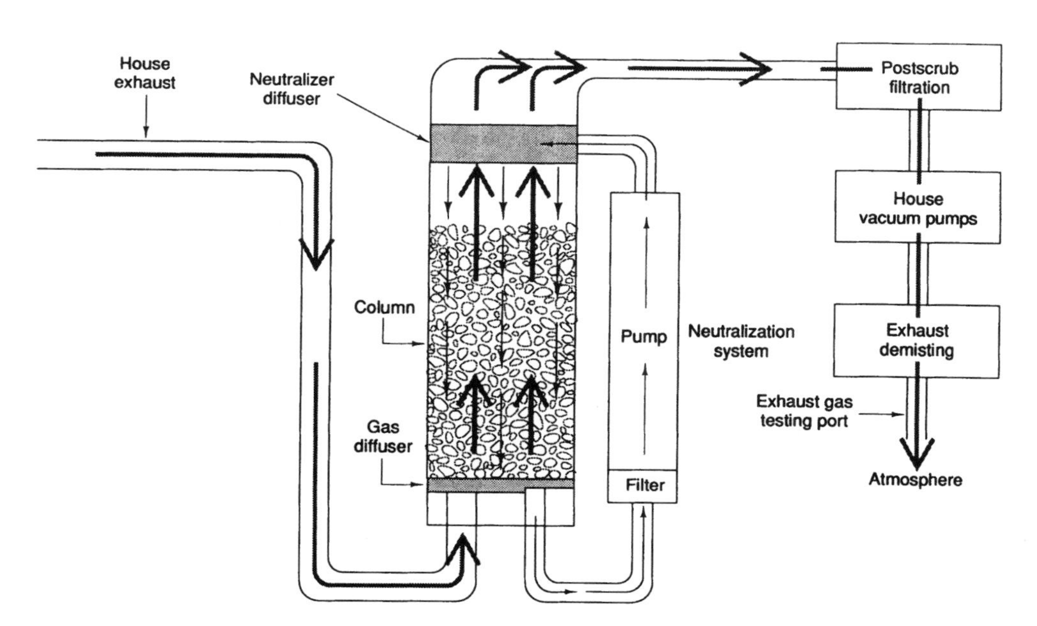

Figure 13-3. Exhaust Gas Reclamation Facility Toxic gases are exhausted from the fab through the neutralization column, a variety of filters, and a set of vacuum pumps before being expelled to the atmosphere. The final effluent is monitored for purity.

chemicals that are found in the gaseous exhausts are usually even more highly toxic than the liquid effluent. The chemicals involved can include arsine, phosphine, organic solvent fumes, silane, hydrogen fluoride, hydrogen chloride, plus any combination of unknown plasma by-products. They must be removed entirely from the exhaust. Since the entire exhaust system must be driven by large vacuum pumps, which can themselves produce effluent (especially as oil ages and small quantities of chemicals escape from the scrubbing system and concentrate in the old oil), a small exhaust scrubbing system must be installed after the main vacuum exhaust pumps. Some installations require less elaborate schemes. For instance, bottle exhausts can often be vented directly to the atmosphere, for example, the addition of ultrapure nitrogen or oxygen to the atmosphere poses no danger whatsoever. Some gas bottles such as silane may be vented to special explosion-proof containers in which the gas ignites to burn into a powder (with water vapor as the primary by-product) or is diluted with huge quantities of air. Obviously, this procedure should be performed only for nontoxic chemicals.

It is imperative that these environmental concerns be implemented properly, since the small effluent from a large number of wafer fabs can seriously damage the environment. A quick look at the number of Environmental Superfund sites in the Silicon Valley will give you an idea of the magnitude of the problem (although most of that contamination occurred before people were aware of the severe consequences of this contamination). It is my firm belief that virtually no contaminating effluent should be allowed to leave any semiconductor facility, as the technology to prevent this escape exists and is available. The long-term consequences of the contamination of our planet are just now becoming evident. The semiconductor industry has always prided itself on being a clean industry, so the precedent for "Zero Toxic Effluent" has been set. In most cases, the ability to reach this goal is largely dependent on the maintenance of the toxic waste elimination systems.

An issue related to environmental safety is that of worker safety. The maintenance organization must maintain the safety procedures in the fab area, that is, make sure that all emergency equipment is available and up to date. This also means that these are usually the people from whom Emergency Response Teams are selected, although virtually all of the groups within the fab have some representation. In addition, the actions of the ERTs are coordinated through the maintenance group because these personnel know the equipment well enough to prevent further damage or safety concerns. The members of the ERT are given training in first aid, handling of air tanks and other emergency equipment, and in testing the environment to determine the type and extent of the emergency. In most cases, the ERT will not attempt to fix the problem although, if the source of the problem is known and everything else is under control, they may isolate the source (for example, if a gas valve has been inadvertently opened they may turn it off; they should never attempt to put out a fire, unless it is very small). The primary responsibility of the ERT is to evacuate the fab area, verify that all personnel are safe, get help for any stricken employees, and verify the extent of the problem. Serious problems are to be left to the professionals (firefighters, for instance).

The final task of the maintenance department is to maintain adequate inventories of supplies at all times, so that most contingencies can be covered. To accomplish this, careful management of the inventories with computerized parts lists is absolutely critical. The number of spare parts required to run a fab can be staggering and the cost of all of these parts even more incredible. In addition, the high quality of the parts must be maintained, so the parts in inventory must be handled carefully and tested prior to installation. Controlling the spare parts in a modern fab is a complex and full-time job, and is a primary reason for the push from many companies for a "ship to stock" guarantee for all parts. This means that the vendor of a part guarantees that it will meet all specifications and absorb the cost of failed parts. Some companies, notably in Japan, impose penalties on vendors for failing to produce the promised quality of goods.

In conclusion, then, the maintenance group is primarily responsible for maintaining a high-quality, operational environment which the manufacturing group can use to produce the wafers, and for keeping the equipment and facilities safe for all personnel. They are also responsible to the community at large for maintaining the facility's environmental protection devices. While there may be appear to be fewer specific tasks for the maintenance group than for the manufacturing group, the level of importance of the maintenance group cannot be underestimated. Without good maintenance, the fab will not reach its ultimate capacity.

13.3 THE ENGINEERING GROUPS

Since the integrated circuit manufacturing operation is so complex and highly technical, there is a high demand for engineers in the semiconductor industry. These personnel provide a wide variety of services throughout the lifespan of an integrated circuit, and are available at essentially every step along the way, from conception to production. In general, the engineering groups can be broken up into several categories, including the design, research and development, process development, process sustaining, product or yield, reliability, and equipment engineering groups. Clearly, the responsibilities of the engineering groups are varied, but several broad outlines can be laid before discussing the individual groups. They are shown in Figure 13-4.

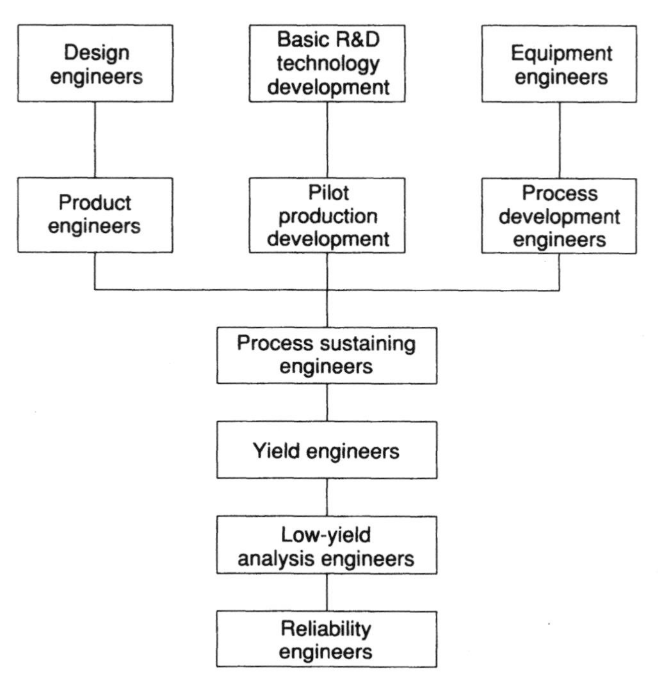

Figure 13-4. Flow of Product Responsibility through the Engineering Groups Integrated circuits go through various phases of development, as do the basic technologies that drive them. The equipment and processes are also undergoing constant change. Ultimately, this development effort must withstand the tests of yield and reliability for the product to succeed.

The main responsibility for all engineering groups is yield. While a few companies are more concerned with line yield than die yield, overall yield is the number one issue at hand for most engineers. As a result, the engineering groups are usually concerned with building in the quality up front. This is done through careful design work, careful process development, and effective implementation of new ideas into processing equipment. Reliability of the chips and of the process equipment is a major concern for the engineering staff, since failures cause reduced total yields. Thus, one of the prime responsibilities for the engineering staff is that of quality control watchdogs. To that end, experiments are carried out, changes are implemented, and procedures are observed and improved, all in an effort to improve yields, even slightly.

Tied into yield improvements are the wholesale process and design improvements that can make or break the wafer manufacturer. These significant changes are developed, tested, and then implemented by the engineer in his or her respective area. The engineer must possess good scientific and statistical analysis skills. The relatively crude skills learned in college will be honed rapidly in the high-pressure environment of the wafer fabrication area. The success of the major development projects and the implementation of new technologies in a wafer fab are usually instrumental to the success of the manufacturing line.

To carry out these responsibilities, a certain amount of salesmanship is necessary for an effective engineer. The engineer is often the primary go-between for several groups, being forced to convince each group that some idea that he or she has come up with should be implemented. Since the engineers have only a certain amount of control over the operations of the fab (the majority being maintained by the manufacturing and maintenance organizations), they must convince every organization of almost every detail. Production control must be sold on giving up wafer start allotments for an engineering test. Manufacturing must be convinced that this will not impact short-term production and will benefit it in the long term. Product engineering must agree that the integrated circuits can perform with the new process or process change. Process engineering must be convinced that there will be no contamination or other problems impacting yields. This adds up to a lot of legwork, and ineffective salesmanship can kill a project before it even starts, even if the project is a good one.

Another responsibility of almost all the engineers is that of teacher. The engineers spend a tremendous amount of time studying the processes and procedures that go into the chips. To make sure that the devices are properly manufactured, it is important that they transmit that information to other pertinent groups so that they can integrate that knowledge with that of their own areas to determine when problems can occur. This type of interaction takes place at all levels, with process engineers teaching operators why they are doing certain operations, and with design engineers teaching process engineers about design features that are critical to the IC's performance. The ability to transmit this information effectively is a good trait to have in an engineer.

A specific task required of almost all engineers is that of specification writing. Since they are the repositories of the technical knowledge of the factory, it is up to them to clearly document that knowledge so that others can benefit from it. The specifications should be written in as clear and coherent a style and should as consistent in format as possible. The ultimate specification should be written so that someone with minimal knowledge about a subject can perform the tasks outlined in the specification without further explanation. This is a very difficult goal to obtain, especially since most processes either change or become obsolete through time. Therefore, the specification writer is faced with describing a moving target. Obviously, part of the charter of the engineers is to make sure that the specifications that have been developed are, in fact, followed.

It is important that the engineers for a manufacturing line all understand as much as possible about the overall operation of the manufacturing process, not just that small area within which they work. This knowledge is invaluable in permitting the development of new procedures and processes. It also prevents decisions being made with insufficient knowledge of the consequences. It is critical, then, to make sure that all engineering personnel have the same information, that it is up to date, and that everyone agrees on responses to problems, so that they will remain consistent. All activities should be recorded in order to facilitate the further dissemination of information as it becomes available.

Finally, it is important in general for the engineer to know the ultimate application for the parts. If the chips are DRAMs or EPROMS, the engineer will know that these devices are under heavy competitive pressure and will push for every little bit of yield possible. If the devices are to be used in vehicles, failures may cause someone to be injured; therefore, reliability and quality are of utmost importance. Without this knowledge, the chips lessen in significance to just "widgets," and in the long run some aspect of the chips will suffer.

Thus, we can see that the engineering staff has many responsibilities. They cover a lot of ground, but can be summarized by saying that the engineer must obtain and retain as much information about the process as possible so that he or she can make major decisions that will significantly impact yield in a reasonable, rapid time frame. The engineer must make these decisions while placed under heavy demands on time for other projects. It is a tribute to the checks and balances of the manufacturing environment that this process runs as smoothly as it does in most wafer fabs.

We will now discuss some the individual responsibilities of the various engineering groups. We will start with the design engineers. These engineers are highly trained in circuit theory and solid state physics and in general are considered among the most valuable of the engineers. They create the designs using the most advanced techniques and tricks that they know of, so that the chip will perform certain functions. It is also up to the designer to place proper test capabilities onto the wafers. This may mean building test structures on the wafer or it may involve designing special test modes for the chip itself (many ICs contain these undocumented modes), or some combination of both. The designers then run their designs through simulation programs which try to test each component of the device in every significant condition expected. This verification process takes a very long time, even on fast computers. Unfortunately, the continuing increase in transistor density means ever-increasing simulation times, far outstripping the increased capabilities of computer technology. Simulation of all functions on a modern IC, such as the 80486 or 68040, continue to be very difficult and time-consuming. The upside to this is that, as the newer techniques are applied to older or simpler technologies, the simulation times for these designs improve dramatically.

After the integrated circuit has successfully passed all of its simulation tests, the set of reticles or masks is made up and used on a controlled batch of real wafers (called "pilot runs") to verify that the design works properly. If it works the first time, the reticles are handed over to production or product engineering, and the next design is started. More often than not, some small bugs are found which must be ironed out before a full product release can be attempted. In this case, the designers go back to their CAD systems, develop and simulate new components, have more reticles or masks manufactured, and run pilot runs a second time. At this point, it is expected that all serious problems will be worked out of the design. Occasionally, subsequent design cycles are required. This often results in severe damage to the marketability of the device, as technology may have progressed to the point where the design is using less than state-of-the-art techniques, or a competitor may have released a similar device and captured most of the market.

The designers must stay involved with the new product until the full qualification has been approved, including reliability aspects, because he or she may be able to provide insight into the operation of the device that will allow the process and product engineering staffs come up with reasonable solutions to processing problems. After final qualification of a product, the design engineer may remain involved in designing upgrades or design shrinks of the device.

The next engineering group that will receive responsibility for a new design is usually the Research and Development group (also called the Technology Development group). These engineers may work on a variety of projects that are only remotely involved with any one product, but in general the largest number of R&D engineers work on developing new techniques and running pilot production of devices. (Pilot production refers to small-volume production of a new design to allow chips to be constructed for test and sampling purposes only. Designs in pilot production are seldom considered to be completely "frozen" in design or process.) The research and development engineers also work with colleges, equipment vendors, other fab personnel, or with other resources to find ways to resolve specific key technical problems. For instance, they may focus on new ways to recrystallize silicon to permit the manufacture of very tiny transistors. They will not typically deal with day-to-day manufacturing issues such as, for example, developing new methods for reducing particles in a process reactor.

The R&D group will take wafers that are in pilot production and evaluate their performance and recommend further actions. If the design is unstable or not manufacturable, a new design may be required. If the fab's performance is below the standards required for successful fabrication of the wafers, the process engineers are given the task of making the required improvements. The R&D engineers will also set rough guidelines for specifications, and will perform tests on specific areas of weakness in the devices to try to home-in on the best processing conditions. It is important that the feedback on device performance is transmitted clearly to all of the other engineering groups, so that problems in computer models (in the design simulation or layout programs) and manufacturing techniques can be ironed out.

After a design is out of pilot production, it goes into full production. At this point, control of engineering of the design is transferred from the R&D engineers to the product, yield, and process engineers. Many companies integrate the position of product engineer and yield engineer, although there are some differences between the roles. For the remainder of this discussion, we will use the term product engineer for these roles.

The product engineer is responsible for most of the die yield issues in the manufacturing process and is typically the one who does the in-depth computer analysis of yield and electrical test trends. The product engineer will sometimes perform engineering experiments on the devices or on the process. Process changes that are developed by either the product engineer or the process engineer should be coordinated by both to verify that no known problems will crop up as a result of the process change.

The product engineer must also verify that new mask or reticle sets work correctly. This is done by using the new reticles on half of a lot of wafers, and the standard reticle on the other half. If there is no difference (or if the new reticle shows a yield improvement), the new reticle is placed into service. If yield degradation is seen, the new reticle is inspected, and, if no problem is found, the test is performed again. If there is still a problem with yield degradation, the reticle is rejected. Considering their prices ($5000+), reticles are rejected or damaged and need to be replaced frequently. Product engineers also have to verify that the tests performed on the wafers are appropriate for the problems and parameters that are under observation.

In general, the product engineer is the primary person who goes between the process engineering staff and the design engineering staff (there is, unfortunately, not that much contact between the designers of the chips and the process engineers who maintain the manufacturing processes). As a result, he or she must have a broad background in a number of fields. For instance, a strong processing background is usually preferable, simply because process engineering is not a skill that is easily taught, and because the importance of the key processing problems will be clearer. However, a strong background in circuit design, even at a component level (designing or constructing printed circuits, for instance), is also very important. Understanding the relationship between these two otherwise divergent fields is very valuable to becoming a successful product engineer.

An engineering group that is somewhat related to the product engineering group is reliability engineering. These are the engineers who study the performance of a chip throughout its lifetime. Increasing the reliability of a chip means that the total lifetime has been increased, or that the mean time between failures (such as occasional lost bits, or other hardware-induced system crashes) has been increased. Reliability concerns also include ensuring that the power consumption and heat dissipation of the chip stay constant through time. Finally, studies are usually made to determine the long-term consequences of radiation damage, sensitivity to radio frequency interference and unstable power-supply voltages, and many other effects. Most of the experiments in the reliability test area require a long time to carry out, although accelerated tests are sometimes performed. Interpreting the test results is fairly tricky since devices in the lab seldom perform under conditions similar to those encountered in the real world. Like the product engineer, the reliability engineer must have command of a wide variety of disciplines.

The next major engineering organization is the process engineering group. These are the engineers who are in the wafer fab on a regular basis, making the day-to-day decisions on a variety of technical issues. The process engineering group is usually broken up into a number of subcomponents, as shown in the organization chart in Figure 13-5. Typically, the group is divided by functional area, such as etch, diffusion, lithography, implant, and thin films. Each of these subgroups will have a few engineers, usually 2 to 10, depending on the size and structure of the organization. Each engineer will be responsible for the

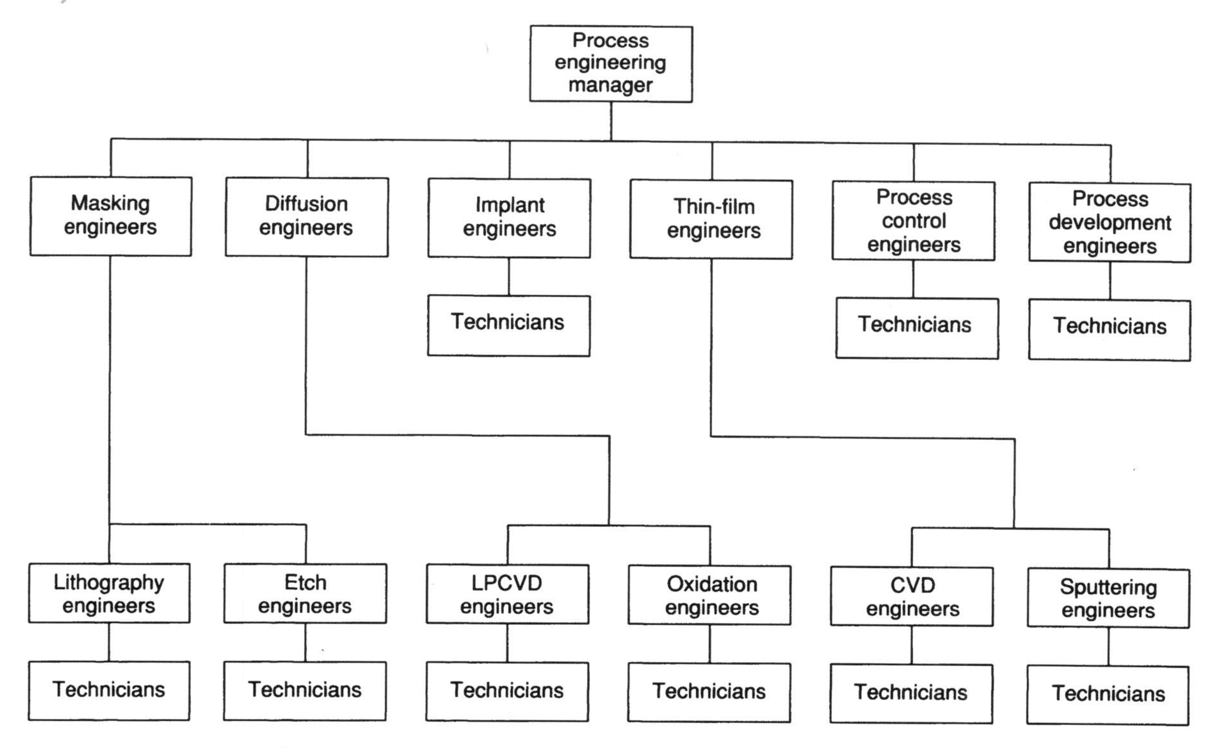

Figure 13-5. Organization of the Process Engineering Group The process engineering staff is broken up in a more horizontal fashion than other organizations. This is due to the large number of different processes that must be maintained by the engineers.

control of a number of systems or processes. For instance, in the diffusion area, an engineer will be assigned to the LPCVD processes, another to the thin oxidation processes, while another will be assigned to the thicker oxidations and anneal steps. While there are significant differences in the details of the jobs performed by the process engineers, there are a substantial number of similarities.

In general, the first priority of the process engineer is to maintain and control yields. This includes virtually all line yield and many of the die yield issues. The process engineer uses the process parametric information to make adjustments or request maintenance procedures. He or she must make use of every bit of available information on a particular piece of equipment before making decisions about its operation. To this end, the system should be thoroughly studied, and the operation and maintenance manuals read and understood. In addition, the past history of the system's operation and control, as well as an in-depth knowledge of the theoretical expectations for the process, greatly improve the effectiveness of the process engineer. In general, the tighter the control on the process, the tighter the control on the yields. This process sustaining work usually consumes 50% or more of the engineer's time.

Solving problems rapidly is one of the primary assets required in a process engineer. Every minute of downtime costs money, so quick decisions are necessary. Obviously, though, the decisions that are made must be correct. An incorrect decision made by a process engineer can cost thousands of dollars in just a few seconds of misprocessing time. As a result, a very logical, systematic approach to problem solving must be followed. Since so many things can go wrong at each step in the wafer manufacturing process, the first step is to verify that the evidence indicating a problem has been interpreted correctly. This could involve recalibration of the test instruments or verification that the discrepant readings are present on the production wafers. Many of the problems encountered in the day-to-day operation of the fab are test instrument- or test-wafer-related, and many hours of frustration can be avoided by checking them first.

If the problem still exists, two primary decisions must be made. The first is what to do with the wafers that have the problems; the second is whether to run the system again or what procedures should be followed before production can be resumed in the system. In both cases, the problem must be resolved in a logical sequence of events.

The first step in solving a problem is to clearly define it. This requires that the problem be broken down into its component parts. If a wafer is hazy, the cause must be determined. The definition of a problem includes determining the source of the problem, the number of wafers that are affected, and the amount of impact to each wafer. After the problem has been defined, it should be fairly easy to isolate the variables involved. If the pertinent variables cannot be discerned, you know that the problem has not been sufficiently defined, and the key questions must be asked in a different manner. For instance, if an automated inspection system is sensitive to focus and intensity of incident light, as well as other details with respect to substrate material, it is important to isolate the effects that occur due to problems in the focusing mechanisms or light sources, so that the source of a variation in the inspection parameters can be isolated.

After the variables have been isolated, the available options are evaluated. For instance, if the BPSG process is producing hazy films, it may be determined that the source of the problem is an improper clean (or poor drying cycle after the clean), contaminants brought in with the wafers, or contaminants generated within the system. A series of tests can be performed to isolate the exact source and to allow you to determine what actions are available to eliminate the problem. For instance, if the problem is with the cleaning process, the options will include shutting the sink down for a chemical replacement, or having the drier cleaned or otherwise adjusted.

Before initiating any actions, the consequences of the action must be estimated. In some cases, such as the clean sink decision above, the consequences are obvious, in that there will be a short period of downtime to clean the system. Further operation of the process would cause further yield loss, so there really is little choice in this decision. However, there are many more marginal decisions that must be made on a daily basis. For instance, if a certain product line is very profitable, minor defects on the wafers will not be considered grounds for wafer rejection. However, major defects on the wafers will be grounds for rejection. Stuck in the middle of these extremes are the judgment calls that must be made when there are no clear criteria for discarding wafers. Therefore, the consequences to fab output and overall profitability must be considered when making these decisions.

Ultimately, the decision to act must be made. Putting off a decision will result only in lost time, although an incorrect decision can result in lost yield. As a result, both improper action and inaction will cost money. Clearly, if an engineer does not have enough experience to make certain decisions, the operation may have to be halted temporarily until a clear decision can be made. Otherwise, if a decision will result in a clear yield benefit, the action should be undertaken. If no decision can be clearly determined, or if all the options have risks associated with them, testing should be carried out to determine the best course of action.

The process engineering job often consists of two segments. The first is the sustaining role, in which the techniques described above are applied to controlling the fabrication area. The other major segment is the process development role. In this role, the process engineer tries new techniques for processing wafers to improve yields with existing equipment, evaluates new equipment that could improve yields, and recommends purchases. Usually, process development is a long, drawn-out task. In most cases, the new process or new piece of equipment must be tested thoroughly, with full split-lot tests (running half of a lot through the old process and half through the new process). These wafers are then followed through the sort operation and sometimes all the way through reliability testing. This effort is necessary to prevent major yield loss due to untested but damaging new processes.

In general, the implementation time for a new process in existing equipment can range from two to six months from conception to completion, depending on the amount of change in the process, the testing procedures required, and the priority (read: yield impact) of the anticipated process change. A year might be required to bring in a new piece of equipment utilizing existing technologies for a new process, especially for a major process step. A totally new technology for providing a manufacturing process can take from two to five years to be fully integrated into the fab environment. Of course, it does not take as long to bring in new equipment that duplicates an existing process. While the new equipment must be tested for operation, and split-lot tests must be run through the system for full qualification, the time required to bring the new system on-line can range from a few weeks to two to three months (that is, after the equipment is installed. Purchasing and ordering considerations can totally distort these time frames). It should be noted that these values are only rules of thumb determined by observation of the semiconductor manufacturing equipment industry.

The process development engineering role is sometimes taken by someone who is devoted only to the new processes. This process development engineer will work side-by-side with the process engineering group to help coordinate the wide variety of improvements or changes that must be made. The process development engineer also works closely with the equipment engineers. These engineers typically work closely with the semiconductor equipment vendors, the facilities design groups, and other groups, consultants, and so on, as required.

The equipment engineers must make sure that the needs of the manufacturing organization are met by the equipment that is being purchased. The job also entails verifying that all facilitization requirements for the new equipment are met prior to installation. These issues include power-supply requirements, air flow requirements, temperature and humidity control requirements, exhaust, gas supply, and cooling water hookups. Other details such as floor vibration, the existence of stray RF fields within the processing area, and other environmental considerations should all be checked out. The new equipment will be evaluated by the equipment and process development engineers for correct operation (often at a vendor's site), and any details or special orders are requested. The equipment engineer also needs to verify that all of the space requirements are met, so that the personnel in the fab will have adequate space in which to work, and so that the facility does not become cramped, increasing contamination densities.

Larger wafer fabrication areas may have a variety of other engineering personnel. These other types of engineers can include process control engineers, metrology engineers, and occasional engineers working for the manufacturing, facilities, or maintenance groups. In smaller fab areas, the responsibilities for these disciplines are distributed among the main engineering groups; often the process engineer assumes many of these roles.

As we can see, there are a wide variety of areas within the engineering departments of a wafer fabrication area. Every aspect of the design, manufacturing, and testing of the integrated circuits must be carefully monitored and controlled. Extreme precision and care are required at all steps of the process, and the costs of *not* providing this care are so high that the expense of a heavy investment in engineering easily pays for itself.

13.4 THE MANAGEMENT TEAM

The management team of the wafer fab includes each of the various group managers, as well as managers from other associated groups, such as accounting, safety, marketing, and so on, who work under a common plant manager. The plant manager may be at the vice presidential level in some of the smaller IC manufacturers, but more often the plant manager reports to a vice president. The organization chart of the management team is shown in Figure 13-6. We can see that the responsibilities of the organization are distributed in a horizontal

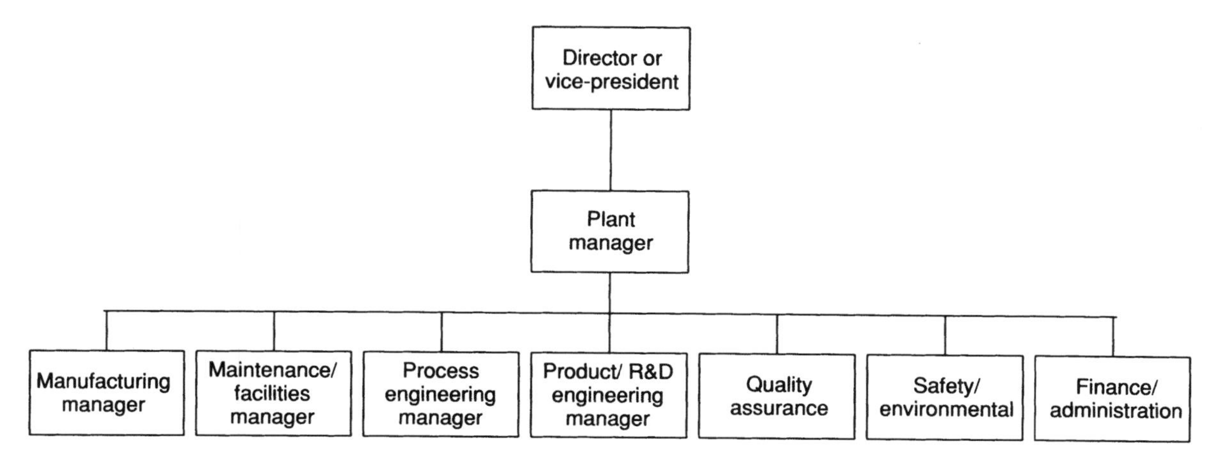

Figure 13-6. The Members of the Management Team While the exact management structure for each company varies, this represents the most typical type of responsibilities that are distributed among the members of the management team.

fashion so that there is a system of checks and balances to keep the company in control. In other words, the quality assurance and manufacturing groups are required to work together as equals in order to produce a sufficient number of chips within the required quality limits.

While the semiconductor industry is still in its youth, there has been enough history that some generalizations can be made about the management styles of the more successful companies. These companies tend to share some common features and practices which I believe are key to making them successful. Unfortunately, these features take a significant amount of effort to develop and many companies have attempted to ignore these practices and take the fast road to quick profits. In most cases, this has cost them in the long run and, as a result, many IC companies have gone out of business or have been reduced in size to bare subsistence. We will discuss these features and practices before discussing the various specific tasks of the fab management team.

Probably the most fundamental requirement for a solid management team is the ability to provide a sound, solid base of operations from which the rest of the personnel can take their cues. This requires that all members of the team be decisive, knowledgeable, and cooperative in providing solutions to the problems that inevitably occur in the day-to-day operation of the fab. Many organizations have been thrown into utter chaos when ineffective managers have tried to impose illogical solutions on manufacturing areas. This problem is amplified many times if other managers in the team are also misinformed. At the costs of operation of a wafer fab, the company cannot afford this kind of chaos and confusion for very long. The most effective teams have been those who do not get rattled easily in the high-stress manufacturing environment, and who can make rational, logical decisions in a consistent manner. A good, long-term plan will make short-term problems easier to solve by allowing options for advancement. Lack of a long-term plan will often result in a "band-aid" solution to a problem, which may cost the company in the long run.

The second most important duty of the management team is to provide clear direction for the operation. This requires having a common vision of where the fab should be going and a positive path to reach the goal. Once a fab has lost its sense of direction, it will no longer be able to keep up in the rapidly changing semiconductor marketplace and can cause it to eventually grind to a halt. The decisiveness of the fab management team becomes very important here, because the advancement of technology requires an aggressive approach. The direction for the fab must not be a timid one, but should still remain consistent with current operations in order to provide a smooth transition to the next level of technology.

At the same time, the most successful companies, in order to convince customers of the quality of devices, publish only reasonably conservative specifications. This means, in reality, that the products must be of even higher quality than would normally be expected for the products to be competitive. For instance, it is well known that Intel microprocessors can operate at clock frequencies as much as 25% higher than specified, with little impact on their reliability other than higher power consumption. Many less-quality-conscious companies may have sold chips specified at higher operating speeds and accepted the slightly higher possibility of chip failure or other problems. The impact on Intel is that they must produce chips of a quality level much greater than that already required to build their advanced ICs, but also gives them a reputation for very high-quality parts.

Another important aspect that is usually found in the more successful semiconductor fabs is a people-oriented attitude and policy. This is based on the simple premise that a worker who is respected and is content with working conditions will be more quality-conscious and productive, although the policy must be genuine, or morale will degrade very rapidly. In general, this people-oriented philosophy is manifested by an environment that stimulates personal growth and free thought and expression. The "open door" policy, wherein employees can visit managers at any reasonable time, is a visible and important component of the environment. Some companies have brought in consultants to teach their personnel how to be assertive and to express themselves without becoming emotionally involved. It is important for everyone to feel that criticism is not directed at individuals but at specific problems, so that the problems can be resolved. This positive attitude of accomplishment will help the management team reach the goals that they have set for the organization, and will help the managers recognize what goals are reasonable and what goals are overaggressive, given the resources available. A people-oriented attitude also reduces the inevitable finger pointing that occurs when there are problems.

Overall, therefore, the companies that have a corporate culture that includes the above attributes have up to now shown performance that is superior to those that have not shown these attributes. This superiority is shown as larger market acceptance for the products, which shows up in the profitability of the companies over the long run. This long-term growth has permitted the companies to make the jumps to successive levels of technology, both financially and technologically due to the fact that the people who are performing well can more easily pick up the nuances of manufacturing the new, more complex products.

These issues are all more issues of style than substance, and even though the implementation of these procedures does take up a significant amount of each person's day, there are many other tasks and responsibilities that must be accomplished by the management team. In general, the members of this team should not attempt to make the day-to-day decisions of fab operations. Obviously, management must sometimes be closely involved, as decisions or problems will arise that may cost the company a lot of money. However, in most cases, the managers should be able to trust the competency and judgment of the fab support personnel. Instead, the management team should be focusing on the more global issues of fab operations, which include maintaining the output level of the manufacturing area and overseeing the quality of the material that is being produced. The safety of the personnel and the protection of the environment, both within and outside of the plant, should also be primary concerns.

In general, the top priority of the factory is safety. Only the most hardened of managers would expect anyone to risk his or her life or health to manufacture some wafers. The safety team usually reports directly to the fab or plant manager, thus preventing conflicts of interest, allowing all of the management team to have knowledge of safety-related issues, and allowing each member of the team to be assigned clear responsibilities and tasks to perform in event of an emergency.

Another critical priority of the management team is to ensure that the environment does not become polluted through the factory's operation. The San Jose area in California is dotted with sites on the Environmental Protection Agency's Superfund list for toxic waste contamination. This list ranges from cancer-causing solvents in the groundwater to soil contaminated with these solvents plus heavy metals and other combinations of toxic waste. In the early days of the industry, these chemicals were not always recognized as hazardous, but there is no longer an excuse for significant amounts of toxic waste to escape from the fab. Waste disposal technologies exist to remove the vast majority of the most toxic pollutants, and there is a push to start removing some of the other dangerous chemicals from the fab areas, such as the replacement of ozone-damaging CFCs with inert perflourocarbons and other related compounds, and to replace chemicals such as HCl and H_2SO_4 with less-dangerous compounds, such as choline. A sufficient number of toxic gas detection and alarm systems must be installed to detect all of these chemicals. It will take some time before the hazardous chemicals are all weeded out of the fabs, but eventually the change will occur, and my guess is that it will be sooner rather than later. In some cities, local ordinances are forcing companies to prevent toxic effluent, a benefit to all of us in the long run.

Obviously, one of the most time-consuming and complex of the tasks facing the management team is forecasting the output of the fab and then obtaining that output. These forecasts require careful planning and good follow-up to verify that procedures necessary to reach the fab's goals are used. Inventory must be controlled and production bottlenecks permanently removed. Conflicts and disagreements must be resolved and all of the various groups sold on making the policies and procedures work for the profit of the fab. Implementation of the fab output plan is complex, requiring a combination of skills, from juggling to salesmanship. Making a forecast and then meeting it is difficult and takes much experience. Many companies stumble when attempting to meet forecasts.

We can see that the management organization must be able to provide a stable, broad, effective base from which other groups can operate. They must be able to provide the leadership and vision that are necessary to excite the employees about their jobs so that they can produce the extremely complex devices more efficiently. They must provide the stability and competency that will command respect in order to make sure that all of the personnel will be willing to follow their leads. The management team must provide internal checks and balances to prevent internal conflicts of interest from causing an impact on reliability, yield, or morale. The managers provide the planning and direction to allow the employees to best carry out the tasks assigned. Finally, they must ensure a relatively safe, pollution-free environment in which everyone can work. The successful companies of the semiconductor industry have managed to merge all of the various positive aspects of these issues into unique, but in many respects similar, cultures.

13.5 THE WAFER FABRICATION TEAM

The most obvious thing that should become apparent in the preceding discussions is that the success of a wafer fab depends strongly on a team effort. While it would be nearly impossible for one person to run a fab alone, it is very possible for one person to slow a fab almost to a halt. A concentrated effort by all parties on building and maintaining a consistent approach, with a great deal of focus on quality workmanship by all parties, will result in a wafer manufacturing area that is both profitable and enjoyable to work in.

Clearly, fab areas have many problems from day to day. They can be very frustrating and can occur over and over until the underlying causes are rooted out and corrected. The wafer fab team needs to have guidelines in place for how to proceed through these difficult regions. Just as a Himalayan mountain climber requires a map and a Sherpa to reach his goal, the combined experience of the fab team must be harnessed to resolve issues as quickly and satisfactorily as possible. Having these procedures and structures in place will help prevent morale loss, yield loss, and other difficult problems, and in the long run will allow the fab to run in a reasonably predictable fashion.

Fab-wide intergroup teams should be formed to carry out the global tasks that are required to maintain the proper level of support for the fab. These types of issues can include identification, definition, and implementation of global specifications, like the cleanroom procedures specification. These teams can then verify that the procedures are followed, identify defects or inconsistencies within the new procedures, and report the progress of the group back to other members of the fab team.

This team approach is much more effective than the traditional multigroup setup usually used. For this structure to survive, a certain amount of discipline and ego reduction is required. However, in the long run, policies are more cohesive and consistently applied to all personnel when teams, not individuals, develop the rules. It is always much easier to convince the other members of the fab team to cooperate when they have a stake in the decision-making progress. There are also a number of pitfalls avoided when the team approach is used for fab management such as finger pointing when problems occur, missed schedules, and other problems that are easy to stumble over in the high-stress and high-cost environment of the fab. The main problem with this approach is the tendency to do everything by committee, becoming, in effect, a little Congress, with all of the associated delays, and bureaucracy.

While not all wafer fabrication areas run under this concept of team effort and concerted joint efforts at problem solving between the otherwise divergent groups, most of the successful companies in the IC business implemented these types of policies early in their existence. While it is difficult to turn around a situation that has had distorted values for a long time, the attempt can be made and, if sincere and consistent, can be successful even in those environments.

Thus, we can see that the wafer manufacturers that have been the most successful are those who have managed to find a distinct balance between the needs of their employees and the surrounding community with the needs of high-volume, high-quality mass production. Internally, it has meant that the leaders of the fab must be aggressive, but able to state their goals clearly, so that other members of the fab team are willing to put forth the extra effort to follow through. The employees of the company are part of the decision-making apparatus so that the aggressive leader does not send the company down a path on which it cannot succeed. Finally, a perfectionist, "We can do it right" attitude often prevails, with quality workmanship viewed as one of the most important attributes of any worker. A company that combines these various skills at a variety of levels within the organization will have ultimate success in the semiconductor industry.

CHAPTER 14

FUTURE TRENDS AND CONCLUSIONS

Now that we have examined the semiconductor manufacturing process in some depth, we can see that there are a number of themes that are repeated throughout the process. They include the importance of maximizing yields, reducing defect densities, and creating a high-quality manufacturing environment. The improvements in these areas must all be accomplished while the semiconductor designs evolve and become more complex at a rapid rate. The general trends of the industry up to now can be used as a guide to allow us to see what type of trends there may be in the next few years. As with any type of prediction, the probability is low that any exact details will be revealed, much less in the volatile semiconductor industry. As a result, we will look mainly at the major trends that appear likely to make a significant impact on the industry.

Looking back over the past 10 to 15 years, we can see that the device performance trends have been toward smaller, faster parts that consume less electricity. In addition, the functional power of the chips has been increased with increasingly high-density components. In general, the increase in device density and functionality has been well over an order of magnitude over the time period. Typical device linewidths have gone from 5 to 10 μm in the late 1970s to about 1-μm levels today (on production devices). The devices that were constructed then were considered to be high-density devices with 10,000 transistors on each chip, while today advanced production ICs have over one million transistors on each. Clock speeds have also improved from 1 to 2 MHz to 25 to 30 MHz. In addition, the functions of hundreds of chips have been integrated into single chips or chip sets that take care of most of the primary operations of the computer systems. For instance, the original IBM PC consisted of around 100 ICs. The same system can now be designed using one CPU, some memory chips, and a system-control IC. The simplest design could consist of four or five components (using modern memory modules). Several showing the improvements in technology over the last several years are shown in Table 14-1.

TABLE 14-1
The Recent Improvements in IC Technology

Systems	CPU	Bits	Clock (MHz)	RAM	Disk	Relative performance[a]	Year[b]
Apple	6502	8	2.5	16k	70k	1	1979
IBM PC	8080	8	4.77	64–640k	360k	4	1982
IBM AT	80286	16	6–10	1M	20M	15	1984
IBM PS/2	80386	32	16–20	1–4M	40M	35	1988
Latest PCs	80486	32	25–33	4–16M	80–160M	50+	1990

CPU[c]	Bits	Clock (MHz)	RAM	Size	PL[d]	C[e]	Functions	Year[b]
6502	8	2.5	64k	35k	0	0	CPU	1975
8080	8	2.5–4	64k	50k	0	0	CPU	1975
8086/8	16/8	4.7–8	1M	100k	0	0	Adds coprocessors	1978
80286	16	6–16	16M	200k	3	0	Add MMU[f]	1981
80386SX	16/32	16	16M	300k	3	0		1989
80386	32	16–33	4G	500k	3	0		1988
80486	32	25–50	4G	1.2M	5	8	Add cache, FPU[g]	1990
i860	64	50	4G	1.4M	3	12	Add parallel process	1990

Memory chip	Bits	Speed (ns)	Year[b]
4116	16k	300–450	1970s
4164	64k	200–300	1982
41256	256k	120–250	1987
(numerous)	1M	70–120	1989
	4M	60–100	1991+

[a] Relative performance is the relative system speed and capabilities based on a number of criteria and is for reference only.
[b] Year of general acceptance, not always year of release.
[c] Except 6502, Intel family used for comparison. Other chip families show similar improvements.
[d] PL = Pipeline queue.
[e] C = On-chip cache sizes.
[f] MMU = Memory management unit.
[g] FPU = Floating point unit.

It is clear that we are reaching the point where another level of technological prowess will be required for the next great boost of power. The standard MOS processes run out of steam at around 50 MHz. Even if improvements can be made in this area, the improvements will continue to be incremental and not of the orders of magnitude seen before. In addition, as the linewidths of the devices shrink, the ability to resolve and produce the devices becomes technologically more difficult and expensive. In addition, other effects (such as quantum effects and maximum current-carrying capacity) start to come into play at the ultramicro miniaturization level. Of course, these effects give rise to phenomena that can be used to create the next generations of devices.

There are quite a few candidates for the technology that will "replace" silicon, and we will discuss some of them. One trend that is becoming quite clear is that there will not be a wholesale replacement for silicon any time in the near future. Instead, the it appears that new, more-complex and expensive technologies are being mated with the older, far-less-expensive silicon devices for many functions. Only the most critical functions of the system will be manufactured with the higher speed parts. Since a significant portion of a computer system does not require these higher speed components, there has been little incentive to provide these devices in the most advanced technologies.

Another factor that will affect the future of the semiconductor market is that the development of new hardware (integrated circuits) has greatly outstripped that of new software. Most of the operating systems on the market today do not operate on the new hardware to anywhere near the optimum efficiencies available. Many of these operating systems were written in the 1970s, or have been created to remain compatible with (and limited by) central processors designed in the 1970s. As a result, there should be a major boom in the software market over the next 5 to 10 years, as the consequences of the immense computing power we now have available become apparent. Unfortunately for the chip market, the same forces may tend to reduce the market demand for some of the newer types of devices. This is not to say that technical advancement will not occur, but it could mean that the future market for these new devices could be smaller than expected, and that growth into these areas will be reduced until the software industry catches up.

From a manufacturing point of view, there will continue to be further advances, although the methods by which new equipment is procured will need some adjustment. The desire to move toward narrower lines and thinner films on semiconductor wafers has the side effect that progressively cleaner, more-stable, and less-damaging processes are required to manufacture the chips. As a result, the costs of equipment development have been increasing rapidly, with very small total available markets for any particular design. The high capital investment required to develop this new equipment has caused most American investors to shy away from new technologies, allowing the Japanese equipment manufacturers to obtain yet another stranglehold on an industry once dominated by Americans. The responsibility of equipment development and procurement processing should be actively supported by the semiconductor manufacturers. While I am not recommending how this should be done, the close relationships between the Japanese equipment vendors and IC manufacturing customer should be looked at closely.

As we can see, a number of divergent forces are at work to drive the semiconductor industry. There is a continued drive for more technological improvement, higher density devices, higher speeds, and smaller systems. This situation is not likely to change any time in the near future. There are still many 8-bit microcomputers on the market running old software. By the end of the century, however, we will have seen a heavy turnover of these machines to 32- and 64-bit computers. New innovations, such as verbally controlled operating systems, integrated televideo/teletex systems, and home control networks will be permitted only through the use of these new systems and, as IC prices come down, the market pressures for these more powerful systems will increase and the older computers will be replaced. Future systems will be able to have gigabytes of storage at any time; most operating systems today are limited to a few megabytes, while the most common—IBM/MS-DOS—directly addresses only 640 kilobytes of RAM. Telecommunications systems will also see great advancements into areas that have been explored but in which many problems, both technical and political, have yet to be resolved. The new technology of high-definition TV (HDTV) will also push the limits of technology, by requiring faster image processing and systems control hardware, and huge amounts of memory.

Forces that are in conflict with this technological drive are the lack of software to adequately implement all of the new features, and the costs to develop and install the new manufacturing equipment and cleanrooms required for the new chips. The risks that are required to keep the United States in the technological race seem to be too high for many investors in this country to face, so that the road ahead appears to be a tough one. However, with the advent of the consortium approach to technology development, the individual risks of the push for the next level of technology are spread between the various chip manufacturers, semiconductor equipment vendors, and chip consumers. This approach may help reverse the trend of American industry giving up technological leadership.

Thus, we can see that the semiconductor industry is in for much of the same type of business climate that it has experienced for much of the last two decades. There will be fairly drastic ups and downs and further consolidation of the various companies, due to the demands from the marketplace for even more functionality. The marketplace and the "book-to-bill" ratios may change, and shakeouts may occur, but in the long run, the industry will continue to grow, driven by the need for higher technology, with the spoils going to those companies who are able to manage the complex task of wafer manufacturing effectively.

14.1 BROAD MARKET TRENDS

Future integrated circuits will have to be manufactured with a number of exotic processes in order to attain the performance and density requirements that the marketplace will be demanding. The chips must be able to attain higher levels of integration on each device, and will be required to process information at much greater speeds than today's chips. In some cases, this increase will be attained through design improvements and tricks, while other improvements will be obtained only through material improvements. We will discuss some of the possible directions for these improvements here. Some of the techniques are already being applied to the highest technology chips.

First, we need to take a glance at what the broad market trends will be over for the computer industry the next 5 to 10 years. Clearly, the integration of multiple systems through networks and modems (through mainframe databases or through the smaller "bulletin board" systems) is one of the most significant trends of the computer scene. Once computers and televisions become completely interlinked (which should occur in the mid-1990s), the ability to obtain large amounts of information from all types of sources (video, audio, data, and other) will be available. Most home computer systems will have complete audio/visual and communications capabilities. For instance, the fax revolution has already started in earnest, and is the forerunner of this new era. For local messages, there are few more cost-effective means of communicating, and for long distance there is nothing more efficient than a fax machine. The integration of video and audio capabilities in general-purpose consumer products is already shown in some systems, including the Amiga and some Apple computer systems. Most computers can be set up to handle combinations of audio and video source material, if enough money and expertise are applied. For example, there are computer CD-ROM players available that can be used as audio CD players, although this is a very expensive way to play CDs.

The next area that will probably see significant changes is in the user/consumer interface and the basic operating systems of computers. Current systems of keyboard data entry are too slow and cumbersome for the human, while icon-driven software tends to limit the user to the choices available and is more complex for the computer to display and work with. There are proponents for both of these interfaces (which are epitomized by the Apple Macintosh and IBM-DOS systems), and there are technical and legal limitations to both. There are other limitations for the many hybrid and competing operating systems (such as UNIX, OS/2), that have prevented their widespread acceptance. The ultimate operating system of the future will probably be able to read and operate code from almost any system on any other system. The reality is that competitive pressure will prevent widespread acceptance of this common operation of programs until there is a major change in the user interface, which is still basically the hand.

Therefore, the next major strides will occur as operating systems become vocalized and intelligent, allowing the operator to control the operation of the system by speaking in normal English (or language of one's nationality) and having the system understand the instructions, search through references to similar occurrences, and execute the appropriate tasks. For instance, the computer systems should be able to monitor the progress of a sales order a few times, determine which items change (exact quote for materials desired and customer), and which come from databases (price lists, availability, and so on), and be able to produce future sales orders itself on command.

Some of these abilities will come from having even higher speed processors available. There has been much discussion about the merits of reduced instruction set (or RISC) technology. In these CPUs, the instruction set is limited to certain key instructions, and all other instructions must then be constructed from them. In the ideal case, each of the basic instructions is programmed to take exactly one clock cycle. In the typical complicated instruction set system (CISC, such as the 680xx or 80x86 microprocessors) some of the instructions may take as long as 15 clock cycles to execute. Therefore, in theory the most basic instructions can be said to be many times faster using RISC technology. The disadvantage is that more of these tiny instructions are required to perform the same tasks that a CISC computer system does in one, which is likely to occur in typical situations. This eats up a lot of time, enough so that for most real computer systems the RISC computer runs about one and one-half to two times the speed of the equivalent CISC computer. Some of the most modern CPUs, such as the 80486, reinterpret many of the multiple clock-cycle instructions from the CISC 80386 CPU as one or two clock-cycle instructions, in a RISC-like fashion. This makes the 80486 execute the same set of instructions as an 80386 about twice as fast.

In addition to the faster CPU speed, the new technological demands will probably require the use of neural network technology. This new type of integrated circuit works on a principle similar to a network of real neurons, in that certain patterns of responses to inputs result in predictable solutions. After several test runs, the chips "memorize" the path to this solution, and when presented with this same set of inputs will produce the correct output. In another remarkable skill, the chip can learn on its own to determine which features of the inputs are critical for a solution. As a result, this technology is expected to be extremely useful in the area of pattern recognition, which will lend itself to video image interpretation, speech recognition and generation, and certain artificial intelligence applications. Although neural networks can solve a variety of simpler tasks also, they do so in a slower, less-efficient way than a standard CPU (at least at this time). We can expect to see the high-speed CPU technology linked with the pattern-recognition technology available with neural networks.

In general, it is safe to say that the 32-bit processor will dominate most of the general-purpose computer world by 1993 to 1995. By then, most of the 8-bit microcomputers will have been phased out, with the most logical trade-up being from 8-bit to 32-bit if it is affordable (the wide variety of available chips is not likely to shrink, as there are still many applications that smaller CPUs can handle, controlling appliances, for instance). The CPUs dominating the marketplace in five years are probably going to be variants of the primary CPUs on the market today, namely the Intel 80386 line (of which the 80486 is a descendent, the CPU containing an enhanced 80386 instruction set), the Motorola 68030 line (of which the 68040 is a direct descendant), and the newer Sun Computer SPARC CPU. This type of chip is more than ample to drive most applications used in a home or business environment, which include things like word processing, spreadsheet, database, desktop publishing, and the other audio and video presentation capabilities noted above. Most of the graphics and communications tasks can be handled by these CPUs. More intensive scientific, high-resolution graphics and multiuser systems will be more efficient with the high-speed 64-bit processors that are now being developed. Many of these CPUs will be able to process data as fast as the mainframe computers of the 1970s and 1980s. As a result of the technology gains from microprocessors, the newer personal computer market has grown to compete with and sometimes even supplant territory that was considered minicomputer or engineering workstation territory only. Within five years, this line will be blurred further, so that the definition of these types of systems will be primarily one of software installation (the traditional "minicomputer" tasks would be taken up by a business software package, the "work-station" tasks set up by a scientific/artificial intelligence package).

As a result of this massive increase in power, the supercomputers of today will be supplanted by even faster supercomputers. These computers will be driven by continuations of a number of current trends. They include the use of a "massively parallel" computer, in which hundreds of CPU chips are placed into a matrix to allow the steps of a process to be carried out simultaneously. Unfortunately, only certain problems are amenable to this type of problem solving (for instance, weather forecasting). The superhigh-speed computers such as the Cray (which uses up to 6 to 12 main processors) use circuits which are made up of extremely high-speed components, such as gallium arsenide chips. These supercomputers are being used in an increasingly large number of research and development centers as well as the more common aerodynamic investigation, weather phenomena, and other very computation-intensive tasks. The use of the supercomputer will spread widely throughout the 1990s.

Thus, we may see that the markets break up into three major areas of competition, as opposed to the very fragmented marketplace today. They will start with the personal or consumer computer, which is similar to today's PC. The second level will be that of the work station. They will be outgrowths of today's minicomputers and work stations, and will probably be the venue for the first of the voice-activated, artificially intelligent systems. These systems will have the power and memory (as well as the price tag) to implement the latest in technology while remaining within the range of most companies. Finally, supercomputers will continue to be used for the most advanced problems. The distinction between the supercomputers and mainframe computers will continue to blur, as the performance of microprocessors pushes the mainframe's board-level CPU out of the market. Ultimately, all of these computers will have interchangeable software and interfaces. This may result in some consolidation and industry shakeout, but should strengthen the industry overall, and should provide a basis for agreements on reasonable standards for future growth. (Think of where the entertainment industry would be now if TV manufacturers had never agreed on one standard broadcast format.)

One technology that is unlikely to be replaced anytime soon, despite repeated attempts to pronounce its ultimate demise, is that of rotating disc media (disk drives). While the semiconductor manufacturers continue to find ways to increase the memory density of their chips, they will not be truly competitive to rotating disk until nonvolatile chips can be manufactured with the bit densities available on the disks. Even then, to actually replace the drives on the market will require that the prices on these chips are reduced significantly to be competitive. Chip manufacturers say that they will eventually meet these goals, but do not give time tables. The fact of the matter is that the disk drive manufacturers are not standing still either, and have already produced optical read/write CDs that can store hundreds of megabytes, and can be purchased for under $1000. However, the support circuitry for these disk drives does allow the chip manufacturers to have some visibility into this marketplace. They will have to remain content with the main system RAM, which must be fast and easily accessible.

In another area, as the applications-specific IC industry evolves, the increasing density of these devices will continue to lead to more and more functional modularity in the designs of the chips. Ultimately, since the computer system manufacturers are interested in solutions to particular system problems, the details of the internal logic of the device will become less important. We can see this evolution occurring already, with chip sets available to make a PC or AT with five or six chips plus memory and CPU, as opposed to the 100 or so required on the original IBM PCs and ATs. This will have a significant impact on the chip manufacturer as the designers are forced to stuff more functionality onto the chips, while at the same time integrating otherwise independent modules into a manufacturable (and profitable) size. Ultimately, the use of these modular designs is leading companies into more specialized market niches, while reducing the number of manufacturers who make large numbers of different types of chips.

So, finally, we can see that there is still more change to come from the point of view of marketplace demands. The new technologies and market requirements that will be emerging in the near future will result in increase of memory density (to handle the large volumes of data, and video or audio information), much higher CPU speeds and parallel processing, as well as more radical technologies such as neural networks to allow pattern recognition, and related "intelligent" activities. Many of these improvements can be largely termed design or implementation improvements, and can be accomplished with current technologies to enhance the operation of the computer systems. The ultimate limitations of these designs will still reside with the basic technology of the chips used in the system.

14.2 TECHNOLOGY ENHANCEMENTS

In general, it can be seen that all of the IC manufacturing technologies have certain limitations. Table 14-2 compares the attributes of each of the major IC technologies. As we can see, each of the technologies has

TABLE 14-2
Attributes of Various IC Technologies[a]

	Power/ gate	Maximum gate speed	Device density	Price/ gate
Bipolar	4	4	6	1
PMOS	5	7	4	2
NMOS	3	6	3	3
CMOS	1	5	1	4
ECL	6	2	5	6
GaAs	7	1	7	7
Bi-CMOS	2	3	2	5

[a] The relative performance of the various IC technologies is shown here. Due to the many variations and rapid improvement in designs, it is almost impossible to show more than a rough comparison in performance. After the top several technology contenders in each category, top performance devices become fairly compatible. Note: lower numbers are better.

certain advantages which determines the types of systems in which the chips can be used. Thus, chips made with NMOS overall are slower than their CMOS equivalents, which are slower than their bipolar equivalents. Faster yet are the emitter-coupled logic chips, which are exceeded by gallium arsenide chips. Naturally, the performance of these technologies is matched by their prices.

One of the primary reasons that the various technologies have different attributes is the electron mobility of the substrate (transistor base) material. For instance, one of the reasons that gallium arsenide is touted as such a fast material is that the electron mobility is about two and one-half times faster than that of silicon. However, design tricks are used to permit slower devices to operate more efficiently, and as a result some of the most advanced silicon devices can keep pace with many average gallium arsenide chips. The ability to squeeze very high performance from the silicon structures indicates that similar gains are possible in the speed of the GaAs devices available at the present time.

Obviously, as has been discussed repeatedly through the book, the race toward increasing circuit densities and shrinking linewidths has not abated. Current near-one-micron design rules will not permit the development of chips of this next generation planned for manufacturing. These chips include the 16- to 64-megabit RAM chips, and CPU chips that exceed 5 million transistors (even today, the 68040 contains about 1.2 million transistors, while the 80486 contains about 1.1 million). As a result, the plans of the majority of the semiconductor industry, and certainly one of the stated goals of SEMATECH and its member companies, is to reach a designed minimum linewidth of 0.35 μm. Japanese companies have stated similar goals with similar timetables, and the experience of the U.S. semiconductor business is a clear indication of the Japanese ability to reach their goals. The European Community, as well as several European companies, have also started to band together to try to reach these difficult, but highly lucrative levels of integration.

There has been continual discussion over when and where GaAs technology or ECL technology or something else will replace silicon in the manufacture of computer systems. Ultimately, it will probably turn out that most computer systems will not need to have parts made of one technology exclusively. While there are always some limitations to mixing the chip technologies, the most likely probability is that the computer systems of the 1990s will have mixed material technologies. For instance, the higher speed parts that are required for the video interface will be constructed in gallium arsenide, while the lower speed parts required for printer, keyboard, and other similar interfaces will be made from silicon. In fact, it may prove to be most profitable to manufacture the basics of the chip on silicon for structural and cost reasons, and then deposit a layer of gallium arsenide on top of that surface to provide the high-speed processing capabilities. A

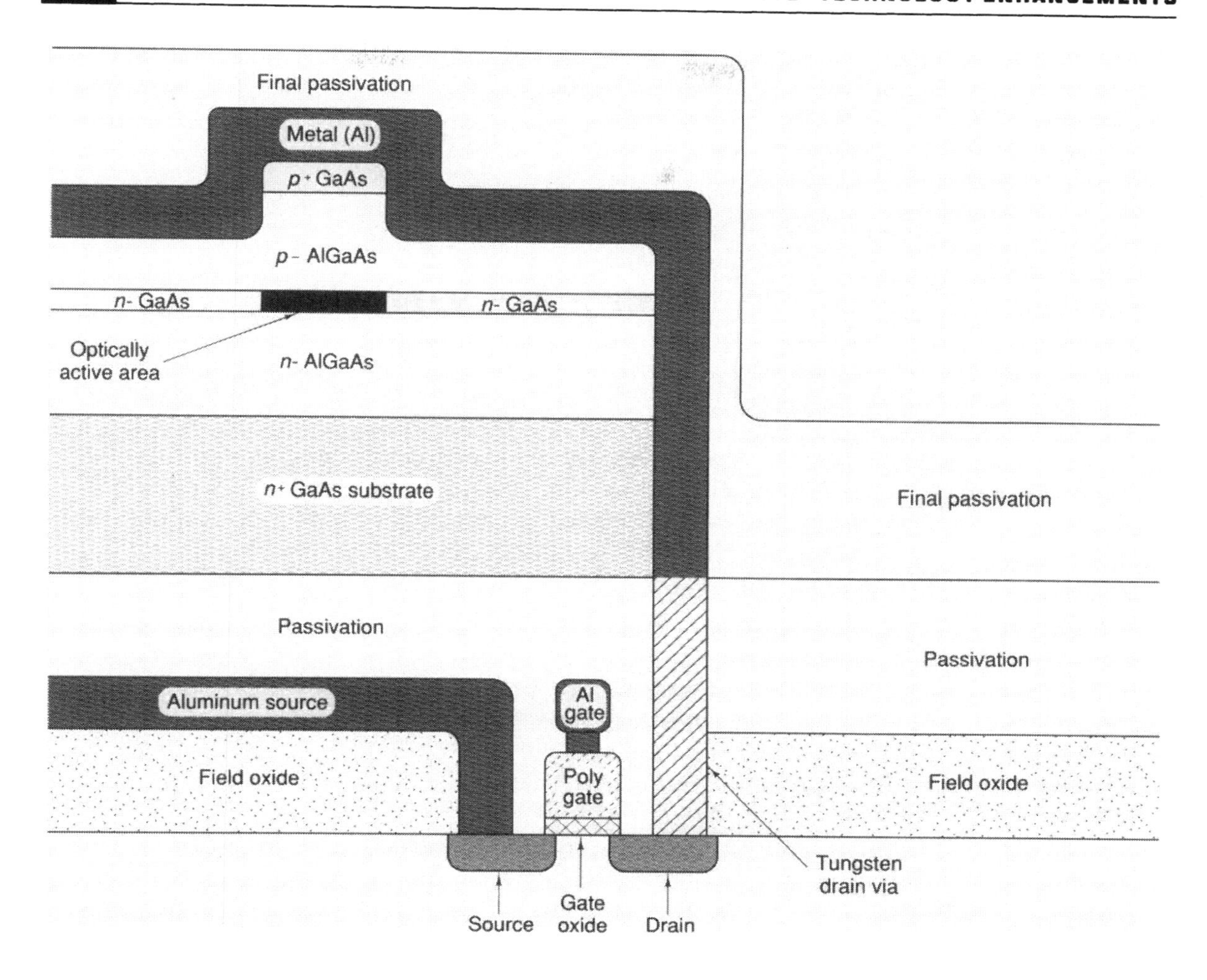

Figure 14-1. A Composite GaAs Device Future developments will include the mixing of various technologies in order to obtain maximum density and performance. Here, a simple GaAs laser is coupled with a silicon MOS transistor for on-chip beam control.

number of techniques have been tested in the laboratory to verify the feasibility of this approach. This structure is shown in Figure 14-1.

Associated with the linkage of the various types of technologies will be the slow but steady growth and integration of optoelectronic parts. At first limited to data transmission and fiber optic networks, the potential for optoelectronic integrated circuits is immense. A CPU could handle instructions of different colors simultaneously, and in variable ways, allowing for fast processing, inherent parallelism, and many self-testing capabilities. Memory chips could likewise be able to handle many times the amount of information per cell, while keeping memory cell size at the same sizes as technologically acceptable now.

If this memory is made to be nonvolatile, power and other requirements for the system will be minimized. Ultimately, far down the road, the use of holographic memory is possible; even with today's technology it is possible to construct crude devices. This technology would allow immense data structures to be stored and manipulated. Another application for optoelectronics is in the design of neural networks. The current designs of the electronics networks themselves revolve around varying the amount of resistance that is required to activate the various "synapses." This type of operation can be simulated using varying intensities of light traveling through the synapse.

Another technology that has been developed in the laboratories to produce ultrahigh-speed transistors is called the quantum electron transistor. In this process, only very thin films of material (usually GaAlAs, or associated materials) are deposited onto a substrate. These films are usually only a few nanometers thick, consisting of only a few monolayers of crystal. The devices are so small that, once activated, the electron flow from source to drain is essentially unimpeded. This flow of electrons is also called "ballistic" electron flow, since the electrons are not affected by collisions within the crystal structure. These devices are many times faster than typical integrated circuits, are much smaller, and have far lower power requirements. This means that they are ideally suited for the manufacture of high-density, high-performance semiconductors. However, the quantum electron devices are very hard to manufacture and, to be manufactured profitably, require mass production quality levels beyond those in use now. However, in the next few years, a significant amount of effort will go into the development of this type of device in order to permit its use in standard integrated circuits.

One of the problems that is encountered in the race to shrink integrated circuit devices is the difficulty of carrying current through the very small lines. Clearly, a line with 0.35 μm design rules cannot carry as much current as a line of under 1.0 μm design rules. In fact, as shown by Figure 14-2, since the height as well as the width of the line must be reduced, and since the total current-carrying capacity is a function of the cross-sectional area of the line, the current-carrying capacity of the 0.35 μm line is only about 12% that of the 1.0 μm line. When the total resistance and relative lengths of the lines are calculated, it is seen that for some devices the lines themselves would melt under maximum current loads if made of aluminum. However, the use of metal silicides, titanium, and other metallic substances has given the chip designer much more latitude in line size and current-carrying capacity. Even some of these materials are hard pressed to carry enough current to operate the devices at the minimum feature sizes.

Research has been ongoing in the field of superconductive thin-film research. A number of companies have developed methods for depositing the new high-T_c superconductive films and making them

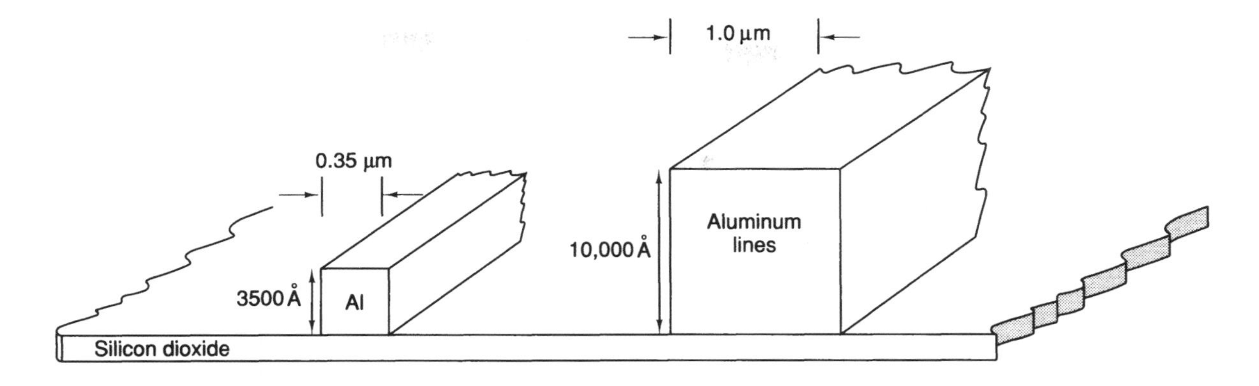

Figure 14-2. Relative Line Sizes As we can see, the reduction in minimum linewidths will result in significant loss of metal volume for each line, thus reducing current-carrying capacity. In this example, maximum current load is reduced by 90%.

operate properly. There are still many problems associated with these devices, such as the ability to produce a uniform crystal (and therefore uniform superconductivity) across the surface of the wafers. Additional problems include the usefulness of a device that must be kept immersed in liquid nitrogen at all times. Also, localized temperature gradients must be considered if the chip itself consumes very much power, as localized regions of the film may have high temperatures and can even cause drastic changes in resistivity. However, considering the extremely high speeds, higher current-carrying capacity, and otherwise lower power consumption of a chip utilizing superconductive materials, research will continue until these issues are resolved. Hopefully, a true room-temperature superconductive film will soon be developed which will permit the operation of these films in a standard integrated circuit.

I have touched on only a few of the more promising technologies that will arise in the 1990s. Without a doubt, many more techniques will be discovered in the next few years. However, the time required between inventing something in a research lab and manufacturing it in a product can be many years, so that the techniques that I have described may be making their impact on the world in the next 5 to 10 years. Developments in the 1990s will probably not reach the common marketplace until early in the next century. Without a doubt, however, the ever-intensifying efforts to develop newer and faster chips will require that these new technologies are studied and exploited in an effective way.

14.3 FUTURE MANUFACTURING PROCESSES

To keep up with the increasing complexity of the semiconductor processes, many new techniques will have to be developed to manufacture wafers. We can see the broad outlines of these trends also. Reduction of the feature sizes of the semiconductor devices requires that all of the various processes receive major upgrades. The precision levels, the absolute process quality levels attainable, the particulate control requirements, and the mechanical handling requirements that will be needed for manufacturing these new technologies will be well beyond those available today. We will discuss some of the key aspects of the next-generation processing techniques in this section. By no means is this a complete list of all possible directions of the wafer fab. However, the majority of the long-term goals should become apparent here.

First, and most obvious, is the issue of lithography. This is by no means trivial, as there are only a few step-and-repeat imaging systems that produce the required lines with any degree of control. In general, visible wavelengths of light cannot be used for submicron lithography, primarily because the wavelength of the light being used is approximately the same as the size of the lines. This leads to severe resolution and focus problems. As a result, the use of deep UV or X-ray wavelengths, or the use of electron beams is typically involved in the development of submicron lithography. There are a number of problems with all of these techniques at this time, that have limited the success of any one system or solution. However, these issues will be resolved over the next few years and the ability to produce lines down to about 0.1 μm will be as commonplace by the end of the century as 1.0 μm lines are now. We can expect to see steppers that can provide better throughput and resolution than was ever thought possible, but it will come at a steep price. The most modern steppers already cost over $1 million. The next generation of these machines could well exceed that price by a significant amount.

After the lines have been developed on the surface of the wafers, they will have to be etched into the appropriate films. Clearly, wet chemical etching has already been ruled out for the critical steps in a wafer fabrication process, and for future technologies may not be allowable under any circumstances. The more advanced plasma etch systems will also provide new problems with radiation and physical damage to the surface of the wafer through the fairly high-energy, reactive collisions. As a result, the etchant plasmas will be generated using remote plasma generators. The afterglow etch process is an example of this remote plasma generator. Another type of reactor is called the reactive ion etcher. All of these various systems are being designed to allow the ions to react with the films in a rapid and controlled fashion without accelerating the ions into the wafer surface and without the damage caused by the radiation developed in the glow discharge (generation) chamber.

The complexity of the etch process is increased by the fact that not only must the films be etched with much greater precision, but some of the films themselves are becoming harder to etch. For example, titanium and certain refractory metal silicides are not as reactive as the more common metals used in semiconductor applications. This can lead to etch selectivity issues. The selectivity of the etch process becomes increasingly more important as linewidths shrink, since very tiny variations in the films can result in significant yield degradation.

Various changes will be seen in the implant process area. There are several areas of development that are currently being explored. Two are in the high-energy and the very low-energy implant regimes. The high-energy systems are often used to implant buried layers of dopants or oxygen, which are then used to provide an insulting layer within the silicon substrate. This structure, called a SIMOX structure, is used to fabricate very high-speed devices, with a minimum of interaction from the substrate, while still maintaining the structural strength of the silicon wafer. The ability to use low-energy implants will permit the very shallow junctions that the devices on top of SIMOX structures use. The low-energy implants also allow greater relative concentrations of the implant species on the surface of the wafers and minimize crystal structure damage.

The diffusion processes will also see some changes, although these processes are more stable, and will not see as much change. The main differences are that the processing temperatures will be reduced as much as possible. Standard thicker oxidation and anneal processes on silicon will continue to remain at high temperatures. However, processes that use GaAs or various other structures will not be able to go much above 400°C, or serious damage will occur within the materials themselves. As a result, oxidation at high pressure (several hundred atmospheres) may become more prominent. Thin oxidation and anneal processes will be carried out on the rapid thermal anneal (or oxidation) systems which permit the surface to be heated to high enough temperatures to obtain high diffusion rates, while keeping the bulk substrate temperature down to reasonable levels. Combining the rapid thermal system with a high-pressure oxidation system minimizes the amount of impact to anything but the top surface of the wafer.

The new designs for integrated circuits lean toward thinner dielectric films. There are designs in the works that use gate oxidation thicknesses of 50 to 90 Å. Some of the films are even being grown in the near-monolayer range. These thickness ranges are very difficult to produce repeatedly with sufficient quality. Even very tiny amounts of chemical or particle contamination can cause complete failure of these films. As a result, special techniques are required for their production.

While the systems in which they are manufactured are similar to the standard diffusion furnace, we will see that more of the vertical diffusion systems will be used, and again we will see a niche for rapid thermal oxidation equipment.

Finally, we will see further changes in chemical vapor deposition processing. The older style, diffusion furnace-type CVD reactor will be replaced, as the control problems and inherent high processing temperatures associated with these systems finally become unbearable. There are a number of contenders for primary type of deposition system. Currently, the plasma-enhanced CVD systems are the processes of choice, but they do have some inherent drawbacks, including difficulties in process control and increased plasma radiation damage. One of the first possibilities is for the use of downstream plasma deposition processing, utilizing a technique similar to that of the downstream etch process. This permits a reduction in total radiation damage.

Another technique that has been developed does not use plasma, but instead uses a technique similar to standard CVD processing. In this system, the process gases are preheated and sprayed onto the heated wafer surface. Uniformity of the system is limited by the ability of the gas control system to provide a uniform gas flow across the wafers. The substrate is usually heated during this type of process, but a case could be made for developing a cross breed between this type of LPCVD reactor and the rapid thermal process so that the film is heated as it is deposited, producing sufficient adhesion and density but not heating the bulk of the silicon substrate.

Another type of process that will gain acceptance through the 1990s is the use of lasers to create films. The ability to use the very high energy available with the laser to dissociate chemical bonds allows the reactions to be very specific and produces molecules in a very high-energy state which allows the film to have some intrinsic planarization benefits. Eximer lasers can be used in a number of different modes, from a parallel-beam setup for wide-area depositions to a normally incident beam setup, where gases stream across the surface of the wafers and the laser writes the pattern directly onto the wafer, completely eliminating the lithography and etching steps. Eximer lasers have also been used in photolithography areas for photoresist exposure, with excellent resolution capabilities, due to the short wavelengths used (around 200 nm) and the high power available. Lasers can also be used in pyrolytic modes in which the laser heats the surface for reaction.

One of the main thrusts for the dielectric CVD area is the push for lower processing temperatures. In most cases, the CVD films are deposited over metal films that have been etched into very fine and sensitive lines on the wafers. High temperatures can cause the lines to distort or crack, or can cause the metals (especially aluminum and its alloys) to form hillocks on the surface of the metal. All of these occurrences are potentially lethal to the chips, but the effects can be reduced through reduced-temperature processing. Thermal LPCVD films have deposition temperatures in the range of 400 to 430°C, while the more common plasma-enhanced processes allow deposition at 350 to 380°C. The more advanced plasma-based systems can obtain deposition temperature at around 300°C, while the laser-enhanced systems have produced films of reasonable quality at 100 to 250°C. The spin-on glass operations develop relatively high processing temperatures with the highest temperatures of around 375 to 400°C obtained during the curing process.

Finally, the CVD films themselves will undergo a significant amount of improvement over the next decade. For instance, both silicon oxide and silicon nitride films are being replaced with silicon oxynitride, a composite material that has the best attributes of both of the other films. Another material that is likely to become prominent in the semiconductor industry is that of diamond or diamondlike carbon films. Diamond films have remarkable properties, and can be used for a number of different tasks. In addition to the obvious extreme hardness and imperviousness to chemical attack, the film normally acts as a very strong dielectric. If doped with the proper chemicals, however, it is possible to force the crystal to become conductive and semiconductive. In theory, a semiconductor chip could be manufactured using only diamondlike materials and various dopants. In reality, at least at this time, the methods for producing diamond films are limited to small substrates with slow deposition rates. This is likely to change soon, as this is a rapidly evolving field.

One more factor that must be kept in mind when trying to discern what future processes will be like is the need for improved metrology. At this time, the ability to measure the very thin film thicknesses and very narrow lines is limited. The process of measuring the narrow lines is usually performed by SEM analysis. While very precise (in that measurements can be made that give repeatable results), the absolute accuracy of these measurements is not known. There is no reproducible way to create and measure standardized linewidths with enough independent verification for absolute certainty of measurement. For example, at the time of this writing, there are no NIST-approved calibration standards for submicron SEM measurements. The creation of this standard will occur within the next five years.

In addition, the techniques for measuring thin films (under 100 Å) are known, but have certain technological limitations (spot size, total amount of light available for measurement, resolution of the detectors, etc.). In addition, there is, once more, no NIST-approved calibration standard under 500 to 1000 Å available at this time. The complexity of producing and storing a sample such as that becomes very critical. We will see significant changes in film thickness measurement devices over the next several years. Finally, the metrology equipment will include more functionality. The film thickness systems will also be able to perform other analysis on the films. For example, the thickness, index of refraction, and doping characteristics of glass films could be determined in one metrology system. The reduced effort that this kind of integration will permit will make the job of producing high-quality wafers that much easier.

Through all of this we can see that there is an increasing need for factory automation to reduce handling, throughput times, and wafer loss through errors. This will be a very complex task requiring all of the equipment vendors to allow themselves to communicate over computer links. The protocols for such an interaction have been defined in the SECS specification (for semiconductor equipment communications standard). Version 1 of this spec lays out how the hardware interface should work. Part 2 of the specification relates to the actual transactions that occur between the processing equipment, the cluster or cell control computer, and the host computer.

Thus, we can see that there are many areas which will see major advances through the next several years. Almost every section in the manufacturing areas will have to be upgraded before 0.35 μm technologies can be produced profitably. These technological improvements will be coordinated with new methods of circuit design, which will lead to ever higher performance chips.

14.4 THE FUTURE OF FAB MANAGEMENT

We can see several manufacturing trends starting to emerge now, largely as a result of the incredible cost of doing business in the semiconductor industry. Some of them are extensions to current trends, although some of the options that have been suggested are more radical departures from current directions. We can see that high-quality manufacturing will be even more important with each succeeding level of technology. This implies that more fab automation will be required. In addition, due to the quite reasonable outcry of the environmentalists and the population as a whole, who do not want the groundwater polluted with HF, heavy metals, or solvents, or the air polluted with phosphine, chlorofluorocarbons, and other deadly fumes, there will be a significant push by the IC manufacturers to clean up and reduce or eliminate contaminants that now escape into the environment.

Clearly, the quality of manufacture will be critical to making sure that a company will remain profitable. The Japanese companies have shown beyond a shadow of a doubt that most users of computer systems want high-quality parts (as anyone who experienced a serious computer system failure can attest, the results are always messy, and often very painful when you discover that data have been lost or were garbled in some way before the system crashed or was backed up). The insistence on high-quality manufacturing will become critical in maintaining profitability as well as market share. The higher quality companies, with their drastically higher yields, will have the luxury to cut prices, placing intense pressure on the companies that cannot produce chips of equivalent quality. This pressure for high quality will result in ever wider use of statistical process control, along with the Just-in-Time and Management-by-Exception methods described in earlier chapters. Only by attaining an appropriate level of inventory and process control can the chip manufacturers expect to maintain profitability.

To monitor these improvements, and to require tighter controls on the process, the statistical quality control rules will start moving toward the establishment of specification limits using $\pm 6\sigma$ statistical rules. This allows each process to vary by as much as $\pm 1.5\sigma$ while keeping a virtually defect-free product. The end effect on the process is to tighten its actual repeatability by about two times. The effect on the product reliability is to reduce the number of failures by some 700 times (with a process of over 1000 separate process steps, a typical number when all of the cleans, the manufacturing processes, inspections, packaging, and so on are counted). By reducing discrepant processes in manufacturing the chips, the yields will become higher, and the profitability of the fab will increase.

Another likely avenue for change in wafer fab management will be the transition from essentially manually or semiautomatically operated equipment to complete fab automation. In these fab areas, the wafers will (ideally) never be handled by humans. This hands-off automation will permit the manufacturer to maintain a complete current database on the product wafers at all points through the manufacturing process. With automated test equipment to provide feedback on the operation of the equipment, the computer will be able to report back to the engineers when a problem has occurred. In fact, with the installation of sufficiently powerful host computers and appropriate expert system software (none of which has yet been developed thoroughly for generalized wafer fab use), the intervention of engineering or maintenance personnel would be required only under the most extreme conditions. The improvement in throughput, quality, and overall factory output that will be seen will more than make up for what promises to be an incredibly expensive proposition.

A recent innovation that has been pioneered by several chip manufacturers is the concept of a cluster tool. This is a module of a process, for instance, an etch module, in which all the components of that particular step are integrated through robots. This allows a lot of wafers to come into an operation, get logged in automatically, cleaned, spun with resist if required, then etched. Measurements are taken on the wafers immediately after the process, and the wafers receive any postprocessing, such as resist removal. The wafers then leave the cluster tool and proceed to the next for the next process. Each cluster tool will consist of components from a variety of manufacturers. Each component must conform to a common set of specifications to allow the automation systems to work properly. The use of cluster tools is expected to reduce the costs and labor requirements for many operations.

One method of reducing the huge start-up costs of these large automated fabs is already being practiced by a number of small foundry fabs in the Far East. These factories typically produce only limited runs of specific types of chips. In these "mini-fabs" or "foundries" there is a minimum amount of equipment, with heavy emphasis on the use of cluster tools, as described above. One of the goals of this type of operation is to achieve nearly full automation. Cleanroom space is kept to a bare minimum, and the wafers are not removed from their protective cases when people are around. By reducing the capital investment required, and then selling fixed lot sizes of chips under contract to key customers, the fab's output and profitability can be much more closely maintained than in the typical large-volume fab. Usually, these types of foundries do not provide much in the way of design services, but may allow cell-based architectures, or other types of ASIC circuits to be designed in-house. At other times, the foundries will merely contract the design of the chip to a third party or will require the customer to supply the designs or even the reticles.

Finally, a major drive over the next 5 to 10 years will be to reduce the effect that the industry has on the environment. While the amounts of waste produced in the semiconductor industry are small compared to many other industries, the effects of the toxins released tend to be very severe. For instance, hydrofluoric acid will remain in the soil and will enter into the drinking water supplies and other parts of the life cycle. Also, the semiconductor (and computer industry, in general) industry is among the heaviest users of CFCs, which have been shown to be extremely damaging to the ozone layer. Over the next several years, a number of processes will have to be modified so that they can operate with other types of chemicals. For example, perflourocarbons can replace CFCs for many of the cleaning processes. It will not be as easy to replace the CFCs in the various plasma-etch processes that use them. Biodegradable chemicals, such as choline, will also replace many of the more toxic and hazardous chemicals such as HCl and sulfuric acid, and will be used as photoresist developer solutions.

Thus, we can see that the ultimate goal of the semiconductor industry, as far as environmental hazards are concerned, should be the policy of "Zero Toxic Effluent." This goal may never be precisely possible, but it is possible, even with today's technology, to remove nearly all of the toxins from the exhausts of a fab. The real question is one of cost. I have little doubt that the semiconductor companies that are being held responsible for digging up old industrial sites and cleaning every ounce of dirt will believe that the goal of Zero Toxic Effluent is a very cost-effective solution to a serious problem. It will be much simpler for the semiconductor industry to move toward that goal, than to wait until the federal or state governments impose these standards. Many cities have enacted very strict effluent regulations to protect themselves from the possibility of contamination.

We can see that most of the fab management issues will be continuations of the concerns of today. They will include heavy foreign competition, high capital investment costs, timid investors, high labor costs, and technological and environmental risks. Success will come only through the imposition of carefully chosen sets of checks and balances, and the use of the most advanced techniques for controlling and predicting the process.

14.5 FINAL COMMENTS AND WRAP-UP

We have covered a large amount of territory in this book. As we have seen, the wafer manufacturing process is complex, has many steps, many potential areas for defects to creep in, and almost unbelievably tight tolerances. Most of the techniques in use today have been developed within the last 10 years or so, so using that as a model, we can rest assured that the next 10 years will result in another massive turnover in processing techniques. As a result, the successful process engineer or wafer fab employee will not be the one who can most quickly learn or follow new specifications and requirements, but instead will be the one who studies and understands the nuances of the equipment technology and the chip manufacturing technology as a whole. This ability will permit the fab employee or engineer to have sufficient flexibility to deal with the rapidly changing situation in the fab. This book has been an attempt to permit these types of individuals a glimpse at each of the various operations in the fab.

We have seen that there are certain key concepts that are stressed throughout the wafer manufacturing operation, regardless of the type of process. These concepts include maximized die yields, maximized line yields, and the highest chip reliability possible. These general goals must be translated into actions, however, and we have discussed what some of these actions are. In general, however, they all boil down to one basic concept: the construction of these items using *quality craftsmanship*. The installation of process control charts in a fab is simply a manifestation of this goal from a corporate level. Each and every individual who handles the wafers or who handles equipment that handles the wafers must feel that same commitment if the factory is to be as successful as possible. The integration of this high-quality, positive attitude with the easy acceptance of new techniques (which comes with the confidence that is gained when one has understanding of the process as a whole) will give a company an almost unbeatable edge in almost every marketplace, regardless of the chip designs produced.

Management techniques such as Just-in-Time and management-by-exception using statistical process control methods will allow the management team to have clear visibility into fab operations. This permits quick action on any problems that may arise. The reaction to and solution of the problems that arise will be greatly affected by the ability of all of the fab personnel to work together as a team. Once again, the most successful companies will be the ones who can effectively resolve the inevitable problems that will occur in the fabs, while bringing on new products at a rapid pace.

Ultimately, then, in 10 years, just as now, and even when the equipment is totally automated and computers provide more control over the process than is available today, the ultimate responsibility for the success or failure of a wafer manufacturing operation will largely hinge on the people who make up the organization. Therefore, all of the individuals in the wafer fab must learn as much as possible about the processes they run and the products they produce.

Remember, the quality of the integrated circuits that are manufactured will reflect the quality of the work that was performed on them, and the quality of the chips will be reflected in the reputation of the company, and therefore on all of the individuals who are making the chips. In the future, the companies with the reputation of making excellent chips will prosper, draw in the best talent, and then grow and continue to prosper. May you all strive to make those companies yours.

APPENDIX A

HAZARDOUS CHEMICAL TOXICITY TABLE

Chemical[a]	Phase[b]	Type	Use	Formula	TLV[c] (ppm)	Attacks	Reaction	Flammable[d]	UEL[e]	LEL[f]
Acetic acid	G/L	Corrosive	Etch buffer	CH_3COOH	10	Eyes, sinus, skin	Inflammation, nausea	Y	16.0%	5.4%
Acetone	G/L	Flammable	Solvent	CH_3COCH_3	750	Eyes, nose, throat	Nausea	Y	13.3%	2.0%
Ammonia	G/L	Nonflammable	Cleaning	H_3/NH_4OH	25	Eyes, skin, respiratory	Severe irritant	Y	28.0%	15.0%
Ammonium fluoride	L	Acid	SiO_2 etch	NH_4F	2.5	Eyes, skin, internal respiratory	Strong irritant	N		
Aqua regia	L	Oxidizer/ corrosive	Gold etch	$HCl:HNO_3$	2	Eyes, skin, respiratory	Severe irritant	N	Forms H_2 gas with metal	
Argon	G	Nonflammable	Purge gas	Ar	—	Displaces O_2	O_2 deprivation	N		
Arsenic	S	Poison	Dopant	As	10	Cancer, internals, skin	Extreme poison	N		
Arsenic (spin-on)	L	Flammable	Dopant	As with polymers, etc.	10	Cancer, internals, skin	Extreme poison	Y	19.0%	3.3%
Arsine	G	Poison/ flammable	Dopant	AsH_3	0.05	Red blood cells	Severe poison	Y		
Boron tribromide	L	Corrosive	Dopant	BBr_3	1	Skin, eyes, mucus	Nausea, shock, coma	Explosive if heated		
Boron trichloride	L	Corrosive	Dopant	BCl_3	1	Skin, eyes, mucus	Severe poison	N		
Buffered oxide etch	L	Corrosive	SiO_2 etch	$HF:NH_4F$	3	Eyes, skin, internals	Severe toxicity	N		
Carbon dioxide	G	Nonflammable	Plasma etch	CO_2	5000	Displaces O_2	O_2 deprivation	N		
Chromic acid	L	Corrosive	Silicon etch	$CrO_3 + H_2SO_4$	0.05	Respiratory, teeth, eyes	Heavy-metal poisoning	N		
Chromium trioxide	S	Oxidizer	Silicon etch	CrO_3	0.05	Eyes, skin	Heavy-metal poisoning	N		
Dichlorosilane	G	Pyrophoric	Si_3N_4 deposition	H_2SiCl_2	5	Respiratory	High toxicity	Y	98.8%	4.1%

(continues)

Chemical[a]	Phase[b]	Type	Use	Formula	TLV[c] (ppm)	Attacks	Reaction	Flammable[d]	UEL[e]	LEL[f]
Ethylene glycol	L	Nonflammable	Coolant	CH_2OHCH_2OH	50	Eyes, respiratory, digestive	Irritant	Y	15.3%	3.2%
Freon 14	G	Nonflammable	Plasma process	Tetra-fluoromethane	1000	Eyes, mucus, displaces air	Irritant, replaces air	N		
Freon 113	L		Refrigerant	CCl_2FCClF_2	1000	Eyes, mucus, displaces air	Irritant, replaces air	N		
Helium	G	Nonflammable	Purge, buffer	He	—	Displaces O_2		N		
HMDS	L	Flammable	Surface primer	Complex	1	Skin, eyes	Irritant, burns	Y	16.3%	0.8%
Hydrochloric acid	L/G	Corrosive	Cleaning	HCl	5	Respiratory, teeth	Severe burns	N	Forms H_2 with metals	
Hydrofluoric acid	L	Corrosive	SiO_2 etch	HF	3	Eyes, skin, internals, respiratory	Severe burns—may not be painful immediately	N		
Hydrogen	G	Flammable	SiO_2 growth	H_2	—	Replaces O_2		Y	75.0%	4.0%
Hydrogen peroxide	L	Oxidizer	Wafer cleans	H_2O_2	1	Skin, eyes, nose, respiratory	Serious irritant	N		
Isopropyl alcohol	L	Flammable	Cleaning	$CH_3CHOHCH_3$	400	Eyes, skin, upper respiratory	Drying, burns eyes	Y	12.0%	2.0%
Methyl alcohol	L	Flammable	Cleaning	CH_3OH	200	Skins, eyes, organs	Poisonous	Y	36.5%	6.0%
Nitric acid	L	Oxidant/ corrosive	Metal etch	HNO_3	2	Eyes, skin, respiratory	Severe burns	N		
Nitrogen	L/G	Nonflammable	Purge gas	N_2	—	Displaces O_2		N		
Oxygen	G	Oxidizer	Many uses	O_2	—			N		
Phenol	L	Poison	Resist process	C_6H_5OH	5	Respiratory, internal	Muscle weakness, collapse	N		
Phosphine	L	Poison	Dopant	PH_3	0.3	Lungs, kidneys, internals	Convulsions, nausea	Y	—	—
Phosphoric acid	L	Corrosive	Nitride etch	H_3PO_4	1	Skin, eyes, respiratory	Severe burns, bronchitis	N		
Phosphorus oxychloride	L	Corrosive/ flammable	Dopant	$POCl_3$	0.1	Eyes, skin, internal	Burns, bronchitis	N	Burns in water	
Photoresist	L	Flammable	Masking	Many	5–100	Eyes, nose, throat	Irritant, long-term hazards	Y	70.0%	1.0%
Potassium hydroxide	L	Corrosive	Developer, cleaning	KOH	2	All body tissues	Serious irritant	N		
Silane	G	Flammable pyrophoric	Silicon source	SiH_4	5	Skin, tissue, internal	Severe fires, silicosis	Y	Any, when confined	
Silicon tetrachloride	L	Corrosive	Etch, cleaning	$SiCl_4$	—	Eyes, respiratory	Serious irritant	N		
Sodium hydroxide	L	Corrosive	Developer cleaning	NaOH	2	Skin, eyes, respiratory	Severe burns	N		
Sulfur hexafluoride	G	Nonflammable	Plasma etch	SF_6	—	Displaces O_2	—	N		
Sulfuric acid	L	Corrosive	Wafer cleans	H_2SO_4	1	All tissue	Severe burns	N		
Trichloroethane (TCA)	L	Possible carcinogen	Chlorine source	CH_3CCl_3	350	Internal	Dizziness, fainting	Y	16.0%	7.0%
Trichloroethylene (TCE)	L	Carcinogen	Cleaning	Cl_2C_2HCl	100	Nervous system, internal, respiratory	Dizziness, nausea	N		
Xylene	L	Flammable	Solvent, resists	$C_6H_4(Ch_3)_2$	100	Skin, eyes, internal, respiratory	Irritant, drowsiness	Y	7.0%	1.0%

[a] There are a number of compounds which have been introduced into the workplace in the last several years. The exact nature of these compounds and their effects on humans are not well understood. Until they are characterized, it would be prudent to treat these chemicals with the proper respect required for toxic chemicals. The chemicals to be aware of include perflourocarbon compounds (replaces CFCs), trimethyl boride, trimethyl arsenide, tetraethylorthosilicate (TEOS), TMCTS, choline, tertiary butyl arsine (TPA), and tertiary butyl phosphine. We have not discussed the environmental effects of these materials, which in many cases can be severe. Many of the toxic compounds, especially the heavy metals and the flourine compounds, cannot be easily removed from the environment once introduced. Concentrations of the compounds will be found in environmental niches, such as the wildlife population. Local laws concerning the handling of these chemicals vary greatly, although federal laws do place many overall restrictions on toxic emissions.
[b] S = solid, L = liquid, G = gas.
[c] TLV = Threshold limit value.
[d] Y = yes, N = no.
[e] UEL = Upper explosive limit.
[f] LEL = Lower explosive limit.

EXPERIMENT DESIGN FOR PROCESS AND PRODUCT DEVELOPMENT

Wafer fabrication is an evolving art, with many processes in a nearly constant state of flux. As a result, a great deal of experimentation must be performed to maintain control of the various pieces of equipment, improve device yield and quality, and enhance the performance characteristics of the devices. This experimentation must be performed in a way that is both statistically valid and economically reasonable. This appendix will go through basic experimental design strategy for developing practical tests.

The costs of experiments in the wafer fab area are considerable when you count in the price of the raw wafers, materials used for processing, lost time and capacity, and possible impact on the time to market for a particular product. In addition, many changes in processing conditions and/or specifications will require customer requalification, especially if the part is of premium quality. This can result in lengthy delays and other associated problems. The best way to reduce these problems is to clearly document the test procedures so that they can be produced for validation of results during customer requalification. With any device, there is a significant risk that changes in one parameter will cause subtle changes in other parameters. This is especially true in the typical multiprocess sequences used in modern IC manufacture. Finally, the standard deviation of the typical process may be so wide that small changes will be hard to discern without carefully controlled experimentation.

Adequate experimental analysis is a requirement for effective decision making during wafer fabrication. Making decisions based on insufficient or improperly presented data can create costly delays when steps have to be retraced and experiments have to be performed again. Therefore, product and process development experiments must be designed in the most practical way. While colleges are excellent sources for theoretical information on statistics, there is seldom enough emphasis placed on the day-to-day details that are important in a manufacturing environment. Therefore, this appendix will not go into great detail about the mathematics of experimental design, presuming that these discussions are covered better elsewhere, and that the reader has some basic knowledge of statistics.

There is no single experimental strategy that will work in all cases for all types of experiments. It is possible, however, to define broad guidelines for how to approach the problem, determine relevant sample sizes, and decide on the best way to analyze the final data. There are many experiment types that have been devised over the years to solve various classes of problems. While any of them can be used, many are very complex and difficult to perform in the wafer production environment. The technique outlined here uses controlled experiment matrices, along with carefully chosen real-time decision-making points, to yield a maximum amount of information from a minimal amount of material. While we will discuss how to obtain high confidence levels for experiments, it is rare that enough material will be freed to permit them. What is important in most cases is verification that an expected result will take place when a certain variable is changed, and that no unwanted interactions occur. This is fundamentally different in both overall direction and scope than theoretical scientific experimentation.

The types of experiments that we will be discussing can be broken down into two categories, "process development" and "product development." In process development experimentation, the goals are to demonstrate improvements in process stability and uniformity, reductions in defects, and other related parameters. Line yield and die yield improvements follow as by-products of improved process control. In product development experimentation, the goals are to improve device performance or to directly impact the yields of the wafers. This type of experimentation is usually accomplished by adjusting parameters, such as changing the implant dose or the gate oxide thickness. The actual tests for these two cases are carried out in similar fashion, and the development efforts are often linked. It is important to design experiments that can discern the differences in yield improvements that are due to process improvements and those that are due to device improvements.

Basic experimental design involves a number of key steps. These are outlined as follows:

1. State the goal of the experiment. If the wrong question is asked, the correct answer cannot be obtained.
2. Define the variables that are to be adjusted, and which parameters can be affected.
3. The general course should be laid out and matrices designed.
4. Define an acceptable goal for the experiment: When is the experiment done?
5. Define what uncertainty is acceptable and calculate required sample sizes and selection criteria.
6. Execute the tests.
7. Analyze the results of the experiments.
8. Make a decision on the results.

Stating the goal of the experiment often sounds very simple, but can lead to some interesting issues. It is at this stage that a significant number of blind alleys can be eliminated. For instance, if a CVD furnace has a particle problem, the real question is not "What causes the particles?" (too broad), or "How do the particles get into the tube?" (probably not relevant), but is something on the order of "What is the source of the particles, and how do we correct that?." For example, many long hours and days have been wasted while CVD engineers the world over have pulled quartzware or otherwise attempted to fix a symptom of a problem instead of the source. This, of course, is a temporary fix that is likely to become less effective through time, as the source of the problem continues to deteriorate. It is also important at this point to identify the scope of the tests. Are these tests being used to further basic research, to solve a particular problem, or to achieve an incremental improvement in device performance or yield?

Next, the variables must be identified. This will reduce the number of potential experiments that could be run. It will also help prevent unwanted side effects from occurring, which is especially important in the event that multistep processes are being investigated. Care must be taken to prevent the confusion of results that can occur in these cases. This is helped by the identification and reduction of all variables to just the pertinent ones. Next, any variables that are thought to be involved, and could interact in some way with the changes proposed, must be identified and measurement procedures defined in order to reduce the possibility of unexpected deleterious effects.

At this point, a general course should be laid out, so that the impact of the experiments proposed can be made clear: How many tests will be needed, how many hours of time and other resources, and so on. At this point, other aspects will often emerge, requiring further detailing of the experiment. It is critical that this "homework" be done prior to running the experiment, even if does not always seem productive. Often, the course of the experiment will change during the test sequence. This is not a particular problem, however, as long as the original experiment design builds this in, and has points defined where decisions can be made for further progress. Not building these decision points into the structure of the tests is a near guarantee that the prescribed schedules will not be met, simply because reality seldom conforms exactly to expectation and adjustments must be made. Time for this should be allotted at the start of the project.

It should be agreed as to what constitutes a completed experiment. Since we have defined what the various components of the experiment will be, we need to make sure that the project does not evolve into an on-going development effort without achieving the purpose originally defined. (It is surprising how often projects go astray in mid-course because conditions were not properly defined at the beginning of the experiment.) In some cases, basic research projects may have open-ended goals, so that a completed project may be defined as a published paper on some aspect of reality. A completed project in product development may be something like a 5% yield improvement due to a change in the field implant.

The next step is to calculate the sample sizes required to obtain reasonable results. In most cases, the sample sizes can be calculated fairly easily by knowing what uncertainty in results is acceptable. In many cases, intuition must be used to tell you what the probabilities of success should be, which will have an effect on sample size. This is because the high cost of experimentation seldom permits the use of enough wafers to obtain good, statistically sound data.

To determine sample size, acceptable risks must be defined. There are two types of risks that are usually described. These are called α and β errors, respectively. For our purposes, they can be defined as follows:

α error—change is expected to have result, but none seen.
β error—change is not expected to have effect, but effect is seen.

Selection of these values is quite subjective, but some criteria can be set by thinking about the effects of the types of error. For example, an α error may not be particularly harmful to the device; instead, the new process will simply not produce the desired results. As a result, there is less risk there and a higher value for the error margin can be accepted. On the other hand, a β error represents an unexpected active change, which could be very hazardous to device yield or performance. As a result, only much smaller β errors are accepted. While little help can be provided as to what are appropriate values for α and β errors, typical values might be 5% (0.05) and 1% (0.01), respectively. This means that there is a 95% chance that the process change will produce the expected results, and a 99% chance that no unexpected results will appear.

Once the acceptable error is determined, the sample size can be calculated. This requires two other pieces of data from the existing wafer manufacturing process. They are the maximum variance allowed (which is the same as the standard deviation of the optimum yield distribution squared) and the increment of change required. For example, the change of a gate oxide of 10 Å may make a significant change in the yield of a semiconductor device.

In calculating of the sample size, the α and β errors are used to calculate probability points for the typical normal distribution. Values for these constants (U_α and U_β) are given in tables, and are available in many of the textbooks on the subject, including the books listed in the bibliography. For $\alpha = 0.05$, $U_\alpha = 1.645$ for a single-variable experiment and 2.326 for a two-variable experiment. $U_\beta = 2.326$ for both cases.

The sample size is calculated using the equation:

$$N = (U_\alpha + U_\beta)^2(\sigma^2/\delta^2)$$

For example, if an oxidation process is critical to device performance and an experiment needs to be run (a split lot between two oxide thicknesses), the sample size must be calculated to give an adequate understanding of the change. If the variance ($= \sigma^2$) of the process is 5 Å, and significant increments of 2 Å ($= \delta^2$) are acceptable, the total number of wafers that must be run with the new process is 40:

$$N = (1.645 + 2.326)^2 \cdot (5/2) = 39.4 \text{ or rounding: } 40$$

Since the experiment must undergo a control, the test wafer must be matched with control wafers that run through the normal process. To achieve the same confidence, the same number of wafers should be processed. However, since this is a working assembly line, it can be assumed that there is some control inherent in the process. As a result, it may be possible to use a much smaller sample (perhaps 10 wafers) as a control, as long as those control wafers all fall within the normal, specified ranges. Therefore, an experiment may be able to be run with two lots of 25 wafers, as long as there is no line yield loss. Assuming that there will be some loss, a full sample size would probably consist of about 75 wafers. Smaller sample sizes will result in reduced certainty of results.

Further preparation of the experiment can include calculation of the selection criteria. This value can be used to help define when a change has made a significant difference or has achieved the desired results. These calculations can reduce the chance of making an incorrect decision. There are a variety of formulas used for this calculation, a typical one being:

$$S = \mu_0 + \sigma U_\alpha/\sqrt{N}$$

where S = selection criterion

μ_0 = process mean

σ = standard deviation of process

U_α = constant described in the α-error discussion above

N = number of samples

In this equation, the solution S describes what the expected process mean should be if the experiment is to be declared a success. If the actual value is less than S, the results have not made a statistical difference. If the actual value is greater than S, the results can be expected to be correct within the α error limits described above (95% probability of success). Many variations on this theme exist, depending on the type of experiment and the number of variables in use.

All of this said, selection criteria are not always formally calculated. Usually, specific goals are targeted; in other words, a certain parametric change in the fab will be expected to produce a certain effect. When this occurs and is verified, the procedure is complete. This, again, is a difference from a purist's view of experiment design, and again is due to the goal-directed nature of the development effort.

Now that the preparatory work has been completed, the next step is to actually start the execution of the test matrices. First, the wafers must be selected. In some cases, wafers are dedicated to the test from the start of the process, while in other cases, the wafers may be chosen from those available in inventory near the step where the test will occur. In most cases, this second method is used only when the problems are serious and the tests must be executed immediately. Otherwise, standard production material would soon be polluted with a variety of wafers that had received nonstandard processing, and would therefore be suspect. If it is a requirement to run a test using material that is in current inventory, it is important to select lots that have not been misprocessed or have had other tests performed on them or had other problems associated with them.

In all cases, the lots should be processed in standard conditions up to the point of the tests, and should then be split in a logical manner. Since tests are run sequentially, or in separate pieces of equipment, wafers from each lot are placed in each of the runs. The wafers are removed one at a time from the original lot and placed into a mixed lot. Each wafer is placed into the next lot, so that two wafers next to each other in the original cassette will end up in separate portions of the experiment. An example of this procedure is shown in Figure B-1. In this example, three lots of 25 wafers are to be run through an experiment to qualify two new processing systems. The lots must be separated into three groups, one for each of the new systems, and one to be processed through a standard production system. A wafer is taken out of a lot, identified, and placed into one of the sample groups. The second wafer is taken, the wafer ID recorded, and it is placed in the next sample group. The third wafer is removed from the lot, identified, and placed in the control group. This procedure is repeated for all three batches of wafers until three new lots have been created, each one of which contains one-third of the wafers from the original lots. The location of each wafer should also be known at this point.

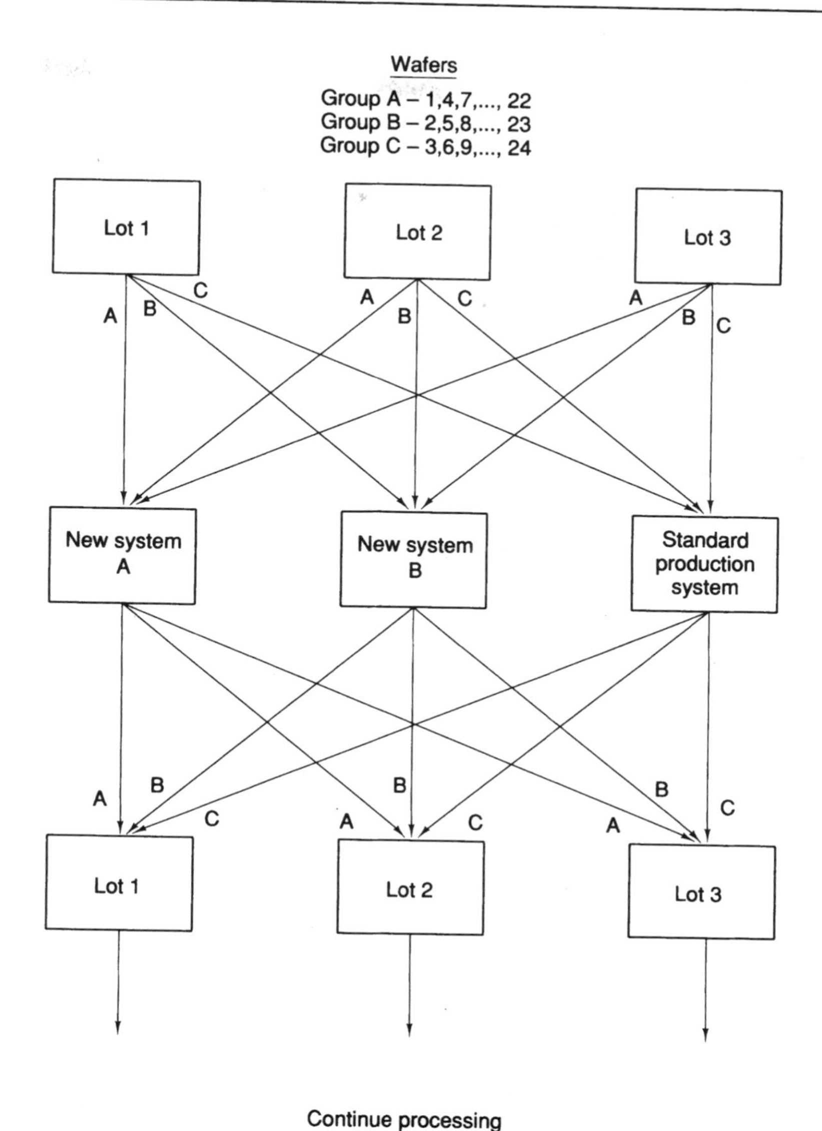

Figure B-1. Separation of Lots for Production Equipment Evaluation Experiments should be carefully designed to maximize the information collected and to maintain control over the results. The wafers are cross-mixed to obtain the most random results, in case other processes have imparted subtle trends in the parameters of the wafers.

Similar strategies should be followed for other experiments. While time-consuming, this randomization step is the only way to ensure that systematic problems that could exist at other process steps are averaged out during the experiment, increasing the chance of a successful test. If the experiment must spread out over numerous days or must be executed in the middle of a complex process sequence, additional mixing may be required to guarantee statistical validity.

Each test run is then processed with the experimental condition changed, and all other parameters kept constant. The actual sequence of events for each process should have been defined in the preliminary work and should be executed in the sequence planned. The actual matrices are often quite easy to define, and depend largely on the subject to be studied or aspect modified.

Typically, one or two variables may be adjusted and any number of parameters evaluated. For instance, the gate oxidation temperature and gas ratios may be varied in order to investigate the impact on film thickness, pinhole density, and substrate crystal damage. Single-variable experiments are the easiest to run, but can result in secondary experiment requirements, as side effects of the testing are discovered. This can add significantly to the length of a test sequence. The typical type of test that is performed here are process characterizations for specific parameters; for example, CVD deposition pressure versus deposition rate.

More often two variables are compared against one another in a matrix, as shown in Figure B-2. In this case, the standard process is represented by the matrix cell in the middle of the matrix. For both of the variables, a point both significantly above and below the standard value are chosen, and tests are run at these points. For example, the pressure specification may be 200 mTorr for a process. The test would then define pressure variations of 180, 200, and 220 mTorr. For two variables, then, the minimum number of runs to complete the test properly is nine. This type of layout allows interactions to be observed. For instance, total photoresist exposure is affected by interactions between light intensity and exposure time. The results of this test matrix can then be used as a basis for further experimental work.

In more aggressive approaches, which are often used in fabs due to the increasing costs of testing and to time constraints, decisions are periodically made as to which particular set of experiments to run, and which will be of particular interest. While reducing the total number of experiments, this method will increase the uncertainty of the results. However, since the development experiment is trying to achieve specific effects, not attempting to make new discoveries, this approach can be very fruitful. It should be noted here that a solution that is found in this manner may not be the most effective method, since some avenues remain unexplored, some of which could be very interesting and fruitful.

To illustrate this procedure, if we look at Figure B-2, we might know that an increase in deposition rate will give better yield results. As a result, since we would expect that increasing pressure and temperature will increase deposition rate, we would run tests from matrix cells 5 (the middle cell), 6, 8, and 9, or the lower right corner of the matrix. If it is then noted that yields stay the same or are reduced as pressure is increased, but improve if the temperature is increased, it may be supposed that the temperature increase is the proper route,

Temperature (°C) \ Pressure (mTorr)	Standard pressure -20 mTorr	Standard pressure	Standard pressure +20 mTorr
Standard temperature -10°C	Cell 1 Temperature, pressure interaction (-) (-) T P	Cell 2 Tests temperature response (-)	Cell 3 Temperature, pressure interaction (-) (+) T P
Standard temperature	Cell 4 Tests pressure response (-)	Cell 5 Standard process; standard temperature and standard pressure	Cell 6 Tests pressure response (+)
Standard temperature +10°C	Cell 7 Temperature, pressure interaction (-) (+) P T	Cell 8 Tests temperature response (+)	Cell 9 Temperature, pressure interaction (+) (+) T P

Figure B-2. A Two-Variable Process Experiment Experiments in two variables should be designed to test both individual and interactive responses. This is done by producing experiment matrices, such as the one shown above (an LPCVD experiment in pressure and temperature). This kind of matrix can be extended into three dimensions, with up to three variables or more, with careful design.

and further experimentation can be planned. Usually, this type of aggressive test strategy is less useful on tests that have many unknowns or take a very long time to complete, because of the possibility of a wrong turn and the (time) consequences of an incorrect decision. In these cases, the more formalized methods of experimentation are more effective.

Once each matrix is completed, the wafers must be remixed with their original lots. It is important while doing this to verify that all wafers are identified to prevent any cross-mixing of lots. Not only does this bring into question the validity of the results obtained on the wafers, but also can reduce the overall quality of the wafers. Lot integrity has a surprisingly strong effect on die yields. Typically, more-stringent metrology requirements are placed on the wafers at this point, for example, requiring inspections and parametric measurements of every wafer. All of the data should be recorded carefully, along with all processing conditions, making special note of all variances in procedure, no matter how small. The wafers should then go through the rest of the process using standard procedures.

To verify that the test will have the required validity, it is important that proper controls be in place at all times. Loss of test validity will result in a significant loss of time, effort, money, and resources when tests must be repeated or, worse yet, yield is lost as a result of a poor decision. The logs of all tests should be carefully and thoroughly analyzed for discrepancies, especially in the event of an unexpected result. To guarantee validity of the test results, the following items must be observed:

1. Before actually executing the tests, verify that they will actually produce the desired results. While this sounds very obvious, it is amazing how many tests are performed that result in little or no information gain.
2. The results should have a good degree of accuracy and precision, so that they can be reproduced independently within a reasonable degree of uncertainty.
3. When designing the test, all procedures should be defined in as much detail as possible to prevent confusion of these details in the future. If possible, the reasons why a particular course is being taken should also be recorded. When the tests are underway, all activities should be recorded, and the procedures followed as closely as possibly. When the inevitable unexpected occurrences come along, the decisions that are made should be recorded for future reference.
4. Extraneous variables must be controlled as much as possible, which can be very difficult if one is trying to run the tests within a manufacturing environment. These types of controls include verification and calibration of all metrology systems involved in the tests, attempting to run all of the tests through a single piece of equipment, timing maintenance to prevent equipment failures as much as possible, and so on.

Now that the tests have been completed and validated, the results must be sorted, and the mean and standard deviation of each group calculated. At this point, the results are compared to the selection criteria defined at the start of the test. If the actual results exceed the target values, the experiment is deemed a success and the process change procedure is initiated. If the results do not exceed the target values, but give an indication of a way to achieve the goal, that new method will be tried. Otherwise, it's back to the drawing board. As always, significant process changes should be verified with yield and reliability analysis in sufficient quantity before committing the fab to the change.

Finally, I cannot emphasize enough the importance of documentation. Write up the experiment, even if only in a cursory form for internal use only. Store a copy or two somewhere. This way, the results will exist for future reference when similar problems come up. A particularly good idea, and increasingly available with the advent of computer LANs, is the documentation library, where documentation and studies on various aspects of the technology are catalogued and stored. After the time, expense, and effort involved in a wafer fab experiment, loss of the valuable information gained is distressing and costly.

Thus, we can see that wafer fab experimentation can be accomplished through fairly straightforward means, as long as the details of the experiments are kept in mind from the very start. The large volume and large number of chips available make the experiments inherently probabilistic, thus permitting the natural use of statistical experiment design. Careful study of a problem followed with the careful implementation of a clear plan of action will usually result in the performance improvements desired. Even though the costs of these experiments can be fairly high, the cost of stagnation from the lack of experimentation is much higher, and easily justifies most well-thought-out experiments.

BIBLIOGRAPHY

As with most works of this type, no progress could be made without the invaluable reference material available. While I could not identify every instance of the use or integration of a report's data (in order to make the text readable), it does not mean that I am not deeply indebted to it all.

It is suggested that anyone looking for more depth in a subject area examine these texts and articles, and read their bibliographies. All of the topics covered in the book are covered in a wealth of detail in the published literature, with much more substance than was possible in this context. This is especially true of the more technical aspects of the IC processing science. By no means should it be assumed that these are the only sources for this information. However, there are many more titles than I can reasonably list here.

The following represents the material that I have in some way touched upon, or discussed in the book, or feel is sufficiently closely related that it warrants further study.

TEXTBOOKS

"An Introduction to Engineering and Engineering Design;" Krick; John Wiley & Sons; 1969

"Chemical Hazards of the Workplace;" Procter and Hughes; J. B. Lippincott Company; 1978

"Data Reduction and Error Analysis for the Physical Sciences;" Bevington; McGraw-Hill, Inc.; 1969

"Deposition Technologies for Films and Coatings;" Bunshah *et al.*; Noyes Publications; 1982

"Device Electronics for Integrated Circuits;" Muller and Kamins; John Wiley & Sons; 1977
"Handbook of Thin Film Technology;" Maissal and Glang; McGraw-Hill, Inc.; 1970
"Microchip Fabrication;" Van Zant; Semiconductor Services; 1986
"Physics of Semiconductor Devices;" Sze; John Wiley & Sons; 1981
"Practical Experiment Designs for Engineers and Scientists;" Diamond; Lifetime Learning Publications; 1981
"Semiconductor and Integrated Circuit Fabrication Techniques;" Gise and Blanchard; Reston Publishing Company, Inc.; 1979
"Silicon Processing for the VLSI Era", Vols. 1 and 2; Wolf and Tauber; Lattice Press; 1986, 1988
"Statistical Quality Control;" Grant and Leavenworth; McGraw-Hill, Inc.; 1980
"Tungsten and Other Refractory Metals for VLSI Applications II;" Broadbent; MRS Proceedings 1986; 1987
"Tungsten and Other Refractory Metals for VLSI Applications;" Blewer; MRS Proceedings 1984–1985; 1986
"VLSI Technology;" Sze; McGraw-Hill, Inc.; 1983

PERIODICALS

There are a number of periodicals written specifically for the semiconductor industry or that include solid state devices as major components of the periodical. They range from the highly technical to the more business-oriented. Some of the more popular are listed here.

Semiconductor International; Cahners Publishing Company. This magazine is the most widely distributed semiconductor publication. It covers the semiconductor business world in detail and also includes selected technical issues relating to wafer or equipment manufacturing.

Solid State Technology; PennWell Publishing Company. This is a more technical publication, often covering technical conferences and other technological updates, as well as general news, in detail.

Microcontamination; Canon Communications, Inc.. This publication covers cleanroom and contamination source issues in depth. It typically contains articles on new systems or techniques for controlling and reducing contamination.

Microelectronic Manufacturing and Testing; Lake Publishing Company. This publication is designed to cover the manufacturing and test equipment marketplace.

Cleanrooms; Witter Publishing Company. This, as the name implies, covers issues relating to cleanroom control, although it is primarily an advertising medium for new cleanroom products.

Journal of the Electrochemical Society; Electrochemical Society. This is a very technical publication covering a wide variety of topics and containing several dozen articles per issue. The Solid State Electronics section covers many interesting articles of interest to the process engineer.

Journal of Applied Physics; Japanese Journal of Applied Physics; IEEE Transactions on Electron Devices. These are also highly technical publications containing articles of interest in a wide variety of fields, including solid state electronics.

ARTICLES

It is impossible to list every paper that has been written on the subject of semiconductor manufacturing. In the listing that follows, I have included a variety of papers that were researched in the preparation of this text as well as a wide variety of others that may be of interest. They can be used as a starting point for research in any of the specific fields desired. Abbreviations used for some journal titles are as follows

APL	*Applied Physics Letters*
CR	*Cleanrooms*
IEEE	*IEEE Transactions on Electron Devices*
JAP	*Journal of Applied Physics*
JECS	*Journal of the Electrochemical Society*
JJAP	*Japanese Journal of Applied Physics*
JVST	*Journal of Vacuum Science Technology*
MC	*Microcontamination*
MMT	*Microelectronic Manufacturing and Technology*
SI	*Semiconductor International*
SSE	*Solid State Electronics*
SST	*Solid State Technology*

A Close Look at Hydrogen Peroxide Purity; Tan and Balazs; SI; November 1989

A Model for Oxidation of Silicon by Oxygen; Cristy and Condon; JECS; October 1981

A New Intrinsic Gettering Technique Using Micro Defects in Czochralski Silicon Crystal: A New Double Preannealing Technique; Kagasawa, Matsushita and Kishino; APL; October 1980

A Method of Forming Thin and Highly Reliable Gate Oxides: Two-Step HCl Oxidation; Hashimoto *et al.*; JECS; January 1980

A Low Pressure BPSG Deposition Process; Foster, Hoeye, and Oldman; JECS; February 1985

A Suspended Boat Loader Based on the Cantilever Principle; Lambert and Bayne; SI; April 1984

A New Preferential Etch for Defects in Silicon Crystals; Jenkins; JECS; September 1977
Advances in Instrumentation for CV Measurements; Nicollian; SST; August 1989
Afterglow Chemical Vapor Deposition of SiO_2; Jackson *et al.*; SST; April 1987
Alkali-Developable Silicon Containing Positive Photoresist for a Two-Layer Resist System; Toriumi *et al.*; JECS; April 1987
Aluminum Alloy Bonding Wires in Corrosive Environments; Ramsey; SI; March 1987
An Electron Microscope Investigation of the Effect of Phosphorus Doping on the Plasma Etching of Polycrystalline Silicon; Irene *et al.*; JECS; September 1981
Analysis of Human Contaminants Pinpoint Sources of IC Defects; Lowry *et al.*; SI; July 1987
Analysis of Polysilicon Diffusion Sources; Probst *et al.*; JECS; March 1988
Analytical Modeling of Non-Linear Diffusion of As in Si; Jeppson, Anderson *et al.*; JECS; September 1987
Anomalous Arsenic Diffusion in Silicon Dioxide; Wada and Antoniadis; JECS; June 1981
Arsine and Phosphine Replacements for Semiconductor Processing; Miller; SST; August 1989
Breakdown Characteristics of Gallium Arsenide; Baliga *et al.*; IEEE; November 1981
Calibration of Ion Implantation Systems; Wittkower; SST; November 1978
Carbon and Oxygen Role for Thermally Induced Microdefect Formation in Silicon Crystals; Kishino *et al.*; APL; August 1979
Certifying a Class 10 Cleanroom Using Federal Standard 209C; Helander; MC; December 1987
Characterization of Surface Treated 316L Stainless Steel Tubings to Prevent Contamination in Gas Distribution; Kumar and Dyar; SST; February 1987
Characterization of LPCVD Thin Silicon Nitride Films; Popova *et al.*; *Thin Solid Films* **122**; December 1984
Characterization of Plasma Enhanced Chemical Vapor Deposition of Silicon Oxynitride; Van den Hoek; *MRS Proceedings*; April 1986
Characterization of Oxygen Doped Plasma-Deposited Si_3N_4; Knolle, Osenbach, and Elia, JECS; May 1988
Charge Trapping in SiO_2 Grown in Polycrystalline Si Layers; Auni *et al.*; JECS; January 1988
Chemical and Radiation Hazards in Semiconductor Manufacturing; Baldwin and Stewart; SST; August 1989
Chlorine Levels in SiO_2 Formed Using TCV and LPCVD at Low Temperatures; Banbolzer and Ghezzo; JECS; February 1987

Class 10 Robots in PECVD Processing; Wu and Carone; SI; February 1988
Cleaning Techniques for Wafer Surfaces; Skidmore; SI; August 1987
Contactless Probing; Singer; SI; May 1989
Controlling Bacterial Growth in an Ultrapure DI Water System; Carmody and Maityak; MC; January 1989
Controlling Static in Wafer Fabrication; Murray; SI; August 1985
Corporate Culture Determines Productivity; *Editorial Industry Week;* May 1981
Correlation of Post-etch Residues to Deposition Temperature in Plasma Etched Al Alloys; JECS; November 1987
Counting and Identifying Particles in High Purity Water; Balazs and Walker; SI; April 1982
Critical Issues in SEM Metrology; Zorich and Holmes; MMT; February 1989
CVD Tungsten and Tungsten Silicide for VLSI Applications; Sachdav and Castellano; SI; May 1985
CVD Dielectric Films for VLSI; Pramanik; SI; June 1988
Deep Submicrometer MOS Device Fabrication Using a Photoresist Ashing Technique; Chung *et al.*; IEEE; January 1986
Defect Etch for <100> Silicon Evaluation; Schimmel; JECS; March 1979
Deposition Kinetics of SiO_2 Film; Maeda and Nakamura; JAP; November 1981
Determination of Exposure Dose Via PDR; Bruce and Lin; SI; June 1988
Development of Positive Photoresists; Mach; JECS; January 1987
Diffusion of Metals in SiO_2; McBrayer, Swanson, and Sigmon; JECS; June 1986
Dislocation Etch for <100> Planes in Silicon; d'Aragona; JECS; September 1977
Downstream Plasma Etching and Stripping; Cook; SST; April 1987
Dry Development Using an Olefinic Polymer and Ozone Developing; Abe, Fujino, and Ban; JECS; August 1987
Dry Etch Chemical Safety; Ohlsen; SST; July 1986
Dry Etching Technology for One Micron VLSI Fabrication; Hirata, Ozaki, Oda, and Kimizuka; IEEE; 1981
ECR Finds Applications in CVD; Kearney *et al.*; SI; March 1989
Effect of Discharge Frequency on Plasma Condition in PECVD; Takagi, Konuma, and Nagasaka; *Japan Science Forum;* December 1981
Effects of Several Parameters on the Corrosion Rates of Al Conductors in Integrated Circuits; Lerner and Eldridge; JECS; October 1982
Effects of Plasma Cleaning on the Dielectric Breakdown in SiO_2 Film on Silicon; Iwamatsu; JECS; January 1982
Electrical Properties of Composite Evaporated Silicide/Polysilicon Electrodes; Koburger, Ishaq and Geipel; JECS; June 1982
Electromigration-Induced Failures in VLSI Interconnects; Ghate; SST; March 1983

Electromigration in Aluminum/Polysilicon Composites; Vaidya; APL; December 1981
Electromigration in Multilayer Metallization Systems; Hoang and McDavid; SST; October 1987
Electron Beam Assisted CVD of SiO_2 and Si_3N_4 Films; Emery *et al.*; *Proceedings of SPIE—The International Society for Optical Engineering* **459**; 1984
Electrostatic Discharge; Thurber *et al.*; *National Bureau of Standards Handbook;* May 1981
Ellipsometry Measurements of Polycrystalline Silicon Films; Irene and Dong; JECS; June 1982
Ellipsometry: A Century Old New Technique; Spanier; *Industrial Research;* September 1975
Evaluating the Effect of Various Process Gases on Filter Performance; Chahine *et al.*; August 1989
Evaluating Voids and Microcracks in Al Metallization; Smith, Welles, and Bivas; SI; January 1990
Evaluation of Ti as a Diffusion Barrier Between Al and Si for 1.2μ CMOS ICs; Faraboni, Turner, and Barnes; JECS; November 1987
Exhaust Gas Incineration and the Combustion of Arsine and Phosphine; Elliott; SST; January 1990
Eximer Lasers: An Emerging Technology in Semiconductor Processing; Znotins; SST; September 1986
Flow Restricting Devices Used with Semiconductor Process Gases; Quinn and Rainer; SST; July 1986
Flows in Sidewal Oxides Grown on Polysilicon Gates; Brown, Hu, and Morrissey; JECS; May 1982
Gallium Arsenide Hazards, Assessment, and Control; McIntyre and Sherin; SST; September 1989
Gate Oxide Charge to Breakdown Correlation to MOSFET Hot-Electron Degradation; Davis and Lahri; IEEE; June 1988
Gettering in Processed Silicon; Sadana; SI; May 1985
Growth Kinematics of a Polysilicon Trench Refill Process; Silvestri; JECS; November 1986
Hazard Potential of Dichlorosilane; Sharp, Arvidson, and Elvey; JECS; October 1982
Hazardous Gas Safety and the Role of Monitoring; Burggraaf; SI, November 1987
Heavy Metal Contamination from Resists During Plasma Stripping; Furimura and Yano; JECS; May 1988
High-Pressure Oxidation NMOS Technology; Baussman; SI; April 1983
High-Rate Photoresist Stripping in an O_2 Afterglow; Spencer, Borel, and Hoff; JECS; September 1986
Hydrogen Plasma Etching of Semiconductors and Their Oxides; Chang, Chang, and Darack; JVST; January 1982

Impact of DI Water Rinses on Silicon Surface Cleaning; Beyer and Kastl; JECS; May 1982
Improving Gas Handling Safety; Murray; SI; August 1986
Influence of Stoichiometry and Hydrogen Bonding on the Insulating Properties of PECVD Silicon Nitride; Chaussat *et al.*; *Physica (Amsterdam)*; 1985
Interaction Between Point Defects and Oxygen in Silicon; Monkowski *et al.*; *ASTM Journal;* 1984
Intermediate Oxide Formation in Double Polysilicon Gate MOS Structure; Sunami, Koyangi, and Hashimoto; JECS; November 1980
Ionic Contamination and Transport of Mobile Ions in MOS Structures; Kuhn and Silversmith; JECS; June 1971
Japan: Quality Control and Innovation; Deming; *Business Week*; July 1981
Junction Depth Measuring Methods: The Pros and Cons; Aley and Turner; SI; May 1980
Laser Activated Chemical Vapor Deposition of SiO_2; Roche and Zorich; *Technical Proceedings of Semicon West;* May 1988
Laser Planarization; Tuckerman and Weisberg; SST; April 1986
Laser Induced Chemical Vapor Deposition of SiO_2; Boyer, Roche, Ritchie, and Collins; APL; April 1982
Lateral Epitaxial Overgrowth of Silicon on SiO_2; Rathman, Silversmith, and Burns; JECS; October 1982
LDD MOSFET's Using Disposable Sidewall Spacer Technology; Pfeister; IEEE; April 1988
Lift-off Techniques for Fine Line Metal Patterning; Frary and Seese; SI; December 1981
Low-Temperature Annealing and Hydrogenization of Defects at the Si–SiO_2 Interface; Johnson *et al.;* JVST; September/October 1981
Low-Temperature Differential Oxidation for Double Polysilicon VLSI Devices; Barnes, Belasi, and Deal; JECS; October 1979
LPCVD of BPSG from Organic Reactions; Williams and Dein; JECS; March 1987
LPCVD of Tantalum Silicide; Reynolds; JECS; June 1988
LPCVD of SI_3N_4; Roenigkm and Jensen; JECS; July 1987
LPCVD of Aluminum and Al–Si Alloys for Semiconductor Metallization; Cooke *et al.;* SST; December 1982
LPCVD of BPSG Film Produced by Injection of Miscible DADBS-TMB-TMP Liquid Source; Levy, Gallagher, and Schrey; JECS; July 1987
Manager's Guide to Cleanroom Operators; Northcraft; MC; May 1986
Measurement of the Depth of Diffused Layers in Silicon by the Grooving Method; McDonald and Goetzberger; JECS; February 1962
Measurement of Electrical Resistivity of CVD-BN Passivation Film from Gate Charge Decay Time of a MOS Transistor; Kim and Shono; JJAP; October 1981
Measurement Tools for Overlay Registration; Murray; SI; February 1987

Mechanical Damage Gettering Effect in Si; Sawada, Kabaki and Watanabe; JJAP; November 1981
Megasonic Particle Removal from Solid State Wafers; Shwartzman, Mayer, and Kern; *RCA Review;* March 1985
Microelectronics Dimensional Metrology in the Scanning Electron Microscope; Postek and Joy; SST; December 1986
MOCVD of III-V Compound Epitaxial Layers; Mawst *et al.;* November 1986
Modeling of LPCVD Processes; Kuiper *et al.;* JECS; October 1982
Modeling of Ambient-Meniscus Melt Interactions Associated with Carbon and Oxygen Transport in EFG of Silicon Ribbon; Kalejs and Chin; JECS; June 1982
MOS Gate Oxide Defects Related to Treatment of Silicon Nitride Coated Wafers Prior to Local Oxidation; Goodwin and Brossman; JECS; May 1982
MOS CV Techniques for IC Process Control; McMillan; SST; September 1972
Nondestructive Analysis of Silicon Nitride/Silicon Oxide/Silicon Structures Using Spectroscopic Ellipsometry; Theetan *et al.;* JAP; November 1981
On the Doping Dependence of Oxidation Induced Stacking Fault Shrinkage in Silicon; Fair and Carim; JECS; October 1982
Optimizing Cleanroom Efficiency; Leavitt; SI; May 1987
Orthogonal Design for Process Optimization and Its Application in Plasma Etching; Yin and Jillie; SST; May 1987
Overlay Accuracy for VLSI Devices; Wakamiya and Nakajima; SI; May 1985
Oxidant Transport during Steam Oxidation of Silicon; Mikkelson; APL, December 1981
Oxidation of Phosphorus Doped Low Pressure and Atmospheric CVD Polycrystalline-Silicon Films; Kamins; JECS; October 1976
Oxidation Process for Double Polysilicon Insulators; Crimi, McDonald, and Montillo; *IBM Technical Publications;* April 1980
Oxidation Induced Stresses and Some Effects on the Behavior of Oxide Films; Hsueh and Evans; JAP; November 1983
Oxide Breakdown Due to Charge Accumulation during Plasma Etching; Ryden, Norstrom, Nender, and Berg; JECS; December 1987
Oxygen, Oxidation Stacking Faults, and Related Phenomena in Silicon; Hu; *Materials Research Society;* April 1981
Particle Control in Process Equipment: A Case Study; Baker and Fishkin; MC; September 1986
Photoinitiated Crosslinking and Image Formation in Thin Polymer Films Containing a Transition Metal Compound; Kutal and Wilson; JECS; September 1987
Planar Plasma Etching of Polysilicon Using CCl_4 and NF_3; Bower; JECS; April 1982

Plasma Processing for Silicon Oxynitride Films; Cavallari and Guslandris; JECS; May 1987
Plasma Etching Methods for the Formation of Planarized Tungsten Plugs Used in Multilevel ULSI Metallization; Saia *et al.;* JECS; April 1988
Plasma TEOS Process for Interlayer Dielectric Applications; Chin and van de Ven; SST; April 1988
Plasma Enhanced Deposition of W, Mo, and WSi_2 Films; Tang, Chu, and Hess; SST; March 1983
Plasma Etching of Aluminum. A Comparison of Chlorinated Etchants; Danner, Daulie, and Hess; JECS; February 1987
Plasma Enhanced Etching of W and WSi_2 in Cl Containing Discharges; Fiscl and Hess; JECS; September 1987
Point Radiation Defects in Boron Doped Silicon; Emtsev, Mashovets, and Nazaryan; *Soviet Physics—Semiconductors (English Translation)*; May 1981
Polycrystalline Silicon Resistors for Use in Integrated Circuits; Hughes; SST; May 1987
Polyimides in Microelectronics; Burggraaf; SI; March 1988
Polysilicon/SiO_2 Interface Microtexture and Dielectric Breakdown; Marcus, Sheng, and Lin; JECS; June 1982
Practical I-line Lithography; Tipton, Marriott, and Fuller; January 1987
Precipitation Process Design for Denuded Zone Formation in CZ-Silicon Wafers; Huber and Reffle; SST; August 1983
Pressure Oxidation of Silicon: An Emerging Technology; Zeto, Korolkoff, and Marshall; SST; July 1979
Process and Film Characterization of PECVD BPSG Films for VLSI Applications; Tang, Scertenleib, and Carpico; SST; January 1984
Process Conditions Affecting Boron and Phosphorus in BPSG Films as Measured by FTIR Spectroscopy; SST; March 1987
Process Modeling and Simulation; Singer; SI; February 1987
Properties of LPCVD WSi_2 as Related to IC Process Requirements; Brors *et al.;* SST; April 1983
Properties of Plasma Enhanced CVD Silicon Films: Films Doped During Deposition; Kamins and Chiang; JECS; October 1982
Properties of Tungsten Silicide Films on Polycrystalline Silicon; Tsai *et al.;* AP; August 1981
Properties of CVD WSi_2 Films Using Reaction of WF_6 and Si_2H_6; Shioya *et al.;* JECS; May 1987
Quantum Yield for Laser Induced CVD of SiO_2 and Si_3N_4; Weber and Anderson; 1988
Reactive Ion Etching of Al/Si in BBr_3/Cl_2 and BCl_3/Cl_2 Mixtures; Bell, Anderson, and Light; JECS; May 1988
Reactive Ion Etching for VLSI; Ephrath; IEEE; 1981
Realtime Inspection of Wafer Surfaces; Blaustein and Hahn; SST; December 1989

Refractive Index Dispersion of Dielectric Films Used in the Semiconductor Industry; Pliskin; JECS; November 1987

Refractory Metals Silicides; McLachlan *et al.*; SI; October 1984

Relaxation Effects at the Semiconductor–Insulator Interface; Zerbst; *German Journal of Applied Physics* **22**; 1966

Role of Chlorine in Silicon Oxidation; Monkowski; SST; July 1979

Segregation of Arsenic to the Grain Boundaries in Polycrystalline Silicon; Swaminathan *et al.*; JECS; October 1980

Monolithic Integration of GaAs and Si; Fan; *Solid State Devices and Materials Conference;* August 1984

Selective Tungsten Processing by Low Pressure CVD; Broadbent and Stacy; SST; December 1985

Selective Dry Etching of Tungsten for VLSI Metallization; Burba *et al.*; JECS; October 1988

Semiconductor Safety Issue; SI; July 1989

Shedding New Light on Multiple Layer Film Thickness Measurements; Pham, Damar, Mallory, and Perloff; MMT; March 1987

Silicon Oxidation Studies: Measurement of the Diffusion of Oxidant in SiO_2 Films; Irene; JECS; February 1982

Silicon Dioxide Removal in Anhydrous HF Gas; Cleavelin and Duranko; SI; November 1987

Silicon on Insulator Films by Oxygen Implantation and Lamp Annealing; Celler; SST; March 1987

Simplified Statistics for Small Numbers of Observations; Dean and Dixon; *Analytical Chemistry* **23;** 1951

SMIF System Performance at 0.22μm Particle Size; Harada and Suzuki; SST; December 1986

Some Problems in Plasma Etching of Al and Al–Si Alloy Films; Hirobe, Kureishi, and Tsuchimoto; JECS; December 1981

Stability of LPCVD Polysilicon Gates on Thin Oxides; Murarka *et al.*; JECS; November 1980

Stacking Fault Generation Suppression and Grown-In Defect Elimination in Dislocation-Free Silicon Wafers by HCl Oxidation; Shiraki; JJAP; January 1976

Statistical Process Control in Photolithography; Pritchard; SI; May 1987

Study on Measurement of Carrier Effective Lifetime in Furnace and Laser Annealed Diodes; Rahman and Furukuwa; JJAP; August 1981

Super Cleanroom Technology: A High Tech Balancing Act; Suzuki, Oikawa, and Sekiguchi; MC; September 1988

Suppression of Stacking Fault Generation in Silicon Wafers by HCl Added to Dry O_2 Oxidation; Shiraki; JAP; January 1976

Survey of Toxic Sensors and Monitoring Systems; Korolkoff; SST; December 1989

System Approach to Low Contamination Wafer Handling; Bantz and Bedini; SST; August 1986

The Comeback of the Vacuum Tube: Will Semiconductor Versions Supplement Transistors?; Skidmore; SI; August 1988
The New Surface Analysis; Singer; SI; November 1989
The LPCVD Polysilicon Phosphorus Doped In Situ as an Industrial Process; Baudrant and Sacilotti; JECS; May 1982
The Oxygen Effect in the Growth Kinetics of Platinum Silicides; Nava, Valeri, and Majni; JAP; November 1981
The Step Coverage of Undoped and Phosphorus Doped SiO_2 Glass Films; Levin and Evans-Lutterodt; JVST; January–March 1983
The Identification and Elimination of Human Contamination in the Manufacture of ICs; Thomas and Calabrese; IEEE; September 1985
The Application of Contour Maps and Statistical Control Charts in Monitoring Dielectric Processes; Smith and Wang; *SPIE Conference;* March 1987
The Shrinkage and Growth of Oxidation Stacking Faults in Silicon and the Influence of Bulk Oxygen; Hu; JAP; July 1980
The Oxidation of Shaped Silicon Surfaces; Marcus and Sheng; JECS; June 1982
The Oxidation Rate Dependence of Oxidation-Enhanced Diffusion of Boron and Phosphorus in Silicon; Lin, Antoniadis, and Dutton; JECS; May 1981
The Nucleation of CVD Silicon on SiO_2 and Si_3N_4 Substrates; Claasen and Bloem; JECS; August 1980 and June 1981
The Relation between Etch Rate and Optical Emission Intensity in Plasma Etching; Kawata, Shibano, Murata, and Nagami; JECS; June 1982
The Structure and Composition of Silicon Oxides Grown in HCl/O_2 Ambients; Monkowski *et al.;* JECS; November 1978
The Growth of Oxidation Stacking Faults and Point Defect Generation at the $Si–SiO_2$ Interface during the Thermal Oxidation of Silicon; Lin *et al.;* JECS; May 1981
The Impact of Stepper Field Registration on Manufacturing; Waldo; SI; May 1986
The CV Technique as an Analytical Tool, Parts 1 and 2; Zaininger and Heiman; SST; May 1970
The Limitation of Short Channel Length N^+-Polysilicon Gate CMOS ICs; Hsu; *RCA Review;* June 1985
The Effects of Processing Conditions on the Outdiffusion of Oxygen from Czochralski Silicon; Heck, Tressler, and Monkowski; JAP; October 1983
The Effect of CMOS Processing on Oxygen Precipitation, Wafer Warpage, and Flatness; Lee and Tobin; JECS; October 1986
Thermal Nitridation of SiO_2 Thin Films on Si at 1150°C; Koba and Tressler; JECS; January 1988
Thermally Enhanced Nd : YAG Laser Annealing of Ion Implanted Silicon; Wilson and Gregory; JVST; January 1982

Thermochemical Calculations on the LPCVD of Si_3N_4 and SiO_2; Spear and Wang; SST; July 1980

Thin SiO_2 Using Rapid Thermal Oxidation Process for Trench Capacitors; Miyni *et al.*; JECS; January 1988

Thin Film Characterization; Smith and Hinson; SST; November 1986

Thin Film Deposition by UV Laser Photolysis; Emery *et al.*; *Proceedings of SPIE—The International Society for Optical Engineering* **459;** 1984

Two-Layer Planarization Process; Schlitz and Pons; JECS; January 1986

Using BPSG as an Interlayer Dielectric; Johnson and Sethna; SI; October 1987

Variable Profile Contact Etching Using Bilayer Planarized Photoresist; Jillier *et al.*; JECS; August 1987

Viscous Behavior of Phosphosilicate and Borophosphosilicate Glasses in VLSI Processing; Levy and Nassau; SST; October 1986

Wafer Annealing Systems; Iscoff; SI; November 1981

Wafer Charging and Beam Interactions in Ion Implantation; White, *et al.*; SST; February 1985

Warpage of Silicon Wafers; Leroy and Plougonven; JECS; April 1980

Water Adsorption in Chemically Vapor-Deposited Borosilicate Glass Films; Arai and Terunema; JECS; May 1974

INDEX